HANDBOOK OF METHODS IN
AQUATIC MICROBIAL ECOLOGY

EDITED BY

PAUL F. KEMP
BARRY F. SHERR
EVELYN B. SHERR
JONATHAN J. COLE

LEWIS PUBLISHERS
Boca Raton   Ann Arbor   London   Tokyo

Library of Congress Cataloging-in-Publication Data

Handbook of methods in aquatic microbial ecology / edited by Paul F.
  Kemp ... [et al.].
      p.   cm.
    Includes bibliographical references and index.
    ISBN 0-87371-564-0
      1. Aquatic microbiology--Methodology--Handbooks, manuals, etc.
    2. Microbial ecology--Methodology--Handbooks, manuals, etc.
    I. Kemp, P. F.
    QR105.H26   1993
    576'.192--dc20   93-18661
                                                                                            CIP

PRINTED IN THE UNITED STATES OF AMERICA
1 2 3 4 5 6 7 8 9 0

Printed on acid-free paper

EDITOR'S FOREWORD

This volume was originally intended to be one section of an edited book entitled *Microbial Interactions in Aquatic Ecosystems,* which is in preparation and will also be published by Lewis Publishers. We soon realized that the suite of techniques available to microbial ecologists was far greater than we had imagined, and that a review of methodology would be far less valuable than an actual compilation of do-it-yourself recipes. Lewis Publishers agreed, and the project began in earnest in January 1991. Originally, we had over 150 methods listed for possible inclusion, an impossible number to deal with in a single volume. By combining variants of methods and dismissing some altogether, we managed to assemble a somewhat shorter list that eventually evolved into the present 86 chapters with 95 contributing authors. Just as well that the final list was shorter: our original goal of five pages per chapter proved to be hopelessly optimistic. At the time of writing this foreword, I am looking at a 10-inch high manuscript and wondering how Lewis Publishers will fit it all into one book.

We invited the originator or a leading proponent of each method to present the procedures that they use. Most of those asked were willing to write a chapter, and most of those came through. Readers should bear in mind that there are many variations to some of these methods, and most chapters include references to alternative procedures that address specific problems.

Our goal in preparing this compilation was to broaden the range of methods that a researcher, whether experienced or novice, might consider applying to a problem. Accordingly, we have deliberately included not only the methods most familar to aquatic microbial ecologists (e.g., AODC and DAPI bacterial enumeration), but also a broad sampling of methods that are relatively new and are still used primarily by specialists. For example, ''molecular ecology'' papers are just beginning to appear regularly, but they have sparked enormous interest among microbial ecologists. We included a sampling of molecular techniques applicable to microbial ecology problems. Some methods in this volume rely on expensive equipment (e.g., flow cytometry and X-ray microanalysis) not available at most institutions. We have included these methods because they are powerful and can be used to address questions not approachable by any other means. It is my hope that this volume will spark greater interest in these techniques and foster collaborations with the lucky few who have access to specialized equipment.

In many cases, the specific procedures included in this volume are intended to represent an entire approach. For example, procedures used to work with anaerobic microbes are given specifically for water column samples, but are representative of those which would be used with anaerobic sediment. The ingenious researcher will use this book not only as a cookbook, but also as an idea book.

The final manuscript was completed almost on schedule, a feat made possible only by the enthusiam and cooperation of the contributors. I would like to express my appreciation to all. I had no idea how much work this book would be, and I can easily imagine how much more work it would have been without the conscientious efforts of the authors.

Paul F. Kemp

Paul Kemp is an oceanographer and microbial ecologist whose work has involved the production and fate of marine bacteria, benthic-pelagic coupling and modeling, and the organization of benthic communities. He received his Ph.D. from Oregon State University in 1985, followed by postdoctoral work at the University of Georgia Marine Institute. He joined the oceanographic group at Brookhaven National Laboratory, originally as a Hollaender Distinguished Postdoctoral Fellow, and later as a staff member. For the past few years he has turned his attention to the application of molecular-biology techniques to problems in marine microbial ecology, especially concerning the structure of bacterial communities and processes controlling microbial growth rates.

Barry and Evelyn Sherr are a husband-wife research team investigating trophic roles of phagotrophic protists (ciliates and flagellates) in both freshwater and marine ecosystems. Barry received his Ph.D. from the University of Georgia in 1977. Evelyn received her doctorate from Duke University in 1974. The Sherrs worked at the Lake Kinneret Limnological Laboratory in Israel and at the University of Georgia Marine Institute before coming to the College of Oceanic and Atmospheric Sciences at Oregon State University, where they share a faculty position. The Sherrs are broadly interested in heterotrophic microbes as facilitators of biogeochemical cycles in estuaries, lakes, and the open ocean. They particularly enjoy developing new methodological approaches for elucidating specific pathways of transfer of carbon and nitrogen within microbial food webs.

Jonathan Cole works at the interface between microbiology and biogeochemistry in aquatic ecosystems. He received his Ph.D. from Cornell University in 1982 and did postdoctoral work at both the Woods Hole Oceanographic Institution and the Marine Biological Laboratory before joining the staff at the Institute of Ecosystem Studies. Cole is particularly interested in the biotic and abiotic regulation of energy flow and nutrient cycles in lakes, rivers, and marine systems, and finds especially useful studies that compare these cycles between diverse systems.

CONTRIBUTOR LIST

Lawrence J. Albright, Department of Biological Sciences, Simon Fraser University, Burnaby, British Columbia, Canada V5A 1S6

Alice L. Alldredge, Department of Biological Sciences, University of California, Santa Barbara, California 93106

Daniel M. Alongi, Australian Institute of Marine Science, P.M.B. No. 3, Townsville M.C., Queensland 4810 Australia

James W. Ammerman, Department of Oceanography, Texas A&M University, College Station, Texas 77843

Amy E. Anderson, Department of Zoology, Oregon State University, Corvallis, Oregon 97331-2914

Ronald M. Atlas, Department of Biology, 139 Life Sciences Building, University of Louisville, Louisville, Kentucky 40292

John A. Baross, School of Oceanography, University of Washington, Seattle, Washington 98195

Russell T. Bell, Department of Limnology, Uppsala University, Norbyvägen 20, 75236 Uppsala, Sweden

Ronald Benner, Marine Science Institute, University of Texas at Austin, Port Aransas, Texas 78373

A. Bianchi, Microbiologie Marine, C.N.R.S. U.P.R. 223, Campus de Luminy, Case 907, F-13288 Marseille cedex 9 France

T.H. Blackburn, Department of Microbial Ecology, Institute of Biology, Aarhus University, DK-8000 Aarhus, Denmark

Beatrice C. Booth, School of Oceanography, University of Washington, Seattle, Washington 98195

Susan L. Bower, Department of Biological Sciences, University of Southern California, Los Angeles, California 90089-0371

Gunnar Bratbak, Department of Microbiology and Plant Physiology, University of Bergen, N-5020 Bergen, Norway

S.A. Brickson, Research Laboratories of the Veterans Administration Medical Center and the Department of Biochemistry and Molecular Biology, University of Miami School of Medicine, Miami, Florida 33125

D.K. Button, Institute of Marine Science, University of Alaska Fairbanks, Fairbanks, Alaska 99775

Juan I. Calderón, Marine Sciences Institute, CSIC, Passeig Nacional s/n, Barcelona, Spain

Lisa Campbell, Department of Oceanography, University of Hawaii, 1000 Pope Road, Honolulu, Hawaii 96822

Josefina García-Cantizano, Department of Genetics and Microbiology, Autonomous University of Barcelona, Bellaterra, Spain

Douglas G. Capone, Center for Environmental and Estuarine Studies, Chesapeake Biological Laboratory, University of Maryland, Solomons, Maryland 20688-0038

Philip G. Carey, Department of Life Sciences, Faculty of Science and Technology, University of Derby, Kedleston Road, Derby DE3 1GB England

Kevin R. Carman, Department of Zoology and Physiology, Louisiana State University, Baton Rouge, Louisiana 70803-1725

David A. Caron, Biology Department, Woods Hole Oceanographic Institution, Woods Hole, Massachusetts 02543

Luis A. Cifuentes, Department of Oceanography, Texas A&M University, College Station, Texas 77843

Richard B. Coffin, U.S. Environmental Protection Agency, Gulf Breeze Environmental Research Laboratory, Sabine Island, Gulf Breeze, Florida 32561

Jonathan J. Cole, Institute of Ecosystem Studies, Box AB, Millbrook, New York 12545

Russell L. Cuhel, Center for Great Lakes Studies, University of Wisconsin-Milwaukee, Milwaukee, Wisconsin 53204

Alan W. Decho, U.S. Geological Survey, Menlo Park, California 94025

Edward F. DeLong, Department of Biological Sciences, University of California-Santa Barbara, Santa Barbara, California 93106

Jody W. Deming, School of Oceanography, University of Washington, Seattle, Washington 98195

Giacomo R. DiTullio, Graduate Program in Ecology, University of Tennessee, Knoxville, Tennessee 37996-1191

Fred C. Dobbs, Department of Oceanography, The University of Hawaii, Honolulu, Hawaii 96822

Susan E. Douglas, Institute for Marine Biosciences, National Research Council, Halifax, Nova Scotia, Canada B3H 3Z1

Hugh W. Ducklow, Horn Point Environmental Laboratory, University of Maryland, Cambridge, Maryland 21613

Michele D. DuRand, Biology Department, Woods Hole Oceanographic Institution, Woods Hole, Massachusetts 02543

M.J. Ferrara-Guerrero, Microbiologie Marine, C.N.R.S. U.P.R. 223, Campus de Luminy, Case 907, F-13288 Marseille cedex 9 France

Robert H. Findlay, Department of Biochemistry, Microbiology, and Molecular Biology and Center for Marine Studies, Darling Marine Center, University of Maine, Walpole, Maine 04573

Stuart Findlay, Institute of Ecosystem Studies, Millbrook, New York 12545

B.J. Finlay, Institute of Freshwater Ecology, Windermere Laboratory, Ambleside, Cumbria, LA22 OLP England

Dian J. Gifford, Graduate School of Oceanography, University of Rhode Island, Narragansett, Rhode Island 02882-1197

Steven C. Hand, Department of EPO Biology, N122 Ramaley Building, University of Colorado, Boulder, Colorado 80309-0334

Mikal Heldal, Department of Microbiology and Plant Physiology, University of Bergen, N-5020 Bergen, Norway

Gerhard J. Herndl, Institute of Zoology, University of Vienna, A-1090 Vienna, Austria

John E. Hobbie, Marine Biological Laboratory, Woods Hole, Massachusetts

Kjell Arne Hoff, BP Nutrition Aquaculture Research Centre, N-4001 Stavanger, Norway

P.A.G. Hofman, Van Hall Institute, Groningen 9721 AA The Netherlands

Hans-Georg Hoppe, Institute für Meereskunde, D-2300 Kiel, Germany

Jung-Ho Hyun, Department of Oceanography and Coastal Sciences, Louisiana State University, Baton Rouge, Louisiana 70803

Rodolfo Iturriaga, Department of Biological Sciences, University of Southern California, Los Angeles, California 90089-0371

S.A. de Jong, Ministry of Transport and Public Works, North Sea Directorate, Rijswijk 2280 The Netherlands

Elisabeth Kaltenböck, Institute of Zoology, University of Vienna, A-1090 Vienna, Austria

David M. Karl, Department of Oceanography SOEST, University of Hawaii, 1000 Pope Road, Honolulu, Hawaii 96822

Paul F. Kemp, Oceanographic and Atmospheric Sciences Division, Brookhaven National Laboratory, Upton, New York 11973

Ronald P. Kiene, Marine Institute, University of Georgia, Sapelo Island, Georgia 31327

David L. Kirchman, College of Marine Studies, University of Delaware, Lewes, Delaware 19958

Bert Klein, GIROQ, Université Laval, Laval P.Q., Québec, Canada G1K 7P4

Carla H. Kuhner, Department of Microbiology, University of Illinois, Urbana, Illinois 61801

Michael R. Landry, Department of Oceanography, University of Hawaii at Manoa, Honolulu, Hawaii 96822

Christopher Langdon, Lamont-Doherty Earth Observatory, Columbia University, Palisades, New York 10964

J. LaRoche, Oceanographic and Atmospheric Sciences Division, Brookhaven National Laboratory, Upton, New York 11973

Paul LaRock, Department of Oceanography and Coastal Sciences, Louisiana State University, Baton Rouge, Louisiana 70803

Barry S.C. Leadbeater, School of Biological Sciences, The University of Birmingham, Edgbaston, Birmingham B15 2TT England

John J. Lee, Department of Biology, The City College of City University of New York, New York 10031

SangHoon Lee, Polar Research Center, Korea Ocean Research and Development Center, Seoul 425-600, Korea

Evelyn J. Lessard, School of Oceanography, University of Washington, Seattle, Washington 98195

Glenn Lopez, Marine Sciences Research Center, State University of New York at Stony Brook, Stony Brook, New York 11794

D.H. Lynn, Department of Zoology, University of Guelph, Guelph, Ontario, Canada N1G 2W1

Erland A. MacIsaac, Department of Fisheries and Oceans, Biological Sciences Branch, West Vancouver Laboratory, West Vancouver, British Columbia, Canada V7V 1N6

D.G. Marty, Microbiologie Marine, C.N.R.S. U.P.R. 223, Campus de Luminy, Case 907, F-13288 Marseille cedex 9 France

Jordi Mas, Department of Genetics and Microbiology, Autonomous University of Barcelona, Barcelona, Spain

George B. McManus, Marine Environmental Sciences Consortium, Dauphin Island, Alabama 36528

Paul A. Montagna, Marine Science Institute, University of Texas at Austin, Port Aransas, Texas 78373

D.J.S. Montagnes, Department of Oceanography, University of British Columbia, Vancouver, British Columbia, Canada V6T 1Z4

Michael T. Montgomery, Center for Biomolecular Science and Engineering, Naval Research Laboratory, Washington, D.C. 20375-5348

Richard Y. Morita, Department of Microbiology, College of Science and College of Oceanography, Oregon State University, Corvallis, Oregon 97331-3804

Gerald Müller-Niklas, Institute of Zoology, University of Vienna, A-1090 Vienna, Austria

Steven Y. Newell, Marine Institute, University of Georgia, Sapelo Island, Georgia 31327

Svein Norland, Institute for Microbiology and Plant Physiology, University of Bergen, Jahnebakken 5, N-5020 Bergen, Norway

Mark D. Ohman, Scripps Institution of Oceanography, La Jolla, California 92093-0227

Robert J. Olson, Biology Department, Woods Hole Oceanographic Institution, Woods Hole, Massachusetts 02543

David E. Padgett, Department of Biological Sciences, The University of North Carolina at Wilmington, Wilmington, North Carolina 28403

Carlos Pedrós-Alió, Marine Sciences Institute, CSIC, Passeig Nacional s/n, Barcelona, Spain

Roland Psenner, Institute of Zoology, University of Innsbruck, A-6020 Innsbruck, Austria

Donald G. Redalje, Center for Marine Science, University of Southern Mississippi, Stennis Space Center, Mississippi 39529

Catherine Riaux-Gobin, CNRS, Laboratoire Arago, Banyuls/mer 66650 France

B.R. Robertson, Institute of Marine Science, University of Alaska Fairbanks, Fairbanks, Alaska 99775

Pierre E. Rouvière, Bacteriology Department, University of Wisconsin-Madison, Madison, Wisconsin 53706

Sybil P. Seitzinger, Academy of Natural Sciences, Division of Environmental Research, Philadelphia, Pennsylvania 19103-1195

Barry F. Sherr, College of Oceanic and Atmospheric Sciences, Oregon State University, Corvallis, Oregon 97331-5503

Evelyn B. Sherr, College of Oceanic and Atmospheric Sciences, Oregon State University, Corvallis, Oregon 97331-5503

Michael E. Sieracki, Bigelow Laboratory for Ocean Sciences, West Boothbay Harbor, Maine 04575

Meinhard Simon, Limnological Institute, University of Konstanz, D-7750 Konstanz, Germany

A.T. Soldo, Research Laboratories of the Veterans Administration Medical Center and the Department of Biochemistry and Molecular Biology, University of Miami School of Medicine, Miami, Florida 33125

John G. Stockner, Department of Fisheries and Oceans, Biological Sciences Branch, West Vancouver Laboratory, West Vancouver, British Columbia, Canada V7V 1N6

Curtis A. Suttle, Marine Science Institute, University of Texas at Austin, Port Aransas, Texas 78373

C.M. Turley, Plymouth Marine Laboratory, Citadel Hill, Plymouth PL1 3PB England

M. Iqubal Velji, Helix Biotech Corporation, Richmond, British Columbia, Canada V6X 1X5

Peter G. Verity, Skidaway Institute of Oceanography, Savannah, Georgia 31416

Johan Wikner, Department of Microbiology, Umeå University, S-90187 Umeå, Sweden

Erik R. Zettler, Biology Department, Woods Hole Oceanographic Institution, Woods Hole, Massachusetts 02543

TABLE OF CONTENTS

Section III. Biomass

Section IV. Activity, Respiration, and Growth

Introduction

John E. Hobbie

This is an exciting book. It reports the latest progress in answering some of the most stubborn and recalcitrant questions in microbiology and ecology, that is, what are microbes actually doing out there in the real world? The progress in the last few years has been both rapid and accelerating. There are suddenly several entire sessions at scientific meetings devoted to these latest methods as the interest swells. There is a real sense of rapid movement in the development of new methods that are actually getting to answers and even to the correct answers. I have long believed that one of the problems in ecological microbiology is that just about any method appears to work; the most artifact-ridden miserable method will produce data with good seasonal cycles; the units of these cycles may be hundreds instead of the millions of bacteria we now know exist in each milliliter of water but by God the data look fine! Now the difference is that we have general agreement about the fundamental characteristics of the answers. For example, we agree on about how many microbes are present or about how fast microbes grow in the plankton. We can now sort out the fine-looking but wrong data from the data that tell us about the real world of microbes. Our challenge is now to improve methods and invent new ones so that the uncertainties will be tens of percents, not hundreds. The exciting progress of the last 5 years in new methods including estimates of viral numbers, molecular probes for bacterial species, bacterial production, and flow cytometry are summarized in the methods presented in this book; many of them point toward new directions in methodology.

The stunning impact that new methods are having on aquatic microbial ecology points out that the field is still methods limited. Despite tremendous progress over the last 25 years, we lag well behind other ecological research areas. We have only a primitive ability to describe the organisms present in nature, what these organisms are actually doing, and what controls their activity and growth. Most of our knowledge describes in great detail the various ways microbes make a living. Fenchel and Blackburn[1] concisely summarize the field by stating that the energy-yielding processes for bacteria are either chemotropic oxidation-reduction reactions or phototrophic reactions. For phototrophic organisms (plants and photosynthetic prokaryotes) there are oxygenic and anoxygenic reactions. For chemotrophic reactions, there are only oxidations of H_2, CHO, CH_4, HS^-, inorganic N compounds, and ferrous iron. However, this understanding of the metabolic pathways tells us only about the microbe's potential activities. For real progress in understanding the ecology of aquatic microbes we must have methods that will allow us to bring together the laboratory studies of the biological capabilities of microbes with measurements of the actual rates of activity in the field. The rate measurements must be reasonably accurate and must be widely accepted. For all these reasons, new methods are a necessity, not a refinement, for modern aquatic microbial ecology.

Why is aquatic microbial ecology still searching for good methods? The easy answer is that the usual methods of laboratory microbiology do not work for most aquatic microbes

0-87371-564-0/93/$0.00 + $.50
© 1993 by Lewis Publishers

and for measuring most of the rates occurring in nature. The microbes from natural waters exist under extremely low concentrations of nutrients and do not grow well on laboratory media. However, even in the future, when we are finally able to put species names on all the microbes found in a milliliter of water, we will have the problem of finding out which of these forms were active in nature; microbes have the ability to survive for long periods with greatly reduced activity. Another problem is that, even when microbes are active in sediments or plankton, the rates of metabolism are low. For example, incubations for respiration have to extend for many hours or days in order to measure any change at all. Unfortunately, when microbes are held in incubation bottles then they soon begin to grow, or die, or become activated. Finally, it is impossible to separate microbes from the abundant nonliving organic debris found in lakes and oceans so that measures of the amount of carbon, nitrogen, or stable isotopes are not possible.

What is a definition of a good method for microbial ecology? I believe strongly that methods that do not tell us what organisms are present in nature or what they are doing are incomplete: not wrong, but not enough. The logic for this statement goes back to basics. Ecology is defined as a "branch of biology dealing with organisms' relations to one another and to their surroundings." (from *The Concise Oxford Dictionary of Current English* 1982; 7th edition). Accordingly, microbial ecology has to ultimately deal with the organisms in their natural surroundings or habitats and this is the part of microbial ecology that is least understood. However, it is also true that for many ecological questions making measurements only in the field is as incomplete as making measurements only in the laboratory. In the long run, both laboratory and field measures are likely necessary. Another part of the definition that may seem strange to outsiders is that a good method must not introduce any artifact into the results. This may seem obvious but artifacts crop up continually in field measurements and are extremely difficult to determine. Finally, a good method must produce data that are ecologically believable. For example, data on microbial growth or the cycling rates of nitrogen must fit within the bigger picture of known cycles for that particular lake, river, or ocean. It is not enough to merely make a measurement; it has to be a reasonable measurement in terms of the energy, carbon, and substrate available.

Therefore, enthusiasm for the astounding progress of the past 3 decades must be tempered with a realization of the distance still to travel. How far have we come? When I first heard about the methods of microbial ecology, in David Frey's 1960 limnology course at Indiana University, Russian scientists led the world in aquatic microbiology but it was difficult to find out details of their methods. Frey translated for us Kuznetsow's[2] book on the role of microorganisms in the cycles of material in lakes which had just been published in Germany. The orientation was geochemical, the information derived from laboratory studies of the cultures and capabilities and from changes in the chemistry of lakes. Numbers of bacteria came from direct counts of stained cells on white cellulose acetate filters. Oxygen changes in incubation flasks were just about the only direct rate measurement.

After Dick Wright and I spent several postdoctoral years developing radioisotope methods for measuring the turnover of organic compounds in lakes, we presented the results in a 1965 Symposium[3] in Plön, Germany. Again, aside from turnover measurements and respiration studies of field samples, most of the methods discussed at the symposium were laboratory studies. A conclusion of the symposium, one of which typified aquatic microbiology in 1965, was that the main obstacle to progress was the lack of standardized media for isolating and counting bacteria.

The advent of the International Biological Program in the early 1970s forced ecologists for almost the first time to account for all of the carbon cycling in soil and freshwater. One result was the general recognition that microbes were responsible for most of the action in the cycling of carbon and nitrogen as well. Quantitative methods were sorely needed; in

1972 Thomas Rosswall organized an International Symposium in Uppsala, Sweden, to assemble information on "Modern Methods in the Study of Microbial Ecology". Among the methods presented[4] were fluorescence microscopy (but not yet an easily usable method), microcalorimetry, kinetic analysis of uptake of organic compounds (but without real knowledge of the concentrations), ATP measurements, autoradiography, cultures on membrane filters, fluorescent antibody methods, and chemostats. Very few of the 80 papers gave information on rates of microbial activity in nature; those that did employed changes in numbers in sterilized soil or artificial chambers or surfaces placed in streams. Virtually none of the methods presented at that symposium have proven to be successful methods, although some were ancestors of those in use today.

These methods of 20 years ago can be characterized as mostly laboratory oriented. Knowledge about the ecology of the microbes came mostly from the performance of cultures on various growth media or in continuous culture, we dealt with the potential ecology of microbes. The methods that were field oriented were either not adequate or not yet well-enough developed to produce reliable data on the presence of various species or their rates of activity.

The years since the early 1970s have produced a flood of new methods for the ecologist. Moreover, most of these are field oriented and make use of a wide variety of techniques from microscopy to molecular biology. One such basic field measure was the direct count of bacteria by epifluorescence microscopy. It was terribly frustrating in the 1960s and early 1970s to measure rates of bacterial activity or total ATP but to have no idea at all how many bacteria were present. A total count would certainly solve most of the problems of ecologists, we thought. The breakthrough came slowly and had to wait until new epiillumination systems were developed for fluorescent antibody use in medicine and until polycarbonate filters with a flat surface and precise pores were invented. These, combined with nuclear stains such as acridine orange and DAPI, allowed the bacteria to be counted as fluorescent spots against a dark background. However, the quantification still did not answer any of the big questions of the role of microbes in natural systems. Numbers per milliliter changed very little over the course of a year in a lake or ocean while the activity changed by several orders of magnitude. The translation of this number per milliliter into organic mass of carbon or nitrogen proved very difficult and laborious when we found that bacteria came in many different sizes and shapes; the tiny bacteria of 0.2 or 0.4 μm were difficult to measure precisely and these uncertain measurements were then cubed to estimate the cell's volume. One solution, DNA content and cell volume of cells by flow cytometry, is presented in this volume.

The answer to all our ecological problems, we thought, would be a measurement of bacterial production analogous to the ^{14}C-bicarbonate incorporation method for planktonic primary production. Most of the bacteria do not use a single substrate so the first idea was to use multiple substrates. While it is possible to measure the concentration and incorporation of all of the most common sugars, amino acids, and fatty acids being used by a bacterial assemblage, it is not practical. We did actually try and were able to make these measurements for 18 amino acids, but the amount of work was so great that only one station every month could be studied. Twelve monthly samples from a North Carolina estuary produced a Ph.D. thesis for one of my students but only over the horrified objections of a terrestrial ecologist on the committee. How could we seriously talk about an entire estuary from one bucketful of water collected once a month?

The first method to give reasonable production values for aquatic microbes utilized the incorporation of ^{3}H-thymidine, a precursor of DNA. From this method we have learned that many bacterial assemblages in natural waters have a generation time as rapid as one day. Some published data even show that there is greater bacterial processing of carbon than algal production. Is this true or is it a calibration artifact? Questions about the method center around several interrelated themes. How much of the total thymidine in the DNA comes from the external pool and how much from internal production? Is there a change in the ratio of external to internal source during incubations? What is a suitable calibration factor for the number of bacteria produced per mole of thymidine incorporated into DNA? These and other questions are still being investigated with the result that this method cannot yet be used without extensive testing in each body of water and at each time of year.

To avoid some of the calibration problems with the thymidine method, the leucine incorporation method was developed for the measurement of bacterial production. This is based upon the finding that leucine makes up a constant percent of bacterial protein and that there is a constant ratio between bacterial protein and bacterial carbon over a wide range of sizes of cells. Because leucine is relatively easy to measure, the actual external substrate concentration may be determined. This concentration times the fractional uptake of ^{14}C- or ^{3}H-leucine gives the incorporation rate; production is calculated from the total protein production. Results of this method are not too different from thymidine results, an encouraging development. Both the thymidine and the leucine methods are producing answers to one of the big questions of aquatic microbiology, that is, the amount of carbon and nitrogen that is processed by microbes.

The first field measurements with thymidine soon showed us that bacterial production rates did not explain the observed numbers of bacteria. Our paradigm that the rate of supply of substrate controlled the bacteria was too simple. Death was the answer and we debated whether it was "natural" death or predation. Again, we thought that just one more method would give us the complete picture.

Epifluorescence proved to work well for counting small (2 to 10 μm) colorless flagellates, the mostly likely predators of bacteria. There are millions per liter of these flagellates in most natural waters, probably enough to control the bacteria by their feeding. However, the feeding and clearance rates of the flagellates are derived from laboratory studies for the most part; they are potential rates only. Measures of rates in nature are few and the methods often controversial. As a result, the search for controls of bacterial numbers has continued, although in my judgment the flagellates are still good candidates.

The latest candidate for control is the viruses. As in the case of the bacteria, early estimates of specific types gave low numbers. For example, estimates of colony forming units (CFUs) ranged from 10^{0} to 10^{4} per milliliter. The studies are in their early stages and show that viruses do have the potential for control; the actual control is still unproven in nature.

The shifts in our paradigm of the control of bacteria in nature is instructive. In 20 years we have gone from an emphasis on substrate supply rate, to flagellates, and then to viruses as the favorite control. Unfortunately, the answers are likely to be complicated interactions of all three factors, sometimes singly and sometimes in concert. Also, we should not neglect the controls exerted by large protozoans that sometimes feed on bacteria and always feed on small flagellates. To sort out when and how these various controls are acting will make the previous difficulties seem simple.

The most exciting developments in recent years have been the application of molecular techniques to ecological questions. Closely related areas of bacterial taxonomy, genetic diversity, adaptation, and evolution have already undergone major changes due to molecular methods. Ecological applications have proven most fruitful in addressing the "what's out there" question. New methods such as PCR, immunofluorescence, and oligonucleotide probes allow the enumeration of specific taxa within a mixed bacterial assemblage. One group of taxa, the "SAR11 cluster" described by Giovannoni et al.[5] may be widespread in the ocean but was previously unknown and uncultured. In a hot spring, none of the 16S rRNA sequences found were identical to previous isolates from hot springs,[6] suggesting that cultures have grossly underrepresented microbial diversity in nature.

Another molecular method, hybridization of total bacterial community DNA,[7] has now been used at several ocean sites to determine that the composition of the community changes seasonally but not weekly. Hybridization has also shown changes in community composition over 10 to 100 m of depth and over kilometers of horizontal distance. At scales of millimeters within an algal mat, molecular probes allow the determination of the thin layer where nitrogen fixation occurs. Molecular techniques are also proving useful to explore the occurrence and physiology of ecological limitations. For example, light adaptation and nutrient limitation in phytoplankton are being examined at the molecular level. It is even possible that nitrogen limitation could be reliably diagnosed by the expression of a particular protein.

In the future activity or rates of microbial processes might even be determined with molecular techniques. There are already promising techniques for determining the relative abundance of RNA and DNA in individual cells as a measure of the growth stage. Another hopeful sign is the small army of medical investigators working on techniques for measuring the activity of the bacteria in urine, or the many microbiologists working on activity in groundwater. If successful, their methods should be transferable to marine and freshwater systems.

The past few decades have produced a revolution in our understanding of the ecology of aquatic microbes. This revolution was possible only because of the rapid improvement in methods and because of the change in emphasis from cultures and laboratory measures of microbial potential to field measures of the actual microbes present, their abundance, activity, and growth. This book includes the old and new methods responsible for this revolution.

REFERENCES

1. Fenchel, T. and Blackburn, T. H., *Bacteria and Mineral Cycling,* Academic Press, New York, 1979.
2. Kusnetzow, S.I., *Die Rolle der Mikroorganismen im Stoffkreislauf der Seen,* Deutscher Verlag der Wissenschaften, Berlin, 1959.
3. Ohle, W., Ed., Stoffhaushalt der Binnengewsser, Chemie und Mikrobiology — Biogenic metabolism of freshwater, chemistry and microbiology, Symposium No. 14, Int. Verein. Theoret. Angewandte Limnologie, 1968.
4. Rosswall, T., Ed., Modern Methods in the Study of Microbial Ecology, Bull. 17 from the Ecological Research Committee of the Swedish Natural Science Research Council, 1973.
5. Giovannoni, S. J., Britschgi, T. B., Moyer, C. L., and Field, K. G., Genetic diversity in Sargasso Sea bacterioplankton, *Nature,* 345, 60, 1990.
6. Ward, D. M., Weller, R., and Bateson, M. M., 16S rRNA sequences reveal numerous uncultured microorganisms in a natural community, *Nature,* 345, 63, 1990.
7. Lee, S. and Fuhrman, J. A., DNA hybridization to compare species compositions of natural bacterioplankton assemblages, *Appl. Environ. Microb.,* 56, 739, 1990.

Section I
Isolation of Living Cells

CHAPTER 1

Isolation and Enumeration of Anaerobic and Microaerophilic Bacteria in Aquatic Habitats

M. J. Ferrara-Guerrero, D. G. Marty, and A. Bianchi

INTRODUCTION

Oxygen availability is an important factor in the regulation of bacterial behavior and distribution in the environment, so it is an important parameter for bacterial cultivation. Low-oxygen or even anaerobic conditions are found below the oxygenated surface waters of lakes and oceans and, except for the deep sea, aquatic sediments are anoxic below a thin oxidized surface layer. Apparently oxygen-saturated environments, such as oxygenated water bodies, have been recently shown to contain anoxic microniches.[1] These environments, in which oxygen access is restricted or absent, are inhabited by oxygen-sensitive bacteria well adapted to these conditions, including

1. Microaerophilic bacteria which use molecular oxygen as electron acceptor, but are not able to grow under oxygen concentration as high as in atmospheric conditions, and proliferate in oxygen-depleted (0.2 to 12.4%) habitats.
2. Anaerobic bacteria which include the facultative anaerobes, which are able to grow in the presence, as well as in the absence of oxygen, and the strict anaerobes, which are inhibited or even killed by oxygen, and require anoxic habitats.

The oxygen-sensitive bacteria must be cultivated under controlled atmosphere. The methods developed for their cultivation are specifically directed toward the total or limited exclusion of molecular oxygen. This exclusion or protection requires the use of closed systems from which oxygen has been removed, and which prevent its reentry. Essentially two types of techniques are used for the isolation and cultivation of these bacteria. With the so-called "conventional" methods, only the incubation step is maintained oxygen-free (anaerobic jar, semisolid-agar culture). These methods are relatively easy and quick to carry out, but are inadequate for the culture of oxygen-sensitive anaerobic bacteria, which may be killed by contact with molecular oxygen during the plating or pouring process. In contrast, the "strictly anaerobic" techniques require that samples, bacteria, and culture media are never exposed to air during the entire process of sample preparation and culturing (Hungate technique, anaerobic glove box). Each of these two anaerobic systems is successfully used for cultivation of extremely anaerobic bacteria, and has advantages and disadvantages. The use of an anaerobic glove box is considerably more expensive and requires more space. The Hungate technique requires more dexterity, but it offers the important advantage of possibility for use in field work. For the same reasons, the utilization of liquid media instead of roll tubes for bacterial enumerations is more practical for field work, since inoculation in solid media involves additional supplies. Glove bags have been recently commercialized: polyethylene bags, equipped with gloves, can be blown up with inert gas. These portable systems are likely useful for field work, but cannot be used for extremely anaerobic bacteria.

MATERIALS REQUIRED

Semisolid Culture

Equipment

- Boiler
- Water bath

Supplies

- Screw-capped test tubes (10 × 200 mm)
- Syringes and spinal-type needles (20G3 Becton Dickinson)
- Ice

Anaerobic Jar Technique

Equipment

- Anaerobic jars (GasPak, BBL; Oxoid, Ltd.)
- For vented jars: vacuum pump; gas mixture tanks

Supplies

- $H_2 + CO_2$ generator sachets (GasPak, BBL)
- Redox indicators (GasPak, BBL)
- Catalyst pellets (GasPak, BBL)

Hungate Technique

Equipment

In its most simple form the Hungate roll-tube technique requires

- Tanks of oxygen-free compressed gasses, and gas regulators.
- Tube-roller equipment (+ boiler, water bath, and ice bath).
- Tube-spinner system. A more sophisticated inoculation and transfer system, including gassing jet movement operated by treadles, a special apparatus of streaking prehardened roll tubes, and a machine for rotating the tubes has been devised by the Virginia Polytechnic Institute, and is commercially available (V.P.I. Anaerobic Culture System, Bellco Glass, Inc.).

Supplies

- Gassing cannula
- Supports, forceps, and clamps to autoclave rubber-stoppered tubes
- Butyl-rubber stopper test tubes (Bellco, 25 × 142 mm)

- Screw-capped tubes with flange-butyl rubber stoppers (Bellco, 16 × 125 mm)
- Sterile needles and syringes

Anaerobic Glove Box Technique

Equipment

- Anaerobic glove boxes. They are commercially available, and vary markedly in size and structure. It appears that the best type is that made of flexible vinyl, as described by Aranki and Freter.[2] The chamber system requires compressed gas with regulator and flow controls; an electric inoculating-loop sterilizer; in some cases, incubator(s).

Supplies

- Catalyst pellets and heater unit to dry the pellets

Media

Anaerobic Bacteria

Various media have been proposed for the study of anaerobic bacteria; there is no single medium that we can recommend unequivocally, and it is obvious that the natural water from which the organisms originated constitutes the most appropriate diluent. The following medium can be recommended for growth of the most frequently studied anaerobes:

- Solution 1 (basal mineral salts, g l^{-1})
    - 0.2 g $CaCl_2 \cdot 2H_2O$
    - 0.5 g KH_2PO_4
    - 0.5 g K_2HPO_4
    - 1.0 g Na_2SO_4
    - 0.5 g NH_4Cl
    - 0.2 g $MgSO_4 \cdot 7H_2O$
    - 0.8 g NaCl (for marine habitats, the amount of NaCl is increased to 25 g)
- Solution 2 (trace elements, g l^{-1})
    - 12.8 g nitrilotriacetic acid (neutralized by NaOH)
    - 0.3 g $FeCl_2 \cdot 4H_2O$
    - 0.02 g $CuCl_2 \cdot 2H_2O$
    - 0.1 g $MgCl_2 \cdot 6H_2O$ (for marine habitats, the amount of $MgCl_2 \cdot 6H_2O$ is increased to 3.0 g)
    - 0.17 g $CoCl_2 \cdot 6H_2O$
    - 0.1 g $ZnCl_2$
    - 0.01 g $H_3BO_3 \cdot 2H_2O$
    - 0.03 g $NiSO_4 \cdot 7H_2O$
    - 0.02 g Na_2SeO_3
- When natural water from the sampling site is available, it may replace solution 1 and solution 2.
- Solution 3 (redox indicator, g l^{-1})
    - Resazurine, 0.1 g, to give a final concentration of 0.0001
- Solutions 4.1 to 4.2 (reducing agents, g l^{-1})
    - Solution 4.1 (for growing fermentative and methanogenic bacteria): 50 g cysteine HCl; 30 g $Na_2S \cdot 9H_2O$; final concentrations of 0.5 and 0.3, respectively.
    - Solution 4.2 (for growing sulfate-reducing bacteria): 60 g $Na_2S \cdot 9H_2O$, final concentration of 0.6.

- Solution 5 (carbonate buffer, g l^{-1})
  80 g NaHCO$_3$, to give a final concentration of 4
- Solutions 6.1 to 6.3 (growth factors and nutrients)
  Solution 6.1 (for growing fermentative bacteria, g l^{-1}): 100 g trypticase; 100 g yeast extract; 500 g glucose; to give final concentrations of 1.0, 1.0, and 5.0, respectively
  Solution 6.2 (for growing sulfate-reducing bacteria, g l^{-1}): 50 g FeSO$_4$·H$_2$O; 100 g trypticase; 100 g yeast extract; 100 g sodium-formate; 200 g sodium-acetate; 300 g sodium-lactate (70%); to give final concentrations of 0.5, 1.0, 1.0, 1.0, 2.0, and 2.1, respectively
  Solution 6.3 (for growing methanogenic bacteria, g l^{-1}): 50 g trypticase; 50 g yeast extract; 250 g sodium-acetate; 300 g sodium-formate; 200 g methanol; 600 g trimethylamine (30%); to give final concentrations of 0.5, 0.5, 2.5, 3.0, 2.0, and 1.8, respectively

Microaerophilic Bacteria

- MBD (mineral basis and diluent): for culturing freshwater or seawater natural populations, use the water of the sampling site. Dilute seawater with distilled water (60:40), to prevent precipitation during autoclaving.
- AMBD (artificial mineral basis and diluent, g l^{-1}): 0.17 g MgCl$_2$ · 6H$_2$O; 0.02 g CaCl$_2$ · 2H$_2$O; 0.002 g Na$_2$MoO$_4$ · H$_2$O; 0.002 g MnSO$_4$ · H$_2$O; 0.002 g H$_3$BO$_3$; for marine bacteria, 20 g sea salts; 1000 ml distilled water; pH 7.5. Substrate: calcium succinate (0.5 g l^{-1}) or another nonfermentable compound.

PROCEDURES

Nonselective Cultivation

Collection and transportation of samples is vital to the success of subsequent examination. Identical sampling procedures can be used for studying both microaerophilic and anaerobic microflora. It is the general rule to prevent any contact of samples with air. A large sample should be taken in order that there is little or no disturbance of the microenvironments in the sample. Water samples may be collected with alcohol-sterilized Niskin bottles, peristaltic pump, or *in situ* pump. Subsamples are transferred into oxygen-free sterilized bottles. Transfer is conducted avoiding any contact with air. Sediment samples may be taken with gravity corer or hand-pushed Plexiglass tubes. The cores are closed, kept in an upright position at a temperature slightly lower than the *in situ* temperature, brought back to the laboratory, and processed as quickly as possible.

Semisolid Culture

Semisolid culture media are used to investigate the growth of oxygen-sensitive bacteria.[3-6] Semisolid media contain a low concentration of agar (0.1 to 0.5%), sufficient to provide a jelly-like consistency and to prevent mixing of the oxygen-rich upper layer with the underlying layers. These media, which are distributed into long and thin glass tubes, provide an oxygen gradient from the agar surface toward the depth. Aerobic bacteria grow at the agar surface, and in the well-oxygenated upmost millimeters; at the opposite, anaerobic bacteria grow in the oxygen-free lower levels. Microaerophilic bacteria are expected to appear in the form of sharp, well-defined growing rings, 5 to 20 mm below the surface, where oxygen concentration (110 to 116 μM) is more convenient for them.[7,8]

Advantages of the method are that no specific equipment is needed, the method can be performed in field conditions, it is very easy to do and very cheap, it provides a unique way to provide a complete oxygen gradient, it offers the best conditions for microaerophiles, and deep layers are sufficiently anoxic to allow the development of facultative anaerobes.

Disadvantages of the method are that this method cannot be used for cultivation of some obligate anaerobes, as the deep layers are not anoxic enough to allow their proliferation, and samples are inoculated at the temperature of agar melting (40 to 45°C) that may be lethal for some bacteria.

1. Use natural MBD or, if unavailable or inappropriate, use AMBD. Supplement with substrate and add 2.0 g l^{-1} bacto-agar as a gelling agent.

2. Dissolve the components by boiling and distribute 9 ml of this medium into 9-ml cotton-plugged glass tubes (10 × 200 mm).

3. To sterilize, autoclave at 120°C for 15 min.

4. Inoculation:

    a. Surface inoculation: stab inoculum into semisolid medium by using platinum loop, and stop the tube with sterile screw cap.

    b. Inoculation of liquid suspension in melted agar: just before utilization, air is removed from the semisolid media by placing the tubes for about 5 min in a boiler. To prevent solidification the tubes are kept in a water bath at 45°C until inoculation. Inoculate with sterile syringe and mix with the melted agar. Immediately after inoculation, place the tube on ice for a few seconds to prevent excessive heating of cells and accelerate agar solidification. Carefully, replace the cotton plug with sterile screw cap.

5. Incubate at appropriate temperature. Caution: temperature must be lower than the agar melting point (45°C).

6. Retrieval or transfer of bacteria is performed by picking colony or growing ring with a sterile plastic syringe equipped with a spinal type needle.

Anaerobic Jar Method

Anaerobic jars are cylindrical vessels, made of metal, glass, or plastic, that can be closed air tight. Media are prepared by standard aerobic methods and used with or without added reducing agents. Anaerobic conditions develop within the jar through generation of gases from chemicals placed in the jar: hydrogen and carbon dioxide are generated by the use of a sachet, containing sodium borohydride, sodium bicarbonate, and citric acid (GasPak, hydrogen + carbon dioxide generator envelopes), to which water is added. Anaerobiosis inside the jars can be controlled by an indicator dye, usually based on methylene blue.

Microaerobic (or even anaerobic) conditions can be obtained through replacement of gases, in the case of vented jars: the air is quickly removed by pumping and the jar is refilled with an appropriate mixture of oxygen-limited (or oxygen-free) gases.

Advantages of the method are that anaerobic jars require relatively little space and are quite simple to use (incubation is anaerobic, but all others operations of subculturing are carried out in air, using normal easy-to-do bacteriological techniques in Petri dishes); anaerobic jars can be used away from the laboratory since no gas supply is required; anaerobic jars are useful for oxygen-tolerant anaerobes, such as endospore-forming clostridia and microaerophilic bacteria.

Disadvantages of the method are that anaerobic jars are inadequate for all anaerobes with high sensitivity to oxygen, because of the exposure of colonies to atmospheric oxygen during handling process; anoxic conditions provided are not good enough for the cultivation of extreme anaerobes.

Hungate Technique

The method, which was devised by Hungate[9,10] for isolation of rumen bacteria, has undergone numerous modifications and improvements[11-13] since the original description. Fundamental to this technique is a supply of oxygen-free gas and the production of pre-reduced, anaerobically sterilized (PRAS) media. Exclusion of air is achieved by simply using rubber-stoppered tubes or flasks for the preparation and tubing media. During transfer of media or bacteria, atmospheric oxygen is kept out of the culture tubes by immediately introducing a steam of oxygen-free gas (CO_2, N_2, H_2) whenever the rubber stopper is removed; CO_2 is the gas of choice, because it is heavier than air.

Solid media are required to isolate and purify bacterial species. In this method, agar plates are replaced by roll tubes: the internal surface of test tube is covered with a thin layer of agar medium, and the interior space of the tube is charged with an anaerobic atmosphere. Bacteria are imbedded in the agar, or are streaked along the surface, making the roll tube analogous to a Petri plate.

Advantages of the method are that these procedures have proved most successful for cultivation of fastidious anaerobes; cultures do not require special incubators; tubes of PRAS can be handled individually, and examined with no anaerobic precaution if kept stoppered (i.e., streak roll tubes can be observed, by using magnification, to count or distinguish colony types); if necessary, aliquots of headspace, and culture medium, can be removed by means of syringe, the needle passing through the rubber stopper; when tubes of PRAS have to be opened to recover colonies, anaerobiosis can be continuously maintained during necessary manipulations with gassing cannula.

A disadvantage of the method is that, for the culture of organisms like the methanogens or other hydrogen-utilizing anaerobes that produce a strong negative pressure within the culture tube, contamination by microbes or oxygen may occur. However, this contamination can be eliminated by using pressurized atmosphere, as described by Balch and Wolfe.[13] When compared to conventional bacteriological procedures, the Hungate method is extremely complex: to manipulate the rubber stopper in an aseptic manner, as well as in a manner to prevent the entrance of oxygen, during inoculation and transfer, requires considerable skill.

1. Bring to boil 919 ml solution 1 + 10 ml solution 2 + 1 ml solution 3, in a round-bottom flask, under a stream of N_2-CO_2 (80:20 mixture).

2. Add 10 ml solution 4 + 50 ml solution 5 + 10 ml solution 6; adjust the pH to 7.5.

3a. For liquid media: distribute 9 ml into screw cap tubes with a rubber seal (Bellco, 16 × 125 mm) outgassed with N_2-CO_2 (80:20) or H_2-CO_2 (80:20). Seal the tubes. Autoclave at 120°C for 15 min.

3b. Alternatively, for roll-tube media: add 2% agar before boiling. Dispense the medium in aliquots of 5 ml into test tubes (Bellco, 25 × 142 mm) outgassed with N_2-CO_2 (80:20) or H_2-CO_2 (80:20). Seal the tubes with butyl rubber stoppers. Place in clamps. Autoclave at 120°C for 15 min.

The entire method of operation and media preparation is outlined in the *Anaerobe Laboratory Manual*.[14]

Anaerobic Glove Box Technique

Anaerobic glove boxes or anaerobic chambers are large work areas which are kept under controlled atmosphere. Usually anaerobic conditions are obtained by an input gas mixture of oxygen-free N_2/CO_2/H_2 (80:10:10) and the presence of a palladinized alumina catalyst.

Other controlled atmospheres can be obtained chainging the gas mixture. The cabinet operates at a small positive pressure, and all manipulations of culturing and subculturing are undertaken with the operator's hands inserted through the glove ports. Tubes, flasks, Petri dishes, and larger equipment can be introduced or removed from the working area via a vacuum-type entry lock.

Advantages of the method are that, in properly designated and operated anaerobic glove boxes, standard Petri plates may be used as easily as in the aerobe laboratory; this method requires less manual dexterity than the Hungate technique, and the method allows for routine genetic procedures, such as replica plating.

Disadvantages of the method are that anaerobic glove boxes are complicated devices, considerably more expensive than the Hungate procedure, and which cannot be used away from the laboratory; the space requirement is greater; time is lost in having to pass through the entry lock all items added to or removed from the chamber; and the cultivation of extremely oxygen-sensitive anaerobes, such as methanogenic bacteria, requires an inner ultra-low oxygen chamber (ULOC).[15]

Selective Cultivation

General Procedures

We describe here general procedures for enumeration of oxygen-sensitive bacteria that utilize PRAS tubes or semisolid media, with an initial inoculum of 1 cm^3 to allow bacterial numbers to be expressed directly by unit volume of sediment or water. The number of bacteria may be determined as CFU (colony forming unit) or as MPN (most probable number).

Serial Dilutions

1. Take 1 ml sample (water with a sterile syringe, and sediment with a sterile syringe with the needle end cut off).

2. Immediately inject into a 16-ml tube (Bellco, 16 × 125 mm) containing 9 ml of sterile dilution solution, remove O_2 by flushing with oxygen-free gas (N_2, CO_2, N_2/CO_2), and seal the tube with butyl rubber stopper and screw cap.

3. Shake using a Vortex mixer or sonicate for 3 min (Bransonic water bath: 48 kz, 100 W), and transfer 1 ml of the mixed contents to another 9 ml sterile dilution solution by means of a sterile syringe. Just before the syringe is used, the air in the dead space should be eliminated by inserting the needle in a stream of oxygen-free gas.

4. In this way, prepare a series of dilutions: 1 ml of this 10^{-1} dilution is transferred into a further 9 ml of diluent and, so, until appropriate dilution has been reached. Dilution series range from 10^{-1} to 10^{-12} depending on the bacterial population density in the original sample.

MPN Enumerations

1. Inoculate 3 sets of tubes (Bellco, 16 × 125 mm) containing 9 ml medium, with 1-ml amounts of the original sample, and with 1 ml of each dilution, using sterile syringes.

2. Incubate the tubes at *in situ* temperature or appropriate temperature for growing bacteria present in the sample.

3. Check for signs of bacterial growth or activity after 1, 2, 3, or 4 weeks.

4. Record the distribution of positive tubes at each dilution level, and use the MPN table for three tubes given in *Standard Methods*[16] to compute the number of bacteria present in the original sample.

CFU Enumerations

1. Inoculate two sets of roll tubes (Bellco, 25 × 142 mm), containing 5 ml of melted agar medium, with 0.5-ml amounts of the original sample and with 0.5-ml volumes from appropriate dilutions; roll tubes are then spun on a roll machine, in an ice bath, until the agar is set.

2. Incubate the tubes vertically at *in situ* temperature or appropriate temperature for growing bacteria present in the sample.

3. For enumeration, count colonies in tubes containing 30 to 300 colonies (with or without magnification), and calculate number of bacteria per milliliter.

Anaerobic Bacteria

Among the numerous anaerobic bacteria inhabiting anoxic aquatic environments, we will discuss three physiologically different types of bacteria which can play an important ecological role: fermentative, sulfate-reducing, and methanogenic bacteria.

Fermentative Bacteria. Fermentative bacteria are facultative or strict anaerobes which hydrolyze polymers and ferment monomers and oligomers. The cultivation medium for fermentative bacteria requires the following constituents: 919 ml solution 1 ± 10 ml solution 2 + 1 ml solution 3 + 10 ml solution 4.1 + 50 ml solution 5 + 10 ml solution 6.1, under a N_2-CO_2 (80:20) atmosphere. Recipes for a variety of media are given by Macy,[17] Hespell and Bryant,[18] and Gottschalk et al.[19]

After 1 and 2 weeks of incubation at the appropriate temperature, fermentative bacteria may be counted according to the MPN technique (liquid media) or the CFU technique (solid media).

1. MPN: bacterial numbers are determined by visual scoring of growth in the medium.

2. CFU: count the number of colonies developed on the agar film in roll tubes containing 30 to 300 colonies, until the number of colonies shows no further increase.

Sulfate-Reducing Bacteria. Sulfate-reducing bacteria are strict anaerobes which use organic substrates as carbon sources and sulfate as terminal electron acceptor; as a consequence, they reduce sulfate (SO_4^{2-}) to hydrogen sulfide (H_2S). The cultivation medium for "total" sulfate-reducing bacteria requires the following constituents: 919 ml solution 1 + 10 ml solution 2 + 1 ml solution 3 + 10 ml solution 4.2 + 50 ml solution 5 + 10 ml solution 6.2, under a H_2-CO_2 (80:20) atmosphere.

In addition, the physiologically different groups of sulfate-reducing bacteria can be cultured in selective media containing restricted carbon sources: sodium-lactate under N_2-CO_2 (80:20), for *Desulfovibrio, Desulfotomaculum,* and *Desulfomonas*; sodium-acetate under H_2-CO_2 (80:20), for *Desulfovibrio*.

Recipes for a variety of media are given by Pfennig et al.[20] and Postgate.[21] After 1 and 2 weeks of incubation at the appropriate temperature, sulfate-reducing bacteria may be counted according to the MPN technique in liquid media, or the CFU technique in solid media: sulfide produced from sulfate forms a black precipitate of FeS, blackening the liquid medium as a whole, or the zone around a colony in solid media. For the MPN method, tubes in which a black precipitate of FeS developed are scored as positive. With the CFU method, count the number of black colonies.

Methanogenic Bacteria. Methanogenic bacteria are very strict anaerobes which produce methane (CH_4) from carbon dioxide reduction and/or from the fermentation of a restricted range of organic compounds. The cultivation medium for "total" methanogenic bacteria requires the following constituents: 919 ml solution 1 + 10 ml solution 2 + 1 ml solution 3 + 10 ml solution 4.1 + 50 ml solution 5 + 10 ml solution 6.3, under a H_2-CO_2 (80:20) atmosphere.

In addition, the physiologically different groups of methanogens can be cultured in selective media containing restricted carbon sources: formate under H_2-CO_2 (80:20), for hydrogen-utilizing methanogens; acetate under N_2-CO_2 (80:20), for aceticlastic methanogens; methanol and trimethylamine under N_2-CO_2 (80:20), for methylotrophic methanogens. Recipes for a variety of media are given by Mah and Smith.[22]

After 2 and 4 weeks of incubation at appropriate temperature, methanogenic bacteria are counted by the MPN technique, on the basis of the presence of CH_4 in the headspace of the tubes. Each tube is analyzed for CH_4 in the headspace by gas chromatography. CH_4 can be accurately identified and quantified by flame ionization detection (FID), as well as thermal conductivity detection (TCD); the former detection is more sensitive than the latter.

1. FID: A 100-μl volume of the gas phase of each tube is removed with a pressure-lock syringe and injected into a gas chromatograph equipped with a flame ionization detector, and a 2 m $\times$ 3 mm (outer diameter) stainless steel column filled with 80- to 100-mesh Porapak R (or Porapak Q). Nitrogen is used as the carrier gas at 30 ml min^{-1}, and the operating temperatures of the injector, column, and detector are 100, 105, and 220°C, respectively.

2. TCD: A 500-μl volume of the gas phase of each tube is removed with a pressure-lock syringe and injected into a gas chromatograph equipped with a thermal conductivity detector, and a 2 m $\times$ 3 mm (outer diameter) stainless steel column filled with 60- to 80-mesh Carbosieve B. Helium is used as the carrier gas at 30 ml min^{-1}, the filament current is 200 mA, and the operating temperatures of the injector, column, and detector are 130, 80, and 170°C, respectively.

Microaerophilic Bacteria

Most of the known microaerophiles are oligocarbonophilic bacteria. Culture media with limited concentrations in organics favor these bacteria. Oligonitrophily or even nitrogen fixation are also very common in this group; these organisms use an exterior combined nitrogen source or dissolved N_2, so they do not require addition of a nitrogen source. To prevent the development of anaerobes, fermentable substrates (i.e., carbohydrates) and electron acceptors (i.e., NO_3^-, SO_4^{2-} . . .) are avoided.

The cultivation medium for microaerophilic bacteria requires the following constituents: MBD or AMBD solution + substrate (0.5 g l^{-1} of calcium succinate or another nonfermentable compound) + 2.0 g l^{-1} bactoagar as gelling agent.

For serial dilution, use natural tap water or seawater, or AMBD, supplemented with 0.10% (w/v) ascorbic acid as reducing agent. Dispense 9 ml of this solution into screw cap tubes with a rubber seal (Bellco 16 $\times$ 125 mm). Microaerobic conditions are obtained by bubbling a N_2/O_2 (94:6) gas mixture through the rubber seal. Recipes for a variety of media are given by Krieg.[23] After 2 weeks of incubation at appropriate temperature, the culture is considered positive when microaerophilic growing rings appear between 5- and 20-mm depth.[7] The number of bacteria present in the original sample is determined using the MPN table for three tubes given in *Standard Methods.*[16]

REFERENCES

1. Sieburth, J. McN., Contrary habitats for redox-specific processes: methanogenesis in oxic waters and oxidation in anoxic waters, in *Microbes in the Sea,* Sleigh, M. A., Ed., Ellis Horwood, Chichester, England, 1987, 11.
2. Aranki, A. and Freter, R., Use of anaerobic glove boxes for the cultivation of strictly anaerobic bacteria, *Am. J. Clin. Nutr.,* 25, 1329, 1972.
3. De Vries, W. and Stouthamer, A. H., Factors determining the degree of anaerobiosis of *Bifidobacterium* strains, *Arch. Mikrobiol.,* 65, 275, 1969.
4. Wimpenny, J. W. T., Coombs, J. P., Lovitt, R. W., and Whittaker, S. G., A gel-stabilized model ecosystem for investigating microbial growth in spatially ordered solute gradients, *J. Gen. Microbiol.,* 127, 277, 1981.
5. Wimpenny, J. W. T., Lovitt, R. W., and Coombs, J. P., Laboratory model systems for the investigation of spatially and temporally organized microbial ecosystems, in *Microbes in their Natural Environments,* Slater, J. H., Wittenbury, R., and Wimpenny, J. W. T., Eds., Cambridge University Press, Cambridge, 1983, 67.
6. Kikuchi, H. E. and Suzuki, T., Quantitative method for measurement of aerotolerance of bacteria and its application to oral indigenous anaerobes, *Appl. Environ. Microbiol.,* 52, 971, 1986.
7. Ferrara-Guerrero, M. J. and Bianchi, A., Comparison of culture methods for enumeration of microaerophilic bacteria in marine sediments, *Res. Microbiol.,* 140, 255, 1989.
8. Ferrara-Guerrero, M. J. and Bianchi, A., Distribution of microaerophilic bacteria through the oxic-anoxic transition zone of lagoon sediments, *Hydrobiologia,* 207, 147, 1990.
9. Hungate, R. E., The anaerobic mesophilic cellulolytic bacteria, *Bacteriol. Rev.,* 14, 1, 1950.
10. Hungate, R. E., A roll tube method for cultivation of strict anaerobes, in *Methods in Microbiology,* Vol. 3B, Norris, J. R. and Ribbon, D. W., Eds., Academic Press, London, 1969, 117.
11. Bryant, M. P., Commentary on the Hungate technique for culture of anaerobic bacteria, *Am. J. Clin. Nutr.,* 25, 1324, 1972.
12. Macy, J. M., Snellen, J. E., and Hungate, R. E., Use of syringe methods for anaerobiosis, *Am. J. Clin. Nutr.,* 25, 1318, 1972.
13. Balch, W. E. and Wolfe, R. S., New approach to the cultivation of methanogenic bacteria: 2-mercaptoethanesulfonic acid (HS-CoM)-dependent growth of *Methanobacterium ruminantium* in a pressurized atmosphere, *Appl. Environ. Microbiol.,* 32, 781, 1976.
14. Holdeman, L. V., Cato, E. P., and Moore, W. E. C., Eds., *Anaerobe Laboratory Manual,* 4th ed., Virginia Polytechnic Institute and State University, Blacksburg, VA, 1977.
15. Edwards, T. and McBride, B. C., New method for the isolation and identification of methanogenic bacteria, *Appl. Microbiol.,* 29, 540, 1975.
16. American Public Health Association, *Standard Methods for the Examination of Water and Wastewater,* 15th ed., American Public Health Association, New York, 1980.
17. Macy, J. M., Nonpathogenic members of the genus *Bacteroides,* in *The Prokaryotes, a Handbook on Habitats, Isolation and Identification of Bacteria,* Vol. 2, Starr, M. P., Stolp, H., Trüper, H. G., Balows, A., and Schlegel, H. G., Eds., Springer-Verlag, Berlin, 1981, chap. 117.
18. Hespell, R. B. and Bryant, M. P., The genera *Butyvibrio, Succinivibrio, Succinimonas, Lachnospira,* and *Selenomonas,* in *The Prokaryotes, a Handbook on Habitats, Isolation and Identification of Bacteria,* Vol. 2, Starr, M. P., Stolp, H., Trüper, H. G., Balows, A., and Schlegel, H. G., Eds., Springer-Verlag, Berlin, 1981, chap. 120.
19. Gottschalk, G., Andreesen, J. R., and Hippe, H., The genus *Clostridium* (nonmedical aspects), in *The Prokaryotes, a Handbook on Habitats, Isolation and Identification of Bacteria,* Vol. 2, Starr, M. P., Stolp, H., Trüper, H. G., Balows, A., and Schlegel, H. G., Eds., Springer-Verlag, Berlin, 1981, chap. 138.

20. Pfennig, N., Widdel, F., and Trüper, H. G., The dissimilatory sulfate-reducing bacteria, in *The Prokaryotes, a Handbook on Habitats, Isolation and Identification of Bacteria*, Vol. 1, Starr, M. P., Stolp, H., Trüper, H. G., Balows, A., and Schlegel, H. G., Eds., Springer-Verlag, Berlin, 1981, chap. 74.

21. Postgate, J. R., *The Sulphate-Reducing Bacteria*, 2nd ed., Cambridge University Press, Cambridge, 1984.

22. Mah, R. A. and Smith, M. R., The methanogenic bacteria, in *The Prokaryotes, a Handbook on Habitats, Isolation and Identification of Bacteria*, Vol. 1, Starr, M. P., Stolp, H., Trüper, H. G., Balows, A., and Schlegel, H. G., Eds., Springer-Verlag, Berlin, 1981, chap. 76.

23. Krieg, N. R., The genera *Spirillum, Aquaspirillum,* and *Oceanospirillum*, in *The Prokaryotes, a Handbook on Habitats, Isolation and Identification of Bacteria*, Vol. 1, Starr, M. P., Stolp, H., Trüper, H. G., Balows, A., and Schlegel, H. G., Eds., Springer-Verlag, Berlin, 1981, chap. 52.

CHAPTER 2

Isolation and Cultivation of Hyperthermophilic Bacteria from Marine and Freshwater Habitats

John A. Baross

INTRODUCTION

The isolation and cultivation of bacteria capable of growing at temperatures greater than 70°C present some unique problems not encountered with other thermal groups of bacteria. This is particularly true for the hyperthermophiles, which grow at 80°C or higher. Except for a few strains of *Sulfolobus* spp. all hyperthermophiles are obligately anaerobic and most have unusual trace nutrient as well as electron acceptor requirements. Most known heterotrophic and chemolithotrophic species require elemental sulfur as a source of energy and/or for the removal of metabolically produced H_2 which is toxic to some strains. Since many hyperthermophiles have minimum growth temperatures above 70°C, the usual isolation and purification procedures of streaking agar media either in Petri plates or roll tubes is not possible: agar evaporates to dryness at such high temperatures. Consequently, the ability of most of the isolated strains of hyperthermophilic bacteria to form colonies on or in solid media is not known.

Most of the approximately 40 species of known hyperthermophilic bacteria have been described only recently. For the purposes of this chapter these organisms fit into three convenient groups: (1) neutrophiles that grow at neutral to slightly acidic conditions, (2) acidophiles that grow optimally at pH 3 or lower, and (3) (neutrophilic) methanogens.[1] This grouping has little to do with nutritional characteristics or phylogeny (a nutritional delineation of hyperthermophiles has little to do with phylogenetic relatedness). Rather it reflects the acidity of different continental and marine thermal environments, which translates directly into the properties of isolation and culturing procedures and media. Attempts to group hyperthermophiles nutritionally are confounded by the examples of some species of *Thermoproteus* and *Sulfolobus* which are capable of growing both chemolithotrophically and heterotrophically. Some *Sulfolobus* species are facultatively anaerobic, whereas all other genera of hyperthermophilic heterotrophs are obligate anaerobes. Representative species from each nutritional group are capable of growing to 110°C. Ultimately, culture conditions for specific hyperthermophilic organisms must reflect the physical and geochemical characteristics of their habitat. This proves particularly true for pH and salinity.

Analyses of 16S rRNA nucleotide sequences, unusual cell wall chemical composition, and presence of ether membrane lipids from representative species of hyperthermophiles have shown that most isolates cluster together in a group distinct from all other bacteria and eukaryotes. Only one group of hyperthermophiles, the order Thermotogales, are eubacteria.

0-87371-564-0/93/$0.00 + $.50

Initially, these species of this order were designated archaebacteria, since they exhibited characteristics consistent with perceptions of the characteristics of Archaean ecosystems. Recently, however, Woese and associates[2] have modified this phylogenetic scheme to reflect the apparently closer relationship of the archaebacteria to eukaryotes than to eubacteria. Their new proposal separates all living organisms into three domains (Bacteria, Archaea, and Eucarya) so as not to perpetuate the concept that the Archaea, formally the archaebacteria, are closely related to bacteria (formerly eubacteria). Although still undergoing scrutiny, other information, including amino acid sequences of proteins and small subunit ribosomal RNA sequences, adds credibility to this proposal.[3,4] The new terminology of Archaea and Bacteria are used in this chapter.

MATERIALS REQUIRED

Equipment

Water bath, incubator, or heating block: No special equipment is necessary for routine incubation of hyperthermophiles up to temperatures of 110°C and 500 kPa (pressure achieved in sealed Bellco tubes). Tubes or serum bottles can be incubated in commercial water baths filled with oil. Dimethyl-silicone oil (Thomas Scientific Silicone Fluid SF 96/50) and related oils are stable to 300°C; they are recommended over organic oils or solvents for their thermal stability. Bellco tube cultures can also be incubated in heating blocks (Reacti-Therm heating modules) filled with oil.[8] Air incubators can be used for culturing hyperthermophiles at temperatures below 85°C, but the temperature will fluctuate considerably if the cultures must be accessed frequently. Standard water baths filled with sand are useful for isolation of hyperthermophiles on shipboard. It is important to note temperature gradients will establish vertically and horizontally in the sand bath. It is recommended that the temperature be measured at the location and depth in the sand where culture tubes are placed.

Supplies

- Crimp-seal "Balch" tubes, 18 × 150 mm (Bellco Glass, Inc.) or serum bottles
- Black butyl rubber septum-type stoppers, gas impermeable (Bellco Glass, Inc. or Geo-Microbial Technologies, Inc.)
- Aluminum crimp seals
- Crimper and decapper for aluminum seals
- Gases and gas mixtures: argon (ultra-high purity grade), H_2, H_2:CO_2 (80:20), N_2:CO_2 (80:20)
- Gassing manifold (Balch and Wolfe;[5] also see Noll[6] for details on construction of a gassing manifold)

Solutions

The following solutions are made with Milli-Q deionized water (Millipore Corp.) or equivalent.

- Synthetic sea water (per liter) (modified from "SME" medium)[7,8]
    NaCl, 19.6 g
    Na_2SO_4, 3.3 g
    KCl, 0.5 g
    KBr, 0.05 g
    H_3BO_3, 0.02 g
    $MgCl_2 \cdot 6H_2O$, 8.8 g

- *Sulfolobus* medium (per liter)[9]
  $(NH_4)_2SO_4$, 1.3 g
  KH_2PO_4, 0.38 g
  $MgSO_4 \cdot 7H_2O$, 0.25 g
  $CaCl_2 \cdot 2H_2O$, 0.07 g
  $FeCl_3 \cdot 6H_2O$, 0.02 g
  $MnCl_2 \cdot 4H_2O$, 1.8 mg
  $Na_2B_4O_7 \cdot 10H_2O$, 4.5 mg
  $ZnSO_4 \cdot 7H_2O$, 0.22 mg
  $CuCl_2 \cdot 2H_2O$, 0.05 mg
  $Na_2MoO_4 \cdot 2H_2O$, 0.03 mg
  $VOSO_4 \cdot 2H_2O$, 0.03 mg
  $CoSO_4$, 0.01 mg
  Yeast extract (Difco), 1.0 g
  Adjust pH to 2.0 with 10 N H_2SO_4 at room temperature
- Trace element solutions (per liter)
  Solution A[8]
    $CuSO_4 \cdot 5H_2O$, 0.01 g
    $ZnSO_4 \cdot 7H_2O$, 0.1 g
    $CoCl_2 \cdot 6H_2O$, 0.005 g
    $MnCl_2 \cdot 4H_2O$, 0.2 g
    $Na_2MoO_4 \cdot 2H_2O$, 0.1 g
    KBr, 0.05 g
    KI, 0.05 g
    H_3BO_3, 0.1 g
    LiCl, 0.05 g
    $Al_2(SO_4)_3$, 0.05 g
    $NiCl_2 \cdot 6H_2O$, 0.01 g
  Solution B[8]
    $VOSO_4 \cdot 2H_2O$, 0.05 g
    H_2WO_4, 0.05 g
    Na_2SeO_4, 0.05 g
    $NiCl_2 \cdot 6H_2O$, 0.05 g
    $SrCl \cdot 6H_2O$, 0.05 g
    $BaCl_2$, 0.05 g
  Vitamin solution[10] (per liter) (store in dark at 2 to 4°C)
    p-Aminobenzoic acid, 5 mg
    Biotin, 2 mg
    DL-Calcium pantothenate, 5 mg
    Cyanocobalamine, 0.1 mg
    Folic acid, 2 mg
    Nicotinic acid, 5 mg
    Pyridoxine-HCl, 10 mg
    Riboflavin, 5 mg
    Thiamine-HCl, 5 mg
    Lipoic acid, 5 mg
  Methanothermus mineral solution 1[10] (per liter)
    K_2HP_4O, 6 g
  Methanothermus mineral solution 2[10] (per liter)
    KH_2PO_4, 6 g
    $(NH_4)_2SO_4$, 6 g
    NaCl, 12 g
    $MgSO_4 \cdot 7H_2O$, 2.6 g
    $CaCl_2 \cdot 2H_2O$, 0.16 g

- Other reagents
    $CaCl_2$, 1% solution
    $NaHCO_3$, 10% solution
    FeEDTA, consisting of $FeSO_4 \cdot 7H_2O$, 1.54 g l^{-1} and Na_2EDTA, 2.06 g l^{-1}
    $(NH_4)_2SO_4$, 43.0 g l^{-1} and KH_2PO_4, 3.6 g
    $Na_2S \cdot 9H_2O$, 3% solution
    Resazuran, 0.1% solution

PROCEDURES

Sampling Methods

Sampling methods vary according to the specific environment of interest. Papers that describe the isolation of hyperthermophiles from terrestrial or shallow marine sulfotaric environments rarely include a detailed description of the sampling protocol or container used for collection. When described, terrestrial hot springs have been sampled by collecting water or sediment into sterile jars or sterile plastic syringes. Plastic syringes with the needle end cut off can be used as coring devices for hot sediments. Shallow marine solfotara sites have been sampled either by scuba divers or surface boat hydrocasts using different coring devices and grab samplers.

Thermal sites associated with submarine hydrothermal vents include hot water, solids from sulfidic smokers and flanges, sediments, and animals; all must be sampled by submersible. The 755-ml titanium syringe described by Von Damm et al.[11] is used frequently to obtain hot water samples from smokers. Under ideal conditions it can collect pure hydrothermal fluids. In general, however, there is some degree of mixing with ambient seawater so that confirming the source of thermophilic bacteria subsequently isolated or detected is rarely a simple matter. Recently, Massoth et al.[12] developed an improved manifold syringe system called the "submersible-coupled *in situ* sensing and sampling system" (SIS^3). This sampling manifold accommodates three titanium syringes and offers the advantages of being able to flush the syringes with hot water before sampling and to monitor the water temperature both at the source of the sample and at the intake of the sampling syringe. We have had considerable success with the SIS^3 in obtaining hydrothermal fluids minimally compromised by seawater.[13] Smoker sulfides and flange solids are usually sampled by breaking apart larger structures with the submersible arm. Solid samples as well as animal specimens should be placed in an insulated "live box" on the submersible basket so as to lessen the chance of losing the sample upon surfacing and help prevent the sample from reaching warm surface temperatures.

Since most of the known hyperthermophiles are anaerobic, it is advisable to inoculate samples directly into anaerobic Balch tubes capped with metal-crimped rubber stoppers (see media section) or to store samples in anaerobic jars until they can be inoculated into suitable media. Anaerobic jars with Gas-Pac's (BBL) are convenient for creating anaerobic conditions in the field. In general, however, most strains of anaerobic hyperthermophiles are stable for at least a short time in the presence of oxygen as long as they are held at temperatures below their minimum for growth. It is recommended that an anaerobic hood be used for initial enrichments for hyperthermophiles from soild samples such as sediments, sulfides, and vent animals, and when transferring cultures inoculated in the field. Inexpensive, disposable plastic glove bags (I^2R Inc., Cheltenham, PA) are suitable for shipboard use. Use either nitrogen or argon gases with these glove bags.

Caution should be taken to prevent environmental samples or cultures from reaching near-freezing temperatures since some strains of hyperthermophiles are lysed at near-freezing temperatures.[14] However, several genera of hyperthermophiles were recently isolated from the plume of a submarine volcanic eruption at MacDonald Seamount after exposure to cold seawater for 1 to 4 d.[15]

In extremely acidic environments (pH <3) it is probable that only species of *Sulfolobus* and *Acidianus* will be isolated. Special care must be taken to adjust the pH of the sample to 5.5 to 6.0 until the sample is prepared for culturing. *Sulfolobus* spp. can grow over a pH range of 1 to 4; while growing they maintain a higher internal pH. The internal pH will quickly reach surrounding values if the cells are not growing, which will result in cell lysis.

So far, no strains of obligately barophilic or halophilic species of hyperthermophiles have been isolated.

Isolation and Maintenance of Hyperthermophiles

The following materials and solutions can be used in the formulation of isolation and growth media for most known species of hyperthermophiles. See sections on neutrophiles, acidophiles, and methanogens for specific formulation and procedures for making media. It should be kept in mind that it is likely that there are other species of hyperthermophiles not yet isolated that may have unusual nutrient, electron acceptor, trace metal, and organic compound requirements. It should be expected that some modifications of media formulations will need to be made to reflect the physical and geochemical properties of specific environments.

Neutrophiles (Nonmethanogenic)

Liquid Media. Most of the neutrophilic, anaerobic, heterotrophic hyperthermophiles can grow on media containing various hydrolyzed protein preparations, such as peptones and trypticase soy, and yeast extract supplemented with trace minerals and elemental sulfur. In general, marine isolates have a seawater requirement. The most commonly used medium formulation for isolation of both sulfur- and nonsulfur-dependent marine heterotrophs consists of synthetic sea water, a source of complex organic carbon, and sulfur.

1. The following is a useful medium formulation for initial isolation of a variety of hyperthermophiles from diverse marine environments (per liter of synthetic seawater):

   10.0 ml trace elements A
   1 ml trace elements B
   10 ml of the $(NH_4)_2SO_4$ and KH_2PO_4 solution
   2 ml FeEDTA solution
   1 g yeast extract (Difco)
   1 g Bacto peptone (Difco)
   1 ml resazurin solution (0.001 g final concentration).

2. The pH is adjusted to 6.5.

3. The medium is filter sterilized and transferred to tubes or serum bottles. When appropriate 5 g l^{-1} sterile elemental sulfur (S°) is added to each tube or bottle (S° is sterilized separately by steaming at 100°C for 1 h on three successive days).

4. The tubes or serum bottles are capped and air is removed and replaced by argon or a blend of $H_2:CO_2$ if appropriate by alternately pulling a vacuum and saturating with gas using a degassing manifold.

5. After degassing any traces of oxygen are removed by addition of sterile 3% Na_2S solution to a final concentration of 0.03%. A head space pressure of approximately 100 kPa is applied to all tubes to prevent air intrusion.

This medium has been used successfully to grow marine *Pyrococcus* spp., *Thermococcus* spp., *Staphylothermus* spp., and related heterotrophic cocci such as ES-1 and ES-4[8,16] and *Pyrodictium brockii*.

Many strains of these organisms also have been shown to grow in Bacto Marine Broth 2216 (Difco). This medium contains synthetic sea salts, peptone, and beef extract. There are also several media formulations for culturing *Pyrococcus furiosus* and related organisms since these species are the focus of considerable research attention on purification and characterization of specific enzymes and physiological studies that require high-volume culturing facilities.[17,18] *Desulfurococcus* spp. isolated from submarine hydrothermal vents were enriched initially on a similar medium but with 0.2% yeast extract as the sole carbon source.[19] Similarly, *Hyperthermus butylicus* was initially cultured with 0.6% tryptone and subsequently found to require peptides for growth.[20] Marine species of *Thermotoga* (Bacteria domain) have been cultured with 0.1% yeast extract and 0.2% glucose. Species of *Archaeoglobus* recently isolated from submarine hydrothermal vent environments reduce SO_4^{2-}, $S_2O_3^{2-}$, or SO_3^{2-} to H_2S and can utilize a variety of carbon sources including proteins, sugars, lactate, formate, and pyruvate.[21-23] One species, *A. fulgidus*, can grow chemolithotrophically only with $S_2O_3^{2-}$ as the electron acceptor and H_2 as the energy source.

Non-salt requiring heterotrophic species of both Archaea and Bacteria can be cultured in the *Sulfolobus* medium adjusted to pH 5.5 to 7.0 and supplemented with appropriate carbon sources and electron acceptors. The $S°$-requiring *Thermoproteus* species will grow heterotrophically on a variety of carbon sources, including glucose, starch, and amino acids, or chemolithotrophically by using $S°$ and H_2 as energy sources and CO_2 as the carbon source.[24] *Thermoproteus* species are slightly acidophilic; the pH of the medium should be adjusted to between 5 and 6.5 depending on the species.

Most of the heterotrophic Archaea have minimum growth temperatures between 60 and 70°C and maximum growth temperatures to 110°C. Choosing incubation temperatures will depend on the species. However, for the initial isolation of hyperthermophilic Archaea, 85 to 90°C incubation is recommended in order to lessen the chance of mixed cultures of organisms that grow in the 80°C range. This is particularly true with submarine hydrothermal vent samples in which we have frequently encountered organisms resembling *Clostridium* spp. (Bacteria) with maximum growth temperatures in the low 80°C range. Initial isolation temperatures for *Thermotoga* spp. should be at 80 to 85°C.

Solid Media. There are few published procedures for culturing heterotrophic hyperthermophiles on solid medium (see review in Wiegel).[14] In general, most solidifying agents are too unstable at temperatures much above 70°C for extended periods. However, there are reports of some species of hyperthermophilic heterotrophs, such as *Pyrococcus furiosus*, forming colonies on agar media at 70°C.[25] In my experience, none of the heterotrophic hyperthermophiles isolated from submarine hydrothermal vent environments were found to form colonies on agar plates. This may be because some species cannot form colonies on surfaces or because 70°C is too close to their minimum growth temperature. Agar media evaporate extremely rapidly at temperatures above 70°C.

Some success has been achieved using the gellan gum, Gelrite (Kelco Division of Merck and Co., Inc.), in shake tubes with environmental samples incubated at 90 to 120°C[26] and as a plating medium at 99°C for the isolation of pure cultures of *Hyperthermus butylicus*.[20] The Gelrite gum, which is prepared from a *Pseudomonas* sp., solidifies rapidly when divalent cations such as calcium and magnesium are added. When preparing the medium described above, it is important to omit the calcium salts from the mineral medium to be autoclaved.

1. After autoclaving, the melted Gelrite solution (8 g gellan gum l^{-1}) should be transferred to an oil bath at approximately 100°C. Additional sterile ingredients preheated to 100°C can be added to the Gelrite. The calcium salts and trace elements should be added last just prior to distribution of medium into shake tubes containing environmental samples[26] or into Petri plates.[14,24]

2. The Gelrite plating medium described by Zillig et al.[24] contained colloidal sulfur for culturing sulfur-dependent hyperthermophiles. The Gelrite was dissolved in boiling water and solidified by addition of $CaSO_4$ (1 g l^{-1}), poured into Petri plates, allowed to solidify, and then soaked with 5 ml of a solution containing saturated levels of S° in 1 M $(NH_4)_2S$ for 2 min. After rinsing with water, the colloidal sulfur was precipitated by addition of 5 to 10 ml of 1 M H_2SO_4 for 2 min and then thoroughly washed with water. These authors soaked the plates with liquid culture medium overnight, dried them at 37°C and stored them in an anaerobic chamber. Dilutions of liquid cultures were plated directly onto these plates and incubated at 99°C.

Acidophiles

The thermoacidophilic Archaea include two genera in the family Sulfolobaceae, *Sulfolobus* and *Acidianus*. Known species include strict and facultative aerobes and obligate and facultative chemolithotrophs. All grow optimally at pH of 2 to 3. *Sulfolobus* spp. are enriched in *Sulfolobus* medium and incubated aerobically at 70 to 80°C. *Sulfolobus* spp. have also been isolated in the field by adding 0.1% yeast extract (using a 1% solution) to thermal spring water and acidifying to pH 2. Both *Sulfolobus* and the related genus *Acidianus* will grow chemolithotrophically with S° and O_2. *A. infernus* is an obligate chemolithotroph capable of growing aerobically or anaerobically. Anaerobic growth involves the reduction of S° by H_2 using a H_2/CO_2 gas mixture.

There are reports of *Sulfolobus* and *Acidianus* forming colonies aerobically and anaerobically on medium solidified by 10% starch. Other solidifying agents would be too unstable at the required acidic pH.

Methanogens

Most of the species of hyperthermophilic methanogens thus far described belong to three genera. All of the marine species belong to the genera *Methanococcus* and *Methanopyrus*, whereas *Methanothermus* spp. have been isolated only from terrestrial solfatara environments. There are other thermophilic methanogens such as *Methanobacterium thermoautotrophicum* that grow at temperatures above 55°C but generally not above 70°C. All of the hyperthermophilic methanogens can grow with H_2 and CO_2, but the growth of many species is stimulated by addition of complex organic material such as yeast extract, trypticase, and peptone or requires vitamins. Some species of *Methanococcus* will grow on formate.

Marine Methanogens

1. A useful medium for initial isolation of marine hyperthermophilic methanogens includes (per liter synthetic seawater):

   10 ml trace elements A
   1 ml trace elements B
   10 ml of the $(NH_4)_2SO_4$ and KH_2PO_4 solution
   10 ml vitamin solution
   10 ml $NaHCO_3$ solution
   1 ml resazurin solution (0.001 g final concentration).

2. The pH is adjusted to 7.0.

3. The medium is filter sterilized and transferred to tubes or serum bottles and capped.

4. Air is removed as described for neutrophiles and replaced by $H_2:CO_2$ leaving a pressure in the tubes or bottles of 200 KPa. After degassing any traces of oxygen are removed by addition of sterile 3% Na_2S solution to a final concentration of 0.03%.

5. Other reducing agents that have been used with methanogens and other anaerobes include cysteine sulfide, titanium citrate, dithiothreitol, and thioglycollate; they are generally employed at a final concentration of 0.025 to 0.05%. In general, it is recommended that various carbon sources be added to the medium for initial isolation of methanogens. Usually, 2 g of both yeast extract and trypticase are added per liter to the medium. Some species of hyperthermophilic *Methanococcus* will grow in a medium containing 2.5 to 5 g l^{-1} formate as a carbon source. Selenium stimulates growth of most species of *Methanococcus*.

Nonmarine Methanogens

1. The recommended medium for isolation of hyperthermophilic nonmarine species of *Methanothermus* is described by Balch et al.[10] and includes (per 905 ml of distilled water):
    37.5 ml each of *Methanothermus* mineral solutions 1 and 2

    10 ml trace elements A
    1 ml trace elements B
    10 ml vitamin solution
    2 g yeast extract (Difco)
    2 g trypticase (BBL)
    5 g sodium acetate
    1.6 mg $NiCl_2 \cdot 6H_2O$
    2 mg $FeSO_4 \cdot 7H_2O$
    1.6 µg coenzyme M
    6 g $NaHCO_3$
    0.25 g cysteine hydrochloride
    0.25 g $Na_2S \cdot 9H_2O$
    1 ml resazurine solution (0.001 g l^{-1}).

2. The pH is adjusted to 6.7 with acetic acid.

3. Medium preparation and sterilization is as described for marine methanogens. For additional media formulations and procedures for isolation and maintenance of methanogens see Balch et al.[10]

Solid Media. There are few published reports of hyperthermophilic methanogens forming colonies on solid media. Stetter[7] indicates that *M. fervidus* can be cultured at 85°C on plated medium solidified by polysilicate and incubated with H_2 and CO_2 in a pressure cylinder.[5] *M. fervidus* will not grow on agar.[7]

NOTES AND COMMENTS

1. Some species of hyperthermophiles grow optimally with doubling times of less than 1 h and reach densities of greater than 10^9 cell ml^{-1}. Other species grow extremely slowly (>10 h doubling times) and will only reach densities of 1 to 5 × 10^7 cells ml^{-1}. This is particularly true for some species of *Thermoproteus* cultured either heterotrophically or chemolithotrophically. It is advisable to use microscopic methods to assess growth in enrichment cultures in instances where there is no evidence of visible turbidity.

2. Many species of hyperthermophilic heterotrophs will lyse relatively rapidly when left at high growth temperatures beyond the early stationary growth phase. Most isolates will remain stable for weeks at room or refrigerated temperatures under anaerobic conditions.

3. Adams[17] has reported that *Pyrococcus furiosus* and other species of heterotrophic hyperthermophiles have been shown to have unique tungsten enzymes and therefore to require a higher concentration of tungsten for optimal growth than is normally found in most trace element mixtures. Adams[17] recommends adding 10 μM Na_2WO_4 l^{-1} of medium. Other species of hyperthermophiles have different metal and electron acceptor requirements; it is important to consult reports on individual species for specific nutrient requirements.

4. Some progress has been made toward the growth of hyperthermophiles in continuous culture to obtain high cell biomass for biochemical and molecular biology studies. Brown and Kelly[18] describe an inexpensive and easily constructed reactor for the continuous culture of hyperthermophilic heterotrophs at temperatures to 100°C. All of the components for this reactor can be purchased "off the shelf". See Kelly and Deming[27] for description of incubation chambers for culturing hyperthermophiles at high temperature and high pressure.

REFERENCES

1. Stetter, K. O., Fiala, G., Huber, G., and Segerer, A., Hyperthermophilic microorganisms, *FEMS Microbiol. Rev.*, 75, 117, 1990.
2. Woese, C. R., Kandler, O., and Wheelis, M. L., Towards a natural system of organisms: proposal for the domains Archaea, Bacteria, and Eucarya, *Proc. Natl. Acad. Sci. U.S.A.*, 87, 4576, 1990.
3. Auer, J., Spicker, G., and Bock, A., Phylogenetic positioning of archaebacteria on the basis of ribosomal protein sequences, *Syst. Appl. Microbiol.*, 13, 354, 1990.
4. Winker, S. and Woese, C. R., A definition of the domains Archaea, Bacteria and Eucarya in terms of small subunit ribosomal RNA characteristics, *Syst. Appl. Microbiol.*, 14, 305, 1991.
5. Balch, W. E. and Wolfe, R. S., New approach to the cultivation of methanogenic bacteria; 2-mercaptoethane-sulfonic acid (HS-CoM)-dependent growth of *Methanobacterium ruminantium* in a pressurized atmosphere, *Appl. Environ. Microbiol.*, 32, 781, 1976.
6. Noll, K. M., Cultivation of strictly anaerobic Archaea: methanogens, in *Protocols for Archaebacterial Research*, Flieschmann, E. M., Place, A. R., Robb, F. T., and Schreier, H. J., Eds., The Center of Marine Biotechnology, Baltimore, 1991, 1.3.1.
7. Stetter, K. O., Methanothermaceae, in *Bergey's Manual of Systematic Bacteriology*, Vol. 3, Staley, J. T., Bryant, M. P., Pfennig, N., and Holt, J. G., Eds., Williams & Wilkins, Baltimore, 1989, 2183.
8. Pledger, R. J. and Baross, J. A., Preliminary description and nutritional characterization of a chemoorganotrophic archaeobacterium growing at temperatures of up to 110°C isolated from a submarine hydrothermal vent environment, *J. Gen. Microbiol.*, 137, 203, 1991.
9. Brock, T. D., Brock, K. M., Belly, R. T., and Weiss, R. L., *Sulfolobus:* a new genus of sulfur-oxidizing bacteria living at low pH and high temperature, *Arch. Microbiol.*, 84, 54, 1972.
10. Balch, W. E., Fox, E. E., Magrum, L. J., Woese, C. R., and Wolfe, R. S., Methanogens: a reevaluation of a unique biological group, *Microbiol. Rev.*, 43, 260, 1979.
11. Von Damm, K. L., Edmond, J. M., Grant, B. C., Measures, C. I., Walden, B., and Weiss, R. F., Chemistry of submarine hydrothermal solutions at 21°N, East Pacific Rise, *Geochim. Cosmochim. Acta*, 49, 2197, 1985.
12. Massoth, G. J., Milburn, H. B., Hammond, S. R., Butterfield, D. A., McDuff, R. E., and Lupton, J. E., The geochemistry of submarine venting fluids at Axial Volcano, Juan de Fuca Ridge: new sampling methods and VENTS Program rationale, in *Global Venting, Midwater, and Benthic Ecological Processes*, De Luca, M. P. and Babb, I., Eds., National Oceanic and Atmospheric Administration, Rockville, MD, 1989, 29.

13. Straube, W. L., Deming, J. W., Somerville, C. C., Colwell, R. R., and Baross, J. A., Particulate DNA in smoker fluids: evidence for the existence of microbial populations in hot hydrothermal systems, *Appl. Environ. Microbiol.,* 56, 1440, 1990.

14. Wiegel, J., Methods for isolation and study of thermophiles, in *Thermophiles: General, Molecular, and Applied Microbiology,* Brock, T. D., Ed., John Wiley & Sons, New York, 1986, 17.

15. Huber, R., Stoffers, P., Cheminee, J. L., Richnow, H. H., and Stetter, K. O., Hyperthermophilic archaebacteria within the crater of erupting Macdonald Seamount, *Nature,* 345, 179, 1990.

16. Pledger, R. J. and Baross, J. A., Characterization of an extremely thermophilic archaebacterium isolated from a black smoker polychaete (*Paralvinella* sp.) at the Juan de Fuca Ridge, *Syst. Appl. Microbiol.,* 12, 249, 1989.

17. Adams, M. W. W., Large scale growth of *Pyrococcus furiosus,* in *Protocols for Archaebacterial Research,* Fleischmann, E. M., Place, A. R., Robb, F. T., and Schreier, H. J., Eds., The Center of Marine Biotechnology, Baltimore, MD, 1991, 1.1.1.

18. Brown, S. H. and Kelly, R. M., Cultivation techniques for hyperthermophilic archaebacteria: continuous culture of *Pyrococcus furiosus* at temperatures near 100°C, *Appl. Environ. Microbiol.,* 55, 2086, 1989.

19. Jannasch, H. W., Wirsen, C. O., Molyneaux, S. J., and Langworthy, T. A., Extremely thermophilic fermentative archaebacteria of the genus *Desulfurococcus* from deep-sea hydrothermal vents, *Appl. Environ. Microbiol.,* 54, 1203, 1988.

20. Zillig, W., Holz, I., Janekovic, D., Klenk, H. P., Imsel, E., Trent, J., Wunderl, S., Forjaz, V. H., Coutinho, R., and Ferreira, T., *Hyperthermus butylicus,* a hyperthermophilic sulfur-reducing archaebacterium that ferments peptides, *J. Bacteriol.,* 172, 3959, 1990.

21. Stetter, K. O., *Archaeoglobus fulgidus* gen., sp. nov.: a new taxon of extremely thermophilic archaebacteria, *Syst. Appl. Microbiol.,* 10, 172, 1988.

22. Burggraf, S., Jannasch, H. W., Nicolaus, B., and Stetter, K. O., *Archaeoglobus profundus* sp. nov., represents a new species within the sulfate-reducing archaebacteria, *Syst. Appl. Microbiol.,* 13, 24, 1990.

23. Zellner, G., Stackebrandt, E., Kneifel, H., Messner, P., Sleytr, U. B., DeMacario, E. C., Zabel, H. P., Stetter, K. O., and Winter, J., Isolation and characterization of a thermophilic, sulfate reducing archaebacterium, *Archaeoglobus fulgidus, Z. Syst. Appl. Microbiol.,* 11, 151, 1989.

24. Zillig, W., Thermoproteales, in *Bergey's Manual of Systematic Bacteriology,* Vol. 3, Staley, J. T., Bryant, M. P., Pfennig, N., and Holt, J. G., Eds., Williams & Wilkins, Baltimore, 1989, 2240.

25. Fiala, G. and Stetter, K. O., *Pyrococcus,* in *Bergey's Manual of Systematic Bacteriology,* Vol. 3, Staley, J. T., Bryant, M. P., Pfennig, N., and Holt, J. G., Eds., Williams & Wilkins, Baltimore, 1989, 2237.

26. Deming, J. W. and Baross, J. A., Solid medium for culturing black smoker bacteria at temperatures to 120°C, *Appl. Environ. Microbiol.,* 51, 238, 1986.

27. Kelly, R. M. and Deming, J. W., Extremely thermophilic archaebacteria: biological and engineering considerations, *Biotechnol. Prog.,* 4, 47, 1988.

CHAPTER 3

Isolation of Psychrophilic Bacteria

Richard Y. Morita

INTRODUCTION

Most of the biosphere is cold. Of the earth's surface, 14% is in the polar regions and approximately 90% (by volume) of the ocean is 5°C or colder. The depth of the discontinuity layer of the oceans varies, mainly in relation to the latitude. The cold portions of the oceans has been referred to as the psychrosphere. Other cold ecosystems are at the top of high mountains, the upper atmosphere, and rivers and lakes at high altitudes.

In spite of the dominance of cold environments, there has been very little ecological or physiological research performed on the true cold-loving bacteria.

Bacteria were known to be capable of growing on foods that required refrigeration, but none of these organisms were truly cold loving. This also applied to bacteria isolated from cold ecosystems. As a result, such terms as Glaciale Bakterien, rhigophile, psychrotolerant, psychrocartericus, psychrobe, rhigophobic bacteria, cryophile, facultative psychrophile, obligate psychrophile, and psychrotrophic bacteria have been used. The first good scientific documentation of a true psychrophile was made by Morita and Haight.[1] The most often used old definition of psychrophiles was that they were organisms capable of producing a visible growth at 0°C in 1 week, and subsequently divided into facultative and obligate psychrophiles: The former having an optimal growth temperature above 20°C and the latter having optimal growth temperature below 20°C. The current definition states that psychrophiles are those bacteria that have an optimum growth temperature at 15°C or lower, a maximum growth at 20°C or lower, and a minimum growth temperature at 0°C or lower.[2] Those organisms that do not fit the foregoing definition but capable of growth at low temperatures are termed psychrotrophs.

The thermosensitivity of psychrophiles has well been demonstrated. Depending on the organism's maximal growth temperature, a few degrees above its maximum growth temperature can bring about thermal death of the cell. Naturally, depending upon the exposure to temperature above its maximal growth temperature, it is time dependent. Temperature above the maximal growth temperatures brings about leakage of cellular material, disruption of the integrity of the membrane, which then, in turn, result in a loss the transport control of the membrane. In addition, some of the enzymes of the cells are also thermally denatured.[2]

0-87371-564-0/93/$0.00 + $.50
© 1993 by Lewis Publishers

MATERIALS REQUIRED

Equipment

- Refrigerators or water baths set at 5 and 21°C.

Solution, Cultures, and Media

- Medium will depend on what physiological type of psychrophile desired.
- Source material for cultures must come from a permanently cold environment. If the material receives solar radiation, the surface temperature could easily exceed 20°C. The inoculum should never be exposed to room temperature for any length of time.

PROCEDURES

It is strongly recommended that the enrichment culture technique be employed first using the desired medium in relation to the physiological type of bacteria desired. The inoculated enrichment culture should be incubated at 5°C for at least 2 weeks, mainly because the lag phase of growth for psychrophiles may be quite long. After incubation, the usual techniques for the isolation of a pure culture can be carried on. After picking colonies, test each isolate for the ability to grow above 20°C. If growth occurs, the organism should be classified as a psychrotroph. If no growth occurs, a psychrophile has been successfully isolated. Many of the psychrophiles have a maximum growth temperature of 10 to 12°C.

The inoculum, all media, including agar plates, should be kept at 5°C. The inoculating loop must also be cold before use. It is preferable that all procedures be conducted in a cold-temperature room.

REFERENCES

1. Morita, R. Y. and Haight, R. D., Temperature effects on the growth of an obligate psychrophilic bacterium, *Limnol. Oceanogr.,* 9, 103, 1964.
2. Morita, R. Y., Psychrophilic bacteria, *Bacteriol. Rev.,* 30, 144, 1975.

CHAPTER 4

Isolation and Characterization of Bacteriocytes from a Bivalve-Sulfur Bacterium Symbiosis

Steven C. Hand and Amy E. Anderson

INTRODUCTION

The occurrence of marine invertebrates harboring intracellular sulfur bacteria is fairly common in marine environments where sulfide and molecular oxygen are simultaneously present (soft-bottom habitats, highly polluted reducing environments, hydrothermal vents).[1] These types of biological associations are thought to be millions of years old, suggesting a high degree of metabolic integration and codependence of the separate components of such a symbiosis. Not surprisingly, many features of the symbiosis are critically dependent on the cellular integrity of both the bacterium and the eukaryotic cell (the bacteriocyte) of the invertebrate host in which the endosymbiont resides. We have developed a procedure for isolating intact bacteriocytes from gill tissue of the shallow-water bivalve *Lucina floridana*, an inhabitant of seagrass beds. The bacteriocyte represents the simplest level of biological organization at which the functional symbiosis can be defined, and consequently, isolated bacteriocytes allow one to study the metabolic interplay between the invertebrate host and its symbionts under conditions where the number of biological variables are kept to a minimum.

The primary advantage of using isolated bacteriocytes for metabolic and biochemical studies is that the symbiotic bacteria are retained in their natural microenvironment. As a consequence, the bacteria receive chemical signals (sulfur compounds, dissolved gases, etc.) via the cytoplasm of the host cell. Furthermore, all bacteriocyte surfaces are in direct contact with medium constituents, so that the effects of slowly exchanging compartments such as connective tissues are minimized. Thus, the individual bacteriocytes are considered functional symbiotic units, and their response to various stimuli quantified on a cellular basis.[2]

Depending on the experimental protocols anticipated, there are certain disadvantages of a practical nature that should be considered when using bacteriocyte preparations. Several hours are required to isolate the bacteriocytes, and experiments must be performed immediately thereafter because of the limited variability of the cells thus far attained (several hours). While several million bacteriocytes can be isolated from 1 to 2 g of host tissue, this number of cells probably represents only a few milligrams of wet mass. Thus, experiments requiring gram quantities of cells are not feasible with bacteriocytes.

Among other metabolic features, we have successfully used bacteriocytes from *L. floridana* to quantify the ability of various sulfur compounds to stimulate carbon fixation, to identify and quantify the radiolabeled synthate produced by the cells and released into the medium, and to measure sulfur-stimulated heat dissipation from the bacteriocytes using microcalorimetry. The release of carbon compounds from bacteriocytes should be analogous to export of carbon to the host tissues.

We suggest that experimental preparations that maintain the endosymbiotic bacteria in more biologically realistic surroundings offer new opportunities for assessing their metabolic potential. Resulting data should improve our understanding of the factors governing the functional and structural unity of the symbiosis, the prokaryotic vs. eukaryotic contributions to sulfur detoxification and ATP synthesis, and carbon assimilation by bacteriocytes and its subsequent export to the host.

MATERIALS REQUIRED

Experimental Animals

In contrast to the hydrothermal vent species like the clam *Calyptogena magnifica*, the research animal used in our studies, *L. floridana*, is a bivalve that is inexpensively obtained. Shiptime is not required for collection. These clams are abundant (up to 90 per square meter) in the shallow seagrass beds of northern Florida.[3] Live specimens are easily shipped at room temperature in ziplock containers and subsequently can be kept for months in laboratory aquaria containing artificial seawater fortified with sulfide.[3] Clam condition appears to be improved if 4 to 6 in. of habitat substrate (sand or mud) is added to the bottom of the aquarium.[4] The addition of a recirculating pump and filters (carbon and biological) helps to maintain the condition of the aquarium (Amy Anderson, personal observations). The bacteriocytes from *L. floridana* gill tissue are hardy in comparison to bacteriocytes of other species, and a higher yield of bacteriocytes per gram gill tissue can be obtained.[2]

Equipment

- Temperature-controlled orbital shaker
- Low-speed, temperature-controlled centrifuge with swinging bucket rotor
- Fluorescence microscope with UV filter cube
- Light microscope and hemocytometer for cell-counting
- Refractometer or other salinity-measuring device

Supplies

- Siliconized glassware (small beakers, pipettes), treated with Sigmacote; 250-μm nylon mesh Tetko, Inc., Elmsford, NY)

Solutions and Reagents

- High potassium artificial seawater (high-K^+ ASW): 275 mM NaCl, 115 mM KCl, 8.5 mM NaHCO$_3$, 18 mM NaSO$_4$, 0.1 mM EGTA, 4.3 mM glucose, 45 mM MgCl$_2$, 0.85 mM dithiothreitol, 8.0 mM CaCl$_2$, pH 8.0.
- Ca^{++}-Mg^{++} free salt solution (CMF solution): 20 mM Hepes, 440 mM NaCl, 9.7 mM KCl, 2.7 mM NaHCO$_3$, 27.8 mM Na$_2$SO$_4$, 0.1 mM EDTA, 4.3 mM glucose, 190 mM maltose, pH 8.0. The total salt concentration was 34 ppt, as measured with a refractometer.
- Artificial seawater (ASW): 20 mM Hepes, 375 mM NaCl, 9.4 mM KCl, 37.7 mM MgCl$_2$, 8 mM CaCl$_2$, 2.7 mM NaHCO$_3$, 17.9 mM Na$_2$SO$_4$, 7.6 mM (NH$_4$)$_2$SO$_4$, 4.3 mM glucose, 190 mM maltose, pH 8.0. Total salt concentration was 34 ppt.
- Digestive enzyme solution: hyaluronidase (type 1-S), 670 units/ml; collagenase (type IV), 830 units ml^{-1}; chymotrypsin (type II), 670 units ml^{-1}; prepared in ASW. The enzyme

concentrations (units ml^{-1}) given above are for use with 1.0 g of gill tissue; these concentrations are changed proportionally with the amount of tissue used. Routinely, 5 ml of digestive enzyme solution are prepared for up to 0.75 g of tissue, while 10 ml are used for 0.75 to 1.5 g of tissue.

- Enzyme inhibitor solution: trypsin-chymotrypsin inhibitor (0.2 mg ml^{-1}) in pre-Percoll suspension solution.
- Pre-Percoll suspension medium: DNase I (type IV), 100 units ml^{-1}; bovine serum albumin (BSA), 1 mg ml^{-1}; in CMF solution.
- Post-Percoll rinse solution: bovine serum albumin (BSA), 1 mg ml^{-1}; DNase I (type IV), 80 units ml^{-1}; prepared in CMF solution.
- Percoll gradient solutions: 30, 50, 60, 70, and 90% solutions (v/v). To prepare 10 ml each of the 30 to 70% solutions, appropriate amounts of Percoll and deionized water are added to 2 ml of a concentrated CMF stock solution (5×). The 90% solution is prepared by adding 9 ml Percoll to 1 ml of 10× concentrated CMF stock.
- Hoechst stain solution: 0.5% of #33258 Hoechst dye (membrane impermeant) prepared in ASW.

All enzymes, trypsin-chymotrypsin inhibitor, Percoll, Sigmacote, and Hoechst stain were purchased from Sigma Chemical Co. (St. Louis, MO).

PROCEDURES

Isolation of Bacteriocytes

1. At the beginning of the day, several 21-ml discontinuous gradients are prepared in 30-ml siliconized glass centrifuge tubes; 5 ml each of the 90, 70, and 60% solutions are layered sequentially into the centrifuge tubes. Then, 3 ml each of the 50 and 30% solutions are added. The gradients are transferred to a 4°C refrigerator and kept chilled until needed.

2. Fresh gill tissue (1 to 2 g) from *L. floridana* is placed on a chilled glass plate and minced with a razor blade into cubes varying in size from 0.5 to 2 mm on a side. The tissue is incubated in high-K$^+$ ASW (to decrease mucus production) for 10 min and then briefly washed in CMF solution.

3. The tissue is transferred to 25-ml beakers and incubated for 20 min at 37°C in 20 ml of CMF solution on a rotary shaker (110 cycles min^{-1}).

4. The tissue is rinsed with ASW. Then 5 to 10 ml (depending on the amount of tissue) of digestive enzyme solution is added, and the tissue is incubated for 1 h at 37°C with shaking. This solution weakens and digests the extracellular matrix of the tissue.

5. At the end of this period, the tissue is rinsed with CMF solution and incubated with shaking for 20 min at 37°C in 10 ml of enzyme inhibitor solution. This step prevents further digestive activity of the enzymes used for tissue dissociation. Additionally, the DNase I degrades large DNA molecules released from broken cells that would tend to cause the intact cells to clump during centrifugation; the albumin also aids in preventing cell aggregation.

6. The tissue is allowed to settle and the medium is carefully discarded. The tissue is rinsed with 2 to 3 ml of pre-Percoll suspension medium and the wash solution is discarded.

7. The tissue is resuspended in 2 ml of the pre-Percoll suspension medium and gently flushed two times through a 5-ml glass pipet or a 2-ml plastic disposable pipet, a procedure that releases large numbers of free cells. The free cells are transferred to a clean tube, and the complete procedure is repeated twice on the remaining tissue cubes.

8. The combined cellular suspension is then filtered through a 250-μm nylon mesh to remove any undissociated tissue.

9. The cellular suspension is subdivided into equal portions and layered onto Percoll gradients at 4°C. The number of cells obtained from about 0.3 to 0.4 g of gill tissue is the maximum to apply to a single gradient tube. The gradients are centrifuged at 75 × g for 15 min at 4°C. This procedure separates the bacteriocytes from other cell types and from acellular debris.

10. At the end of the centrifugation run, the cells accumulating at the desired interfaces are collected with a plastic disposable pipet. Typically, the majority of single, intact bacteriocytes accumulate at the 70 to 90% Percoll interface. The density of bacteriocytes slowly decreases as a function of the time that clams are held in the laboratory. Thus, several weeks after specimen collection, the Percoll concentrations may have to be altered to maintain optimal separation and purity of bacteriocytes.

11. The isolated bacteriocytes collected from each gradient are combined and then diluted about fivefold in post-Percoll rinse solution. The cells are allowed to settle (at 0°C, on ice). The overlying solution is removed and the washing procedure repeated twice. This washing removes the Percoll, which can interfere in the assay for CO_2 incorporation.

12. Bacteriocytes are stored on ice until used for experiments (see cautionary note below).

13. A portion of the cells are diluted in post-Percoll rinse solution, and 100 μl of Hoechst stain solution are added to about 200 to 300 μl of diluted cell suspension. The cells are viewed in a concave counting chamber with UV illumination to determine the percentage of intact cells. To avoid breaking cells, load the suspension onto the surface of the counting chamber and then apply the cover slip. If a cell is broken, the dye will permeate the cell and result in intense fluorescence due to staining of cellular DNA. Intact cells produce only a faint epifluorescence. The use of trypan blue as an indicator of cellular integrity does not work well with bacteriocytes. It is exceedingly difficult to detect internal staining of broken cells with trypan blue, presumably due to the high concentrations of bacteria that fill the internal cytoplasmic space of the bacteriocytes. We found about 75 to 80% of the bacteriocytes were intact based upon the Hoechst staining technique.

14. In order to make cell counts, bacteriocytes are diluted as above and loaded beneath the cover slip of a hemocytometer. Counts are made with light microscopy using a scanning objective. Because of their fragility and size, some cells will be broken during the loading process. Yields of 2 to 3 million bacteriocytes per gram tissue can be achieved.

15. The distinguishing features used to identify bacteriocytes under light microscopy are their granular appearance (due to intracellular bacteria), presence of large pigment granules, lack of cilia, and relatively large size (20 to 70 μm).[2,3] Based on these criteria, the final cell preparation contains approximately 90 to 95% bacteriocytes.

Suggested Applications

After intact bacteriocytes have been isolated, a variety of experimental approaches can be taken to characterize the morphological and functional features of these cells.

Morphology

Techniques for preparing the isolated bacteriocytes for viewing with electron microscopy have been developed.[2] Briefly, isolated cells to be fixed for electron microscopy are transferred to Beem capsules and centrifuged at low speed (500 × g) to concentrate the cells. Upon removing the supernatant, glutaraldehyde (4% in 0.3 M PIPES buffer, pH 7.2, at room temperature) is layered over the cells. After 30 to 60 min, the glutaraldehyde solution is removed, and the Beem capsules are filled with warmed agar (1.5% in 0.3 M PIPES, pH 7.2). When the agar solidifies, the capsules are given three 15-min washes in buffer and placed in 1% osmium tetroxide (prepared in 0.2 M potassium phosphate buffer, pH 7.4) for 2 to 3 h. The preparations are washed thoroughly with deionized water, dehydrated in a graded acetone series, and embedded in Spurr's low-viscosity media. Thin sections are cut and stained with 2% uranyl acetate followed by 0.2% lead citrate.

Carbon Fixation and Transport

It is now generally accepted that sulfur-driven CO_2 fixation by the bacteria may be of significant nutritional importance to the integrity/health of these symbioses. Activities of the enzymes of the Calvin-Benson cycle and the CO_2 fixation and translocation from the symbiont-containing tissues have been reported for many of these species (see Somero et al.[1] for review). In the case of *L. floridana*, previous work has demonstrated the autotrophic nature of this symbiosis by measuring the activities of the diagnostic enzymes of the Calvin-Benson cycle and the bacterial enzymes of sulfide oxidation in the symbiont-containing tissues.[3] However, we contend the metabolic interplay between an invertebrate host and sulfur-bacteria symbionts can be clarified best by studying isolated bacteriocytes. For example, while evidence supports the idea that assimilated carbon is transported from the bacteria to the host tissues,[5] virtually nothing is currently known about the nature of this translocation process. By adding sulfur compounds to stimulate carbon fixation in bacteriocyte suspensions, it should be possible to identify the radiolabeled carbon compounds released into the medium by the cellular symbiosis, thereby revealing the identity of the carbon compounds transported to the host organism.

Sulfur Metabolism

Similarly, much is yet to be explained about the sulfur metabolism occurring in these symbioses. Bacterial enzymes of the sulfur cycle, and the production of partially oxidized sulfur species, including elemental sulfur, have been documented in a multitude of these animals (for review see Somero et al.[1]). Thus far, sulfide oxidation has been shown to occur in the animal tissues of some species,[6-8] or implicated as a bacterial activity in symbioses that contain sulfide-binding proteins, which prevent sulfide poisoning and the oxidation of sulfide by the host tissues.[9,10] Recent work with the bivalve *Solemya reidi*[4] has begun to separate the roles of the bacteria and host tissues in the metabolism of sulfide. However, the elucidation of the separate metabolic roles of the individual partners in these symbioses can only be partially identified by studies using whole animals, tissues, or body fluids. Again, we suggest that bacteriocytes may offer many experimental advantages for these questions as well. For example, differential capacities for sulfur oxidation could easily be compared among cell types isolated from the gills of *L. floridana*. If bacteria are responsible for a large portion of this activity, bacteriocytes should have higher sulfur oxidation capacity than ciliated epithelial cells. This pattern would likely be reversed if mitochondrial oxidation of sulfide predominated,[7] because high densities of mitochondria exist in the epithelial cells.[3] (Supported by NSF grant OCE-8900107.)

NOTES AND COMMENTS

1. All pieces of glassware (including pipettes) that come in contact with the bacteriocytes must be siliconized before use. Tips of Pasteur pipettes should be firepolished as well. Alternatively, plasticware can be substituted effectively.

2. The most common difficulty in isolating intact bacteriocytes is handling the cells too harshly at various steps in the isolation procedure. For example, concentrating and washing the cells by pelleting them multiple times with centrifugation or transferring the cells in narrow bore pipets (or pipets that are not siliconized and firepolished) can result in reduced numbers of intact cells.

3. Based on assays of CO_2 incorporation, isolated bacteriocytes show little decline in viability at room temperature over a 4-h period. Beyond this period of time, a decrease in CO_2 fixation capacity is observed. Thus, experiments should be initiated immediately upon completion of the isolation protocol.

REFERENCES

1. Somero, G. N., Childress, J. J., and Anderson, A. E., Transport, metabolism, and detoxification of hydrogen sulfide in animals from sulfide-rich marine environments, *Rev. Aquatic Sci.,* 1(4), 591, 1989.
2. Hand, S. C., Trophosome ultrastructure and the characterization of isolated bacteriocytes from invertebrate-sulfur bacteria symbioses, *Biol. Bull.,* 173, 260, 1987.
3. Fisher, M. R. and Hand, S. C., Chemoautotrophic symbionts in the bivalve *Lucina floridana* from seagrass beds, *Biol. Bull.,* 167, 445, 1984.
4. Anderson, A. E., Childress, J. J., and Favuzzi, J. A., Net uptake of CO_2 driven by sulphide and thiosulphate oxidation in the bacterial symbiont-containing clam *Solemva reidi, J. Exp. Biol.,* 133, 1, 1987.
5. Fisher, C. R., Jr. and Childress, J. J., Translocation of fixed carbon from symbiotic bacteria to host tissues in the gutless bivalve, *Solemya reidi, Mar. Biol.,* 93, 59, 1986.
6. Powell, M. A. and Somero, G. N., Sulfide oxidation occurs in the animal tissue of the gutless clam, *Solemya reidi, Biol. Bull.,* 169, 164, 1985.
7. Powell, M. A. and Somero, G., Hydrogen sulfide oxidation is coupled to oxidative phosphorylation in mitochondria of *Solemya reidi, Science,* 233, 563, 1986.
8. Powell, M. A. and Somero, G. N., Adaptations to sulfide by hydrothermal vent animals: sites and mechanisms of detoxification and metabolism, *Biol. Bull.,* 171, 274, 1986.
9. Childress, J. J., Felbeck, H., and Somero, G. N., Symbiosis in the deep sea, *Sci. Am.,* 256(5), 114, 1987.
10. Wilmot, D. B. and Vetter, R. D., The bacterial symbiont from the hydrothermal vent tubeworm *Riftia pachyptila* is a sulfide specialist, *Mar. Biol.,* 106, 273, 1990.

Additional References

Anderson, A. E., Felbeck, H., and Childress, J. J., Aerobic metabolism is maintained in animal tissue during rapid sulfide oxidation in the symbiont-containing clam *Solemya reidi, J. Exp. Biol.,* 256, 130, 1990.

Belkin, S., Nelson, D. C., and Jannasch, H. W., Symbiotic assimilation of CO_2 in two hydrothermal vent animals, the mussel *Bathymodiolus thermophilus* and the tube worm *Riftia pachyptila, Biol. Bull.,* 170, 110, 1986.

Cary, S. C., Vetter, R. D., and Felbeck, H., Habitat characteristics and nutritional strategies of the endosymbiont-bearing bivalve *Lucinoma aequizonata, Mar. Ecol. Prog. Ser.,* 55, 31, 1989.

Cavanaugh, C. M., Symbiotic chemoautotrophic bacteria in marine invertebrates from sulphide-rich habitats, *Nature,* 302, 58, 1983.

Cavanaugh, C. M., Levering, P. R., Maki, J. S., Mitchell, R., and Lidstrom, M. E., Symbiosis of methylotrophic bacteria and deep-sea mussels, *Nature,* 325, 346, 1987.

Childress, J. J., Fisher, C. R., Brooks, J. M., Kennicutt, M. C., II, Bidigare, R., and Anderson, A. E., A methanotrophic marine molluscan (Bivalve, Mytilidae) symbiosis: mussels fueled by gas, *Science,* 233, 1306, 1986.

Dando, P. R., Southward, A. J., and Southward, E. C., Chemoautotrophic symbionts in the gills of the bivalve mollusc *Lucinoma borealis* and the sediment chemistry of its habitat, *Proc. R. Soc. Ser. B,* 227, 227, 1986.

Felbeck, H., Sulfide oxidation and carbon fixation by the gutless clam *Solemya reidi*: an animal-bacterial symbiosis, *J. Comp. Physiol.,* 152, 3, 1983.

Felbeck, H., Childress, J. J., and Somero, G. N., Calvin-Benson cycle and sulphide oxidation enzymes in animals from sulphide-rich habitats, *Nature,* 293, 291, 1981.

Felbeck, H., Powell, M. A., Hand, S. C., and Somero, G. N., Metabolic adaptations of hydrothermal vent animals, *Biol. Soc. Wash. Bull.,* No. 6, 261, 1985.

Felbeck, H. and Somero, G. N., Primary production in deep-sea hydrothermal vent organisms: roles of sulfide-oxidizing bacteria, *Trends Biochem. Sci.,* 7, 201, 1982.

Fiala-Medioni, A. and Metivier, C., Ultrastructure of the gill of the hydrothermal vent bivalve *Calyptogena magnifica*, with a discussion of its nutrition, *Mar. Biol.,* 90, 215, 1986.

Fisher, C. R., Jr., Chemoautotrophic and methanotrophic symbioses in marine invertebrates, *Rev. Aquatic Sci.,* 2, 399, 1990.

Fisher, C. R., Childress, J. J., Oremland, R. S., and Bidigare, R. R., The importance of methane and thiosulfate in the metabolism of the bacterial symbionts of two deepsea mussels, *Mar. Biol.,* 96, 59, 1987.

Giere, O., Structure and position of bacterial endosymbionts in the gill filaments of Lucinidae from Bermuda (Mollusca, Bivalvia), *Zoomorphology,* 105, 296, 1985.

Hand, S. C. and Somero, G. N., Energy metabolism pathways of hydrothermal vent animals: adaptations to a food-rich and sulfide-rich deep-sea environment, *Biol. Bull.,* 165, 167, 1983.

O'Brien, J. and Vetter, R. D., Production of thiosulfate during sulphide oxidation by mitochondria of the symbiont-containing bivalve *Solemya reidi, J. Exp. Biol.,* 149, 133, 1990.

Southward, E. C., Gill symbionts in thyasirids and other bivalve molluscs, *J. Mar. Biol. Assoc. U.K.,* 66, 889, 1986.

Spiro, B., Greenwood, P. B., Southward, A. J., and Dando, P. R., $^{13}C/^{12}C$ ratios in marine invertebrates from reducing sediments: confirmation of nutritional importance of chemoautotrophic endosymbiotic bacteria, *Mar. Ecol. Prog. Ser.,* 28, 233, 1986.

Terwilliger, R. C., Terwilliger, N. B., and Arp, A., Thermal vent clam (*Calyptogena magnifica*) hemoglobin, *Science,* 219, 981, 1983.

Vetter, R. D., Elemental sulfur in the gills of three species of clams containing chemoautotrophic symbiotic bacteria: a possible inorganic energy storage compound, *Mar. Biol.,* 88, 33, 1985.

Wood, A. P. and Kelly, D. P., Methylotrophic and autotrophic bacteria isolated from lucinid and thyasirid bivalves containing symbiotic bacteria in their gills, *J. Mar. Biol. Assoc. U.K.,* 69, 165, 1989.

CHAPTER 5

General Techniques for the Isolation and Culture of Marine Protists from Estuarine, Littoral, Psammolittoral, and Sublittoral Waters

John J. Lee

INTRODUCTION

Anyone who has cupped their hands together and tried to hold a parcel of water in them, for any length of time, knows the problem. Sea water is ephemeral. It is very difficult to hold, except for a brief moment in space and time. At the microphytic and micro- and meiofaunal levels of trophic dynamics thousands of individuals and hundreds of species interact in very localized patches or assemblages to make up a community mosaic. We really have very few specifics on the subtle biotic and chemical factors that regulate a dynamically changing population and community structure. Because of the ephemeral qualities of samples of seawater containing marine microbes and protists, some workers have chosen "black box" approaches to study functional problems. This approach has the advantage of simplicity and it can generate numbers. However, it does miss many qualities of biogeocoenoses. At the other extreme there are those who isolate bacteria, protists, or meiofauna in axenic culture to study the physicochemical and intraspecific biological aspects of their niches. I have contended elsewhere[1-4] that, while we can learn much from the autecological approach, it also falls short of giving us a realistic understanding of small-scale biotic interactions. There is a middle ground, the study of gnotobiotic cultures (note: gnotobiotic = all organisms are known to observer; agnotobiotic = not all organisms present are known). These have given us great insights into qualitative biotic interactions of protists and meiofauna.[5-8]

There are many different successful approaches to growing marine protists in the laboratory. These can be divided as follows

A. Axenic/monoorganismic
  (i) Phototrophs
  (ii) Osmotrophs
B. Fortuitous/agnotobiotic
  (i) No intervention
  (ii) Enriched and controlled
  (iii) Fed natural mixtures
C. Fortuitous/partially gnotobiotic batch, or continuous, cultures of carnivorous, bactivorous, or herbivorous protists
  (i) Fed live food

    (ii) Fed heat killed food
    (iii) Food organisms preinoculated or inoculated and encouraged to grow with consumer
  D. Inductive methods
    (i) Isolation of food organisms from the habitat for use in (iii) below
    (ii) Tracer feeding to identify likely candidates for (iii) below
    (iii) Synxenic (i.e., coculture of two or more organisms) or gnotobiotic cultures

Each of the above has advantages and disadvantages. Fortuitous/agnotobiotic cultures have the advantages of ease of setup and low-cost maintenance, but they are often unreliable. Some photosynthetic and osmotrophic marine protists are easily isolated in axenic culture, but with others considerable effort is required to axenize them or set up synxenic cultures. Aseptic technique is required to maintain them gnotobiotic.

MATERIALS REQUIRED

Equipment

- Illuminated incubator, light bank, or shelf with controllable light. Intensity varies depending on organisms
- Good quality dissection microscope with darkfield and transillumination base.
- For inductive methods: scintillation counter

Supplies

Axenic/Monoorganismic

Autotrophs

- Sterile pasteur pipettes
- Petri dishes with solidified differential media and antibiotic mixture (Tables 1 and 2)
- Microspatula (e.g., Fisher Scientific Co., Cat. no. 21-401-10; ground to 1 to 2 mm at tip)
- Plastic bags (Zip-lock if possible).

Osmotrophs

- Sterile nine-hole spot plates (Corning cat. no. 7220) inside sterile petri dishes (20 × 150 mm)
- Sterile capillary pipettes attached to a mechanical pipetting device
- Waste receptacle

Fortuitous/Agnotobiotic

- Stacking culture dishes (Carolina Supply Co., Wheaton cat. no. 41684, tissue culture flasks, or deep petri dishes (Corning cat. nos. 3160-152, 324, or 3140)
- Plastic wrap
- Freshly collected net plankton or benthic algal sample
- Graduated cylinder
- Pipettes

Table 1. Media for the Isolation of Unicellular Algae[32]

Natural Media

1. Unenriched seawater
2. Erdschrieber medium (95 ml seawater, 5 ml soil extract, 10 mg KNO_3, 50 mg $NaH_2PO_4 \cdot H_2O$)
3. Seawater + 100 mg l^{-1} KNO_3
4. Seawater + 50 mg l^{-1} NaH_2PO_4
5. Seawater + vitamin mixture I (= 10 μg B_{12}, 34 μg thiamine, 24 μg biotin per 100 ml)

Synthetic Media

6. Basal medium (g l^{-1})

NaCl	25
$MgSO_4 \cdot 7H_2O$	9
KCL	0.7
$CaCl_2$	0.3
$NaSiO_3 \cdot 9H_2O$	0.07
NaH_2CO_3	0.1
NaH_2PO_4	0.01
NTA	0.07
Tris	1
P II metals[32]	30 ml L^{-1}

7. Basal medium + 0.1 g L^{-1} $NaNO_3$
8. Basal medium + 0.1 g L^{-1} NH_4Cl (see also ref. 15)
9. Basal medium + 0.1 g L^{-1} NH_4NO_3
10. Basal medium + 0.1 g L^{-1} NH_4NO_3 + 0.05 g L^{-1} $NaH_2PO_4 \cdot H_2O$
11. #10 + the following amino acid mixture (all at 1 mM)

Alanine	0.0089
Na glutamate	0.044
Glycine	0.0075
Na glutamate	0.0147
Lysine	0.0146
Arginine	0.0174
Na aspartate	0.0133

12. #10 + vitamin B_{12} 0.00001 mM
13. #10 + vitamin mixture as in #5 above
14. #10 + 50 ml l^{-1} soil extract
15. #13 and the following

Na lactate	0.034 mM
Na acetate	0.025 mM
Glucose	0.054 mM

16. #13 + Na glutamate 0.044 mM
17. #13 + asparagine 0.04 mM
18. #13 + glycerol 0.028 mM
19. #13 + acetone-extracted or autoclaved natural material (e.g., *Enteromorpha*)

Note: Based on media published earlier by others and our group.[9,15,16,29-32] May be solidified by the addition of a good quality agar. Antibiotic mixtures (Table 2 and References 10, 16, 33, and 34) can be very effective against bacteria without inhibiting algae, protists, or meiofauna.

Fortuitous/Partially Agnotobiotic

- Tissue culture flasks, test tubes, or other incubation vessels
- Axenic cultures of microphytes (available from American Type Culture Collection, 12301 Parklawn Drive, Rocklawn, MD 20852 or Provasoli-Guillard Center for Culture of Marine Phytoplankton, Bigelow Laboratory for Ocean Sciences, West Boothbay Harbor, ME 04575). Cultures that have worked well in the past:

  Cylindrotheca closterium
  Nitzschia acicularis
  Amphora spp.
  Fragilaria construens
  Achnanthes hauckiana
  Navicula spp.
  Dunaliella salina

Table 2. Antibiotic-Antimycotic Mixes that Inhibit Many Marine Bacteria and Fungi but not Most Protists or Meiofauna

Component	Mix 1[a] (100×)	Mix 2[a]	Mix 3[b]
Penicillin	10,000 units	5 mg	12,000 units
Streptomycin	10,000 µg	5 mg	
Chloramphenicol			50 µg
Fungizone (or mycostatin)[c]	25 µg		
Polymyxin B			50 µg
Neomycin sulfate		10 mg	60 µg

Note: Based on the experience of others and our own group.[15,16,33,34]

[a] Available commercially from GIBCO Laboratories, 3175 Staley Road, Grand Island, NY 14072, Cat #600-5240 and #600-5640, respectively.
[b] Available commercially from Sigma Chemical Co., P.O. Box 14508, St. Louis, MO 63178-9915, Cat. #P9681.
[c] Fungizone (amphotericin B) and mycostatin (nystatin), available from E.R. Squibb & Sons, New York).

Inductive Methods

- Radiation safety supplies (e.g., lab coat, gloves, tray mats, radionuclide waste containers, plastic shields)
- Liquid scintillation vials and cocktail
- ^{32}P or $NaH_2^{14}CO_3$ for labeling
- Nine-hole spot plates
- Clinical centrifuge
- Sterile seawater

Solutions

- See Tables 1 and 2.

PROCEDURES

Isolation of Axenic Monoorganismic Cultures

Isolation of Axenic Phototrophs

Axenic cultures of nonmotile and gliding algae (e.g., chlorophytes, unicellular rhodophytes, chrysophytes, and diatoms) are often easily obtained by picking the colonies formed on agar.[9,10] A now-classical nutritional analysis of an epiphytic (on *Enteropmorpha intestinalis*) sublittoral diatom assemblage from a salt marsh gives us a model for the approach to isolation of phototrophs from complex assemblages.[9] Briefly, the study, which used 24 varieties of unenriched and enriched natural and artificial sea water media, showed that no one medium was suitable for the isolation of all colony types. However, taken as a whole spectrum of media, the success rate for the isolation of the most common diatoms was as high as 90% and averaged >66% in weekly samples. More than twice the number of colony types were isolated on an agar prepared with sea water from the collection site than on Erdschrieber, a very popular medium enriched with soil extract, nitrate, and phosphate. The greatest variety of colonies grew on a medium enriched with an acetone extract of the host plant. The various media differed in the numbers of different colony types which grew on them over the course of the summer (e.g., there was a steady increase in the number of colonies and colony types which grew on media enriched with thiamine, biotin, and B12 from the 8th week to the end of the summer). With various modifications the media that are listed in Table 1 have been used to isolate axenic clones of endosymbiotic diatoms, chlorophytes, and rhodophytes[10-13] and epiphytic and water column free-living diatoms, chlorophytes, and chrysophytes from semitropical and tropical seas.[9,14,15]

It does not seem to matter whether the media are prepared from sea water from the local sampling site or from an artificial sea water base. The media are solidified with 1.5% agar of a high quality (one which does not have orange flakes when cooled, e.g., Difco Noble agar, Cole Labs agar #2). It is difficult to view colonies on plates made with lower quality agar. Filter-sterilized heat-labile enrichments (e.g., vitamins) and antibiotic mixtures (see Table 2) are aseptically added to cooled (60°C in a water bath), autoclaved media just before they are poured into petri dishes. In some habitats the addition of an antibiotic mixture is strongly recommended.

1. An inoculum of 0.1-ml sample added to each dish is spread with an alcohol-sterilized bent glass rod.

2. Since the plates must be incubated in moderate light (~60 μE) for 1 to 2 weeks, they must be sealed in plastic bags. Some condensation usually occurs, so it is important to make sure the plates are inverted (agar side up) during incubation.

3. After 1 week, colonies are observable through the inverted plates in an ordinary transilluminated, epiilluminated, or dark field dissection microscope. With experience, more than 60 different diatom colony types can be recognized under a dissection microscope.[9]

4. Colonies for picking are marked on the bottom of the dish. If there is moisture condensed on the lid of the dish, it must be replaced with a dry one, before the dish is turned agar side down on the microscope stage.

5. Colonies are picked with an alcohol-sterilized spatula.

6. The colonies are transferred to a liquid version of the same medium. After the isolates grow they can be restreaked on agar to confirm their purity.

Isolation of Axenic Osmotrophs

Two methods are commonly used for the isolation of axenic osmotrophs (e.g., ciliates and flagellates): (1) migration and washing techniques popularized by Guillard, Provasoli, and co-workers[16-19] and (2) the silicone oil dilution method developed by Soldo (described in Chapter 11). The former works particularly well with motile species.

1. Sterile sea water containing an antibiotic mixture is aseptically pipetted into the wells of the spot plate.

2. The sample containing the organism of interest is introduced into the first well of the spot plate. The organisms in the well are found under a dissection microscope. When one of the species of interest moves so that it is sufficiently separated from others, a capillary pipette is lowered to capture it.

3. The organism is gently discharged into the next well in the plate.

4. After 5 min, it is picked up again in a sterile capillary pipette and transferred to the next well.

5.   This procedure is repeated. After the 12th to 18th serial passage bacteria-free protists can usually be obtained.

Fortuitous/Agnotobiotic Cultures

No Intervention

These are the easiest to set up and maintain. They are always worth trying because they require minimal effort.

1.   As soon as possible after collection, samples are diluted and placed in sterile sea water at ambient temperature. Stacking aquaria are useful culture vessels because of their large surface area and ease of access, but they have the disadvantage of high evaporation; tissue culture flasks have almost the opposite characteristics: lower evaporation rate but more difficult access to the protists.

2.   Incubation can be either in front of daylight fluorescent tubes ($\sim$25 to 60 μE) or near a window, but out of direct sunlight. This approach works well with many benthic marine protists (e.g., amoebae and foraminifera).[20]

3.   The trick to success is to change the sea water daily for the 1st week and then change the water at least weekly after that. For benthic protists the change is accomplished by careful decanting. Plastic wrap or another stacking culture dish can be placed as a lid on top of the stacking culture dish to retard evaporation.

Enriched and Controlled

Often it is possible to grow bactivorous and herbivorous protists and their food in the same vessel. The growth of the latter can often be encouraged by the addition of nitrate (KNO_3 or NH_4NO_3, 100 mg l^{-1}), phosphate ($NaH_2PO_4 \cdot H_2O$, 50 mg l^{-1}), and soil extract (50 mg l^{-1}). The trick is to encourage growth of food organisms without the overgrowth of other organisms. Sometimes this works, by chance, for a week to a month and at other times it may not. Sometimes transfers are successful, but, at other times they are not. Trial and error can help zero in on the amount of enrichment and/or light needed to balance the growth of the protists and their food.

Fed Natural Mixtures

Planktonic and benthic foraminifera have been maintained, and sometimes cultured, by feeding them natural mixtures or organisms selected from natural mixtures.[21,22]

1.   Planktonic foraminifera can be harvested by SCUBA divers or gentle net tows. Benthic foraminifera can be harvested by sieving the upper 1 cm of sediment or by dislodging them from seaweeds. The latter can be accomplished by gently harvesting the seaweeds into wide-mouth plastic bottles with seawater and agitating them vigorously. The seaweeds can be removed with the aid of forceps. After a few moments to allow the foraminifera to settle, the seawater with the epiphytes can be decanted, leaving the foraminifera and the sediment.

2.   The foraminifera are placed in the stacking culture dishes (or other culture vessel).

3.   Epiphytes obtained during the collection of the foraminifera (above), or prepared for the purpose, can be pipetted into the culture dishes. The epiphytes can be concentrated by

pouring them into a graduated cylinder. After they settle the excess water can be decanted. Some workers have "heat killed" the epiphytes first by placing them in a 50°C water bath for 10 to 30 min before adding them to the foraminifera. Nauplei or copepods from natural collections can be pipetted gently into the rhizopodal nets of planktonic foraminifera.

Fortuitous/Partially Agnotobiotic Batch or Continuous Cultures of Carnivorous, Bactivorous, or Herbivorous Protists

More reliable (or stable) maintenance or cultures can be obtained by beginning with known species of potential food organisms.

1. Inoculate a culture dish with a dense supply ($>10^6$ cells mm^{-3}) of food organisms, e.g., *Dunaliella salina* for *Euplotes vannus*, *Navicula* sp. for *Pontiflex maximus.*, *Escherichia coli* for *Vexillifera minutissima*.

2. Inoculate small numbers (<100 cells) of protists into a clean culture vessel. Warning! Some protists are voracious consumers and reproduce rapidly in the presence of particular food species. For example, *Euplotes vannus* will reduce a population of *Dunaliella salina* from 10^6 to $<10^3$ in 3 d or less; *Pontifex* will clear a petri dish covered with a lawn of *Navicula* in less than a week. Thus, food organisms must be grown in advance if the culture is to be successful. The protocol can be varied so that the food organisms are heat killed or inactivated before inoculation into the protist culture.

3. Heat kill food organisms by immersing test tubes containing them in a 50°C water bath for 10 to 30 min.

4. Inoculate a known volume of the heat-killed organisms into the cultures with the protists.

5. Observe the cultures to see if food is in excess or limiting.

6. Adjust feeding regime accordingly.

Inductive Methods

Inductive methods require more time, experience, and skills, but they offer greater potential for success than fortuitous methods. They are particularly useful in the study of protists which have partitioned their niches along food resource gradients (e.g., foraminifera).[14,22] The following protocol has been very useful with foraminifera and could be modified to be applied to other protist groups):

1. Isolate potential food organisms from the community associated with the protist of interest. The protocol outlined above (Axonic Phototrophs) is appropriate for this purpose.

2. Grow the axenic food organism in liquid media in a test tube until it reaches mid-log phase. For most salt marsh and littoral algae this is approximately 5 d.

3. Aseptically add filter-sterilized radionuclide (^{14}C or ^{32}P) to the growing algae. The concentration of the labeling medium should be approximately 1 to 10 μCi ml^{-1}. Each of these labels has advantages and disadvantages. The ^{14}C label has the advantage that it is a weak β emitter and does not bind to glass. The disadvantage is that some of it is rapidly metabolized and lost in respiration. The ^{32}P is more tightly bound in the alga and its consumer. It has the disadvantage that it is a strong β emitter and binds strongly to glassware.

4. Incubate culture for an additional 3 d.

5. About 4 h before beginning the next step inoculate the foraminifera and seawater into the wells of nine-hole spot plates. Place the spot plates into a large petri plate with a moist paper towel on the bottom.

6. Transfer culture to conical bottom centrifuge tube.

7. Centrifuge algae out of labeling medium ($\sim$1000 $\times$ g, 10 min); decant supernatant into radioactive waste.

8. Resuspend in sterile seawater; agitate with a vortex mixer.

9. Repeat steps 6 and 7 (above) until there is no label left in the supernatant medium (usually five to eight washes).

10. Resuspend cells in a small known volume of seawater.

11. Take aliquots to assay radioactivity and to enumerate algal population.

12. With the aid of a dissection microscope examine the spot plate prepared in step 5. Add aliquots of tracer-labeled algal cells to wells containing foraminifera with extended reticulopodia.

13. Incubate for 3 to 6 h.

14. With the aid of forceps or brushes, transfer the foraminifera to another well with sterile seawater. Brush the foraminifera vigorously and change the seawater three times.

15. Incubate foraminifera in spot plates inside large petri dishes for an additional 12 to 24 h. (While uptake of food could be measured at this point, experience has shown that a varying fraction of what is ingested by foraminifera is egested within the first 24 h[14]).

16. Repeat step 14.

17. Withdraw all the water from the well with a pasteur pipette and add 0.1 ml of 1 mM EDTA. The tests of the foraminifera should dissolve. Since this varies with age and species, check with the aid of a dissection microscope.

18. With the aid of a pipette remove the partially dissolved foraminifera and transfer them to a liquid scintillation vial with cocktail.

19. After subtracting background, and compensating for efficiency and quench, the DPM of the foraminifera in the vial are divided by the DPM per algal cell and the weight of the foraminifera to get the amount of algae consumed and retained per milligram foraminifera.

20. Examine the data and choose two or more food organisms to be used to set up synxenic cultures.

21. Inoculate the foraminifera and a mixture of food organisms into a tissue culture flask. Add a mixture of antibiotics to suppress bacterial growth.

22. The next steps in the study of synxenic (gnotobiotic) or axenic cultures of protists or meiofauna should be tailored to the aims of the investigator. Three different approaches have generally been used in the past: (1) substitution of one organism for another, (2) substitution of nonliving organism-derived materials for the organism itself, and (3) substitution of essential amino acids, lipids, vitamins, or other metabolites for live food. Each of these approaches has made major contributions to our present understanding of the factors which underly quality related aspects of microbial (and meiofaunal) food webs.[2,3,18,19,23-28] Generation time, reproduction (or not), and longevity are some of the criteria which have been used to make comparative judgments on the value of the substitutions.

REFERENCES

1. Lee, J. J., Brief perspective on the autecology of marine protozoa, in *Protozoa and their Role in Marine Processes,* NATO ASI Series, Vol. G 25, Springer-Verlag, Berlin, 1991, 181.
2. Rubin, H. A. and Lee, J. J., Informational energy flow as an aspect of the ecological efficiency of marine ciliates, *J. Theor. Biol.,* 62, 69, 1976.
3. Lee, J. J., Physical, chemical and biological quality related food-web interactions as factors in the realized niches of microzooplankton, *Ann. Inst. Oceanogr.,* 58, 19, 1982.
4. Lee, J. J., Informational energy flow as an aspect of protozoan nutrition, *J. Protozool.,* 27, 5, 1980.
5. Lee, J. J., Tietjen, J. H., Saks, N. M., Ross, G. G., Rubin, H., and Muller, W. A., Educing and modeling the functional relationships within sublittoral salt marsh aufwuchs communities — inside one of the black boxes, *Estuarine Res.,* 1, 710, 1975.
6. Lee, J. J., Tietjen, J. H., Mastropaolo, C., and Rubin, H., Food quality and the heterogenous spatial distribution of meiofauna, *Helgolander Wiss. Meeresunters.,* 30, 272, 1977.
7. Rubin, H. A. and Lee, J. J., Food quality and microzooplankton patchiness, in *Ecology of Marine Protozoa,* Capriulo, G. M., Ed., Oxford University Press, New York, 1990, 271.
8. Muller, W. A. and Lee, J. J., Biological interactions and the realized niche of *Euplotes vannus* from the salt marsh Aufwuchs, *J. Protozool.,* 24, 523, 1977.
9. Lee, J. J., McEnery, M. E., Kennedy, E. M., and Rubin, H., A nutritional analysis of a sublittoral diatom assemblage epiphytic on *Enteromorpha* from a Long Island salt marsh, *J. Phycol.,* 11, 14, 1975.
10. Hoshaw, R. W. and Rosowski, J. R., Methods for microscopic algae, in *Handbook of Phycological Methods, Culture Methods and Growth Methods,* Stein, J. R., Ed., Cambridge University Press, Cambridge, 1979, chap. 3.
11. Reimer, C. W. and Lee, J. J., New species of endosymbiotic diatoms (Bacillariophyceae) inhabiting larger foraminifera in the Gulf of Elat (Red Sea), Israel, *Proc. Nat. Acad. Sci. Phil.,* 140, 339, 1988.
12. Lee, J. J., Crockett, L. J., Hagen, J., and Stone, R., The taxonomic identity and physiological ecology of *Chlamydomas hedleyi* sp. nov., algal flagellate symbiont from the foraminifer *Archaias angulatus,* Br. J. Phycol., 9, 407, 1974.
13. Lee, J. J., Fine structure of the rhodophycean *Porphyridium purpureum in situ* in *Peneroplis pertusus (Forskol)* and *P. acicularis (Batsch)* and in axenic culture, *J. Foram. Res.,* 20, 162, 1990.
14. Lee, J. J., Erez, J., ter Kuile, B., Lagziel, A., and Burgos, S., Feeding rates of two species of larger foraminifera, *Amphistegina lobifera* and *Amphisorus hemprichi,* from the Gulf of Elat (Red Sea), *Symbiosis,* 5, 61, 1988.
15. Lee, J. J., Tietjen, J. H., Stone, R. J., Muller, W. A., Rullman, J., and McEnery, M., The cultivation and physiological ecology of members of salt marsh epiphytic communities, *Helgolander Wiss. Meeresunters.,* 2, 136, 1970.
16. Guillard, R. L., Methods for microflagellates and nannoplankton, in *Handbook of Phycological Methods, Culture Methods and Growth Methods,* Stein, J. R., Ed., Cambridge University Press, Cambridge, 1979, chap. 4.
17. Provasoli, L., Cultivation of animals, in *Marine Ecology,* Vol. 3, Kinne, O., Ed., John Wiley & Sons, New York, 1977, chap. 5.
18. Shiraishi, K. and Provasoli, L., Growth factors as supplements to inadequate algal foods for *Tigriopus japanicus, Tohuku. J. Agric. Res.,* 10, 89, 1959.
19. Provasoli, L., Shiraishi, K., and Lance, J. R., Nutritional idiosyncrasies of *Artemia* and *Tigriopus* in monoxenic culture, *Ann. N.Y. Acad. Sci.,* 77, 250, 1959.
20. Anderson, O. R., Lee, J. J., and Faber, W. W., Collection, maintenance and culture methods for the study of living foraminifera, in *Biology of Foraminifera,* Lee, J. J. and Anderson, O. R., Eds., Academic Press, London, 1991, chap. 10.
21. Hemleben, C., Spindler, M., and Anderson, O. R., *Modern Planktonic Foraminifera,* Springer-Verlag, New York, 1989.

22. Bradshaw, J. S., Preliminary laboratory experiments on ecology of foraminiferal populations, *Micropaleontology,* 1, 351, 1957.

23. Lee, J. J., McEnery, M. E., Pierce, S., Freudenthal, H. D., and Muller, W. A., Tracer feeding in littoral foraminifera, *J. Protozool.,* 13, 659, 1966.

24. Lee, J. J., Tietjen, J. H., and Garrison, J. R., Seasonal switching in the nutritional requirements of *Nitocra typica,* a harpacticoid copepod from salt march aufwuchs communities, *Trans. Am. Micros. Soc.,* 95, 628, 1976.

25. Muller, W. A. and Lee, J. J., Apparent indispensability of bacteria in foraminiferan nutrition, *J. Protozool.,* 16, 471, 1969.

26. Dougherty, E. C., Ed., Symposium on Axenic Culture of Invertebrate Metazoa: A Goal, *Ann. N.Y. Acad. Sci.,* Vol. 77, 1959.

27. Conklin, D. E. and Provasoli, L., Nutritional requirements of the water flea *Moinia macrocopa, Biol. Bull.,* 152, 337, 1977.

28. Lee, J. J. and Pierce, S., Growth and physiology of foraminifera in the laboratory. IV. Monoxenic culture of an allogromiid with notes on its morphology, *J. Protozool.,* 10, 404, 1963.

29. Provasoli, L., McLaughlin, J. J. A., and Droop, M. R., Development of artificial media for marine algae, *Arch. Microbiol.,* 25, 1957.

30. Lee, J. J. and Muller, W. A., Culture of salt marsh microorganisms and micrometazoa, in *Culture of Marine Invertebrate Animals,* Smith, W. L. and Chanley, M. N., Eds., Plenum Press, New York, 1975, 87.

31. McLachlan, J., Growth media-marine, in *Handbook of Phycological Methods, Culture Methods and Growth Methods,* Stein, J. R., Ed., Cambridge University Press, Cambridge, 1979, chap. 2.

32. Pintner, I. J. and Provasoli L., Artificial cultivation of a red-pigmented marine blue-green alga *Phormidium persicinum, J. Gen. Microbiol.,* 18, 190, 1958.

33. Droop, M. R., A procedure for routine purification of algal cultures with antibiotics, *Br. Phycol. Bull.,* 3, 295, 1967.

34. Provasoli, L., Effect of plant hormones on *Ulva, Biol. Bull.,* 114, 375, 1958.

CHAPTER 6

Long-Term Culture of Marine Benthic Protists

Philip G. Carey

INTRODUCTION

The interstitial spaces of marine sediments are known to harbor a rich diversity of protistan groups that appear to play a key role in the passage of energy through food webs. Two groups have received substantial attention from biologists: ciliates tend to dominate the protistan assemblage in porous sands of the intertidal zone and foraminifera in deeper water, where interstices are blocked by fine particles. Other important groups such as the naked amoebae and the smaller zooflagellates, euglenids, and bodonids have only recently been the subject of detailed study.

Experimental studies on organisms maintained in laboratory culture have involved nutrition, growth rate, physiological ecology, ultrastructure, and development. The maintenance and long-term culture of four groups of marine benthic protists is described here: the ciliates, foraminifera, naked amoebae, and the heterotrophic flagellates. Representatives of other benthic protist groups such as the heliozoa, testate amoebae, labyrinthulids, mycetozoa, and the enigmatic Xenophyophorea and Komokiacea may be encountered in sediment samples. These organisms have not yet been cultured and whether they play an important role in sediment ecology is still unknown.

Our knowledge of the marine benthic ciliates is extensive. Summaries of this ecologically important group can be found in References 1 to 6. Benthic ciliates fall broadly into two groups: obligate forms displaying morphological adaptations to the benthic environment and those that are not restricted solely to sediments. The first group is of great interest as it contains the "karyological relict" ciliates and anerobic ciliates that form intimate associations with sulfur bacteria. Food materials consumed by ciliates include diatoms, flagellates, other ciliates, and bacteria. The development of the "maintenance unit" has enabled these protists to be maintained in the laboratory for periods of up to 6 months, making them freely available for detailed study at sites far removed from the sea.[7] The maintenance unit is a small-scale aquarium system designed to circulate clean, oxygenated seawater around, but not directly through, a cube of sand retained within a mesh bag. Interstitial water is gently oxygenated by moving water over the sand surface.

Foraminifera are phagotrophic amoeboid protists that dwell within a test and capture food using anastomosing pseudopods. Trophic mechanisms of foraminifera in their natural habitats involve uptake of dissolved organic matter, herbivory, carnivory, omnivory, suspension feeding, scavenging, parasitism, and cannibalism. Symbiosis with algae commonly occurs.[8] The number of laboratory studies on the nutrition of these organisms has been considerable (see References 9 and 10). They are selective in their feeding habits; however, many benthic genera are easily maintained in the laboratory. The simplest method is to place live organisms in a circulating seawater aquarium in which the water is gently flowing. Nevertheless, most research studies have utilized organisms cultured in small dishes of seawater to which mineral grains, bacterial supplements, and algae have been added. Success is dependent on the maintenance of a range of algal species to act as a food source for the establishment of di- or polyxenic cultures; commercially available cultures are usually employed. In general the small, weakly silicified pennate diatoms and small unicellular chlorophytes appear to be good nutritional sources.

Naked amoebae are now recognized to play an important role in the open ocean ecosystem (reviewed by Anderson[1]) and their importance in benthic sediments is also being appreciated to an increasing extent.[11] They are often overlooked in sediment samples; nevertheless, they are common in all types of shallow marine sediments.

Flagellate protists of the open ocean have been shown to be very important grazers of bacteria and they clearly play a significant role in nutrient cycling.[6,12,13] A substantial fauna has also been found in shallow and deep water sediments; however, the importance of these organisms to the benthic ecosystem has not been clearly elucidated.[14-18] The photosynthesizing flagellates of the benthos have been more widely studied than the heterotrophic flagellate biota. Many of the smaller flagellates do not appear to possess food vacuoles and are considered to be osmotrophic, feeding on dissolved organic matter. Bacteriophagous forms and diatom feeders are common. Pedinellalean (actinomonad) helicoflagellates mainly consume bacteria whereas some euglenids and bodonids are able to browse on diatoms attached to sediment particles. General questions concerning microflagellate bactivory and size selective grazing have been addressed by Anderson and Fenchel.[12]

MATERIALS REQUIRED

Maintenance Unit

The advantage of this system is that it uses inexpensive and readily obtainable equipment designed for the aquarist, and a series of stand-alone units can be arranged for separate samples requiring different environmental conditions. The main disadvantage is one of volume. With its small dimensions, only a limited quantity of sediment can be held.

- Polypropylene holding tank, 30 × 30 × 15 cm
- Polypropylene reservoir tank, 30 × 30 × 60 cm
- Polypropylene sample container, 22 × 15 × 8 cm, drilled with numerous holes
- Tank fittings and top in nontoxic plastic
- Polythene tubing, various sizes
- Nylon cloth, 140-μm mesh size
- Aquarium undergravel filter plates
- Nontoxic plastic strips, various sizes
- Undergravel filter "power head" pump, giving 180 l^{-1}
- Aquarium canister pump, giving 420 l^{-1}
- Crushed cockle shell or clean coarse coral sand
- Seawater of correct salinity, natural or good quality artificial
- Air pump and diffusers, preferably of wood
- Protein skimmer, air or power operated
- Nitrate and nitrite aquarium test kits
- Polypropylene corers 150 mm long × 30 mm diameter
- Constant temperature room

Tanks for the sand sample, bench-top holding tank, and underbench reservoir should be made from food-grade polypropylene. Many of the component parts including tanks, polyethylene tubing, and nontoxic tank fittings can be purchased from hardware stores; all other parts can be purchased from aquarium suppliers. Nylon cloth, the mesh size chosen to retain the collected sand, can be obtained from a specialist supplier (Nitex, from Tobler, Ernst and Traber Inc., or Nylon Bolting Cloth, from Henry Simon, Stockport, U.K.). Undergravel filter plates are made by a number of companies but those manufactured by Rena, Annecy, France, fit the floor of the holding tank exactly. A point of attachment is provided for the fitting of an undergravel filter "power head" pump (Eheim 1007, G. Eheim Inc., Germany). A canister pump (Fluval 202, R.C. Hagen Corp.) filled with crushed cockle shell or coarse coral sand is used to buffer pH. Clean seawater of the correct salinity collected at the sampling site can be used, but, as the apparatus takes a working volume of 50 l, a good grade of artificial seawater (Tropic Marin, Biener GmbH Aquarientechnik, Germany or Instant Ocean, Aquarium Systems) can be substituted. An air pump is used to provide air to two diffusers, one situated in each tank as well as to the air-operated protein skimmer (Sander WT250, E. Sander Elecktronapparat, Germany) which is employed to clean seawater by foam fractionation. High-efficiency, power-operated skimmers are manufactured by Tunze GmbH, Germany. Nitrate and nitrite seawater aquarium test kits are supplied by a number of companies.

Clonal Culture

Equipment

- Constant temperature environment with artificial illumination
- Low-power microscope

Supplies

- Culture dishes with high surface:volume ratio
- Membrane-filtered seawater
- Micropipettes made from pasteur pipettes

Solutions, Media, and Cultures

- Mineral grains obtained from sterilized beach sand, selected for size by sieving and selective sedimentation techniques.
- For Foraminifera, cultures of suitable marine algae can be obtained from the American Type Culture Collection, 12301 Parklawn Drive, Rockville, MD 20852 and the Culture Centre of Algae and Protozoa, Scottish Marine Biological Association, Dunstaffnage Marine Research Laboratory, P.O. Box 3, Oban, Argyll, PA34 4AD, U.K. Suitable organisms are listed below:

Diatoms	**Chlorophytes**
Amphora spp. (*Helamphora*)	*Dunaliella* spp.
Nitzschia hungarica	*Nannochloris* sp.
N. brevirostris	*Chlorococcum* sp.
N. acicularis	*Chlorella* spp.
N. ovalis	
Phaeodactylum tricornutum	
Cylindrotheca closterium	

- Egg yolk medium
    Seawater, 0.45 μm Millipore filtered
    Streptomycin sulfate, 100 mg^{-1} final concentration
    Pencillin-g, 100 mg^{-1} final concentration
    Freshly boiled egg yolk
- Foyn's Erdschreiber Medium
    Stock Solutions
        A. NaNo$_3$ 2.35 M
        B. Na$_2$HPO$_4$ 0.07 M
        C. Soil extract prepared by mixing 1 kg unmanured garden soil and 1 l distilled water;
            bring to pH 8 with NaOH; autoclave for 1 h at 15 lb. pressure; decant or preferably
            filter supernatant liquid which is the soil extract; store under refrigeration.
- Final composition of medium
    Membrane-filtered seawater   950 ml
    Soil extract   50 ml
    Stock solutions A and B   1 ml each
- Cerophyl
    Dried cereal grass leaf, obtainable from Cerophyl Laboratories Inc., Kansas City, MO;
    supplied by International Marketing Corp. 36, Lenexa Business Center, 9900 Pflumm
    Road, Lenexa, KS 66215.

PROCEDURES

Maintenance Unit

1. The unit should be situated in a constant temperature environment. The apparatus is
    arranged with the holding tank (A) situated on the bench top and the reservoir (B) un-
    derneath. The filter plates (UG) are placed on the floor of the holding tank and the power
    head filter (PH) attached. Tubes for the high-level drain (HLD) and low-level drain (LLD)
    with tap are fitted. The canister filter or circulating pump (CP) connects the reservior to
    the holding tank. The protein skimmer (PS) is installed.

2. The apparatus is filled with seawater and all the pumps started. The water is vigorously
    aerated. The power head is so situated that it pumps water from beneath the filter plates
    and additionally around the sample container. Any gaps over the plates are filled with
    strips of plastic to ensure good all-around flow. The canister filter takes water from the
    reservoir, pumps it through the buffering material, and into the top tank. After its passage
    around but not directly through the sample, the seawater is removed from the top tank
    via the high-level drain. Water in the reservoir tank is purified by protein skimming.

3. At the collecting site, the sample container, lined with nylon mesh, is filled with thixotropic
    sand to ensure the presence of interstitial water and its associated ciliate fauna. A single
    block of sand shaped to fit the sample container can be taken or a number of scoops of
    sand can be combined. Transport and handling should be carried out at low temperature
    to avoid depletion of oxygen in transit.

4. In the laboratory the sample container is quickly placed in the holding tank, on top of the
    filter plates. The holding tank can be covered to reduce evaporation.

5. Routine measurements are made on a number of parameters. Temperature should be
    checked every day in addition to the concentration of nitrites. A maximum level of 0.1
    mg l^{-1} nitrite should be maintained throughout. The use of the protein skimmer is essential
    to maintain water quality. Nitrate levels are determined weekly and should never rise
    about 10 mg l^{-1} for littoral sand communities and 20 mg l^{-1} for estuarine sand communities.

Weekly partial water changes can reduce the level of nitrites considerably. Water is kept fully oxygenated, as measured by weekly readings of dissolved oxygen (DO).

6. Small samples of sand can be taken directly for examination. Sampling with a corer is effected by reducing the water level in the holding tank, then opening the tap on the low-level drain until the level is reduced to the surface of the sand. All pumps are kept running during this procedure. As the core is taken, the tap is closed and the normal water level is restored within 60 s. A vertical zonation is usually established within the sample and areas of the sand may tend toward anoxia.

Clonal Culture

The advantage of establishing single cell cultures is self-evident. With care such cultures can be maintained for many years and now form the bulk of material in international culture collections. The disadvantage is that, when stocks are built up, the number of cultures that need to be handled tends to rise exponentially.

Ciliates

A range of benthic ciliates can be cultured in dishes by simple enrichment; either solid or liquid nutrients can be added to these cultures to increase numbers of bacteria and algae. Carnivorous species will require the separate cultivation of suitable ciliates to be added as food. Cereal grains or Cerophyl will initiate short-term cultures of many genera.

Clones are established by transferring selected specimens, from growing isolates, by micropipette into fresh media. Deep form, 50-mm diameter, plastic disposal petri dishes are suitable containers. Selection of ciliates from sand samples is preferable to obtaining organisms after seawater-ice extraction, where they may have been subject to osmotic stress. Long-term clonal culture involves rigorous techniques to ensure that no contaminating organisms are transferred, as some ciliates, especially the ubiquitous *Euplotes*, can rapidly overgrow other species. Many interstitial ciliate genera have proved difficult to establish in culture. Fenchel[19] studied a range of ciliates feeding on natural food materials, but only one genus of karyorelictid ciliates, *Tracheloraphis*, has been successfully established in clonal culture using an egg yolk medium,[20] subculturing every 2 to 4 weeks.

As many of the interstitial ciliates are microaerophillic, living around the redox potential discontinuity layer, greater success with these demanding organisms may be obtained by cultivation in seawater containing low levels of dissolved oxygen. Laboratory modeling of sulfide-rich sand communites was undertaken by Fenchel[21] using microaquaria containing seawater and 1 mM sodium sulfide.[22] Recent studies have involved the use of stoppered vials containing wheat grain medium and lettuce medium in the presence of 1 mM sulfide (HS$^-$).[23,24]

Foraminifera

Algal strains are established on suitable media to act as food materials for foraminiferan culture. Live foraminifera and seawater samples are then collected for establishment of suitable gnotobiotic cultures. Primary cultures of collected material are established in Erd-schreiber medium. Live, healthy cells should be picked out from benthic samples under low-power magnification and are used to establish cultures with the rigorous exclusion of other organisms. Cultures are fed with suitable algae, single strains, or mixtures. The concentration is adjusted so that overgrowth of algae does not occur. Cultures thus established

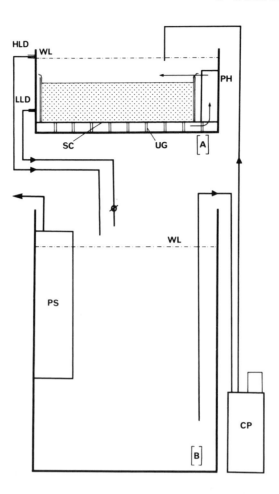

Figure 1.   Maintenance unit. A, holding tank; B, reservoir tank; CP, circulating pump; HLD, high-level drain; LLD, low-level drain with tap; PH, undergravel power head pump; PS, protein skimmer; SC, sand container lined with mesh; UG, undergravel filter plates; WL, water level. (From Carey, P. G., *J. Protozool.*, 33(3), 442, 1986. With permission.)

are maintained at a suitable temperature, 10 to 15°C, under artificial light of approximately 250 W cm^{-2}. The dishes are always covered to prevent evaporation and changes in salinity. Fresh, heat-killed algae may also be tried. Culture tubes of algae are killed by plunging for a few seconds into a hot water bath at 60°C. The algae are then centrifuged for a short time and one drop of this concentrated culture is added to each dish. Mineral grains of the correct size may be prepared from fine beach sand and a small quantity may be added to those cultures that require foreign matter to be incorporated into their test construction. Subcultures can be made from single cell isolates at regular intervals. Cultures are observed every 7 d to determine frequency of subculture.

Naked Amoebae

Despite the difficulty in identifying the presence of naked amoebae in field samples, most species will adapt readily to culture conditions using simple tried and trusted microbiological techniques. A detailed summary of collecting methods, initial handling, culture conditions, clonal isolation, media, and an illustrated key to identification, the indispensable guide to marine gymnamoebae by Page,[25] is essential reading. The key to freshwater and soil amoebae by Page[26] contains useful information and should also be consulted. The technique which has yielded greatest success is the enrichment of field samples, followed by the establishment of clonal cultures on an agar substrate or in liquid media. By inoculating freshly collected material into several media, a full range of amoebae and their food materials can be revealed. Some amoebae will only grow in a liquid medium, but many will grow on both solid and in liquid media. Some genera such as *Platyamoeba* thrive at lower salinities, so one set of cultures would be prepared in full-strength seawater, i.e., 35 ppt salinity and one set at a lower (26 ppt) salinity. After mixed cultures of amoebae and their accompanying food organisms are established, clonal cultures utilizing the same food organisms are prepared. Further refinement of clonal cultures can be undertaken by using known food materials. Suitable food organisms successfully employed with nonbenthic amoebae for this stage are the diatom *Amphiprora hyalina* for the amoeba *Parameoba eilhardi*, the bacterium *Escherichia coli* for *Heteramoeba clara*, and *Ocenanospirillum* for various marine genera.

Cultures without a liquid overlay have clear advantages over ''wet'' cultures. They are easier to handle and they do not spill their contents when tipped. However, debris can build up in such cultures, making observation difficult. Single species may tend to dominate in enriched samples and less numerous species may easily be overlooked.

Flagellates

Culture methods have been fully documented elsewhere.[27-29] The older works of Droop[30] and Lewis[31] should also be consulted. The method usually involves four basic steps: (1) enrichment of freshly collected material, (2) dilution of the culture thus established, (3) isolation of single cells or colonies to act as clone founders, and (4) migration or application of antibiotics to establish axenic cultures. This methodology has proved its worth with many species; nevertheless, some organisms may require specialized techniques. The bacterivorous flagellate isolated from deep-sea sediment by Turley et al.[32] did not grow at 2°C and 1 atm; however, it grew prolifically in a simulated deep-sea environment at 2°C and 450 atm.

Cultures once established can be maintained for years and cultures of many common species are available from culture collections. Flagellates that are more demanding in their requirements, e.g., those from the deep sea, require specialized equipment.

REFERENCES

1. Anderson, O. R., *Comparative Protozoology, Ecology, Physiology, Life History,* Springer-Verlag, New York, 1988, chap. 7.
2. Carey, P. G., *Marine Interstitial Ciliates — An Illustrated Key,* Chapman and Hall, London, 1991.
3. Corliss, J. O., Hartwig, E., and Lenk, S. E., Ciliophora, in *Introduction to the Study of Meiofauna,* Higgins, R. P. and Thiel, H., Eds., Smithsonian Institution Press, Washington, DC, 1988, 258.
4. Fenchel, T., *Ecology of Protozoa: The Biology of Free-Living Phagotrophic Protists,* Springer-Verlag, Berlin, 1987, chap. 8.
5. Lee, J. J., The ecology of marine protozoa: an overview, in *Ecology of Marine Protozoa,* Capriulo, G. M., Ed., Oxford University Press, New York, 1990, chap. 1.
6. Patterson, D. J., Larson, J., and Corliss, J. O., The ecology of heterotrophic flagellates and ciliates living in marine sediments, *Prog. Protistol.,* 3, 185, 1989.

7. Carey, P. G., A method for long-term laboratory maintenance of marine psammophilic ciliates, *J. Protozool.*, 33, 422, 1986.

8. Lipps, J. H., Biology/paleobiology of foraminifera, in *Foraminifera: Notes for a Short Course,* Vol. 6, Broadhead, T. W., Ed., University of Tennessee, 1982, 1.

9. Lee, J. J., Towards understanding the niche of foraminifera, in *Foraminifera,* Vol. 1, Headly, R. H. and Adams, C. G., Eds., Academic Press, London, 1974, 207.

10. Lee, J. J., Nutrition and physiology of the foraminifera, in *Biochemistry and Physiology of the Protozoa,* Vol. 3, 2nd ed., Levendowsky, M. and Hutner, S. H., Eds., Academic Press, New York, 1980, 43.

11. Gooday, A. J., Sarcomastigophora, in *Introduction to the Study of Meiofauna,* Higgins, R. P. and Thiel, H., Eds., Smithsonian Institution Press, Washington, DC, 1988, chap. 17.

12. Andersen, P. and Fenchel, T., Bactivory by microheterotrophic flagellates in seawater samples, *Limnol. Oceanogr.,* 30, 198, 1985.

13. Fenchel, T., The ecology of heterotrophic microflagellates, *Adv. Microb. Ecol.,* 9, 57, 1986.

14. Fenchel, T., Studies on the decomposition of organic detritus derived from the turtle-grass *Thalassia testudinum, Limnol. Oceanogr.,* 15, 14, 1970.

15. Fenchel, T., The quantitative importance of the benthic microfauna of an arctic tundra pond, *Hydrobiologica,* 46, 445, 1975.

16. Fenchel, T. and Harrison, P., The significance of bacterial grazing and mineral cycling for the decomposition of particulate detritus, in *The Role of Terrestrial and Aquatic Organisms in Decomposition Processes,* Anderson, J. M. and McFadyen, A., Eds., Blackwell Scientific, Oxford, 1976, 285.

17. Fenchel, T. and Jorgensen, B. B., Detritus food chains of aquatic ecosystems: the role of bacteria, *Adv. Microb. Ecol.,* 1, 1, 1977.

18. Robertson, M. L., Mills, A. L., and Ziemann, J. C., Microbial synthesis of detritus like particulates from dissolved organic carbon released by tropical seagrasses, *Mar. Ecol. Prog. Ser.,* 7, 279, 1982.

19. Fenchel, T., The ecology of marine microbenthos. II. The food of marine benthic ciliates, *Ophelia,* 5, 73, 1968.

20. Lenk, S. E., Small, E. B., and Gunderson, R., Preliminary observations of feeding in the psammobiotic ciliate *Tracheloraphis, Origins Life,* 13, 229, 1984.

21. Fenchel, T., The ecology of marine microbenthos. IV. Structure and function of the benthic ecosystem, its chemical and physical factors and the microfauna communities with special reference to the ciliated protozoa, *Ophelia,* 6, 1, 1969.

22. Fenchel, T., Perry, T., and Thane, A., Anaerobiosis and symbiosis with bacteria in free-living ciliates, *J. Protozool.,* 24, 154, 1977.

23. Fenchel, T. and Finlay, B. J., Endosybiotic methanogenic bacteria in anaerobic ciliates. Significance for the growth efficiency of the host, *J. Protozool.,* 38, 18, 1991.

24. Fenchel, T. and Finlay, B. J., Synchronous division of an endosymbiotic methanogenic bacterium in the anaeorobic ciliate *Plagiopyla frontata,* Kahl, *J. Protozool.,* 38, 22, 1991.

25. Page, F. C., *Marine Gymnamoebae,* National Environment Research Council, Institute of Terrestrial Ecology, Cambridge, 1983.

26. Page, F. C., *A New Key to Freshwater and Soil Gymnamoebae,* Culture Centre of Algae and Protozoa, Freshwater Biological Association, U.K., 1988.

27. Guillard, R. L. L., Methods for microflagellates and nannoplankton, in *Handbook of Phycological Methods,* Stein, R. J., Ed., Cambridge University Press, London, 1979, chap. 4.

28. Hoshaw, R. W. and Rosowski, J. R., Methods for microscopic algae, in *Handbook of Phycological Methods,* Stein, J. R., Ed., Cambridge University Press, London, 1979, chap. 3.

29. McLachlan, J., Growth media-marine, in *Handbook of Phycological Methods,* Stein, J. R., Ed., Cambridge University Press, London, 1979, chap. 2.

30. Droop, M. R., A note on the isolation of small marine algae and flagellates for pure culture, *J. Mar. Biol. Assoc. U.K.,* 33, 511, 1954.

31. Lewis, R. A., The isolation of algae, *Rev. Algol.,* 3, 181, 1959.

32. Turley, C. M., Lochte, K., and Patterson, D. J., A barophilic flagellate isolated from 4500 m in the mid-North Atlantic, *Deep-Sea Res.,* 35, 1079, 1988.

CHAPTER 7

Behavior and Bioenergetics of Anaerobic and Microaerobic Protists

B. J. Finlay

INTRODUCTION

A single chemical factor — the dissolved oxygen tension — plays a cardinal role in controlling the spatial distribution of many protists living in sediments and stratified water columns.[1-3] Protists living in the mixed open water of oceans and lakes will rarely sense an oxygen gradient, but this is not so for the great diversity of protists living at greater depths. Many of the latter are microaerophilic,[4-9] preferring an oxygen tension around 5% of the atmospheric saturation value, and a surprising variety of species are anaerobic.[10]

It has been shown that microaerobic species grow best at low oxygen tensions,[5] and that anaerobes requires the almost complete absence of oxygen.[11] It is clear therefore that many of the methods commonly employed to study the behavior and various bioenergetic parameters of protists, especially those methods which provide homogeneously high oxygen tensions, are unsuitable.

It is relatively straightforward to produce natural oxygen gradients (i.e., produced by the oxygen consumption of microorganisms) in simple glass chambers in the laboratory, to introduce protists, and to allow them to demonstrate their preferred oxygen regime: when offered a choice, their normal chemosensory behavior will reveal their natural preference. As shown below, it is also possible to extend the use of these chambers to the measurement of some basic bioenergetic parameters, e.g., respiration and cell yield. With a little skill and some patience, a complete energy budget for a phagotrophic protist can be determined using a single capillary microslide.

Most of our experience is with anaerobic and microaerobic ciliates, but there is no reason why the methods described below should not be extended to other protists. Some of the methods have already been adapted for photosensitive and phototrophic species.[4,7,12,13]

The main advantages of using the types of glass chambers described are (1) that the behavior and health of the protists can be continuously monitored, (2) that microaerobic species are able to choose their optimal oxygen tension, and (3) that the behaviors observed and the values measured for bioenergetic parameters (e.g., respiration) are likely to be representative of the natural condition. The main disadvantages are (1) that some of the techniques, especially those involving capillary microslides, do require some dexterity, (2) that oxygen gradients may take a long time (e.g., a day) to become established if organism density is low, and (3) the processes within these chambers never reach a steady state, e.g., the oxygen gradient may be maintained by a dense population of microaerophilic bacteria which may in turn serve as food for microaerophilic protozoa: so the protozoa gradually consume the biological cause of the steep oxygen gradient which determines their own spatial distribution.

MATERIALS REQUIRED

Equipment

- Light microscope (final magnification 40 to 100×)
- Low-speed bench centrifuge

Supplies

- Microslides (width 4 mm; height 0.4 mm)
- Glass Pasteur pipettes and micropipettes
- Conical pyrex centrifuge tubes (about 15-ml capacity)
- Filtration rig for 25-mm filters
- 0.22-μm pore size, 25-mm membranes
- Vaseline
- Paraffin oil
- Anaerobic handling cabinet or atmos bag (Aldrich Chemical Co.)
- Cylinder of oxygen-free nitrogen
- Glass chambers: A variety of types of small, flat, glass chambers are commercially available, and several of these are ideal for examining the behavior and bioenergetics of protists in the presence and absence of oxygen. We have used three types of chamber, which have the following useful properties:

  1. The "Sedgewick-Rafter chamber" is 1 mm deep and it holds 1 ml. The base area is calibrated in 1-mm squares. A double oxygen gradient is easily established by arranging the cover slip perpendicular to the long axis of the chamber and by pipetting in a volume of approximately 0.7 ml, which remains open on two sides to the air or other gas phase (Figure 1).[8] If the liquid contains high-enough numbers of microaerobic bacteria and protists, clear superimposed bands of both will appear without further assistance, within a few hours (Figure 2). One band can then be kept as a control while the other is manipulated, for example, by the injection of small volumes of respiratory inhibitors.[8]
  2. "Microslides" are flat glass capillaries. They are produced with various dimensions: the most useful ones are 50 mm long, with an internal width of 4 mm and an internal depth of 0.4 mm. The shallow depth of these chambers means that it is possible to keep all protists in focus, especially when using low-power objectives. The ends of microslides are easily sealed with paraffin wax or oil. Microslides are particularly suitable for anaerobic preparations,[14] in which case they are manipulated within an anaerobic handling cabinet or nitrogen filled "atmos bag", and filled with anoxic medium.
  3. Spectrophotometer cells are commercially available with pathlengths down to 1 mm. With one end open to the atmosphere, these chambers can be used to establish a single oxygen gradient. They are prepared by pipetting a centrifuged pellet of protists into the spectrophotometer cell, then placing the latter inside a 20-ml centrifuge tube and recentrifuging the protists onto the base of the spectrophotometer cell. Most of the respiratory oxygen consumption will be associated with the protists, bacterial aggregates, and other sedimented particles, and an anoxic boundary will spread from the base of the chamber upward. Note that 1-mm spectrophotometer cells can be laid horizontally and uncapped. When held upright, these chambers (particularly those with a 2-mm path length) are ideal for observing and photographing the effects of gravity on free-swimming protists (Figure 3).[7,12]

PROCEDURES

Observing Microaerobic and Anaerobic Behavior of Protists

A population of microaerobic protists will, if left undisturbed and free to swim in an oxygen gradient, develop a pronounced aggregation close to the oxic-anoxic boundary[11,13] (Figure 4). This will usually take a few hours. Anaerobic protists also have a chemosensory response to oxygen and their motility is minimal in the complete absence of oxygen[11,15] (Figures 5 and 6). Therefore, given the choice, they will accumulate on the anoxic side of the oxic-anoxic boundary. Most crude cultures of microaerobic and anaerobic protists also contain high numbers of bacteria which are microaerobic or facultatively so (the most active and abundant bacteria are often *Spirillum* species). These bacteria often form a well-defined dense band at the oxic-anoxic boundary and act as a visual marker (Figure 2).

It is quite easy to manipulate the behavior of protists in Sedgewick-Rafter chambers. Most importantly, the microaerobic or anaerobic character of the protists can be confirmed by changing the composition of the gas phase which is in contact with the liquid supporting the protists, for example, by placing the chamber in a perspex sandwich box adapted to admit N_2, N_2/CO_2 (to check the role of pH), or pure O_2.[8]

It can be particularly rewarding to record the behavior of protists in oxygen gradients, using video, or long-exposure still photography. The details of the methods are outside the scope of this brief discussion but relevant methods have been described elsewhere.[7,8,12,13] Using such methods it is possible to separate taxic, transient, and kinetic swimming behaviors.

Measuring Respiration Rate in Glass Capillaries

The oxygen uptake rate of protists is usually a function of the oxygen tension, and the data can invariably be fitted with a Monod-type equation.[5] Since microaerobic ciliates are adapted for life at relatively low-oxygen tensions, "normal" respiration rates are obtained only if the ambient oxygen tension is maintained at some typically low-level. This can be done using any small respiration chamber fitted with an oxygen electrode: but a large number of cells will need to be introduced into the chamber and the dissolved oxygen tension must be maintained at a suitably low level. An alternative method, which needs fewer cells and which provides the cells with a self-sustaining constant oxygen tension, makes use of microslides.

Microaerobic protozoa introduced into air-saturated water will usually clump together.[5,7] As the clump consumes oxygen, it creates a microaerobic microenvironment, and the microaerobic cells soon aggregate at the oxic-anoxic boundary (Figure 7). If the population is physically constrained within the cylindrical chamber of a microslide, most of the protozoa will then migrate as a dense band along the slide. The rate of movement of this band will be proportional to the rate of oxygen consumption by the ciliate population. The following procedure was used with the marine microaerophilic ciliate *Euplotes*.[5] The method can be used with other protists, substituting water, mineral medium, or sterile culture medium for seawater.

1. Centrifuge cells into membrane-filtered culture medium (repeat low-speed centrifugation several times to separate protozoa from contaminating bacteria). Remove the supernatant, filter it (0.22-μm pore size membranes), bubble it with air, and use it for the final suspension of washed protozoa.

2. Using micropipettes, fill a microslide with air-saturated seawater and seal one end with paraffin oil.

3. To the other end of the microslide add a dense suspension of ciliates (about 1000), then seal the end with paraffin oil.

4. Place the microslide on a microscope stage and observe with weak illumination.

5. After a few minutes the ciliates will form a band perpendicular to the length of the microslide. This band will slowly move in the direction of air-saturated water. Measure the migration rate of the band of cells (use an eyepiece graticule, or graph paper on the microscope stage).

6. At the termination of the experiment, count the number of ciliates in the microslide.

7. Calculate the rate of oxygen consumption as the volume of water depleted of oxygen per unit time, e.g., in the case shown (a marine *Euplotes*),[5] the microslide contained 1250

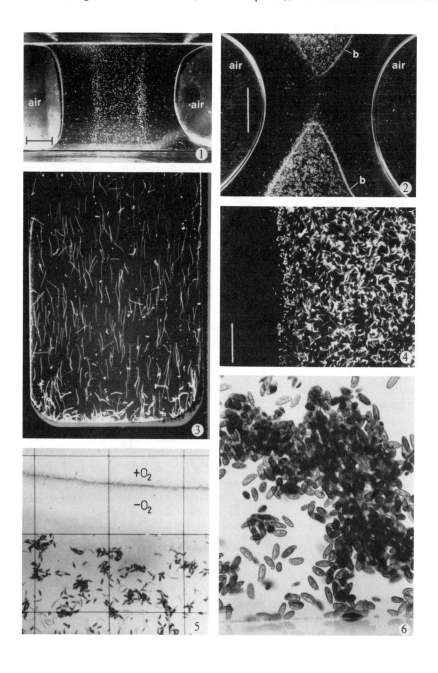

ciliates, each millimeter of microslide contained 9.3 nl dissolved O_2, the band moved 4.3 mm h^{-1} so oxygen consumption was 40 nl O_2 per 1250 ciliates per hour, or 0.032 nl O_2 ciliate^{-1} h^{-1}.

Measuring Population Growth and Yield

Glass capillaries can be used to directly record both consumption of food by individual ciliates and the subsequent growth of the population. The example given below is of the anaerobic marine ciliate *Metopus contortus*.[14] The method may also be used with microaerobic and aerobic ciliates but it should be remembered that, over the time scale of population growth, the ciliates in the microslide will require access to oxygen. In the example given, both the ciliate and the food organisms are fairly large (length 120 μm and several microns, respectively): this helps greatly in obtaining accurate counts. It would be difficult to use this method to determine the rate of consumption by a filter-feeding ciliate, although feeding by some flagellates (e.g., chrysomonads feeding on large particles) can be quantified. In both cases microslides can of course be used to determine population growth of the protozoa.

With the assumption that the food organism does not grow in the microslide, three bioenergetic parameters of the ciliate can be determined: the growth rate constant, (μm), the rate of food consumption (U), and the yield (Y) of the ciliates (on a volume basis).

1. Fill the microslide with a suspension of baker's yeast (about 10^6 cells ml^{-1}) in oxygen-free seawater, inject a few microliters of concentrated cells from a micropipette and seal the ends of the capillary with vaseline. The ciliates should be concentrated by gentle centrifugation, the seawater should be outgassed with oxygen-free nitrogen, and the filling of the microslide should be carried out with the latter enclosed in an anaerobic handling cabinet or an atmos bag filled with nitrogen.

2. Attach the microslide to a normal glass slide and place the latter on the microscope stage.

3. Count the number of ciliates in the microslide at least once per day.

4. Determine the rate of ingestion of yeast cells by direct observation (observe 10 to 15 ciliates for at least 2 to 5 min).

Figure 1. A population of the microaerobic ciliate *Loxodes striatus* stratified in a double oxygen gradient (oxygen is diffusing from the left and from the right) in a Sedgewick-Rafter chamber. The center of the population is probably anoxic. Cell density is higher in the region of each of the two oxic-anoxic boundaries. Scale bar = 5 mm.

Figure 2. A population of *Loxodes striatus*, divided into two parts, in a double horizontal oxygen gradient in a Sedgewick-Rafter chamber (some of the water has evaporated). Tight bands of microaerobic bacteria (b) mark the approximate positions of the oxic-anoxic boundary; 2-s exposure. Scale bar = 5 mm. (Adapted from Finlay, B. J., Fenchel, T., and Gardener, S., *J. Protozool.,* 33, 157, 1986.)

Figure 3. Cells of *Loxodes striatus* showing positive geotaxis in oxygenated water; 5-s exposure. The chamber is a spectrophotometer cell (1 cm wide).

Figure 4. A 5-s exposure of a population of the ciliate *Loxodes striatus*. The cells toward the left are demonstrating a kinetic response (reduced motility) to their preferred low-oxygen tension. The more motile cells toward the right are swimming in anoxic water. Scale bar = 2 mm. (Adapted from Finlay, B. J., Fenchel, T., and Gardener, S., *J. Protozool.,* 33, 157, 1986.)

Figure 5. Typical aggregation of an anaerobic ciliate (*Metopus paleformis*) in anoxic water. A band of microaerobic bacteria indicates the oxic-anoxic boundary. The ciliates restrict themselves to a region deep inside the anoxic zone. Background squares have sides of 1 mm.

Figure 6. Typical clumping of anaerobic ciliates (*Plagiopyla frontata*) suspended in oxygenated water. Each ciliate is approximately 100 μm long.

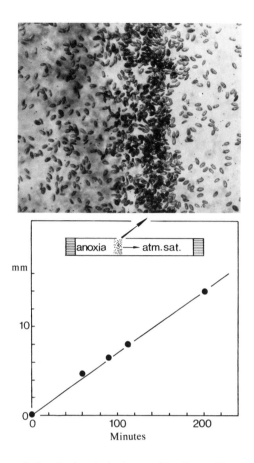

Figure 7.   The migration velocity of a band of microaerobic ciliates (*Euplotes* sp.) through a capillary previously filled with aerated water. Migration velocity is proportional to respiration rate. (Adapted from Fenchel, T., Finlay, B. J., and Gianni, A., *Arch. Protistenk.*, 137, 317, 1989.)

Calculation — In this example, the anaerobic ciliate has a doubling time in microslides of 75 h [μ = (ln 2)/doubling time = 0.0092 h^{-1}]. The rate of uptake (U) of yeast cells is 98 ciliate^{-1} h^{-1}, the average ciliate cell volume (V_c) is 1.3 $\times$ 10^5 μm^3, and the average yeast cell volume (V_y) is 135 μm^3. Thus, the ciliate yield (also known as gross growth efficiency) is

$$Y = [\mu \times V_c]/[U \times V_y]$$
$$= (0.0092 \times 1.3 \times 10^5)/(98 \times 135)$$
$$= 0.09, \text{ or } 9\% \text{ in terms of volume.}$$

NOTES AND COMMENTS

1.   In order to obtain sufficiently high densities of organisms, it is often necessary to concentrate them by centrifugation. There are no firm guidelines on how to do this. We always use 20-ml conical pyrex centrifuge tubes and a swing-out head. Most large living ciliates will sediment into a loose pellet after 100 $\times$ g for a little less than one minute: smaller protists may require up to 1000 $\times$ g. It is particularly useful to have a fast brake on the centrifuge, especially if, as is often the case, the ciliates quickly swim out of their loose-sedimented pellet.

2.   Sedgewick-Rafter chambers are made in glass and in plastic; the glass ones are more durable, but more expensive. Both types can be purchased from Graticules, Ltd., Morley Rd., Tonbridge, Kent, U.K. Microslides are sold by Camlab, Ltd., Nuffield Rd., Cambridge, U.K.

3.   The methods described examine the respiration rate of protozoa which are separated from their food. When starved, the respiration rate of protozoa quickly declines,[16] so it is advisable to complete the experiment as quickly as possible. In practice it takes 2 to 3 h to get reasonable results, even with high densities of cells. An alternative solution is to introduce heat-killed or nonliving food into the microslides.

4.   Population growth rate and yield: in this example the ciliate is relatively large and anaerobic, so its generation time is relatively long. Small, aerobic protozoa will need to be counted every 2 to 15 h, depending on cell size and temperature.[17,18]

5.   There are many ways to measure the cell volume of protists. A simple, quick, and accurate method requires fixation (e.g., 4% formalin, final concentration) of cells followed by either (1) squeezing cells between glass slide and cover slip until they have plane parallel sides or (2) filtration onto black membrane filters (0.22-μm pore size), with or without staining of cytoplasm. In both cases the area of the compressed cell is determined most easily by still or video photography. Care must be exercised in measuring as accurately as possible the (average) depth of the compressed specimen. This can be done using the focusing wheel (this *must* be calibrated) on the microscope and high-power objectives (e.g., 54 to 100×).

REFERENCES

1. Fenchel, T., Kristensen, L. D., and Rasmussen, L., Water column anoxia: vertical zonation of planktonic protozoa, *Mar. Ecol. Prog. Ser.,* 62, 1, 1990.
2. Finlay, B. J., Oxygen availability and seasonal migrations of ciliated protozoa in a freshwater lake, *J. Gen. Microbiol.,* 123, 173, 1981.
3. Finlay, B. J., Physiological ecology of free-living protozoa, *Adv. Microb. Ecol.,* 11, 1, 1990.
4. Berninger, U.-G., Finlay, B. J., and Canter, H. M., The spatial distribution and ecology of zoochlorellae-bearing ciliates in a productive pond, *J. Protozool.,* 33, 557, 1986.
5. Fenchel, T., Finlay, B. J., and Gianni, A., Microaerophily in ciliates: responses of an *Euplotes* species (Hypotrichida) to oxygen tension, *Arch. Protistenk.,* 137, 317, 1989.
6. Finlay, B. J., Berninger, U.-G., Clarke, K. J., Cowling, A. J., Hindle, R. M., and Rogerson, A., On the abundance and distribution of protozoa and their food in a productive freshwater pond, *Eur. J. Protistol.,* 23, 205, 1988.
7. Finlay, B. J., Berninger, U.-G., Stewart, L. J., Hindle, R. M., and Davison, W., Some factors controlling the distribution of two pond-dwelling ciliates with algal symbionts (*Frontonia vernalis* and *Euplotes daidaleos*), *J. Protozool.,* 34, 349, 1987.
8. Finlay, B. J., Fenchel, T., and Gardener, S., Oxygen perception and O_2 toxicity in the freshwater ciliated protozoon *Loxodes, J. Protozool.,* 33, 157, 1986.
9. Finlay, B. J., Span, A. S. W., and Harman, J. M. P., Nitrate respiration in primitive eukaryotes, *Nature,* 303, 333, 1983.
10. Fenchel, T. and Finlay, B. J., The biology of free-living anaerobic ciliates, *Eur. J. Protistol.,* 26, 201, 1991.
11. Fenchel, T. and Finlay, B. J., Oxygen toxicity, respiration and behavioural responses to oxygen in free-living anaerobic ciliates, *J. Gen. Microbiol.,* 136, 1953, 1990.
12. Fenchel, T. and Finlay, B. J., Photobehavior of the ciliated protozoon *Loxodes*: taxic, transient and kinetic responses in the presence and absence of oxygen, *J. Protozool.,* 33, 139, 1986.
13. Finlay, B. J. and Fenchel, T., Photosensitivity in the ciliated protozoon *Loxodes*: pigment granules, absorption and action spectra, blue light perception, and ecological significance, *J. Protozool.,* 33, 534, 1986.
14. Fenchel, T. and Finlay, B. J., Anaerobic free-living protozoa: growth efficiencies and the structure of anaerobic communities, *FEMS Microbiol. Ecol.,* 74, 269, 1990.

15. Finlay, B. J. and Fenchel, T., An anaerobic protozoon, with symbiotic methanogens, living in municipal landfill material, *FEMS Microb. Ecol.,* 85, 169, 1991.
16. Fenchel, T. and Finlay, B. J., Respiration rates in heterotrophic, free-living protozoa, *Microb. Ecol.,* 9, 99, 1983.
17. Fenchel, T., The ecology of marine microbenthos. III. The reproductive potential of ciliates, *Ophelia,* 5, 123, 1968.
18. Finlay, B. J., The dependence of reproductive rate on cell size and temperature in freshwater ciliated protozoa, *Oecologia,* 30, 75, 1977.

CHAPTER 8

Culturing Free-Living Marine Phagotrophic Dinoflagellates

Evelyn J. Lessard

INTRODUCTION

Dinoflagellates are a nutritionally diverse group of protists with phototrophic, hetero-trophic, and mixotrophic members. Although the majority of dinoflagellate species probably are heterotrophs,[1,2] most culturing efforts in the past century have focused on the photo-synthetic forms (an excellent review of the history of these efforts can be found in Reference 3). Interestingly, some of the first dinoflagellates to be brought into culture were heterotrophs[4,5] and two heterotrophic species, *Crypthecodinium cohnii* and *Oxyrrhis marina*, have become the protistan equivalent of the laboratory rat for numerous studies, e.g., genetics,[6] nutri-tion,[7-9] ultrastructure,[10,11] digestion,[12] and grazing.[13] With the recent discovery that heter-otrophic dinoflagellates are abundant and widespread in aquatic systems[14-18] and have im-portant trophic roles in aquatic food webs,[16,19-21] more concerted efforts are being made to culture heterotrophic dinoflagellates.[18,22-24]

Heterotrophy is a nutritional mode in which an organism acquires some or all of its nutritional requirements from organic sources. Osmotrophs utilize dissolved organic matter, whereas phagotrophs feed on particulate organic matter. Dinoflagellates capable of growing solely on dissolved organic matter in culture include *C. cohnii*,[25] *O. marina*,[26] *Noctiluca scintillans*,[27] and *Gyrodinium lebourii*.[28] These dinoflagellates are also capable of phago-trophy. In the dilute natural environment, phagotrophy is undoubtedly the predominant nutritional mode of these and other heterotrophic dinoflagellates. For certain purposes, axenic culture is desirable;[26,29] specific media and approaches for axenic culture can be found in the above references. This chapter will focus on methods for culturing heterotrophic dino-flagellates as phagotrophs. It should be noted that many phototrophic dinoflagellates have phagotrophic capabilities,[30-32] and the inclusion of particulate food sources also may enhance or make possible the culturing of these mixotrophs.

As dinoflagellates feed by a variety of mechanisms, this will be one factor to consider in providing appropriate culture conditions and prey items. Athecate dinoflagellates, which lack a rigid outer cell covering, can engulf their food[2,33] or use a specialized appendage ("peduncle") to pierce their prey, then suck out their cell contents.[34] Some thecate dino-flagellates also feed with a peduncle (*Dinophysis* spp.[35]). Other thecate dinoflagellates feed by engulfing their prey with a pseudopodium, then digest it extracellularly.[22,36] These mech-anisms of feeding allow dinoflagellates to utilize a wide range of prey sizes (including prey larger than themselves), and both motile and nonmotile prey. Thus, potential food items for culturing include bacteria, cyanobacteria, flagellates, ciliates, and diatoms.[2,14,19,35]

MATERIALS REQUIRED

Equipment

- Dissecting microscope with darkfield illumination

Supplies

- Micropipettes (75-μl capacity recommended)
- Aspirator tubing with mouthpiece
- Nine-well Pyrex depression plates, petri dishes, slides
- Culture vessels (see discussion below)

Solutions and Media

- See Table 1

PROCEDURES

General Principles

Attempts to isolate and culture phagotrophic dinoflagellates have followed many of the same principles and procedures used in culturing other protists.[3,29,37-38] As only a few phagotrophic dinoflagellates have been successfully cultured, little is known about their specific nutritional needs or their degree of sensitivity to toxic compounds, light cycles, temperature changes, etc. While some dinoflagellates (e.g., *O. marina*) tolerate neglect (even abuse!) and will grow on a wide variety of foods, others, especially oceanic and thecate types, are very sensitive to culture conditions. As a general guideline, environmental conditions (temperature, light, turbulence) during isolation and culturing should be held as close to ambient as possible to avoid trauma to the organisms.

Culture Vessels

Dinoflagellates have been grown successfully in many types of culture vessels. The best choices for both media preparation and incubations are those made of plastic: polycarbonate, polysulfone, or Teflon flasks, bottles (Nalgene or Corning), and centrifuge tubes (10 to 50 ml, Nalgene Oak Ridge type). The use of glass is not recommended due to the difficulty in completely removing trace metals[39] and the potential for releasing silicic acid from the glass during autoclaving. We also have successfully grown dinoflagellates in polystyrene and polypropylene Falcon tubes, although the success rate was lower. A longer lag period also often occurred than when using polycarbonate containers (in addition, they do not withstand autoclaving). For initial isolations and short-term incubations, we have found that polystyrene 12-well tissue culture plates work quite well for some dinoflagellates. Whatever culture vessels are used, it is important that they be thoroughly clean. Many protists are sensitive to trace metals[39,40] as well as to plasticizers and rubber. Closures for flasks and tubes should be nontoxic, either glass caps, polypropylene screw-type or friction-fit caps.

Table 1. Composition of Culture Media

Media for phototrophs — f/2[37] and K[42]

Nutrients	Final concentration in medium (mol 1⁻¹)	
	f/2	K
Major nutrients		
NaNo₃	8.83×10^{-4}	8.83×10^{-4}
NaH₂PO₄ · H₂O	3.36×10^{-5}	—
Na₂glycero PO₄	—	1×10^{-5}
Na₂SiO₃ · 9H₂O	5.54×10^{-5}	5.54×10^{-5}
Trace elements		
Na₂EDTA	1.17×10^{-5}	1×10^{-4}
FeCl₃ · 6H₂O	1.17×10^{-5}	—
FeEDTA	—	1.17×10^{-5}
MnCl₂ · 4H₂O	9×10^{-7}	9×10^{-7}
ZnSO₄ · 7H₂O	8×10^{-8}	8×10^{-8}
CoSO₄ · 7H₂O	5×10^{-8}	5×10^{-8}
Na₂MoO₄ · 2H₂O	3×10^{-8}	3×10^{-8}
CuSO₄ · 5H₂O	4×10^{-8}	1×10^{-8}
Vitamins		
Thiamin-HCl	3×10^{-7}	3×10^{-7}
Vitamin B12	3.7×10^{-10}	3.7×10^{-10}
Biotin	2.1×10^{-9}	2.1×10^{-9}
Other additions		
H₂SeO₃	—	1×10^{-8}
NH₄Cl	—	5×10^{-5}

Note: Make separate stock solutions for each of the major nutrients, trace metals and other additions; for f/2, FeCl₃, and Na₂EDTA can be combined in one stock solution. A single working stock of trace metals can be made by combining trace metal stock solutions, (and for K media) the extra Na₂EDTA, and selenium. Teflon bottles should be used for stock solutions. The stock solutions can be sterilized by autoclaving or filtering (0.22-μm Millipore MF membrane), then stored in a refrigerator. Vitamins can be combined into one stock, filtered-sterilized (to maintain full potency) in small aliquots and stored in the freezer. TRIS buffer (1 m*M*) also was included in K media.[42] However, as TRIS has been noted to be toxic to some protists, it is best omitted when culturing phagotrophs. For growing stock algal cultures for food, use f/2 or K at full strength. Seawater for media should be filtered through glass fiber filters (Whatman GF/F) to remove particulates and autoclaved in Teflon or polycarbonate bottles. To avoid problems with precipitation and changes in concentration or activity of the additions, it is recommended that all nutrients, trace metals, and vitamins be added aseptically after autoclaving. This can be accomplished by adding the nutrient stocks with sterile pipettes under a sterile hood. Silicate should only be added for diatoms. If seawater cannot be obtained from natural sources, there are several formulations available for making artificial seawater.[3,55] Media for initial isolations of phagotrophs and for subsequent transfers of established frequently fed phagotroph cultures can be made by adding trace metal mixtures and vitamins from either f/2 or K at reduced levels, typically 1/100.

Black phenolic screw caps are best avoided as they can be toxic. If used, they should be autoclaved separately in deionized water and be fitted with Teflon liners.[3] Cultureware first should be washed in Micro detergent (Baxter Scientific Products), rinsed with deionized water, soaked in 2 *N* HCL overnight, and rinsed thoroughly with deionized water. Before autoclaving, empty vessels should be filled with deionized water to minimize the introduction of trace contaminants from the autoclave steam.[41]

Isolation and Culture Methods

Isolating heterotrophic dinoflagellates for culture can be accomplished in several ways.

Enriching a Water Sample

Simply incubating an unaltered water sample for 1 to 3 d often will result in a "bloom" of heterotrophic dinoflagellates. However, enrichments with f/2[37] or K media[42] at 1/100 strength will stimulate the growth of both phytoplankton and their grazers. To stimulate the growth of bacterivores, enrich with a sterile rice grain and incubate in the dark. Incubating subsamples (100 to 150 ml) of the enriched seawater in 250 ml polycarbonate flasks will increase the chances for one or more dinoflagellate cultures. The enrichments should be checked daily, for the dinoflagellate bloom may be short lived. Desirable cells then can be isolated in selected media and fed a variety of food items to determine the optimal combination.

Micropipetting Individual Cells Directly from a Water Sample

While enrichments are convenient as they often result in many cells to work with, the species that multiply in enrichments are not necessarily typical of the ambient community. One way to obtain a more representative selection of protists is to isolate single cells directly from a water sample. During the isolation procedure, it is extremely important to proceed as quickly as possible while maintaining the ambient temperature of the water sample and media. Before isolating cells, a supply of micropipettes should be drawn out to a fine point over a flame. It is also helpful to survey a sample with a dissecting microscope so as to become familiar with the appearance and swimming patterns of heterotrophic dinoflagellates before starting the isolation procedure. Isolated cells can be examined at a higher magnification on a compound microscope by placing them in a drop of water on a slide. To confirm whether the isolates are heterotrophs, epifluorescence microscopy is useful, as many heterotrophic dinoflagellates autofluoresce green when exposed to blue light.[15,18,43,44] However, as not all heterotrophic dinoflagellates autofluoresce green, an absence of red or orange autofluorescence (characteristic of phototrophs) will confirm a heterotroph as well. Darkfield illumination for the dissecting microscope is critical for detecting and isolating small cells (with practice, isolating cells down to 5 μm is possible). If a darkfield illumination system is not available, the mirror can be adjusted to achieve a similar affect. Each cell then is drawn up into the micropipette using suction by mouth (or rubber bulb) through the aspirator tubing, then aspirated into a depression well containing filtered seawater or media. Successive transfers of each cell through filtered seawater or media will help remove contaminants, then the cell is aspirated into a culture vessel with media and food organisms. Other approaches to single cell isolations can be found in Reference 3 and references therein.

Diluting a Water Sample

A third method of isolation is to dilute a sample with filtered seawater to obtain 0.1- or 1.0-ml samples containing only one cell. This aliquot is added to many tubes of media with food organisms and incubated.

Incubation Methods

For successful cultures, it is important to mimic natural conditions (temperature, light) as much as possible, especially during the initial isolation. For herbivores, maintaining a light-dark cycle (e.g., 14:10 or 12:12 light:dark hours) is desirable. Maintaining relatively low light levels (<30 μE m^{-2} s^{-1}) is also helpful to prevent "overgrowth" of the food organisms. Bacterivores can be grown in the dark. While many dinoflagellates grow very well in still cultures, some, notably many of the larger thecate types, require gentle motion (not agitation) to be grown successfully. This is necessary to keep the typically nonmotile food (diatoms) in suspension where they can be engulfed by the dinoflagellate pseudopodial feeding apparatus. Slowly rotating (approximately 1 RPM) tubes or flasks end over end[14] achieves the best results.

Culturing Herbivores

The search for optimum food organisms and essential nutrients required for the growth of the more common phagotrophic dinoflagellates is relatively recent. The search has been a combination of trial and error, knowledge of nutrient requirements of other protists (both phototrophs and heterotrophs), and careful observations of their natural prey, as observed in live samples[22,45] or in the food vacuoles of preserved specimens (reviewed in Reference 2).

The choice of food organisms depends on the species of phagotrophic dinoflagellate to be isolated. Table 2 lists food organisms that have been used to successfully culture some phagotrophic dinoflagellates. The vast majority of phagotrophic dinoflagellates have not as yet been cultured, nor have optimum prey items been determined for those that have been cultured. Some generalizations can be made as a guide in deciding which prey to use. In general, small athecate dinoflagellates (<20 μm, *Gymnodinium, Gyrodinium, Katodinium, Amphidinium* spp.) can be cultured on small (1 to 5 μm) flagellates and cyanobacteria, whereas larger athecate dinoflagellates (e.g., *Gymnodinium* and *Gyrodinium* spp., *Polykrikos kofoidi*) can use photosynthetic dinoflagellates (e.g., *Scrippsiella trochoidea, Gymnodinium simplex*), diatoms, and ciliates. Thecate dinoflagellates (*Oblea, Zygabikidium, Protoperidinium*) can be cultured on a variety of diatoms and/or dinoflagellates. Jacobson and Anderson[22] list the specific prey species that they observed being ingested by 18 species of thecate dinoflagellates in short-term cultures. Mixtures of food species often result in better growth than single species.

Trace metals, essential for protist growth, are not always sufficient in prey alone. Depending on their concentration, trace metals also can be toxic to protists.[40,46] The degree of need for, or sensitivity to, trace metals has not been extensively examined in heterotrophic protists[46] and, until more is known, exact recommendations cannot be made. The addition of chelated trace metal mixtures that have been formulated for phototrophs (e.g., medium f/2 Guillard[47]) has been found to enhance the growth of phagotrophs. According to Gifford,[47] their addition at 1/10 to 1/100 f/2 levels (corresponding to Na_2EDTA levels of 10^{-6} and 10^{-7} M, with a ratio of chelator:total metals of 1:1) resulted in optimal growth rates for a coastal oligotrich ciliate. We have cultured successfully both coastal and oceanic phagotrophic dinoflagellates in 1/100 levels of f/2 trace metals plus vitamins (Table 2). A newer formulation, medium K,[42] has resulted in much better growth of oceanic phototrophs. Medium K has a higher level of chelation (Na_2EDTA levels of 10^{-4} M, ratio of chelator:metals of 10:1), lower copper concentration, and added selenite. Shapiro et al.[18] used medium K to culture successfully two phagotrophic species, *Gymnodinium* sp. and *Gyrodinium* sp.,

Table 2. Food Organisms of Phagotrophic Dinoflagellates in Culture

Species	Prey species	Type	Media	Source
Athecate				
Oxyrrhis marina	Nitzchia sp.	dia	SW+soil	Tidepool[5]
	14 algal spp. in 6 orders	—	FSW+soil	Tidepool[48]
	Saccharomyces exiguus	yeast	ASW+soil	Tidepool[7]
	Rhodomonas sp.	cryp	f/2	n.a.[49]
	Phaeodactylum tricornutum	dia	f/2	n.a.[50]
	Isochrysis galbana	prym	f/2	n.a.[50]
	Dunaliella tertiolecta	chlor	ASW+nuts	Culture contaminant[13]
	Dunaliella primolecta	chlor	ASW+nuts	CCAP[12]
Noctiluca miliaris	Dunaliella parva	chlor	Erdsch.	—[51]
	Dunaliella tertiolecta	chlor	ASW+nuts	North Sea[52]
	Dunaliella sp.	chlor	FSW+nuts	Long Island Sound[27,29]
	Platymonas tetrethale	pras	ASW+nuts	—[27,29]
Gyrodinium fucorum	Mixed bacteria	bact	FSW+Fucus	Rotting Fucus[5]
Polykrikos kofoidi	Scrippsiella trochoidea	dino	FSW	L.A. Harbor[53]
Katodinium fungiforme	Dunaliella salina	chlor	Grund ESW	Culture contaminant[34]
Gymnodinium sp.	Gymnodinium sp., unid.	dino	K	Gulf of Maine[18]
Gymnodinium sp.	2–3 μm phototrophs, bact.			
Gymnodinium sp.	I. galbana, Synechococcus, Micromonas pusilla, Emiliania huxleyi	cyano, pras, prym	IMR	N. Pacific (Sta. P)[23]
Gymnodinium sp.A,B	M. pusilla, Synechococcus sp.	pras, cyan	f/200tm+vit	Georges Bank[24]
Gymnodinium sp.A	Prymnesium, M. pusilla	pyrm, pras	f/200tm+vit	Sargasso Sea[24]
Gymnodinium sp.B	Heterocapsa triquetra	dino	f/200tm+vit	Chesapeake Bay[24]
Gymnodinium sp.C	Bacteria, hetero. flag.	—	Rice grain	Dabob Bay[24]
Thecate				
Protoperidinium hiroboris	Leptocylindricus danicus	dia	F/2	Perch Pond[14]
Dinophysis rotundata	Tiarina fusus	cil	F/2	Kattegat[35]
Oblea rotunda	Scrippsiella trochoidea	dino	F/200tm+vit	Chesapeake Bay[24]
	Heterocapsa triquetra	dino		

Note: dia: diatom; cryp: cryptophyte; prym: prymnesiophyte; chlor: chlorophyte; pras: prasinophyte; dino: dinoflagellate; cyano: cyanobacteria; bact: heterotrophic bacteria; cil: ciliate; SW: seawater; FSW: filtered seawater; ASW: artificial seawater; soil: soil extract; nuts: inorganic nutrients; vit: vitamins; Erdsch.: Erdschreiber (sw + nuts + soil); f/2: Reference 35; K: Reference 40; IMR: Reference 51; n.a.: not available.

from the Gulf of Maine (Table 2). The efficacy of the K for other phagotrophs, however, has not been examined systematically. High levels of EDTA have been shown to be detrimental to some protists.[47] To optimize future culturing efforts, as well as increase our knowledge of protist physiology, the effects of chelator and trace metal concentrations on protist growth needs to be studied further.

With the exception of *O. marina*, the vitamin requirements of phagotrophic dinoflagellates have not been determined. *O. marina* requires vitamin B12 and thiamine, and is stimulated by biotin.[8] F/2 vitamins can be added to the growth media to ensure an adequate vitamin supply, although the need for vitamins in addition to food organisms has not been rigorously tested.

Two approaches to growing herbivorous dinoflagellates have proven successful. The frequent-feeding method is preferred for initial isolations of most dinoflagellates and for long-term culture of the larger thecate dinoflagellates. The cotransfer method is convenient for maintenance of some types. In the frequent-feeding method, aliquots of food from exponentially growing algal cultures are added to an herbivore culture approximately once a week in phagotroph media (containing only trace metals and vitamins, f/200 recipe, Table 1). Frequent feeding (one or two times per week) with relatively low prey concentrations (<100 µg C l^{-1}) seems to be a key to success for many dinoflagellates. Transfers to fresh media should be made every 2 to 4 weeks. In the cotransfer method, herbivore and algae are transferred together into complete phototroph media (containing all the necessary nutrients for algal growth, e.g., f/2 or K media, Table 1). Herbivore and prey grow together in a relatively stable association. Many small gymnodinoids [Katodinium ($=$ *Gymnodinium*) *fungiforme*;[34] see also References 18 and 24] and some of the "weed" species (e.g., *O. marina*) can be cultured easily this way.

Culturing Bacterivores

We have grown several species of dinoflagellates (estuarine *Gymnodinium* sp. and *O. marina*) on bacteria alone, or on bacterivorous flagellates and ciliates. Cells are isolated in wells of 12-well tissue culture plates or in 10- or 50-ml polycarbonate centrifuge tubes filled with filtered seawater. To provide a mixed bacterial inoculum, the seawater from which the cells are isolated is filtered through an 0.8-µm filter and a drop is added to each well or tube. A sterile rice grain or 0.01% (W/W) yeast extract (Difco) can be added to provide a substrate to stimulate the growth of bacteria. Cultures can be incubated in the dark.

REFERENCES

1. Kofoid, C. A. and Swezy, O., *The Free-Living Unarmoured Dinoflagellata*, Vol. 5, University of California Press, Berkeley, 1921.
2. Gaines, G. and Elbrächter, M., Heterotrophic nutrition, in *The Biology of Dinoflagellates*, Blackwell Scientific, Oxford, 1987, 224.
3. Guillard, R. R. L. and Keller, M. D., Culturing dinoflagellates, in *Dinoflagellates*, Spector, D. L., Ed., Academic Press, New York, 1984, chap. 12.
4. Küster, E., Ein kultivierbare Peridinee, *Arch. Protistenk.*, 11, 351, 1908.
5. Barker, H. A., The culture and physiology of the marine dinoflagellates, *Arch. Mikrobiol.*, 6, 157, 1935.
6. Beam, C. A. and Himes, M., Dinoflagellate genetics, in *Dinoflagellates*, Spector, D. L., Ed., Academic Press, New York, 1984, chap. 8.
7. Droop, M. R., A note on some physical conditions for cultivating *Oxyrrhis marina*, *J. Mar. Biol. Assoc. U.K.*, 38, 599, 1959.

8. Droop, M. R., Water-soluble factors in the nutrition of *Oxyrrhis marina*, *J. Mar. Biol. Assoc. U.K.,* 38, 605, 1959.

9. Provosoli, L. and Gold, K., Nutrition of the American strain of *Gyrodinium cohnii,* *Arch. Mikrobiol.,* 42, 196, 1962.

10. Dodge, J. and Crawford, R. M., Fine structure of the dinoflagellate *Oxyrrhis marina.* I. The general structure of the cell, *Protistolgica,* 7, 295, 1971.

11. Dodge, J. and Crawford, R. M., Fine structure of the dinoflagellate *Oxyyrhis marina.* III. Phagotrophy, *Protistologica,* 10, 239, 1974.

12. Opik, H. and Flynn, K. J., The digestive process of the dinoflagellate *Oxyrrhis marina* Dujardiin, feeding on the chlorophyte, *Dunaliella primolecta* Butcher: a combined study of ultrastructure and free amino acids, *New Phytol.,* 113, 143, 1989.

13. Goldman, J. C., Dennett, M. R., and Gordin, H., Dynamics of herbivorous grazing by the heterotrophic dinoflagellate *Oxyrrhis marina, J. Plankt. Res.,* 11, 391, 1989.

14. Jacobson, D. M., The Ecology and Feeding Behavior of Thecate Heterotrophic Dinoflagellates, Ph.D. thesis, Massachusetts Institute of Technology and Woods Hole Oceanographic Institution, 1987.

15. Lessard, E. J., Oceanic Heterotrophic Dinoflagellates: Distribution, Abundance and Role as Microzooplankton, Ph.D. thesis, University of Rhode Island, 1984.

16. Lessard, E. J., The trophic role of heterotrophic dinoflagellates in marine ecosystems, *Mar. Microb. Food Webs,* 5, 49, 1991.

17. Smetacek, V., The annual cycle of protozooplankton in the Kiel Bight, *Mar. Biol.,* 63, 1, 1981.

18. Shapiro, L. P., Haugen, E. M., and Carpenter, E. J., Occurrence and abundance of green-fluorescing dinoflagellates in surface waters of the Northwest Atlantic and Northeast Pacific, *J. Phycol.,* 25, 189, 1989.

19. Lessard, E. J. and Swift, E., Species-specific grazing rates of heterotrophic dinoflagellates in oceanic water, *Mar. Biol.,* 87, 289, 1985.

20. Bjornsen, P. K. and Kuparin, J., Growth and herbivory by heterotrophic dinoflagellates in the Southern Ocean, studied by microcosm experiments, *Mar. Biol.,* 109, 397, 1991.

21. Hansen, P. J., Quantitative importance and trophic role of heterotrophic dinoflagellates in a coastal pelagial food web, *Mar. Ecol. Prog. Ser.,* 73, 253, 1991.

22. Jacobson, D. and Anderson, D. M., Thecate heterotrophic dinoflagellates: feeding behavior and mechanism, *J. Phycol.,* 22, 249, 1986.

23. Strom, S., Growth and grazing rates of an herbivorous dinoflagellate (*Gymnodinium* sp.) from the open subarctic Pacific Ocean, *Mar. Ecol. Prog. Ser.,* 78, 103, 1991.

24. Lessard, E., unpublished observations.

25. Tuttle, R. C. and Loeblich, A. R., III, An optimal growth medium for the dinoflagellate *Crypthecodinium cohnii, Phycologia,* 14, 1, 1975.

26. Droop, M. R., Nutritional investigation of phagotrophic protozoa under axenic conditions, *Helgolander Wiss. Meeresunters.,* 20, 272, 1970.

27. McGinn, M. P. and Gold, K., Axenic cultivation of *Noctiluca scintillans, J. Protozool.,* 16(suppl.), 13, 1969.

28. Lee, R. E., Saprophytic and phagocytic isolates of the colourless heterotrophic dinoflagellate *Gyrodinium lebouriae* Herdman, *J. Mar. Biol. Assoc. U.K.,* 57, 303, 1977.

29. Provosoli, L., Axenic cultivation, in *Marine Ecology,* Vol. 3, Kinne, O., Ed., John Wiley & Sons, New York, 1977, 1295.

30. Norris, D., Possible phagotrophic feeding in *Ceratium lunula, Limnol. Oceanogr.,* 14, 448, 1969.

31. Dodge, J. and Crawford, R. M., The morphology and fine structure of *Ceratium hirundinella* (Dinophyceae), *J. Phycol.,* 6, 137, 1970.

32. Schnepf, E. and Winter, S., A microtubular basket in the armoured dinoflagellate *Prorocentrum micans* (Dinophyseae), *Protistenk.,* 138, 89, 1990.

33. Buck, K., Bolt, P. A., and Garrison, D. L., Phagotrophy and fecal pellet production by an athecate dinoflagellate in Antarctic sea ice, *Mar. Ecol. Prog. Ser.,* 60, 75, 1990.

34. Spero, H. J., Phagotrophy in *Gymnodinium fungiforme* (Pyrrhophyta): the peduncle as an organelle of ingestion, *J. Phycol.,* 18, 356, 1982.

35. Hansen, P. J., Dinophysis — a planktonic dinoflagellate genus which can act as both prey and predator of a ciliate, *Mar. Ecol. Prog. Ser.,* 69, 201, 1991.
36. Gaines, G. and Taylor, F. J. R., Extracellular digestion in marine dinoflagellates, *J. Plankt. Res.,* 6, 1057, 1984.
37. Guillard, R. R. L., Culture of phytoplankton for feeding marine invertebrates, in *Culture of Marine Invertebrate Animals,* Smithe, W. L. and Chanley, M. H., Eds., Plenum Press, New York, 1975, 29.
38. Lee, J. J., Hutner, S. H., and Bovee, E. C., *An Illustrated Guide to the Protozoa,* Allen Press, Kansas, 1985.
39. Fitzwater, S. E., Knaur, G. A., and Martin, J. H., Metal contamination and its effect on primary production measurements, *Limnol. Oceanogr.,* 27, 544, 1982.
40. Brand, L. E., Sunda, W. G., and Guillard, R. R. L., Reduction of phytoplankton reproduction rates by copper and cadmium, *J. Exp. Mar. Biol. Ecol.,* 96, 225, 1986.
41. Brand, L. E., Guillard, R. R. L., and Murphy, L. S., A method for the rapid and precise determination of acclimated phytoplankton reproduction rates, *J. Plankt. Res.,* 3, 193, 1981.
42. Keller, M. D., Selvin, R. C., Claus, W., and Guillard, R. R. L., Media for the culture of oceanic ultraphytoplankton, *J. Phycol.,* 23, 633, 1987.
43. Carpenter, E. J., Chang, J., and Shapiro, L. P., Green and blue fluorescing dinoflagellates in Bahamanian waters, *Mar. Biol.,* 108, 145, 1991.
44. Lessard, E. J. and Swift, E., Dinoflagellates from the North Atlantic classified as phototrophic or heterotrophic by epifluorescence microscopy, *J. Plankton Res.,* 8, 1209, 1986.
45. Spero, H. J. and Moree, M. D., Phagotrophic feeding and its importance to the life cycle of the holozoic dinoflagellate, *Gymnodinium fungiforme, J. Phycol.,* 17, 43, 1981.
46. Stoecker, D., Sunda, W. H., and Davis, L. H., Effects of copper and zinc on two planktonic ciliates, *Mar. Biol.,* 92, 21, 1986.
47. Gifford, D., Laboratory culture of marine planktonic oligotrichs (Ciliophora, Oligotrichida), *Mar. Ecol. Prog. Ser.,* 23, 257, 1985.
48. Droop, M. R., The role of algae in the nutrition of *Heterameoba clara* Droop, with notes on *Oxyrrhis marina* Dujardin and *Philodina roseola* Ehrenberg, in *Some Contemporary Studies in Marine Science,* Barnes, H., Ed., Allen and Unwin, London, 1966, 269.
49. Klein, B., Gieskes, W. W. C., and Kraay, G. G., Digestion of chlorophylls and carotenoids by the marine protozoan *Oxyrrhis marina* studied by h.p.l.c. analysis of algal pigments, *J. Plankton Res.,* 8, 827, 1986.
50. Barlow, R. G., Burkill, P. H., and Mantoura, R. F. C., Grazing and degradation of algal pigments by marine protozoan *Oxyrhhis marina, J. Exp. Mar. Biol. Ecol.,* 119, 119, 1988.
51. Gross, F., Zur Biologie und Entwicklungsgeschichte von *Noctiluca miliaris, Arch. Protistenk.,* 83, 178, 1934.
52. Eckert, R., Bioelectric control of bioluminescence in the dinoflagellate *Noctiluca, Science,* 147, 1140, 1965.
53. Morey-Gaines, G. and Ruse, R. H., Encystment and reproduction of the predatory dinoflagellate, *Polykrikos kofoidi* Chatton (Gymnodiniales), *Phycologia,* 19, 230, 1980.
54. Eppley, R. W., Holmes, R. W., and Strickland, J. D. H., Sinking rates of marine phytoplankton measured with a fluorometer, *J. Exp. Mar. Biol. Ecol.,* 1, 161, 1967.
55. Harrison, P. J., Waters, R. E., and Taylor, F. J. R., A broad spectrum artificial seawater medium for coastal and open ocean phytoplankton, *J. Phycol.,* 1, 28, 1980.

CHAPTER 9

Enrichment, Isolation, and Culture of Free-Living Heterotrophic Flagellates

David A. Caron

INTRODUCTION

Microbiologists traditionally have relied on the study of pure cultures of microorganisms to obtain basic information on the physiology of important species. The culture of a species under carefully controlled conditions provides an experimental framework within which to examine its biology in the absence of potentially confounding interactions with other living organisms. Heterotrophic flagellates are no exception to this maxim, and much of our understanding of flagellate biology has come from laboratory-based studies of cultured species.

Laboratory studies of heterotrophic flagellates have provided baseline ecological data on a wide variety of species. Based on these data we have begun to place limits on the activities of these microorganisms in nature. Studies of cultured flagellates have included examinations of the nutritional modes, feeding rates, growth rates, growth efficiencies, and nutrient remineralization of heterotrophic flagellates.[1-8] These investigations are far from complete, and our understanding of their ecological role in nature will undoubtedly improve as more species of heterotrophic flagellates are brought into culture. As a consequence of these studies, a fairly wide diversity of heterotrophic flagellates are presently available from private sources and service collections. A recent summary[9] provides information on the sources of these species.

Free-living zooflagellates (Phylum Sarcomastigophora, Subphylum Mastigophora)[10] constitute several orders within the Class Zoomastigophorea. In addition, many Phytomastigophorean Orders also contain numerous chlorotic and apochlorotic species which feed phagotrophically (and perhaps osmotrophically in some instances). As might be suspected from their diverse taxonomic affinities, heterotrophic flagellates display a number of trophic modes. As a consequence, there are many variations on the basic methods used to culture these microorganisms. Provided below are some general guidelines on culturing these species. It should be noted that these methods might need to be adapted to suit the culture requirements of a particular species.

0-87371-564-0/93/$0.00 + $.50
© 1993 by Lewis Publishers

MATERIALS REQUIRED

Equipment

Very little specialized equipment is needed for enriching and isolating most heterotrophic flagellates. In general, common bacteriological and phycological equipment and media are sufficient. Several large items, however, are necessary in order to provide adequate facilities for climate control and observation.

- Culture facility — Many species of heterotrophic flagellates have fairly wide ranges of tolerable physical/chemical conditions. Therefore, a quite corner of an air-conditioned room often will suffice as a culture room. For a substantial culture collection or for important species, however, a growth incubator equipped with an emergency power supply is highly preferable. Several manufacturers produce reliable models that can be outfitted with preset light and temperature regimes (e.g., Percival Manufacturing Co., P.O. Box 249, Boone, IA 50036 Tel. 515-432-6501).
- Microscopes — A compound microscope, preferably equipped for phase contrast or differential interference contrast, is essential for examining cultures at high magnification. A high-quality dissecting microscope is also essential for micropipetting protista, for rapidly establishing the presence or absence of protista in dilution/extinction experiments, and as a means of rapidly checking the vigor of established cultures.
- Autoclave — Sterilization is necessary for preparing media and glassware used for enriching, isolating, and maintaining cultures of flagellates. For field and seagoing research programs, a large-capacity pressure cooker is adequate. The use of sterile, disposal plasticware (pipettes, tissue culture flasks and dishes, test tubes, etc.) can reduce the need for autoclaving in the field.
- Transfer hood — Aseptic technique is generally preferable but it is not required for all aspects of flagellate culture. It is, however, absolutely necessary for maintaining axenic cultures. Bacterial contamination can be minimized tremendously by using a transfer hood equipped for UV sterilization, or a positive-pressure laminar flow hood. The liberal use of 95% ethanol for sterilizing the transfer area also is recommended.

Supplies

- Inoculating loops
- Petri dishes
- Graduated and Pasteur pipettes
- Carboys
- Erlenmeyer flasks
- Test tubes
- Multi-well tissue culture dishes
- An assortment of micropipettes (10 to 200 μl) and small-diameter tubing (to fit the micropipettes) are used for isolating flagellates by the micropipetting method (see below).

Solutions and Media

Many different types of culture media have been employed to culture heterotrophic flagellates. They range from simple organic substrates added to natural seawater, to rather complex, chemically defined media. The medium required for growing heterotrophic flagellates is dependent primarily on the type of flagellates to be enriched or cultured. Bacterivorous flagellates require media for growing bacteria, while herbivorous flagellates require phytoplankton media for growing the algal prey. In many situations, heterotrophic flagellates are enriched by promoting the growth of the appropriate naturally occurring prey in the water sample. In other instances, direct addition of precultured prey can be advantageous. Formulations for various types of media can be found in References 9 and 11 to 15.

Artificial Seawater

Depending on the fastidiousness of the flagellate strain to be cultured, either natural or artificial seawater can be used. Natural and artificial seawaters have been used extensively to culture protista (both photosynthetic and heterotrophic). Most investigators prefer the use of natural, aged (in the dark) seawater collected from an oligotrophic, oceanic locale, but coastal seawater is also extensively used for robust species. Fluctuations in salinity are an obvious consideration for the use of coastal water, as are the presence of nuisance algae or organic compounds. For these reasons, several artificial seawater formulations have been developed and employed. The formula for a common one is provided below (taken from Reference 18).

1. Solution A (dissolve in 7 l of distilled water)

    - $NaCl$, 205.1 g
    - KCl, 5.8 g
    - $MgCl_2 \cdot 6H_2O$, 92.8 g
    - $CaCl_2 \cdot 2H_2O$, 13.0 g (or anhydrous $CaCl_2$, 9.8 g)

2. Solution B (dissolve in 2.58 l of distilled water)

    - Na_2SO_4, 34.4 g

3. Solution C (dissolve in 1 l of distilled water). Filter-sterilize solution C through a sterile 0.2-μm pore size filter.

    - $NaHCO_3$, 16.8 g
    - $NaHPO_4$, 0.014 g

    To prepare 1 l of nonsterile artificial seawater (30%), mix 730 ml of solution A, 260 ml of solution B, and 10 ml of solution C. It is reported that autoclaving the combined solutions will precipitate excess heavy metals and yield a seawater with an ionic content similar to natural seawater. For enriched microbiological media, add the nutrients (and agar, if used) to solution A, and autoclave solutions A and B separately. After autoclaving, add solution C to solution B, and then combine A and B. Note that $NaCl$, Na_2SO_4, and KCl should be dried (125°C) for at least 4 h and cooled in a dessicator.

f/2 Medium for Algal (Prey) Culture

Algal media typically contain major nutrients (N and P), vitamins, and trace metals. Exact composition varies according to the algal species. A commonly used medium for many algae is the f/2 medium of Guillard,[12] as follows:

Stock Solutions

1. $NaNO_3$, 15 g/200 ml of distilled water (DW)
2. $NaH_2PO_4 \cdot H_2O$, 1 g/200 ml of DW

3. NaSiO$_3$·9H$_2$O, 15 g/500 ml of DW
4. Trace metals: Prepare the working stock solution of trace metals by dissolving 5.0 g of ferric sequestrene in 900 ml of DW. Add 1 ml of each trace metal solution (given below) and bring to 1 l.

    - CuSO$_4$·5H$_2$O, 980 mg/100 ml of DW
    - ZnSO$_4$·5H$_2$O, 2.2 g/100 ml of DW
    - CoCl$_2$·6H$_2$O, 1 g/100 ml of DW
    - MnCl$_2$·4H$_2$O, 18 g/100 ml of DW
    - Na$_2$MoO$_4$·2H$_2$O, 630 mg/100 ml of DW

5. Vitamins: A working stock solution of vitamins consists of the following amounts dissolved in 100 ml [NOTE: primary stock solutions of biotin (0.1 mg ml^{-1}) and B12 (1 mg ml^{-1}) are used to make this solution].

    - Thiamine·HCl, 20 mg
    - Biotin, 1 ml of 1° stock
    - B12, 0.1 ml of 1° stock

    To prepare f/2 medium, add 1 ml of stock solutions (1) to (4) and 0.5 ml of solution (5) to 1 l of natural, aged (preferably filtered) seawater, or to artificial seawater (see above). The vitamin stock solutions should be autoclaved and stored at 4°C. Silicate and trace metals are usually sterilized separately and added after autoclaving to prevent precipitation. Silicate can be omitted for phytoplankton other than diatoms. Usually a buffer can be omitted with full-strength seawater.

Enriched Seawater Media for Bacterial (Prey) Culture

The use of precultured bacteria as prey for bacterivorous flagellates has some advantages over simply enriching the natural bacterial assemblages. Undesirable prey (such as fungi) may be enriched by the addition of organic compounds to natural samples. More importantly, high concentrations of labile organic material may result in the depletion of oxygen in the cultures and elimination of the desired flagellate species. Many bacterial strains can be grown in large quantities in liquid medium containing complex organic compounds. Numerous nutrient media exist. Two "tried and true" formulations are given here. Products are available from a number of sources (e.g., Difco Laboratories, P.O. Box 331058, Detroit, MI, Tel. 800-521-0851).

1. Nutrient broth

    - 3 g beef extract l^{-1}
    - 5 g peptone l^{-1}

2. 2216 medium[16]

    - 5 g peptone l^{-1}
    - 0.01 g FePo$_4$ l^{-1}

3. 2216E Medium:[17] 2216 Medium + 1 g yeast extract l^{-1}

For short-term storage, bacterial strains can be streaked with an inoculating wire onto seawater agar slants enriched with either of the nutrient media above. Prepare medium in a flask $\geq 2 \times$ the volume of the medium. Add 15 g agar l^{-1} of medium. Heat slowly over a flame until agar dissolves (CAUTION: agar foams readily when boiled.) Dispense into screw-top or cotton-plugged test tubes (approximately one third of the tube volume). Autoclave. When tubes are removed from autoclave, lay them on their side on a clean surface (or in a transfer hood), elevating the mouth of the tube about 1 cm (a pencil or marker usually works well) such that the medium forms a slanted surface along one side of the test tube. After the tubes are cooled, bacteria can be streaked along the length of the agar surface and stored immediately at 5°C. To use, simply remove a tube from refrigeration and incubate at optimal growth temperature.

Individual isolates of bacteria can be used as prey or mixed assemblages can be obtained by filtering natural seawater through a sterile 0.8-μm pore size filter, and inoculating the filtrate into an enriched medium (these bacterial enrichments should be examined before use for the presence of minute bacterivorous flagellates that sometimes can pass through the filter). If desired, cultured bacteria can be rinsed free of organics by centrifuging them in 250-ml centrifugation flasks (10 to 20 min at 10,000 rpm) and resuspending the pellet in the desired seawater medium.

PROCEDURES

General Notes on Culturing Heterotrophic Flagellates

Most procedures for culturing free-living, heterotrophic flagellates are not complicated. The most useful tool is common sense and some rudimentary knowledge of the nutrition and chemical/physical requirements of the protista to be cultured. Obviously, it is somewhat problematical trying to anticipate these latter conditions. For that reason, insight may need to be combined with a fair amount of trial and error before the desired species can be brought into culture.

The approach most commonly employed for obtaining cultures of a desired flagellate species is to begin with a wide array of enrichment media and culture approaches with the intent of providing conditions that will support the growth of the desired species (the growth of other species will undoubtedly be stimulated as well). Once an adequate enrichment procedure has been established, efforts can then be focused on gradually eliminating unnecessary or undesired conditions and co-occurring species until the desired culture type is obtained (e.g., clonal culture of the desired species with living prey, or an axenic culture of the desired species).

Several simple methods are described below for the enrichment, isolation, and maintenance of heterotrophic flagellates. In general, these methods also are applicable for culturing many of the ciliates and amoeboid protozoa with the exception of planktonic foraminifera, radiolaria, and acantharia which require special considerations.

Enrichment Cultures

Enrichment cultures are performed with the intent of increasing the abundance (both absolute and relative to other protista) of the desired species of flagellate in a natural sample. Numerical dominance of the species to be cultured is highly advantageous. It favors the selection of the desired species during micropipetting procedures and it is absolutely necessary for the success of isolation procedures based on dilution/extinction. Enrichment cultures also can provide valuable information on the conditions that will support the growth of a flagellate once it is isolated. Failure of a species to grow in a particular type of enrichment culture should not, however, be taken as a clear indication that those conditions will not support growth of the flagellate once it is isolated. Competition with, and predation by, other protista can affect the outcome of enrichment cultures for slow-growing species of protista.

Some natural samples are already numerically dominated by one or a few flagellates, and isolation procedures (see below) can be employed immediately for these species. In the case of flagellate species that are easily identifiable when viewed using a dissecting microscope (e.g., many dinoflagellates), isolation procedures also can be conducted without prior enrichment. In most instances, however, some initial work is required to increase the density of the desired flagellate in the sample. Given below are some general guidelines for accomplishing this objective.

Two basic types of enrichment cultures can be attempted: batch and continuous culture. Batch enrichment cultures are preferable because these cultures are generally easy to prepare and maintain. Any of a variety of vessels including test tubes, small flasks, or petri dishes may be used as culture vessels for batch enrichments. The specific vessel is somewhat a matter of space and convenience, but it is important to consider the rate of oxygen diffusion into the medium in organically enriched samples (utilization of oxygen by bacteria can rapidly deplete oxygen in some containers). Tissue culture flasks (e.g., Costar Corp., Tel. 800-492-1110) filled to less than one quarter of their capacity are particularly convenient because they provide a large air space and they can be examined for protistan growth using a dissecting or inverted microscope without removing a sample.

Batch enrichment cultures can be established easily by inoculating a natural water sample with a substrate(s) that will promote the growth of prey for the flagellate to be cultured. For example, the addition of a complex organic substrate (e.g., yeast extract at 0.01 to 0.1%) will promote bacterial growth which, in turn, will promote the growth of bacterivorous protista. Cultured prey species also can be added directly to water samples. This approach is straightforward if one wishes to enrich a specific nutritional type of flagellate (e.g., bacterivore, herbivore) but it is more problematical if one is attempting to enrich a particular species of heterotrophic flagellate for which the nutritional mode is not known. In the latter situation, a range of additions might be attempted in replicate samples in an attempt to enrich different prey species and thus provide the appropriate prey in some of the containers. Once the cultures have been inoculated, periodic examination (every few days) will determine the success of the various types of enrichments. Depending on the type of enrichment, physical conditions (primarily temperature), and species composition of the water sample, subculturing or the addition of more substrate or prey may be necessary after several days.

Batch cultures have several practical advantages. They are simple to prepare and maintain, and therefore cultures containing several different media types or culture conditions can be attempted simultaneously. Diversity in the culture conditions can be helpful in trying to enrich particularly fastidious microorganisms. Another advantage of batch cultures is that normally a temporal succession of species occurs in the vessels. Different species of flagellates can dominate a single enrichment culture at different times during this succession. Batch cultures sampled periodically can thus yield a number of different flagellate species. This approach is particularly useful for enriching some slow-growing species that would be outcompeted and lost from continuous cultures.

Continuous enrichment cultures are much more complicated and laborious in their construction. This method has been employed routinely to culture phytoplankton species[19] but has only occasionally been used to enrich and/or culture heterotrophic protista.[20-23] Continuous enrichment cultures are composed of a culture vessel which is kept well mixed using magnetic stirring and/or bubbling. A continuous supply of prey organisms (or substrate for prey) is pumped into the culture vessel from a reservoir. Excess medium from the culture vessel is removed by gravity or by air pressure that maintains a slight positive pressure in the vessel. The reservoir may contain substrate for the prey of the heterotrophic flagellates (e.g., organic material to promote bacterial growth) or, more commonly, precultured prey themselves (e.g., a bacterial or phytoplankton population).

Continuous cultures possess a singular advantage over batch cultures by providing a mechanism for selecting protistan species with specific abilities even when the initial densities of these species are quite low. Because there is a constant removal of liquid from the culture vessel (and replacement with reservoir liquid) continuous enrichment cultures strongly select for those species that can grow and reproduce rapidly on the prey species provided. For example, this enrichment method has been used to enrich flagellates that will consume diatoms by pumping a cultured diatom into the vessel.[22] The resulting enrichments were then used to initiate clonal cultures of herbivorous flagellates using a micropipetting technique. Similarly, species can be selected on the basis of their growth rates, tolerance to specific temperatures, etc. This method is generally highly selective and thus yields fewer species for isolation, but obtaining a successful enrichment by this method immediately provides the investigator with a substantial amount of information on the growth requirements of the species. This advantage of continuous culture for enriching species of heterotrophic flagellates is mitigated by the fact that this method typically constitutes a significantly greater effort to conduct than batch enrichment cultures. In addition, many flagellates (particularly dinoflagellates) do not respond well to the rigorous mixing regime of continuous culture systems.

Food

The type of protistan species that will grow in an enrichment culture is, to a large degree, dependent on the food provided in the culture. Enriching the bacterial assemblage of a water sample will result, in most cases, in a rather dramatic increase in the number of small bacterivorous flagellates (and possibly ciliates) in the sample within a few days. High densities of bacterivorous protista in these same cultures may, after several days to a few weeks, result in the enrichment of carnivorous protista. Herbivorous flagellates often can be enriched by adding an aliquot of a microalgal culture to a water sample, or by adding nutrients to the sample and providing light for the growth of the endogenous microalgae. The specific type of prey species provided or enriched in a water sample, therefore, will dictate the species of heterotrophic flagellates that will bloom in the culture.

Ideally, acceptable prey (or dissolved organic compounds) should be supplied continuously in order to keep pace with flagellate population growth. This objective is a specific advantage of the continuous culture method, but it is not easily attained in batch cultures. Many phagotrophic protistan species increase their population abundances rapidly in response to high prey concentrations but then die rapidly (or are consumed) when the prey density declines. This "boom and bust" growth cycle results in a succession of different species of heterotrophic flagellates in most batch enrichment cultures. Because of this species succession, timing is important in the application of isolation procedures to batch enrichment cultures. The cultures should be examined often at high magnification in order to provide the best chance of obtaining an enrichment at the desired stage of species succession.

A practical solution to preventing the loss of desired species in batch enrichments is to occasionally supplement the cultures with more prey. This solution requires that the prey density be monitored fairly regularly. An easier solution is to provide a mechanism that will maintain a slow-growing prey population in the cultures. For example, a common trick to prolong healthy growth of bacterivorous flagellates is the addition of bacterial substrate that can only be metabolized slowly (sterile, uncooked rice grains work nicely). Slow growth of the bacteria provides a constant supply of prey thereby maintaining vigorous growth of bacterivorous protista. For herbivorous protista, the addition of phytoplankton nutrients and the presence of light can help maintain a photosynthetic prey population.

Culture Conditions

Food is a major determinant of the types of flagellates that will grow in enrichment cultures but the importance of this factor is always moderated by chemical and physical parameters. Temperature, pressure, light, and oxygen are all important considerations for successfully enriching heterotrophic flagellates. Even the type and severity of mixing may affect the enrichment of some species of flagellates (e.g., many dinoflagellates are particularly sensitive to stirring). Most protista grow over a fairly wide range of temperatures, but optimal temperatures exist for all species. Therefore, temperature can be used as a selection factor in enrichment techniques.

Aerobic and anaerobic species of free-living heterotrophic flagellates exist, and therefore oxygen concentration is an important determinant of the outcome of enrichment cultures. For the enrichment and isolation of aerobic flagellates, aeration is important particularly when organic compounds are added to promote the growth of bacterial prey for bacterivorous flagellates. On the other hand, enrichments of anaerobic flagellates have been accomplished using adaptations of the Hungate technique for culturing anaerobic bacteria.[24] Cowling[9] described the use of a microscope slide ring chamber for isolating these species.

Light can have a direct effect on the successful enrichment of some flagellates. Most apochlorotic flagellates do not have an obligate requirement for light, but some phagotrophic (mixotrophic) algae are incapable of growth in continuous darkness. Light may also have an indirect effect on the growth of herbivorous flagellates by supporting the growth of suitable algal prey in enrichment cultures.

Most flagellates cultured up to this time have been isolated from surface waters (≤ 100 m) and cultured at 1 atmosphere pressure. For the few species of flagellates cultured from the deep sea, however, the growth of some of these species has been shown to be enhanced at high pressure.[25] In general, it may be necessary to go to considerable lengths to mimic environmental conditions in order to culture flagellate species from severe environments.

Predators

It may prove necessary in some situations to take steps to exclude potential predators of desired flagellate species. If the abundance of a small (e.g., <10 μm) flagellate has been enriched in a water sample it is likely that ciliates and larger flagellates that can prey on the smaller flagellate will subsequently become enriched. In order to avoid the removal of the desired flagellate by its predators it may be necessary to prescreen water samples to remove most potential predators. I have found it useful to use the filtrate of 10- or even 5-μm Nitex screening as the inoculum for enrichment cultures for small bacterivorous flagellates. Alternatively, differences in the growth rates of a small flagellate and its predators may be used to one's advantage. Larger protozoa typically have longer generations times than smaller ones and it may be possible to conduct isolation procedures (see below) for the smaller flagellate prior to a substantial increase in the abundance of its consumers. The enrichment of metazoan populations is generally not a problem, although I have occasionally experienced problems with rotifers and copepods in long-term enrichment cultures with volumes ≥ 1 l.

Establishing Clonal Cultures

The establishment of a clonal culture of a heterotrophic flagellate can be accomplished in several ways. The most appropriate method for this task depends on the abundance of the species in a sample, its size, and its ability to grow on various types of media.

Micromanipulation

The micromanipulative method appears to be the most commonly used means of establishing clonal cultures of flagellates. In situations where the desired flagellate is not numerically dominant, this technique may be the only means of isolating a particular species. Highly sophisticated (and expensive) micromanipulators can be purchased for isolating microorganisms, but considerable success can be achieved with a very simple setup and some patience.

A simple arrangement that I have used repeatedly for isolating small bacterivorous and herbivorous flagellates involves the use of glass microcapillary tubes as micropipettes. The tubes are sterilized, individually heated in a flame, pulled to a very narrow diameter, and then broken in the middle to yield two micropipettes. The pipettes are carefully stored in a sterile container for future use. The micropipettes are used with a small-diameter tubing that can be controlled by a hand-held bulb or by mouth (equipped with a filter cartridge to prevent the ingestion of material). For isolating protista, a small drop of the sample or enrichment culture is placed on a sterile microscope slide and viewed with a dissecting microscope equipped for dark-field illumination. We often add the drop containing the protista to several drops of sterile medium that will be used for rinsing and subsequent culturing. This process dilutes bacteria and other protista in the sample or enrichment and allows the flagellates to adjust to slight differences in osmotic tension between the liquids. Three separate drops of sterile medium are placed on a second slide to be used as rinses. The tip of the micropipette is viewed through the microscope and manipulated into the culture medium and close to a desired protistan cell (NOTE: it is important to "charge" the pipette with sterile medium before entering the drop containing live microorganisms in order to avoid liquid being drawn into the pipette by capillary action). The protist is gently sucked into the micropipette, transferred to one of the drops of sterile medium, and gently expelled (these operations should be viewed continuously under the dissecting microscope. The pipette is removed and the cell is relocated. If the flagellate is active, its mobility will move the protist away from any contaminating bacteria and other protista in the drop. Motility of the protist also is critical for relocating particularly small protista. Once the protist is relocated, a new micropipette is then used to recapture the cell (taking as little surrounding water as possible) and again transferred to a new, sterile drop of medium. This process is repeated once more, and the recaptured cell is then transferred to a container with sterile medium (containing the appropriate food).

This method can be used to obtain clonal cultures of most robust flagellates. With practice and a high-quality dissecting microscope equipped with dark-field illumination, even very small (<5 μm) flagellates can be isolated. This method also is useful for obtaining axenic cultures of some flagellates if aseptic technique is maintained and care is taken to allow the flagellates to swim out of the water carried over in the micropipette at each transfer.

A number of adaptations on this method exist. Cowling[9] described an "agar cavity-channel" technique in which hemispherical cavities in an agar dish can be used as receptacles and culture vessels for isolated flagellates.

Dilution/Extinction

A number of variations have been employed to isolate numerically dominant species in an enrichment culture. The principle of all of these methods is to transfer smaller and smaller aliquots of a sample into a suitable growth medium until protistan growth is no longer observed. This process is accomplished by serially diluting an aliquot of a sample in a protist-free diluent. The series is examined regularly for the growth of protista. Theoretically, the last dilution in which the protist grows is the result of the growth of a single cell (i.e., it is a clonal culture). In practice, of course, one cannot be certain that this culture is actually the result of a single cell but the ease of this method (relative to most micromanipulative methods) makes it a popular one. The dilution/extinction method has been employed not only for the establishment of clonal cultures of heterotrophic flagellates but also as a method of enumerating protista in soil,[26] freshwater,[27] and marine samples.[28]

Dilution/extinction can be applied to solid media for some species of flagellates, although this is not a commonly used procedure. A sample or enrichment is diluted in sterile liquid medium (or ''soft'' liquid agar for heat-tolerant species) and spread over a lawn of prey microorganisms (usually bacteria). Under these conditions some flagellates can form small colonies that can be observed with a dissecting microscope. This method also is effective with some amoebae and ciliates, and is analogous to methods used to identify plaque-forming bacteriophage.

Cowling[9] used a ''dipping wire'' method in which a metal-inoculating wire is dipped into a sample or enrichment-containing protista and then sequentially transferred through a liquid medium series. As with other dilution/extinction methods, the last positive culture in the dilution series is the result of one (or a few) protistan cells.

Axenizing Flagellate Strains

Phagotrophy is clearly the nutritional mode of most heterotrophic flagellates in nature, and the usual method for culturing these species in the laboratory is to provide living prey for them. A number of these species, however, can be cultured in the absence of other living microorganisms. The establishment of axenic cultures of flagellates is desirous even though it represents a highly artificial mode of existence for these protista. Many physiological processes are difficult to study in mixed cultures of microorganisms. For example investigations of nutrient remineralization by heterotrophic protista are difficult to conduct when live bacteria, which can contribute to nutrient uptake and remineralization, are present in the culture vessel.

The establishment of axenic cultures requires a fair amount of work (and a reasonably hardy protist). The most problematical part of this process is the removal of bacteria which are almost always present at higher densities than the protist. Usually, a clonal culture of the flagellate is first established with prey that are clearly distinguishable from the flagellate. It is possible to micropipette individual cells (described above for establishing clonal cultures) directly into medium that contains no other living organisms but, in practice, this shortcut is fruitful only if there is some assurance that a suitable nonliving food has been provided in the culture vessel.

A variety of methods exist for reducing, limiting, or in some cases eliminating the bacteria in protistan cultures, but antibiotics are most commonly employed. Several antibiotics have been used for axenizing microalgae (e.g., 50 to 100 mg l^{-1} penicillin/30 to 60 mg l^{-1} streptomycin is a commonly used mixture). The use of antibiotics for ridding cultures of bacteria must be tailored to each species. Some protist species are susceptible to the effects of prokaryotic metabolic inhibitors, and will cease their growth even though ample food is present.

Antibiotics occasionally can be effective in eliminating live bacteria from cultures of protista, but problems can arise in trying to completely rid them from cultures of heterotrophic flagellates. Unlike algal cultures, high concentrations of nonliving organic material must be added to cultures of heterotrophic flagellates as the bacteria are eliminated (to provide food for the protista). If bacteria in the cultures acquire antibiotic resistance, therefore, these latter bacteria can grow readily in these media (one reason for using an antibiotic mixture). For this reason, the use of antibiotics to eliminate bacteria may be supplemented with the use of a dilution/extinction method by first establishing numerical dominance of the protist using antibiotics and then performing a dilution series to obtain a bacteria-free suspension of the protist. Alternatively, micromanipulation (as noted above) can be used to isolate flagellates from the remaining bacterial assemblage.

As living microorganisms are removed from protistan cultures, they must be replaced with nonliving food. This food may be in the form of dissolved or particulate organic material, or killed prey. High concentrations (grams per liter) of complex organic compounds can be used to support some flagellates in axenic cultures, although osmotrophic growth often is not as rapid as phagotrophic growth. In some cases, however, even very high concentrations of dissolved organic compounds cannot sustain protistan growth.[1] For those species in which phagotrophy is an obligate or preferred mode of nutrition, heat-killed prey can sometimes be used to promote their growth.[29]

Culture Maintenance

Short-Term Maintenance

Maintaining healthy cultures of protista is necessary when these cultures are involved in day-to-day use in the laboratory. Once desired species have been established in monoxenic or axenic cultures, it is imperative to maintain these cultures on a transfer (subculturing) schedule that is commensurate with the growth and feeding rates of the species. The routine maintenance of a culture collection, even a modest one, is not a trivial matter. Particularly fastidious species (e.g., many of the heterotrophic dinoflagellates) can constitute a considerable commitment of time. One efficient method is to subculture groups of species (usually according to their type of culture media and their optimal transfer time) and to maintain a notebook indicating the schedule for each species. I have found that most clonal cultures of bacterivorous flagellates grown in bacterized cultures with rice grains can withstand 3 to 4 weeks before subculturing is necessary. There is, however, considerable variability in the hardiness of each species, and some experimentation is required to find an acceptable schedule.

Any attempt to maintain live cultures of protista should include a system for maintaining at least two, and preferably three, replicate sets of each culture in order to prevent accidental loss. We routinely keep three sets of each culture that we are investigating; one active stock culture, and the two most recent subcultures. The subcultures are kept in a separate incubator to minimize the chances for loss of a strain.

Long-Term Maintenance

Long-term maintenance and storage of flagellate cultures is desirable for several reasons. In a very practical sense, long-term storage minimizes the amount of time that must be spent tending to cultures. Also, storage of cultures provides a mechanism to prevent accidental loss of live cultures that often have taken considerable time and effort to establish. Nearly everyone that routinely cultures microorganisms has experienced the anguish of having cultures destroyed in a failed incubator. More importantly, however, the proper storage of isolated species will maintain the robustness of the clone for a longer period of time. It is not uncommon for protistan cultures to become "fatigued" over long period of vegetative growth, lose vigor, and eventually stop growing. Long-term preservation methods (if successful) can maintain the viability of a stock culture while circumventing the need for constant attention to live cultures during periods when the cultures are not in use.

It is possible to freeze or lyophilize (freeze dry) many flagellate species. A recent manual of these procedures has been published by The American Type Culture Collection [ATCC Preservation Methods: Freezing and Freeze-Drying, 2nd edition, ATCC Marketing (NR84), 12301 Parklawn Dr., Rockville, MD 20852; Tel. 301-881-2600]. In addition, a number of depositories exist that will accept cultures of heterotrophic flagellates. These collections provide an efficient mechanism for investigators to maintain a species in culture without expending space and effort for its upkeep. These facilities also provide a mechanism for investigators to obtain species for their studies. Cowling[9] provides a list of several culture collections that maintain heterotrophic flagellates.

ACKNOWLEDGMENTS

This work was supported by National Science Foundation Grant OCE-8901005. Woods Hole Oceanographic Institution Contribution Number 8036.

REFERENCES

1. Haas, L. W. and Webb, K. L., Nutritional modes of several non-pigmented microflagellates from the York River Estuary, Virginia, *J. Exp. Mar. Biol. Ecol.*, 39, 125, 1979.
2. Fenchel, T., Ecology of heterotrophic microflagellates. II. Bioenergetics and growth, *Mar. Ecol. Prog. Ser.*, 8, 225, 1982.
3. Sherr, B. F., Sherr, E. B., and Berman, T., Grazing, growth, and ammonium excretion rates of a heterotrophic microflagellate fed with four species of bacteria, *Appl. Environ. Microbiol.*, 45, 1196, 1983.
4. Goldman, J. C., Caron, D. A., Andersen, O. K., and Dennett, M. R., Nutrient cycling in a microflagellate food chain. I. Nitrogen dynamics, *Mar. Ecol. Prog. Ser.*, 24, 231, 1985.
5. Fenchel, T., The ecology of heterotrophic microflagellates, *Adv. Microb. Ecol.*, 9, 57, 1986.
6. Jacobson, D. M. and Anderson, D. M., Thecate heterotrophic dinoflagellates: feeding behavior and mechanisms, *J. Phycol.*, 22, 249, 1986.
7. Caron, D. A., Grazing of attached bacteria by heterotrophic microflagellates, *Microb. Ecol.*, 13, 203, 1987.
8. Geider, R. J. and Leadbeater, B. S. C., Kinetics and energetics of growth of the marine choanoflagellate *Stephanoeca diplocostata*, *Mar. Ecol. Prog. Ser.*, 47, 169, 1988.
9. Cowling, A. J., Free-living heterotrophic flagellates: methods of isolation and maintenance, including sources of strains in culture, in *The Biology of Free-Living Heterotrophic Flagellates*, Patterson, D. J. and Larsen, J., Eds., Clarendon Press, Oxford, 1991, 45, 477.
10. Lee, J. J., Hutner, S. H., and Bovee, E. C., *An Illustrated Guide to the Protozoa*, Society of Protozoologists, Lawrence, Kansas, 1985, 629.
11. Guillard, R. R., Methods for microflagellates and nannoplankton, in *Handbook of Phycological Methods — Culture Methods and Growth Measurements*, Stein, J. R., Ed., Cambridge University Press, Cambridge, 1973, 70.

12. Guillard, R. R. L., Culture of phytoplankton for feeding marine invertebrates, in *Culture of Marine Invertebrate Animals,* Smith, W. L. and Chanley, M. H., Eds., Plenum Press, New York, 1975, 29.

13. Provasoli, L., A catalogue of laboratory strains of free-living and parasitic protozoa (with sources from which they may be obtained and directions for their maintenance), *J. Protozool.,* 5, 1, 1958.

14. Daggett, P. M. and Nerad, T., *American Type Culture Collection Catalogue of Protists — Algae and Protozoa,* ATCC, Rockville, MD, 1985.

15. Thompson, A. S., Rhodes, J. C., and Pettman, I., *Culture Collection of Algae and Protozoa: Catalogue of Strains,* CCAP, Cumbria, 1988.

16. Oppenheimer, C. H. and Zobell, C. E., The growth and viability of sixty-three species of marine bacteria as influenced by hydrostatic pressure, *J. Mar. Res.,* 11, 10, 1952.

17. Zobell, C. E., Studies on marine bacteria. I. The cultural requirements of heterotrophic aerobes, *J. Mar. Res.,* 4, 42, 1941.

18. Sieburth, J. M., *Sea Microbes,* Oxford University Press, New York, 1979, 491.

19. Rhee, G.-Y., Continuous culture in phytoplankton ecology, *Adv. Aq. Microbiol.,* 2, 151, 1980.

20. Hamilton, R. D. and Preslan, J. E., Observations on the continuous culture of a planktonic phagotrophic protozoan, *J. Exp. Mar. Biol. Ecol.,* 5, 94, 1970.

21. Gold, K., Methods for growing Tintinnida in continuous culture, *Am. Zool.,* 13, 203, 1973.

22. Goldman, J. C. and Caron, D. A., Experimental studies on an omnivorous microflagellate: implications for grazing and nutrient regeneration in the marine microbial food chain, *Deep-Sea Res.,* 32, 899, 1985.

23. Caron, D. A., Growth of two species of bacterivorous nanoflagellates in batch and continuous culture, and implications for their planktonic existence, *Mar. Microb. Food Webs,* 4, 143, 1990.

24. Miller, T. L. and Wolin, M. J., A serum bottle modification of the Hungate technique for cultivating obligate anaerobes, *Appl. Microbiol.,* 27, 985, 1974.

25. Turley, C. M., Lochte, K., and Patterson, D. J., A barophilic flagellate isolated from 4500 m in the mid-North Atlantic, *Deep-Sea Res.,* 35, 1079, 1988.

26. Singh, B. N., A method of estimating the numbers of soil protozoa, especially amoebae, based on their differential feeding on bacteria, *Ann. Appl. Biol.,* 33, 112, 1946.

27. Baldock, B. M., A method for enumerating protozoa in a variety of freshwater habitats, *Microb. Ecol.,* 12, 187, 1986.

28. Caron, D. A., Davis, P. G., and Sieburth, J. M., Factors responsible for the differences in cultural estimates and direct microscopical counts of populations of bacterivorous nanoflagellates, *Microb. Ecol.,* 18, 89, 1989.

29. Caron, D. A., Porter, K. G., and Sanders, R. W., Carbon, nitrogen and phosphorus budgets for the mixotrophic phytoflagellate *Poterioochromonas malhamensis* (Chrysophyseae) during bacterial ingestion, *Limnol. Oceanogr.,* 35, 433, 1990.

CHAPTER 10

Determination of Pressure Effects on Flagellates Isolated from Surface Waters

C. M. Turley

INTRODUCTION

In oceanic waters small heterotrophic flagellates are important components of the pelagic microbial food web and are a major link in the detrital decomposer pathway. Aggregates of phytodetritus of various sizes fall from the euphotic zone and form the main energy source for the deep sea. Layers and pockets of phytodetritus have been found on the mid-North Atlantic sea bed as deep as 4500 m. The phytodetritus had a distinct flora of bacteria and cyanobacteria; deep-sea detrital feeders such as foraminifera and nematodes colonized the detritus subsequent to its arrival on the sea bed. Reports of small flagellates associated with this material indicate that they may also play an important role in the deep-sea food web. Two apparently barophilic flagellates have been isolated from such an environment indicating that there may be a population of flagellates adapted to growth in the deep sea.

Aggregates generally originate in the upper mixed layer of the oceans and contain high concentrations of bacteria and bacterivorous protozoa. The protozoa may be transported to the sea bed by their association with rapidly sinking aggregates (30 to 300 m d^{-1}). However, protozoa from euphotic waters can function under high-pressure and low-temperature conditions to varying degrees and this seems to depend on the species. Sedimenting particles may therefore undergo a succession of bacterivorous flagellates influenced by changes of pressure incurred during their descent. The ability of some species to grow and continue to graze over a wide pressure range will enhance remineralization of the aggregate throughout its journey and may influence aggregate concentration. This process releases carbon dioxide and nutrients and consumes oxygen and, as aggregates contribute a major flux from surface waters, about 10% of primary production, it may significantly influence geochemical cycles in the oceans.

Bacterivorous flagellates can be cultured under *in situ* pressure and temperature from most water depths by enriching seawater with a source of particulate material in order to enhance the growth of bacterial prey. The source of enrichment material and water sample depth depends on the questions you are asking. For example, Lochte and Turley[1] incubated phytodetritus which had reached the sea bed at 4500 m at *in situ* pressure and temperature in order to look at the growth of flagellates on the naturally occurring material, while in another investigation sterilized detritus was added to water in contact with the sediment to see if there was a benthic reaction by the deep-sea flagellate population to a simulated detrital

0-87371-564-0/93/$0.00 + $.50
© 1993 by Lewis Publishers

fall.[2] In another investigation, the question of what happens to the growth rates of flagellates, associated with aggregates from the euphotic zone when they sink through the water column and undergo substantial pressure changes, was approached by incubating flagellates isolated from particles collected at 40 m under a range of different pressures.[3] It is the latter type of experiment which will be addressed here while reference can be made to the above publications for methods of sampling and isolating deep-sea flagellates.

MATERIALS REQUIRED

Equipment

- Large volume *in situ* particle sampling pump
- Oceanographic sampler (e.g., ''Go-Flo'' sampler) attached to a CTD
- Pressure vessels fitted with pressure gauge and pressure pump (see Notes and Comments — 6)
- Heat sealer for polyethylene bags
- Autoclave
- TEM
- Epifluorescent microscope
- Filtration manifold and equipment for counting bacteria and flagellates (see relevant chapters in this issue)
- Phase contrast microscope for observing live flagellates
- Cold room or cold stage for phase microscope

Supplies

- Sterile volumetric cylinders
- Sterile pipettes
- 0.6-μm pore size, 293-mm diameter Nuclepore filters
- Sterile container
- Sterile pipettes
- Sterile 80-mm flattened, 0.1-mm thick polyethylene bags
- Slides and coverslips

Solutions and Reagents

- Sterile seawater
- 25% SEM grade glutaraldehyde
- Fluorochrome stain for enumerating bacteria and flagellates (see Chapter 25)

PROCEDURES

Sampling

1. Collect naturally occurring suspended particulate material from the required water depth on a 0.6-μm pore size, 293-mm diameter Nuclepore filter in a large volume *in situ* particle sampling pump.

2. Rinse the particulates from the filter with autoclaved seawater collected from the same depth, using an oceanographic ''Go-Flo'' sampler, and cooled to 2 to 5°C.

3. Incubate approximately 50 ml of the detrital/sterile seawater mixture in a sterile tissue culture flask in the dark at 2 to 5°C.

Subculturing and Barotolerance Experiments

1. A mixed population of flagellates will develop within 4 to 8 weeks (depending on the initial population concentration) and can be maintained by adding 5 ml of the original culture to 200 ml seawater enriched with five barley grains (both having been sterilized by microwave treatment).[4] The barley acts as a detrital source for the gradual growth of bacteria which in turn act as a food source for the bacteriverous flagellates. The cultures can be maintained by subculturing every 2 to 3 weeks.

2. Once a substantial population of flagellates develop (around 4 to 8×10^5 ml^{-1}), replicate 5-ml subsamples can be placed aseptically into 10-ml volume, 80-mm flattened diameter, 0.1-mm thick polyethylene bags that have been sterilized with ethylene oxide gas or γ-irradiation (2.5 Mrad cobalt 60).

3. Fix replicate subsamples in 2.5% glutaraldehyde (final concentration) for determination of numbers of bacteria and flagellates at the start of the experiment (N_{t1}).

4. Heat seal the bags making sure that as little air as possible is enclosed in the bag and place them in pressure vessels.

5. Fill the pressure vessels with aerated tap water at 2 to 5°C to allow a limited exchange of gases across the bag membrane.

6. Pump pressure vessels to a range of pressures (for example, 1, 50, 100, 200, 300, 400, and 500 atm) in order to examine barotolerance of the species you have isolated. Incubate the pressure vessels at a constant temperature between 2 and 5°C, that of the deep sea.

7. After the appropriate incubation period (around 3 d), slowly decompress the pressure vessels at a rate of <100 atm min^{-1} to avoid intra- and extracellular bubble formation which may disrupt some protozoa (see Notes and Comments — 1).

Determination of Motility, Growth, and Growth Efficiency

1. Remove subsamples for live examination using phase contrast microscopy of motility and flagellar movement (these can aid in identification, see Notes and Comments — 3) and species diversity. This should be carried out in either a cold room or using a cooled microscope.

2. Fix a further subsample in 2.5% glutaraldehyde for confirmation of identity by whole mount transmission electron microscopy (TEM) and for determination of flagellate and bacterial numbers (see Notes and Comments — 4) and biomass (see Notes and Comments — 5) by epifluorescence direct count techniques.

3. Calculate growth rates (μ_n) from total flagellate or species numbers (N) by

$$\mu_n(d^{-1}) = (\ln N_{t2} - \ln N_{t1})/(t_2 - t_1) \tag{1}$$

where t_1 (d) is the start and t_2 (d) is the end of the incubation period.

4. Calculate doubling times (Dn) by

$$D_n (d) = \ln(2)/\mu_n \tag{2}$$

5. Calculate apparent flagellate carbon conversion efficiency (CeF) on the assumption that decreases in bacterial carbon are only attributable to consumption by flagellates (see Notes and Comments — 2):

$$\text{CeF (\%)} = \frac{\text{carbon incorporated into flagellate biomass}}{\text{bacterial carbon used}} \cdot 100 \qquad (3)$$

6. Compare the growth rates you find under high pressure and low temperature with surface water flagellates populations. For example, doubling times of *Bodo* sp. isolated by Turley et al.[2] of 1.6 to 3.2 d at 450 atm are generally longer than those of 0.12 to 0.96 d generally reported for shallow water flagellates,[5-9] but similar to the doubling times of 0.96 to 1.44 d recorded by other groups of workers.[10,11]

7. By assuming a Q_{10} of two, correct for the temperature difference between your experiment and the above experiments (carried out at around 20°C). A mean doubling time of 2.11 d for the barophilic *Bodo* sp. was corrected to 0.59 d and is within the lower range for shallow water flagellates.[2] By doing this you can investigate whether low temperature rather than high pressure is the important influence on the doubling time of the flagellates in your experiment.

8. Compare your value for carbon conversion efficiency with estimates of 24 to 62% for shallow water flagellates[6,12,13] and 17 to 25% for a barophilic flagellate.[2]

NOTES AND COMMENTS

1. In studies of gas supersaturation tolerances of *Tetrahymena* and *Euglena* cell disruption, due to the formation of intra- and/or extracellular bubbles, occurred with rapid decompression (1 to 2 s return to atmospheric pressure from >175 atm). However, both species were unaffected by slow decompression (over several minutes) from 225 and 300 atm.[14,15] Turley and Carstens[3] found decompression from 450 to 1 atm over 5 to 7 min did not result in visible bubble formation and cell rupture.

2. The carbon conversion efficiencies assume that the flagellates only ingest bacteria. Should the flagellates also graze directly on detrital particles, the carbon conversion efficiency would be lower. Estimates of conversion efficiency using this method are therefore upper estimates. The carbon not converted into flagellate biomass is an estimate of the ingested carbon dissipated through respiration and excretion.

3. Despite the importance now attached to marine flagellates,[16] very few are identified with any degree of confidence. This is because very few studies have been conducted, and complete identification often requires electron-microscopical examination of the material. Turley and Carstens[3] show that an investigation purely at the population level could have resulted in a misinterpretation of the influence of pressure and point out the importance of investigations at the species level as well as the population level. Identification is substantially aided by prior familiarization of the species characteristics of live specimen using phase microscopy which generally remain active for 40 min at 2 to 5°C after decompression.

4. It is useful to monitor the bacterial food supply during your experiments so that they remained within the range of those usually found necessary to support a growing flagellate population.

5. The width and length or area of 70 flagellates can be measured and the mean cell volume calculated for the population by assuming basic geometric shapes. The carbon equivalent of flagellate volume can be estimated assuming a conversion factor of 1.1×10^{-13} g C μm^{-3}.[17]

6. The construction and design of pressure vessels, including specialized high-pressure sampling and transfer vessels, are described by Jannasch,[18] Gilchrist and MacDonald,[19] and Tabor et al.[20]

REFERENCES

1. Lochte, K. and Turley, C. M., Significance of bacteria and cyanobacteria associated with phytodetritus and its decomposition in the deep-sea, *Nature,* 333, 67, 1988.
2. Turley, C. M., Lochte, K., and Patterson, D. J., A barophilic flagellate isolated from 4500 m in the mid-North Atlantic, *Deep-Sea Res.,* 35, 1079, 1988.
3. Turley, C. M. and Carstens, M., Pressure tolerance of oceanic flagellates: implications for remineralization of organic matter, *Deep-Sea Res.,* 38, 403, 1991.
4. Keller, M. D., Bellows, W. K., and Guillard, R. R. L., Microwave treatment for sterilization of phytoplankton culture medium, *J. Exp. Mar. Biol. Ecol.,* 117, 279, 1988.
5. Fenchel, T., Ecology of heterotrophic microflagellates. I. Some important forms and their functional morphology, *Mar. Ecol. Prog. Ser.,* 8, 211, 1982.
6. Fenchel, T., Ecology of heterotrophic microflagellates. II. Bioenergetics and growth, *Mar. Ecol. Prog. Ser.,* 8, 225, 1982.
7. Linley, E. A. S. and Newell, R. C., Estimates of bacterial growth yields based on plant detritus, *Bull. Mar. Sci.,* 35, 409, 1984.
8. Andersen, P. and Fenchel, T., Bacterivory by microheterotrophic flagellates in seawater samples, *Limnol. Oceanogr.,* 30, 198, 1985.
9. Goldman, J. C. and Caron, D. A., Experimental studies on a omnivorous microflagellate: implications for grazing and nutrient regeneration in the marine microbial food chain, *Deep-Sea Res.,* 32, 899, 1985.
10. Sherr, B. F. and Sherr, E. B., Role of heterotrophic protozoa in carbon and energy flow in aquatic ecosystems, in *Current Perspectives in Microbial Ecology,* Klug, M. J. and Reddy, C. A., Eds., American Society for Microbiology, Washington, DC, 1984, 412.
11. Andersen, P. and Sorensen, H. M., Population dynamics and trophic coupling in pelagic microorganisms in eutrophic coastal waters, *Mar. Ecol. Prog. Ser.,* 33, 99, 1986.
12. Caron, D. A., Goldman, J. C., Andersen, O. K., and Dennett, M. R., Nutrient cycling in a marine microflagellate food chain. II. Population dynamics and carbon cycling, *Mar. Ecol. Prog. Ser.,* 24, 243, 1985.
13. Sherr, B. F., Sherr, E. B., and Berman, T., Grazing, growth, and ammonium excretion rates of a heterotrophic microflagellate fed with four species of bacteria, *Appl. Environ. Microb.,* 45, 1196, 1983.
14. Hammingsen, E. A., Cinephotomicrographic observations of intracellular bubble formation in *Tetrahymena, J. Exp. Zool.,* 200, 43, 1982.
15. Hemmingsen, B. B. and Hemmingsen, E. A., Intracellular bubble formation: differences in gas supersaturation tolerances between *Tetrahymena* and *Euglena, J. Protozool.,* 30, 608, 1983.
16. Fenchel, T., *Ecology of Protozoa,* Brock-Springer, Madison, 1986, 197.
17. Turley, C. M., Newell, R. C., and Robins, D. B., Survival strategies of two small marine ciliates and their role in regulating bacterial community structure under experimental conditions, *Mar. Ecol. Prog. Ser.,* 33, 59, 1986.
18. Jannasch, H. W., Experiments in deep-sea microbiology, *Oceanus,* 21, 50, 1978.
19. Gilchrist, I. and MacDonald, A. G., Techniques for experiments with deep-sea organisms at high pressure, in *Experimental Biology at Sea,* MacDonald, A. G. and Priede, I. G., Eds., Academic Press, London, 1983, 239.
20. Tabor, P. S., Deming, J. W., Ohwada, K., Davis, H., Waxman, M., and Colwell, R. R., A pressure-retaining ocean sampler and transfer system for measurement of microbial activity in the deep-sea, *Microb. Ecol.,* 7, 51, 1981.

Additional References

Alongi, D. M., The distribution and composition of deep-sea microbenthos in a bathyal region of the western Coral Sea, *Deep-Sea Res.,* 34, 1245, 1988.

Caron, D. A., Grazing of attached bacteria by heterotrophic microflagellates, *Microb. Ecol.,* 13, 203, 1987.

Fenchel, T., The significance of bacterivorous Protozoa in the microbial community of detrital particles, in *Aquatic Microbial Communities,* Cairns, J., Ed., Garland, 1977, 529.

Patterson, D. J., Larsen, J., and Corliss, J. O., The ecology of heterotrophic flagellates and ciliates living in marine sediments, *Prog. Protist.,* 3, 185, 1989.

Rice, A. L., Billet, D. S. M., Fry, J., John, A. W. G., Lampitt, R. S., Mantoura, R. F. C., and Morris, R. J., Seasonal deposition of phytodetritus to the deep-sea floor, *Proc. R. Soc. Edinburgh,* 88B, 265, 1986.

Silver, M. W. and Alldredge, A. L., Bathypelagic marine snow: deep-sea algal and detrital community, *J. Mar. Res.,* 39, 501, 1981.

Thiel, H., Pfannkuche, O., Shriever, G., Lochte, K., Gooday, A. J., Hemleben, C., Mantoura, R. F. C., Turley, C. M., Patching, J. W., and Riemann, F., Phytodetritus on the deep-sea floor in a central oceanic region of the Northeastern Atlantic, *Biol. Oceanogr.,* 6, 203, 1990.

Turley, C. M., Protozoan association with marine 'snow' and 'fluff' — a session summary, in *Protozoa and Their Role in Marine Processes,* Reid, P. C., Turley, C. M., and Burkill, P. H., Eds., (NATO-ASI Series, Vol. G25), Springer-Verlag, Berlin, 1991, 309.

CHAPTER 11

Isolation, Cloning, and Axenic Cultivation of Marine Ciliates

A. T. Soldo and S. A. Brickson

INTRODUCTION

Compared to the large numbers of bacterial species and related microorganisms which have been maintained in pure culture in the laboratory, few protozoan species have been placed in axenic culture.[1] The reasons for this are manifold: (1) most protozoa are highly fastidious creatures and often manifest exacting nutritional requirements; consequently, they are difficult to culture and maintain in the laboratory, (2) except for a relatively few species, most protozoa (especially ciliates) do not readily form discrete colonies on solid surfaces (agar, agarose, gelatin, etc.),[2-6] making it necessary to adopt other more tedious procedures to obtain clones, and (3) many are fragile and easily damaged in micropipettes used for transfer through numerous washes needed to free them of contaminating microorganisms. For these reasons we developed an overall approach for the collection, isolation, cloning, and cultivation of these organisms, which greatly reduces the need to manipulate them individually and may be carried out with materials and equipment easily attainable from commercial sources. Briefly, samples are taken from the "wild" and enriched to encourage the growth of the desired species. Individuals are then cloned using the silicone oil plating procedure (SOPP)[7] and maintained in the laboratory as nonaxenic (bacterized) cultures. The bacterized isolates are then placed in a suitable growth medium and maintained as axenic cultures. Initially, we were interested in the isolation, cloning, and cultivation of a group of marine philasterine ciliates. However, the procedures we developed may be applied to other protozoa as well.

MATERIALS REQUIRED

Equipment

- Dissecting microscope

Supplies

- Scintillation vials (30-ml capacity) (New England Nuclear, Inc.)
- Polyethylene screw-cap bottles (Cole-Parmer, Inc.)
- Petri dishes (150 × 20 mm, pyrex) (Corning, Inc.)

0-87371-564-0/93/$0.00 + $.50

- Nine-spot depression slides (Corning, Inc.)
- Pasteur transfer pipettes (Corning, Inc.)
- Glass beads (Corning, Inc.)
- Culture tubes and stainless steel, fingerless caps (Bellco, Inc.)
- Sterile plastic petri dishes (35 × 10 mm, #1008) (Falcon Plastics)
- Graduated plastic tubes (1.7 × 10 cm) (Falcon Plastics)
- Nitrogen gas

Solutions, Reagents, and Media

- Proteose peptone (Difco, Inc.)
- Nutrient broth (Difco, Inc.)
- Bacto agar (Difco, Inc.)
- Trypticase-peptone (#11921) (BBL)
- Ribonucleic acid (Torula yeast type VI) (Sigma)
- Silicone oil, a viscosity standard (9.6 centipoises) (Brookfield Engineering Laboratories, Inc., Stoughton, MA)
- Percoll (Pharmacia)
- Instant Sea Water (ISW): Dissolve 1 lb. of Aquamarine salts (Aquatrol Inc., Anaheim, CA) in about 8 l of distilled water under constant stirring at room temperature. After about 1 h, determine the density of the water with a suitable hydrometer and adjust, by the addition of distilled water to a final density of 1.030 g ml^{-1}. Filter through glass wool to remove any undissolved materials and store at 4°C in a large plastic container. Dilute to 1.015 g ml^{-1} with an equal volume of distilled water as needed.
- Cerophyl Medium (CM): Add 0.5 g of Cerophyl (Cerophyl Laboratories, Inc., Kansas City, MO) to 100 ml of ISW (d = 1.015 g ml^{-1}) and heat to a boil. Filter three times while hot through glass wool. Sterilize at 121°C for 15 min. Filtered source water may also be used. Bacterized Cerophyl Medium (BCM) is prepared by inoculating CM with a bacterium obtained from source water and incubating overnight at 30 to 37°C.
- Axenic Culture Medium (ACM):[8] To prepare 100 ml ACM, add 0.5 g of Cerophyl to 90 ml of ISW (d = 1.015 g ml^{-1}) and heat to a boil. Filter three times while hot through glass wool. To the hot solution dissolve, under constant stirring:

  > 1.0 g proteose peptone
  > 1.0 g trypticase
  > 0.1 g ribonucleic acid
  > 0.1 ml vitamins (1000 ×)

- Adjust to pH 7.2 with 1 N NaOH and bring to volume with ISW (1.015 g ml^{-1}). Dispense 5 ml per tube and sterilize at 121°C for 15 min.
- Vitamins (1000 ×) mg; (source: Calbiochem)

Biotin	0.01
Folic acid	50
Nicotinamide	50
D-Pantothenate, Ca	100
Pyridoxal HCl	50
Riboflavin	50
Thiamine HCl	150
DL-Thioctic acid	1

- Suspend the vitamins in 100 ml double-distilled water, and, under constant stirring, dispense 1-, 2-, or 5-ml portions in screw-cap tubes (16 × 125 mm). Flush with nitrogen and store at −20°C. Thaw as needed only once. Discard any unused portion.

PROCEDURES

Collection of Samples

Small samples of seawater are taken from shallow pools, inlets, near submerged vegetation, decaying logs, tree roots, etc. by immersing 30-ml capacity screw-cap scintillation vials under the water at the desired location and allowing them to fill to about two thirds their capacity. Larger samples may be taken in a similar manner using 500-ml capacity screw-cap plastic bottles. Alternatively, small quantities of decaying materials (fish, meat, etc.) may be placed in perforated containers. Plastic tissue holders used in histology laboratories are excellent for this purpose, or holes may be drilled in plastic screw-caps and mounted on glass scintillation vials. These containers are then immersed at various locations, secured by means of monofilament fishing line to a tree or any other permanent fixture near the shoreline, and allowed to remain for periods ranging from several hours to several days before retrieval. When taking samples, be certain to record information such as location, time of day, water temperature, pH, salinity, etc. In transporting the samples back to the laboratory, make certain to avoid excessive exposure to sunlight and heat.

Processing Samples in the Laboratory

Upon return to the laboratory, small dip samples may be examined immediately for the presence of protozoa. After shaking to distribute the organisms, 0.5 ml of each sample is transferred by Pasteur pipette to individual wells of nine-spot depression slides and observed with the aid of a 10 to 30 × binocular dissecting microscope. Large samples are first filtered through cheesecloth or glass wool to remove coarse debris and then concentrated by gravity through a millipore filter (pore size = 0.45 μm in diameter) to about 50 ml or so. A portion of the filtrate may be saved for use in preparing culture medium. Both the small dip samples and the concentrated large samples may be enriched by adding small quantities (1 to 2 mg ml^{-1}) of dry cerophyl, hay, wheat germ, dried egg yolk, tuna fish, etc., followed by incubation in the dark at 22 to 27°C for a few days. It is advisable to heat the dried additives at 180°C for an hour to kill any contaminating protozoa that may be present. We routinely plate out an unfiltered sample on nutrient or cerophyl agar prepared with source water to isolate and maintain one or more bacterial species for use as a food source for bacterial feeders.

Cloning by the Silicone Oil Plating Procedure (SOPP) and Cultivation in Bacterized Medium

After incubating the enriched samples for a few days, there are usually a sufficient number of ciliates for plating by the SOPP. A model protocol is as follows:

1. Ciliates are suspended in instant sea water, d = 1.015 g ml^{-1} (ISW) at a concentration of about 1000 ml^{-1} or less and diluted with an equal volume of sterile cerophyl medium (CM).

2. 200 μl of the diluted suspension (about 100 ciliates) are added to 3 ml of sterile silicone oil in a screw-cap tube.

3. The tube is then vortexed vigorously for 15 to 20 s and its contents immediately poured into a 35 × 10 mm sterile plastic petri dish. Under these conditions microdroplets form, some containing individual ciliates, and become fixed between the hydrophobic surfaces of the silicone oil and the bottom of the petri dish. Usually the diluted CM permits the growth of a sufficient number of bacteria to serve as a food source for the ciliates trapped in the microdroplets and enables them to multiply and form a clone.

4. The plates are allowed to remain undistributed for about 30 min to insure complete formation of the microdroplets before placing them in the incubator.

5. The plates may be examined at suitable intervals with the aid of a dissecting microscope at 10 to 30 × magnification.

6. Microdroplets containing ciliate populations are subsequently transferred to 0.5-ml volumes of fresh bacterized culture medium contained in the wells of nine-spot depression slides or in culture tubes with the aid of sterile micropipettes, "pulled" to 100 to 200 μm in diameter in a gas flame from Pasteur transfer pipettes.

Maintenance of Isolates in Bacterized Medium

1. After individual species have been placed in bacterized medium, they are examined periodically for growth.

2. Isolates that produce clones are transferred to sterile culture tubes (16 × 125 mm) equipped with fingerless stainless steel caps and maintained by doubling the volume every few days or so by the addition of fresh culture medium depending upon the rate of growth of the organisms.

3. When the volume reaches 5 ml or so, 0.5ml of the culture is transferred to a fresh tube and the process repeated. It is necessary to decide on a transfer regimen that is optimal for a given isolate. In most cases, the bacteria present in the original culture will serve as an adequate food source for the protozoa. In some cases, it may be necessary to rejuvenate a culture by adding freshly prepared bacterized medium (BCM). It is important to examine the cultures very carefully to insure that each culture contains a single protozoan species. If there is any doubt, "plate out" the culture using the SOPP and reestablish it in bacterized culture.

Sterilization and Maintenance in Axenic Medium

1. Double the volume of the desired bacterized protozoan strain until a volume of 5 or 10 ml is obtained.

2. Add 250 μl of Percoll for each 5 ml of culture to bring the concentration to 5% (v/v).

3. Into a capped, sterile, graduated plastic tube, add sterile glass beads (4 mm diameter) to the 9-ml mark.

4. Then, by means of a sterile pipette, add the ciliate culture containing the Percoll to a level near the upper surface of the glass beads taking care to avoid touching the insides of the tube with the pipette.

5. Add more beads to the 12-ml mark followed by the addition of sterile ISW containing 2.5% Percoll (v/v) to just cover the glass beads.

6. Carefully overlayer with 2 ml of sterile ISW and allow the tube to remain upright and undisturbed for a period long enough to enable the ciliates to migrate into the overlay (10 min to 1 h or so).

7. At this point, remove 1 ml of the overlay containing the migrated ciliates, transfer to a fresh sterile tube, and stand for 2 to 3 h. This permits sufficient time for the ciliates to either digest or expel any microorganisms, spores, etc. on which the ciliates may have been feeding.

8. To the suspension of ciliates that have been allowed to stand for 2 to 3 h, add sufficient Percoll to bring the concentration to 5% (v/v).

9. To a fresh sterile plastic tube, add sterile glass beads to the 3-ml mark followed by the suspension of freshly migrated ciliates.

10. Then add glass beads to the 12-ml mark and add ISW containing 2.5% Percoll (v/v) to just cover the glass beads.

11. Carefully overlayer 2 ml ISW and allow the ciliates to migrate once again. This time, however, monitor the migration by removing 5- to 10-µl samples from the overlay at intervals. Place these on a glass slide and observe in the binocular microscope until the ciliates first appear.

12. At this point and at short intervals thereafter, transfer 100-µl portions to tubes containing 3-ml sterile silicone oil; add 100-µl fresh axenic culture medium (ACM) and vortex and plate as described above. Incubate at 27°C and observe on a daily basis. Microdroplets that contain any bacteria that may have comigrated with the ciliates will grow and multiply and are easily detected as white "plaques" after a few days. These, of course, are to be avoided. Ciliates that develop clones are usually observable in the microdroplets after 4 to 6 d. These are picked up in micropipettes and transferred to fresh 0.5-ml portions of fresh full-strength ACM medium contained in wells of depression slides.

13. After incubation at 27°C for 5 to 7 d, clones are transferred to 5 ml ACM and maintained by transferring 0.1-ml portions of 7-d-old cultures to 5 ml of fresh ACM on a weekly basis. These are tested to insure against the presence of any slow-growing microbial contaminants that may be present by plating on agar containing the crude culture medium and/or standard nutrient agar and incubating at 37°C for periods of up to a week.

14. This procedure usually produces axenized cultures on the first try with only a small portion of the microdroplets showing contamination. On occasion, we have observed that a substantial portion of the microdroplets may be contaminated. In this case, it is advisable to repeat the procedure using antibiotics in ISW and culture medium. Neomycin and gentamicin are excellent for this purpose. Avoid the use of penicillin, which is unstable in sea water.

REFERENCES

1. Provasoli, L., A catalogue of laboratory strains of free-living and parasitic protozoa (sources from which they may be obtained and directions for their maintenance), *J. Protozool.*, 5, 1, 1958.
2. Asami, K., Nodake, Y., and Ueono, T., Cultivation of *Trichomonas vaginalis* on solid medium, *Exp. Parasitol.*, 4, 43, 1955.
3. Neff, R. J., Mechanisms of purifying amoebae by migration on agar surfaces, *J. Protozool.*, 5, 226, 1958.
4. Padan, E., Ginzberg, D., and Shilo, M., Growth and colony formation of the phytoflagellate *Pyremnesium parvum* Carter on solid medium, *J. Protozool.*, 14, 477, 1967.
5. Samuels, R., Agar techniques for colonizing and cloning trichomonads, *J. Protozool.*, 9, 103, 1962.
6. West, R. A., Barbera, P. W., Kolar, J. R., and Murrel, C. B., The agar layer method for determining the activity of diverse materials against selected protozoa, *J. Protozool.*, 9, 65, 1962.
7. Soldo, A. T. and Brickson, S. A., A simple method for plating and cloning ciliates and other protozoa, *J. Protozool.*, 27, 328, 1980.
8. Soldo, A. T. and Merlin, E. J., The cultivation of symbiote-free marine ciliates in axenic medium, *J. Protozool.*, 19, 519, 1972.

CHAPTER 12

Isolation and Laboratory Culture of Marine Oligotrichous Ciliates

Dian J. Gifford

INTRODUCTION

Tintinnid ciliates have been cultured in the laboratory by a number of investigators during the past 2 decades.[1-10] The generally more abundant aloricate oligotrich ciliates have been less amenable to laboratory culture. This may be due in part to their greater fragility: lack of a lorica may render them more susceptible to damage during collection and handling. Bacterivorous forms of these organisms were isolated from interstitial[11,12] and pelagic[13] environments in the early 1980s. A limited number of species of the phagotrophic, pelagic forms which are the target of the protocol which follows were grown in the laboratory prior to 1985[14] when these organisms were first cultured on a large scale by Gifford.[15] They have since been reared routinely in the laboratory.[16-19]

This paper describes procedures for the nonclonal serial culture of herbivorous, phagotrophic marine planktonic oligotrichs. The method, which is primarily that of Gifford,[15] should apply equally well to the culture of tintinnid ciliates,[20] to bacterivorous forms,[13,21] to applications requiring continuous culture,[22] and to applications requiring clonal cultures.[10] The protocol presented below is not axenic, but may be made so by addition of appropriate antibiotics.[16,17,19] To date, the method has been used primarily to culture estuarine and coastal forms. However, several oligotrichous ciliates from the oceanic subarctic Pacific have been cultured successfully using the protocol (D. Gifford, unpublished data), and it is likely that other oceanic forms can be reared as well.

In addition to careful attention to the chemical composition of the culture medium and to physical and biological culture conditions (i.e., light, temperature, food composition, food concentration), successful culture of oligotrichous ciliates in the laboratory requires attention to details of the processes of collection, concentration, and isolation of the target organisms. These topics are discussed separately below. The major factors in successful laboratory culture of oligotrich ciliates appear to be avoidance of metal and solvent contamination, chelation of the culture medium, avoidance of turbulence, and provision of appropriate, i.e., not overly dense, levels of suitable food organisms.

The culture of marine planktonic ciliates is an art as much as a science. Hence, the protocol which follows is not a foolproof recipe. A number of nonquantitative anecdotal comments are included throughout the text. A great deal of trial and error is involved, particularly with respect to determining suitable food items and suitable levels of the food. The advantage of this method is that, given patience and reasonably clean laboratory conditions, it usually works. The disadvantage is that the isolation and maintenance of oligotrich cultures is extremely labor intensive.

MATERIALS REQUIRED

Equipment

- See comments on sampling apparatus below
- Dark-field dissecting microscope
- Lighted, temperature-controlled incubators
- Autoclave

Supplies

- Pasteur pipettes
- Teflon bottles
- Micro detergent (Cole-Parmer Instrument Company; Chicago)
- Culture vessels — Polycarbonate and teflon have the advantage that they can be cleaned thoroughly, but clean glass vessels may also be used for some forms. A general cleaning protocol is to wash with Micro detergent, rinse well with deionized water, soak in 5% HCl, and rinse three times with deionized water. When sterility is required, clean vessels are autoclaved with distilled water in them and the water is subsequently discarded prior to use.
- Drawn pipettes — Drawn pipettes are made by rotating the mid-tip of a Pasteur pipette in a gas flame and pulling the two sections apart. The oral end of the pipette is plugged with cotton, and the pipettes are autoclaved in paper wrappers. For use, the drawn pipette tip is clipped off with a pair of clean forceps.

Solutions, Media, and Cultures

- EDTA
- HCl
- Metals for f/2 medium
- The factors common to successful oligotrich culture media are use of seawater collected from the same source as the oligotrichs; clean, metal-free preparation of sterile medium; chelation of the culture medium; and addition of trace metals to the culture medium.[15-19] Although the overall culture method is not axenic, media preparation, handling, and storage are. If seawater from the same source is not available, aged oligotrophic ocean water, such as Sargasso Sea water, can sometimes be substituted if adjusted to the appropriate salinity with deionized water. The seawater is passed through a small porosity glass fiber or membrane filter to remove particles. A nonmetal filtration manifold is used. The filtered water is autoclaved in teflon bottles, and the iron-EDTA trace metal component of phytoplankton f/2 medium[23] is autoclaved separately and added to the medium axenically for a final concentration of 10^{-8} to 10^{-7} M. For storage, the hot medium is transferred to sterile polycarbonate bottles.
- Suitable algal cultures for maintenance of oligotrichs, e.g.,

  Isochrysis
  Heterocapsa
  Scrippsiella
  Tetraselmis
  Chroomonas

PROCEDURES

Field Collection of Microplankton

Because collection by nets destroys aloricate ciliates,[15] target organisms are isolated from bulk seawater which is collected using clean methods. How "clean" the collection conditions are is dictated by the specific environment: for example, estuarine forms should not be as susceptible to metal, or other toxic, contamination as are oceanic forms. A rigorous metal-free cleaning protocol is described by Fitzwater et al.[24] This cleaning method is optimal for target organisms from oceanic environments, and may be applied less rigorously for collection of near-shore and estuarine organisms.

To collect bulk seawater, a clean plastic surface bucket may suffice. It is important to note that even brief exposure to the standard latex tubing supplied on Niskin bottles is extremely toxic to marine organisms.[25] Hence, in oceanic environments, clean teflon-coated Go-Flo bottles, which do not contact the surface film of the water and do not contain interior metal parts, are preferred. If these are not available, Niskin bottles can be rerigged with silicone tubing and o-rings and teflon-coated springs, then cleaned. If Go-Flo or Niskin bottles are used, the water is drained from them through silicone tubing into clean receptacles. Turbulence destroys aloricate ciliates and should be avoided while draining: the siphon tube should contact the bottom of the receiving vessel so that water is transferred into water rather than into air. If large grazers that might prey on the protozoans are present, the seawater may be siphoned gently through a submerged 200-μm mesh to remove them, although losses of aloricate ciliates will occur.

Isolation of Monospecific Serial Cultures

Target organisms may be dilute (<1 ciliate ml^{-1}) in seawater. It is considerably easier to sort them if they are concentrated. Methods which employ nets or meshes may be used to concentrate tintinnids, but are extremely destructive to aloricate ciliates, and have the effect of diluting them further.[15]

Pipetting

The ciliates are manipulated individually under a dissecting microscope using a drawn pipette. They are most easily viewed using dark-field illumination. Because the oligotrichs cannot, in general, be identified unambiguously under these conditions, preliminary sorting is done on the basis of size, color, and swimming pattern: some forms swim in tight circles, others swim in loose spirals; plastidic ciliates appear yellow under dark-field illumination, whereas aplastidic forms are white. The sorted ciliates are transferred into incubation vessels containing culture medium and food. The incubation vessels may be covered plastic multiwell plates, or crystallization dishes, beakers, or flasks of various volumes, depending on the abundance of available ciliates and the goals of the particular investigation. Multiwell plates and crystallization dishes have the advantage that their contents can be scanned rapidly under the dissecting microscope with minimal disturbance to the ciliates, whereas larger beakers and flasks must be subsampled. Multiwell plates offer the additional advantage that ciliate growth under a variety of food conditions can be compared easily. The number of ciliates sorted into the culture vessel depends on the vessel volume. If clonal cultures are required, the inoculum is a single ciliate added to each vessel. This general procedure is applied to successive subsamples of the bulk seawater, or to subsamples of the bulk seawater in which the protozoan assemblage has been encouraged to grow for a period of time before processing, described below.

Enrichment

Bulk seawater may be enriched with cultured phytoplankton and incubated for a period of days in order to obtain more dense concentrations of ciliates. In this procedure, bulk seawater is transferred to a number of vessels. Appropriate densities of various phytoplankton species are added singly or in combination to each vessel and the vessels are incubated at environmental temperature. Subsamples of the water are examined under the dissecting microscope at 1- to 2-d intervals to assess ciliate growth. When the ciliates are sufficiently dense, they are sorted as described above. This procedure has the advantage that it is not labor intensive and involves minimal handling of the ciliates.

Culture Conditions

Food Type

The choice of appropriate phytoplankton food for planktonic oligotrichs is dictated by the organisms' *in situ* environment, by their mouth dimensions and feeding mechanics, and by the availability of suitable phytoplankton cultures. Phytoplankton cells should be motile and able to remain in suspension in the oligotrich cultures, which should not be shaken or stirred. Phytoplankton cells having complex or spinose morphology that is likely to cause handling difficulties to the consumer should be avoided,[26] as should phytoplankton, such as *Olisthodiscus lueteus*, known to be toxic to ciliates,[27] and phytoplankton cultures, such as *Dunaliella tertiolecta*, which are probably not nutritionally replete.[28] Common microalgal clones which promote oligotrich growth in culture include species of *Isochrysis, Heterocapsa, Scrippsiella, Tetraselmis,* and *Chroomonas.*[15]

Food Density

As a rule of thumb, too little food is better for oligotrich cultures than too much food. As a first approximation, one can attempt to match the ratio of consumers to phytoplankton cells to the organisms' *in situ* environmental conditions. As a practical matter, it is usually more convenient to initiate replicate cultures at several food levels, e.g., ''low'', ''medium'', and ''high'' in order to determine the optimal level.

Physical Factors

Light. Oligotrichs which are strictly heterotrophic can be grown in either the light or the dark. Forms which retain chloroplasts must be grown in the light.[16,17] The advantage of maintaining cultures in the light is that the phytoplankton food will also grow, and it is possible to maintain the oligotrichs in a quasi-steady state with their food. Cultures maintained in the dark must be fed more frequently.

Temperature. The oligotrich cultures and the bulk seawater collected to isolate them are maintained at a temperature as close as possible to that of the environment from which the water was collected. Abrupt changes in temperature are usually fatal to oligotrichs. It is generally easier to acclimate oligotrichs collected from a warm environment to colder temperatures than the reverse. If cultures adapted to different temperatures are required, the temperature should be changed slowly ($\sim 0.5°C$ d^{-1}) over a protracted period of time.

Maintenance of Cultures

During the early stages of isolation, the cultures are examined frequently under the dissecting microscope to ensure that a single oligotrich type is isolated. If organisms other than the target are present, they are removed with a drawn pipette and discarded. Phytoplankton food is added as necessary. Debris is removed from the bottoms of multiwell plates and crystallization dishes with a pasteur pipette, as necessary. Cultures are transferred at regular intervals by pouring or pipetting a portion of the culture volume into a clean vessel containing culture medium and phytoplankton food.

Oligotrich cultures commonly have a finite lifetime in the laboratory, surviving from weeks to many months. Most cultures eventually decline, cease to reproduce, and die off. This decline, which has also been observed in tintinnid cultures,[4,9] is attributed to cessation of conjugation. To postpone what appears to be an inevitable decline, separate strains of the same oligotrich species may be mixed together periodically in order to encourage conjugation and genetic diversity.

NOTES AND COMMENTS

1. The quality of ambient air in the laboratory is critical to the success of oligotrich culture. The fumes of volatile solvents, tobacco smoke, and histological fixatives are capable of destroying oligotrichs in an open culture vessel. As a general practice, persons who smoke should not handle culture vessels or supplies because of the toxic residues on their hands and clothing. Laboratory personnel should avoid wearing or using substances containing perfume or fragrance.
2. Anecdotally, better growth occurs with a minimum innoculum of several ciliates ml^{-1}.

ACKNOWLEDGMENTS

Manuscript preparation was supported by grant number N00014-90-J-1437 from the U.S. Office of Naval Research.

REFERENCES

1. Gold, K., The role of ciliates in marine ecology. I. Isolation and cultivation of a member of the order Tintinnida, *Am. Zool.,* 6, 513, 1966.
2. Gold, K., Some observations on the biology of *Tintinnopsis* sp., *J. Protozool.,* 15, 193, 1968.
3. Gold, K., Tintinnids: feeding experiments and lorica development, *J. Protozool.,* 16, 507, 1969.
4. Gold, K., Cultivation of marine ciliates and heterotrophic flagellates, *Helgolander Wiss. Meeresunters.,* 20, 264, 1970.
5. Gold, K., Growth characteristics of the mass-reared tintinnid *Tintinnopsis beroida, Mar. Biol.,* 8, 105, 1971.
6. Johansen, P. L., A Study of Tintinnids and other Protozoa in Eastern Canadian Waters with Special Reference to Tintinnid Feeding, Nitrogen Excretion, and Reproduction, Ph.D. thesis, Dalhousie University, Halifax, Nova Scotia, 1976.

7. Laval-Peuto, M., Reconstruction d'une lorica de forme *Coxliella* par le trophont nu de *Favella ehrenbergii* (Ciliate, Tintinnina), *C. R. Hebd. Seanc. Acad. Sci.*, 284D, 547, 1977.

8. Heinbokel, J. F., Studies on the functional role of tintinnids in the Southern California Bight. I. Grazing and growth rates in laboratory cultures, *Mar. Biol.*, 47, 177, 1978.

9. Paranjape, M., Occurrence and significance of resting cysts in a hyaline tintinnid, *Helicostomella subulata* (Ehre.) Jorgensen, *J. Exp. Mar. Biol. Ecol.*, 48, 23, 1980.

10. Verity, P. G., Grazing, respiration, excretion and growth rates of tintinids, *Limnol. Oceanogr.*, 30, 1268, 1985.

11. Martinez, E. A., Sensitivity of marine ciliates (Protozoa, Ciliophora) to high thermal stress, *Estuar. Coast. Mar. Sci.*, 10, 369, 1980.

12. Martinez, E. A., Effects of temperature and food concentration on competitive interaction in three species of marine Ciliophora, *Carib. J. Sci.*, 19, 3, 1983.

13. Gold, K., Simplified procedures for maintaining *Strombidium sulcatum*, and SEM observations on its morphology, *J. Protozool.*, 29, 297, 1982.

14. Rassoulzadegan, F., Dependence of grazing rate, gross growth efficiency and food size on temperature in a pelagic oligotrichous ciliate *Lohmaniella spiralis* Leeg., fed on naturally occurring particulate matter, *Ann. Inst. Oceanogr.*, 58, 177, 1982.

15. Gifford, D. J., Laboratory culture of marine planktonic oligotrichs (Ciliophore, Oligotrichida), *Mar. Ecol. Prog. Ser.*, 23, 257, 1985.

16. Stoecker, D. K., Silver, M. W., Michaels, A. E., and Davis, L. H., Enslavement of algal chloroplasts by four *Strombidium* spp. (Ciliophora, Oligotrichida), *Mar. Microb. Food Webs*, 3, 79, 1988.

17. Stoecker, D. K., Silver, M. W., Michaels, A. E., and Davis, L. H., Obligate mixotrophy in *Laboea strobila*, a ciliate which retains chloroplasts, *Mar. Biol.*, 99, 415, 1988.

18. Putt, M. and Stoecker, D. K., An experimentally determined carbon:volume ratio for marine "oligotrichous" ciliates from estuarine and coastal waters, *Limnol. Oceanogr.*, 34, 1097, 1989.

19. Putt, M., Metabolism of photosynthate in the chloroplast-retaining ciliate *Laboea strobila*, *Mar. Ecol. Prog. Ser.*, 60, 271, 1990.

20. Stoecker, D. K., Guillard, R. R. L., and Kavee, R. M., Selective predation by *Favella ehrenbergii* (Tintinnina) on and among dinoflagellates, *Biol. Bull.*, 160, 136, 1981.

21. Ohman, M. D. and Snyder, R. A., Growth kinetics of the omniverous oligotrich ciliate *Strombidium sulcatum*, *Limnol. Oceanogr.*, 36, 922, 1991.

22. Scott, J. M., The feeding rates and efficiencies of a marine ciliate, *Strombidium* sp., grown under chemostat steady-state conditions, *J. Exp. Mar. Biol. Ecol.*, 90, 81, 1985.

23. Guillard, R. R. L., Culture of phytoplankton for feeding marine invertebrates, in *Culture of Marine Invertebrate Animals*, Smith, W. L. and Chanley, M. H., Eds., Plenum Press, New York, 1972, 29.

24. Fitzwater, S. E., Knauer, G. A., and Martin, J. H., Metal contamination and its effect on primary production, *Limnol. Oceanogr.*, 27, 544, 1982.

25. Price, N. M., Harrison, P. J., Landry, M. R., Azam, F., and Hall, K. J. F., Toxic effects of latex and tygon tubing on marine phytoplankton, zooplankton and bacteria, *Mar. Ecol. Prog. Ser.*, 34, 41, 1986.

26. Verity, P. G. and Villareal, T. A., The relative food value of diatoms, dinoflagellates, flagellates, and cyanobacteria for tintinnid ciliates, *Arch. Protistenkd.*, 131, 71, 1986.

27. Verity, P. G. and Stoecker, D. K., The effects of *Olisthodiscus luteus* on the growth and abundance of tintinnids, *Mar. Biol.*, 72, 79, 1982.

28. Langdon, C. J. and Waldock, M. J., The effect of algal and artificial diets on the growth and fatty acid composition of *Crassostrea gigas* spat, *J. Mar. Biol. Assoc. U.K.*, 61, 431, 1981.

CHAPTER 13

Extraction of Protists in Aquatic Sediments via Density Gradient Centrifugation

Daniel M. Alongi

INTRODUCTION

Protozoans are numerically and functionally important in most benthic habitats.[1-3] Unfortunately, taxonomic and ecological studies of these organisms have lagged behind those of larger organisms because of difficulties in extraction and a lack of taxonomic expertise. A wide variety of extraction methods have been used to harvest cells from fine sediments and organic detritus, testifying to the persistence of the harvesting problem. Innumerable variations of at least eight methods have been used: decantation/preservation with formaldehyde,[4,5] cultivation and serial dilution,[6,7] pipetting of diluted mud,[8-10] adhesion onto coverslips,[11] the seawater ice method,[12,13] flushing with a narcotizing agent,[14] flushing with a vital stain,[15] and density centrifugation with a silica gel-sorbitol mixture.[16] Three of the techniques appear most promising, yielding superior recovery of cells: dilution, flushing with a vital stain, and density centrifugation.

The silica gel-sorbitol technique has successfully extracted a wide variety of protists (ciliates, flagellates, amoebae, foraminifera) from tropical muds, detrital carbonates, and quartz, and carbonate sands.[16-18] Advantages of the method are (1) cells can be extracted either dead or alive, (2) harvested cells are usually undamaged to aid in identification, (3) extraction efficiency can be easily estimated and is generally high, (4) cells of different density (e.g., testate vs. nontestate protists) can be separated within the same samples, and (5) very fragile, nanoprotozoa can be harvested and concentrated. Disadvantages of the method are (1) the preferred silica sol used, Percoll, is expensive outside the U.S. (>$250 per liter), (2) the procedure is initially time consuming as the optimum density of the solution needs to be determined, (3) the solution is difficult to filter, (4) the solution gels if preservatives are added above a particular concentration (e.g., >1% formalin), (5) the solution must be kept sterile and out of direct light as much as possible as it is a good growth medium for bacteria and microalgae, and (6) the optimum density of solution (i.e., the amount of sorbitol added) is dependent on sediment type. This last shortcoming appears to be the most serious problem but, once preliminary samples are taken to test for an optimum density, the procedure is rapid, more efficient, and less time consuming than the other methods. Other advantages of using the silica gel-sorbitol techniques are listed by Price et al.[19] and Schwinghamer.[20]

MATERIALS REQUIRED

Equipment

- Filtration apparatus (pump, receiving flask, filter tower and base, forceps)
- Balance (accurate to 0.01 g)
- Autoclave
- Vortex mixer
- Petri dishes
- Centrifuge
- Dissecting microscope (to 500 ×)
- Compound microscope (to 1000 ×)
- Hemocytometer

Supplies

- Autoclaved glassware (graduated cylinders and beakers)
- 0.2-μm pore-size membrane filters
- Screw-capped 1-l container
- Scintillation vials
- Syringes
- 10- and 50-μl micropipettes
- Centrifuge tubes (preferably 30-ml size)
- Dialysis tubing (for Ludox)

Solutions and Reagents

- 50% (w/v) $MgCl_2$ solution
- 5% sodium tetraborate-buffered formalin solution
- 1 to 2% (w/v) nigrosin black or methyl green solution[23]

PROCEDURES

Preparation

It is best to start with the Percoll-sorbitol mixture of Price et al.[19] as the density of this solution (1.15 g/cm³) should be sufficient to extract most cells lacking tests or shells. The mixture is prepared by first dissolving 91.1 g D-Sorbitol, 2.64 g Tris-HCl, and 4.03 g Tris reagent base in 0.5 l Percoll (all chemicals can be obtained from Sigma Chemical Co., MO). Then, 1.43 g $MgCl_2 \cdot 6H_2O$ is dissolved in 100 ml of filtered (0.2 μm), autoclaved seawater [salt content must match habitat salinity; for freshwater organisms, use physiological (0.85% NaCl) saline]. This solution is added to the Percoll mixture, which is made up to a final volume of 1 l with additional Percoll. All glassware used must be clean and sterile. It is preferable that the container used to store the Percoll-sorbitol mixture be darkened and have a screw-cap (stoppered containers may be difficult to reopen because the mixture is sticky and will dry, sealing the container; parafilm may be used to avoid this problem). Schwinghamer[20] recommends that the mixture be filtered before each use but I have found this time consuming and unnecessary if all labware is properly sterilized. The mixture must be stored refrigerated and the density of the medium should be checked regularly with density beads which can also be obtained from Sigma Chemical Co.

The silica gel Ludox can be used instead of Percoll. Ludox is stored in NaOH and must be detoxified by overnight (12 to 16 h) dialysis against running water. Dialyzed Ludox gels during this process but can be liquified by adding sterile water. Both density and pH must be checked. The NaOH must be extracted even if extracting preserved organisms because retention of the NaOH will cause elevated osmotic potential that may lead to cell plasmolysis.[19]

Salinity will also affect osmotic potential so it may be necessary to use water of either lower or higher salinity depending upon habitat salinity and the result of trial runs. As noted earlier, the density of the solution may have to be changed depending on sediment type. In my experience, tropical sediments require a less dense mixture (75 g D-sorbitol instead of 91.1 g) than temperate muds and sands. The reason for this is not clear but it may be due to the different concentrations of $CaCO_3$ and differences in clay type (illite vs. kaolinite) between sediments of different latitudes. The amount of sorbitol added will have to be determined by preliminary trial runs to test extraction efficiency.

Extraction

1. Obtain replicate cores from the study site. A large number of small cores (<1 cm^2) is preferable to a smaller number of larger cores; the actual number of samples should be determined by the required level of precision about the mean.[21] I use a plastic medical syringe (1 or 2 cc) with the needle end cut off as a coring device. The actual volume of sediment used for extraction should be small because extraction efficiency usually drops rapidly for volume ratios greater than one part sediment to four parts Percoll-sorbitol.[20]

2. In the laboratory, gently extrude each core into a separate centrifuge tube (30 ml is best) containing 5 ml of the silica gel mixture (volume added will depend on sediment volume).

3. Shake or vortex each sample at slow speed for 1 to 2 min to mix the solution thoroughly into the sediment.

4. Allow to stand for 1 h. This allows for the development of a density gradient (minimizing centrifugation time) and permits acclimatization by protists to the solution.

5. Centrifuge at 1000 to 3000 rpm for 10 to 40 min. Both speed and time of centrifugation will vary with sediment type. They must be sufficient to form a compact slug at the bottom of the tube and a sediment-free supernatant, but allow easy resuspension of the slug for additional extractions.

6. Subsequent procedures will depend upon whether or not there is interest in cells <20 μm. If only the larger (>20 μm) cells are to be enumerated (most ciliates, larger pigmented flagellates, amoebae, some yeasts, foraminifera), two or three further extractions are needed; the number for optimum recovery being determined by the recovery efficiency procedures. The first supernatant is poured into a 5-cm glass Petri dish that has the bottom gridded into 0.5-cm^2 squares. To facilitate counting under a dissecting microscope, a 50% (W/V) $MgCl_2$ solution may be added dropwise into the dish to slow the movements of the protists. A fixed number of grids are counted randomly and subsequent cell counts (obtained by examining the supernatant from the second or third extraction) are added to obtain total numbers per total grids counted per core.

    If cells <20 μm are of interest, step 7 below should be carried out in addition to steps 1 to 6:

7. A 1-ml aliquot of the first supernatant (from step 6) is placed into an acid-washed scintillation vial containing 50 μl of a 5% sodium tetraborate-buffered formalin solution together with either a 1 to 2% solution of either methyl green or negrosin black.[18] These stains mix well in silica gel and formalin. Some stains such as Lugol's solution and acridine

orange tend to clump in the gel mixture limiting identification. It should be noted here that some (but not all) possible variations on the procedure have been attempted. I have tried acridine orange staining followed by filtration onto blackened membrane filters (epifluorescence microscopy), but the solution does not filter well and the AO clumps onto the filter. Other possible, untested, variations to be tried include use of DAPI or FITC and filtration of a greatly diluted aliquot. For enumeration of nanoprotozoans (5 to 20 μm), two aliquots (5 to 10 μl) from the first 1 ml of formalin-preserved sample are counted on duplicate 0.1-mm^3 blocks on a hemocytometer at 400 ×.[22] As mentioned earlier, this procedure has a comparatively high detection limit ($>1.0 \times 10^4$ cells ml^{-1}) at a low magnification level. It does have the advantage of better image clarity than some other epifluorescence procedures. For instance, in my experience, counts using the epifluorescence technique of Bak and Nieuwland[15] can be unreliable because of the fluorescence of detrital particles.[18]

Extraction Efficiency

Either of two procedures may be used to determine recovery efficiency: the cumulative yield method[20] or the cell addition technique.[16] The cumulative yield method is the simplest: repeat the extraction-centrifugation procedure (adding the cumulative yield of all counts) until no further cells are harvested. This method assumes that the total yield equals the actual number of recoverable cells per sediment sample. This assumption may not be valid because it is probable that some cells will be lost or destroyed by repeated sample handling and by addition of the mixture.

The alternative procedure[16] is to add a known number of various protists grown in laboratory culture. A great variety of culturing procedures may be used (see References 23 and 24 for the simplest techniques) particularly for large ciliates, foraminifera, amoebae, and flagellates. A listing of research centers from where cultures may be obtained is provided by Cowling.[25]

In my experience, the simplest method is to scrape the upper few mm of surface sediment from the study habitat, add it to a series of 25-cm^2 plastic tissue culture flasks (Dow Corning Co.), and inoculate with 10 to 15 ml of aged (1 to 2 weeks) seawater plus a few grains of rice or some mixed cereal (baby pablum) or oatmeal flakes. After 2 to 3 weeks incubation (usually in the dark at 20 to 25°C), the cultures may be examined for growth of cell populations under a dissecting microscope. Cells can be picked out individually using a sterile glass pipette and collected into a sterile Syracuse dish. The cells can then be added to a small (1 to 2 cm^3) volume of sediment from the study habitat, previously defaunated by alternate freezing and thawing. Once subjected to the extraction procedure, the recovery efficiency can be calculated for each extraction. This method is more time consuming than the cumulative yield method but is advantageous in that the absolute number of cells per sample is known.

Calculations

Large (>20 μm) Protists in Petri Dish

Assume that a 1-cm^3 sediment sample was extracted once with 5 ml Percoll-sorbitol mixture and counted in a 5-cm diameter Petri dish gridded into 0.5-cm^2 squares. Assume cells were counted in 15 of these squares. Calculate total number of cells per core:

$$N = nA/a \qquad (1)$$

where   N = number of cells in petri dish
        A = total number of 0.5-cm squares in dish
        a = number of squares counted
        n = cumulative number of cells counted in a square

If all 5 ml of the solution is added to the dish, numbers from Equation 1 = total numbers per core (less extraction efficiency). If additional extractions are required, count each extract with the protocol given above and simply add the counts together to get the number of cells per core. Correct for extraction efficiency (e.g., if recovery efficiency is 75%, multiply total numbers per core by 1.33).

Nanoprotozoa by Hemocytometer

Assume that one 10-μl aliquot of the first ml of the 5-ml supernatant is counted on duplicate 0.1-mm^3 blocks on a hemocytometer. Assume that we find one cell per duplicate block (each block = 0.1 μl volume). Calculate total number of cells per core (assuming 100% recovery efficiency) as follows

$$N = C \times d \tag{2}$$

where   N = number of cells per 10-μl aliquot
        C = mean number of cells per duplicate block ($= 0.2$ μl)
        d = volume correction ($= 50$)

and

$$N_c = V \times N \times e \tag{3}$$

where N_c = numbers of cells per core
        V   = volume of supernatant
        N   = number of cells per 10-μl aliquot
        e   = volume correction ($= 500$)

If more extractions are required, repeat the calculation on each extract and add the results after correction for extraction efficiency.

ACKNOWLEDGMENT

I thank Dr. Peter Doherty for clarifying and improving the text. Contribution No. 618 from the Australian Institute of Marine Science.

REFERENCES

1. Alongi, D. M., Flagellates of benthic communities: characteristics and methods of study, in *The Biology of Free-Living Heterotrophic Flagellates,* Patterson, D. J. and Larsen, J., Eds., Clarendon Press, Oxford, 1991, chap. 5.
2. Patterson, D. J., Larsen, J., and Corliss, J. O., The ecology of heterotrophic flagellates and ciliates living in marine sediments, *Prog. Protist.,* 3, 185, 1989.

3. Fenchel, T., *Ecology of Protozoa,* Science Tech, Madison, 1987, 197.

4. Dye, A. H., Quantitative estimation of protozoa from sandy substrates, *Est. Coast. Mar. Sci.,* 8, 199, 1979.

5. Dye, A. H., Composition and seasonal fluctuations of meiofauna in a Southern Africa mangrove estuary, *Mar. Biol.,* 73, 165, 1983.

6. Mare, M. F., A study of a marine benthic community with special reference to the microorganisms, *J. Mar. Biol. Assoc. U.K.,* 25, 517, 1942.

7. Lighthart, B., Planktonic and benthic bacteriovorous protozoa at eleven stations in Puget Sound and adjacent Pacific Ocean, *J. Fish. Res. Board Can.,* 26, 299, 1969.

8. Kemp, P. F., Bacterivory by benthic ciliates: significance as a carbon source and impact on sediment bacteria, *Mar. Ecol. Prog. Ser.,* 49, 163, 1988.

9. Wyatt, C. E. and Pearson, T. H., The Loch Eil project: population characteristics of ciliate protozoans from organically enriched sea-loch sediments, *J. Exp. Mar. Biol. Ecol.,* 56, 279, 1982.

10. Bark, A. W., The temporal and spatial distribution of planktonic and benthic protozoan communities in a small productive lake, *Hydrobiology,* 85, 239, 1981.

11. Webb, M. G., An ecological study of brackish water ciliates, *J. Anim. Ecol.,* 25, 148, 1956.

12. Uhlig, G., Eine einfache methode zur extraktion der vagilen, mesopsammalen mikrofauna, *Helgo. Wiss. Meeres.,* 11, 78, 1964.

13. Fenchel, T., The ecology of marine microbenthos. I. The quantitative importance of ciliates as compared with metazoans in various types of sediments, *Ophelia,* 4, 121, 1967.

14. Alongi, D. M. and Hanson, R. B., Effect of detritus supply on trophic relationships within experimental benthic food webs. II. Microbial responses, fate and composition of decomposing detritus, *J. Exp. Mar. Biol. Ecol.,* 88, 167, 1985.

15. Bak, R. P. M. and Nieuwland, G., Seasonal fluctuations in benthic protozoan populations at different depths in marine sediments, *Neth. J. Sea Res.,* 24, 37, 1989.

16. Alongi, D. M., Quantitative estimates of benthic protozoa in tropical marine system using silica gel: a comparison of methods, *Est. Coast. Shelf Sci.,* 23, 443, 1986.

17. Alongi, D. M., The distribution and composition of deep-sea microbenthos in a bathyal region of the western Coral Sea, *Deep-Sea Res.,* 34, 1245, 1987.

18. Alongi, D. M., Abundances of benthic microfauna in relation to outwelling of mangrove detritus in a tropical coastal region, *Mar. Ecol. Prog. Ser.,* 63, 53, 1990.

19. Price, C. A., Reardon, E. M., and Guillard, R. R. L., Collection of dinoflagellates and other marine microalgae by centrifugation in density gradients of a modified silica gel, *Limnol. Oceanogr.,* 23, 548, 1978.

20. Schwinghamer, P., Extraction of living meiofauna from marine sediments by centrifugation in a silica sol-sorbitol mixture, *Can. J. Fish. Aquat. Sci.,* 38, 476, 1981.

21. Sokal, R. R. and Rohlf, F. J., *Biometry,* W. H. Freeman, San Francisco, 1981, 859.

22. Collins, C. H. and Lyne, P. M., *Microbiological Methods,* Butterworths, London, 1976, 521.

23. Kudo, R. R., *Protozoology,* Charles C. Thomas, Springfield, IL, 1977, 1174.

24. Lee, J. J., Hunter, S., and Bovee, E. C., *An Illustrated Guide to the Protozoa,* Allen, Kansas, 1985, 629.

25. Cowling, A. J., Free-living heterotrophic flagellates: methods of isolation and maintenance, including sources of strains in culture, in *The Biology of Free-Living Heterotrophic Flagellates,* Patterson, D. J. and Larsen, J., Eds., Clarendon Press, Oxford, 1991, chap. 27.

Section II
Identification, Enumeration, and Diversity

CHAPTER 14

Statistical Analysis of Direct Counts of Microbial Abundance

David L. Kirchman

INTRODUCTION

The problem of counting microorganisms has been examined by statisticians since the turn of the century (e.g., Student[1]). After nearly a century, there is still some confusion. The statistical analysis of microbial abundance determined by epifluorescence microscopy may appear to be more problematic than other measurements of microbial assemblages because there are several levels at which subsamples can be taken. These levels include the following:

1. River, lake, or ocean being studied
2. Large volume (100 ml to >1 l) sampling container, e.g., Nisken bottle
3. Subsample from large volume container which is preserved in a small container (<20 to 100 ml) until staining
4. Subsample (<2 to 50 ml) from preserved water, which is stained with a fluorochrome, and filtered onto a filter
5. Microscopic fields ("subsample" of stained water) examined to count number of cells.

Two questions arise: how many replicates should be taken at each of these levels? Once the data are collected, how are they analyzed statistically? This entire discussion applies to enumeration of all microorganisms by epifluorescence microscopy, ranging from bacteria to ciliates. Recent studies also have reported epifluorescence direct counts of viruses.[2]

To answer the first question, we need estimates of the actual variance contributed by the different levels and some estimate of the costs of performing replicates at each level.

Kirchman et al.[3] examined the variability contributed by the lowest three levels (levels 3, 4, and 5) in counting heterotrophic bacteria after staining with acridine orange. They found that variation among microscopic fields (level 5) was highest, contributing 62 to 80% of total variance. Variance contributed by the filters (level 4) was 16 to 27% of the total, whereas the preserved subsamples (level 3) contributed 0 to 10%. Examining variance at level 2 is equivalent to examining the spatial and temporal variation in microbial abundance; that is, if Niskin casts are taken repeatedly over time, differences in microbial abundance over space and time will contribute to the total variance. By determining the variance at level 1, we are essentially examining the intersystem differences in microbial abundance.

0-87371-564-0/93/$0.00 + $.50

If costs are ignored, the most efficient sampling scheme is to replicate only at the highest level (actually level 2 in the above list, the large volume sampler) and have only one sample at the lowest level.[4] This is the most efficient scheme because we can calculate an error with replication only at the highest level; that level will include the variance of the lower levels. In contrast, we do not know the total error if only the lowest level is replicated because the variance contributed by the higher levels is not included. An important rule is that replication at the highest level is usually most useful, even if that level is not the main source of variance.

This most efficient scheme, however, is not the optimal one. That is determined by including the costs. The equations to determine the optimal number of replicates at levels 4 and 5 are

$$N_f = \sqrt{[(C_g S_f^2)/(C_f S_g^2)]} \tag{1}$$

$$N_g = \sqrt{[(C_v S_g^2)/(C_g S_v^2)]} \tag{2}$$

where N_f and N_g are the optimal number of fields and filters, respectively; C_v, C_g, and C_f are the marginal costs; and S_v^2, S_g^2, and S_f^2 are the variance components of a replicate subsample (v) (level 3), filter (g), and field (f), respectively. The total cost (C) of doing one count is

$$C = C_v N_v + C_g N_g N_v + C_f N_v N_g N_f \tag{3}$$

After calculating N_f and N_g, the optimal number of subsamples (N_v) is calculated with Equation 3. To use this equation, the investigator must decide on the total cost he or she is willing to spend on the sample, i.e., assign a value to C.

PROCEDURE: SUGGESTED OPTIMAL SAMPLING STRATEGY

An investigator could calculate an optimal sampling strategy for his or her specific problem by going through the above equations. Alternatively, we give the following sampling strategy, based on the cost and error analysis of Kirchman et al.:[3]

Number of preserved subsamples: 2 (1 is sometimes acceptable)
Number of filters per subsample: 1
Number of microscopic fields per filter: >7, probably 10

The volume filtered per subsample should result in approximately 30 cells per field. Parts of a microscopic field can be counted such that the average is approximately 30. Of course, the fraction of the field counted is included in figuring the number of cells per volume of water. Kirchman et al.[3] showed that the coefficient of variation (CV) for bacterial cells per field was minimized at approximately 30. The CV is higher for fewer cells because cells per field follows a Poisson distribution, but when cells per field exceeded 30 the CV increased, probably because of operator fatigue.

The above sampling scheme is probably not different from those commonly used without a formal statistical analysis. The subtle but important difference concerns the number of preserved subsamples. Most investigators probably only have one preserved subsample. That is acceptable when the scientific question concerns variation at level 2, e.g., the temporal variation in abundance could be examined by taking only one sample per time point. The changes over time could be examined by a regression analysis, but the investigator would not be able to determine if the difference between two specific time points (for example) was statistically significant because the variance contributed by subsampling would be unknown.

A simpler question perhaps is, how is the error of a single microbial count determined? The common approach would be to simply calculate the error associated with level 5, i.e., the variation contributed by the microscope fields. If this is the only level that is replicated, which is often the case, the error calculated with these data will underestimate the true error by as much as 30%. The best estimate of the true error includes the variance contributed by the other levels, either actually measured or taken from the literature. The true error is then calculated by nested analysis of variance (ANOVA).

Here is another common problem. While comparing the means from two filters (level 4), the investigator senses that more data should be taken to show whether or not a difference is statistically significant. Should more fields be counted or should new filters be prepared? At the very least, new filters should be prepared (assuming that the preserved sample is less than a few days old) because this increases replication at a higher level than simply counting more fields.

Several investigators base their counting strategies on the number of microorganisms per microscopic field, e.g., they continue until 300 cells are counted, and a graph is used to calculate the error, which is based solely on the number of cells per field. This was reasonable, particularly before the advent of personal computers, because cells per field follow a Poisson distribution and this level does contribute the most variance. However, it is now possible to calculate directly the standard error of a direct count estimate with the aid of statistical packages which are easy to use on personal computers. This is preferable because it provides the true estimate of the error independent of any assumed cell distribution. Furthermore, as already pointed out, the total error of the estimate will be more than the error at this lowest level, the microscopic field. In any case, if one follows the scheme suggested above (1 subsample, 10 microscopic field with roughly 30 cells per field), both strategies result in the same amount of effort.

Perhaps the largest problem with the direct count method is between-operator variation. Unless care is taken to "standardize" the recognition of cells, differences in counts among operators can be quite large. The problem is especially large for the smallest microorganisms. This is well known for bacteria[3,5] and undoubtedly will be true for viruses. The problem can be difficult if small colloidal particles are stained.[6] This is one more source of variation, which can be included in the nested ANOVA if necessary.

REFERENCES

1. Student, On the error of counting with a haemocytometer, *Biometrika,* 5, 351, 1907.
2. Hara, S., Terauchi, K., and Koike, I., Abundance of viruses in marine waters: assessment by epifluorescence and transmission electron microscopy, *Appl. Environ. Microbiol.,* 57, 2731, 1991.
3. Kirchman, D. L., Sigda, J., Kapuscinski, R., and Mitchell, R., Statistical analysis of the direct count method for enumerating bacteria, *Appl. Environ. Microbiol.,* 43, 769, 1982.
4. Sokal, R. and Rohlf, F. J., *Biometry: the Principles and Practice of Statistics in Biological Research,* 2nd ed., W.H. Freeman, San Francisco, 1981.
5. Nagata, T., Someya, T., Konda, T., Yamamoto, M., Morikawa, K., Fukui, M., Kuroda, N., Takahashi, K., Oh, S.-W., Mori, M., Araki, S., and Kato, K., Intercalibration of the acridine orange direct count method of aquatic bacteria, *Bull. Jpn. Soc. Microbial Ecol.,* 4, 89, 1989.
6. Koike, I., Hara, S., Terauchi, K., and Kogure, K., Role of sub-micrometre particles in the ocean, *Nature,* 345, 242, 1990.

CHAPTER 15

Enumeration and Isolation of Viruses

Curtis A. Suttle

INTRODUCTION

The methods described here are designed to be used for the enumeration and isolation of viral pathogens that infect aquatic bacteria, cyanobacteria, and phytoplankton. However, many of the procedures are similar to virological methods that have been in place for many years in other scientific and medical disciplines. This contribution brings together in a convenient forum some of the more suitable methods.

Basically, four methods are currently used for enumerating viruses in aquatic environments. These are plaque assays, most-probable-number assays (MPNs), transmission electron microscopy (TEM), and epifluorescence microscopy. The methods tell us different things. Plaque assays and MPNs are for quantifying the abundance of infectious units which cause lysis of a particular host. Obviously these assays require the host of interest to be culturable. TEM is typically used for enumerating the number of viral-like particles (VLPs), either in whole water or in culture medium, while epifluorescence microscopy has been used for quantifying viral-sized particles containing double-stranded DNA. The procedure used depends on the question being addressed and the accuracy and sensitivity required. The advantages and disadvantages associated with each of the methods are addressed in the following protocols.

Successful isolation of a virus depends on bringing host and pathogen together so that infection of the host and amplification of the virus occurs. Unless viruses infecting the host are abundant, the chances of isolating one directly by plaque or MPN assays are slim; hence, the probability of finding a virus increases with the volume of water screened. Also, keep in mind that host density is the major factor dictating the rate of viral propagation. Screening a relatively large volume of water can be accomplished either by growing the host to high density in the water to be assayed or by concentrating the virus community using ultrafiltration and adding some of the concentrate to an exponentially growing culture of the host. Additional information on recovering viruses from the natural environment can be found in Berg.[1]

0-87371-564-0/93/$0.00 + $.50
© 1993 by Lewis Publishers

MATERIALS REQUIRED

Enumeration of Viruses

Plaque Assays

Equipment

- Autoclave
- Microwave oven
- 25-mm polysulfone filter funnel (Gelman)
- 1- and 5-ml adjustable pipettes
- Constant-temperature dry bath or water bath
- Centrifuge
- Microcentrifuge
- Vortex mixer
- Culture facilities for bacteria and/or algae

Supplies

- Polypropylene or polycarbonate sampling bottles
- 25-mm, 0.22-, and 0.45-μm pore-size polyvinylidine difluoride (PVD) "Durapore" membrane filters (Millipore)
- 25 mm, 0.2-, and 0.4-μm pore-size polycarbonate membrane filters (Nucleopore or Poretics)
- Petri plates
- Pipet tips
- 250-ml screw-cap borosilicate-glass media bottle or Erlenmeyer flask
- 1.5-ml microfuge tubes
- 13 $\times$ 100 mm disposable glass culture tubes

Solutions, Reagents, and Media

- Purified agar (e.g., Fisher Scientific)
- Liquid culture medium
- Bottom agar — In a 1-l container add 5 g of agar to 500 ml (1% w/v) of an appropriate growth medium. Partially melt the agar by microwaving (e.g., 3 min at 700 W) and autoclave for 20 min at 121°C. Allow the medium to cool to about 60°C and under sterile conditions pour 15 to 20 ml into each of 20 to 30 plates. Leave the petri plate lids slightly ajar while the plates are cooling to prevent condensation accumulating on the lids. Once the agar has solidified, stack the plates and invert them to prevent condensation dripping on the surface of the agar. These plates can be used about 12 h after pouring if the surface of the agar is not wet. The plates can be kept at room temperature for a week or more if they are sealed in a plastic bag.
- Top agar — Prepare 0.6% (w/v) top agar by adding 0.6 g of agar to 100 ml of medium in a 250-ml screw-cap Erlenmeyer flask or media bottle and sterilize, with the lid off, as described above. If sealed after sterilization the medium can be stored at room temperature for many months and remelted in a microwave oven before use. Repeated melting may result in precipitates forming. For each water sample that is to be assayed pour 2 to 2.5 ml of molten top agar into each of three 7-ml 13 $\times$ 100 mm disposable glass culture tubes and keep at 45 to 47°C. For each experiment also include three tubes of agar that will be used for controls.
- Host cells — For microalgae such as *Synechococcus* or *Chlorella*, cells are grown in liquid culture to about 10^7 cells ml^{-1}. While still in exponential growth the cultures are harvested

by gentle centrifugation and resuspended in liquid medium at a concentration of about 10^9 ml^{-1}. For bacteria grown in nutrient-rich broth it is generally not necessary to concentrate the cells.

Most-Probable-Number (MPN) Assays

Equipment

- 5-ml adjustable pipette
- Fluorometer with filter set for chlorophyll a determination (420_{ex}, $> 640_{em}$; e.g., Corning CS 5-60 and CS 2-64)
- Culture facilities for phytoplankton
- 25-mm polysulfone filter funnel (Gelman)

Supplies

- 1-l polypropylene or polycarbonate container
- 25-mm, 1.0-μm pore-size polycarbonate membrane filters (Nuclepore or Poretics)
- 7- and 50-ml glass culture tubes (13 $\times$ 100 and 25 $\times$ 150 mm, respectively) with polypropylene screw caps
- Pipet tips
- Liquid culture of host organism in exponential growth (10^7 cells ml^{-1})

Transmission Electron Microscopy (TEM)

Equipment

- Ultracentrifuge with swinging-bucket rotor
- Transmission electron microscope
- Tweezers for handling EM grids
- Centrifuge support platforms for EM grids

Supplies

- 200 to 400 mesh copper TEM grids coated with carbon and Formvar
- Ultracentrifuge tubes
- Parafilm
- TEM grids and grid platforms — The grid platforms must be strong but not so dense that the safety of the rotor is compromised. The simplest approach is to mold perfectly balanced platforms out of epoxy resin or machine them out of hard plastic such as Plexiglass.[2] The platforms can be easily removed if a threaded hole is included in the platform into which a small rod can be inserted. Preferably, the grids should be held in place within grid-size shallow depressions on the surface of each platform. A small notch beside each depression will facilitate removal of the grids.

For this technique, carbon-coated Formvar films on copper grids are probably the best compromise between strength and electron transparency. These are commercially available (e.g., Ted Pella, Inc.), but expensive, or they can be prepared if one has access to a carbon evaporator.[3] A problem with support films is that uneven charge on the surface can affect the distribution of VLPs, thereby resulting in inaccurate estimates of virus density. Consequently, the surface of the films should be made evenly hydrophylic by UV irradiation, by exposure to a strong electrical field in a reduced atmosphere (glow discharge), or by chemical treatment.[3] A simple treatment is to float the grids for 1 min on a drop of 4000 MW poly-L-Lysine and wick away the excess solution immediately before using the grids.

Solutions and Reagents

- 25% (v/v) EM-grade glutaraldehyde
- 1% (w/v) uranyl acetate solution
- Uranyl-acetate stain — Add 0.25 g of uranyl acetate ($C_4H_6O_6U$) to 25 ml of distilled water and gently stir for about 30 min to make up a 1% stain solution. Be careful when handling uranyl acetate as it is toxic and radioactive. Store it at 4°C in a tightly capped dark bottle as it is light sensitive and prone to oxidation. The stain can be stored for several weeks, but as precipiatate may form it should be centrifuged or filtered prior to use. This is nasty stuff so always dispose of unwanted stain as toxic waste.

Epifluorescence Microscopy of DAPI-Positive Particles

Equipment

- Epifluorescent microscope with a 100-W Hg bulb, 334- to 365-nm excitation and >420 nm emission filter sets, and an ocular quadricule divided into 100-grid squares.
- 20-μl and 2-ml adjustable pipettes
- Microcentrifuge for 500-μl centrifuge tubes
- 25-mm polysulfone filter funnel (Gelman)

Supplies

- Polypropylene or polycarbonate bottles for sampling
- 0.03- and 0.2-μm pore-size, 25 mm, polycarbonate membrane filters (Poretics)
- 0.45-μm pore-size, 25-mm diameter, nitrocellulose membrane filters
- 500-μl centrifuge tubes
- Microscope slides
- 22 × 22 mm cover slips
- Low-fluorescence immersion oil

Solutions and Reagents

- 5 μg ml^{-1} solution of DAPI (4′, 6-diamidino-2-phenylindole) (Sigma D 1388). Make up a 5 μg ml^{-1} solution of DAPI in McIlvaine's buffer (pH = 4.4). To make the buffer dissolve 3.561 g of $Na_2HPO_4 \cdot 2H_2O$ in 100 ml of distilled water (solution A) and 2.101 g of citric acid·H_2O in 100 ml of distilled water (solution B) and combine 8.82 ml of solution A with 11.18 ml of solution B. Store at 4°C in the dark as DAPI is very light sensitive.
- Irgalan black (Ciba-Geigy). Make the 0.03-μm pore-size filters nonfluorescent by soaking them for several hours at 90°C in a solution of 2 g Irgalan black, dissolved in 1 l of 2% acetic acid. After staining, rinse the filters in filtered-distilled water. For larger orders, stained 0.03-μm pore-size filters are currently available from Poretics Corp. (Livermore California).
- 0.2 M solution of $Na_2HPO_4 \cdot 2H_2O$ (M.W. 178.05)
- 0.1 M solution of Citric acid. H_2O (M.W. 210.14)
- 0.015 M solution of NaCl (M.W. 58.44)
- Acetic acid ($C_2H_4O_2$; M.W. 60.05)
- Formaldehyde or 25% EM-grade glutaraldehyde
- DNase (DNase I, Sigma D 4527) Solution — The proportion of DAPI-positive particles that are DNase sensitive is usually small; however, the use of DNase will often make counting

easier by reducing background fluorescence. To make a stock solution of DNase dissolve 10,000 Kunitz units in 1.0 ml of ice-cold 0.15 M NaCl. Aliquot 50-μl subsamples into 500-μl microfuge tubes and freeze at $-80°C$. This should keep for many months, but it should be assayed periodically for activity per the manufacturer's instructions.

Isolation of Viruses

Equipment (AP = Amplification Procedure, CP = Concentration Procedure)

- Fluorometer (Turner Designs) to monitor phytoplankton growth or spectrophotometer to measure bacteria growth (AP and CP)
- Filtration apparatus with plastic or stainless-steel filter supports for 47-mm diameter filters (AP)
- 1- and 5-ml adjustable pipettes (AP and CP)
- Lighted incubator or other system for culturing phytoplankton and bacteria (AP and CP)
- Pump for vacuum and pressure filtration (AP and CP)
- Two 142-mm stainless-steel filter holders (CP)
- Stainless-steel reservoir for pressure filtration (CP)
- Peristaltic pump (several liters per min) for ultrafiltration (CP)
- 30,000 MW-cutoff spiral ultrafiltration cartridge (Amicon S1Y30) (CP)

Supplies (AP = Amplification Procedure, CP = Concentration Procedure)

- 0.2- and 1.0-μm pore-size, 47 mm, polycarbonate membrane filters (Nuclepore, Poretics) (AP)
- Culture flasks for phytoplankton and bacteria (AP)
- Nutrients for enriched sample water and media (AP and CP)
- 13 $\times$ 100 mm borosilicate culture tubes (AP and CP)
- 142-mm diameter glass-fiber filters (MFS GC50, 1.2-μm nominal pore size) (CP)
- 142-mm diameter, low protein-binding "Durapore" membrane filters (Millipore, 0.22- or 0.45-μm pore size) (CP)

PROCEDURES

Enumeration of Viruses

Plaque Assays

Plaque assays are routinely used to estimate titers of viruses that cause lysis of bacteria, cyanobacteria, and algae that can be grown on solid media.[4-6] The basis of the method is that each infective unit will form a clearing (plaque) on a lawn of host cells. Lawns are typically made by mixing host cells and viruses in molten agar and quickly and evenly pouring this mixture over a bottom layer made with a higher percentage of agar. The number of plaque-forming units (PFUs) in a given volume of water can be estimated using this method. One must be cautious, however, as a number of aquatic bacteria will form plaques on lawns of bacteria and algae.[7,8] Filtration or pretreatment of the water sample with chloroform are the methods traditionally used to distinguish viral and bacterial pathogens. However, as some algal viruses may be 0.4 μm or larger in diameter,[9] and some are chloroform sensitive, these approaches may be selective.

Advantages of the plaque-assay method include the following: (1) it is relatively sensitive with a practical detection limit of about 5 PFUs ml^{-1}. However, in natural seawater samples viruses infecting specific hosts are often present at even lower concentrations. (2) Accurate results can be achieved with a modest amount of effort. Plaque counts can even be done by image analysis. (3) Pathogens can be easily purified by cloning individual plaques picked from a plate. (4) Only infective viruses are enumerated.

Disadvantages include that (1) the host must be culturable on solid medium, (2) different pathogens cannot easily be distinguished, (3) natural bacteria can interfere with lawn formation, particularly of slower-growing hosts, and (4) filtration may remove some viruses.

1. Sample collection and filtration — Ideally, samples should be collected immediately before plaque assaying. Use a clean container that has relatively low protein binding, such as a polycarbonate or polypropylene bottle. Although prefiltration of the sample selects against some viruses, it is often necessary in order to prevent bacteria overgrowing the lawn. Filtration, through a 0.45-μm PVD or 0.4-μm polycarbonate filter is often adequate, although in some instances 0.2-μm pore-size filters may be required. Also, many marine viruses show little decrease in titer for several weeks if samples are prefiltered and stored in the dark at 4°C (unpublished data).

2. In a sterile microfuge tube combine 500 μl of cell concentrate with 500 μl of the sample or control water to be assayed. Mix quickly by gentle vortexing to prevent multiple infections of single cells. Briefly spin in a microcentrifuge to remove liquid from the top and side of the tube, and allow adsorption of the viruses to the host cells. If virus titers are high, decrease the volume of sample added or do a dilution series using sterile medium, prior to adsorption. A 30- to 45-min incubation should be adequate if the adsorption kinetics are unknown.

3. Following adsorption add the sample to a tube containing 2 to 2.5 ml of molten top agar, vortex and quickly pour the contents onto the bottom agar.

4. After 30 min invert the plates and incubate them under appropriate conditions. The plaques are counted after an even lawn appears.

Most-Probable-Number Assays (MPNs)

MPNs are usually done to determine the concentration of lytic viruses that infect hosts which cannot be grown on solid medium but which can be grown in liquid.[10] This is particularly a problem with delicate phytoplankton. The assay is simple but labor intensive. It is also very sensitive; a detection limit of 0.01 viruses ml^{-1} is not unreasonable. However, MPNs lack the accuracy and precision of a plaque assay. Traditionally, MPN assays have been done using a series of tenfold dilutions with three to ten replicates at each dilution. The replicates in which no growth or growth followed by cell lysis occurs are assumed to contain at least one infective virus. By comparing the number of replicates that contain viruses, to a MPN table, the number of infective units can be estimated. With the wide availability of computers and other technologies such as microplate readers, it is not necessary to be constrained by a specific number of replicates or order-of-magnitude dilution regimes. The first step is to determine the detection limit desired. For example, to achieve a detection limit of 1 virus ml^{-1}, several milliliters of water needs to be screened. In lieu of information on viral titer it is necessary to do a broad dilution series. A series of five, tenfold dilutions, starting with 1 ml of sample should span the range of concentrations typically found, although larger volumes can be screened if desired. Changes in phytoplankton biomass are monitored by measuring *in vivo* chlorophyll fluorescence, which allows hundreds of cultures to be nondestructively monitored on a daily basis. The culture tubes specified fit directly into a fluorometer. Alternatively, the procedure can be modified to substitute microtiter plates and a plate reader for the culture tubes and fluorometer. The MPN, standard error, and confidence interval are calculated using a program written in BASIC.[11]

1. Transfer 50 ml of an exponentially growing batch culture of the host to be screened into 450 ml of fresh medium. As soon as exponential growth of the culture begins, transfer 4 ml into each of 55 7-ml culture tubes. Immediately add 1 ml to each tube from the dilution series outlined below (see steps 2d to f).

2. Collect sample water in a clean polypropylene or polycarbonate container and prepare the dilution series as follows:

   a. Filter 25 ml of sample water through a 1-μm pore size polycarbonate filter into a filter flask that has been prerinsed with filtrate.

   b. Into each of four 50-ml culture (dilution) tubes place 18 ml of medium. Into the first dilution tube add 2 ml of 1-μm-filtered sample water. Mix well.

   c. Take 2 ml from the first dilution tube (10^{-1}) and add it to the second (10^{-2}). Complete the dilution series for the 10^{-3} and 10^{-4} dilutions.

   d. Add 1 ml from the undiluted sample to each of ten replicate 7-ml culture tubes, each containing 4 ml of exponentially growing host-cell culture (from step 1).

   e. For each of the four dilutions, add 1 ml to each of ten replicate 7-ml culture tubes.

   f. Reserve the five remaining tubes as controls.

3. Gently mix each tube and measure the fluorescence on a daily basis. Cultures which have not cleared after 7 d in stationary phase are assumed not to contain a virus.

4. For each dilution, record the number of tubes that have cleared and use these data to calculate the MPN for the concentration of viruses in the sample. The results are generally expressed as the number of tubes from each dilution in which lysis occurs (e.g., 10, 10, 3, 0, 0). The MPN can be determined from published tables of values[12] or preferably using a computer program which also provides confidence intervals and standard errors for the MPN estimates.[11]

Transmission Electron Microscopy (TEM)

There are a variety of protocols for determining the concentration of viral-like particles (VLPs) in aqueous solutions using TEM.[13,14] Currently, only two of these have been adapted for use in enumerating viruses in natural aquatic communities. Both methods require that the viruses be concentrated before they can be enumerated because of the high magnifications used in electron microscopy. The viruses can be concentrated using ultrafiltration and subsamples of the concentrate spotted onto EM grids[15] or the viruses can be pelleted directly onto grids by ultracentrifugation.[2,16] The grids are then stained and viewed using TEM.

Both protocols share several disadvantages. (1) They require access to a transmission electron microscope, an expensive facility that is generally only available at large research centers. (2) There are no definitive criteria for what constitutes a VLP in a natural viral community. Although many viruses are easy to recognize because of the presence of phage-like tails, many nontailed particles are difficult to categorize. (3) Particulates and aggregates can interfere with counting by obscuring large areas of the EM grid. (4) At the present time no information can be obtained on whether the VLPs are infective and if so whether the host is a bacterium, phytoplankton, or other organism. As well, if ultrafiltration is used the concentration efficiency must be determined using a tracer. Inert 50-nm polystyrene beads (Polyscience) can be easily enumerated in concentrates using epifluorescence microscopy and can be used as internal standards for determining the concentration of VLPs on EM grids. Alternatively, concentration efficiencies can be ascertained using virus tracers and plaque assays. If viruses are being concentrated from seawater there is the added difficulty that the salts must be removed before the sample can be spotted and dried onto a grid. Proctor and Fuhrman[15] accomplished this by pipetting a 2-μl sample onto a formvar-coated grid and floating this on a drop of distilled water to dialyse the salts. Care must be taken to ensure that there are no holes in the film through which viruses could diffuse. If an ultracentrifuge and swinging-bucket rotor are available pelleting of the viruses is simpler and more rapid.

The significant advantages of using TEM to count viruses are that information is obtained on the morphological diversity and abundance of the total virus community, and because infectivity is not an issue unfiltered samples can be preserved immediately after collection, thereby eliminating problems of virus decay and amplification.

1. Sampling and fixation — Collect duplicate water samples and if possible centrifuge them immediately without fixatives or preservatives, which can cause artifacts such as clumping. However, keep in mind that in unpreserved samples continuing biological processes may influence the concentration of VLPs. If necessary the samples can be stored for several weeks in 1 to 2% EM grade glutaraldehyde, in the dark, at 4°C. Viruses can be dispersed from aggregates in preserved samples by adding 10 μg ml^{-1} of a surfactant such as polyoxyethylene sorbitan monooleate (i.e., Tween 80) to the sample prior to centrifugation (Cochlan, personal communication), similar to the method used by Yoon and Rosson[17] for dispersing bacteria attached to particulate material.

2. Ultracentrifugation — Place, carbon-coated side up, duplicate hydrophylic grids onto each platform and carefully fill and balance each centrifuge tube. A variety of swinging-bucket rotors can be used, although those holding approximately 100-mm centrifuge tubes are suitable for most water samples. In order to pellet the smallest viruses, centrifuge the samples until particles of 80S are sedimented with 100% efficiency. For samples in distilled water at 20°C this can be calculated using the following formula:

$$T = (1/s) * (ln[r_{max}/r_{min}]/[w^2 * 60]) \tag{1}$$

where T = time in minutes; s = sedimentation coefficient in seconds (e.g., an 80S particle has a sedimentation coefficient of 80×10^{-13} s); r_{min} = distance (cm) from the center of the rotor to the top of the sample in the centrifuge tube; r_{max} = distance (cm) from the center of the rotor to the surface of the EM grid; w (angular velocity in radians) = 0.10472 * rpm.

Example for a Beckman SW 40 rotor at 30,000 rpm with a full centrifuge tube and a 0.5 cm-high grid platform:

$$\frac{1}{80 \times 10^{-13}} * \frac{ln\,(15.38/6.67)}{(0.10472 * 30000)^2 * 60} = 176.4 \text{ min} \tag{2}$$

Sedimentation times will vary as a function of temperature and salinity because of the associated changes in the density and viscosity of water. Increase centrifugation times about 12% if pelleting viruses out of 3.5% salt vs. distilled water. Lower temperatures have a more dramatic effect on sedimentation times because of the associated changes in the viscosity of water. Increasing sedimentation times by 25% for each 5°C decrease in temperature below 20°C will adequately compensate for the increased viscosity. Decreases in the viscosity of water at temperatures above 20°C are much less pronounced.

3. Staining — VLPs are most easily enumerated if they are positively stained. In negatively stained preparations the stain deposition is often uneven making it difficult to count all

of the VLPs. Some experimentation may be necessary to achieve optimum results for a given sample, although the following protocol is generally adequate.

    a. After centrifugation remove the grids, wick off the excess water with a piece of filter paper, and remove the remaining salts by floating the grids, sample-side down, through several drops of distilled water. Wick off the excess water at each step but make sure the surface of the grid is never dry.

    b. To stain, float the grid face-down on a drop of 1% uranyl acetate for about 30 s. Wick off the excess stain, rinse through two or three drops of distilled water, briefly allow the grids to dry on filter paper, and view with TEM. One advantage of using uranyl acetate over other stains is that the grids can be stored in a desiccator for several weeks without appreciable loss of detail. Hayat and Miller[3] present an authoritative review on staining and visualization of viruses by TEM.

4.   Enumeration — Establish objective criteria for recognizing VLPs and scan each grid to make sure that the VLPs are distributed randomly and that larger particles will not interfere with counting. VLPs can be counted directly off the phosphorescent screen but for better resolution should be counted from photographic negatives. Mathews and Buthala[18] obtained a precision of $\pm 6\%$ at the 95% confidence interval by counting a minimum of 100 VLPs from five random fields at a magnification of $10,000\times$. For most applications, counting a minimum of 20 fields containing at least 200 VLPs represents a reasonable compromise between accuracy and effort. Assuming a Poisson distribution this yields upper and lower 95% confidence limits of 174 and 230. As increases in accuracy are a function of the square root of the number counted it is necessary to count much larger numbers of viruses to get any appreciable increase in accuracy. If desired, 95% confidence intervals can be estimated from the following formulae:

$$\text{Upper} = n + 1.96 * \sqrt{(n + 1.5)} + 2.42 \tag{3}$$

$$\text{Lower} = n - 1.96 * \sqrt{(n + 0.5)} + 1.42 \tag{4}$$

where n is the number of viruses counted.

    In order to determine the VLP concentration, first calculate a taper correction factor (S_f), because the particles do not sediment in parallel paths. This can be approximated as follows:[18]

$$S_f = [0.5\, r_{max} - (0.5\, r_{min}^2 / r_{max})] / (r_{max} - r_{min}) \tag{5}$$

Use the following formula to convert the number of VLPs in a field, to a concentration:

$$\text{VLPs ml}^{-1} = P_f / (A_f * H * S_f) \tag{6}$$

where P_f = number of particles counted per field; A_f = area of grid represented in the microscope field or by the photographic negative (cm^2); H = height of sample (cm) (i.e., $r_{max} - r_{min}$).

Epifluorescence Microscopy of DAPI-Positive Particles

    The protocol has the advantages that DAPI positive is an objective criterion, a modest amount of equipment is required compared to TEM, and the method is amenable to use at sea. The most notable disadvantages are that (1) the viruses must be concentrated on a filter or by ultrafiltration before being counted, (2) no information is obtained on the composition of the viral community or the organisms that the viruses infect, (3) DAPI is specific for double-stranded DNA (dsDNA), so only dsDNA viruses are DAPI positive (DAPI binds to dsRNA as well, but the binding constant is an order of magnitude less than that of DNA,[19] (4) some small dsDNA viruses are difficult to visualize using DAPI, (5) samples must be prefiltered to remove bacteria which can also remove a large proportion of the viruses, (6) small bacteria cannot be distinguished from large viruses, (7) individual viruses cannot be distinguished from small clumps of viruses (although we have not found clumping of viruses to be a problem).

Viruses can be concentrated and enumerated in a variety of ways. The simplest way is by filtration onto stained 25-mm polycarbonate filters.[20] Alternatively, viruses can be concentrated to $>10^8$ ml^{-1} using ultrafiltration or vortex flow filtration and enumerated directly on glass slides.[21-23] A variety of ultrafiltration systems can be used to concentrate viruses from tens or hundreds of liters of water (see virus isolation procedures), but as this generally takes several hours it is not practical for routine virus counts. Small-volume centrifugation concentrators potentially provide an alternative, but more research is required to increase the efficiency of virus recovery from natural water samples.

Filtration Method

1. Filter a freshly collected water sample through a 0.2-μm pore size polycarbonate filter.

2. To a 2-ml subsample add 50 μl of stock DNase solution (500 Kunitz units) and incubate for 30 min at room temperature. After DNase treatment the sample can be fixed in 1% glutaraldehyde or formalin.

3. Add 0.2 ml of the DAPI stock solution (1 μg ml^{-1} final concentration) and incubate in the dark for 30 min. The fluorescence is less if higher concentrations of DAPI are used.[24]

4. Filter the entire sample through a 25-mm, 0.03-μm pore size, irgalan-black-stained polycarbonate filter (vacuum 200 mmHg); this filtration may take >30 min. Alternatively, use an unstained 0.02-μm Anodisc (Anotec, Banbury, U.K.) membrane. Do not rinse the filter as the DAPI binding is reversible. The 0.03-μm filter should be laid over a prewetted 0.45-μm pore size nitrocellulose filter for even filtration. These filters are very fragile and must be handled with care. Turn the vacuum off as soon as the filter drys. If the filter is too wet or dry the slide will be poor.

5. Lay the filter over a small drop of low-fluorescence immersion oil on a microscope slide. Place three tiny drops of oil on a cover slip and gently lay this over the filter, trapping the oil between the cover slip and filter. The filter should be of even color, with no wrinkles, dry spots, or emulsion of water and oil.

6. At 1000 × magnification count 20 to 100 DAPI-positive particles in each of 20 random fields. Calculate the number of viruses per milliliter (N_v) from

$$N_v = P_f * (A_s/A_f) * (1/V_s) * ([V_s + V_e + V_g + V_D]/V_s) \qquad (7)$$

where P_f = number of fluorescent particles per field; A_s = filtration area of filter (μm^2); A_f = area of field (μm^2); V_s = volume of sample (ml); V_e = volume of DNase added (μl); V_g = volume of glutaraldehyde or formalin added (μl); V_D = volume of DAPI added (μl).

Concentration Method. In order to count DAPI-stained viruses directly on glass slides the abundance must be about 10^{10} ml^{-1}. The method is particularly well suited for counting viruses in laboratory experiments where further concentration is frequently not required.

However, 10^{10} ml^{-1} is considerably greater than in most natural water samples; hence, the viruses must be concentrated before they can be counted.

The efficiency of concentration should be checked using an internal standard such as 50-nm fluorescent beads, labeled with fluorescein isothiocyanate (FITC, excitation/emission = 458/540 nm), or by adding a trace addition of a marine bacteriophage to one of a pair of duplicate samples and titering the replicates by plaque assay before and after concentration.

1. Concentrate the viruses from a natural water sample as is outlined below in the virus isolation procedures. Transfer 20 µl of the concentrate to a microfuge tube.

2. Dilute 10 µl of stock DNase solution with 90 µl of ice-cold 0.15 M NaCl, and for each 20-µl sample place 5 µl of DNase I (five Kunitz units) on the inside of each microfuge tube.

3. Briefly centrifuge the sample (approximately 15,000 × g for 3 s), vortex, recentrifuge, and allow to incubate for 30 min at room temperature. DNase will dissolve free DNA that could interfere with the counting method, but will not interfere with DNA protected by a protein coat. If required the samples can be fixed at this point with 1% formalin or glutaraldehyde. Do not add excess glutaraldehyde as it imparts background fluorescence and makes viruses difficult to count.

4. To stain, add 5 µl of DAPI stock solution (1 µg ml^{-1} final concentration), centrifuge and vortex, as above, and incubate in the dark at 4°C for at least 30 min.

5. Pipet 3 to 5 µl on a clean glass slide and cover with a clean 22 × 22 mm cover slip. At 1000 × magnification count 20 to 100 DAPI-positive particles in each of 20 random fields. Use the appropriate number of grid squares in the quadricule. The viruses will tend to adsorb to the surfaces of the cover slip and slide, so be careful to count the DAPI-positive particles in both planes of focus. Calculate the concentration of viruses per milliliter (N_v) using the following formula:

$$N_v = P_f * (A_c/A_f) * (1/[V_c * C_f]) * ([V_s + V_e + V_g + V_D]/V_s) * 1000 \qquad (8)$$

where P_f = number of fluorescent particles per field; A_c = area of cover slip (µm²); A_f = area of field (µm²); V_c = volume of sample under the cover slip (µl); C_f = concentration factor of the sample; V_s = volume of sample to which DNase, fixative, and DAPI are added (µl); V_e = volume of DNase added (µl); V_g = volume of glutaraldehyde or formalin added (µl); V_D = volume of DAPI added (µl).

Isolation of Viruses

Amplification Method

This protocol is extremely simple, requires little equipment, and allows relatively large volumes of water to be assayed for the presence of lytic viruses. The major disadvantages are that separate amplifications must be done for each host that is screened and the culture volumes can be large.

1. Filter several liters of sample water through a 0.2- or 1.0-µm pore size polycarbonate filter to remove the natural bacteria or phytoplankton community, depending on whether one is screening for bacteriophages or algal viruses.

2. Enrich seawater with inorganic (e.g., ESNW[25]) or organic (0.05% Bacto-Peptone and 0.05% Casamino Acids, Difco) nutrients as appropriate and a 10% inoculum of an exponentially growing culture of the potential host added.

3. Measure *in vivo* fluorescence or optical density (600 nm) to monitor the growth of the phytoplankton or bacteria, respectively. Within a few days after growth of the host has ceased the water is screened for the presence of viruses.

4. For each enriched water sample to be screened add a 10% inoculum from an exponentially growing host-cell culture into each of ten 13 × 100 mm borosilicate culture tubes (with polypropylene caps) containing fresh medium (0.4 ml of culture to 3.6 ml of medium).

5. When exponential growth is observed add 1 ml of 0.2- or 1.0-μm filtered water from the enriched sample into five of the tubes. As a control add 1.0-μm filtered water from a stationary culture to the other five tubes.

6. If a decrease in fluorescence or optical density is observed relative to that of the control cultures, repeat the experiment by transferring 5% inocula from a control and a potentially infected culture into ten additional tubes containing exponentially growing cultures that had never been exposed to the suspected pathogen. Repeat the process numerous times to dilute nonreplicating viruses from the original concentrate.

For hosts that will grow on solid substrate, screening of the enriched water sample can be done by plaque assay as described in the above section on enumeration.

Concentration Method

The other approach of increasing the probability of host-virus encounter is to increase the concentration of the natural virus community by ultrafiltration, and add aliquots of the concentrate to potential hosts. Using this method a large number of different bacteria and phytoplankton can be screened expeditiously. As well, the concentrates serve as a library of natural virus communities that can be screened for other pathogens. The disadvantages of the approach are that prefiltration removes viruses[23] and that ultrafiltration is somewhat expensive and time consuming.

1. Dispense 20 to 100 l of water into a pressure vessel. Stainless steel pressure vessels can often be obtained from soft drink wholesalers for a modest deposit.

2. Pressure filter the water (<130 mmHg) through 142-mm diameter glass fiber and Durapore membrane filters, connected in series and held in place by stainless-steel filter holders. The 0.45-μm pore size filters let bacteria through and make ultrafiltration slower, but the 0.22-μm pore size likely excludes many of the large algal viruses. The 0.22-μm pore size is suitable for smaller viruses.

3. Concentrate the filtrates 100 to 1000-fold using a 30,000 MW ultrafiltration cartridge (Amicon). Flow rates at a back pressure of 1000 mmHg are about 850 ml min^{-1}. The cartridge is cleaned after use by flushing with 2 l of 0.1 N NaOH heated to 45°C and is then stored refrigerated in 50 mM phosphoric acid. Despite the manufacturer's recommendation we have found that long-term storage in 0.01 N NaOH can result in failure of the cartridge. Prior to reuse, flush the cartridge with 7 l of deionized-distilled water.

4. The concentrate is screened in the same manner as is used for the enriched water samples. Inocula from a concentrate are introduced into exponentially growing cultures of phytoplankton which are monitored fluorometrically. If there is evidence of a pathogen the effect is propagated. Alternatively, the concentrate can be screened by plaque assay if the host will grow on solid medium.

Cloning of the Virus

If the pathogen can be propagated the first step is to demonstrate that it is viral, as a number of bacteria are predatory on other bacteria and phytoplankton.[7,8,26] Test if the infective agent can be removed by filtration, whether it is sensitive to antibiotics and whether it is host specific. Finally, use TEM to confirm the presence of VLPs.

Cloning of the virus is also accomplished by a modification of the above methods. If the host is amenable to growing on agar the simplest way to clone the virus is through plaque purification. Serially diluted culture lysate from a virus amplification is plaque assayed, and a single well-separated plaque removed from the lawn. The plaque is eluted in medium overnight and the eluent used for another dilution series and plaque assay. This procedure should be repeated several more times to be sure that the virus has been cloned.

For hosts that will not grow on solid medium cloning must be accomplished by amplifying a single infectious unit in a liquid culture. The first step is to determine the titer of a culture lysate by an MPN assay. Based on these results a dilution series is set up and 0.2 of an infectious unit is added into each of 20 exponentially growing cultures of the host. The probability of a culture receiving a single virus is 0.164 and therefore would be expected to occur with a frequency of 3.27 out of 20 cultures. The probability of a culture receiving two or more viruses is <0.02.[10] Hence, by repeating this procedure twice one can be very confident that the virus has been cloned.

ACKNOWLEDGMENTS

I am deeply grateful to a number of people who familiarized me with much of what is included here. They include Dr. D. T. Brown, A. M. Chan, M. T. Cottrell, F. Chen, R. Mitchell, and Dr. L. M. Proctor. Comments by Dr. W. P. Cochlan were instrumental in improving the manuscript. Support for much of the work reported here was provided by the Office of Naval Research (N00014-90-J-1280), National Science Foundation (OCE-9018833), and the Texas A&M University Sea Grant Program (NA-16RG0457-01). Contribution No. 92-001 of the Marine Science Institute.

REFERENCES

1. Berg, G., *Methods for Recovering Viruses from the Environment,* CRC Press, Boca Raton, FL, 1987.
2. Borsheim, K. Y., Bratbak, G., and Heldal, M., Enumeration and biomass estimation of planktonic bacteria and viruses by electron microscopy, *Appl. Environ. Microbiol.,* 56, 352, 1990.
3. Hayat, M. A. and Miller, S. E., *Negative Staining,* McGraw-Hill, New York, 1990.
4. Adams, M. H., *Bacteriophages,* Wiley-Interscience, New York, 1959.
5. Safferman, R. S. and Morris, M. E., Algal virus: isolation, *Science,* 140, 679, 1963.
6. Van Etten, J. L., Burbank, D. E., Kuczmarski, D., and Meints, R. H., Virus infection of culturable *Chlorella*-like algae and development of a plaque assay, *Science,* 219, 994, 1986.
7. Stewart, W. D. P. and Daft, M. J., Microbial pathogens of cyanophycean blooms, *Adv. Aquat. Microbiol.,* 1, 177, 1977.
8. Sakata, T., Fujita, Y., and Yasumoto, H., Plaque formation by algicidal *Saprospira* sp. on a lawn of *Chaetoceros ceratosporum, Nippon Suisan Gakkaishi,* 57, 1147, 1991.
9. Van Etten, J. L., Lane, L. C., and Meints, R. H., Virus and viral-like particles of eukaryotic algae, *Microbiol. Rev.,* 55, 586, 1991.
10. Cottrell, M. T. and Suttle, C. A., Wide spread occurrence and clonal variation in viruses which cause lysis of a cosmopolitan, eukaryotic marine phytoplankter, *Micromonas pusilla, Mar. Ecol. Prog. Ser.,* 78, 1, 1991.
11. Hurley, M. A. and Roscoe, M. E., Automated statistical analysis of microbial enumeration by dilution series, *J. Appl. Bacteriol.,* 55, 159, 1983.
12. Koch, A. L., Growth measurement, in *Manual of Methods for General Bacteriology,* Gerhardt, P., Ed., American Society of Microbiologists, Washington, DC, 1981, 179.

13. Sharp, D. G., Quantitative use of the electron microscope in virus research, *Lab. Invest.*, 14, 831, 1965.

14. Miller, M. F. II, Particle counting of viruses, in *Principle and Techniques of Electron Microscopy: Biological Applications,* Vol. 4, Hayat, M. A., Ed., Van Nostrand Reinhold, New York, 1974, 79.

15. Proctor, L. M. and Fuhrman, J. A., Viral mortality of marine bacteria and cyanobacteria, *Nature,* 343, 60, 1990.

16. Bergh, O., Borsheim, K. Y., Bratbak, G., and Heldal, M., High abundance of viruses found in aquatic environments, *Nature,* 340, 476, 1989.

17. Yoon, W. B. and Rosson, R. A., Improved method of enumeration of attached bacteria for study of fluctuation in the abundance of attached and free-living bacteria in response to diel variation in seawater turbidity, *Appl. Environ. Microbiol.,* 56, 595, 1990.

18. Mathews, J. and Buthala, D. A., Centrifugal sedimentation of virus particles for electron microscopic counting, *J. Virol.,* 5, 598, 1970.

19. Manzini, G., Xodo, L., Barcellona, L., and Quadrifoglio, F., Interaction of DAPI with double-stranded ribonucleic acids, *Nucleic Acids Res.,* 13, 8955, 1985.

20. Hara, S., Terauchi, K., and Koike, I., Abundance of viruses in marine waters: assessment by epifluorescence and transmission electron microscopy, *Appl. Environ. Microbiol.,* 57, 2731, 1991.

21. Suttle, C. A., Chan, A. M., and Cottrell, M. T., Infection of phytoplankton by viruses and reduction of primary productivity, *Nature,* 347, 467, 1990.

22. Suttle, C. A., Chan, A. M., and Cottrell, M. T., Use of ultrafiltration to isolate viruses from seawater which are pathogens of marine phytoplankton, *Appl. Environ. Microbiol.,* 57, 721, 1991.

23. Paul, J. H., Jiang, S. C., and Rose, J. B., Concentration of viruses and dissolved DNA from aquatic environments by vortex flow filtration, *Appl. Environ. Microbiol.,* 57, 2197, 1991.

24. Coleman, A. W., Maguire, M. J., and Coleman, J. R., Mithramycin and 4'-6-diamidino-2-phenylindole (DAPI)-DNA staining for fluorescence microspectrophotometric measurement of DNA in nuclei, plastids and virus particles, *J. Histochem. Cytochem.,* 29, 959, 1981.

25. Harrison, P. J., Waters, R., and Taylor, F. J., A broad spectrum artificial seawater medium for coastal and open ocean phytoplankton, *J. Phycol.,* 16, 28, 1980.

26. Taylor, V. I., Baumann, P., Reichelt, J. L., and Allen, R. D., Isolation, enumeration and host range of marine bdellovibrios, *Arch. Microbiol.,* 98, 101, 1974.

CHAPTER 16

Total Count of Viruses in Aquatic Environments

Gunnar Bratbak and Mikal Heldal

INTRODUCTION

The ecological significance of virus has recently been focused in relation to the relative high numbers of viruses or virus-like particles (VLPs) found in aquatic ecosystems. New methods for direct count of viral particles[1] and indirect estimation of viral total count,[2] combined with the rapid decay of free viruses,[3,4] have shown that bacterial communities are strongly influenced by viruses. Proctor and Fuhrman[2] reported high frequencies of bacterial cells to contain mature phages by thin sectioning of the cells. Heldal and Bratbak[3] were able to include total count of bacterial cells which contained mature phages and they could also estimate burst sizes for cells in lysis. Both these studies indicated that high fractions of the bacterial community were in a lytic cycle. Moebus[5] used a number of host strains and were by plaque counts of specific phages able to show that maximal concentrations of various phages lasted only few days in the water around Helgoland. These observations indicate dynamic and rapidly changing microbial communities in aquatic environments.

Most marine bacteria apparently do not thrive on agar plates and plate counts of bacteria in samples from natural waters will usually make up less than 1% of the total bacterial counts. The majority of marine bacteria can therefore not be isolated for subsequent use as hosts in counting of phages with the traditional plaque count technique (plaque forming units, PFU). Moreover, if isolated bacteria are lysogenic they will not be useful as host bacteria for counting of PFU of the phage they carry as they will be immune to reinfection and lysis by this phage.

Moebus[5] found that for most indicator bacteria the number of PFU varied in the range of <1 to 20 to 30 PFU ml^{-1}. Cumulative PFU counts employing several bacterial strains was in this study four to five orders of magnitudes less than what is usually found by direct total counts of viral particles in TEM. Nevertheless, we conclude that both these methods are needed, in addition to other methods, to reveal the significance of viruses in natural microbial communities. The method used for total virus counts in TEM is described in the following.

The main advantage of the method is the ability to obtain total counts of viral particles. It is also possible to count the fractions of bacteria which contain mature viral particles, and to make estimates of burst sizes from cells in lysis. Further advantages are related to estimation of size and morphology of VLPs. Total counts and size distribution of bacteria and flagellates can be estimated from similar preparations as used for virus total counts. The method is rapid as preparation and counting of six samples can be performed in about 2 h.

The disadvantages of the method is the need for equipment like ultracentrifuge and transmission electron microscope and equipment for making EM grids (evaporation of carbon on the grids and for glow discharging). It is also a disadvantage that no clear-cut criteria can be used (so far) to distinguish viral particles from other particles of similar size, morphology, and staining properties.

MATERIALS REQUIRED

Equipment

- TEM microscope
- Ultracentrifuge
- Swing-out rotor
- Ultracentrifuge tubes (e.g., polyallomer tubes, Beckman Instruments Inc. Spinco Division, 1050 Page Mill Road, Palo Alto, CA 94304) with flat bottoms molded in epoxy glue (e.g., Araldit, Ciba-Geigy)
- Balance
- Tweezers
- Scalpel
- Grid storage box

Supplies

- EM grids (nickel or copper grids, 400 mesh) with carbon-coated formvar or colloidon film; glow-discharged grids are recommended, but not necessary (see Chapter 44)
- Filters for grid support, e.g., Sartorius cellulose nitrate filters (or any other kind with good wetting properties)
- Kodak Projection Print Scale (Eastman Kodak Company, Rochester, NY 14650)
- Double-sided tape
- Filter paper

Solutions

- 2% Uranyl acetate in distilled water (or another electron microscope stain)
- 37% Formaldehyde (p.a.) or 25% glutardialdehyde (electron microscope grade)

PROCEDURES

Sample Preparation

1. Preserve 100-ml water samples immediately after sampling with 2% formaldehyde or 1% glutardialdehyde (final concentrations).

2. Cut the filters to fit on the flat bottom of the centrifuge tubes and attach a small piece (2 × 5 mm) of double-sided tape to the center of the filter.

3. Tape the EM grids to the filter but make sure that it is only the edge of the grids that touches the tape.

4. Fill the centrifuge tubes to the brim with sample water. Slide the filter with attached grids into the tubes at an angle of about 80° to avoid trapping air bubbles underneath the grids and let it sink to the bottom.

5. Balance the tubes to a constant weight so that the water is no more than 2 to 3 mm from the top of the tubes.

6. Centrifuge the samples at 100,000 × g for 2.5 h or 200,000 × g for 1 h (depending on the rotor head used). Other speeds and centrifuge times that will pellet particles with a sedimenting coefficients of 100 Svedbergs may also be used.

7. Remove the supernatant with a pipette and pick out the filters. Loosen the grids from the tape with a scalpel and place them in a grid storage box.

8. The grids should be locked in tweezers during staining and rinsing. Stain the grids with 2% uranyl acetate (UAc) for 30 s.

9. Drain off the UAc by touching the grid with a piece of filter paper and rinse once by dipping the grid in distilled water.

10. Drain off the water and air dry.

Counting in TEM

1. Total counts of viruses should be done at a magnification of 100,000 ×. Larger viruses (>150 nm) may with some experience be counted at lower magnification (20,000 ×).

2. Adjust the condenser and the objective apertures to give good contrast. With our Jeol 100CX and an accelerating voltage 80 kV we use the condenser aperture no. 1 and the objective aperture no. 3.

3. Count viruses in 10 to 100 view fields, depending on the number of viruses. Counting may be done on the whole fluorescent screen or on a smaller field with known area. Make sure that the whole screen is illuminated and not masked by the aperture.

The main reason for our positive staining with UAc (i.e., staining followed by rinsing) is to obtain uniform-stained particles on the whole grid. For studies of morphology and size negative-stained samples may be more suitable as more details may be observed. VLPs will in general show a variable degree of staining and thus a whole range of gray levels. Large viruses (100 to 300 nm) are densely stained and easy to recognize, but for large fractions of the smaller viruses we have to decide a limit in gray level for the particles to include in total counts. The reason for this may be that larger viruses have more nucleic acid in the head structure and therefore bind more stain than do smaller viruses. In addition, spontaneous release of genetic material from viruses will result in a reduced nucleic acid content and an accordingly variable staining. With a gray level scale like Kodak Projection Print Scale we use level 16 as a minimum level for including a particle in the total viral counts, provided that the background gray level is 32 or lower.

Particles which may interfere with our total counts are protein crystals, particles from octopus ink, scales from flagellates, or mineral particles. Most of these groups of particles should be easy to sort out, and they may interfere only occasionally.

Calculations

The water volume (V) from which particles are harvested during centrifugation may be calculated as follows:

$$V = A * (R^2 - r^2)/2R \tag{1}$$

where R is the maximum and r the minimum radius of the water in the centrifuge tubes during centrifugation (units are centimeters), and A is the area of the view fields counted on the grids (units in centimeters squared).

R and r are calculated from maximum radius of the rotor employed (see manufacturer's data sheet), the thickness of the molded flat bottoms, and the height of the water column in the centrifuge tube. The area of the view fields may be calculated as

$$A = \pi * (r_{view\ field}/mag)^2 \tag{2}$$

where $\pi = 3.14$, $r_{view\ field}$ is the radius of the microscopes view field, i.e., the fluorescent screen (unit is centimeters), and mag is the magnification.

The concentration of particles in the sample (unit is milliliters) is calculated by multiplying the average particle counts per view field with V^{-1}.

REFERENCES

1. Bergh, Ø., Børscheim, K. Y., Bratbak, G., and Heldal, M., High abundance of viruses found in aquatic environments, *Nature,* 340, 467, 1989.
2. Proctor, L. M. and Fuhrman, J. A., Viral mortality of marine bacteria and cyanobacteria, *Nature,* 343, 60, 1990.
3. Heldal, M. and Bratbak, G., Production and decay of viruses in marine waters, *Mar. Ecol. Prog. Ser.,* 72, 205, 1991.
4. Bratbak, G., Heldal, M., Thingstad, T. F., Riemann, B., and Haslund, O. H., Incorporation of viruses into the budget of microbial C-transfer. A first approach, *Mar. Ecol. Prog. Ser.,* 83, 273, 1992.
5. Moebus, K., Preliminary observation on the concentration of marine bacteriophages in the water around Helgoland, *Helgolander Meeresunters.,* 45, 411, 1991.

Additional References

Børscheim, Y., Bratbak, G., and Heldal, M., Enumeration and biomass estimation of planktonic bacteria and viruses by transmission electron microscopy, *Appl. Environ. Microbiol.,* 56, 352, 1990.
Bratbak, G., Heldal, M., Norland, S., and Thingstad, T. F., Viruses as partners in spring bloom microbial trophodynamics, *Appl. Environ. Microbiol.,* 56, 1400, 1990.

CHAPTER 17

Improved Sample Preparation for Enumeration of Aggregated Aquatic Substrate Bacteria

M. Iqubal Velji and Lawrence J. Albright

INTRODUCTION

Within the aquatic environment, bacteria can be found either attached to various substrates or existing as free-living single or aggregated cells in the water column. Substrates to which bacteria are likely to attach include marine snow, fecal and detrital particles in the water column, sediment particles, and aquatic plants.

Quantitation of free-living bacteria in the water column can be carried out easily and accurately by staining the bacterial cells with a fluorochrome followed by the direct epifluorescent microscopic count method.[1-3] However, enumeration of bacteria attached to the various substrates can be problematic using standard epifluorescent methods because of uneven distribution, aggregation and layering of bacterial and other microbial organisms embedded in colloidal matrices, or the presence of opaque particulate matter which causes interference with the observation of bacteria.

The sample preparation method delineated here overcomes these problems by utilizing a combination of chemical and physical procedures for dispersing the bacteria from their attached sites or aggregated forms. The method involves initial fixation and strengthening of bacteria in the sample with formaldehyde, followed by addition of dispersant, tetrasodium pyrophosphate (PPi), and, subsequently, an application of ultrasound for physical separation of the bacterial cells.[4,5] This is followed by the standard epifluorescence microscopic count method for enumeration of the dispersed bacteria.

There are several advantages to this method of sample preparation. First, it allows for bacterial enumeration from aquatic substrates that are difficult or impossible to enumerate via the use of standard methods; second, the sample preparation method disperses bacteria evenly and thereby decreases variance in bacterial counts between microscopic fields. This dispersal reduces the time and tedium required for enumeration. Finally, the method allows for the determination of both total and free bacteria under certain conditions.

As with other methods of sample preparation, there are certain disadvantages inherent in this method. These include the fact that bacteria are no longer viable after being subjected to the treatment regime. Further, the original distribution of bacteria on the substrate will be changed after the treatment. Finally, the use of PPi in some samples with high concentration of multivalent cations in water can cause precipitation which would interfere with the observation of bacteria.

0-87371-564-0/93/$0.00 + $.50

MATERIALS REQUIRED

Equipment

- Epifluorescence light microscope with UV and light source
- 4',6-diamidino-2-phenylindole (DAPI) filter set for blue light (e.g., Zeiss BP365/10, FT 310, and LP 395)
- Sonicator with microtip
- Vortex mixer
- Filtration rig for 25-mm filters
- Macro- and micropipettes

Supplies

- 10-ml sterile disposable polypropylene tubes
- Ice bath
- 0.2 μm pore size, 25 mm diameter Nuclepore filters dyed with Irgalan black or Nuclepore-black filters
- Cork borer 4 mm internal diameter

Solutions and Reagents

- 37% formaldehyde
- Fresh tetrasodium pyrophosphate (PPi) stock solution (0.22-μm Millipore filter sterilized) for water samples
- NaCl solution (0.22-μm filter sterilized) 0.85% for fresh water samples or made up to the same salinity (parts per thousand) as seawater
- PPi-NaCl (concentration of PPi and NaCl depend on concentration of divalent cations and salinity of sample)
- Stock DAPI (0.5 mg ml^{-1}), filter sterilized prior to use
- Irgalan Black (unless black filters are purchased)
- Gelatin glycerol

PROCEDURES

Sample Preparation

1. Preserve bacterial samples after collection with formaldehyde (3.7%, final concentration) for water samples or in NaCl-buffered water for solid or semisolid samples. For marine sediments or plants, the formaldehyde is made up in NaCl solution of the same salinity (parts per thousand) as the seawater, and for fresh water samples the formaldehyde is made up in 0.85% NaCl solution. Store samples at 4°C until analysis.

2. Add stock solution of PPi to the water sample to give final concentration in the range of 0.5 to 10 mM, depending on concentration of multivalent cations present, and a final volume of 5 ml in polypropylene tubes. For semisolid or solid samples of known weight or volume, the sample is suspended in 5 ml of PPi and concentration of up to 0.5 M (the optimum concentration of PPi is one above which precipitation of the cations is observed visually in the sample or there is interference in observation of bacteria when the sample is examined under oil immersion). Mix by using a vortex mixer and incubate the sample at room temperature (22°C) for 15 to 30 m with shaking.

3. Precool sample to 4°C and pack the sample tube with crushed ice. Sonicate the sample at a power level of 100 W for 30 to 60 s with the ultrasound probe immersed in the water sample to 1 to 2 cm above the bottom of the tube. The sonication may have to be carried in shorter bursts if the heat cannot be dissipated adequately.

4. After mixing briefly, the whole sample or subsample can be stained with the fluorochrome of choice and filtered onto a predyed 0.20-μm Nuclepore filter.

Enumeration of Bacteria

Bacteria can be stained using the fluorochromes acridine orange,[2] DAPI,[3] Hoechst 33258,[6] or 4′,5-(4,6-dichlorotriazin-2-yl) aminofluorescein (DTAF).[7] DAPI is preferred for most environmental samples because of lower nonspecific background staining. The final DAPI concentration in the sample should be in the range of 1.0 to 5.0 μg ml^{-1}, rather than 0.01 μg ml^{-1} as suggested by Porter and Feig[3] for planktonic bacteria.[4,5,8,9]

1. Add 10 to 50 μl of stock DAPI to the 5-ml sample, mix for 15 s, incubate for 10 to 20 min in the dark, and then mix again prior to filtration onto 0.2-μm pore size filter.

2. Adjust sample volume so that there are between 10 and 20 fluorescing bacterial cells per grid field, and 20 to 30 randomly chosen fields are counted per filter.

3. Total number of bacteria per sample can be calculated by taking into consideration the mean number per grid, area of filtration, any dilution or concentration of the sample, and the initial volume, weight, or area of the sample.

NOTES AND COMMENTS

1. Do not use commercially available artificial seawater mix for suspending marine samples because they may contain multivalent cations which will precipitate on addition of PPi.

2. If there is interference in viewing bacteria from sediment particles following sonication, separate detached bacteria by addition of glycerol followed by centrifugation to pellet the sediment particles.[10]

3. Alternatively, Tween 80 at concentration of 10 μg ml^{-1} can be used to reduce hydrophobic interactions.[11] We suggest a combination of PPi (0.5 to 500 mM) and Tween 80 (5 to 10 μg ml^{-1}) for reduction of both ionic and nonionic interactions, followed by ultrasound.

4. Aquatic plant leaves of fronds can be sampled with a sterile cork borer. The epiphytic bacteria are then stained by placing the disc in DAPI solution (1 μg ml^{-1}) for 10 min, followed by three rinses with NaCl-buffered water to remove any excess stain. Each disc is then placed on a glass slide and coated with warm glycerol gelatin. A glass coverslip should be placed immediately on top and the gelatin allowed to solidify before examining the slide under oil immersion for epiphytic bacteria.[12]

5. The following are selected references for use of the method in different conditions:

    a. Seawater column[13]
    b. Marine snow[14,15]
    c. Fresh water sediments and artificial substrates[10]
    d. Marine sediments[16]
    e. Seawater, marine sediments, fecal matter, lesser giant kelp blades[5]
    f. Humic particles[17]
    g. Free and attached bacteria under culture conditions for grazing experiments[18]
    h. Fluorescently labeled bacteria (FLB) used for grazing experiments[7,8,19]

REFERENCES

1. Daley, R., Direct epifluorescence enumeration of native aquatic bacteria: uses, limitation and comparative accuracy, in *Native Aquatic Bacteria Enumeration, Activity, and Ecology,* Costerton, J. W. and Colwell, R. R., Eds., American Society for Testing and Materials, Philadelphia, 1979, 29.

2. Hobbie, J. E., Daley, R., and Jasper, S., Use of Nuclepore filters for counting bacteria by fluorencence microscopy, *Appl. Environ. Microbiol.,* 33, 1225, 1977.

3. Porter, K. G. and Feig, Y. S., The use of DAPI for identifying and counting aquatic microflora, *Limnol. Oceanogr.,* 25, 943, 1980.

4. Velji, M. I. and Albright, L. J., The dispersion of adhered marine bacteria by pyrophophate and ultrasound prior to direct counting, in *Intrenational Colloquium on Marine Bacteriology,* Second Centre National de la Recherche Scientifique, IFREMER, Actes de Colloques, 3, Brest, France, 1986.

5. Velji, M. I. and Albright, L. J., Microscopic enumeration of attached marine bacteria of seawater, marine sediment, fecal matter and kelp blade samples following pyrophosphate and ultrasound treatments, *Can. J. Microbiol.,* 32, 121, 1986.

6. Paul, J. H., Use of Hoechst dyes 33258 and 33342 for enumeration of attached and planktonic bacteria, *Appl. Environ. Microbiol.,* 3, 939, 1982.

7. Sherr, B. F., Sherr, E. B., and Fallon, R. D., Use of monodispersed, fluorescently labeled bacteria to estimate *in situ* protozoan bacterivory, *Appl. Environ. Microbiol.,* 53, 958, 1987.

8. Sieracki, M. E., Haas, L. W., Caron, D. A., and Lessard, E. J., Effect of fixation on particle retention by microflagellates: underestimation of grazing rates, *Mar. Ecol. Prog. Ser.,* 38, 251, 1987.

9. Schallenberg, M., Kalff, J., and Rasmussen, J. B., Solutions to problems in enumerating sediment bacteria by direct counts, *Appl. Environ. Microbiol.,* 55, 1214, 1989.

10. Kaplan, L. A. and Bott, T. L., Diel fluctuations in bacterial activity on streambed substrata during algal blooms: effects of temperature, water chemistry, and habitat, *Limnol. Oceanogr.,* 34, 718, 1989.

11. Yoon, W. B. and Rosson, R. A., Improved method of enumeration of attached bacteria for study of fluctuation in the abundance of attached and free-living bacteria in response to diel variation in seawater turbidity, *Appl. Environ. Microbiol.,* 56, 595, 1990.

12. Velji, M. I., Improved Microscopic Enumeration of Attached Aquatic Bacteria following Deflocculant and Ultrasound Treatment, M.Sc. thesis, Simon Fraser University, Burnaby, British Columbia, 1985.

13. Jensen, L. M., Sand-Jensen, K., Marcher, S., and Hansen, M., Plankton community respiration along a nutrient gradient in a shallow Danish estuary, *Mar. Ecol. Prog. Ser.,* 61, 75, 1990.

14. Amy, P. S., Caldwell, B. A., Soeldner, A. H., Morita, R. Y., and Albright, L. J., Microbial activity and ultrastructure of mineral-based marine snow from Howe Sound, British Columbia, *Can. J. Fish. Aquat. Sci.,* 44, 1135, 1987.

15. Simon, M., Alldredge, A. L., and Azam, F., Bacterial carbon dynamics on marine snow, *Mar. Ecol. Prog. Ser.,* 65, 205, 1990.

16. van Duyl, F. C. and Kop, A. J., Benthic bacterial biomass supported by streamwater dissolved organic matter, *Mar. Ecol. Prog. Ser.,* 59, 249, 1990.

17. Tranvik, L. J. and Sieburth, J. McN., Effects of flocculated humic matter on free and attached pelagic microrganisms, *Limnol. Oceanogr.,* 34, 688, 1989.

18. Sibbald, M. J. and Albright, L. J., Aggregated and free bacteria as food sources for heterotrophic microflagellates, *Appl. Environ. Microbiol.,* 54, 613, 1988.

19. Albright, L. J., Sherr, E. B., and Fallon, R. D., Grazing of ciliated protozoa on free and particle-attached bacteria, *Mar. Ecol. Prog. Ser.,* 38, 125, 1987.

CHAPTER 18

Direct Estimates of Bacterial Numbers in Seawater Samples Without Incurring Cell Loss Due to Sample Storage

C. M. Turley

INTRODUCTION

Direct microscopic counts of samples stained with nucleic acid fluorochromes such as Acridine Orange (AO) or 4'6-diamidino-2-phenylindole (DAPI) are essential for studying and enumerating populations of bacterioplankton and have become routine over the last 10 to 15 years.

Numbers vary from around 0.2 to 3.0 $\times$ 10^9 l^{-1} in the upper mixed layer of the oceans. Apart from being an important carbon pool in its own right, with the exception of phytoplankton, often achieving more biomass than any other group of organisms, bacteria are the main agents making some of the extensive DOC pool available to the rest of the food web via bacterivorous organisms, particularly flagellates. Bacteria are now recognized as a large and active part of the marine ecosystem and are, therefore, an important component of investigations or models of oceanic carbon and nitrogen flow.

In the oceans bacterial concentrations can vary considerably both on the micro- and macroscale so intense sampling frequency in time and space, closely coordinated with measurements of the biology, chemistry, and physics of the water masses is generally attempted in oceanographic investigations of bacterial dynamics. With one investigator able to analyze a maximum of 15 to 20 samples per day, the impracticability of counting at sea, and the development of ambitious field programs, many microbial oceanographers have resorted to taking large numbers of preserved seawater samples for later analysis, over a period of weeks or months, back in the laboratory. However, the effect of long-term storage of preserved seawater samples has recently been shown to dramatically effect the concentration of bacteria such that counts from stored, preserved seawater samples may have grossly underestimated bacterial numbers and their importance in the marine ecosystem.

Fortunately, frozen storage of freshly preserved and prepared (stained, filtered, and mounted on slides) seawater samples for up to 70 d results in no cell decrease. This means that more shipboard time is required to process (to stain, to filter, to mount on slides, and to microscopic check for statistically correct numbers of bacteria per quadrat) the preserved seawater samples in order to have accurate estimates of bacterial numbers. Samples prepared in this way may then be stored frozen for counting on land.

0-87371-564-0/93/$0.00 + $.50
© 1993 by Lewis Publishers

MATERIALS REQUIRED

Equipment

- 25- and 47-mm diameter filtration apparatus and vacuum pump.
- An epifluorescent microscope fitted with filters for excitation of cells stained with Acridine Orange; for example, we use an Olympus BH-2 microscope with a reflected light fluorescence attachment equipped with a 100-W high-pressure mercury burner (HPO 100W/2). Excitation of cells which had taken up the AO fluorochrome was achieved using an exciter filter (BP-490), a blue excitation filter combination (Dichroic mirror-500 with built in barrier filter 0-515), and an additional barrier filter (0-530). The sample was viewed through a 100 × oil immersion objective (S Plan Apo 100, 1.40 oil), a 1.25 × column magnifier, and 10 × WHK eyepieces giving a total magnification of 1250 ×. One of the eyepieces was equipped with an indexed-squared graticule E11A (Graticules, Ltd.) (width 82 μm) which is divided into 100 small quadrats which facilitates accurate counting. The graticule was calibrated with a stage micrometer S8 (Graticules, Ltd.)
- Freezer and refrigerator.
- Dispenser for 25% filtered glutaraldehyde.

Supplies

- Sterile, particle free sample bottles.
- Micropipette and sterile tips for dispensing 200-μl volumes.
- 0.2-μm pore size 25- and 47-mm diameter Nuclepore filters (see Notes and Comments — 9).
- Particle-free containers to store solutions.

Solutions

- Irgalan black (IB): 2 g l^{-1} 2% acetic acid in 2.5% glutaraldehyde.
- Particle-free water (PFW): 0.2-μm pore size filtered 2.5% glutaraldehyde made up in freshly distilled water.
- Particle-free acridine orange (AO) solution: 1 mg ml^{-1} 2.5% glutaraldehyde) (0.2-μm pore size filtered). Store in the dark.

All three solutions should be stored at 5°C. The AO solution may need to be made up monthly as loss of staining ability has been observed by some workers.

PROCEDURES

Sample Preparation and Protocol

1. Preserve the samples within 3 h of sampling, with particle-free (0.2-μm pore size filtered) 25% SEM grade glutaraldehyde (2.5% final concentration) (see Notes and Comments — 8).

2. Within 12 h of fixation, pour a known volume of the sample into a 25-mm filter unit holding 0.2-μm pore size, 25-mm diameter, Nuclepore polycarbonate filter stained for a minimum of 2 h in Irgalan Black solution (see Notes and Comments — 9). Sample volumes should be adjusted for location and depth to give the required number of cells per microscope field (see Notes and Comments — 5 and 6).

3. Filter the samples under a maximum vacuum of 80 mm-Hg until approximately 2 ml remains and stop filtration or, if your sample volume is under 2 ml, to ensure random distribution of cells on the filter, make up to 2 ml with particle-free water (PFW).

4. Add 200 μl of AO solution for 5 min to enable staining of the DNA. (See Notes and Comments — 1.)

5. Filter the remaining 2 ml down on to the filter. Care at this stage should be taken not to allow the filter to dry out as this may result in an underestimation of cell numbers.

6. Place the damp filter (bacteria uppermost) onto a clean slide with a fine smear of non-fluorescent immersion oil (e.g., Nikon sd = 1.515).

7. Add a small drop of the immersion oil to the center of the filter. Note that too much oil can result in "floating" of the cells off the filter so that there are two plains of focus.

8. Mount a coverslip on top of the filter and oil drop and firmly press down until the oil moves out to the edge of the filter and forms a seal. The pressure on the coverslip must ensure no lateral movement of the coverslip as this can result in unequal distribution of cells on the filter.

9. The edges of the coverslip can be sealed with nail varnish.

10. The filter funnels should be thoroughly rinsed with PFW to ensure no contamination between samples.

11. Background counts should be carried out to ensure that the PFW, IB, and AO solutions and glassware are bacteria free.

12. Prior to acceptance of the preparation for counting, examine the prepared slide using an epifluorescent microscope to ensure that there is

    a. One focal plane (see 7 above)
    b. Equal distribution of cells on the filter (see 8 above)
    c. Approximately 20 to 50 cells/quadrat (see Notes and Comments — 6).

Sample Storage and Enumeration

1. If the preparation passes all the above requirements immediately freeze it at −20°C until ready for counting on land (see Notes and Comments — 3 to 5). Samples have been stored frozen for up to 70 d without any loss in cells.

2. Thaw sample and using an epifluorescent microscope count 20 to 40 quadrats and a minimum of 600 cells which, according to the tables in Reference 1, give a precision of ±10% at the 95% confidence interval (see Notes and Comments — 10).

3. Calculate bacterial cell numbers from

$$\text{cells/ml} = [(SC - BC) \times CF \times F]/V \tag{1}$$

where SC = mean of sample counts/quadrat, BC = mean of background counts/quadrat, CF = effective filter area/quadrat area, F = (volume preservative/volume sample preserved) + 1, and V = volume preserved sample filtered (milliliters). (See Notes and Comments — 2.)

NOTES AND COMMENTS

1. DAPI can also be used to fluorescently stain bacteria for their enumeration[2] and is particularly useful for samples with a high particulate loading as the blue fluorescence from the stained cells makes them more easily distinguishable, e.g., from detritus, than AO-stained cells.

2. Biomass estimates of bacterial populations are often made by measuring epifluorescent images with an ocular microscope, from enlarged photographic images projected onto a digitizing board interfaced with a computer, or by image analyzer[3,4] in order to determine biovolume. An estimate of cell numbers and a conversion factor from biovolume to cellular carbon can be used to estimate bacterial biomass (see Reference 5 and Chapters 35–37 and 44).

3. There is a dramatic decrease, on average 39% during the first 40 d, in bacterial cell numbers with storage time of glutaraldehyde preserved seawater samples using the AO fluorochrome method of estimating bacterial numbers commonly used in most laboratories.[6] Similar losses were found in formalin-fixed samples, samples stored in brown glass bottles, and when using the DNA-specific fluorochrome DAPI. This indicates that the apparent cell loss with storage time was not peculiar to the preservative (glutaraldehyde), the storage bottles (polystyrene tissue culture flasks), or the fluorochrome (AO).

4. Attachment to the inner surfaces of sample bottles containing preserved seawater samples stored for several hundred days did account for 41 to 48% of the original seawater bacterial concentration before storage. Sonication of stored preserved seawater samples increased the concentration of unattached bacteria by 5 to 199%. Such inconsistency of cell release indicates that treatment by sonication cannot be used as a method to ensure full recovery of cells and an accurate estimate of the original concentration of bacteria in the preserved seawater prior to storage. Furthermore, as 29 to 91% of the original concentration remained unaccounted for in either the attached or unattached population of bacteria, there seems to be a genuine loss of cells within the sample bottle that cannot be explained solely by attachment. The effectiveness of the fixatives to preserve or cross link the nucleic acids, dissolution of the protein, and cell lysis[7] are possible explanations.

5. Counts carried out at sea and counts carried out on the same samples stored frozen give the same estimates of bacterial numbers.[6] Since counting at sea can be difficult and unpleasant I recommend fixation, staining, and filtration of the sample as soon after sampling as possible. After viewing under the epifluorescent microscope, in order to check the correct volume of sample has been filtered to give a statistically significant number of cells per quadrat, the prepared slide can be immediately frozen for later counting on land.

6. Sample volumes should be adjusted for location and depth to give the required number of cells per quadrat. For example, during the summer in the NE Atlantic 0.5- and 5-ml volumes of seawater from 2 and 300 m are needed to be filtered to give approximately 25 cells per quadrat and a final concentration of 3.0×10^6 ml^{-1} and 3.0×10^5 ml^{-1}, respectively, while 55 ml is required from a 2750-m seawater sample to give the same cells per quadrat and a final concentration of 2.5×10^4 ml^{-1}.

7. A decrease in the mean cell size and shift in size frequency distribution was also observed with storage time of preserved seawater samples. Estimates of bacterial numbers, biovolume, and biomass from preserved seawater samples stored for weeks and months will substantially underestimate the importance of bacteria in oceanic ecosystems.[6]

8. Some researchers find 2% buffered formaldehyde an acceptable alternative preservative to glutaraldehyde.

9. Black filters are available from Nuclepore but have, on occasions, had high background fluorescence making it difficult to count fluorescing cells. Although I have not tried them recently the filters now receive more favorable reports of low background fluorescence.

10. Researchers often have different criteria for determining the numbers of cells and quadrats to count. If less precision is acceptable 200 cells can be counted. If higher precision is required, as in this protocol, 600 should be counted. Indeed, you may wish to consider the recommendation by Kirchman et al.[8] (see Chapter 14) of counting more than one filter from each sample.

REFERENCES

1. Cassell, E. A., Rapid graphical method for estimating the precision of direct microscopic counting data, *Appl. Microb.*, 13, 293, 1965.
2. Porter, K. G. and Feig, Y. G., The use of DAPI for identifying and counting aquatic microflora, *Limnol. Oceanogr.*, 25, 943, 1980.
3. Bjørnsen, P. K., Automatic determination of bacterioplankton biomass by image analysis, *Appl. Environ. Microb.*, 51, 1199, 1986.
4. Sieracki, M. E., Johnson, P. W., and Sieburth, J. McN., Detection, enumeration, and sizing of planktonic bacteria by image analysed epifluorescence microscopy, *Appl. Environ. Microb.*, 49, 799, 1985.
5. Fry, J. C., Direct methods and biomass estimation, *Meth. Microb.*, 22, 41, 1990.
6. Turley, C. M. and Hughes, D. J., Effects of storage on direct estimates of bacterial numbers of preserved seawater samples, *Deep-Sea Res.*, 39, 375, 1992.
7. Hopwood, D., Theoretical and practical aspects of glutaraldehyde fixation, in *Fixation Chemistry,* Stoward, P. J., Ed., Chapman and Hall, London, 1973, 47.
8. Kirchman, D., Sigda, J., Kapuscinski, R., and Mitchell, R., Statistical analysis of the direct count method for enumerating bacteria, *Appl. Environ. Microb.*, 44, 376, 1982.

Additional References

Daley, R. J. and Hobbie, J. E., Direct counts of aquatic bacteria by a modified epifluorescence technique, *Limnol. Oceanogr.*, 20, 875, 1975.

Jones, J. G. and Simon, B. M., An investigation of errors in direct counts of aquatic bacteria by epifluorescence microscopy, with reference to a new method for dyeing membrane filters, *J. Appl. Bacteriol.*, 39, 317, 1975.

Zimmerman, R. and Mayer-Reil, L. A., A new method for fluorescent staining of bacterial populations on membrane filters, *Kiel. Meeresforsch.*, 30, 24, 1974.

CHAPTER 19

Total and Specific Bacterial Counts by Simultaneous Staining with DAPI and Fluorochrome-Labeled Antibodies

Kjell Arne Hoff

INTRODUCTION

Coons et al.[1] introduced the use of fluorochrome-conjugated antibodies, and in 1942 introduced the fluorescein isocyanate-labeled antibodies.[2] These labeled antibodies were used to stain tissues of mice, heavily infected by pneumococci. The potential of the method in autecological work was examined by Hobson and Mann working with rumen bacteria.[3] Schmidt and Bankole adapted the method to terrestrial environments, working with the fungus *Aspergillus flavus*.[4] A review of the immunofluorescence technique in ecological work has been given by Bohlool and Schmidt.[5]

There are two ways of making the target antigen "visible" by use of fluorochrome-labeled antibodies: the direct and the indirect method. Both methods rely on a primary antibody. The primary antibody may be polyclonal or monoclonal, depending on the wanted specificity. For many purposes in ecological work, polyclonal antibodies are sufficient. Rabbits are quite popular as experimental animals for producing antibodies since they are easy to handle and the serum volume obtained (approximately 100 ml) is theoretically sufficient to stain 5000 to 10,000 samples, assuming that the serum can be diluted at least 100-fold.

Having obtained the serum after an immunization program and final bleeding, the researcher has to decide whether to conjugate the antibodies in the serum with a fluorochrome or to use a conjugated secondary antibody. The secondary antibody is directed toward antigens on the molecules of the primary antibody. Secondary antibodies are obtained by immunizing one species with antibodies of a second species (e.g., immunizing swine or sheep with rabbit antibodies).

The advantage of the direct method is greater simplicity in the staining procedure. Savings in time relative to the indirect method can outweigh the greater initial work if there are many samples to stain. A disadvantage with the direct method is that conjugation with the fluorochrome must be repeated several times during the initial setup of the procedure. At the end of the immunization program, a small volume of blood is collected from the animal and tested to see if it contains the desired antibodies and to determine the optimal dilution of the serum. It is normal procedure to trigger the animal's immune response by administrating a booster dose of the antigen, thereby elevating the level of the desired antibodies. However, the final bleeding cannot be performed before the level and specificity of the antibodies have been determined to be satisfactory. During conjugation of the antibodies

0-87371-564-0/93/$0.00 + $.50

with fluorochromes, inevitably, some of the antibodies will be lost, the amount being highly dependent on the skill and the experience of the worker.

The advantage of the indirect method is that the secondary antibody is commercially available with a wide range of conjugated molecules: fluorochromes, enzymes, metals, and radioisotopes. When stored in a refrigerator, the shelf life is good. In autecological work, fluorescein isothiocyanate (FITC)-labeled antibodies are most commonly used. As there are several epitopes on the primary antibody (thus the possibility that several secondary antibodies may bind to the primary one), use of secondary antibodies may enhance the fluorescence intensity compared to using labeled primary antibodies. However, this may also be a disadvantage in cases where the primary antibody and possibly also the secondary antibody binds nonspecifically to surfaces in the samples. This may cause a high background fluorescence.

In well-prepared specimens using both the direct and the indirect method, the background is homogenous and no defined structures besides antibody stained bacteria are visible. Even when the bacterioplankton is concentrated on membrane filters the number of specific bacteria stained with antibodies may be low (i.e., less than one bacteria per observed microscopic field). This means that there are few or no structures to observe, and this may cause difficulties, particularly with establishing that the microscope is properly focused. This may lead to a tedious search for stained particles of any kind to ensure focus, and thereafter a search for stained bacteria in the vicinity of such particles. This is easily overcome if both the total number of cells and the specific stained cells could be viewed in the same field. In this case, focus is secured by viewing the numerous bacteria present by staining with a fluorochrome such as 4',6-diamidino-2-phenylindole (DAPI), then switching the optical filters of the microscope to view antibody-stained cells. With this refinement the search for antibody-stained bacteria is greatly facilitated.

A second disadvantage using labeled antibodies is false-positive readings. An antibody-stained cell will appear with a bright corona, the size will be enlarged, and it is difficult to observe any details, for instance, whether or not the cell is dividing.[6] Especially in environmental samples, structures similar to stained bacteria may occur (e.g., where the specific bacteria to which the antibodies are directed against is rod shaped, but an apparent positive finding appears circular and approximately the same size). In such cases it is highly desirable to have a look on what is "really present" in the field.

A third disadvantage using labeled antibodies is the increased uncertainty in quantifying the ratio of total and specific counts when this has to be done with two separate preparations. In many cases bacteria do not distribute evenly on the filters. Normally the specific bacteria in question constitute only a small fraction of the total counts in natural samples and it is easy to bias the counts. For example, if several specific bacterial cells are attached to a particle, this may bias the count severely. This bias will be diminished if the total bacterial number on the same particle could be counted.

MATERIALS REQUIRED

Equipment

- Filtration rig for 25-mm filters
- Vacuum pump
- Epifluorescence microscope, including filter assemblies for rapid switching between DAPI and FITC

Supplies

- Polypropylene support filters (25 mm) (Gelman Science, Ann Arbor, MI)
- Polycarbonate membrane filters (25 mm; 0.2- or 0.45-μm pore size) (Nuclepore Corp., Pleasanton, CA)
- Disposable plastic syringes (10 ml or larger)
- Disposable filter units for plastic syringes (e.g., Millex-GV; Millipore Corp., Bedford, MA)
- Pipettes

Solutions

- DAPI (4′,6-diamidino-2-phenylindole) (Sigma Chemical Co., St. Louis, MO)
- Irgalan black BGL (Ciba-Geigy)
- Acetic acid, 2% (vol/vol)
- Formaldehyde solution, 0.35% (vol/vol)
- Rabbit serum containing primary antibodies
- FITC-conjugated swine anti-rabbit immunoglobulins (DAKO, Copenhagen, Denmark)
- 0.01 M phosphate buffer, pH 7.2
  Solution A 3.56 g $Na_2HPO_4 \cdot 2H_2O$ in 1 l distilled water
  Solution B 2.76 g $NaH_2PO_4 \cdot H_2O$ in 1 l distilled water
  Mix 410 ml of solution A and 90 ml of solution B and add 500 ml of distilled water.
  Adjust pH to 7.2 if necessary.
- 0.01 M phosphate-buffered saline, pH 7.2
  As above but supplemented with 20 g NaCl and 4.3 g $MgCl_2 \cdot 6H_2O$
- Buffered glycerol, pH 7.5
  1 volume 0.5 M carbonate buffer, pH 9.5
  9 volumes glycerol
  Adjust pH
- 0.5 M carbonate buffer, pH 9.5

$Na_2CO_3 \cdot 10H_2O$	42.93 g
$NaHCO_3$	29.40 g
Distilled water	1 l

PROCEDURES

It is important that all solutions used in the preparation are filtered through filters with pore size of 0.2 μm to ensure that no bacterial contamination will occur. All equipment should be cleaned thoroughly.

Preparation of Filters

1. The polypropylene support filters are stored in a DAPI solution. The solution is made up of 10 μg DAPI ml^{-1} of distilled water. The polypropylene filters act as support for the Nuclepore filters and their basic role is to distribute the vacuum more evenly when evacuating the water from the Nuclepore filters. DAPI will diffuse upwards, staining bacteria on the Nuclepore filter. After use, the polypropylene filters are transferred back to the DAPI solution. Another advantage of soaking the filters in DAPI solution is that the filters are moist when placed on the filtration funnel; a dry support filter and a moist Nuclepore filter will immediately stick to each other and it is more cumbersome to position the filters correctly.

2. Untreated polycarbonate membrane filters will show a high background when used with an epifluorescence microscope. This can be avoided by staining them with the dye Irgalan black BGL which will also enhance contrast.

   a. Polycarbonate membrane filters with pore size of 0.2 or 0.45 μm are covered with stain overnight.

   b. Thereafter the staining solution is decanted off and the filters rinsed three times with distilled water.

   c. Finally, the filters are stored in a 0.35% formaldehyde solution. After some time more dye is released and the formaldehyde solution should be renewed.

   d. Alternatively, prestained black polycarbonate filters are also available commercially (Nuclepore).

Staining

The present method was developed for seawater samples, and this will be described here.

1. The support filter is first placed on the filter rig with a pair of forceps.

2. The Nuclepore filter is removed from its storage solution, placed upon the supporting filter, and the top of the filtering unit is fastened. The sample volume is typically 10 ml. (The arguments regarding optimal sample volume is the same for this method as for staining with acridine orange or DAPI, and this has been covered elsewhere.) The filter unit is connected to a vacuum pump with a facility to adjust the vacuum and a manometer measuring the vacuum. A pressure of -0.2 bar works well. In the hands of the author, it did not appear to matter whether the filter runs dry after filtering the water sample or if the vacuum is released just after the samples are completely filtered. When operating a filtering rig without a separate valve for each funnel, it is nearly impossible to prevent some of the filters from running dry.

3. After filtering the samples, fluorescent-antibody staining is performed by prewashing each filter three times with 2 ml of distilled water. In many reports, washing is performed by using 0.01 M phosphate buffer, pH 7.2.[7] However, no improvement in fluorescence of marine samples has been noticed by this author using phosphate buffer.

4. The working solution of the primary antibodies may conveniently be stored in 10- or 50-ml plastic syringes equipped with a disposable 0.22-μm pore size filter unit. Each filter is covered with the freshly filtered serum solution, washed three times with 2 ml distilled water, covered with the working dilution of secondary antibodies (conjugate), and finally washed three times. Staining time for both serum and conjugate is 20 min. If the administered volume of serum or conjugate is too small, more is added during the staining period to avoid drying out of the filters.

5. After staining the specimens are mounted in filtered (0.2 μm) buffered glycerol, pH 7.5.[7]

6. When working with marine samples, the primary antibodies are diluted in 0.01 M phosphate-buffered saline, pH 7.2. The dilution ratio must be determined individually for each sample of serum, but is typically in the range 1:100 to 1:800 for polyclonal sera. This author has used FITC-conjugated swine anti-rabbit immunoglobulins. These secondary antibodies have been diluted 1:100 in the same buffer as described for the primary antibodies and stored the same way.

Microscopy

The microscope must be equipped with two optical filter packages to view both the DAPI-stained bacteria and the antibody-labeled bacteria. It must be easy to switch between the two packages. As an example for the Leitz Orthoplan microscope, a combination of the filter packages Ploemopak A (BP 340-380 exciting filter, RKP 400 beam-splitting mirror, LP 430 suppression filter) was used for viewing DAPI staining and Ploemopak I 2 (BP 450-490 exciting filter, RKP 510 beam-splitting mirror, LP 515 suppression filter) was used for viewing FITC staining. The Ploemopak I 2 works for acridine orange as well. A $63 \times$ or $100 \times$ oil-immersion objective is appropriate.

Development, Use, and Refinement

Separate Staining Step with DAPI

It is fully possible to stain the bacteria with DAPI before or after the staining with primary antibodies and labeled secondary antibodies. After filtering the sample, the filters are covered with DAPI (1 to 10 μg ml^{-1}) for 1 to 5 min. The DAPI solution is sucked off; thereafter, staining with antibodies continues as described above.

Staining of Imprints

Staining bacteria with the combination of DAPI and FITC-labeled antibodies has also been tested on imprints, as follows.

1. Material from moribund or newly dead Atlantic salmon suffering from cold-water vibriosis were collected and examined. Bleeding areas on the skin or the trans-sectioned milt were laid gently on glass slides and then removed to produce imprints. Smears of gut content were also prepared on glass slides.

2. Imprints and smears on glass slides were heat fixed before staining.

3. The imprints were stained by using the same solutions and dilution rates as previous described.

   a. First the imprint area is covered by adding drops of rabbit antiserum in phosphate buffer.
   b. After 20 min the glass slide is gently washed with distilled water and allowed to drain and dry.
   c. Thereafter the imprint area is covered with FITC-conjugated anti-rabbit immunoglobulins (1:100) for 20 min, washed, and drained.
   d. Finally, the area is covered with freshly filtered DAPI (1 μg ml^{-1}) for 5 min, washed, and drained.
   e. The specimen is mounted in filtered buffered glycerol, pH 7.5.

Distilled water is preferred for washing the slides because buffered phosphate saline will cover the slides with a thin layer of salt precipitates when the slides are drained and allowed to dry. It must be checked if the bacteria in question do not lyse during washing. Normally most bacteria, even of marine origin, should withstand this washing procedure since the cells are heat fixed.

NOTES AND COMMENTS

1. The method has proved to be very robust. Students and colleagues introduced to the method have made excellent preparations in their first trial. We have not made any effort to investigate how the method will work with filters and reagents of other brands than

described here. If DAPI staining does not perform well according to the present method, it is possible to stain with DAPI in a more traditional way as described above (Separate Staining Step with DAPI).

2.   If the bacteria in question lyse due to washing with distilled water, the washing steps should be carried out using phosphate-buffered saline.

3.   A general problem when using labeled antibodies is the question of specificity and cross reactions. Therefore, the primary antibody must be tested against several strains of closely related bacteria to demonstrate the absence of cross-reactions. For further discussions of controls to be performed, see Reference 8.

4.   A problem encountered with some optical filter packages of certain brands is fading and excitation problems. It has been described that, if the microscopic field firstly is observed for DAPI-stained bacteria, these bacteria will also be visible when switching to the FITC filter package, thereby interfering with the observation of specifically stained bacteria. This is overcome by first looking for FITC and thereafter for DAPI.

5.   The method was originally developed for general staining of planktonic bacteria in seawater and specific staining of the fish pathogen *Vibrio salmonicida* in the same sample. Later the method has been applied successfully for the staining of general and specific bacteria in marine sediments. It has also proven itself in specific staining of populations of sulfate-reducing bacteria in samples collected from water and waste water from oil wells in the North Sea.

REFERENCES

1. Coons, A. H., Creech, H. J., and Jones, R. N., Immunological properties of an antibody containing a fluorescent group, *Proc. Soc. Exp. Biol. Med.*, 47, 200, 1941.
2. Coons, A. H., Creech, H. J., Jones, R. N., and Berliner, E., The demonstration of pneumococca antigen in tissues by the use of fluorescent antibody, *J. Immunol.*, 45, 159, 1942.
3. Hobson, P. N. and Mann, S. O., Some studies on the identification of rumen bacteria with fluorescent antibodies, *J. Gen. Microbiol.*, 16, 463, 1957.
4. Schmidt, E. L. and Bankole, R. O., Detection of *Aspergillus flavus* in soil by immunofluorescent staining, *Science,* 136, 776, 1962.
5. Bohlool, B. B. and Schmidt, E. L., The immunofluorescence approach in microbial ecology, in *Advances in Microbial Ecology,* Vol. 4, Alexander, M., Ed., Plenum Press, New York, 1980, 203.
6. Hoff, K. A., Rapid and simple method for double staining bacteria with 4',6-diamidino-2-phenylindole and fluorescein isothiocyanate-labeled antibodies, *Appl. Environ. Microbiol.*, 54, 2949, 1988.
7. Kawamura, A., Jr., *Fluorescent Antibody Techniques and Their Applications,* University Park Press, Baltimore, 1969.
8. Schmidt, E. L., Fluorescent antibody techniques for the study of microbial ecology, *Bull. Ecol. Res. Comm.*, 17, 67, 1973.

CHAPTER 20

Use of RFLPs for the Comparison of Marine Cyanobacteria

Susan E. Douglas

INTRODUCTION

Since the discovery of very small unicellular cyanobacteria in the oceans,[1,2] numerous methods have been used to identify them in various ecological systems. These have included comparisons of such characteristics as phycobiliprotein profiles[3] and fluorescence,[4] G:C composition,[5] and immunofluorescence.[6] However, these methods have had limited use in determining the genetic relatedness of these organisms.

Restriction fragment length polymorphisms (RFLPs) have been shown to be good markers of taxonomic relatedness in a number of groups of organisms (see Reference 7). The method is relatively simple and can be performed by anyone familiar with basic molecular biology techniques. DNA is prepared by a simple lysis procedure, subsequent removal of proteins by organic extraction, and purification on a cesium chloride density gradient. The DNA is digested with restriction endonucleases, the fragments separated electrophoretically on agarose gels, and transferred to nylon membranes by Southern blotting. Specific genes are located by hybridization with labeled probes and the fragments localized by autoradiography. The patterns of hybridizing fragments are indicative of the variability of the genomes under study and are distinctive in each organism or group of closely related organisms.

Advantages of the method are (1) simple to perform, (2) large number of potential scorable markers makes statistics and phylogenetic analysis possible, (3) availability of many heterologous probes and many different restriction endonucleases allows discrimination between broadly divergent or closely related organisms, (4) small size of genome of marine cyanobacteria[8] means the autoradiographic pattern of bands is usually simple, (5) availability of nonradioactive detection systems, (6) nylon membranes may be stripped and reprobed several times, and (7) no cloning is required.

Disadvantages of the method are (1) DNA is generally purified over cesium chloride gradients which adds to the expense of the procedure, (2) axenic cultures are usually required, and (3) DNA may be methylated,[9,10] making complete digestion of the DNA difficult and thus obscuring the autoradiographic pattern of bands.

MATERIALS REQUIRED

All general chemicals and solvents should be of analytical grade and carbon-filtered deionized water should be used in all solutions and media. Restriction endonucleases, Klenow fragment, and [α-^{32}P]-dCTP (3000 Ci mmol^{-1}) from various sources are used according to the manufacturer's instructions except that 4 mM spermidine is incorporated into all restriction endonuclease incubations. Nylon membranes (Zeta-Probe) are from BioRad (Richmond, CA). Type I agarose (low EEO), low melting point agarose, RNase A, sarkosyl, and lysozyme are from Sigma Chemical Co. (St. Louis, MO). Pentamers for labeling probes, bacteriophage lambda DNA, and Sephadex G50 are from Pharmacia (Uppsala, Sweden). Proteinase K is from Boehringer Mannheim (Germany). Carnation-brand skim milk powder is used for BLOTTO in hybridization (see below).

Equipment

- Refrigerated centrifuge for harvesting cells
- Preparative ultracentrifuge for cesium chloride gradient
- Clinical centrifuge
- Water Bath (37, 65, 100°C)
- Heat sealer and bags
- -70°C freezer
- Short-wave ultraviolet lamp
- Horizontal gel electrophoresis apparatus
- Camera with Wratten #23A filter
- Short-wave ultraviolet transilluminator
- Dark room
- Vacuum oven (80°C)
- Protective shielding for radioisotope work (1/2 inch plexiglass)
- Geiger counter
- Liquid scintillation counter

Supplies

- Pasteur pipettes
- Glass flasks
- Polyallomer ultracentrifuge tubes
- 16-gauge hypodermic needles
- X-ray film (Fuji, Kodak)
- Intensifying screens (Dupont Cronex Lightning Plus)
- Autoradiography cassette

Solutions and Reagents

- TE buffer: 10 mM Tris, 1 mM EDTA, pH 8.0
- 1$\times$ TBE: 0.09 M Tris borate, 0.002 M EDTA
- Wash buffer: 120 mM NaCl, 10 mM Tris, 1 mM EDTA, pH 8.0
- Lysis buffer: 25% sucrose (w/v), 10 mM Tris, 1 mM EDTA, pH 8.0
- 6$\times$ Loading Buffer: 0.25% bromophenol blue, 0.25% xylene cyanol, 15% Ficoll (Type 400)
- 10$\times$ SSC: 1.5 M NaCl, 0.15 M Na citrate, pH 7.0
- 10$\times$ BLOTTO: 2.5% powdered milk in water
- 20$\times$ SSPE: 3.6 M NaCl, 0.2 M NaH$_2$PO$_4$, 0.02 M EDTA, pH 7.4
- Buffer A: 0.1$\times$ SSC, 0.5% SDS
- Buffer B: 1% SDS, 0.25% BLOTTO, and 6$\times$ SSPE
- 0.25 M EDTA, pH 8.0
- RNAse A, 10 mg ml^{-1}, boiled as per Sambrook et al.[11]
- Proteinase K

- Phenol/chloroform/isoamyl alcohol (25:24:1) (see Reference 11)
- Chloroform/isoamyl alcohol (24:1)
- Isopropanol equilibrated with a saturated NaCl solution
- Ethidium bromide (10 mg ml^{-1} stock)
- Restriction endonucleases
- Materials for gel electrophoresis: a complete description of agarose gel electrophoresis is given in Reference 11
- Denaturant: 0.25 M HCl
- Neutralization solution: 1.5 M NaCl, 0.5 M NaOH
- Transfer solution: 1 M NH$_4$OAc, 0.02 M NaOH
- Developer and fixer (Kodak LX24 developer and FX40 fixer)

PROCEDURES

RFLP

First, DNA is isolated by gentle lysis of freshly grown mid-logarithmic phase cells from a 1 l culture or from frozen cell pastes, organic extraction, and cesium chloride density centrifugation.

1. Collect fresh cells by centrifugation for 10 minutes at 3000 × g and 4°C.

2. Resuspend the cell pellet in 100 ml of wash buffer and recentrifuge as above.

3. Resuspend the cell pellet or thawed cell paste in 20 ml of lysis buffer.

4. Add 5 ml of 0.25 M EDTA, pH 8.0 and 0.3 ml of RNase A solution (10 mg ml^{-1}; boiled as per Sambrook et al.)[11] and incubate at 37°C for 50 min.

5. Add proteinase K to 50 μg ml^{-1} and incubate at 65°C for 60 min.

6. Remove protein and cell debris by two phenol/chloroform/isoamyl alcohol extractions and one chloroform/isoamyl alcohol extraction, taking care not to shear the DNA by excessive agitation. The cell lysate is gently swirled with an equal volume of organic phase in glass flasks, as plastic is attacked by chloroform. The emulsion is transferred to glass centrifuge tubes and the phases separated by centrifuging at low speed for 10 min. The upper aqueous phase is removed from the lower organic phase using a Pasteur pipette, which has had the end broken off to make the bore larger, and transferred to a clean flask for reextraction.

7. To the final aqueous phase, add cesium chloride (1 g ml^{-1} lysate) and ethidium bromide (20 μl) and centrifuge at 184,000 × g for 18 h and 66,200 × g for 2 h at 20°C in polyallomer tubes in an ultracentrifuge.

8. Carefully remove the tubes from the centrifuge rotor and view from the side with a hand-held ultraviolet lamp. Only one band should be evident in the tube. If others are present, they may represent DNA from contaminants, and the purity of the culture should be checked. The DNA is collected by puncturing the side of the tube using a 16-gauge needle and slowly drawing off the band. A description of this procedure is illustrated in Reference 11. The remainder of the solution in the tube should be put in a bottle reserved for ethidium bromide-contaminated material, and disposed of appropriately.

9. Remove the ethidium bromide by extracting several times with isopropanol equilibrated with a saturated NaCl solution. The upper organic (isopropanol) layer will be pink for the first two or three extractions and should be removed using a Pasteur pipette into a container reserved for ethidium bromide-contaminated material. When the upper layer is no longer pink, all of the ethidium bromide has been removed.

10. Dilute the DNA twofold with water, add two volumes of 95% ethanol, and incubate at $-20°C$ for 2 h. Precipitate the DNA by centrifugation at 12,000 $\times$ g for 30 min at 4°C.

11. Rinse the DNA pellet in 70% ethanol, dry under vacuum, and resuspend in approximately 1 ml of TE.

12. Precipitate the DNA a second time by the addition of 0.5 volumes of 7.5 M ammonium acetate and 0.56 volumes of isopropanol to enhance restrictability.[12] The final precipitate is collected by centrifugation as described above and dissolved in TE to a final concentration of approximately 50 μg ml^{-1}.

13. Incubate the DNA (1 μg) with restriction endonuclease (approximately 10 units) in a final volume of 20 μl of the manufacturer's recommended buffer for at least 3 h at 37°C. Certain DNAs may require an overnight incubation with excess enzyme in order to achieve digestion.

14. Terminate the reaction by incubation at 65°C for 10 min and addition of 4 μl of 6$\times$ loading buffer.

15. Load samples into wells of 0.7% agarose gels and electrophorese for 6 h at 50 V in 1$\times$ TBE containing 0.5 μg ml^{-1} ethidium bromide. Lambda DNA digested with a restriction endonuclease such as HinDIII, EcoRI, or StyI is included in at least one well of each gell for size markers.

16. Visualize restricted DNA in the gel using short-wave ultraviolet light and photograph.

17. A complete description of the next procedure, Southern transfer, is given in Reference 11.

18. Soak gel in denaturant for 15 min, neutralization solution for 30 min, and transfer solution for 30 min.[13]

19. Transfer the DNA by capillary action to a nylon membrane (presoaked in water and then transfer solution) for at least 4 h.

20. Rinse the membrane in 2$\times$ SSC to remove any adhering agarose, air dry, and bake *in vacuo* at 80°C for 30 min.

Next, the fragment of DNA containing the probe sequence is released from the vector sequence by restriction endonuclease digestion, separated by electrophoresis and labeled using the random-prime method of Feinberg and Vogelstein.[14] A sample (25 ng) of bacteriophage lambda DNA is also labeled for identifying the size markers.

21. Digest recombinant plasmid containing the probe sequence with the appropriate restriction endonuclease.

22. Separate the probe sequence from the vector sequence by electrophoresis in 0.7% low-melting point agarose and 1$\times$ TBE.

23. Excise the gel slice containing the probe sequence, add water (1.5 ml g^{-1} agarose), boil the sample for 7 min, and store at $-20°C$.

24. Aliquots containing 25 ng of DNA are labeled for 4 h at 37°C with 25 μCi of [α-^{32}P]-dCTP using any one of a number of commercially available kits.

25. Purify probe from unincorporated nucleotides by passage over a 1-ml column of Sephadex 650 (see Reference 11) and measure radioactivity by counting 1 μl of probe solution in 10 ml of water in a liquid scintillation counter.

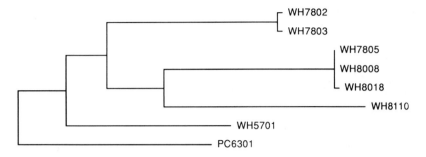

Figure 1. Phylogenetic analysis of RFLP data from murine cyanobacteria studied by Douglas and Carr[9] using PAUP 3.0n.[16] A total of 192 characters were analyzed and a single most parsimonious tree of 206 steps was obtained. PC6301 was used as the outgroup and the exhaustive search option was used.

The next step is hybridization of isolated DNA with probe.

26. Prewash nylon membranes in buffer A at 65°C for 1 h and prehybridize at 65°C for 4 h in a solution containing buffer B.

27. Remove prehybridization solution and replace with fresh buffer B to which 10^6 cpm ml^{-1} of the purified probe have been added. Labeled bacteriophage lambda DNA, which will hybridize to the size markers, is also added.

28. Perform hybridization in heat-sealed bags for at least 12 h at 65°C.

29. Wash membranes in $2\times$ SSC/0.1% SDS twice for 30 min each, and then check for radioactivity using a Geiger counter. Washes in SSC of decreasing ionic strength and at increased temperatures are performed when necessary to remove background.

30. Autoradiography is conducted for 1 to 7 d at $-70°C$ with intensifying screens.

31. Membranes are stripped of probe by incubating at 65°C in 0.4 N NaOH for 30 min, followed by washing in buffer A at 65°C for 60 min. Additional hybridizations with different probes are then performed as described above.

Data Analysis

Although it is sometimes easy to group small numbers of organisms by visual comparison of the RFLP patterns,[9] larger numbers may require a more detailed analysis.[15] Phylogenetic analysis may be performed using either parsimony or distance methods. In the former, the presence and absence of the fragments detected by autoradiography in each organism is tabulated and shared homologous characters are used to derive the tree. In the latter, a similarity matrix is constructed and the molecular evolutionary distance between organisms calculated. (For an example of the data tables used to construct the trees, please refer to Reference 15). It is useful to include one or two organisms (which are distantly related to the organisms under investigation) which can serve as outgroups and be used to root the tree. Parsimony analysis of the RFLPs obtained using three probes[9] with the program PAUP[16] yielded the phylogenetic tree shown in Figure 1. In a similar study, Wood and Townsend[15] used the MUPG distance method of Li[17] and the Dollo parsimony method of Felsenstein[18] to produce phylogenetic trees. The two sets of data and three different analysis programs yielded similar trees, indicating the robustness of the derived phylogenies.

NOTES AND COMMENTS

1. It may be possible to obtain good-quality DNA without cesium chloride density gradient centrifugation, using procedures based on CTAB extractions such as that described by Doyle and Doyle.[19]
2. The choice of probe is important. One should choose a probe which will give a strong hybridization signal, and which will detect genes present in low copy number. Genes involved in photosynthesis are usually present in one or a few copies and are highly conserved, thus giving RFLP patterns with marine cyanobacteria which are easy to interpret. Usually, heterologous probes are used, although a few homologous probes from marine cyanobacteria are available. In addition to the phycoerythrin genes cloned from WH7803 (Newman and Carr, unpublished), the phycocyanin genes recently cloned from WH7803[20] should prove very useful for the phycocyanin-containing cyanobacteria.
3. The recent advent of polymerase chain reaction technology (PCR)[21] has made possible rapid comparisons of related organisms. The gene of interest is amplified from organisms of interest and the size of the product determined on agarose gels. The product may also be cloned and sequenced. This technique is sensitive to trace amounts of contaminating DNA from other sources. Sequence information from the genes to be amplified is required for designing primers and this may not be available for many of the genes in the marine cyanobacteria yet. Primers based on highly conserved genes from other organisms may be used successfully, however.
4. The RAPD method[22] also shows promise for comparisons of marine cyanobacteria. It does not rely on the availability of sequence information for cyanobacterial genes since random primers are used for the amplification. However, it is very sensitive to small changes in the amplification conditions and requires the best thermocycler available (a large expense). As with PCR, the DNA must not contain trace amounts of contamination from other sources.

ACKNOWLEDGMENT

This is NRCC publication number 33777.

REFERENCES

1. Waterbury, J. B., Watson, S. W., Guillard, R. R. L., and Brand, L., Widespread occurrence of a unicellular, marine, planktonic, cyanobacterium, *Nature,* 277, 293, 1979.
2. Johnson, P. W. and Sieburth, J. McN., Chroococcoid cyanobacteria in the sea: a ubiquitous and diverse phototropic biomass, *Limnol. Oceanogr.,* 24, 928, 1979.
3. Alberte, R. S., Wood, M., Kursar, T. A., and Guillard, R. R. L., Novel phycoerythrins in marine *Synechoccus* spp.: characterization and evolutionary and ecological implications, *Plant Physiol.,* 75, 732, 1984.
4. Wood, A. M., Horan, P. K., Muirhead, K., Phinney, D. A., Yentsch, C. M., and Waterbury, J. B., Discrimination between types of pigments in marine *Synechococcus* spp. by scanning spectroscopy, epifluorescence microscopy, and flow cytometry, *Limnol. Oceanogr.,* 30, 1303, 1985.
5. Waterbury, J. B., Watson, S. W., Valois, F. W., and Franks, D. G., Biological and ecological characterization of the marine unicellular cyanobacterium *Synechococcus,* in *Photosynthetic Picoplankton,* Platt, T. and Li, W. W. K., Eds., *Can. Bull. Fish. Aquat. Sci.,* 214, 71, 1986.
6. Campbell, L. and Carpenter, E. J., Characterization of phycoerythrin-containing *Synechococcus* spp. populations by immunofluorescence, *J. Plankton Res.,* 9, 1167, 1987.
7. Banks, J. A. and Birky, C. W., Jr., Chloroplast DNA diversity is low in a wild plant, *Lupinus texensis, Proc. Natl. Acad. Sci. U.S.A.,* 82, 6950, 1985.
8. Cuhel, R. L. and Waterbury, J. B., Biochemical composition and short-term nutrient incorporation patterns in a unicellular marine cyanobacterium *Synechococcus* (WH7803), *Limnol. Oceanogr.,* 29, 370, 1984.

9. Douglas, S. E. and Carr, N. G., Examination of genetic relatedness of marine *Synechococcus* spp. by using restriction fragment length polymorphisms, *Appl. Environ. Microbiol.*, 54, 3071, 1988.

10. Lambert, G. R. and Carr, N. G., Resistance of DNA from filamentous and unicellular cyanobacteria to restriction endonuclease cleavage, *Biochim. Biophys. Acta*, 781, 45, 1984.

11. Sambrook, J., Fritsch, E. F., and Maniatis, T., *Molecular Cloning: A Laboratory Manual*, 2nd ed., Cold Spring Harbor Laboratory, Cold Spring Harbor, NY, 1989.

12. Owen, R. J. and Borman, P., A rapid biochemical method for purifying high molecular weight bacterial chromosomal DNA for restriction enzyme analysis, *Nucleic Acids Res.*, 15, 363, 1987.

13. Rigaud, G., Grange, T., and Pictet, R., The use of NaOH as transfer solution of DNA onto nylon membrane decreases the hybridization efficiency, *Nucleic Acids Res.*, 15(2), 857, 1987.

14. Feinberg, A. P. and Vogelstein, B., A technique for radiolabelling DNA restriction endonuclease fragments to high specific activity, *Anal. Biochem.*, 132, 6, 1983.

15. Wood, A. M. and Townsend, D., DNA polymorphism within the WH7803 serogroup of marine *Synechococcus* spp. (cyanobacteria), *J. Phycol.*, 26, 576, 1990.

16. Swofford, D. L., PAUP Version 3.0n, Illinois Natural History Survey, Champagne, IL, 1990, 136 pp.

17. Li, W.-H., Simple method for constructing phylogenetic trees from distance matrices, *Proc. Natl. Acad. Sci. U.S.A.*, 78, 1085, 1981.

18. Felsenstein, J., Parsimony in systematics: biological and statistical issues, *Annu. Rev. Ecol. Syst.*, 14, 313, 1983.

19. Doyle, J. J. and Doyle, J. L., Isolation of plant DNA from fresh tissue, *BRL Focus*, 12, 13, 1990.

20. Wilson, W. H., Newman, J., Mann, N. H., and Carr, N. G., Cloning and sequence analysis of the phycocyanin genes of the marine cyanobacterium *Synechococcus* sp. WH7803, *Plant Mol. Biol.*, 17, 931, 1991.

21. Saiki, R. K., Scharf, S., Faloona, F., Mullis, K. B., Horn, G. T., Erlich, H. A., and Arnheim, N., Enzymatic amplification of β-globin genomic sequences and restriction site analysis for diagnosis of sickle cell anemia, *Science*, 37, 170, 1985.

22. Williams, J. G. K., Kubelik, A. R., Livak, K. J., Rafalski, J. A., and Tingey, S. V., DNA polymorphisms amplified by arbitrary primers are useful as genetic markers, *Nucleic Acids Res.*, 18, 6531, 1990.

CHAPTER 21

Use of High-Resolution Flow Cytometry to Determine the Activity and Distribution of Aquatic Bacteria

D. K. Button and B. R. Robertson

INTRODUCTION

Populations, cell size, DNA, and biomass-distributions of pelagic bacteria can be obtained by flow cytometry from event frequency, forward scatter and fluorescence intensity of cells stained with DAPI. Subpopulations resolved by multiparameter analysis can also be characterized, sorted, and collected for further study. Bacterial biomass associated with sediments is computed from that of the bacteria shaken from the particles times the ratio of radiolabeled nutrient uptake rates for the two groups of organisms. Kinetic measurements are related to cell mass so that nutrient flux measurements reflect substrate accumulation ability which can be translated to growth rate.

Anticipated uses include evaluating organism activity from simultaneous analysis of cell size, DNA content, and RNA content as well as evaluating growth and selective predation rates from rates of change of bacterial populations and their size distributions in systems of limited and controlled populations of predators.

Advantages of the method are (1) statistical accuracy in individual cell parameters due to the large sample size (100 to 100,000 cells per analysis), (2) speed where populations can be characterized in 5 to 30 min depending on size and requirements, (3) accuracy in population determination giving reproducibility of $\pm 5\%$, (4) sensitivity for measuring ultramicrobacteria whose mean organism size is below resolution by light microscopy, and resistance to cultivation further erodes confidence, (5) preservability of samples which allows convenient analyses of stored samples, (6) minimal interference from other particles since photosynthetic organisms can be identified from fluorescent pigments, debris fails to give a DNA signal, and DNA viruses are below the size range of bacteria, (7) resolution of specific subpopulations during culture development, (8) identification of genome numbers as an aid to determining activity and purity of subpopulations, (9) sorting capabilities where sufficient material from subpopulations can be delivered to characterize cells by electron microscopy or to determine radioactivity previously incorporated, and (10) multiparameter analysis capabilities that potentially allow the analysis of at least three parameters on each cell such as size, DNA content, and RNA content along with total population.

Disadvantages include (1) background fluorescence from brownwater lakes and from some mineral sediments, (2) an absence of convenient biological standards for size and DNA content, and (3) limitations of DAPI, an AT-sensitive DNA stain, due to variation in the GC content which ranges from 35% for *Bacillus* to 69% for *Pseudomonas*.[1]

0-87371-564-0/93/$0.00 + $.50
© 1993 by Lewis Publishers

MATERIALS REQUIRED

Equipment

- Flow cytometer — We use a Cytofluorograf IIS equipped with a model 2151 computer system, high-sensitivity photomultiplier tubes, 3.5-decade logarithmic circuitry, quartz flow-cell with 75-μm orifice, adjustable beam-shaping lens assembly, temperature controller, sorter (now maintained by Becton Dickinson, Mountain View, CA), and two 5-W argon lasers (Coherent, Inc., Palo Alto, CA). One laser is sufficient for size/DNA/total-population assays.
- Coulter counter — A model Z_{BI} Coulter counter with a 30-μm orifice and channelyzer (Coulter Electronics, Hialeah, FL) and calibrated with human erythrocytes of known volume, about 80 μm^3, is used to quantify fluorescent latex beads added to samples as an internal population and fluorescence standard. Electrical impedance of the cells passing through the orifice is taken as a measure of cell volume; bacteria >0.2 μ^3 are sized for calibration of the forward scatter signal to obtain cell volume by flow cytometry.
- Filtration devices — A high flow-rate unit with autoclavable cartridges rated at 0.1-μm pore size (Millidisk; Millipore Corp., Bedford, MA) is used for vacuum filtration of sheath fluid, and a unit for 13-mm diameter membrane filters is used for sample filtration.
- Epifluorescence microscope — Ours is a Leitz Dialux 20EB equipped with a 100-W Hg lamp and filter cube "A" for resolving blue fluorescent bacteria from UV excitation of DAPI-DNA and cube "I2" for green fluorescence of internal standard microspheres excited at 488 nm. Although the spheres are clearly resolved by flow cytometry with UV excitation, resolution by epifluorescence microscopy is much better with 488-nm illumination.
- Cold storage — Stains and staining solutions are stored frozen, and preserved samples are stored at 0 to 5°C in the dark.

Supplies

- Sample vials; sterile 1.5-ml microcentrifuge tubes and 20-ml scintillation vials
- Filters; 1-μm pore size, 13-mm diameter polycarbonate membranes for water samples; small, formaldehyde-resistant, disposable units such as Acro13 filters (Gelman Sciences, Ann Arbor, MI)
- Syringes; two 1-ml glass syringes, one dedicated to formaldehyde use and the other to the staining solution, are rinsed after each use
- Fluoresbrite Plain Microspheres, 0.96, 0.60, and 0.90-μm diameter (Polysciences, Inc., Warrington, PA)
- Red blood cells (Count-A-Part; Diagnostic Technology, Inc., Hauppauge, NY)
- Micropipets for adding preservative and stain

Solutions

- Formaldehyde; 37% as commercially supplied, filtered just before use
- Triton X-100; 5% aqueous stock solution, stored dark, frozen in 1-ml aliquots
- DAPI (4′,6-diamidino-2-phenylindole); 0.5 mg ml^{-1} aqueous stock solution, stored dark, frozen in 0.5-ml aliquots
- Staining solution; a mixture of 50 μl DAPI stock and 1 ml Triton X-100 stock solutions, filtered (0.2 μm, Acro13) just before use
- Sheath fluid, distilled water filtered through autoclaved 0.1-μm pore size Millidisk cartridges (Millipore Corp., Bedford, MA)
- Sheath fluid for sorting cells. Distilled water solution of sodium chloride at 0.85%, filtered as above

PROCEDURES

General Comments

Essential to the analysis of aquatic bacteria is good laser performance, clean optics, a well-aligned optical path, and stable sample flow. Flow cells are selected by inspection under 50 × magnification for minimal defects and a centered orifice, routinely observed for cleanliness, and changed as necessary. After each day's use, the entire sample line, including the flow cell, is rinsed with filtered chlorine bleach (0.05% sodium hypochlorite) followed by distilled water to eliminate accumulated stain and prevent bacterial growth. The flow cell is cleaned every few days with dichromate solution.

Excitation of the bacteria is by UV light (351.1 and 363.8 nm), at 100 mW of power, with the beam focused to a 5 × 44 μm wide ellipse and aligned to impinge on the center of the flow stream and maximize observed fluorescence and scatter from 0.96-μm fluorescent spheres. Sample flow is set at 5 μl min^{-1} to give a sample stream that is <11 μm in diameter for consistent illumination by the Gaussian distributed beam from the laser.

Light scattered in the forward direction, 0.5 to 20° from the path of the beam (obscured by a 1.5-mm blocker bar), is directed by a short-wavelength-reflecting 424-nm long-pass dichroic through a 310- to 370-nm bandpass filter to remove fluorescence signal. Side scatter, taken from 72 to 108°, is obtained by a similar filter arrangement. Blue fluorescence from UV-excited DAPI emerging from the side scatter dichroic is isolated by a 450- to 490-nm bandpass filter. Scatter and fluorescence are directed by fiber optics to photomultipliers. Processing of the signal, integrated over time (area mode), is triggered by fluorescence to reduce interference from nonfluorescent debris.

Although instrument alignment is performed with linear scaling, logarithmic amplification is used for bacterial analyses to give a multidecade dynamic range. Instrument sensitivity and threshold settings under conditions for bacterial analyses limit the useful range to 2.5 decades for both forward scatter and fluorescence. Scaling in both linear and log modes is calibrated for each photomultiplier-analog board pair[2] using fluorescent microspheres of similar size but different fluorescent intensities (Polysciences 0.90- and 0.96-μ spheres); with logarithmic processing, intensity, displayed as channel number, is converted to its linear equivalent channel according to the calibration curves.

Samples are typically analyzed for total bacterial population, size distribution, biomass (the product of population, size, and density, at 1.04 pg μm^{-3}), DNA content of the cells, and subpopulations based on bivariate histograms of size and DNA content. Such histograms give recognizable patterns characteristic of the water systems from which the samples are taken. Mean size, reported as colume or mass, and mean DNA content of the cells for the whole population and subpopulations can be tabulated to further characterize the sample.

Cytometer Operation

Preparation of Internal Standards

Internal standards are used to calculate the analyzed volume for population determinations and to set the photomultiplier gains for forward scatter and fluorescence. A mixture of 0.90- and 0.60-μm diameter fluorescent microspheres in filtered (0.2-μ pore size), distilled water is used. The larger particles can be counted by Coulter counter for accurate concentration at 1 × 10^8 ml^{-1} in the stock suspension. They allow the concentration to be monitored by Coulter counter over several days' use, produce a clear side-scatter signal with minimal interference from biological particles for calculating sample volume,[3] and provide a measure of fluorescence stability which is sometimes useful in addition to that derived from the smaller spheres. However, at instrument settings used for bacteria, the 0.90-μm spheres are above scale for forward scatter. The 0.60-μm spheres, at about 1×10^8 ml^{-1} in the stock suspension, produce a signal within the scaling range for both forward scatter and fluorescence and allow monitoring of instrument stability throughout analysis as well as comparison among samples analyzed at different times.

Preparation of Samples

1. Water samples are preserved with formaldehyde (to 0.5%; 13.5-μl filtered stock ml^{-1} sample) which also improves stain permeability. Preserved samples may be stored cold in the dark for weeks to months.

2. Samples are filtered (1-μm pore size) to remove large particles and clumps of bacteria.

3. A 1-ml volume is treated with 20 μl of the staining solution of Triton X-100 (to 0.1% for further cell permeabilization) and DAPI (to 0.5 μg ml^{-1}) for 1 h at 10°C and amended with 1 μl of the internal standard microsphere mixture to give a concentration of 1×10^5 ml^{-1} for the 0.90-μm diameter spheres.

Sample Analysis

1. Photomultiplier gains are set so that the signal from unstained 0.60-μm standard microspheres appears in channel 850 (out of 1000) for forward scatter and 660 for DAPI-DNA fluorescence, and the signal from the 0.90-μm microspheres appears in channel 200 for side scatter.

2. A control sample of stained sheath fluid containing internal standard microspheres is used to equilibrate the sample line with stain and to monitor the staining solution for contamination by bacteria and other particles. Absorption of stain by the spheres will increase their intensity by about 20 channels.

3. Signal processing is triggered on fluorescence so that only those events associated with fluorescence above an adjustable threshold will be analyzed; in a stain-free system, signal from nonfluorescent debris and unstained bacteria is below the threshold and ignored by the processor.

4. Sample flow rate, determined from the accumulation rate of side scatter signal from the 0.90-μm spheres, is monitored during sample analysis to ensure constant illumination across the sample stream.

5. Data acquisition is in list mode and normally continues for 5 to 30 min depending upon bacterial population. Analysis of 10,000 cells is often convenient and sufficient to produce smooth histograms.

6. Stored list data are subsequently reaccumulated for display, manipulation, and production of the hard copies.

7. Usual analysis rate is between 2 to 20 samples per hour depending on the cell density and the information and accuracy required. Size distribution, DNA content, and total population data can be obtained and recorded from a population of 10^6 cells ml^{-1} in about 10 min.

Sorting

Two subpopulations selected during analysis can be sorted and collected at a time for further study. To facilitate the process, thresholds are increased for both forward scatter and fluorescence to eliminate as much unwanted signal as possible. For >99% purity of sorted fraction, the sorter is set for deflection of a single drop containing a cell with desired characteristics for collection. Internal standard 0.90-μm spheres are sorted along with one of the subpopulations to evaluate the sorting process by epifluorescence microscopy. A portion of each sorted fraction is resuspended in stain solution and analyzed to confirm sorting success.

Data Analysis

Bacterial populations are computed from the number of events recorded on the bivariate histogram of log forward scatter vs. log DAPI-DNA fluorescence in the area bounded by forward scatter channels 1 to 1000 and fluorescence channels 600 to 1000 and from the volume of the sample processed as determined from the number of standard 0.90-μm spheres detected by side scatter as compared to the concentration added.

Mean values for forward scatter and fluorescence are computed by integrating histograms of frequency on intensity[4] with a digitizer system having spreadsheet capability (SigmaScan, Jandel Scientific; Corte Madera, CA) to convert logarithmic intensity to linear equivalent intensity according to analog board calibration curves. Newer computer systems allow direct transfer of list data to spreadsheets for statistical manipulations. For assigning quantitative values, signal intensities are normalized with respect to signal from internal standard 0.60-μm microspheres and then related to standard curves prepared for cell size and DNA content.

Standard Curve

Cell Size. Cell volume is obtained from a standard curve prepared using various bacteria sized by Coulter counter; it relates forward scatter intensity to Coulter cell volume.[3] Current efforts involve use of oligobacteria which average 0.06 μm^3 in volume and have been sized by scanning cryogenic electron microscopy. For particles of diameter (d) less than one fifth the wavelength of the light, scattering increases according to the Rayleigh-Gans equation with d^6, with experiments giving a d^4 dependency.[5] Cell volume measurements show agreement with Mie theory using scatter integrated over the appropriate collection angle. Values for cell volume remain in question due to lack of biological standards in the appropriate 0.1-μm^3 size range. While there has been recent success in cultivating ultramicrobacteria for this purpose, their absolute size has not yet been resolved.

DNA. Standard curves for DNA are prepared from replication-inhibited *Escherichia coli*.[6] A culture in log growth phase is treated with rifampin at 150 $\mu g\ ml^{-1}$ to inhibit initiation of DNA synthesis and produce cells containing two, four, or eight genome copies. The chromosome of *E. coli* has 4.1 fg of DNA with a GC content of 0.5. Subpopulations containing integral genome copies are resolved by flow cytometry and their DAPI-DNA fluorescence intensity is related to DNA content for the standard curve. Correlation coefficients of 0.999 typically result but slopes vary among preparations so standards should be included for each run. A stock of rifampin-treated cells, preserved and stored as described above, makes a convenient source of the standard.

Interpretation of Data

Flow cytometric data can be used as the primary analytical tool in conjunction with a variety of techniques to evaluate the characteristics of natural populations and identify their role in ecosystem function. Procedures for interpreting data from several of these combinations are as follows.

In Situ Activity. Activity toward dissolved substrates is often measured by exposing water samples to radiolabeled substrate and measuring the rate of label incorporation or production of $^{14}CO_2$ or ^3H_2O. Procedures are well known.[7] For comparisons among systems, rates are related to biomass and substrate concentration according to the relationship

$$v_A = a^\circ_A \, X \, A \tag{1}$$

where v_A is the uptake rate of substrate A in g l^{-1} h^{-1}, X is biomass in g l^{-1}, and a°_A is the specific affinity of the cells for the substrate in l g^{-1} cell^{-1} h^{-1}. Specific affinity is the best indication of the ability of an organism to accumulate a substrate.[8] Use of Equation 1 involves an assumption that $A < K_A$ where K_A is the Michaelis constant for substrate transport. Background substrate dilutes added radiolabeled substrate by the same amount as that taken up, so it does not affect specific affinity unless quantities are large enough to cause saturation.[9]

Ambient substrate concentrations can approach Michaelis constants for uptake[10] and some can be measured directly. A more accurate value for the specific affinity is then given by

$$a^\circ_A = V_{max}/K_A \tag{2}$$

V_{max} is the maximal rate of substrate uptake as computed from the asymptote of the uptake rate on concentration and K_A is the associated Michaelis concentration at half-maximal rate. Specific affinity values from the literature have been organized according to substrate type for comparison.[11]

Cell Composition and Bacterial Activity. Bacterial activity is reflected by DNA and RNA content and cell size.[12] When observed subpopulations have DNA contents in the ratio of 1:2, as is the case for slow-growing dilution cultures, the pattern is taken to reflect 1n and 2n chromosomes.[13] Factors which increase the proportion of 2n- to 1n-organisms can be regarded as either stimulating growth rate or retarding cell division. Few ribosomes and small amounts of RNA are also characteristic of slow-growing bacteria.[12] Analysis of cells stained with DAPI and Pyronin Y (for RNA) and excited with UV and 517 nm in a dual-laser system can give correlated measurements of cellular DNA, RNA, and size. Although the method has only been applied to eukaryotes,[14] it has great potential for assessing bacterial activity in aquatic systems since six pairs of correlated data can be obtained for each cell.

Characterizing Subpopulations. Activity in various subpopulations can be evaluated following physical separation. Activity has been related to cell size by measuring radiolabel uptake rates and determining the mean cell volume in filter-fractionated samples analyzed by flow cytometry.[15] By this method we found that the smallest bacteria were the most active.[4]

Another method involves physical sorting by flow cytometry. Following radiolabel uptake, bacteria can be sorted according to subpopulations with unique combinations of size and DNA content. Fidelity is excellent; reanalyzed populations contain only members of the desired population. ^3H-substrates are used because their high specific activity allows small amounts of collected sample to be counted by scintillation spectroscopy.

A water sample is exposed to ^3H (mixed)-amino acids (4×10^8 dpm μg^{-1}) at 1 μg l^{-1} for 3 h at 10°C; 5 ml of the incubation mixture is immediately preserved with formaldehyde. Prior to analysis and sorting the sample is stained as described above. At a flow rate of 5 μl min^{-1}, about 1 ml of sample is analyzed during a 3-h sorting period. Volumes of the collected fractions vary depending on the number of sorting events (population), but can be greater than the initial volume due to dilution with sheath fluid; 1 to 2 ml at about 1000 dpm ml^{-1} can be expected. A portion of each collected fraction is restained, amended with internal standard microspheres, and analyzed to determine population recovery and confirm sorting success. The remainder is filtered to collect sorted radiolabeled bacteria on a 0.1-μm membrane for scintillation counting. Biomass associated with each selected fraction is determined from population and cell size data provided by flow cytometry and related to cell radioactivity for that fraction.

Evaluating Bacterial Populations in Particles and Sediments. Bacterial activity in the nonpelagic fraction or particle fraction of aquatic systems can be assessed in two ways. One is to filter-fractionate the system with a 1-μm pore size filter. That most of the pelagic bacteria pass through the filter can be verified by analyzing a 5-μm filtrate by flow cytometry; populations and the distributions should be the same for both filtrates.

A material balance gives the total biomass X_t as that in the particles X_p plus the free-living fraction X_f:

$$X_t = X_p + X_f \qquad (3)$$

Radiolabeled amino acids are added to both a filtrate fraction and an unfiltered sample, and the rates of radioactivity uptake determined. From Equation 1 the uptake rate in the total population v_t is given by

$$v_t = a_A X_t A \qquad (4)$$

For the pelagic or free-living bacteria, that rate is

$$v_f = a_A X_f A \qquad (5)$$

By assuming that the specific affinity of the free-living bacteria is the same as that of bacteria in the particles, one has

$$v_p = a_A X_p A \qquad (6)$$

which is related to equation 4 by a rate balance

$$v_p = v_t - v_f \qquad (7)$$

Then, from Equation 6,

$$X_p = v_p/(a_A A) \qquad (8)$$

and substituting in Equation 7 gives the mass of organisms in the particles,

$$X_p = (v_t - v_f)/(a_A A) \qquad (9)$$

While there may be a difference in the specific affinities of pelagic and particle-associated bacteria, this gives an estimate of populations which may not otherwise be known. As the kinetic characteristics of these organisms become better established, appropriate corrections may be applied.

The second method is similar, but the bacteria are mostly attached to sediment particles such as in seashore sediments. It has been used to evaluate bacterial activity in oil spill-impacted beaches.[16] Sample material is suspended in filtered seawater (0.2 μm), vigorously shaken to dislodge some bacteria from the particles, and allowed to settle. A portion of the liquid fraction is removed, radiolabeled substrate is added to both it and the remaining slurry, and uptake rates are determined. Biomass associated with sediment is computed from the ratio of uptake rates in the decanted portion and the slurry times the biomass of the decantate analyzed by flow cytometry. Again, specific affinities of the bacteria in both fractions are assumed to be similar.

Determining Rates of In Situ Growth. Average specific growth rates may be determined by a dilution technique[17] where predators are diluted to restrict their activity. A sample of volume V_f is filtered through 0.2-μm pore size filters and added to an ulfiltered volume V_u so that the ratio R_u of unfiltered to total volume also reflects the ratio of organisms used in the experiment as compared to those originally present. Bacterial growth rate μ is assumed to be dependent on an unaltered value for substrate concentration, and to be fast with respect to the growth rate of the predators P, so grazing is taken as constant over time t (when t is small). Loss to predation is second order in predators and bacteria as specified by a grazing rate constant g, and

$$dX/dt = \mu X - g\,(P_0\,R_u)(X_0\,R_u) \tag{10}$$

where P_0 and X_0 are the original populations of predators and bacteria, respectively. If the growth rate of the predators is small so that change in its population over the course of the experiment is negligible, the change in bacterial population X_1 to X_2 integrated over time is

$$\ln X_2/X_1 = (\mu - g\,P_0 R_u)t \tag{11}$$

The right-hand side is an apparent growth rate μ_a for the bacteria

$$\mu_a = \mu - g\,P_0\,R_u \tag{12}$$

so that a plot of μ_a vs. R_u extrapolates to μ, the real *in situ* growth rate of the organisms in the system, as R_u approaches 0. Moreover, size-selective grazing becomes apparent if, for example, atypically large size distributions appear (as they do with intermediate inoculum sizes in dilution cultures; see below) at small values of R_u but not large where grazing becomes important. Microflagellate predators can be discriminated from phytoplankton by an absence of chlorophyll fluorescence and counted as well, so long as several hundred organisms per milliliter remain.

This method has not previously enjoyed great success for measuring bacterial growth rates. However, the improved speed and accuracy of flow cytometry together with its ability to resolve subpopulations renders this method worthy of reexamination. For example, the amount of predation by slow-growing predators can be affected by reducing either of the populations added. Thus, predation effects on size distributions of bacterial subpopulations developing can be evaluated.

Growth of Oligobacteria in Dilution Culture. To obtain typical marine bacteria for laboratory examination it appears useful to avoid solid media and added nutrients which can discourage growth, large initial populations in which allelopathic interactions can occur, and overgrowth by commonly cultured organisms which represent 1 to 2% of the total in surface water. A technique called dilution culture consistently shows between 2 and 100% viability for near-surface seawater which means that many or most marine bacteria are culturable given proper conditions. To obtain these cultures one dilutes the original population to a small and known number of organisms in sterile seawater as determined by epifluorescence microscopy and follows the developing population. Viability is computed from multiple inoculations and the binomial theorem (in preparation).

Flow cytometry is especially useful for evaluating growth in dilution cultures which typically begin with a statistical single organism (0.02 viable cells ml^{-1} in 50 ml). It allows early detection of growth since populations as low as 100 to 1000 ml^{-1} can be resolved, and progress of population development can be followed. In cultures with minimal nutrient supplements, populations often truncate at about 10^6 ml^{-1}, so meaningful data can be obtained long before growth cessation. Capability of screening 50 to 100 samples per day facilitates handling large numbers of cultures. Culture purity is suggested as well because pure cultures of growing oligobacteria produce characteristic histograms. Normally one sees two clusters of DNA fluorescence, those from organisms with 1n and 2n chromosomes. These correspond to 1n and 2n amounts of biomass (forward scatter). Sometimes these smear together as a characteristic streak oriented at 45° indicating partly replicated DNA and incompletely grown cells.

For dilution cultures, growth rates cn be estimated from growth curves according to the relationship

$$X_t = X_0 e^{\mu t} \tag{13}$$

where X_t and X_0 are bacterial populations at the time of measurement and at the time of inoculation. This technology is seen as providing a large number of common new and important marine organisms for investigation.

Theoretical Considerations

Interpretation of flow cytometry data in terms of microbial dynamics often involves application of principles peculiar to the kinetics of microbial processes.

Kinetic Constants. The primary kinetic constants are maximal rate which gives substrate uptake rate per unit biomass at saturation with substrate and specific affinity which gives maximal rate of substrate accumulation when normalized to small substrate concentrations where flow of the individual molecules into the cell is at maximal rate. The relationship between the specific growth rate and the specific affinity of organisms for a particular substrate (A) is given by

$$\mu_A = a°_A \, A \, Y_{XA} \tag{14}$$

where Y_{XA} is cell yield, g cells produced per g substrate A consumed. Specific affinities need to be several thousand liters g^{-1} cell^{-1} h^{-1} for substantive growth rates and usual ambient concentrations. These values should be examined because organisms are sometimes affected by manipulation due to measurements and also because the small nutrient concentrations are subject to analytical error. These errors are usually cumulative giving specific affinities too low to account for observed growth in the environment. By considering the values of the kinetic constants, one can often locate such errors. Correct kinetic constants are essential in generating useful mathematic models of environmental processes.

By changing the substrate A (B, C, . . .) concentration with added carrier, one may relate its flux to concentration, thereby accounting for saturation. Hyperbolic kinetics are usually observed. In such cases,

$$V_{\text{max A}} = a^\circ_A K_A \tag{15}$$

where V_{max} is the maximal uptake rate for substrate A and K_A is the associated Michaelis constant.[18] An oligotrophic capacity plot, or plot of log a°_A vs. $-$log K_A, then gives a straight line with the best oligotrophs at the extreme. Osmotrophic marine pelagic organisms growing on their usual substrates should then lie near the extreme. This is because large specific affinities may be achieved with either high-affinity transporters or large amounts of them, but a combination of both reflects the most efficient distribution of protein. More significant than affinity of transporters in setting K_A is the ratio of the amount of cytoplasmic rate-limiting enzyme to transporters which can increase K_A dramatically. This relationship between protein distribution and kinetic constants, whether organisms favor substrate accumulation or rapid utilization of that accumulated, can be used to understand the ecological strategy of the organisms involved from their molecular composition.

Multiple Substrates. Where the rate of growth depends on simultaneous accumulation of multiple substrates

$$\mu = \mu_A + \mu_B + \mu_C \ldots \tag{16}$$

specific growth rates can be estimated from specific affinities, ambient substrate concentrations, and yields of each according to Equation 14. It gives growth rate from specific affinity data and ambient nutrient concentrations and must be compatible with data from the dilution technique above.

CONCLUSIONS

All of the above techniques indicate the power of flow cytometry in microbial ecology. It gives a convenient way of estimating biomass for a wide range of organisms without which kinetic measurements and associated interpretation of activity is severely compromised. Moreover, the quality of the biomass can be simultaneously evaluated as well.

REFERENCES

1. Betz, J. W., Aretz, W., and Hartel, W., Use of flow cytometry in industrial microbiology for strain improvement programs, *Cytometry*, 5, 145, 1984.
2. Schmid, I., Schmid, P., and Giorgi, J. V., Conversion of logarithmic channel numbers into relative linear fluorescence intensity, *Cytometry*, 9, 533, 1988.
3. Robertson, B. R. and Button, D. K., Characterizing aquatic bacteria according to population, cell size and apparent DNA content by flow cytometry, *Cytometry*, 10, 70, 1989.
4. Button, D. K. and Robertson, B. R., Kinetics of bacterial processes in natural aquatic systems based on biomass as determined by high-resolution flow cytometry, *Cytometry*, 10, 558, 1989.
5. Morel, A. and Yu-Hwan, A., Optical efficiency factors for free-living marine bacteria, *J. Mar. Res.*, 48, 145, 1990.
6. von Meyenburg, K., Boye, E., Skarstad, K., Koppes, L., and Kogoma, T., Mode of initiation of constitutive stable DNA replication in RNase h-defective mutants of *Escherichia coli* K-12, *J. Bacteriol.*, 169, 2650, 1987.
7. Button, D. K., Schell, D. M., and Robertson, B. R., Sensitive and accurate methodology for measuring the kinetics of concentration-dependent hydrocarbon metabolism rates in seawater by microbial communities, *Appl. Environ. Microbiol.*, 41, 936, 1981.

8. Gottschal, J. C., Some reflections on microbial competitiveness among heterotrophic bacteria, *Antonie Leeuwenhoek J. Microbiol.*, 51, 473, 1985.

9. Button, D. K., Robertson, B. R., and Craig, K. S., Dissolved hydrocarbons and related microflora in a fjordal seaport: sources, sinks, concentrations, and kinetics, *Appl. Environ. Microbiol.*, 42, 708, 1981.

10. Button, D. K. and Robertson, B. R., Dissolved hydrocarbon metabolism: the concentration-dependent kinetics of toluene oxidation in some North American estuaries, *Limnol. Oceanogr.*, 31, 101, 1986.

11. Button, D. K., Kinetics of nutrient-limited transport and microbial growth, *Microbiol. Rev.*, 49, 270, 1985.

12. Dennis, P. P. and Bremer, H., Macromolecular composition during steady-state growth of *Escherichia coli* b/r, *J. Bacteriol.*, 119, 270, 1974.

13. Binder, B. J. and Chisholm, S. W., Relationship between DNA cycle and growth rate in *Synechococcus* sp. strain PCC 6301, *J. Bacteriol.*, 172, 2313, 1990.

14. Miller, R. M. and Bartha, R., Evidence from liposome encapsulation for transport-limited microbial metabolism of solid alkanes, *Appl. Environ. Microbiol.*, 55, 269, 1989.

15. Robertson, B. R. and Button, D. K., Refinement of standard curves for cell size and DNA content of aquatic bacteria analyzed by flow cytometry, Abstracts, *XVII Congress Int. Soc. Anal. Cytol.*, Colorado Springs, CO, 1993.

16. Button, D. K., Robertson, B. R., McIntosh, D., and Juttner, F., Interactions between marine bacteria and dissolved-phase and beached hydrocarbons after the Exxon Valdez oil spill, *Appl. Environ. Microbiol.*, 58, 243–251.

17. Li, W. K. W. and Dickie, P. M., Growth of bacteria in seawater filtered through 0.2 μm Nuclepore membranes: implications for dilution experiments, *Mar. Ecol. Prog. Ser.*, 26, 245, 1985.

18. Button, D. K., Biochemical basis for whole-cell uptake kinetics: specific affinity, oligotrophic capacity, and the meaning of the Michaelis constant, *Appl. Environ. Microbiol.*, 57, 2033, 1991.

Additional Reference

Button, D. K. and Robertson, B. R., Methodology for analysis of a small marine bacterium by flow cytometry, *Cytometry,* suppl. 1, 103, 1987.

CHAPTER 22

Phytoplankton Analysis Using Flow Cytometry

Robert J. Olson, Erik R. Zettler, and Michele D. DuRand

INTRODUCTION

Phytoplankton, as autofluorescent particles suspended in a fluid medium, are well suited for analysis by flow cytometry; this technique can be used to rapidly count, measure, and isolate individual cells. Because flow cytometry is a rapidly evolving field with a steady stream of new instruments and applications, and because the phytoplankton comprise a wide range of species that cannot all be analyzed with any one flow cytometer or protocol, our treatment will cover only a limited number of analyses. We will present a specific protocol for analyzing the picophytoplankton (the group most easily studied by flow cytometry at present), references to protocols for some other phytoplankton groups, and general practical suggestions for flow cytometric analyses of natural samples. We will attempt to avoid instrument-specific analyses, so that investigators using any flow cytometer can benefit from the material presented. Specialized instrumentation or methodology will be only briefly discussed, with references to the pertinent literature.

A flow cytometer examines individual particles flowing through a sensing region (Figure 1). Typically, the suspension of sample particles (here, phytoplankton cells) is introduced into the center of a flowing "sheath" of particle-free water in a flow cell. The sheath guides the sample through a focused laser beam and constricts the sample stream so that only a single cell is present in the beam at any one time. As each cell passes through the beam, it scatters and absorbs light. The light scattered at various angles can be measured, and the absorbed light may be emitted as fluorescence of various colors, which can be measured by separate detectors. Some instruments can measure the volume of each cell by the Coulter principle. For each cell, the correlated measurements of the size of each signal can be stored in a computer for later analysis. In addition, some flow cytometers can physically sort particles that satisfy predetermined criteria.

The optical measurements from flow cytometry contain information which can be used to identify and enumerate groups of phytoplankton such as cyanobacteria, prochlorophytes, coccolithophorids, pennate diatoms, and cryptophytes. At a finer scale, such characteristics as the size and pigment content of the cells within each group can also be examined. The light scatter signals are related to particle size, shape, and refractive index; the magnitude and color of the autofluorescence signals contain information about the quantities and types of pigments present. In addition to measuring these intrinsic optical properties, fluorescently labeled antibodies and molecular probes can be used to mark specific groups, and compounds

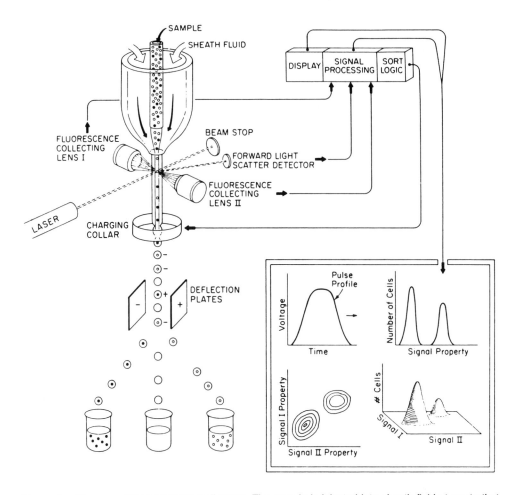

Figure 1.   Diagram of flow cytometer/cell sorter. The sample is injected into sheath fluid at a rate that permits analysis of properties of individual cells as they pass through a laser beam. The pulse profiles of fluorescence emission and light scatter signals of each cell are processed and displayed as one-dimensional histograms, bivariate two-dimensional contour plots, or three-dimensional plots. Based on a predesigned sort logic, particles of interest can be physically sorted from the emerging droplet stream by electrostatic deflection. (From Chisholm, S. W., Armbrust, E. V., and Olson, R. J., in *Photosynthetic Picoplankton*, Platt, T. and Li, W. K. W., Eds., *Can. Bull. Fish. Aquat. Sci.*, 214, 343, 1986. With permission.).

such as DNA, protein, and lipids can be stained with specific dyes and quantified;[1-7] these techniques will not be discussed here.

Advantages of flow cytometry include

- Speed: The technique is rapid; it has the potential to examine thousands of individual cells per second.
- Sensitivity: The fluorescence of cells too dim for manual epifluorescent microscopy can be detected and measured.
- Precision: The coefficients of variation (CV = standard deviation/mean) for light scatter and fluorescence measurements of uniform microspheres (which may be included in samples as internal standards) can be as low as 1%.
- Sorting: One of the most powerful and unique advantages of flow cytometry is the ability to physically isolate cells based on any combination of the optical characteristics measured. This makes it possible to obtain pure samples of specific cells of interest for further analysis.

The major disadvantages of flow cytometry are

- Limited resolution: Since a flow cytometry typically measures only peak or integrated signals (as opposed to ultrastructural details), phytoplankters can rarely be identified to genus or species. Instead, cells are classified on the basis of optical characteristics *(ataxonomy)*.[8] Other methods such as epifluorescence microscopy and image analysis allow far greater resolution of the heterogeneity in a plankton sample.
- Small sample size: Most flow cytometers analyze very small volumes (<0.5 ml) so that cells which occur at less than about 10^3 ml^{-1} cannot be conveniently analyzed without preconcentration or instrument modification (see Notes and Comments — 1). A related problem (not unique to flow cytometry) is the detection of rare cells in the presence of large numbers of interfering particles such as detritus (see Notes and Comments — 2).

MATERIALS REQUIRED

Equipment

- Flow cytometer (see Notes and Comments — 3)
- Ultrasonic cleaning bath
- Epifluorescence microscope with filter sets for chlorophyll (CHL) and phycoerythrin (PE) to monitor sorting
- Vortex mixer
- Liquid nitrogen refrigerator

Supplies

- Micropipets: 10 to 100 μl for beads, 100 to 1000 μl for sample preparation
- Cryovials
- Beads (uniform fluorescent microspheres);[13] we find Polysciences calibration grade "Fluoresbrite" beads in the 0.5- to 2-μm size range to be particularly useful

Solutions and Reagents

- Glutaraldehyde
- Liquid nitrogen for sample preservation

PROCEDURES

This protocol is written for an investigator with access to a flow cytometer and an experienced operator. Instrument setup and alignment involves a number of steps which vary from instrument to instrument. Hence, only procedures that may be specific to analyses of picophytoplankton (*Synechococcus, Prochlorococcus,* and eukaryotes) are covered here in detail.

Data Acquisition

Instrument Setup

Sheath Fluid. For best precision, the refractive index of the sheath fluid should be matched to the sample (e.g., for seawater samples, use 0.22-μm filtered seawater of similar salinity for sheath).

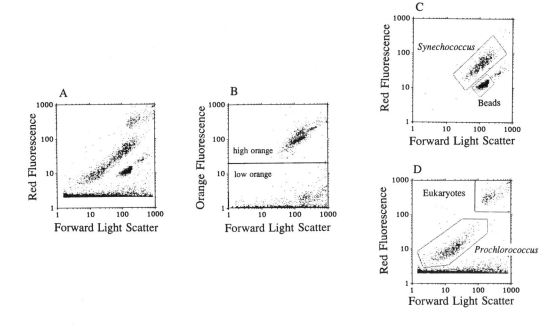

Figure 2. Examples of flow cytometric signatures from a sample taken at Hydro Station S (32°10'N, 64°30'W), southeast of Bermuda, in February 1989. Beads (0.57 μm) were added to the sample. (A) In this scatterplot of forward light scatter (FLS) vs. red fluorescence (CHL), it is difficult to separate the populations from each other. This scatterplot was generated from a list mode file which was gated on red fluorescence to exclude most particles without chlorophyll. (B) If we examine the same data on a scatterplot of FLS vs. orange fluorescence (PE), we can draw a gate separating particles with high orange fluorescence from those with low orange fluorescence. (C and D) Replotting these two subsets of the data on the original axes of FLS vs. red fluorescence allows us to resolve all four populations of interest. *Synechococcus* and beads are marked with windows in (C). Particles above and to the right of the bead singlets are bead doublets and triplets. Eukaryotes and *Prochlorococcus* are separated from each other and from nonfluorescent particles near the baseline in (D).

Initial Settings. A test sample (e.g., a culture of *Synechococcus*) is useful for a rough adjustment of settings: the cells of interest should appear near the center of the scale for each parameter (here, light scatter, chlorophyll fluorescence, and phycoerythrin fluorescence; see Notes and Comments — 3.b.i.). Ideally, *Prochlorococcus* cells (which are smaller and less brightly fluorescent than *Synechococcus* and are thus usually the limiting factor in terms of sensitivity) would also be used as test samples, but these cells are not yet as easily cultured as *Synechococcus* (see Notes and Comments — 4).

Alternatively, beads alone can be used to estimate initial settings. For example, we have found that 0.57-μm Polysciences YG (yellow-green) "Fluoresbrite" beads have about the same amount of "red" fluorescence (when excited with 488-nm light and measured through a 660- to 700-nm bandpass interference filter) as the *Prochlorococcus* in many seawater samples (e.g., Figure 2). (These beads scatter light much more effectively than *Prochlorococcus* cells, so they should be well above the scatter detection limit).

Beads provide an instrument quality control check at the approximate settings for the actual analysis. The CV of the beads should be measured at this point, to ensure that precision is acceptable. The bead CV will depend on the bead quality and the instrument configuration (especially the sample flow rate: in general, for a given laser spot size, precision is inversely related to sample analysis rate). Note that, even with a bead CV for forward light scatter (FLS) as high as 20%, useful data can be obtained (e.g., Figure 2).

Data Storage Format. We collect list mode files of logarithmic signals for all parameters, for maximum flexibility in data analysis. However, this method has the disadvantage of large storage requirements, so the list mode files are "gated" on CHL by specifying a minimum threshold CHL fluorescence which the cell must exceed in order for its measurements to be stored (e.g., Figure 2A). This prevents detritus or noise pulses from filling up the list mode file. To retain information about particles below the gate, ungated data can also be stored as two-parameters scatterplots which require a finite amount of storage space regardless of the number of cells (e.g., we save ungated scatterplots of FLS vs. CHL and FLS vs. PE).

Sample Preparation

Sample Collection and Storage. Water samples are collected using PVC Niskin or GoFlo bottles and should be transferred to acid-cleaned dark plastic bottles (taking care to shield samples from subsurface depths from sunlight) as soon as possible. Surface samples can be obtained with a plastic bucket. The samples can be kept in the dark at room temperature for short periods while awaiting analysis on the flow cytometer. For periods longer than a few hours, they should be preserved. Prokaryotic picoplankton cells can be preserved for several days by refrigeration, but eukaryotic phytoplankton cells deteriorate more quickly under these conditions. For longer-term storage, samples can be fixed with glutaraldehyde (1% final concentration) for 10 min, then frozen in liquid nitrogen.[14] Before analysis, frozen samples can be thawed in a room temperature water bath and should then be run immediately, as the background fluorescence from glutaraldehyde increases quickly. We use 0.1% glutaraldehyde, with results comparable to those obtained with higher concentrations. Paraformaldehyde is a possible alternative to glutaraldehyde since it does not fluoresce.[15]

Prescreening. In productive waters, or with cultures that may clump, samples should be prescreened to remove large aggregates which may clog the flow cell (see Notes and Comments — 5). In oligotrophic waters, or when using a large flow cell (e.g., 250 μm), it is usually not necessary to prescreen the samples.

Bead Addition. We routinely add fluorescently labeled microbeads (calibration grade "Fluoresbrite," Polysciences) as internal standards to our samples (but see Notes and Comments — 6). This allows us to continuously monitor instrument performance during a run, and makes it easy to compare samples run at different instrument settings.

Prepare a working stock of beads for use as an internal standard (see Notes and Comments — 7). Beads should be chosen such that they are on scale on each parameter; this allows comparison of the light scatter and fluorescence values in different samples, by normalizing cell values to bead signals. If beads are not on scale at the settings used, it is possible to calibrate between PMT settings.[16]

Sonicate the working stock of beads to reduce clumping. We normally do this for about 5 min at the beginning of each day, but, since bead stocks may differ, the need for more frequent sonication can be evaluated by noting the relative frequency of bead singlets, doublets, and triplets. Before each use, vortex the tube for 5 s to suspend the beads uniformly. Add a known volume of calibrated bead stock (see Notes and Comments — 8) to the sample and mix by swirling the sample tube.

Flow Cytometric Analysis

Run the sample until you have collected data from a sufficient number (e.g., 10,000) of cells (not just signals, since many could be from detritus you are not interested in). Store the accumulated data to disk.

Data Analysis

After the samples are run on the flow cytometer, the data can be analyzed on the instrument computer or transferred to another computer so that the instrument is available for running more samples. Data analysis techniques will vary depending on the type of sample, the analysis software, and what information is required from the data. Although progress is being made toward automated data analysis,[17] most data analyses still incorporate operator interaction and judgment. If the data were saved as list modes, analysis involves observing multiple two-parameter scatterplots of the data and drawing windows around the populations of interest. The computer software then counts the number of events within each window and calculates statistics on the population (see Notes and Comments — 9).

Here we provide an example of how we analyze a typical open ocean picoplankton sample. We generally classify the cells in a sample into one of three groups: *Synechococcus, Prochlorococcus,* and "eukaryotic picophytoplankton" (see Notes and Comments — 10). If beads were included as an internal standard, there is a total of four categories that we wish to analyze in a typical sample. Since these populations cannot all be distinguished using one or two parameters (e.g., Figure 2), multiple scatterplots or three-dimensional plots are necessary.

Example Analysis

One way to analyze the typical sample is as follows. In a scatterplot of FLS vs. CHL (Figure 2A) the beads (whose position is known, see Initial Settings) can be discriminated from all the phytoplankton and could therefore be isolated at this point. Alternatively, the analysis can begin with a gating based on presence or absence of orange fluorescence (Figure 2B). In a FLS vs. PE scatterplot, *Synechococcus* should appear as a distinct subpopulation (though they may overlap the beads). Once *Synechococcus* and beads have been separated from *Prochlorococcus* and eukaryotic phytoplankton in this manner, all four populations can be isolated on FLS vs. CHL scatterplots (Figure 2C and D).

Now that the four groups have been defined, you can cycle through all the selected plots to refine or check the discrimination. Most software windows use "AND" logic so that the particle must be in the window on all of the plots to be included in the category. Once you are satisfied with the discrimination, have the computer generate the statistics for each category and go on to the next sample. The statistics we use most frequently are number of particles in the category, the mean, the mode, and the CV for each parameter.

The summary statistics can then be loaded into a spreadsheet program for rough plotting and manipulation. Calculate cells per milliliter for each category using Equation 1 (below). Light scatter and fluorescence values should be normalized to the bead standard to allow comparison between settings and between samples.

Normalization and Calibration

Although flow cytometers are capable of making very precise measurements, these measurements are relative ones and their magnitudes are to some degree instrument specific. Comparisons of flow cytometric results between instruments, and indeed between samples, are usually done by referring sample values to signals obtained from reference beads. Note that this only applies to changing the laser power, PMT high voltage, or linear amplifier gain; changing the optical filters may make it very difficult to compare different samples since beads and phytoplankton can have very different emission spectra.[20] Procedures for calibrating the hardware, such as linear amplifiers, PMTs, and logarithmic amplifiers, are described elsewhere[16,21,22] and will not be discussed here. Linearity and sensitivity can also be conveniently tested with commercial sets of precalibrated beads.[23]

Determination of Cell Concentration. Measurement of the sample volume analyzed is accomplished using a stock of beads of known concentration (see Notes and Comments — 8). Even instruments with metered flow rates should be checked with a calibrated stock of beads to correct for coincidence (particles missed because more than one is present in the laser beam) and electronic dead time. At high acquisition rates (as when there are many nontarget particles in the samples), the electronics may miss target cells while they are processing other signal pulses. Having an internal standard of calibrated beads allows you to correct for these missed cells, since the beads should be missed in the same proportion as cells. By adding a known volume of calibrated beads to a sample, the cell concentrations can be calculated as

$$\text{cells ml}^{-1} = (\text{cells counted/beads counted}) * \text{beads ml}^{-1} \tag{1}$$

Determination of Cell Optical Properties. In addition to enumeration, information related to cell size (i.e., from forward light scatter) and pigment content (i.e., from fluorescence) can be normalized for each population, by dividing the cell signal by the bead signal for each parameter of interest. Note that, if logarithmic amplifiers are used, the data are generally linearized before this normalization is made (otherwise, for example, the geometric mean rather than the arithmetic mean will be obtained).

Sorting. The mechanics of sorting are best left to the instrument operator.[24] What is important to understand is that any population which can be defined using gates, as described in the preceding paragraphs, can be physically sorted away from the other particles in the sample. This pure sample of a single class of cell can then be studied further. Sorting can be used to isolate live cells into culture,[18] but be aware that many cells suffer some damage from the stresses of flow cytometry and sorting.[25]

Other Applications. In productive waters, it is sometimes possible to distinguish a variety of larger eukaryotic phytoplankton even with a small sample volume. For example, pennate diatoms and cryptophytes in a sample from Vineyard Sound, MA were separated based on their unique light-scattering and fluorescence characteristics (Figure 3). These groups and coccolithophorids, which depolarize scattered light, are probably the best examples of eukaryotic phytoplankton that can be identified by flow cytometry on the basis of their inherent optical properties.[26]

NOTES AND COMMENTS

1. See Introduction (Disadvantages of flow cytometry) — Since the volume analyzed on most flow cytometers is very small, cells which occur at less than about 10^3 ml^{-1} must be concentrated before analysis. Centrifugation, tangential flow filtration, or gravity filtration onto polycarbonate membrane filters (e.g., Nuclepore) may be satisfactory for qualitative work or preparation of concentrates for sorting. We have concentrated up to 300 l of Sargasso Sea water down to about 15 ml using this latter method (using several sets of 3-μm pore size filters with 47-mm diameter on a set of eight filter rigs). For quantitative work with large rare cells such as coccolithophorids, we have found it necessary to modify the flow cytometer to accommodate larger volumes.[9] Populations of only 25 cells ml^{-1} were analyzed using a modified EPICS V flow cytometer. Sample pickup

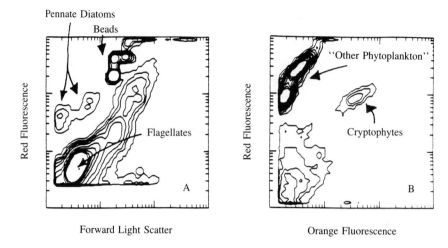

Figure 3.   Flow cytometric signatures of distinctive populations of eukaryotic phytoplankton; 2-ml water samples from Vineyard Sound, MA were analyzed after the addition of 3.79-μm fluorescent beads as standards, and the population identities verified by cell sorting. (A) Small flagellated phytoplankton with low amounts of red (chlorophyll) fluorescence were the most abundant cell type, but there were also two populations of pennate diatoms whose forward light scatter signals were smaller than for other cells with the same chlorophyll. Other populations of larger phytoplankton were present but not identified here. (B) Cryptophytes, with both red and orange (phycoerythrin) fluorescence, were easily distinguished from the other phytoplankton.

and injection needles and tubing of about 1.2-mm inner diameter were used. A standard flow cell body was bored out to accommodate the larger injection needle and the 250-μm square quartz biosense tip was used without restrictor tubing. A 150-mm lens block was used to spread out the laser beam to 750 μm. These modifications allowed sample flow rates of about 10 ml min⁻¹.

2.   See Introduction (Disadvantages of flow cytometry) — Interference by high concentrations of nonfluorescent or autofluorescent nonphytoplankton particles occurs with some samples from the particle maximum in the ocean and with samples (such as old cultures) containing precipitates and/or debris. In cultures, the problem can be minimized by sterile filtration of medium. If the problem is autofluorescence of particles which are not of interest to you, it may be worth trying various combinations of emission filters to optimize the signal-to-noise ratio. If most of the interfering particles are small and dim, try turning up discriminator triggers on forward light scatter (FLS) or the fluorescence signal so that the instrument does not detect those particles. Otherwise, the only solution is to run the sample slowly enough so that coincidence is minimized (see also Notes and Comments — 6).

High count rates (particularly on FLS) can be caused by electronic noise, dirt on the optics or flow cell, or misalignment of the laser or flow cell. Electronic noise can usually only be eliminated by reducing the gain (although operating other equipment such as a sonicator on the same circuit as the flow cytometer can also cause noise). Stray light noise can be minimized by keeping the instrument laboratory darkened during operation. Scattered light is often the main problem during high-sensitivity operation. It can be caused by dirty or misaligned optics or flow cells and can be minimized by a regular maintenance schedule. Sample pickup and injection needles can be cleaned by sonicating in a detergent solution, or changed altogether, as can sample and sheath tubing. The whole fluidics system can be cleaned with laboratory detergent or 10% commercial bleach by filling the sheath tank and flushing, followed by rinsing with particle-free distilled water. Optics should be cleaned sparingly with optical grade methanol and lens paper or cotton swabs. Since many optics are coated, do not clean them unless they need it, to avoid inadvertently scratching the coating.

3. See Equipment — Flow cytometer: There are several commercial flow cytometers available that are being used without any modifications for oceanographic research (price range $50,000 to $250,000). For more information on sources of commercial instruments, see Reference 10. It is also possible to construct a suitable flow cytometer from an epifluorescence microscope[10,11] or a laser and optical table.[12] Building your own instrument may be less expensive, but requires a significant input of time and effort. Considerations in choosing an instrument include:

   a. Light source — The light source must provide light of suitable wavelength and power to excite detectable fluorescence from the photosynthetic pigments of a single picoplankter. In most flow cytometers, excitation light is supplied by a laser. Air-cooled argon ion lasers, the most commonly used, provide 488-nm light in the range of tens of milliwatts, while more expensive water-cooled lasers can provide several other wavelengths, as well as higher power (up to several watts). A few instruments use arc lamps, but the power density of arc lamps is low even compared to low-power lasers, and the fluorescence sensitivity of these instruments is usually considered inadequate for studying picoplankton. The 488-nm line of argon ion lasers common to many commercial instruments is strongly absorbed by both carotenoid and chlorophyl *b* accessory pigments in eukaryotes and by the phycourobilin chromophores in the phycoerythrin of most marine *Synechococcus*. Other pigments such as chlorophyll *a* itself, phycoerythrobilin-rich phycoerythrin, and phycocyanin are less effectively excited by this wavelength, but in many cases suboptimum absorption can be compensated by high excitation intensity, so that, for example, even *Synechococcus* without phycourobilin can be detected with 488 nm excitation.

   b. Detectors —

      i. Fluorescence: The flow cytometer should be able to simultaneously measure at least two colors of fluorescence. The ability to change the optical filters in front of the photomultiplier tubes (PMTs) allows greater versatility in applications and in optimizing the signal-to-noise ratio. However, instruments with fixed filters configured to measure fluorescence in the red (e.g., 650-nm longpass or 660 to 700 bandpass, for chlorophyll) and orange (e.g., 560 to 600 bandpass, for phycoerythrin) regions of the spectrum have proved satisfactory for picoplankton discrimination.

      ii. Light scattering: The ability to measure light scattering allows one to examine nonfluorescent as well as fluorescent particles. Particle size can be related to measurements of laser light scattered in the near-forward direction (FLS); light scattered at larger angles (90° light scatter, 90LS), while influenced by size, also provides information about refractive index and/or surface roughness of the particles. Forward light scatter detectors in commercial instruments are often photodiodes rather than PMTs and, if low power lasers are used, these may be inadequate for detecting light scattered by picoplankton-sized cells. Depending on the flow cytometer, 90LS may be used in addition to or in place of FLS.

   c. Data collection — The use of logarithmic rather than linear amplifiers on all signals extends the dynamic range of measurements to 3 to 4 decades, as opposed to 2 decades or less with typical linear amplifiers. This is important in dealing with natural populations where signal sizes encountered span several orders of magnitude. "List mode" data collection (in which every signal from every cell is stored separately) is more versatile than one-parameter histograms or two-parameter scatterplots, and is especially valuable in exploratory analysis and when increasing numbers of parameters are measured.

4. See Initial Settings — Live standards: Cultures with known characteristics (for example, *Synechococcus* with or without phycourobilin, *Prochlorococcus*, coccolithophorids with

or without liths, cryptophytes, and pennate diatoms) should be run to verify that the system is discriminating the cell types in which you are interested. Cultures can be obtained from the Provasoli-Guillard Center for Culture of Marine Phytoplankton, Bigelow Laboratory for Ocean Science, West Boothbay Harbor, ME 04575.

5. See Prescreening — Prescreening can be accomplished using a 3-ml syringe and an 18-gauge needle whose point has been filed or cut off and covered with Nitex mesh (we use 53-μm mesh when using the 76-μm jet-in-air Coulter flow cell tip). The mesh is held in place with a short piece of plastic tubing slipped over the mesh and needle. The sample is drawn up through the mesh, the needle removed, and the sample ejected into a clean tube.

6. See Bead Addition — Nonfluorescent particles: If you are concerned with analyzing nonfluorescent particles in a sample, be aware that there are often nonfluorescent contaminant particles in bead stocks. Beads can be cleaned up by centrifuging and removing the supernatant along with the small contaminants, then resuspending the beads. To avoid any added particles, samples can be run without internal standards. In this case, check the instrument frequently by running beads as external standards between samples, to ensure it is operating properly, and each time any settings are changed. If you are not using a metered flow instrument, it will be necessary to measure the volume (gravimetrically or volumetrically) of each sample. In this case, underestimation due to cells missed during electronic ''dead time'' must be considered; the analysis rate should be kept low relative to the instrument's electronic processing capability.

7. See Bead Addition — Preparation of bead working stock: Prepare a working stock of beads by diluting the bead stock with distilled water (note that some kinds of beads will disintegrate over a period of hours to days if resuspended in seawater). The dilution factor will vary with bead manufacture and bead size (they are generally sold as percent solids so that stocks of large beads have fewer beads per milliliter). Working stock concentration should ensure that the number of beads analyzed with each sample is sufficient to obtain good counting statistics, but not so high that the beads are the most common particles in the sample (i.e., causing limitation of disk space or computer memory). A good number to aim for is several thousand beads analyzed when you add 5- to 10-μl beads to a 1-ml sample.

8. See Bead Addition — Calibration of bead working stock:

    a. Make up a bead calibration sample exactly as your experimental samples, using filtered seawater. To minimize coincidence so that every bead is counted, minimize noise pulses by turning down the laser power, amplifiers, or PMT high voltages (but make sure you can count the beads on scale).

    b. Establish sample flow and allow it to stabilize (this fills the instrument ''dead volume'' with sample).

    c. Remove the calibration sample and weigh it on a balance that measures to 1 mg.

    d. Replace the sample on the instrument and collect data for a suitable length of time (approximately 10 min).

    e. Stop the data collection and sample flow at the same time, and immediately reweigh the sample. The difference in the weights divided by the density of the medium gives the volume run. (Slightly less accurate volumes can be determined by withdrawing the remaining sample volume with a series of calibrated micropipets; this is useful at sea where milligram balances are impractical).

    f. Analyze the sample file to count the beads run and determine the concentration of beads in the sample.

Working stock beads should be calibrated daily. It is possible to obtain CVs of <5% for successive bead calibrations. Beads are kept refrigerated; diluted working stocks in particular seem to degrade more rapidly at room temperature. Bacteria can also grow in diluted stocks. If stored refrigerated and kept on ice during use, working stocks can last a year or more, although the calibration may change with time. Note that multiple pipeting with air displacement pipets from cold liquids can cause noticeable inaccuracies; thus, it is important to use a consistent method of pipeting.

9. See Data Analysis — The software which is supplied with most instruments and much commercial software is geared toward clinical needs, but it can be used for analyzing phytoplankton samples as well. It is convenient to be able to use nonrectangular windows for defining the populations, and list modes allow much greater flexibility in using multiparameter data to separate populations from each other and background noise.

10. See Data Analysis — Any of the three main groups that we classify can have subpopulations. A good example is the presence of bright and dim *Synechococcus* cells in many coastal water stations. These can be analyzed by dual beam flow cytometry.[18,19]

ACKNOWLEDGMENTS

We thank Sallie Chisholm, Sheila Frankel, Brian Binder, and Jeff Dusenberry for valuable discussions and advice. This work was supported in part by NSF grants OCE-8316616, OCE-8421041, OCE-8508032, OCE-8814332, OCE-9012147, OCE-9014724, OCE-9024380, and by ONR contracts N00014-83-K-0661, 84-C-0278, 87-K-0007, and 89-J-1110.

REFERENCES

1. Chisholm, S. W., Armbrust, E. V., and Olson, R. J., The individual cell in phytoplankton ecology: cell cycles and flow cytometry, in *Photosynthetic Picoplankton,* Platt, T. and Li, W. K. W., Eds., *Can. Bull. Fish. Aquat. Sci.,* 214, 343, 1986.
2. Amaral, L. A., Lim, E. L., Caron, D. A., and DeLong, E. F., The Use of Ribosomal RNA-Based Oligonucleotide Probes for Examining Natural Assemblages of Nanoplanktonic Protists, ASLO Aquatic Sciences Meeting Program Abstracts, Santa Fe, NM, February 9 to 14, 1992.
3. Amman, R. I., Binder, B. J., Olson, R. J., Chisholm, S. W., Devereux, R., and Stahl, D. A., Combination of 16S rRNA-targeted oligonucleotide probes with flow cytometry for analyzing mixed microbial populations, *Appl. Environ. Microbiol.,* 56, 1919, 1990.
4. Olson, R. J., Vaulot, D., and Chisholm, S. W., Effects of environmental stresses on the cell cycle of two marine phytoplankton species, *Plant Physiol.,* 80, 918, 1986.
5. Vaulot, D., Olson, R. J.,, and Chisholm, S. W., Light and dark control of the cell cycle in two marine phytoplankton species, *Exp. Cell Res.,* 167, 38, 1986.
6. Vaulot, D. and Partensky, F., Cell cycle distributions of prochlorophytes in the Northwestern Mediterranean Sea, *Deep-Sea Res.,* 39, 727, 1992.
7. Ward, B. B., Immunology in biological oceanography and marine ecology, *Oceanography,* 3(1), 30, 1990.
8. Yentsch, C. M. and Pomponi, S. A., Automated individual cell analysis in aquatic research, *Int. Rev. Cytol.,* 105, 183, 1986.
9. Olson, R. J., Zettler, E. R., Chisholm, S. W., and Dusenberry, J. A., Advances in oceanography through flow cytometry, in *Particle Analysis in Oceanography,* Demers, S., Ed., Springer-Verlag, Berlin, 1991, 351.
10. Shapiro, H. M., *Practical Flow Cytometry,* 2nd ed., Alan R. Liss, New York, 1988.
11. Olson, R. J., Frankel, S. L., and Chisholm, S. W., An inexpensive flow cytometer for the analysis of fluorescence signals in phytoplankton: chlorophyll and DNA distributions, *J. Exp. Mar. Biol. Ecol.,* 68, 129, 1983.
12. Frankel, S. L., Binder, B. J., Chisholm, S. W., and Shapiro, H. M., A high-sensitivity flow cytometer for studying picoplankton, *Limnol. Oceanogr.,* 35, 1164, 1990.
13. Sherr, E. and Sherr, B., Protistan grazing rates via uptake of fluorescently labeled prey, in *Current Methods in Aquatic Microbial Ecology,* Kemp, P. F., Sherr, B. F., Sherr, E. B., and Cole, J. J., Eds., CRC Press, Boca Raton, FL, 1993, chap. 80.
14. Vaulot, D., Courties, C., and Partensky, F., A simple method to preserve oceanic phytoplankton for flow cytometric analyses, *Cytometry,* 10, 629, 1989.

15. Campbell, L., personal communication, 1991.

16. Durand, R. E., Calibration of flow cytometer detector systems, in *Methods in Cell Biology,* Vol. 33, Darzynkeiwicz, Z. and Crissman, H. A., Eds., Academic Press, New York, 1990, chap. 53.

17. Frankel, D. S., Olson, R. J., Frankel, S. L., and Chisholm, S. W., Use of a neural net computer system for analysis of flow cytometric data of phytoplankton populations, *Cytometry,* 10, 540, 1989.

18. Olson, R. J., Chisholm, S. W., Zettler, E. R., and Armbrust, E. V., Analysis of *Synechococcus* pigment types in the sea using single and dual beam flow cytometry, *Deep-Sea Res.,* 35, 425, 1988.

19. Olson, R. J., Chisholm, S. W., Zettler, E. R., and Armbrust, E. V., Pigments, size, and distribution of *Synechococcus* in the North Atlantic and Pacific Oceans, *Limnol. Oceanogr.,* 34, 45, 1990.

20. Vaulot, D., Filter characteristics and standardization in flow cytometry, *Signal Noise,* 2(2), 1989.

21. Bagwell, C. B., Baker, D., Whetstone, S., Munson, M., Hitchcox, S., Ault, K. A., and Lovett, E. J., A simple and rapid method for determining the linearity of a flow cytometer amplification system, *Cytometry,* 10, 689, 1989.

22. Gandler, W. and Shapiro, H., Logarithmic amplifiers, *Cytometry,* 11, 447, 1990.

23. Vogt, R. F., Cross, D. C., Phillips, D. L., Henderson, L. O., and Hannon, W. H., Interlaboratory study of cellular fluorescence intensity measurements with fluorescein-labeled microbead standards, *Cytometry,* 12, 525, 1991.

24. Dean, P. N., Helpful hints in flow cytometry and sorting, *Cytometry,* 6, 62, 1985.

25. Rivkin, R. B., Phinney, D. A., and Yentsch, C. M., Effects of flow cytometric analysis and cell sorting on photosynthetic carbon uptake by phytoplankton in cultures and from natural populations, *Appl. Environ. Microbiol.,* 52, 935, 1986.

26. Olson, R. J., Zettler, E. R., and Anderson, O. K., Discrimination of eukaryotic phytoplankton cell types from light scatter and autofluorescence properties measured by flow cytometry, *Cytometry,* 10, 636, 1989.

Additional References

Chisholm, S. W., Olson, R. J., Zettler, E. R., Goericke, R., Waterbury, J. B., and Welschmeyer, N. A., A novel free-living prochlorophyte abundant in the oceanic euphotic zone, *Nature,* 334, 340, 1988.

Chisholm, S. W., Frankel, S. L., Goericke, R., Olson, R. J., Palenik, B., Waterbury, J. B., West-Johnsrud, L., and Zettler, E. R., *Prochlorococcus marinus nov.* gen. *nov.* sp.: an oxyphototrophic marine prokaryote containing divinyl chlorophyll *a* and *b, Arch. Microbiol.,* 157, 297, 1992.

Darzynkeiwicz, Z. and Crissman, H. A., Eds., *Methods in Cell Biology,* Vol. 33, Academic Press, New York, 1990.

Dubelaar, G. B. J., Groenewegen, A. C., Stokdijk, W., van den Engh, G. J., and Visser, J. W. M., Optical plankton analyser: A flow cytometer for plankton analysis. II. Specifications, *Cytometry,* 10, 529, 1989.

Li, W. K. W. and Wood, M., Vertical distributions of North Atlantic ultraphytoplankton: analysis by flow cytometry and epifluorescence microscopy, *Deep-Sea Res.,* 35, 1615, 1988.

Neveux, J. D., Vaulot, D., Courties, C., and Fukai, E., Green photosynthetic bacteria associated with the deep chlorophyll maximum of the Sargasso Sea, *C. R. Acad. Sci.,* 308, 9, 1989.

Olson, R. J., Chisholm, S. W., Zettler, E. R., Altabet, M. A. and Dusenberry, J. A., Spatial and temporal distributions of prochlorophyte picoplankton in the North Atlantic Ocean, *Deep-Sea Res.,* 37, 1033, 1990.

Van Dilla, M. A., Dean, P. N., Labrum, O. D., and Melamed, M. R., *Flow Cytometry: Instrumentation and Data Analysis,* Academic Press, New York, 1985.

Veldhuis, M. J. W. and Kray, G. W., Vertical distribution and pigment composition of a picoplanktonic prochlorophyte in the subtropical North Atlantic: a combined study of HPLC-analysis of pigments and flow cytometry, *Mar. Ecol. Prog. Ser.,* 56, 177, 1990.

CHAPTER 23

Enumeration of Phototrophic Picoplankton by Autofluorescence Microscopy

Erland A. MacIsaac and John G. Stockner

INTRODUCTION

Phototrophic picoplankton, or picophytoplankton, were a little known part of the marine and freshwater phytoplankton prior to their discovery as an abundant and ubiquitous component of many plankton communities.[1,2] Prokaryotic cyanobacteria and prochlorophytes and a variety of eukaryotes have now all been identified as members of the picophytoplankton.[3,4] Phototrophic picoplankton are responsible for a significant fraction of total primary production in many pelagic environments and they have been implicated as important conduits for carbon, nutrient, and energy cycling in microbial foodwebs in plankton communities.[5]

Epifluorescence microscopy has evolved into a relatively routine method for quantifying the abundance and biomass of phototrophic picoplankton in both marine and fresh waters. Overlap of cell sizes with heterotrophic bacteria originally made microscopic identification of picophytoplankton difficult before the adoption of epifluorescent techniques for bacteria counting.[6] By utilizing the natural autofluorescence of chlorophyll and phycobilin photosynthetic pigments, phototrophic cells were easily identified among the heterotrophic picoplankton.

Autofluorescence techniques are relatively simple to learn and apply. Sample enumeration is straightforward, lacking much of the subjectivity that afflicts epifluorescence bacteria enumerations using fluorochrome stains. Phototrophic picoplankton are easily separated by their autofluorescence from heterotrophic bacteria, larger phytoplankton, and detritus particles in samples. Different types of picophytoplankton with different photopigment compositions and fluorescence emissions can also be identified by changing the excitation wavebands used for epifluorescence enumeration.

However, the technique requires investment in an epifluorescence microscope and various filter/dichroic-mirror sets, and only picophytoplankton cells with relatively healthy pigment complements can be enumerated. Some picophytoplankton are reportedly difficult to enumerate due to very weak photopigment fluorescence. Autofluorescence also offers few morphological details for further identification or classification of cells. Preservation and long-term storage of samples can be difficult for some fragile picophytoplankton and samples are easily and irreversibly destroyed by mishandling during processing and storage.

The following describes the most common methods used to handle and enumerate picophytoplankton by autofluorescence microscopy. Many of the common pitfalls and technical

0-87371-564-0/93/$0.00 + $.50

refinements are described. For plankton ecologists interested in quantifying total phytoplankton community composition and biomass, autofluorescence picophytoplankton techniques are an essential companion technique to the conventional Utermöhl techniques for nano- and microphytoplankton enumeration.

MATERIALS REQUIRED

- Epifluorescence microscope with
    - 100× high numerical-aperture, flat-field, oil-immersion objective
    - High-pressure 50 to 200 W mercury, or 75 to 150 W xenon lamp
    - 8 to 12.5× eyepiece oculars with micrometer and counting grid
    - Stage micrometer
    - Exchangeable filter/dichroic-mirror sets (specifications follows)
    - 50% neutral density filter
- Glass slides and cover slips
- Nonfluorescent immersion oil (e.g., Cargille type A)
- 25-mm filterholders and filtration manifold

PROCEDURES

Selecting Excitation Filter/Mirror Sets

Different combinations of excitation filters and dichroic mirrors are used to isolate specific wavebands of light and excite the photopigments characteristic of different classes of phytoplankton.[7-9] The selective excitation of any photopigment depends on matching the optimum light absorbtion waveband of the pigment to the intensity and waveband of the excitation light. Shorter wavelengths of light (<600 nm) are used for excitation and the longer fluorescence wavelengths (>550 nm) for microscopic observation. The primary autofluorescent pigments used to enumerate picophytoplankton are the chlorophylls (chlorophyll *a* and divinyl-chlorophyll *a*) and the phycobilin pigments (phycoerythrins and phycocyanins). Although phytoplankton carotenoids are not generally autofluorescent, they do absorb and transfer light energy from longer wavelengths to chlorophyll *a* fluorescence in live cells. Similarly, chlorophylls *b* and *c* are not significantly autofluorescent *in vivo* but they also absorb and transfer light energy to chlorophyll *a* fluorescence. Thus, the light absorption and fluorescence emission spectra of living cells, in fresh preparations for epifluorescence microscopy or flow cytometry fluorescence analysis, can differ markedly from those of preserved cells.[10,11] Chemical fixation or freezing uncouples the transfer of light energy from the accessory pigments to chlorophyll *a*, resulting in light absorption and fluorescence emission spectra for cells that reflect those of the individual autofluorescent pigments. Since fixation and preservation of field samples is usually required for routine autofluorescence picophytoplankton enumeration, the recommended excitation filter/mirror sets and enumeration procedures are specifically tailored toward the excitation and fluorescence characteristics of the individual primary autofluorescent pigments of fixed or dead cells.

Chlorophyll *a* (or divinyl-chlorophyll *a* for prochlorophytes), present in all picophytoplankton, is most strongly excited by violet-blue (420 to 440 nm) wavelengths and fluoresces deep red (670 to 690 nm).[4] In preserved samples, degradation of chlorophylls to weaker fluorescing pheophytins that absorb at shorter wavelengths (400 to 430 nm) can occur during fixation, freezing, or storage.

The phycobilin accessory pigments of cyanobacteria picoplankton encompass a variety of phycoerythrins and phycocyanins.[9] Phycocyanin and allophycocyanin absorb excitation light in the orange-red wavelengths (620 to 650 nm) and overlap their bright red fluorescence emission (640 to 660 nm). The absorbtion and fluorescence emission maxima of phyco-erythrins depend on the type of chromophores in the phycobiliprotein. Phycoerythrin that contains both phycourobilin (PUB) and phycoerythrobilin (PEB) chromophores (type I) has absorption maxima at 490 to 500 nm and 540 to 565 nm due to each respective chromophore, while type II phycoerythrin has only the PEB chromophore and its 540 to 565 nm absorbtion maxima. Both types of phycoerythrins absorb green light and fluoresce yellow-orange (550 to 580 nm) while only type I phycoerythrin also absorbs blue light efficiently, and either may dominate the photopigment autofluorescence of cyanobacteria picophytoplankton. In contrast to chlorophyll, phycobilins may increase in fluorescence immediately after pres-ervation due to the uncoupling of energy transfer to chlorophyll *a* or alteration of their protein complexes.[12]

Two filter sets are recommended for use during picophytoplankton enumeration, one for excitation of the chlorophylls and type I phycoerythrin, and one for excitation of type II phycoerythrin and the phycocyanins. Although various specialized filter sets designed for specific individual pigments can be used, particularly for studying the pigment characteristics of living cells, our focus is on the minimum complement required for a routine autoflu-orescence enumeration technique applicable to all cells. The blue waveband (450 to 500) filter set commonly used for chlorophyll *a* and phycoerythrin excitation[13] is not recom-mended, especially for epifluorescence microscopes equipped with high-pressure mercury lamps. First, the light energy of a mercury lamp is concentrated into high-intensity emission lines at 406, 435, 546, and 578 nm, with relatively low-energy emission in the blue wave-band. A mercury lamp is also less effective for blue-light excitation than a comparable intensity xenon lamp due to the flatter and higher average emission spectrum of the latter in the blue waveband. Second, the optimum wavelengths for maximum chlorophyll *a* ab-sorption and excitation are actually in the violet (400 to 450 nm) waveband. Coupled with the presence of two high-intensity mercury emission lines in the violet at 406 and 435 nm, a violet-blue (400 to 500 nm) excitation filter set is recommended for efficient excitation of chlorophyll *a* (and divinyl-chlorophyll *a*), as well as cyanobacteria with type I phycoer-ythrin. The violet-blue waveband also encompasses the short-wave shifted absorption peaks for the pheopigments that can result from degradation of chlorophylls in preserved samples. Thus, for picophytoplankton samples degraded by preservation or storage, violet-blue ex-citation is more effective at exciting the autofluorescence of the remaining pheophytins. A typical filter set for violet-blue excitation consists of a 395-nm longwave pass filter coupled with a 500-nm shortwave pass filter, a 510-nm dichroic mirror, and a 520-nm longwave pass barrier filter.

The second filter set isolates a green-yellow waveband (520 to 560 nm) for excitation of phycocyanins and type II phycoerythrin. A 520-nm longwave pass filter is coupled with a 560-nm shortwave pass filter, a 580-nm dichroic mirror, and a 590-nm longwave pass barrier filter. Although optimum phycocyanin excitation typically occurs at wavelengths >560 nm, the absorption peaks are broad enough that sufficient excitation light can usually be delivered at the shorter wavelengths for adequate autofluorescence. Similarly, although the optimum fluorescence emission of type II phycoerythrin occurs at wavelengths <590 nm, there is usually sufficient fluorescence at longer wavelengths to allow discrimination. The phycobilin pigments are relatively robust in their excitation and fluorescence characteristics. By switch-ing between the violet-blue and the green-yellow filter sets during enumeration of a micro-scope field, differences in the autofluorescence of cells are used to indicate the predominance of the different types of photopigments.

The intensity of excitation light is a function of the excitation bandwidth and the emission spectra of the lamp. It also depends on the percent light transmission of the excitation filters and dichroic mirror and the various lenses in the microscope. It is important that all components of the microscope light path be selected to optimize the effiency of light transmission. Generally, combinations of sharp cutoff longwave pass and shortwave pass excitation filters are more efficient at light transmission than bandpass filters. Flat-field, high numerical-aperture objectives, specifically designated for fluorescence microscopy, should only be used. Other components in the microscope light path such as coated lenses or internal filters may selectively absorb certain wavelengths of excitation and fluorescent light, interfering in the technique. It is particularly important to consider all components of the microscope when retrofitting an existing compound microscope for epifluorescence illumination.

Selection and Preparation of Membrane Filters

Polycarbonate membrane (e.g., Nuclepore, Unipore) or aluminum oxide (Anotech) filters are normally used for picoplankton enumeration because cells can be concentrated into a flat field of view for easy microscopic analysis. Both filter types are first dyed black to reduce background fluorescence, although predyed black polycarbonate filters are available commercially. Filters are soaked for >48 h in 0.2-μm filtered Irgalan black (acid black 107, Ciba-Geigy) solution (2 g l^{-1} in 2% acetic acid). Dyed filters may be rinsed and used fresh or they can be dried and stored in their original packages. Alternate dyes used for polycarbonate filters includes Sudan Black (type B) dissolved in absolute ethanol and diluted to 1:15000 (w/v) with 50% ethanol or a hot (60 to 70°C) solution of 2.5 g l^{-1} Ebony Black (Dylon #8) dissolved in 95°C 0.25% (w/v) NaCl. Dye solutions should be prefiltered to remove particulates. Some destaining and increased filter background fluorescence may occur when acidic (e.g., humic-stained) waters are filtered or wetting agents are used in sample preparation.

Although 0.2-μm pore size polycarbonate filters are routinely used, significant numbers of picophytoplankton cells have occasionally been found in their filtrates.[14] Smaller pore size (0.1-μm) polycarbonate filters can be used; however, the lower flow and filter loading rates greatly reduce the filterable sample volumes. Alternatively, aluminum oxide filters have been reported as superior to polycarbonate filters for picoplankton retention,[15] and 0.1-μm aluminum oxide filters can be used routinely or as a check. Their high pore density relative to polycarbonate filters allows higher filter loading rates and ensures that adequate sample volumes can be filtered despite the smaller pore size. Filters of cellulose nitrate or mixed ester, although lacking the flat surface of the polycarbonate or aluminum oxide filters, having also been used with apparent success[16] although no quantitative comparisons with the other filters have been reported for picophytoplankton enumeration.

The collection efficiency of any filter for field populations of picophytoplankton is determined not only by the nominal pore size of the filter but also by the types of picophytoplankton cells present (size and flexibility), the surface charge (electrostatic) adsorption of cells by the filter, and the vacuum pressure.[17] Filter adsorption of cells is affected by the pH and ionic strength of the water sample and sampling protocols using aldehyde fixation prior to filtration may reduce cell passage by fixing the cell walls of small, flexible picophytoplankton. Aldehyde fixatives can also affect the pH and ionic characteristics of the sample water; thus, filter bypass of picophytoplankton is a phenomenon dependent on both the sampling environment and the methods used for collection and fixation of picophytoplankton.

Preparation of Slides

1.   Water samples, freshly collected or stored at <4°C in the dark for a few hours, are briefly aldehyde fixed immediately prior to filtration (see Sample Preservation).

2. Aliquots are filtered onto predyed, 25-mm, polycarbonate membrane or aluminum oxide filters in 25-mm filterholders. A support membrane filter (e.g., 0.45-μm cellulose ester) promotes an even vacuum and distribution of cells over the filter surface. Vacuum pressure should be less than 5 kPa. All filtration and subsequent handling is carried out under subdued light to protect the pigments from photobleaching. Although typically 10- to 25-ml sample volumes are used for oligo- and mesotrophic marine and freshwaters, the aliquot filtered should be as high as practically possible and only limited by the loading capacity of the filter for the sampled waters. Picophytoplankton are rarely so abundant in natural waters that cell densities are too high in the microscope field for easy enumeration. High microscope field densities significantly reduce enumeration times and improve counting statistics, particularly for rare cell types.

3. Immediately after all water has drawn through, the moist filter is removed from the filter holder under reduced vacuum and placed on a glass slide.

4. A small drop of nonfluorescent immersion oil is placed on top of the filter and a cover slip added and gently pressed in place with forceps to exclude trapped air. At this point, the filters are prepared for enumeration or may be immediately frozen for long-term storage (see Sample Preservation).

5. Another drop of oil is placed on the cover slip for enumeration under 100× oil immersion.

Identification of Morpho-Types

Three main groups of phototrophs (cyanobacteria, prochlorophytes, and eukaryotes) have been found in marine and freshwater picoplankton communities,[2-4] but they can be difficult to distinguish under the epifluorescence microscope. Only general cell size, shape, and fluorescent color are available to classify cells. The fluorescence characteristics of the cells under different excitation wavebands is also complicated by changes in the complements of autofluorescent pigments due to prior growth conditions of the cells. Although other types of photosynthetic bacteria besides *Synechococcus* (e.g., *Rhodospirillaceae*, Chlorobiaceae) contain autofluorescent photopigments (bacteriochlorophyll, carotenoids) and can be enumerated by autofluorescence techniques,[18] they are usually restricted to specialized anaerobic boundary environments and their relatively large cell sizes would typically exclude them from picophytoplankton enumerations.

Cyanobacteria are the most common reported picophytoplankton, largely due to their ease of enumeration, and they exhibit the greatest diversity of form in field samples (Figure 1). They include unicellular coccoid, ovoid, or rod-shaped *Synechococcus* spp. (0.5 to 1.5-μm diameter); a variety of microcolonial forms of 4 to 50 *Synechococcus*-like cells in mucilaginous or tightly clustered aggregates; and filamentous *Oscillatoria* types ranging from 4 to 30 μm in length. The microcolonial and filamentous forms are more common in freshwater. Depending on their dominant phycobilin photopigment, they fluoresce brightly under either the violet-blue or green-yellow excitation filter sets.

The only planktonic prochlorophytes currently known are small (0.3- to 1.0-μm diameter), unicellular, coccoid, prokaryotic cells distinguishable from *Synechococcus* cells only by their unique pigment complement and distinctive ultrastructure in TEM thin sections.[4] They lack phycobilin pigments but possess unique divinyl-chlorophyll *a* and accessory pigments. Currently, they have only been found in oceanic waters where they were an important component of the picophytoplankton. Fresh cells are reportedly difficult to enumerate by autofluorescence microscopy using blue light excitation; however, excitation of fixed cells, by a violet-blue waveband with optimum emission in the absorbtion maximum for divinyl-chlorophyll *a*, might significantly improve their autofluorescence.

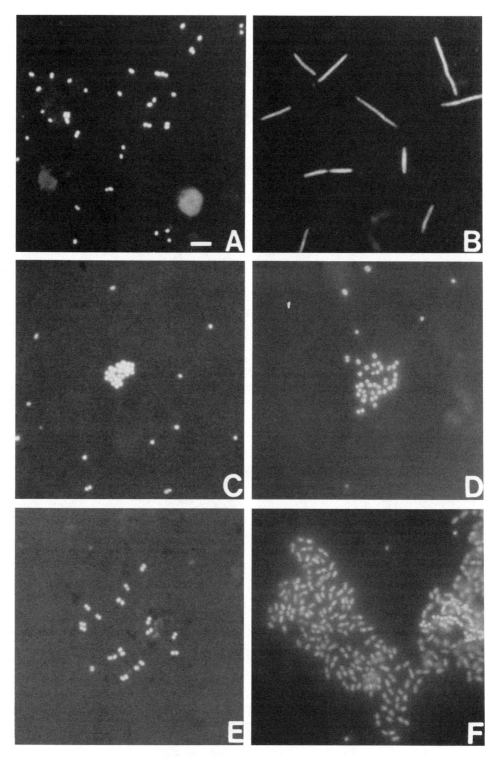

Figure 1. The diversity of forms found in field populations of cyanobacteria picophytoplankton. Bar in (A) equals 5 μm for all. Unicellular *Synechococcus* spp. (A) common to fresh and marine waters, *Oscillatoria*-type filaments (B), tightly clustered microcolonial aggregate of *Synechococcus*-type cells (C), loose microcolonial mucilaginous aggregate (D), paired (*Aphanocapsa*-like) cells in a mucilaginous microcolony (E), and large *Aphanothece*-like colony (F). *Oscillatoria*-type and microcolonial forms are found largely in fresh water.

Eukaryotic picophytoplankton are generally unicellular chlorococcoid or flagellated cells, although flagella are not visible under the epifluorescence microscope without using ancillary fluorochrome stains. Their deep red chlorophyll a fluorescence under violet-blue excitation is usually restricted to regions of distinct chloroplasts within the cell and accurate morphological details are usually lacking. However, the localization of chlorophyll in chloroplasts is a diagnostic feature useful for distinguishing eukaryotes from prokaryotic phototrophs since the photopigments of the latter are typically dispersed in peripheral thylakoids throughout the cell.

A routine enumeration protocol involves first locating the filter surface and picophytoplankton cells under violet-blue excitation. Eukaryotes and prochlorophytes both fluoresce deep red due to chlorophyll a and divinyl-chlorophyll a, but eukaryotic cells are distinguished by localization of chlorophyll fluorescence within the cell. Only cyanobacteria with type I phycoerythrin fluoresce (bright yellow-orange) under violet-blue excitation and they show no cellular localization of fluorescence. Switching to green-yellow excitation for the same field reveals cyanobacteria with phycocyanin (red) or type II phycoerythrin (orange) as their dominant phycobilin pigment. These cells may be weakly visible under violet-blue excitation but show greatly enhanced autofluorescence using the green-yellow waveband.

Enumeration Procedures

Using the above protocol for identifying picophytoplankton types, cells are enumerated by random fields at a magnification of $1000\times$ to $1250\times$ using a $100\times$ objective. Counting statistics are similar to those for fluorochrome enumeration of heterotrophic bacteria (see Reference 19 and Chapter 58). Generally, 100 to 200 cells of each of the dominant picophytoplankton types are counted, with an upper limit set at 20 to 30 random microscope fields when picophytoplankton abundances are very low. The density of cells on the filter is often fixed before enumeration, thus the area of the microscope field is varied by using a counting grid or portion of it for the most abundant cell types and the entire field and additional fields for rare forms. One or two carefully placed transects across the filter can also be used if cell densities are very low or for counting rare, large microcolonial forms. Increasing the field area by reducing the objective magnification also reduces the excitation light and cell fluorescence intensity and can only be used for highly fluorescent cells.

Estimate cell volumes by measuring the dimensions of subsets of cells with the eyepiece micrometer or by scoring cells during enumeration into predetermined cell types and size classes using the counting grid or eyepiece micrometer as a size reference. Photomicrographs or slide projections can also be used to determine cell dimensions. Cell volumes are calculated using formulae for geometric shapes (e.g., spheres, prolate spheroids, cylinders) approximating the shapes of the cells.

Sample Preservation

Picophytoplankton samples that cannot be enumerated immediately require preservation to minimize degradation of photopigment autofluorescence and to protect the cells from lysis during storage. The success of any preservation technique will depend on the types of picophytoplankton present in the sample (cyanobacteria, eukaryotes, and prochlorophytes) since each has its own unique pigment complement and susceptibility to preservation and handling losses. Picocyanobacteria are the easiest types of cells to preserve since these prokaryotic cells can often withstand freezing, filtration, and even desiccation on filters, without prefixation.[12,20,21] However, many eukaryotic picoplankton such as flagellates that lack a rigid cell wall cannot survive filtration or freezing without fixation.[13] Preservation of prochlorophytes for epifluorescence enumeration has not been investigated in detail for this relatively new group or picophytoplankton.[12] In fact, they are reported to be very difficult to enumerate by epifluorescent techniques due to weak photopigment autofluorescence and most of our knowledge about them comes from flow cytometry fluorescence studies.

Various sample preservation methods have been compared for autofluorescence enumeration of picophytoplankton and other phytoplankton, and for closely related flow cytometry studies with similar preservation requirements for phytoplankton cells and autofluorescence.[12,16,20,22] The latter specifically requires whole water samples for analysis. Although fixation with aldehydes protects fragile cells during storage and filtration, they have a concentration- and time-dependent degradation effect on pigment fluorescence. Water samples fixed with formaldehyde or glutaraldehyde and refrigerated in the dark quickly lose chlorophyll and phycobilin autofluoresence within a few days to several weeks. Brief fixation with low concentrations of formaldehyde or glutarldehyde followed by quick freezing of prepared slides or whole water samples generally produces the best preservation of cells and pigment autofluorescence for most picophytoplankton. Prefixation with 0.2% (v/v) formaldehyde and $-40°C$ storage of prepared slides was significantly more effective at preserving picophytoplankton, particularly eukaryotes, than either fixation with 2 to 4% glutaraldehyde/formaldehyde or freezing of prefixed whole water samples.[16] However, quick freezing of whole water samples in liquid nitrogen ($-70°C$) prefaced by a 10-min fixation with 1% glutaraldehyde was very effective at long-term preservation of the photopigment autofluorescence and cell structure of prochlorophytic picoplankton, as well as most cyanobacteria and eukaryotes, for flow cytometry analysis.[12]

Either formaldehyde or glutaraldehyde can be used as fixatives; however, differences in their fixation properties dictate different protocols in their application to samples. Picophytoplankton samples require exposure to either aldehyde at a concentration and duration to both arrest biological activity in the sample and fix picophytoplankton cell membranes and contents, yet minimize their penetration and fixation (degradation) of the fluorescent photopigments. Formaldehyde penetrates cells more quickly than glutaraldehyde, and the concentrations or exposure times used should be lower. Prefixation protocols successfully used for picophytoplankton range from 10 min with 1% glutaraldehyde to 1 h with 0.2% formaldehyde.

Fresh formaldehyde solutions are prepared from paraformaldehyde powder instead of commercially prepared solutions because the latter may have an undesirably low pH and usually contain methanol or other impurities that contribute to degradation of photopigment fluorescence. A 2% paraformaldehyde stock solution is prepared by dissolving 2 g paraformaldehyde in 96 ml of 0.2-μm filtered sample water at 70°C. Low-alkalinity freshwaters may require addition of a few drops of 1 N NaOH to aid dissolution. The pH of the stock solution should be checked and adjusted to 7.0 to 8.0 if necessary with NaOH and stored at 4°C. The stock is then used as is or diluted with filtered sample water prior to addition to the samples to achieve the desired concentration. Glutaraldehyde should be electron-microscope grade, preferably dated as fresh, and stored under nitrogen to ensure stability. Glutaraldehyde stocks are prepared fresh by dilution with filtered sample water to the required concentration.

Buffers, to maintain a neutral sample pH during storage, are not added to either the formaldehyde or glutaraldehyde stocks. They were shown to be of no benefit for preserving photopigment autofluorescence and may actually cause the collapse of fragile cells[12] and interfere in the fixation reactions of aldehydes.[23] Low pH in the samples is not a problem if the aldehyde solutions are prepared with filtered sample water rather than distilled water and used at the low concentrations and for the brief fixation periods recommended.

Sample water collected in opaque glass vials or tubes is mixed with an equal quantity of an aldehyde stock solution, having a concentration twice the required level, prepared using filtered sample water. This minimizes the osmotic shock of introducing small volumes of high concentration fixatives to the sample. The dilution of sample is corrected for in the filtration procedure or enumeration calculations.

Unless filtration facilities are lacking in the field, there is little advantage to fixing and freezing whole water samples for later thawing and filtration. Picophytoplankton are easily prefixed and collected on filters in the field, mounted in immersion oil as described for the enumeration procedures, and frozen for storage. The oil protects the cells from desiccation during storage, and the slides can be quickly thawed in the lab for enumeration. Some researchers have also reported success using gelatine/glycerol mountants to prepare slides for more long-term frozen storage.[16,22]

The light absorbtion and fluorescence emission of freshly collected picophytoplankton can differ markedly from preserved cells because fixation and preservation can uncouple light energy transfer from accessory pigments, chemically alter or degrade the pigments, or shift the wavelengths of their light absorbtion and fluorescence emission peaks. For example, degradation of chlorophyll a to its pheopigment reduces its fluorescence intensity to about half but also shifts its light absorbtion maximum to shorter wavelengths (from 430 to 410 nm).[24] For eukaryotic picophytoplankton, with chlorophyll a as their primary autofluorescent pigment, this shift in the absorbtion spectra is important for evaluating preservation techniques if blue light (450 to 500 nm) is used for excitation and enumeration of preserved samples. However, it is often possible to recover ''lost'' picophytoplankton counts in preserved samples by using the wider wavebands for excitation that encompass the shifted absorbtion peaks and deliver higher total excitation light intensities over the broader waveband.

NOTES AND COMMENTS

1. Rapid fading of cell fluorescence during enumeration can be a problem, particularly in fresh samples. Reducing the excitation light intensity by inserting 50% neutral density filters in the lamp path reduces the intensity of cell fluorescence and slows fading sufficiently to complete microscopic analysis.

2. There are other nonbiological particles that autofluoresce under the wavelengths used to enumerate picophytoplankton. Small particles that fluoresce brightly and do not gradually fade like picophytoplankton cells may be suspended mineral particles commonly found in turbid coastal marine waters or glacially turbid and shallow lakes.

3. Overlap with Utermöhl enumerations: Nano- and microphytoplankton are typically enumerated by Utermöhl settling techniques concurrently with epifluorescence picophytoplankton counts. The danger in using two entirely different methods to enumerate potentially overlapping size classes of phytoplankton is that some cells may be counted twice, inflating estimates of biomass, while others may be missed entirely. This is especially true when a cell appears strikingly different under the two techniques and when size, without a good scale of reference, is used as the sole separating criteria. Although epifluorescence techniques have been used to enumerate nano- and microphytoplankton,[25] the methods are not in general use. Thus, picophytoplankton and larger phytoplankton are not amenable to enumeration on the same sample, making cross comparisons and adherence to clear separation criteria very important.

4. Problem with colonial forms: Colonial aggregates of picoplankton-sized cells are particularly common in freshwaters, and range from tight clusters of 4 to 50 cells to loose aggregates of 10 to 50 cells in mucilaginous colonies. The former can be readily counted by transects or random fields to estimate the number of colonies and the average number

of cells per colony. However, the latter may include forms readily visible and often counted during Utermöhl enumerations (e.g., *Aphanocapsa, Aphanothece*). Although the colonies are microplankton size, the individual cells fit the picoplankton size class. Larger mucilaginous colonies break up when collected on membrane filters for autofluorescence microscopy and it is not always easy to distinguish solitary picophytoplankton from the loose colonial forms.

5. When occasional, nonrandom clumps of picophytoplankton cells are found on the filter surface, the presence of mucilaginous colonial forms is suspect, and can be confirmed under light microscopy in settled Utermöhl chambers. These fields can be excluded during epifluorescence counting and the colonies instead enumerated by Utermöhl techniques, although average cell sizes are best estimated under the epifluorescence microscope. If colonial forms are abundant, gentle prescreening of the sample water through 2-μm polycarbonate filters may be required. Colonial cells can then be enumerated by Utermöhl techniques or by the difference between screened and unscreened counts if cells are dispersed into single cells first using ultrasonics and chemical dispersal aids.

REFERENCES

1. Waterbury, J. B., Watson, S. W., Guillard, R. R. L., and Brand, L. E., Widespread occurrence of a unicellular, marine, planktonic, cyanobacterium, *Nature,* 277, 293, 1979.
2. Johnson, P. W. and Sieburth, J. McN., Chroocooccoid cyanobacteria in the sea: a ubiquitous and diverse phototrophic biomass, *Limnol. Oceanogr.,* 24, 928, 1979.
3. Johnson, P. W. and Sieburth, J. McN., *In situ* morphology and occurrence of eucaryotic phototrophs of bacterial size in the picoplankton of estuarine and oceanic waters, *J. Phycol.,* 18, 318, 1982.
4. Chisholm, S. W., Olson, R. J., Zettler, E. R., Goericke, R., Waterbury, J. B., and Welschmeyer, N. A., A novel free-living prochlorophyte abundant in the oceanic euphotic zone, *Nature,* 334, 340, 1988.
5. Stockner, J. G. and Antia, N. J., Algal picophytoplankton from marine and freshwater ecosystems: a multidisciplinary perspective, *Can. J. Fish. Aquat. Sci.,* 43, 2472, 1986.
6. Daley, R. J. and Hobbie, J. E., Direct counts of aquatic bacteria by a modified epifluorescence technique, *Limnol. Oceanogr.,* 20, 875, 1975.
7. Caldwell, D. E., Accessory pigment fluorescence for quantitation of photosynthetic microbial populations, *Can. J. Microbiol.,* 23, 1594, 1977.
8. Wilde, E. W. and Fliermans, C. B., Fluorescence microscopy for algal studies, *Trans. Am. Microsc. Soc.,* 98, 96, 1979.
9. Wood, A. M., Horan, P. K., Muirhead, K., Phinney, D. A., Yentsch, C. M., and Waterbury, J. B., Discrimination between types of pigments in marine *Synechococcus* spp. by scanning spectroscopy, epifluorescence microscopy, and flow cytometry, *Limnol. Oceanogr.,* 30, 1303, 1985.
10. Tsuji, T., Ohki, K., and Fujita, Y., Determination of photosynthetic pigment composition in an individual phytoplankton cell in seas and lakes using fluorescence microscopy; properties of fluorescence emitted from picophytoplankton cells, *Mar. Biol.,* 93, 343, 1986.
11. Navaluna, N. A., Perry, M. J., and Talbot, M. C., The effect of chemical fixation on some optical properties of phytoplankton, *J. Plankton Res.,* 11, 15, 1989.
12. Vaulot, D., Courties, C., and Partensky, F., A simple method to preserve oceanic phytoplankton for flow cytometric analyses, *Cytometry,* 10, 629, 1989.
13. Davis, P. G. and Sieburth, J. McN., Differentiation of phototrophic and heterotrophic nanoplankton populations in marine waters by epifluorescence microscopy, *Ann. Inst. Oceanogr.,* 58(Suppl.), 249, 1982.
14. Li, W. K. W., Experimental approaches to field measurements: methods and interpretation, in *Photosynthetic Picoplankton,* Platt, T. and Li, W. K. W., Eds., *Can. Bull. Fish. Aquat. Sci.,* 214, 1986, 251.

15. Jones, S. E., Ditner, S. A., Freeman, C., Whitaker, C. J., and Lock, M. A., Comparison of a new inorganic membrane filter (Anopore) with a track-etched polycarbonate membrane filter (Nuclepore) for direct counting of bacteria, *Appl. Environ. Microbiol.*, 55, 529, 1989.

16. Hall, J. A., Long-term preservation of picophytoplankton for counting by fluorescence microscopy, *Br. Phycol. J.*, 26, 169, 1991.

17. Brock, T. D., *Membrane Filtration: A User's Guide and Reference Manual,* Science Technology, Madison, 1983.

18. Mazunder, A. and Dickman, M. D., Factors affecting the spatial and temporal distribution of phototrophic sulfur bacteria, *Arch. Hydrobiol.*, 116, 209, 1989.

19. Kirchman, D., Sigda, J., Kapuscinski, R., and Mitchell, R., Statistical analysis of the direct count method for enumerating bacteria, *Appl. Environ. Microbiol.*, 44, 376, 1982.

20. Kuuppo-Leinikki, P. and Kuosa, H., Preservation of picoplanktonic cyanobacteria and heterotrophic nanoflagellates for epifluorescence microscopy, *Arch. Hydrobiol.*, 114, 631, 1989.

21. MacIsaac, E. A. and Stockner, J. G., Current trophic state and potential impacts of coal mine development on productivity of Middle Quinsam and Long lakes, *Can. Tech. Rep. Fish. Aquat. Sci.*, 1381, 1985.

22. Tsuji, T. and Yanagita, T., Improved fluorescent microscopy for measuring the standing stock of phytoplankton including fragile components, *Mar. Biol.*, 64, 207, 1981.

23. Jones, D., Chemistry of fixation and preservation with aldehydes, in *Zooplankton Fixation and Preservation,* Steedman, H. F., Ed., Unesco Press, Paris, 1976, 155.

24. Baker, K. S., Smith, R. C., and Nelson, J. R., Chlorophyll determinations with filter fluorometer: lamp/filter combination can minimize error, *Limnol. Oceanogr.*, 28, 1037, 1983.

25. Brock, T. D., Use of fluorescence microscopy for quantifying phytoplankton, especially filamentous blue-green algae, *Limnol. Oceanogr.*, 23, 158, 1978.

CHAPTER 24

Estimating Cell Concentration and Biomass of Autotrophic Plankton Using Microscopy

Beatrice C. Booth

INTRODUCTION

Microscopic analysis can be used to determine both the community structure and the biomass of the planktonic autotrophs in the sea. Using traditional light microscopy (LM), the Utermohl method,[1] autotrophs and heterotrophs below about 5 μm usually cannot be distinguished. However, epifluorescence microscopy (EFM) does allow that distinction permitting accurate enumeration and sizing of both autotrophs[2,3] and heterotrophs,[4] but often without identification to species. Electron microscopy (both transmission, TEM, and scanning, SEM) allows such identification of many of the pico- and nanoplankton-sized organisms. If four aliquots of one sample are analyzed, one by each these four methods, a comprehensive understanding of the autotrophic community can be gained. From cell concentrations and dimensions, estimates of biomass (as carbon) can be made using standard conversions from volume to carbon.

As cumbersome as such an approach may seem the alternatives are clearly more limited in scope: for example, biomass (as carbon) of the entire autotroph assemblage can be estimated from concentrations of chlorophyll a in the sample,[5] but only if the carbon:chlorophyll a ratio for that time and place is known. Cell concentrations of size classes can be obtained from electronic counters (e.g., Coulter) but these will not distinguish autotrophs from heterotrophs or detritus, and they do not correlate well with microscopic counts.[6] Cell concentrations of autotrophs can be obtained using flow cytometry but the sizing of organisms is not yet precise (it is based upon side scatter, a function of both size and surface texture). This technique has been used successfully on homogeneous populatons of distinctive size or pigment such as *Synechococcus* or cryptomonads[7-9] but cannot distinguish the wide range of shapes and sizes of the larger eukaryotes in a field sample.[10] Use of pigments other than chlorophyll a by high-pressure liquid chromatography, HPLC, has potential for large-scale studies,[11] but the interpretation of results is presently restricted by our incomplete knowledge of the pigment composition of aquatic species.

Thus, one advantage of the microscopical method is greater biological detail for the whole size spectrum of microbial autotrophs. If care is taken in selecting sample sizes, especially for the larger cells (LM), summation of the biomass of the individual size fractions (from LM and EFM) is an accurate estimate of the total autotroph carbon. This number, traditionally difficult to obtain, is most useful. From it can be estimated the carbon:chlorophyll a ratio of the autotroph assemblage. Together with ^{14}C-uptake rates, it can be used to estimate *in situ* growth rates of the autotroph assemblage.[12]

0-87371-564-0/93/$0.00 + $.50
© 1993 by Lewis Publishers

One further advantage, that of dependability, results from the preparation of samples at sea for analysis on land: there are no complex instruments or *in situ* incubations to go wrong at sea.

Disadvantages of the method are (1) the time it takes to analyze four aliquots per sample and combine the data, (2) higher operator error than purely mechanical analyses, (3) variation between investigators using different conversions from volume to carbon. Partial amelioration of (1) is possible by preparing only selected samples for SEM and TEM in situations where the species composition is not expected to change suddenly.

MATERIALS REQUIRED

Equipment

- Oceanographic water sample gear (e.g., Niskin bottles)
- Fume hood
- Filter rack with waste trap and vacuum pump
- Glass filter holders with effective diameter <20 mm (e.g., Nuclepore Micro 25-mm glass filter holders)
- Stainless steel (not sintered glass) filter supports
- $-20°C$ freezer for storage of slides
- Epifluorescence microscope with blue illumination and a field diaphram; mercury lamp preferable to halogen
- Multikey denominator or computer enumeration system

Supplies

- Black or unstained polycarbonate membrane filters (Nuclepore)
- 4-μm Millipore filters
- Glass slides, cover slips, and slide box (remove any paper liners which would absorb oil from the slides)
- Cargille DF low-fluorescence immersion oil

Solutions

- Glutaraldehyde (50% diluted to 25% with distilled water)
- Buffered formalin

PROCEDURES

Sampling and Fixation

1. Collect one large water sample using subsurface sampler (e.g., Niskin, Van Dorn). For integration with other data this sample should come from the same water bottle that provides water for pigment analysis (chlorophyll a) and ^{14}C-uptake experiments. Immediately sub-sample aliquots for epifluorescence microscopy (EFM), scanning electron microscopy (SEM), light microscopy (LM), and transmission electron microscopy (TEM).

2. Fix LM sample immediately with formalin. Keep remaining samples dark and cold (but do not freeze) until processed. Process immediately in this order: EFM, SEM, TEM.

Adequate fixation is critical to the method, both to preserve the cells from deterioration and to stabilize cell membranes of nonthecate forms in preparation for filtration. Live counts should be made on unconcentrated samples without using a cover slip.[13] In low biomass areas such a procedure is impractical. Buffered formalin is preferable to Lugol's for the thecate forms (LM) because it does not stain the cells. Glutaraldehyde has been the preferred fixative for nonthecate cells.[3,14-16] Reynolds[17] found glutaraldehyde ineffective for 1-μm flagellates. It is possible that paraformaldehyde is better for picoplanktonic eukaryotes,[13,18] although other differences in technique may be at fault in these comparisons.

Preparations for EFM, SEM, and TEM should be made at sea.

Epifluorescence Microscopy (EFM)

1. Slide preparation and storage: It is important that the cells be quite dense on the filter to minimize time scanning the filter, but they should not overlap. For subarctic Pacific waters, where chlorophyll a averages 0.3 μg l^{-1} and rarely exceeds 1 μg l^{-1}, we filtered 50 ml in a funnel with an effective diameter of 15 to 20 mm.[15,19] Fix the sample (1.0% glutaraldehyde, final concentration, 10 min in the dark) in the filter funnel over the filter to avoid loss of organisms from adhesion to walls of an intermediate container. We used blackened Nuclepore filters (either commercially blackened or using Irgalan Black), pore size 0.4 μm, mounted over a stainless steel frit (glutaraldehyde clogged sintered glass frits). Smaller mesh (0.2 μm) did not improve retention of pico- or nanoplankton and frequently clogged. Use a backing filter (Millipore HA) to ensure random distribution of cells on the filter. Nonrandom distribution has been one criticism of filter methods.[20,21]

2. Filter under low vacuum (<100 mm Hg) until the filter is just dry.

3. Mount the filter on a glass slide over a drop of low-fluorescence immersion oil. We use Cargille type DF immersion oil because type B produces too much background fluorescence and type FF so little that the ocular micrometers are obscure. Then add a second drop of immersion oil in the middle of the filter and place a cover slip on top. Do not move the cover slip thereafter.

4. Store the filter flat in the dark at 4°C for a few days while the water in the preparation is replaced by immersion oil, then in a freezer (always with the filters flat, not vertical) until examination. We used a −20°C freezer.

5. Examining the filter: Try to examine samples within 1 year of collection, although they will keep longer. Minimize or eliminate refreezing. Examine under blue light. We used a Zeiss standard microscope equipped for epifluorescence (Zeiss filter set 48 77 09, reflector 510 nm, excitation 450 to 490 nm, barrier filter at 520 nm, 50-W mercury light source). A mercury light is presumably preferable to halogen to elicit the weak autofluorescence of eukaryotic picoplankton. To minimize fading of autofluorescence, illuminate only small portions of the filter very briefly by closing field diaphragm and moving the slide continuously: a transect 24 μm wide of the entire filter is preferable to an equal area composed of large fields.

6. Use a stratified counting program (five increasingly larger segments of the filter examined using magnifications from ×500 down to ×125) to provide accurate counts for different size cells. We used several ocular micrometers (for calibration see Reference 21) to allow a choice of count areas depending on the cell density of each sample (count >100 cells in the major categories).[22] Generally, we counted 10 fields (173 μm × 173 μm) for *Synechococcus* at ×500, 1 transect (25 to 50 μm × 15000 μm) for eukaryotes <5 μm at ×500, 1 transect (279 μm × 15000 μm) at ×312 for cells 5 to 15 μm, 2 transects (437 μm × 15000 μm) at ×200 for cells 16 to 32 μm, and half or all the filter at ×125 for cells over 32 μm. To speed analysis we counted cells <6 μm into 1-μm intervals and lumped larger, less common cells into 5-μm intervals.

7. Measure the effective filter diameter of each sample in the microscope. This number was not consistent even for a given funnel/base combination, varying instead with each sample.

8. Identification: Organisms can be separated into *Synechococcus* (yellow fluorescing), various cryptomonads (orange fluorescing), and various sizes and species of dinoflagellates and other flagellates (all red fluorescing). Diatoms can be distinguished by shape, but their frustules do not fluoresce and LM counts are more specific. Coccolithophorids will be included with other eukaryotes as the coccoliths do not fluoresce; separate LM counts must be made. General dinoflagellate types, such as *Gymnodinium* and *Gyrodinium* can be distinguished. Heterotrophic forms fluoresce pale green (perhaps from the glutaraldehyde) and it is usually best to count them simultaneously with the eukaryotes to avoid overestimates of each group that might result from the inclusion of questionable specimens during independent counts, even on the same slide. Distinguishing heterotrophic cryptomanads from the spindle-shaped dinoflagellate heterotrophs is particularly difficult.

9. Enumeration: Because of the large number of count categories (cell types × size classes), and because work is done in the dark, several tiered numerical denominators are needed, the keys coded for touch. Alternatively, a computer with spreadsheet and a macro program (we used Toshiba T1000, LOTUS 123 and Prokey) can be programmed as a denominator. Each time a specific key is pressed the number in a spreadsheet cell is increased by 1. The keys should be coded for touch. This system has the advantage of simultaneous data entry for later computer reduction. Semiautomated image analysis[23] accomplishes the same objective as well as operator-assisted cell measurement using a camera lucida.

10. Computation: Compute cell concentration in each size-population from:

$$\text{cells } l^{-1} = \text{count} * (\text{Area}_{filter} / \text{Area}_{counted}) * (1000 \text{ ml/subsample volume in ml}) \quad (1)$$

Calculate cell volumes from the cell measurements using formulae for different shapes in Reference 21 and convert to cell carbon: >4 μm;[24] <4 μm,[25] 0.22 pgC μm^{-3}, *Synechococcus* 0.21 pgC/cell.[26] Calculate biomass (as carbon) for the various size populations from cell concentration and carbon/cell.

Light Microscopy (LM)

1. Sample preparation: Choose a large-enough sample such that you can easily count 100 specimens of the most common species. For the subarctic Pacific we collected 250-ml samples, and fixed them using buffered formalin (0.4%, final concentration). In more oligotrophic areas a larger sample is necessary. Two preparations are possible: settling for the inverted microscope[1] (for details see References 21 and 27) or filtration.[21] For the latter technique samples are filtered onto 4-μm Millipore filters, air dried cells down in immersion oil on a cover slip until the filter clears (2 to 4 d), then mounted in immersion oil and examined using a standard light microscope.[28] Such preparations must be examined within 2 weeks because diatoms frustules appear to dissolve with longer storage on the filter.

2. Identification and enumeration: Examine one or more transects for coccolithophorids, and all or half the area for diatoms and plankton >20 μm. Smaller dinoflagellates are best enumerated using EFM because of the difficulty of distinguishing autotrophic and heterotrophic forms. Larger, less-abundant dinoflagellates of known trophic mode are enumerated using LM. Identify all cells to the lowest possible taxon, genus, or species, measure, calculate cell volumes (formulae in Reference 21), and convert to cell carbon.[24] Compute cell concentration and biomass as in EFM.

Transmission Electron Microscopy (TEM)

Selected depths were sampled for TEM work. These samples (250 to 1000 ml) were concentrated by centrifuge or back filtration and whole mounts prepared (method modified from Reference 29; see Reference 30) from the concentrate onto nickel London grids, dried, and stored. Our basic modification was in handling the grids at sea. We settled cells from a drop of concentrate onto the grid by placing the grid into the drop (on a plastic surface) instead of dropping microdrops onto the grid. They were shadow cast in the land lab[30] and examined using a JEOL 100B TEM.

Carbon estimates in various size classes determined from EFM were partitioned according to the relative abundance of various species on the TEM grids to give biomass estimates of different species. Species to be identified are those with unique scales or flagellation such as prymnesiophytes, prasinophytes, chrysophytes, Parmales, and choanoflagellates.

Scanning Electron Microscopy (SEM)

More detailed identification of many of the nanoplankton, especially cryptomonads and dinoflagellates, is obtained using SEM. The method is that of Paerl and Shimp[31] and is seaworthy. We fixed 50-ml samples in 1 to 2.5% glutaraldehyde for 20 min over a 0.4-μm Nuclepore filter, carefully filtered under low vacuum, and transferred the filters while still wet into an alcohol series using small aluminum envelopes.[32] They were stored in 75% ethanol at 4°C until critical point drying with Freon, then viewed using a JEOL 840A SEM with X-ray analysis capability. Relative abundances of the various species are used as in TEM calculations. One limitation of this method is that autotrophs are not distinguished from heterotrophs in SEM and TEM. A time-consuming solution is to view the same grid sequentially using first EFM, then TEM.[33-35]

REFERENCES

1. Utermohl, H., Neue Wege in der quantitativen Erfassung des Planktons, *Verh. Int. Verein. Theor. Angew. Limnol.*, 5, 567, 1931.
2. Brock, T. D., Use of fluorescence microscopy for quantifying phytoplankton, especially filamentous blue-green algae, *Limnol. Oceanogr.*, 23, 158, 1978.
3. Wilde, E. W. and Fliermans, C. B., Fluorescence microscopy for algal studies, *Trans. Am. Micro. Soc.*, 98, 96, 1979.
4. Geider, R. J., Abundance of autotrophic and heterotrophic nanoplankton and the size distribution of microbial biomass in the southwestern North Sea in October 1986, *J. Exp. Mar. Biol. Ecol.*, 123, 127, 1988.
5. Lorenzen, C. J., A method for the continuous measurement of *in vivo* chlorophyll concentration, *Deep-Sea Res.*, 13, 223, 1966.
6. Arvola, L., A comparison of electronic particle counting with microscopic determinations of phytoplankton and chlorophyll a concentrations in three Finnish lakes, *Ann. Bot. Fennici*, 21, 171, 1984.
7. Chisholm, S. W., Armbrust, E. V., and Olson, R. J., The individual cell in phytoplankton ecology: cell cycles and applications of flow cytometry, in *Photosynthetic Picoplankton*, Platt, T. and Li, W. K. W., Eds., *Can. Bull. Fish. Aquat. Sci.*, 214, 343–369, 1986.
8. Li, W. K. W. and Wood, A. M., Vertical distribution of North Atlantic ultraphytoplankton: analysis by flow cytometry and epifluorescence microscopy, *Deep-Sea Res.*, 35, 1815, 1988.
9. Olson, R. J., Chisholm, S. W., Zeittler, E. R., Altabet, M. A., and Dusenbury, J. A., Spatial and temporal distributions of prochlorphyte picoplankton in the North Atlantic Ocean, *Deep-Sea Res.*, 37, 1033, 1990.

10. Legner, M., Phytoplankton quantity assessment by means of flow cytometry, *Mar. Microb. Food Webs,* 4, 161, 1990.

11. Ondrusek, M. E., Bidigare, R. R., Sweet, S. T., Defreitas, D. A., and Brooks, J. M., Distribution of phytoplankton pigments in the North Pacific Ocean in relation to physical and optical variability, *Deep-Sea Res.,* 38, 243, 1990.

12. Booth, B. C., Lewin, J., and Lorenzen, C. J., Spring and summer growth rates of subarctic Pacific phytoplankton assemblages determined from carbon uptake and cell volumes estimated using epifluorescence microscopy, *Mar. Biol.,* 98, 287, 1988.

13. Massana, R. and Gude, H., Comparison between three methods for determining flagellate abundance in natural waters, *Ophelia,* 33, 197, 1991.

14. Coulon, C. and Alexander, V., A sliding-chamber phytoplankton settling technique for making permanent quantitative slides with application in fluorescence microscopy and autoradiography, *Limnol. Oceanogr.,* 17, 149, 1972.

15. Booth, B. C., The use of autofluorescence for analyzing oceanic phytoplankton communities, *Bot. Mar.,* 30, 101, 1987.

16. Kuuppo-Leinikki, P. and Kuosa, H., Preservation of picoplanktonic cyanobacteria and heterotrophic nanoflagellates for epifluorescence microscopy, *Arch. Hydrobiol.,* 114, 631, 1989.

17. Reynolds, N., The estimation of the abundance of ultraplankton, *Br. Phycol. J.,* 8, 135, 1973.

18. Hall, J. A., Long-term preservation of picoplankton for counting by fluorescence microscopy, *Br. Phycol. J.,* 26, 169, 1991.

19. Booth, B. C., Size classes and major taxonomic groups of phytoplankton at two locations in the subarctic Pacific Ocean in May and August, 1984, *Mar. Biol.,* 97, 275, 1988.

20. Holmes, R. W., The Preparation of Marine Phytoplankton for Microscopic Examination and Enumeration on Molecular Filters, Spec. Sci. Rep., U.S. Fish and Wildlife Service, Fisheries, Washington, D.C., 433, 1, 1962.

21. Wetzel, R. G. and Likens, G. E., *Composition and Biomass of Phytoplankton, Limnological Analyses,* Springer-Verlag, New York, 1991.

22. Hobro, R. and Willen, E., Phytoplankton countings. Intercalibration results and recommendations for routine work, *Int. Rev. Ges. Hydrobiol.,* 62, 805, 1977.

23. Roff, J. C. and Hopcroft, R. R., High precision microcomputer based measuring system for ecological research, *Can. J. Aquat. Sci.,* 43, 2044, 1986.

24. Strathmann, R. R., Estimating the organic carbon content of phytoplankton from cell volume or plasma volume, *Limnol. Oceanogr.,* 12, 411, 1967.

25. Mullin, M. M., Sloan, P. R., and Eppley, R. W., Relationship between carbon content, cell volume and area in phytoplankton, *Limnol. Oceanogr.,* 11, 307, 1966.

26. Waterbury, J. B., Watson, S. W., Valois, F. W., and Franks, D. G., Biological and ecological characterization of the marine unicellular cyanobacterium *Synechococcus, Can. Bull. Fish. Aquat. Sci.,* 214, 71, 1986.

27. Hasle, G. R., The inverted microscope method, in *Phytoplankton Manual,* Sournia, A., Ed., UNESCO, Paris, 1978, 88.

28. McNabb, C. D., Enumeration of freshwater phytoplankton concentrated on the membrane filter, *Limnol. Oceanogr.,* 5, 57, 1960.

29. Moestrup, O. and Thomsen, H. A., Preparation of shadow-cast whole mounts, in *Handbook of Phycological Methods: Developmental and Cytological Methods,* Gantt, E., Ed., Cambridge University Press, Cambridge, 1980, chap. 31.

30. Booth, B. C. and Marchant, H. J., Parmales, a new order of marine Chrysophytes, with descriptions of 3 new genera and 7 new species, *J. Phycol.,* 23, 245, 1987.

31. Paerl, H. W. and Shimp, S. L., Preparation of filtered plankton and detritus for study with scanning electron microscopy, *Limnol. Oceanogr.,* 18, 802, 1973.

32. Booth, B. C., Lewin, J., and Norris, R. E., Nanoplankton species predominant in the subarctic Pacific in May and June, 1978, *Deep-Sea Res.,* 29, 185, 1982.

33. Davis, P. G. and Sieburth, J. McN., Differentiation and characterization of individual phototrophic and heterotrophic microflagellates by sequential epifluorescence and electron microscopy, *Trans. Am. Microsc. Soc.,* 103, 221, 1984.

34. Estep, K. W., Davis, P. G., Hargraves, P. E., and Sieburth, J. McN., Chloroplast containing microflagellates in natural populations of North Atlantic nanoplankton, their identification and distribution, including a description of five new species of *Chrysochromulina* (Prymnesiophyceae), *Protistologica*, 20, 613, 1984.
35. Hoepffner, N. and Haas, L. W., Electron microscopy of nanoplankton from the North Pacific Central Gyre, *J. Phycol.*, 26, 421, 1990.

Additional References

Borsheim, K. Y. and Bratbak, G., Cell volume to cell carbon conversion factors for a bactivorous *Monas* sp. enriched from seawater, *Mar. Ecol. Prog. Ser.*, 36, 171, 1987.
Ahlgren, G., Comparison of methods for estimation of phytoplankton carbon, *Arch. Hydrobiol.*, 98, 489, 1983.
Vargo, G. A., Using the fluorescence microscope, in *Phytoplankton Manual*, Sournia, A., Ed., UNESCO, Paris, 1978, 197.

CHAPTER 25

Preservation and Storage of Samples for Enumeration of Heterotrophic Protists

Evelyn B. Sherr and Barry F. Sherr

INTRODUCTION

During the past 15 years increasing attention has been given to roles of protists 2 to 200 μm in size in aquatic ecosystems.[1-3] Earlier, lack of adequate methodology had hampered research on heterotrophic flagellates and ciliates in their natural habitats. In particular, there were no good methods by which numerical abundance/biomass estimates could be made for *in situ* assemblages. Application of epifluorescence staining methodologies overcame this deficiency and has been the basis for much of the new information about aquatic protists. In addition, researchers have become much more sophisticated in the use of sampling and preservation techniques for these organisms, which are basically just "bags of water" that are easily lysed. Below we describe commonly used preservation methods for aquatic protists. First, however, we will list some major problems that have been identified regarding preservation and storage of samples for enumeration of <200-μm-size protists.

1. Loss of naked flagellate and ciliate cells due to sample handling: Passing a sample through any kind of screen inevitably results in lysis of some protists, particularly larger naked ciliates.[4] Therefore, it is recommended that whole water samples be used for estimates of total protist numbers and biomass. Since unpreserved protist cells are readily lysed during filtration, it is imperative that water samples be preserved before being collected on filters.
2. Loss of cells due to preservation: Various workers have reported apparent lysis of a fraction of flagellate or ciliate cells preserved with formaldehyde or glutaraldehyde. Pace and Orcutt[5] found significantly fewer protists with 2% final concentration formalin that with 1% iodide or with mercuric chloride plus bromophenol blue. Bloem et al.[6] reported that freshwater flagellates were lysed when fixed with borate-buffered formaldehyde or glutaraldehyde, but were preserved with constant cell numbers over a period of weeks with unbuffered 1% glutaraldehyde or 5% formaldehyde. We have observed rapid lysis of some species of cultured marine ciliates when fixed with 2% borate-buffered formalin. Preservation with dilute concentrations (<1 to 2%) of acid Lugol's solution has also been reported to result in significant loss of numbers of ciliates[7] and flagellates.[8]
3. Distortion of cell shape by formaldehyde: We as well as others[5,9] have observed that formalin-preserved cells, especially those of ciliates and dinoflagellates, are often disrupted and distorted compared to the shape of the living cell. Ciliates may be partially or completely "blown up" and dinoflagellate thecae may look "stringy", sometimes in such a way that at first glance the dinoflagellate cell may be mistaken for a ciliate.

4. Decrease in cell biovolume upon preservation: Cells of both flagellates[9,10] and ciliates[7,10] show significant shrinkage in biovolume, to about 75 to 45% of the live cell biovolume, after preservation with aldehyde fixatives as well as acid Lugol's solution. However, the cells may first swell when preserved; thus, cell volumes may take some hours to reach a stable value.[7] As a general assumption, the biovolume of fixed cells will be about 50% of that of live cells, and the C:unit biovolume ratio of fixed cells will be approximately 0.16 to 0.20 pg C μm^{-3}.[6,9,10]

5. Loss of cells when stored in liquid samples: Anecdotal information suggests that the abundance of most microbial cells, including bacteria and protists, decreases with time in preserved liquid samples. For this reason, it is recommended that fresh samples be immediately preserved, stained, filtered, and the filters mounted onto slides and kept frozen until ready to be inspected.

MATERIALS REQUIRED

Equipment

- Filtration apparatus
- For storage: desiccator, refrigerator, -20 to $-70°C$ freezer (depends on method)

Supplies

- 0.2-μm cellulosic filters
- 0.2-μm polycarbonate membrane filters
- 0.45-μm Gelman triacetate metricel filters (for Tsuji and Yanagita method)
- Glass slides, cover slips

Solutions and Reagents

The following preservatives are discussed in the next section:

- Formalin (nonbuffered, or buffered with either sodium tetraborate or hexamethylenetriamine)
- Paraformaldehyde (powder)
- Glutaraldehyde
- Cacodylate (used with glutaraldehyde)
- Mercuric chloride
- Bromophenol blue (used with HgCl method)
- Acid Lugol's: 1 g iodine (I) + 2 g potassium iodide (KI) dissolved in 200 ml distilled or deionized water; add 20 ml glacial acetic acid
- Alkaline Lugol's: 10 g I + 20 g KI + 10 g sodium acetate, dissolved in 140 ml of distilled or deionized water
- $Na_2S_2O_3$, 3 g in 100 ml distilled water (used with alkaline Lugol's)

Other reagents used include:

- 1 N NaOH
- HCl
- Dry ice
- Glycerol
- Glycerine mountant: Tsuji and Yanagita[14] used the following recipe for the glycerine mountant: 20 ml of deionized water is added to 10 g gelatin (white extra fine DAB-7, Merck)

and heated to 80°C. After the gelatin is dissolved, 50 ml of 80°C glycerol is mixed into the solution. Slow heating in an incubator prevents the formation of bubbles. The glycerine mountant is stored at 5°C.

PRESERVATIVES

Formalin

Various workers have used 1 to 5% final volume formalin (37% formaldehyde) to preserve protists in water samples.[6,11] The formalin may be buffered by saturating with sodium tetraborate (borax)[6,7] or hexamethylenetriamine[11] to minimize pH shift in the sample. However, Bloem et al.[6] reported that use of buffered formalin resulted in 100% lysis of freshwater flagellates, while 5% final volume-unbuffered formalin preserved flagellate cells in liquid samples for 1 week. Stoecker et al.[11] used 20% formaldehyde buffered with 100 g l^{-1} of hexamethylenetramine mixed with seawater samples to yield a final concentration of 2% formaldehyde, and reported that 80 to 90% of ciliates were preserved with this method for up to 5 weeks when stored at 4°C in the dark.

A solution of formaldehyde may be prepared from paraformaldehyde powder dissolved in distilled or deionized water to minimize the presence of chemicals which may contaminate commercial formalin.

Glutaraldehyde

Glutaraldehyde is another commonly used preservative for microbes in seawater. It has the disadvantages of being more hazardous than formalin, breaking down at room temperature, and having a short shelf life (on the order of months). EM-grade glutaraldehyde should be used in preparation of cells for electron microscopy; otherwise, a cheaper grade may be employed. Glutaraldehyde-fixed cells will have a faint green autofluorescence when excited by blue light during epifluorescence microscopy. Bloem et al.[6] reported that 1% glutaraldehyde (final concentration) preserved chlorophyll autofluorescence better than did 5% formalin. Adding 4% ice cold glutaraldehyde, diluted with 0.2-μm filtered water of the same salinity as the sample, and mixed volume/volume with the sample, is an effective way to prevent egestion of food vacuole contents of protists.[12,13]

Mercuric Chloride

Pace and Orcutt[5] compared three methods of preserving freshwater ciliates for inspection via inverted microscopy: 1% final volume Lugol's solution, 2% formalin, and mercuric chloride + bromophenol blue, and reported that the last method gave the best results. Their method was to preserve 230 ml of water with 10 ml of a saturated solution of HgCl, and then stain with a drop of 0.04% bromophenol blue. Formalin resulted in disintegration of ciliate cells; Lugol's solution preserved the ciliates, but stained detritus also so that the ciliates were difficult to visualize. HgCl is not commonly used to preserve marine samples.

Lugol's Solution

Low concentrations of Lugol's solution (0.5 to 1%, "weak tea color") have been traditionally used to preserve phytoflagellates for counts via Utermohl inverted microscopy.

Some workers recommend higher concentrations (up to 10%) of acid Lugol's solution for fixation and storage of marine ciliates in liquid samples prior to enumeration[15] (see Chapter 12). Lugol's solution should be kept in a dark bottle at room temperature.

Disadvantages of Lugol's solution are that it strongly colors cells so that autotrophic and heterotrophic cells cannot easily be distinguished, and cells cannot be stained for epifluorescence microscopy.

Mixtures

Glutaraldehyde:Paraformaldehyde

Tsuji and Yanagita[14] preserved marine nanoplankton using a mixture of the two aldehyde fixatives to give a final volume concentration in the sample of 1% glutaraldehyde and 0.03 to 1% paraformaldehyde. Their method is as follows:

1. Dissolved 1 g paraformaldehyde powder in 50 ml of distilled water at 65°C by continuous stirring.

2. One to five drops of 1 N NaOH is added to clear the solution.

3. The paraformaldehyde solution is cooled and then mixed with 40 ml of 25% glutaraldehyde, and the pH adjusted to 7 with NaOH or HCl.

4. Distilled water is added to adjust the volume to 100 ml, and the mixture filtered through 0.45 μm and stored at 5°C.

5. To fix a seawater sample, a 1:9 ratio of preservative mixture: sample volume is used. Preserved samples are stored at 5°C for >1 h before filtration.

Cacodylate-Buffered Glutaraldehyde

A recipe for cacodylate-buffered glutaraldehyde to preserve protists for electron microscopy (from S. Nishino, personal communication) is as follows:

1. 160 ml of 25% EM-grade glutaraldehyde + 32 g of cacodylate is dissolved in 840 ml of 0.2-μm filtered artificial seawater of appropriate salinity.

2. Adjust the pH to 8.0 with a concentrated NaOH solution.

3. Filter through a 0.2-μm cellulosic filter to remove any flocculated material.

4. Store at 5°C. Refilter every few weeks before use.

5. Mix 1:1 with seawater samples for a final concentration of 2% glutaraldehyde and 0.1 M cacodylate. Both compounds are highly toxic.

Lugol's Solution:Formalin:Sodium Thiosulfate

We routinely use the following method, suggested to us by F. Rassoulzadegan:

1. Quick-fix 20-ml samples with 10 μl of alkaline Lugol's solution.

2. Follow immediately with addition of 0.4 to 0.5 ml borate-buffered formalin and then 20 μl of 3% sodium thiosulfate (3 g $Na_2S_2O_3$ in 100 ml deionized water).

The sodium thiosulfate bleaches out the iodine color so that cells may be observed via epifluorescence microscopy. Chlorophyll autofluorescence is well preserved with this method. We have found that both flagellates and ciliates appear more intact with the combined method than with aldehyde fixation alone, and that Lugol + formalin preservation prevents egestion of food vacuole contents.[12]

SAMPLE STORAGE

Until recently, it has been common practice to store aldehyde-preserved samples in the dark and cold (5°C) for days to 1 to 2 weeks, or Lugol-preserved samples in the dark at room temperature for weeks to months, prior to inspection. Attention to cell loss with time in liquid samples was not routinely made. Anecdotal evidence is now mounting to suggest that storage of preserved cells in liquid sample can lead to variable, often significant, loss of flagellate and ciliate cells in less than a day. Therefore, it is highly recommended that preserved samples be processed, filtered, mounted on slides, and stored frozen ($-20°C$ or lower temperature) as soon as conveniently possible after sampling, and that control experiments to determine the potential magnitude of cell loss of preserved compared to live samples be made. For later inspection via epifluorescence microscopy, samples can be stained with a fluorochrome prior to, or immediately upon, filtration onto a polycarbonate membrane filter. If it is desirable to look at the sample via transmitted light microscopy, or a combination of transmitted light and epifluorescence microscopy, two procedures may be followed:

Filter-Transfer-Freeze (FTF) and Glycerine Mounting[16]

1. Filter the sample onto a membrane filter, not quite to dryness.

2. Place the filter sample-side down onto a glass slide on which a drop of water has been smeared.

3. Put the slide, filter-side up, onto a piece of dry ice. When the sample is hard frozen to the slide, the filter is peeled away.

4. The sample on the slide can be permanently fixed with glycerine gel. Hewes[16] recommends letting the glycerine-gel preparation dry completely (0.5 to 2 h), then washing the mounted preparation for 10 to 20 min in cold deionized water to remove soluble enzymes, and finally letting the gelatin preparation dry completely again before putting on a coverslip.

Glycerine Mounting Method of Tsuji and Yanagita[14]

1. Samples are filtered onto a Gelman triacetate metricel filter (GA 6, 0.45-μm pore size).

2. A small lump (1 mm³) of glycerine mountant is placed onto a cover glass and melted on a hot plate at 50°C.

3. The sample filter is placed onto the melted glycerine, sample-side down, and the preparation cooled to room temperature.

4. One or two drops of glycerol are placed on the reverse side of the filter to clear it. The filter mounted onto the cover glass can be stored in a dessicator at $-20°C$ for several months. About 3 d of storage in the dessicator is required for the filter to become completely transparent.

5. For microscopic inspection, the filter-cover slip preparation is warmed to room temperature in the dessicator, and then mounted onto a glass slide.

REFERENCES

1. Porter, K. G., Sherr, E. B., Sherr, B. F., Pace, M., and Sanders, R. W., Protozoa in planktonic food webs, *J. Protozool.,* 32, 409, 1985.
2. Capriulo, G. M., Ed., *Ecology of Marine Protozoa,* Oxford University Press, New York, 1990.
3. Reid, P. C., Turley, C. M., and Burkill, P. H., Eds., *Protozoa and Their Role in Marine Processes,* (NATO ASI Series, Ecological Sciences, Vol. 25), Springer-Verlag, New York, 1991.
4. Gifford, D. J., Laboratory culture of marine planktonic oligotrichs (Ciliophora, Oligotrichida), *Mar. Ecol. Prog. Ser.,* 23, 257, 1985.
5. Pace, M. L. and Orcutt, J. D., Jr., The relative importance of protozoans, rotifers, and crustaceans in a freshwater zooplankton community, *Limnol. Oceanogr.,* 26, 822, 1981.
6. Bloem, J., Bar-Gilissen, M.-J., and Cappenberg, T. E., Fixation, counting, and manipulation of heterotrophic nanoflagellates, *Appl. Environ. Microbiol.,* 52, 1266, 1986.
7. Ohman, M. D. and Snyder, R. A., Growth kinetics of the omnivorous oligotrich ciliate, *Strombidium* sp., *Limnol. Oceanogr.,* 36, 922, 1991.
8. Klein Breteler, W. C. M., Fixation artifacts of phytoplankton in zooplankton grazing experiments, *Hydrobiol. Bull.,* 19, 13, 1985.
9. Choi, J. W. and Stoecker, D. K., Effects of fixation on cell volume of marine planktonic protozoa, *Appl. Environ. Microbiol.,* 55, 1761, 1989.
10. Putt, M. and Stoecker, D. K., An experimentally determined carbon-volume ratio for marine oligotrichous ciliates from estuarine and coastal waters, *Limnol. Oceanogr.,* 34, 1097, 1989.
11. Stoecker, D. K., Taniguchi, A., and Michaels, A. E., Abundance of autotrophic, mixotrophic, and heterotrophic planktonic ciliates in shelf and slope waters, *Mar. Ecol. Prog. Ser.,* 50, 241, 1989.
12. Sherr, B. F., Sherr, E. B., and Pedrós-Alió, C., Simultaneous measurements of bacterioplankton production and protozoan bacterivory in estuarine water, *Mar. Ecol. Prog. Ser.,* 54, 209, 1989.
13. Sanders, R. W., Porters, K. G., Bennett, S. J., and DeBiase, A. E., Seasonal patterns of bacterivory by flagellates, ciliates, rotifers, and cladocerans in a freshwater planktonic community, *Limnol. Oceanogr.,* 34, 4, 673, 1989.
14. Tsuji, T. and Yanagita, T., Improved fluorescent microscopy for measuring the standing stock of phytoplankton including fragile components, *Mar. Biol.,* 66, 9, 1982.
15. Gifford, G. J., Impact of grazing by microzooplankton in the Northwest Arm on Halifax Harbour, Nova Scotia, *Mar. Ecol. Prog. Ser.,* 47, 249, 1988.
16. Hewes, C. D. and Holm-Hansen, O., A method of recovery of nanoplankton from filters for identification with the microscope: the Filter-Transfer-Freeze (FTF) technique, *Limnol. Oceanogr.,* 28, 389, 1983.

CHAPTER 26

Staining of Heterotrophic Protists for Visualization via Epifluorescence Microscopy

Evelyn B. Sherr, David A. Caron, and Barry F. Sherr

INTRODUCTION

Studies of the *in situ* abundance and activity of heterotrophic protists, especially cells <20 μm in size, were hampered due to inadequate methods of enumeration of this group of microbes until epifluorescence microscopical techniques began to be applied to the problem in the early 1980s.[1-4] The basic premise of all these methods is that a fluorescent compound is used to visualize protistan cells in natural samples. These compounds bind to various components of the protistan cell (see Table 1) which are then visualized by stimulation of the fluorochrome with high-energy, short-wavelength light, causing it to emit light of lower energy (i.e., longer wavelength). The light emanating from the fluorochrome is then separated from the excitation beam by a series of filters within the microscope and diverted to the ocular lenses. Since photosynthetic pigments fluoresce, epifluorescence microscopy may also be used to distinguish phototrophic protists that can be abundant in the nanoplankton size class. The autofluorescence of photosynthetic pigments varies from yellow to red and is dependent upon the type of pigment and the filter set employed.

Epifluorescence microscopy is accomplished by a series of filters which selects a specific segment of the light spectrum to excite the stained preparation and then allows only excited (fluoresced) light to pass to the ocular lenses. Typically, a bandpass (BP) excitation filter is used in conjunction with epifluorescence microscopy to illuminate the prepared sample containing stained protista with light at wavelengths that causes the fluorochrome to fluoresce. A dichromatic beam splitter is then used to direct the excitation beam through the objective lens to the sample preparation. Fluoresced (and incident) light pass back through the objective lens but the properties of the beam splitter are such that only fluoresced (longer wavelength light) passes through it and on to the ocular lenses. A long pass (LP) emission filter removes unwanted wavelengths before the light passes up to the ocular lenses.

Most companies (Olympus, Zeiss, Leitz, Nikon) which sell research microscopes that can be outfitted for epifluorescence microscopy have a variety of filter sets designed for specific fluorochromes, e.g., blue light filter sets for acridine orange, proflavin, fluorescein isothiocyanate (FITC), and 5-(4,6-dichlorotriazine -2-yl) aminofluorescein (DTAF); UV light filter sets for 4′,6-diamidino-2-phenylindole (DAPI) and primulin; and green light filter sets for rhodamine isothiocyanate (RITC) and Nile Red. Lamps for epifluorescence typically have a broad spectrum of wavelength emissions including UV as well as visible light. The

Table 1. Fluorochromes Commonly Used in Epifluorescence Microscopy

Fluorochrome	Excitation (nm)	Emission peak (nm)	Target
DAPI	365	390–400	DNA
Primulin (Direct Yellow 59)	365	425	Cytoplasm
Proflavin hemisulfate	470–490	500–520	Cytoplasm
FITC isomer 1	470–490	500–520	Cytoplasm
DTAF	470–490	500–520	Cytoplasm
Acridine orange (ZnCl salt)	470–500	550–570	RNA, DNA
Nile Red lipids	488–525	570–600	Neutral lipids
Rhodamine RITC	540–560	620	Cytoplasm

two common types of lamps used in epifluorescence microscopy are 75- or 100-W xenon lamps and 50- or 100-W mercury lamps. Both are expensive (about $150 to $200 each) and are rated for only a few hundred hours of use. In our experience, xenon lamps are difficult to adjust for optimum illumination of the microscope field of view.

A brief description of fluorochrome staining methods currently used to visualize heterotrophic protists is given below. The great advantage of epifluorescence microscopy for counting protistan (and prokaryotic) populations relative to most other techniques is the relative ease of the method. Slides can be prepared at sea or under relatively harsh field conditions and then sealed and stored frozen until they can be returned to the lab and examined. Microscopical examination is laborious for large numbers of samples and still somewhat subjective, but advances in molecular biology, immunology, and flow cytometry hold the promise of higher precision and automated counting techniques.

There are, however, some significant disadvantages that still remain with these methods. A shortcoming of all the epifluorescence microscopical methods presently in use is that the presence of a chloroplast (i.e., chlorophyll autofluorescence) does not necessarily confirm that photosynthesis is the sole source of nutrition for these individuals. A number of algal species are capable of phagotrophy (so-called "mixotrophic" species). Likewise, a number of herbivorous species of ciliates retain the chloroplasts from ingested prey and maintain these chloroplasts in a functional state.[5] Mixotrophic behavior undermines the use of chlorophyll autofluorescence as a feature to distinguish between phototrophic and heterotrophic microorganisms. Another shortcoming is that epifluorescence microscopy provides relatively few taxonomic criteria. Critical ultrastructural features cannot be discerned with this method, and electron microscopy and the observation of live specimens remain important methods for identifying small protists. Future applications of the oligonucleotide and immunological probes to epifluorescence microscopy may provide taxonomic information, as well as abundance data.

Examples of fluorescently stained protists are shown in Figures 1 to 3. For photographing epifluorescence preparations, we recommend ASA 400 film, e.g., Ektachrome 400 or Kodak T-MAX.

MATERIALS REQUIRED

Equipment

- Microscope outfitted for epifluorescence microscopy, with appropriate filter sets for the fluorochromes of interest (see specific comments in staining procedures)
- 25-mm filtration apparatus

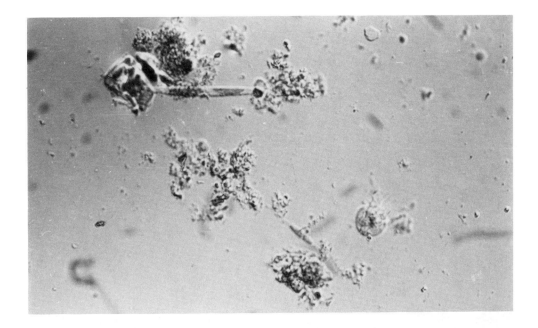

A

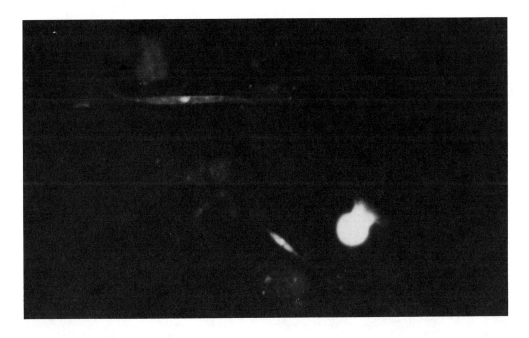

B

Figure 1.   Visualization of microbes via transmitted and epifluorescence microscopy. Two 50-μm pennate diatoms and a 20-μm ciliate seen via (A) transmitted light and (B) epifluorescence after staining with FITC. Original magnification 160 ×.

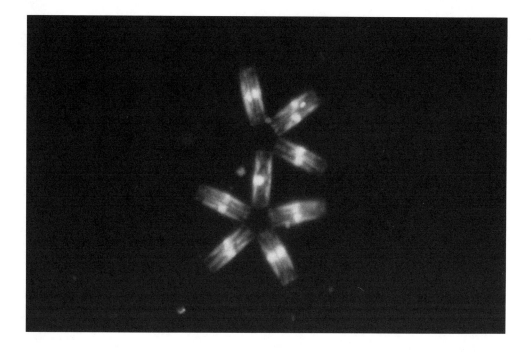

A

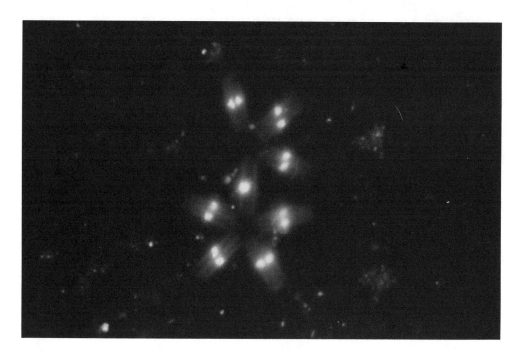

B

Figure 2. Comparison of FITC and DAPI staining. Chain of 10-μm long pennate diatoms: (A) whole cells visualized via FITC and chlorophyll-a autofluorescence and (B) nuclei of cells visualized via DAPI. Original magnification 500×.

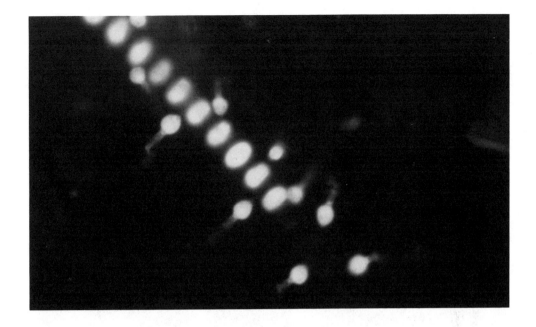

A

B

Figure 3. Heterotrophic protists visualized via FITC staining: (A) 4-μm choanoflagellates attached to a chain-forming diatom, *Skeletonema costatum*; (B) 8-μm heterotrophic flagellate; and (C) 16-μm choreotrichous ciliate. Original magnification 1000 ×.

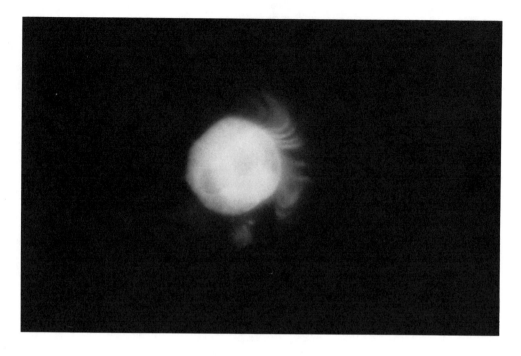

FIGURE 3C.

- Stage and slide micrometers for measuring the widths of the microscope fields of view and/or transect lengths.
- Ocular micrometer for measuring cell size.

Supplies

- Black-stained polycarbonate membrane filters, 0.2 μm or 0.8 μm are typically used. The use of 0.8-μm filters is sometimes preferred because much of the prokaryote assemblage will pass through these filters, thereby resulting in less extraneous material in the field of view. Two suppliers of black-stained membrane filters are available: Nuclepore Corp. (Pleasanton, CA, 1-800-882-7711) and Poretics Corp. (Livermore, CA, 1-800-922-6090).
- Cellulosic backing filters to support the membrane filters and aid in the even dispersion of cells over the membrane filter surface. Commonly used backing filters are 0.45-μm Millipore filters or 0.8-μm Nuclepore Membra-fil filters.
- Slides, cover slips
- Immersion oil (e.g., Cargille type A, Cargille type LF, FITC mounting fluid for FITC staining).

Solutions

Fluorochromes (see Table 1) may be ordered from Sigma Chemical Company, St. Louis, MO, (1-800-325-3010, outside the U.S./Canada phone collect 314-771-5750). Sigma chemicals are sold by several affiliated companies in Europe.

- FITC staining solution; this recipe may be doubled
    0.25 ml of 0.5 M sodium carbonate buffer (pH 9.5)
    1.1 ml of 0.01 M potassium phosphate buffer (pH 7.2)

1.1 ml of 0.85% sodium chloride

1.0 mg of FITC

- Recipes for the FITC buffers
    1. carbonate buffer — mix 100 ml of 4.2 g $NaHCO_3$/100 ml DiW (distilled/deionized water) with ~60 ml of 5.3 g $NaCO_3$/100 ml DiW, adjust to pH 9
    2. phosphate buffer — mix 72 ml of 0.87 g KH_2PO_4/500 ml DiW with ~6.2 ml of 0.65 g K_2HPO_4/500 ml DiW
- Acridine orange solution

    0.1% W/V in distilled water, preserved with 1% formalin or glutaraldehyde
- Primulin

    Working-strength solution is 250 μg of primulin per milliliter in distilled water with 0.1 *M* Trizma-HCl at pH 4.0
- DAPI

    A concentrated stock solution of DAPI (1.0 mg ml^{-1} in 0.2-μm filtered distilled water) is prepared and stored at −20°C

PROCEDURES

Samples should be preserved before filtration. Glutaraldehyde or formalin fixatives are most frequently used. In some of the fluorochrome methods, however, cells may be stained before preservation. One can live-stain protists with dilute acridine orange or DAPI; visualization of living protists via epifluorescence microscopy is an interesting demonstration for students.

A typical problem in the use of epifluorescence microscopy for counting protistan (and prokaryotic) assemblages is the presence of "background fluorescence". This unwanted fluorescence can result from not thoroughly rinsing stained preparations, nonspecific staining of material in the water sample, and the autofluorescence of the membrane filters or the immersion oil. A few cautions on minimizing this problem are noted here.

Most epifluorescence staining methods require a lower background fluorescence than can be obtained with plain membrane filters. Originally, membrane filters were darkened by staining with irgalan black.[6] However, black-stained membrane filters are now available in several pore sizes from both Nuclepore Corp. and Poretics Corp. (see above). A wetted cellulosic backing filter is recommended to support the membrane filter in order to minimize wrinkling and to facilitate even distribution of the sample. The same backing filter may be used for several samples. It is important that both both filters be prewetted with deionized or filtered seawater and that bubbles be excluded when mounting the filters on the filtration assembly in order to obtain an even dispersion of cells onto the membrane filter.

Autofluorescence of some immersion oils also can contribute significantly to background fluorescence. Special low-fluorescence oils are commercially available but are not typically required for routine protistan counts by the methods listed below. Cargille type A is sufficient for most applications. We have found, however, that immersion oils can acquire somewhat higher fluorescence as they age. Care should be taken to keep bottles tightly capped when not in use, and to try a new bottle of oil with preparations for which a high background fluorescence is obtained but cannot be easily explained.

All staining and rinsing solutions should be filtered (0.2 μm) before use. A variety of filter types are suitable but Millipore cellulosic filters are typically used. Filtration will remove undissolved contaminants in stains as well as bacteria that might be present in significant quantities (even in distilled water). Although it is improbable that contaminating bacteria and other materials would be confused with protists, a clean preparation is much easier to examine and count.

One final note concerns the application of any of the following methods to counting protists. No single method is the best for all samples or for all situations. All of the fluorochromes differ somewhat in their ability to stain protists and their organelles, in their specificity for living biomass, and in their ability to allow other features to be distinguished (e.g., the autofluorescence of photosynthetic pigments). The methods also differ in their complexity. The quickest method (e.g., the acridine orange, AO, method, see below) may be the method of choice in situations where a known population is being counted (e.g., in a laboratory culture), but this method may be less appropriate than others for use with most natural samples. The amount of particulate material suspended in the water as well as the size and composition of the nanoplankton assemblage can influence the effectiveness of a method. The most appropriate method for a particular sample or situation is to some degree a matter of trial and error, and some preliminary work to determine the best method for a particular situation is encouraged prior to initiating a large-scale investigation.

Acridine Orange (AO)

The AO method was one of the first epifluorescence microscopical methods to gain wide use for counting aquatic bacteria.[7] Davis and Sieburth[1] formalized the use of this fluorochrome for enumerating heterotrophic nanoplankton (cells 2 to 20 μm in size), although previous studies had also used this fluorochrome for counting protista.[8] Heterotrophic nanoplankton are enumerated using this method by examining the nanoplankton counts from two subsamples, one unstained and one stained with AO. The number of phototrophic cells is counted in the unstained preparation (using the autofluorescence of photosynthetic pigments as a diagnostic tool), and the number of both phototrophic and heterotrophic cells is counted in the AO-stained preparation. The difference between the two counts is the number of heterotrophic cells.

1. A preserved water sample of 2 to 25 ml (depending on the trophic state of the ecosystem) is poured into the filter toward on which backing and membrane filters have been placed.

2. A volume of AO solution (0.1% W/V in distilled water, preserved with 1% formalin or glutaraldehyde) equal to 10% of the sample volume is added to the sample.

3. After staining for 3 min, the sample is gently filtered down, and 2 ml of distilled water is added to the sample as a rinse when the meniscus of the sample reaches the filter surface. Samples should be filtered using low vacuum pressure (<5 to 10 in. Hg). We generally remove the membrane filter from the tower while still under vacuum to minimize water retention by the filter and prevent cells from floating off the filter surface.

4. The filter may be mounted onto a glass slide by smearing a small drop of immersion oil onto the slide, placing the filter onto the area of the slide covered with oil, and then placing another drop of oil onto the top of the filter. An alternative mounting method is to fog the slide with a quick hard breath and then rapidly place the filter onto the slide and a drop of immersion oil onto the filter. This latter procedure must be performed before the moisture disappears from the slide. If performed correctly the filter will adhere and lie very flat against the slide.

5. After a drop of oil is placed on the filter it is covered with a coverslip. At this stage the mounted filter may be examined immediately or frozen for weeks to months before inspection. Sealing the edges of the coverslip with melted paraffin or fingernail polish is recommended for long-term storage to prevent movement of the coverslips and changes in the moisture content of the preparations.

6. After the filter is mounted onto a slide, it is examined via epifluorescence microscopy using a blue light filter set (e.g., Zeiss filter set 487709: BP450-490 exciter filter, FT510

dichromatic beam splitter, and LP520 barrier filter). Eukaryotic cells will appear yellow to orange in color.

The advantages of the AO technique (relative to other epifluorescence microscopical methods) are that the method is simple and quick (3-min staining time) and stains most cells thoroughly. Although AO stains nucleic acids preferentially, the cytoplasm and even flagella can be intensely stained. In addition, the fluorochrome itself is relatively stable. Working solutions can be kept in darkened containers at 4°C for several months. We routinely enumerate heterotrophic flagellates in cultures by adapting the AO direct count method. Several milliliters of a culture are preserved, incubated in the filter tower with 0.5 ml of AO solution (10 mg per 100 ml of filtered seawater + 6 ml of formalin) for 1 min, and collected onto a 0.8-μm black membrane filter. The cells fluoresce brightly and are easily counted. Typically the nucleus (DNA) of the cells fluoresces yellow, and the cytoplasm (presumably RNA) fluoresces reddish orange.

The disadvantages of the AO method primarily stem from the fact that its emission spectrum strongly overlaps with the emission spectrum for chlorophyll. Because of this shortcoming it is necessary to count two preparations to obtain one count of the heterotrophic nanoplankton in a sample. Another consequence of this problem is that it is difficult to obtain an accurate count of heterotrophic nanoplankton when the abundance of phototrophic cells is high relative to the abundance of heterotrophic forms because one must substract two large numbers (total nanoplankton − phototrophic nanoplankton) to obtain a small number (heterotrophic nanoplankton). Nonspecific staining of particulate material and a high degree of carcinogencity of AO are further limitations with this method. Finally, AO adheres to the filter tower and can contaminate subsequent unstained samples. Therefore, it is recommended that a separate tower assembly be used exclusively for AO staining.

Proflavin

The inability of the AO procedure to distinguish autofluorescent (i.e., phototrophic) nanoplanktonic protists from heterotrophic ones in a single preparation stimulated the development of several methods designed to preclude that problem. Haas[3] proposed the use of the protein-binding state proflavin to overcome this problem and allow the enumeration of autotrophs and heterotrophs in the same microscopical field.

1. A freshly collected, unpreserved water sample is added to a filtration tower mounted with black membrane and backing filters.

2. 20 μl ml⁻¹ of sample of a solution of proflavin (0.033% W/V in distilled water) is added to the unpreserved water and the contents gently agitated to mix in the stain.

3. After 2 min, 50 μl ml⁻¹ of glutaraldehyde (6.0% V/V in 0.2 μm prefiltered seawater, final volume 0.3%) is added to the sample and after gentle agitation the contents are allowed to sit for an additional 2 min.

4. The sample is then collected onto the filter using a low vacuum (<5 in. Hg), and mounted onto a slide with low fluorescence immersion oil (e.g., Cargille type LF).

5. The sample is inspected using blue light excitation (band pass filter of 450 to 490 nm) with an emission through a 500- to 530-nm long pass filter. Eukaryotic cells, including flagella, appear bright green on a black background; chlorophyll fluorescence appears bright red to orange-yellow; detritus appears a dull pink. The proflavin working solution can be stored at 4°C for months.

Under appropriate conditions the proflavin method affords the ability to simultaneously visualize nanoplankton cells and examine those cells for autofluorescence of photosynthetic pigments. Caution should be used when following this procedure, however, due to the low concentration of glutaraldehyde fixative which is likely to cause the loss of delicate flagellates and ciliates. We are not aware that the proflavin method has been tried with initially preserved cells or with other preservation techniques.

FITC

The use of fluorescein isothiocyanate (FITC) as proposed by Sherr and Sherr[9] also provides a method for visualizing nanoplanktonic cells in natural samples while simultaneously allowing the differentiation of phototrophic (autofluorescent) and heterotrophic forms. Like the AO technique, it is an adaptation of a method used originally for enumerating bacteria.[10] FITC is a protein stain and thus provides good, general cellular staining including flagellar staining. However, since FITC fluoresces most intensely at alkaline pH (9 to 10), the sample must be rinsed with high pH carbonate buffer solution, and the preparation mounted using FA mounting fluid (immersion oil of pH 9, Difco. Co., Detroit) between filter and cover slip.

1. A formalin-preserved sample is added to a filtration tower mounted with black membrane and backing filters. The sample is filtered at low vacuum pressure (<5 in. of Hg) onto the membrane filter.

2. The sample is then flooded with the FITC staining solution. (Note on incubation of the filter: liquid will gradually drip through a 0.8-μm membrane filter by gravity; this can be prevented by sealing off the tower from the atmosphere to maintain a positive pressure to the back of the sample filter during incubation.)

3. Following a 5- to 10-min incubation with FITC, the sample is filtered and rinsed twice with 10 ml of the cold sodium carbonate buffer, and mounted onto a glass slide as described above, using FA mounting fluid, pH 9.

4. The preparation is examined via blue light excitation (e.g., a Zeiss BG12 excitation filter, FT510 dichromatic beam splitter, and an LP510 barrier filter). Nanoplanktonic protists fluoresce apple green.

FITC has been applied successfully in coastal waters where particle concentrations can be quite high. Because nonspecific staining of particulate material is a problem with some of the fluorochromes (e.g., AO), the FITC method may be preferable over other methods in environments where the concentration of suspended particulate material is high. Additionally, FITC is not as significant a health hazard as nucleic acid fluorochromes (e.g., AO and DAPI), and the staining solution can be stored at 4°C in the dark for a week. However, it is possible to overstain with FITC so that chlorophyll autofluorescence is partially masked; optimum staining times should be empirically determined. The FITC fluorescence fades after about 20 to 30 s of illumination.

Primulin (Direct Yellow 59)

Caron[4] proposed using primulin, a UV-excited, blue-fluorescing protein stain to minimize overlap between the emission spectra of the fluorochrome stain and chlorophyll. This method was specifically developed to aid in the examination of samples from oceanic environments in which phototrophic assemblages are often dominated by very small cells with minute amounts of chlorophyll autofluorescence. A specific advantage of the primulin technique (and DAPI; see below) relative to some of the other fluorochromes is that the excitation and emission spectra of the molecule are lower than some of the other commonly applied fluorochromes. The excitation of the stained preparation with UV light allows visualization of the nanoplanktonic protists (primulin fluorescence), and by quickly switching to blue light excitation the investigator can determine which of these cells are phototrophic (by their chlorophyll autofluorescence). Because primulin fluoresces only minimally with blue light excitation, even small amounts of autofluorescence can be detected. Caron[4] used an HBO 50-W mercury lamp and Zeiss filter sets 487702 (G365 band pass excitation filter, an FT420 beam splitter, and an LP418 barrier filter) for UV excitation and 487709 (see AO method above) for blue light excitation. Binding and fluorescence of primulin is optimal at acid pH (around 4.0).

1. A working-strength solution of the stain is prepared. The solution is filtered through a 0.2-μm filter prior to use.

2. A preserved sample is added to a filtration tower mounted with black membrane and backing filters and gently filtered onto the membrane filter.

3. The sample on the filter is rinsed with two 1-ml aliquots of acid rinse (0.1 M Trizma HCl in distilled water, pH 4.0).

4. After the rinse solution is drawn through, the sample is flooded with the primulin solution and incubated for 15 min (taking care that the stain does not drip through the filter during the incubation, see note above).

5. After staining, the primulin solution is removed by gentle vacuum, the sample is rinsed with two 2-ml aliquots of the pH 4.0 rinse water, and then mounted onto a glass slide using Cargille type A immersion oil.

6. Eukaryotic cells are first visualized using UV excitation and then classed as heterotrophic or phototrophic on the basis of the presence of autofluorescing chloroplasts within the cell by switching to blue light excitation.

Caron[4] found that the primulin method yielded higher counts of phototrophic nanoplankton than did two other staining procedures: proflavin[3] and FITC,[9] and gave total nanoplankton counts equivalent to those determined using the AO enumeration method. Primulin staining has also been applied successfully to the enumeration of freshwater protists.[11-13] These studies have demonstrated that a lower concentration of primulin can be effective in some environments.

The primulin method has the advantage that the fluorochrome fluoresces only minimally with blue light excitation. For this reason it is particularly appropriate for environments dominated by very small phototrophs or in which the autofluorescence of the phototrophs is particularly weak (both situations are common in surface waters of the open ocean). It is, however, a longer staining procedure than the methods mentioned above, and nonspecific staining can lead to problems of high background fluorescence in particle-laden environments. Recent modifications of this method using lower concentrations of primulin in combination with high concentrations of DAPI have improved on this problem (see below). An additional *caveat* is the requirement that the microscope must possess a mechanism for rapidly (and conveniently) changing between the UV and blue light filter sets. Neofluor objective lenses must be used, as other types of lenses (e.g., Zeiss Planachromat) have coatings which absorb in the excitation range of the UV light filter set. A xenon or mercury lamp is required for sufficient UV output.

DAPI

The fluorochrome 4'6'-diamidino-2-phenylindole (DAPI) is highly specific for DNA, and in theory only fluoresces brightly when complexed with DNA (specifically, by intercalating between adenine-thymine pairs). When excited by UV light (365 nm), the DAPI-DNA complex fluoresces blue, while unbound DAPI and DAPI associated with detritus fluoresce yellow.[2] Porter and Feig[2] were among the first to suggest that DAPI could be used to visualize and enumerate aquatic microbes, including bacteria, phytoplankton, and heterotrophic protists. This method has gained wide acceptance as a method for counting bacteria using a relatively low concentration of DAPI (3 to 5 μg ml^{-1}). At this low concentration the nuclei of protistan cells stain intensely. Recently, it has been established that higher concentrations of DAPI (10 to 50 μg ml^{-1}) stain cells rather nonspecifically and can be used to visualize nanoplanktonic protists.

1. A working solution of 0.1 μg of DAPI ml^{-1} of distilled water is prepared from the stock solution and can be stored at 4°C in the dark for several weeks.

2. The working DAPI solution is filtered prior to use, and added at a ratio of approximately 1:10 stain:sample in a filter tower mounted with a black membrane filter and wetted Millipore backing filter. The sample should be kept from strong light during and after staining as DAPI is light sensitive.

3. After standing for >5 min, the sample is concentrated onto the membrane filter with gentle vacuum (<7 in. Hg) and the filter is mounted onto a slide (see above).

4. The preparation can be inspected via epifluorescence microscopy using a UV filter set (e.g., Zeiss filter set 487702 [see primulin method] or 477701: BP365/10 excitation filter, FT390 dichromatic beam splitter, and 395 barrier filter). Neofluor objective lenses must be used. Bacterial cells appear bright blue against a black background; eukaryotic cells are recognized by blue-fluorescing nuclei together with, in the case of phototrophs, dull-red fluorescing chlorophyll. (Note: chlorophyll fluorescence is poorly stimulated by this filter arrangement, and switching to blue light excitation is required for most small nanoplankton).

Although DAPI is a fluorochrome specific for DNA, in practice at high concentrations DAPI staining will cause the entire cell as well as flagella and cilia to fluoresce. We use concentrated DAPI (1.0 mg ml^{-1} of deionized water), adding 50 $\mu l/ml$ of sample, and incubating for 7 min or longer in the dark (up to overnight in sample vials). To conserve DAPI, which is the most expensive of the routinely used fluorochromes, a sample >5 ml can be filtered down to 5-ml volume in the filter tower before adding DAPI. DAPI solutions should be filtered through a 0.2-μm filter prior to use because mats of yellowish fibers (undissolved DAPI) can interfere with inspection of a prepared sample. Cells stained with concentrated DAPI have blue-fluorescing cytoplasm and bright blue-white nuclei.

The use of DAPI for visualizing and counting protists has many of the same advantages and disadvantages as the primulin technique, and is now often used in combination with that fluorochrome (see below). Differentiation of autofluorescence and fluorochrome fluorescence are maximized by the use of different filter sets. DAPI has extremely low fluorescence when excited with blue light, and therefore weakly autofluorescent phototrophs can be identified with this technique. In addition, DAPI-stained cells do not fade during illumination, as do cells stained with other fluorochromes (e.g., AO and FITC); thus, it is easier to make micrographs of DAPI-stained preparations.

The DAPI method, however, does require the use of two filter sets in most instances and so it is not convenient with some microscope designs. Also, DAPI, like AO, should be considered highly carcinogenic because of its ability to intercalate between nucleotide bases. A note of caution with respect to using DAPI for enumerating and sizing bacteria: we have unpublished data suggesting that DAPI-stained bacterial cells appear smaller in size, and in some cases are less numerous, compared to bacterial cells stained with AO.

Dual Staining Procedures

Samples may be stained with two different fluorochromes in order to visualize different parts of protistan cells. Sherr and Sherr[14] developed a double-staining procedure using the protein stain fluorescein isothiocyanate (FITC) and DAPI. In this procedure, the samples are prestained using 200 μl ml^{-1} of a DAPI solution (0.1 mg ml^{-1} distilled water) per 5 ml of sample, before staining with FITC. Eukaryotic cells are visualized via FITC fluorescence using blue light excitation, and then the cell nuclei and DAPI-stained bacteria within food vacuoles can be inspected by switching to UV excitation.

Double staining with primulin and DAPI also has been employed by some investigators in order to aid the visualization of protistan cells and stain both the cytoplasm and the nucleus.[15,16] These fluorochromes are both UV excited, so they present the possibility of thorough staining of the cells with minimal overlap of the emission spectrum of chlorophyll autofluorescence. The use of two filter sets, however, is required in order to properly visualize chlorophyll autofluorescence (as noted above).

The following method was described by Martinussen and Thingstad:[16]

1.   10 to 50 ml of a preserved seawater sample are gently filtered onto a 1.0-μm black membrane filter.

2.   The vacuum is turned off and the filter is flooded with 2 ml of DAP1 solution (50 μg ml^{-1}) and stained for 3 to 5 min.

3.   The filter is then rinsed twice with 1 to 2 ml of 0.1 M Trizma-HCL, pH 4.0, and flooded with 2 ml of primulin solution (250 μg ml^{-1} in 0.1 Trizma-HCl, pH 4.0).

4.   The primulin is filtered through, followed by two additional washes with 0.1 M Trizma-HCl rinse.

5.   The filter is placed specimen side up on a film of liquid paraffin (immersion oil can also be used) on a glass slide, a few drops of liquid paraffin are put in the center of the filter, and a coverslip placed on top.

Other Types of Fluorescent Stains

The concept of using epifluorescence microscopy to inspect microbes originated with biomedical researchers. Cell physiologists have a sophisticated arsenal of fluorochromes used to address specific questions; there are plenty of as-yet undiscovered applications for aquatic microbial ecologists to explore. For instance, there are fluorochromes specific for RNA, for cell membranes, and for mitochondria, as well as pH-sensitive fluorochromes. Klut et al.[17] investigated the use of 25 specific fluorochromes on live cells of 5 phytoplankton species, in order to visualize and identify specific subcellular compartments, storage products, and other metabolites. A variety of cellular stains is sold by Molecular Probes (Eugene, OR, 97402-0413, 503-344-3007), which has an excellent advisory service for individual research problems. An example of a specific stain is Nile Red, 9-diethylamino-5H-benzo[a]-phenoxazine-5-one, which fluoresces brilliant orange-red when in hydrophobic environments such as intracellular lipid droplets.[18-20] Cooksey et al.[19] used Nile Red to evaluate the relative production of lipids by freshwater algae, and Cole et al.[20] used Nile Red to monitor the concentration of lipid droplets within hymenostome ciliates over a cell division cycle. We have used Nile Red to visualize lipids within marine diatoms. The method of Cole et al.[20] for Nile Red is

1. A solution of Nile red dissolved in acetone (500 μg ml^{-1}) is added to preserved water samples to yield a final concentration of 5 μg ml^{-1} of Nile Red.

2. After staining for 30 min, the samples are gently centrifuged and rinsed prior to microscopic observation. There is little background fluorescence because Nile Red fluoresces only in the presence of lipids.

Enumerating Stained Protista

Estimates of the number of protista contained in a sample collected onto a membrane filter are made by enumerating cells of the category of interest (flagellates, ciliates, etc.) on some fraction of the filter. The portion of the filter examined can be either transects of carefully measured length across the filter surface (the length of these transects must be measured accurately with the stage micrometer), or a number of fields of view from various segments of the filter. It is important to note that protistan (and bacterial) cells will not be distributed evenly over the surface of the filter even under the best conditions for sample preparation. For this reason, care should be taken to include fields of view or transects from different segments of the filter (i.e., near the center, near the edges, etc.).

The abundance of protista in the original sample is determined knowing the sample volume filtered, the average number of protista per area of filter counted (e.g., the number of heterotrophic flagellates per transect or per field of view), the area of the filter covered by sample (determined by the filter funnel diameter), the area of the filter examined per transect or per field of view, and the dilution due to the addition of a preservative (a constant factor of 0.9 if a 1:9 preservative:sample ratio is employed). Some of these features are constants if the filtration apparatus, preservation method, and magnification remain constant over a series of counts, and can be combined for convenience into a single conversion factor. Some of these features (e.g., the sample volume), however, may vary between environments or within a sample for different population counts (e.g., flagellates and ciliates). The area of the filter examined per transect or per field of view differs, of course, with the objective lens used for counting the sample. Calibration of each microscope lens with a slide micrometer is necessary to obtain this latter measurement. Some careful bookkeeping is required to keep track of the different variables for each sample and population count.

Once the sample has been counted and all of the pertinent conversion factors have been determined, the abundance of protista per milliliter is calculated by the following equation (this example assumes the area counted is one field of view):

$$\#/\text{ml} = \frac{(\text{cells/field of view}) \times (\text{area of filter covered by sample})}{(\text{field of view area}) \times (\text{DF}) \times (\text{ml filtered})} \tag{1}$$

where DF = preservative dilution factor.

Heterotrophic protists are usually enumerated at magnifications of $630\times$ to $1000\times$. Typical abundances in most marine environment are 10^2 ml^{-1} to 10^3 ml^{-1}. The abundance of larger, rarer cells such as heterotrophic dinoflagellates and ciliates should be determined at a lower magnification (e.g., $200\times$ to $500\times$). The abundance estimates resulting from epifluorescence microscopical counts can be converted to biovolume (usually expressed as μm^3 ml^{-1}) by using either a predetermined average cell biovolume for all cells enumerated or by empirically estimating a biovolume for each cell enumerated in individual samples (by making microscopical measurements of the protista and calculating volume based on their approximate geometric shapes). The biovolume of small protista is often converted to biomass (with some trepidation) using published conversion factors (e.g., Reference 21).

REFERENCES

1. Davis, P. G. and Sieburth, J. McN., Differentiation of phototrophic and heterotrophic nano-plankton populations in marine waters by epifluorescence microscopy, *Ann. Inst. Oceanogr.*, 58(Suppl.), 249, 1982.
2. Porter, K. G. and Feig, Y., The use of DAPI for identifying and counting aquatic microflora, *Limnol. Oceanogr.*, 25, 943, 1980.
3. Haas, L. W., Improved epifluorescence microscopy for observing planktonic micro-organisms, *Ann. Inst. Oceanogr.*, 58(Suppl), 261, 1982.
4. Caron, D. A., Technique for enumeration of heterotrophic and phototrophic nanoplankton, using epifluorescence microscopy, and comparison with other procedures, *Appl. Environ. Microbiol.*, 46, 491, 1983.
5. Stoecker, D., Michaels, A. E., and Davis, L. H., Large proportion of marine planktonic ciliates found to contain functional chloroplasts, *Nature*, 326, 790, 1987.
6. Hobbie, J. E., Daley, R., and Jasper, S., Use of Nuclepore filters for counting bacteria by fluorescence microscopy, *Appl. Environ. Microbiol.*, 33, 1225, 1977.
7. Francisco, D. E., Mah, R. A., and Rabin, A. C., Acridine orange epifluorescence technique for counting bacteria, *Trans. Am. Micros. Soc.*, 92, 416, 1973.
8. Fenchel, T., Studies on the decomposition of organic detritus derived from the turtle grass *Thalassia testudinum*, *Limnol. Oceanogr.*, 15, 14, 1970.
9. Sherr, B. and Sherr, E., Enumeration of heterotrophic microprotozoa by epifluorescence mi-croscopy, *Est. Coast. Shelf Sci.*, 16, 1, 1983.
10. Fliermans, C. B. and Schmidt, E. L., Fluorescence microscopy: direct detection, enumeration, and spatial distribution of bacteria in aquatic systems, *Arch. Hydrobiol.*, 76, 33, 1975.
11. Carrick, H. J., Fahnenstiel, G. L., Stoermer, E. F., and Wetzel, R. G., The importance of zooplankton-protozoan trophic couplings in Lake Michigan, *Limnol. Oceanogr.*, 36, 1335, 1991.
12. Sanders, R. W., Porter, K. G., Bennett, S. J., and DeBiase, A. E., Seasonal patterns of bacterivory by flagellates, ciliates, rotifers, and cladocerans in a freshwater planktonic com-munity, *Limnol. Oceanogr.*, 34, 673, 1989.
13. Bloem, J., Bar-Gilissen, M. J. B., and Cappenberg, T. E., Fixation, counting, and manipulation of heterotrophic nanoflagellates, *Appl. Environ. Microbiol.*, 52, 1266, 1986.
14. Sherr, E. B. and Sherr, B. F., Double-staining epifluorescence technique to assess frequency of dividing cells and bacterivory in natural populations of heterotrophic microprotozoa, *Appl. Environ. Microbiol.*, 46, 1388, 1983.
15. Hobbie, J. E. and Cole, J. J., Response of a detrital food web to eutrophication, *Bull. Mar. Sci.*, 35, 357, 1984.
16. Martinussen, I. and Thingstad, T. F., A simple double staining technique for simultaneous quantification of auto- and heterotrophic nano- and pico-plankton, *Mar. Microb. Food Webs*, 5, 5, 1991.
17. Klut, M. E., Stockner, J., and Bisalputra, T., Further use of fluorochromes in the cytochemical characterization of phytoplankton, *Histochem. J.*, 21, 645, 1989.
18. Greenspan, P., Myer, E. P., and Fauler, S. D., Nile Red: a fluorescent stain for intracellular lipid droplets, *J. Cell Biol.*, 100, 965, 1985.
19. Cooksey, K. E., Guckert, J. B., Williams, S. A., and Callis, P. R., Fluorometric determination of the neutral lipid content of microalgal cells using Nile Red, *J. Microbiol. Methods*, 6, 333, 1987.
20. Cole, T. A., Fok, A. K., Ueno, M. S., and Allen, R. D., Use of Nile Red as a rapid measure of lipid content in ciliates, *Eur. J. Protistol.*, 25, 361, 1990.
21. Borsheim, K. Y. and Bratbak, G., Cell volume to cell carbon conversion factors for a bacter-ivorous *Monas* sp. enriched from seawater, *Mar. Ecol. Prog. Ser.*, 36, 171, 1987.

CHAPTER 27

A Quantitative Protargol Stain (QPS) for Ciliates and Other Protists

D. J. S. Montagnes and D. H. Lynn

INTRODUCTION

Ciliates may be a major component of planktonic food webs.[1] These protists act as heterotrophs, functional autotrophs, and mixotrophs[2-4] and may use resources more rapidly and efficiently than metazoan competitors. Further, they can ingest prey too small to be captured by zooplankton and thus may be important trophic intermediates.[5,6] Consequently, planktonic ciliates influence energy flux both as primary and secondary producers and by contributing to nutrient regeneration.[3,4,7,8]

Typically in food web studies, ciliates have been lumped into a few groups identified by silhouettes (all that can be seen in settling chambers). This has been an ecological and taxonomic oversight since ciliate assemblages can be diverse, occuping a number of trophic levels.[9-14] A major step toward rectifying this situation would be to partition ciliate biomass into trophic categories based on species.

Some recent studies have identified planktonic ciliate assemblages to the genus or species level.[15-18] Concomitantly, there has been an increased appraisal of the taxonomy of planktonic ciliates based primarily on protargol stained specimens;[9-11,17,19-26] we recommend these works for species identification but encourage use of our own work for description guidelines of the naked oligotrichs.[9-11,24-26]

Protargol is a recommended approach to examine ciliates.[27,28] Originally a brand name, protargol is now vernacular for the silver proteinate stain. This method binds silver proteinate to structures such as ciliary and cortical microtubules, basal bodies, and nuclei. The silver is then reduced to reveal the structures. The resulting preparation is virtually comprehensive in staining a mixed population of ciliates and also adequately stains many flagellates, diatoms, sarcodines, and metazoa.

Many ecologists view protargol silver staining of ciliates with trepidation, maintaining that it is difficult and time consuming. The QPS method is not difficult and is quicker than traditional methods (e.g., Reference 29). Settling chambers have been the routine method for estimating planktonic ciliate biomass. However, after accounting for delicate species, which may rupture, QPS is both as accurate and precise as settling chambers.[30] Admittedly, QPS takes longer than other means of enumerating plankton (e.g., settling chambers) but it provides unparalleled delineation of structures essential for identification of ciliates. Further, unlike other methods QPS provides a permanent and easily stored preparation.

0-87371-564-0/93/$0.00 + $.50

Modifications have been developed to increase the reproducibility and enhance the quality of protargol stains.[29,31-38] Below are directions for a slightly modified quantitative protargol stain (QPS),[30] descriptions of the reactions likely to occur during staining, and adjustments that can overcome the variability with which taxa react to staining.

MATERIALS REQUIRED

Fixation

Supplies

- Sampling bottle or gentle pump
- Bottles for storing fixed samples
- Large flasks or graduated cylinders for settling samples
- Siphon for removing upper water in settling vessels

Solutions and Reagents

- Formalin, 37% CH_2O
- Calcium carbonate, $CaCO_3$
- Picric acid, $(NO_2)_3C_6H_2OH$
- Glacial acetic acid, CH_3COOH
- Concentrated Bouin's fluid 5 to 10% (fixative volume/sample volume). To make concentrated Bouin's fluid saturate formalin with calcium carbonate (until sediment appears), decant, saturate with picric acid crystals, and decant. Then, just prior to fixation add 1% glacial acetic acid (fixative volume/sample volume).

The Quantitative Protargol Stain, QPS

Equipment

- 60°C water bath
- Heating plate
- Bunsen burner
- Millipore (or similar) filter holder with a fritted glass base for 22-mm filters

Supplies

- 0.8- to 3.0-μm porosity 22-mm filters comprised of mixed esters of cellulose (e.g., MF-Millipore). Use gridded filters if subsampling is desired.
- 5-μm porosity backing filters
- 22-mm² coverslips
- 22 × 40 mm coverslips
- Microscope slides
- pH paper
- Thin copper wire (1- to 3-mm diameter)
- Columbia staining jars: available from A. H. Thomas, Inc., P. O. Box 779 Philadelphia, PA 19105-0779

Solutions and Reagents

- Protargol powder is available from a number of sources including BDH, Associate of E. Merk, Frankfurter Strasse 250, D-6100 Darmstadt, Germany; Polyscience Inc., 400 Valley Road, Warington, PA 18976; sold as "strong silver Roques for Bodian method" from Roboz Surgical Co. Inc., 810 18th St. Northwest, Washington, DC 20006; Sigma Chemical Company, P.O. Box 14508, St. Louis, MO 63178-9916.
- Agar
- Potassium permanganate, $KMnO_4$
- Oxalix acid, $(COOH)_2 \cdot 2H_2O$
- Hydroquinone, $C_6H_6O_2$
- Sodium sulfite, Na_2SO_3
- Anhydrous sodium carbonate, Na_2CO_3
- Distilled water
- Reagent grade isopropyl alcohol, $CH_3CHOHCH_3$
- Reagent grade xylene, $C_6H_4(CH_3)_2$
- Hydrochloric acid, 0.1 N HCl
- Sodium hydroxide, 0.1 N NaOH
- 0.5% (weight/volume) gold chloride solution, $AgCl_4$
- Sodium thiosulfate, $Na_2S_2O_3 \cdot 5H_2O$
- Permount, available from Fisher Scientific (or another xylene- or toluene-based mounting medium)

PROCEDURES

Fixation

1. Plankton samples should be collected with a bottle or gentle pump, not a net since delicate forms will rupture.

2. Rapidly add the sample to the fixative; samples will stay preserved in this state for months.

3. Concentrate the sample by settling and decanting the upper fluid by siphoning.

4. Samples fixed in acid Lugol's fixative can often be successfully stained after postfixation with concentrated Bouin's fluid.

    a. First, transfer the sample to a tall flask (graduated cylinders work well), allow the sample to settle, and remove ~75% of the fluid by siphoning.

    b. Then, add 5 to 10% concentrated Bouin's fluid, without the glacial acetic acid, to the Lugol's fixed sample, mix gently, and leave for 12 to 24 h

    c. If the sample color has not changed to the characteristic "yellow" color of Bouin's repeat this procedure. Samples may then be stained.

The Quantitative Protargol Stain, QPS

1. Place the sample in a Millipore filter holder with a smooth fritted glass base for supporting filters (Figure 1) and a 5-μm porosity backing filter (for even dispersion). Use 22-mm filters (0.8- to 3.0-μm porosity) composed of mixed cellulose esters (gridded filters if subsampling is desired). NOTE: Do not use cellulose acetate filters as these do not withstand organic solvents.

2. Apply suction at ≤100 mmHg.

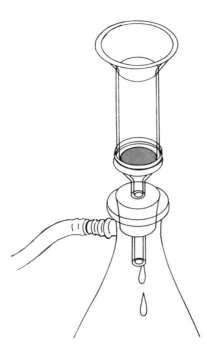

Figure 1.   The sample is placed in a Millipore filter holder with a smooth fritted glass base for supporting filters. NOTE: The filter is placed on a backing filter.

3.   Before all the liquid is removed, add water slowly to the side of the column and apply suction until no Bouin's (yellow color) remains; be careful not to resuspend settled cells since this results in uneven distribution. NOTE: An alternative method is to draw the sample onto the filter and proceed to step 3 without rinsing. If this is done, be sure that all the yellow color is removed during step 10.

4.   Suck water from the column until it is just empty (when the meniscus just disappears); otherwise, cells may rupture.

5.   Remove the filter and place it, residue up, on a piece of warm glass (30 to 40°C) (Figure 2). This removes some excess moisture from the filter and heats the melted agar (see below) so it does not solidify too quickly.

6.   Immediately place a drop of liquid agar (2.5% weight/volume) on a 22-mm² cover slip. Invert the coverslip, and keeping it parallel to the filter, lower it gently but quickly onto the filter (Figure 3). Observe the residue to ensure that it is not resuspended. If resuspension occurs, allow the filters to dry longer on the heated glass prior to embedding. Agar embeds and adheres the organisms to the filter while the coverslip produces a flat surface. Placing the agar on the cover slip, rather than on the filter, prevents resuspension and uneven distribution of the residue. NOTE: Heat the agar in a water bath (~60°C) to maintain a liquid state.

7.   Immediately apply pressure on the coverslip to obtain a thin layer of agar. (Figure 4).

8.   Allow the agar to solidify (2 to 5 min).

9.   Trim two of the four edges of the filter that extend past the cover slip and peel off the filter (Figures 5 and 6). This allows filters to fit in Columbia staining jars. All reagents, unless stated otherwise, are in these jars and the filters are transferred from jar to jar. However, fewer jars may be used; instead of moving the filters, the top of the jar may be partially covered and the contents poured out. NOTE: Use separate jars for the 100% isopropyl, xylene, and Permount steps since these must be void of water.

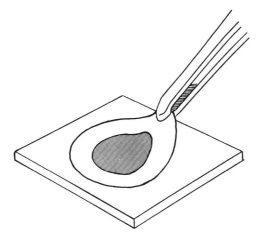

Figure 2.   The filter is placed, residue up, on a piece of warm glass (30 to 40°C).

10. As filters are processed place them back to back in a Columbia jar containing water (Figure 7).

11. Transfer the filters to a 0.5% (weight/volume) solution of potassium permanganate for 5 min. When not in use, store stock potassium permanganate in refrigerator. Discard when solution becomes brown. This step is a preparation for impregnation of stain by bleaching or oxidizing cytoplasmic components that otherwise become stained and obscure the desired structures.

12. Wash in water until the excess potassium permanganate is removed. Specimens should be brown and the water clear after ~10 min. NOTE: For all washes, transfer the filters

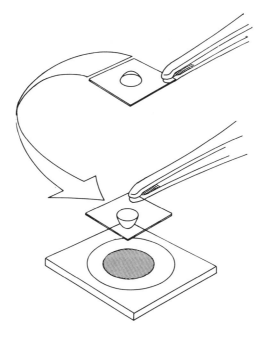

Figure 3.   A drop of liquid agar is placed on a cover slip, the coverslip is inverted, and lowered quickly onto the filter.

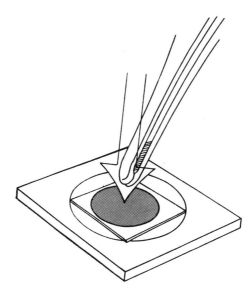

Figure 4. After placing the coverslip on the filter apply pressure on the coverslip to obtain a thin layer of agar.

to a Columbia jar filled with tap water and completely immerse it in a constantly flushed, 18 to 20°C water bath (too high temperatures may dissolve agar). This step is to remove excess potassium permanganate, which would inhibit the stain.

13. Place filters in 5% (weight/volume) oxalic acid for 5 min. This completely removes the potassium permanganate by reduction.

14. Wash and drain, as in step 10, for 10 min to remove the oxalic acid, which inhibits staining.

15. Place filters in 0.5 to 4.0% (weight/volume) protargol solution (the actual strength of the solution depends on the manufacturer, age of the silver proteinate, and species being stained). Increase the concentration of protargol if the cells are understained. To prepare the protargol solution, place a 0.5 g coil of thin copper wire in the bottom of the jar. Alternately, place coverslip-shaped pieces of copper sheet between the filters (flame the copper over a Bunsen burner until it is red hot then plunge it into 100% isopropyl alcohol

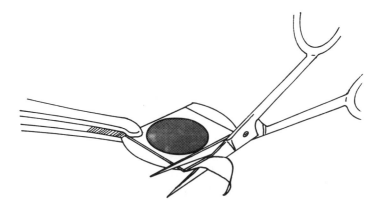

Figure 5. After sliding the cooled filter off the glass, trim two of the four edges of the filter that extend past the coverslip.

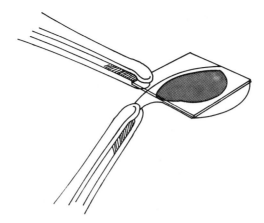

Figure 6. Peel the filter off the coverslip and discard the coverslip.

and rinse it in distilled water. This will expose a nonoxidized "bright" surface). Then sprinkle the protargol powder on the surface of the distilled water in the jar and allow it to dissolve without stirring. Test the pH of the solution using narrow range pH paper and adjust to pH ~8.8 with 0.1 N HCl or 0.1 N NaOH. The protargol solution and copper wire must be made fresh each run. However, the copper sheets may be cleaned and reused. Store protargol powder in a refrigerator.

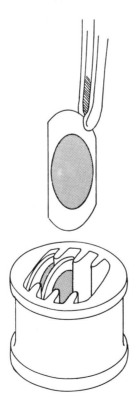

Figure 7. As filters are processed place them back to back in a Columbia jar containing water to await staining.

16. Leave filters in protargol for ~24 h (pH should shift to ~7 but good stains may result even without a pH shift).

17. Remove the filters from the protargol without draining and transfer them to hydroquinone solution: 1% (weight/volume) hydroquinone dissolved in 5% (weight/volume) sodium sulfite solution plus 4% anhydrous sodium carbonate for 5 to 10 min. Replace this solution each run. Store stock solution in the dark and, when it turns brown, discard. Hydroquinone reduces (develops) the silver. Sodium sulfite, an antioxidant, preserves the hydroquinone solution. Sodium carbonate increases the pH, facilitating the silver-reducing activity of hydroquinone.[32]

18. Wash as in step 10 for 5 min to stop the developing process.

19. Place filters in 0.5% (weight/volume) gold chloride. The duration of this step varies with the organisms being stained, and may be a few to several seconds. A suggested time is 1 s. Alternatively, use a dilute gold chloride dip (0.05%) and leave the filters in much longer (up to several minutes). The 0.5% gold chloride solution may be used repeatedly and should be stored in the dark. However, the 0.05% solutions must be replaced each run. The purpose is to "tone" the stain. Gold amalgamates with the silver proteinate and through reduction displaces some silver. This adds contrast and a reddish color to the stain.[37]

20. Place filters in 2% (weight/volume) oxalic acid solution for ~2 min. Overexposure at this stage may over hydrolyze structures and fade the stain. However, underexposure may result in poor toning. This procedure produces a purple-gray tone by reducing the gold-silver protein amalgamate and hydrolyzes obscuring cellular components.

21. Wash in water as in step 10 for 5 min to stop "bluing" process by removing oxalic acid.

22. Place filters in 5% (weight/volume) sodium thiosulfate solution for 5 min. This fixes the stain by reduction. Lengthening this step may degrade the stain.[37]

23. Wash as in step 10 for at least 5 min to remove excess sodium thiosulfate.

24. Dehydrate in isopropyl alcohol series (30-50-70-95-100-100-100%) for 5 min in each. Replace the 100% alcohol rinses every run and periodically change the others. The purpose is to dehydrate, since the clearing agent and mounting medium are not miscible with water.

25. Transfer the filters through three 100% xylene rinses of 5 min each. This makes the filters transparent and clears the cells. Rotate and replace the xylene so the final rinse is always new (i.e., 2 becomes 1, 3 becomes 2, and 1 is replaced and becomes 3). If xylene contains water, the filters may become opaque after mounting and hardening. CAUTION: The filters become very flexible at this stage. Be careful not to bend them.

26. Place the filters in a 20% xylene:80% Permount solution for a minimum of 0.5 h. This replaces xylene with the mounting medium, which prevents opaque air pockets forming when the Permount hardens.

27. Mount each filter with Permount on a slide by placing drops below and on the filter, then cover with a 22 × 40 mm² coverslip. Ensure that no air is trapped under the filter or coverslip. NOTE: If there is too much mounting medium or the filters have become bent during staining, the preparation will be too thick for oil immersion observation. If this is the case apply pressure to the coverslip while the Permount dries (a bent paper clip works well as a clamp or use a small weight on top of the coverslip). Excess mounting medium ensures that air pockets will not form during hardening.

28. If filters become opaque, the Permount has not completely infiltrated the filter or water still remains in the preparation. To salvage the filter, soak the slide in xylene, remove the filter, and repeat process from 100% isopropyl alcohol in step 24.

NOTES AND COMMENTS

1. Estimating ciliate numbers using either settling chamber (SC) or quantitative protargol stained (QPS) significantly underestimated the "true" density, established by direct count (DC), by ~13%. Further, the abundance of "delicate" species was underestimated by QPS since they ruptured on the filter.[30] Since the effect on delicate species will vary with fixation, technique, and species composition, we cannot provide a universal correction factor. Instead, we recommend for a comprehensive appraisal that QPS be initially tested against SC or DC[39,40] to determine the underestimation of species in natural populations.

2. Unlike settling chambers the QPS method (without using backing filters) fails to produce a Poisson distribution of cells on the filters; for natural populations, the coefficient of variation from replicate counts was reduced to near 15% by counting ~300 cells.[30] However, further investigation (unpublished results) indicates that using backing filters does result in a random distribution of cells. Clumping of residue can be a problem. This may be reduced by carefully homogenizing the sample prior to filtration and ensuring there is no air trapped under the filter during concentration.

3. DC provides the best estimate of density.[30,39,40] However, direct counting is not always practical for large collections of natural populations (although see Reference 40). Further, DC and SC cannot be used to identify species, genera, and sometimes even families. Therefore, to estimate the density and composition of a protist assemblage, QPS offers a more useful approach.

4. The quantitative protargol stain offers a number of advantages beyond identification and enumeration. (1) It provides a permanent record while SC does not. (2) It allows inferences of functional morphology (e.g., membranellar spacing, position, and structure of oral region) by observing stained specimens (see Figure 8). (3) It provides estimates of the trophic status of ciliates by observing vacuole contents. (4) It reveals specific division stages of ciliates and thus ontogeny may be described. (5) It provides a means to estimate species' growth rate from an enclosed natural assemblage of ciliates by serial sampling and staining. (6) It is shorter and more efficient than previously described protargol methods (e.g., Reference 29).

5. Some people use tap water to make up the protargol solution. At times this seems to give better results than using distilled water. The reason for this improvement is unclear but may be related to the "hardness" of the water. We have used distilled water that has been saturated with calcium carbonate, boiled, and decanted. This modification improved the quality of the stain by darkening structures that stained lightly when only distilled water was used. Again, the reason for these improvements is unclear. Protargol selectively stains structures with silver. Copper acts as a catalyst[37] and may agglomerate with the stain[38] to produce a blue tinge. The pH adjustment standardizes silver proteinate solutions and likely facilitates the silver proteinate binding to cell structures.

ACKNOWLEDGMENTS

The organization, presentation, and assessment of this method was improved considerably by Alan Martin, Julie Koester, and Colin Field.

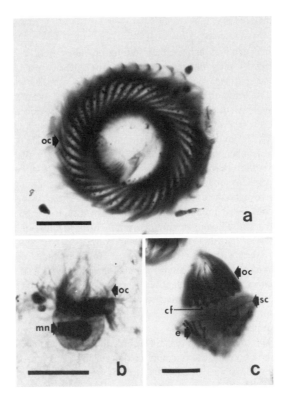

Figure 8. Three QPS stained ciliates. (a) *Strobilidium spiralis* (from Reference 9); (b) *Lohmanniella oviformis* (from Reference 9); and (c) *Strombidium acutum* (from Reference 10). Note the oral ciliature (oc) in all three ciliates, cytoskeletal fibers (cf), somatic ciliature (sc), and extrusomes (e) in *S. acutum* and the macronucleus (mn) in *L. oviformis*. These structures are revealed by QPS and are used to identify ciliates. Scale bars equal 20 μm.

REFERENCES

1. Lynn, D. H. and Montagnes, D. J. S., Global production of heterotrophic marine planktonic ciliates, in *Protozoa and their Role in Marine Processes,* Reid, P. C., Turley, C. M., and Burkill, P. H., Eds., Springer-Verlag, Berlin, 1991, 281.

2. Capriulo, G. M., Sherr, E. B., and Sherr, B. F., Trophic behaviour and related community feeding activities of heterotrophic marine protists, in *Protozoa and their Role in Marine Processes,* Reid, P. C., Turley, C. M., and Burkill, P. H., Eds., Springer-Verlag, Berlin, 1991, 219.

3. Smith, W. O., Jr. and Barber, R. T., A carbon budget for the autotrophic ciliate *Mesodinium rubrum, J. Phycol.,* 15, 27, 1979.

4. Stoecker, D. K., Mixotrophy in marine planktonic ciliates: physiological and ecological aspects of plastid-retention by oligotrichs, in *Protozoa and their Role in Marine Processes,* Reid, P. C., Turley, C. M., and Burkill, P. H., Eds., Springer-Verlag, Berlin, 1991, 161.

5. Porter, K. G., Paceo, M. L., and Battey, J. F., Ciliate protozoans as links in freshwater planktonic food chains, *Nature,* 277, 563, 1979.

6. Sherr, E. B. and Sherr, B. F., Planktonic microbes: Tiny cells at the base of the ocean's food webs, *Trends Ecol. Evol.,* 6, 50, 1991.

7. Azam, F., Fenchel, T., Field, F. G., Gray, J. S., Meyer-Reil, L.-A., and Thingstad, F., The ecological role of water-column microbes in the sea, *Mar. Ecol. Prog. Ser.,* 10, 257, 1983.

8. Gast, V. and Horstmann, U., N-remineralization of phyto- and bacterioplankton by the marine ciliate *Euplotes vannus, Mar. Ecol. Prog. Ser.,* 13, 55, 1983.

9. Lynn, D. H. and Montagnes, D. J. S., Taxonomic description of some conspicuous species of Strobilidiine ciliates (Ciliophora: Choreotrichia) from the Isles of Shoals, Gulf of Maine, *J. Mar. Biol. Assoc. U.K.,* 68, 639, 1988.

10. Lynn, D. H., Montagnes, D. J. S., and Small, E. B., Taxonomic description of some conspicuous species in the family Strombidiidae (Ciliophora: Oligotrichida) from the Isles of Shoals, Gulf of Maine, *J. Mar. Biol. Assoc. U.K.,* 68, 259, 1988.

11. Lynn, D. H., Montagnes, D. J. S., Dale, T., Gilron, G. L., and Strom, S. L., A reassessment of the genus *Strombidinopsis* (Ciliophora, Choreotrichida) with descriptions of four new planktonic species and remarks on its taxonomy and phylogeny, *J. Mar. Biol. Assoc. U.K.,* 71, 597, 1991.

12. Lynn, D. H. and Corliss, J. O., Ciliophora, in *Microscopic Anatomy of Invertebrates,* Vol. 1, Harrison, F. W. and Corliss, J. O., Eds., Wiley-Liss, New York, 1991, 333.

13. Montagnes, D. J. S., Lynn, D. H., Roff, J. C., and Taylor, W. D., The annual cycle of heterotrophic ciliates in the waters surrounding the Isles of Shoals, Gulf of Maine: an estimate of their trophic role, *Mar. Biol.,* 99, 21, 1988.

14. Small, E. B. and Lynn, D. H., Phylum Ciliophora, in *An Illustrated Guide to the Protozoa,* Lee, J. J., Hutner, S. H., and Bovee, E. C., Eds., Allen Press, Lawrence, KS, 1985, 393.

15. Gilron, G. L., Lynn, D. H., and Roff, J. C., The annual cycle of biomass and production of tintinnine ciliates in a tropical neritic region near Kingston, Jamaica, *Mar. Microb. Food Webs,* 5, 95, 1991.

16. Lynn, D. H., Roff, J. C., and Hopcroft, R. R., Annual abundance and biomass of aloricate ciliates in tropical neritic waters off Kingston, Jamaica, *Mar. Biol.,* 106, 437, 1991.

17. Martin, A. and Montagnes, D. J. S., An analysis of the winter assemblage of ciliates in a British Columbia fjord: Biomass, species composition and descriptions, and an assessment of their trophic impact, *J. Euk. Microbiol.,* in press.

18. Montagnes, D. J. S., The Annual Cycle of Planktonic Ciliates in the Waters Surrounding the Isles of Shoals, Gulf of Maine: Estimates of Biomass and Production, M.Sc. thesis, University of Guelph, Ontario, 1986, 114.

19. Krainer, K. H. and Foissner, W., Revision of the genus *Askenasia* Blockmann, 1895, with proposal of two new species, and description of *Rhabdoaskenasia minima* n. g., n. sp. (Ciliophora, Cyclotrichida), *J. Protozool.,* 37, 414, 1990.

20. Krainer, K. H., Contributions to the morphology, infraciliature and ecology of the planktonic ciliates *Sombidium pelagicum* n. sp., *Pelagostrombidium mirabile* (Penard, 1916) n. g., n. comb. and *Pelagostrombidium fallax* (Zacharias, 1896) n. g., n. comb. (Ciliophora, Oligotrichida), *Eur. J. Protistol.,* 27, 60, 1991.

21. Laval-Peuto, M. and Brownlee, D. C., Identification and systematics of the Tintinnina (Ciliophora): evaluation and suggestions for improvement, *Ann. Inst. Oceanogr.,* 62, 69, 1986.

22. Maeda, M. and Carey, P. G., An illustrated guide to the species of the family Strombidiidae (Oligotrichida, Ciliophora) free swimming protozoa common in aquatic environment, *Bull. Ocean Res. Inst., Univ. Tokyo,* 19, 68, 1985.

23. Maeda, M., An illustrated guide to the species of the families Halteriidae and Strobilidiidae (Oligotrichida, Ciliophora) free swimming protozoa common in aquatic environment, *Bull. Ocean Res. Inst., Univ. Tokyo,* 21, 67, 1986.

24. Montagnes, D. J. S., Lynn, D. S., Stoecker, D. K., and Small, E. B., Taxonomic description of one new species and redescription of four species in the family Strombidiidae (Ciliophora, Oligotrichida), *J. Protozool.,* 35, 189, 1988.

25. Montagnes, D. J. S., Taylor, F. J. R., and Lynn, D. H., *Strombidium inclinatum* n. sp. and a reassessment of *Strombidium sulcatum* Claparède and Lachmann (Ciliophora), *J. Protozool.,* 37, 318, 1990.

26. Montagnes, D. J. S. and Lynn, D. H., Taxonomy of choreotrichs, the major planktonic ciliates, with emphasis on the aloricate forms, *Mar. Microb. Food Webs,* 5, 59, 1991.

27. Kirby, H., *Materials and Methods in the Study of Protozoa,* University of California Press, Berkeley, 1950, 72.

28. Dale, T. and Small, E. B., Marine and estuarine strobilidiid oligotrich ciliates studied with protargol, *J. Protozool.,* 30(Suppl.), 16A, 1983.

29. Lee, J. J., Small, E. B., Lynn, D. H., and Bovee, E. C., Some techniques for collecting, cultivating and observing protozoa, in *An Illustrated Guide to the Protozoa,* Lee, J. J., Hutner, S. H., and Bovee, E. C., Eds., Allen Press, Lawrence, KS, 1985, 1.

30. Montagnes, D. J. S. and Lynn, D. H., A quantitative protargol stain (QPS) for ciliates: method description and test of its quantitative nature, *Mar. Microb. Food Webs,* 2, 83, 1987.

31. Antipa, G. A., The parlodion blanket, a method for affixing ciliates and other specimens to slides or coverslips before staining, *Trans. Am. Microc. Soc.,* 104, 287, 1985.

32. Aufderheide, K. J., An improvement in the Protargol technique of Ng and Nelson, *Trans. Am. Microsc. Soc.,* 101, 100, 1982.

33. Brownlee, D. C., Measuring the Secondary Production of Marine Planktonic Tintinnine Ciliates, Ph.D. thesis, University of Maryland, 1982, 202.

34. McArdle, E., Cytological staining of ciliate protozoa cultured on membrane filters, *Trans. Am. Micros. Soc.,* 97, 582, 1978.

35. McCoy, J. W., New features of the tetrahymenid cortex revealed by Protargol staining, *Acta Protozool.,* 8, 155, 1974.

36. Ng, S. F. and Nelsen, E. M., The Protargol staining technique: an improved version for *Tetrahymena pyriformis, Trans. Am. Microsc. Soc.,* 96, 369, 1977.

37. Tuffrau, M., Perfectionnements et pratique de la technique d'impregnation au protargol des infusoires ciliés, *Protistologica,* 3, 91, 1967.

38. Zagon, I. S., *Carchesium polypinum:* cytostructure after Protargol silver deposition, *Trans. Am. Microsc. Sco.,* 89, 450, 1970.

39. Dale, T. and Burkhill, P. H., Live counting — a quick and simple technique for enumerating pelagic ciliates, *Ann. Inst. Oceanogr.,* 58, 267, 1982.

40. Sime-Ngando, T., Hartmann, H. J., and Grolière, C. A., Rapid quantification of planktonic ciliates: comparison of improved live counting with other methods, *Appl. Environ. Microbiol.,* 56, 2234, 1990.

CHAPTER 28

Preparation of Pelagic Protists for Electron Microscopy

B. S. C. Leadbeater

INTRODUCTION

Electron microscopy has made a significant contribution to the study of pelagic protists in three respects: (1) in their identification, taxonomy, and systematics;[1-7] (2) in distribution studies, which in the last 20 years have focused on protists of the nanoplankton[8-11] and more recently on those of the picoplankton;[12,13] and (3) in ultrastructural, physiological, and biochemical studies. For the first two categories, information has generally been obtained from the study of intact whole cells and ultrathin sections, and both transmission electron microscopy (TEM) and scanning electron microscopy (SEM) have made a major contribution. For the third category of studies, information has come principally from TEM studies of sectioned material, although other specialized techniques and electron microscopes have been used. TEM has a resolving power of about 2Å and an effective magnification of $100,000\times$, whereas SEM has a resolving power of about 10Å and an effective magnification of 20,000 to $30,000\times$.

Since many pelagic protists are delicate and subject to distortion or disruption when handled, considerable care must be exercised when collecting and concentrating samples of plankton. Two major constraints associated with EM must always be borne in mind when selecting appropriate preparative techniques. These derive from the fact that, for observation, specimens must be exposed to a high vacuum and must withstand irradiation by a beam of electrons. Fixation of cells followed by dehydration, whether it be air drying or substitution of water with a dehydrating fluid, followed by critical-point drying or embedding in resin are therefore almost universal preparative operations before viewing of specimens is possible. The most favored fixatives used for specimen preparation are (1) glutaraldehyde, a bifunctional cross-linking agent which covalently complexes proteins and other molecules to their neighbors and (2) osmium tetroxide, used as a vapor or in solution, which binds and stabilizes lipid bilayers as well as tissue proteins. Neither fixative is capable of preserving all biological molecules but used in conjunction, either as glutaraldehyde followed by osmium tetroxide or simultaneously, satisfactory preservation can be achieved. To some extent choice of fixative will be dependent on the technique being used.

Although most techniques can be applied equally well to freshwater and marine protists, it is necessary to take account of the original salinity and osmolarity of samples when preparing some reagents. In particular, fixatives should be made up in solutions that are isotonic with the original water sample. This can be achieved by the addition of (1) a synthetic, defined salt solution and (2) nonionic compounds that are osmotically active such as sucrose. Most fixatives are also made up in a buffer, usually 0.1 M sodium cacodylate solution at a pH between 6.0 and 8.0. Standard phosphate buffer at concentrations between 0.05 and 0.1 M may be used with fixatives for freshwater protists. Organic reagents including sucrose should not be added to osmium tetroxide solutions.

All techniques described here have been used in qualitative studies of pelagic protists. Although some have also been applied to quantitative studies,[10,11] considerable difficulties remain over obtaining reliable numerical data. Three major criteria must be satisfied before quantification becomes possible: (1) concentration of samples must be achieved without disruption or loss of cells; (2) the precision with which the final volume of liquid is achieved must be scientifically acceptable; (3) the final distribution of cells, either dried down on a substrate or within thin resin sections, must be random. So far only samples prepared on membrane filters for SEM satisfactorily meet these criteria.

In the following discussion pelagic protists are taken to refer to microscopical pigmented and nonpigmented organisms of the nano- and picoplankton. Excluded from the discussion are the larger protozoa including ciliates, heliozoa, foraminifera, and radiolaria. In the text it is assumed that most workers will have access to a EM laboratory with standard equipment. Details of techniques are only given where they specifically apply to pelagic protists; standard, routine EM procedures will not be described.

MATERIALS REQUIRED

Equipment

- For collection, a 20 μm mesh nylon net
- Cooled incubator
- Top-pan balance
- Centrifuge: Centrifuges may be used for concentrating living or preserved protists on board ship or in the laboratory. Two types of centrifuge are available: (1) the cup-type centrifuge where the sample is enclosed within a tube; there is a choice between swing-out or fixed-angle rotors and (2) continuous centrifuge in which a continuous stream of sample is passed through the centrifuge chamber. Continuous centrifuges have been used for concentrating phytoplankton[14] and nanoplankton[15] and for some flagellates this method can be used satisfactorily for quantitative work. Only the procedure for use with the cup-type centrifuge will be described here.
- Millipore graduated 250-ml funnel (part of a Sterifil 47 mm Aseptic Filter holder)
- Stand with clamp to hold Millipore funnel
- Filtration rig for 25-mm diameter membrane filters
- Critical-point drying apparatus
- Vacuum evaporator
- Oven (60°C)
- Refrigerator

Supplies

- For collection, a collecting bottle, plastic funnel, and 5-l water container
- For cup-type centrifuges, standard tubes for holding samples are made of glass or plastic. Glass tubes have the advantage that cells do not adhere to the sides of the tube; they can be washed and sterilized and are not easily scratched. Plastic tubes have the advantage that they

are robust and do not break but when used for the first time cells may stick to the wall of the tube by electrostatic attraction.

- Pasteur pipettes, some with capillary tips, and teats. To produce capillary tip, hold the larger end of a pipette with the left hand and the narrower end with a pair of forceps. The narrow end is heated in the midregion with a small bunsen flame and when soft the pipette is removed from the flame and the tip pulled with forceps to form a fine capillary. The capillary tip is then carefully broken to give a smooth end.
- 1 m of plastic tubing and screw clip
- Cellulose nitrate membrane filters, 47-mm diameter, 0.45- or 0.8-μm pore size
- Unipore polycarbonate membranes, diameter 25 mm, pore size 0.45 μm
- Forceps for handling filters, 100-ml beaker
- Double-sided cellotape, silver base adhesive
- SEM sample stub, SEM
- Slides
- Petri dishes
- Stainless steel wire mesh
- Gold/palladium (60/40) wire, chromium granules, platinum/palladium (80/20) wire, carbon, or other metal as required
- Formvar/carbon coated copper grids, 3.05-mm diameter. Formvar/carbon coated grids should be exposed for 2 to 3 min to glow discharge ("ion cleaning") in a vacuum evaporator. This treatment enhances spreading of the drop containing suspended material on the grid surface.
- 7-ml glass specimen tubes with snap-on closures
- Embedding capsules, 8-mm diameter with pyramidal tips

Solutions and Reagents

A note of caution: appropriate safety precautions must always be taken when handling fixatives, reagents, and embedding resins because many are toxic and some are powerful oxidizing or reducing agents. All fixatives must be of electron microscope (EM) grade and other reagents of Analar grade.

- 0.1 M cacodylate buffer pH 7.0 prepared by dissolving sodium cacodylate in distilled water.
- Aqueous acetone solutions (10, 25, 50, 75, and 100% acetone)
- Ethanol/distilled water series (10, 25, 50, and 75% absolute ethanol)
- 10 ml aqueous solution of 1.0 mg ml^{-1} poly-L-lysine hydrochloride (mw = 21,000)
- 1 and 2% w/v osmium tetroxide in 0.1 M cacodylate buffer pH 7.0
- 0.5 to 1.0% w/v aqueous solution of uranyl acetate, membrane filtered (membrane pore size 0.22 μm); or 0.5 to 2.0% w/v aqueous solution of phosphotungstic acid, pH adjusted to neutrality with 1 N potassium or sodium hydroxide and membrane filtered before use; or 0.5 to 3.0% w/v aqueous ammonium molybdate solution
- 5% w/v glutaraldehyde solution in 0.1 M cacodylate buffer pH 7.0 (0.25 M sucrose added for marine protists)
- Propylene oxide, Epon resin

PROCEDURES

Collection of Protists

Collection of pelagic protists from marine and freshwater habitats is usually accomplished by use of water bottles. Most standard water collectors, such as Van Dorn, Nansen, Niskin, and Kammerer bottles are satisfactory. The volume of the sampler should be at least 2 l and the material from which the sampler is made should be tested with pelagic protists before use to ensure that no chemical component is released into the sample that might disrupt or kill cells. Net and pump plankton samplers are generally unsatisfactory because of the damage they inflict on delicate microscopical cells. Before use all collecting apparatus should be thoroughly washed with water from the habitat to be sampled to remove chemical contaminants. Under no circumstances should detergents, fixatives or other noxious chemicals be brought into contact with apparatus used for collecting and concentrating protists. For critical quantitative work, bottles should be washed with filtered water or use made of specialist collecting bottles which can be sterilized prior to sampling. After collection, water samples should be filtered through a 20-μm mesh nylon net to remove larger plankton and then processed as rapidly as possible. If water contains suspended particles of silt or debris, it may be necessary to allow samples to stand for a while to allow some sedimentation to occur before further processing can take place. If samples must be stored for any period of time they should be maintained at a temperature similar to that recorded at the time of sampling.

Concentration

Two methods of concentrating pelagic protists are commonly used: (1) centrifugation and (2) filtration. Advantages of centrifugation include (1) a rapid and effective method for concentrating cells for qualitative purposes and (2) can be used on small volumes of sample. Disadvantages of centrifugation include (1) damage to delicate cells and (2) reliable quantification depends on levels of precision that may be difficult to achieve. Filtration is generally less disruptive to delicate cells and if used carefully it can be used for quantitative estimates of cell concentrations.

By Centrifugation

1. Generally a cup centrifuge with swing-out rotor is most suitable for concentrating pelagic protists. Cells are concentrated at the extreme tip of each tube. Depending on the centrifuge, the rotor may contain eight or more cups which may hold 10-, 20-, or 50-ml tubes. In this way it is possible to concentrate between 8 and 400 ml of sample at a time.

2. Tubes should be filled with sample, adjusted to the same weight using a top pan balance, and arranged in pairs as is usual practice.

3. Centrifugation speed should be chosen to pellet cells within 5 to 10 min; the relative gravitational force should range from 200 to 1500 × g. It is desirable but not essential to have some form of temperature control.

4. After centrifugation, the supernatant is carefully removed from the tube with a Pasteur pipette without disturbing the pellet. For larger volumes the supernatant may be siphoned off.

5. For 10-ml samples 0.5 ml of liquid should be left in the tube. Very carefully the pellet of cells is resuspended in the remaining liquid, initially by shaking the tube and then by gently sucking the suspended cells up and down the capillary of a ''pulled-out'' Pasteur pipette. Care must be exercised not to remove liquid or cells from the tube during this process.

By Filtration

Concentration of protists is achieved by removal of water from a plankton sample through a membrane filter.[16] If the sample contains a low concentration of cells then a 47-mm filter, pore size 0.8 or 0.45 μm is recommended.

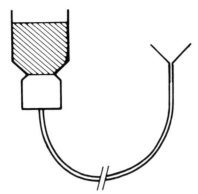

Figure 1. Diagram showing filter funnel and tubing required for concentration of protists by gravity filtration.

1. The simplest method of filtration is to attach a length of plastic tubing to the base of the membrane filter holder (Figure 1), prime the filter with sample allowing the tubing to fill with an unbroken column of water, and then allow gravity filtration. The liquid in the funnel must be continually topped up so that the membrane is not permitted to dry out. To obtain an adequate cell suspension it may be necessary to filter 1 l or more of liquid in this way. The most appropriate sample volume for a particular location must be determined empirically.

2. To terminate filtration the lower end of the plastic tubing is raised to the level of the funnel and a screw clip on the tubing closed.

3. The concentrated cell sample is gently swirled to dislodge cells adhering to the membrane and then subsamples can be removed for whole mount preparations or centrifuged to form a pellet for fixation and embedding.

By Filtration for SEM and Quantitative Studies

This technique differs from that described above in that usually a smaller diameter filter, 25-mm diameter, is used and a smaller volume of sample is filtered.[10,11,17] The object is to deposit all particles in the sample, above a stated pore size, onto the membrane surface. Advantages of this method include (1) with adequate apparatus specimens can be prepared *in situ*; (2) direct counts of component species are possible; and (3) many species can be identified. Disadvantages of the method include (1) it is a very time consuming way of quantifying taxa; (2) SEM has limited resolution and some species with detailed ultrastructural features cannot be identified with certainty; and (3) quantitative methods are not entirely satisfactory; the question of whether or not cells settle onto membranes with a random distribution has not been adequately resolved.

1. The filter funnel is usually attached to a vaccum pump and a gentle vacuum, no more than 60 mmHg, applied.

2. The funnel is loaded with about 20 ml of sample and the liquid drawn down until about 5 ml remain above the membrane surface.

3. 0.5 ml of a 1% aqueous solution of osmium tetroxide with 0.1 M cacodylate buffer at pH 7.0 is added and cells allowed to fix for 10 min.

4. Fixation is followed by addition of 10 ml of cacodylate buffer and this is slowly sucked through the filter to be followed by 10-ml aliquots of a series of acetone solutions in

distilled water (10, 25, 50, 75, and 100%). Each solution is added dropwise over 5 to 10 min and the filtration allowed to proceed slowly never allowing the membrane to dry out.

5. When 100% acetone has been added, the filter is ready for transfer to a critical-point-drying apparatus. This requires considerable care so as to minimize disturbance of cells on the filter surface and to prevent drying of the filter. A 100-ml beaker containing 40 ml of 100% acetone is prepared; the filter funnel is dismantled and the moist membrane containing cells is transferred immediately, using a pair of flattened forceps, and immersed into the acetone in the beaker.

6. The membrane can now be transferred to a critical-point drying apparatus already loaded with 100% acetone. Critical-point drying is by standard means.[17]

7. The dried membrane is mounted on to a SEM sample stub with double-sided cellotape and a silver base adhesive.

8. Membranes mounted on stubs are sputter coated with gold or another metal of choice and viewed with a SEM.

There are numerous variations of this protocol, for instance, a 5% glutaraldehyde solution in 0.1 M cacodylate buffer with 0.25 M sucrose added for marine specimens can be used (one part to four parts of sample) instead of osmium tetroxide solution. However, osmium tetroxide is usually considered to be superior.[17] Desalination of marine specimens can be achieved by passing a decreasing series of seawater and distilled water mixes through the filter prior to dehydration. Other dehydrating agents such as ethanol may be substituted for acetone.

Since all particles in a known volume of liquid are deposited on the membrane surface it should be possible to obtain reasonably accurate quantitative assessments of species concentrations. Booth et al.[10] enumerated cells along random transects of the mounted membrane, where a single transect represented 0.036 ml of the original sample.

Preparation of Whole Mounts of Intact Cells

Advantages of preparing whole mounts for TEM are that it is (1) a rapid and relatively easy technique and (2) excellent for qualitative identification purposes; loaded grids can be stored for years. Disadvantages include (1) cells shrink and collapse during drying and therefore allowances for shrinkage must be made when measuring the dimensions of cells.

In advance of requirement, 3.05-mm diameter copper grids should be coated with a Formvar plastic film and a thin layer of carbon following a standard protocol. When required for use, Formvar/carbon grids may be further coated with polylysine.[18] This enhances the adherence of cells to the plastic film on the grid and permits subsequent addition or removal of liquids to the grid without undue disturbance of the attached cells. To obtain a polylysine coating, a drop of an aqueous solution of 1 mg ml^{-1} poly-L-lysine hydrochloride in distilled water is placed on the plastic/carbon film covering a grid and rinsed off with distilled water after 1 to 10 min.[18] After washing, grids should be stored in a dust-free atmosphere either in a grid box or small petri dish.

Two techniques for the preparation of whole mounts can be recommended. Method 1 involves the processing of a cell suspension on the grid,[19] whereas method 2 involves fixing and washing cells before loading grids. The first method has the advantage of minimizing cell loss during processing but, because the second method does not involve drying down salt crystals onto the grid, cleaner preparations are usually obtained.

Problems commonly encountered in the preparation of whole mounts include (1) overloading the grid with cells which may be caused by having a too concentrated suspension

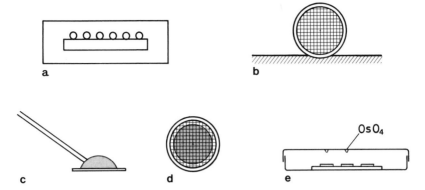

Figure 2. Preparation of whole mounts of intact cells. (a) Slide with strip of double-sided cellotape with attached grids; (b) higher magnification showing attachment of grid to cellotape; (c and d) addition of drop of concentrated cells to grid; (d) shows diameter of drop in relation to diameter of grid; (e) slide with attached grids in Petri dish containing drops of 1% osmium tetroxide solution.

of cells and/or loading a grid with too large a drop of liquid; (2) underloading the grid so that too few cells are present in which case the concentration process should be adjusted; (3) cells aggregated into clumps so that they cannot be recognized individually. This can be overcome by gently shaking the concentrate to resuspend cells; vortexing may become necessary but this must be done with care to avoid damaging cells; (4) contaminating precipitate may spoil clarity of material; contamination can arise from a number of sources including (i) overexposure of sample to osmium tetroxide vapor which results in opaque deposits on grid; (ii) incomplete washing may leave undissolved salt crystals, (iii) excessive debris in original sample may overlie specimens, (iv) oil or other pollutants either in original sample or from dirty equipment may produce film on specimen; and (5) Formvar film may tear during preparation; this may result from too thin a film or an imperfect film already punctured with holes.

Method 1

1. Coated grids should be secured by their under edge to the edge of a strip of double-sided cellotape attached to a slide (Figure 2a and b).

2. A drop, about 5 µl, containing suspended cells is placed on the coated surface of the grid by means of a "drawn out" Pasteur pipette. The drop produced should occupy about three quarters the diameter of the grid (Figure 2c and d). Care must be taken not to (i) damage the grid with the pipette tip whilst transferring the drop, (ii) get liquid onto the under surface of the grid, or (iii) overload the grid with cells.

3. Fixation of cells within the drop is achieved by exposing loaded grids to osmium tetroxide vapor. To achieve this the slide containing the loaded grids is placed in a Petri dish and two or three drops of a 1% w/v aqueous solution of osmium tetroxide are placed on the inner surface of the lid which is then replaced on to the base of the Petri dish (Figure 2e). Osmium tetroxide solution readily volatilizes and after 30 s to 1 min the cells within the drop will be fixed.

4. The slide containing the grids should then be removed from the Petri dish and placed in a clean Petri dish to so that the drops of liquid will dry. Drying can be expedited by placing the Petri dish containing the slide in an incubator at about 40°C.

5. Once dry, grids are carefully removed from the cellotape with fine forceps and washed thoroughly in a jet of distilled water delivered by a wash bottle to remove crystallized salt.

6. After 1 min the grid is dried by bringing a sliver of filter paper into contact with the edge of the grid to remove excess water.

An alternative method of washing is to place grids, Formvar film upward, on a wire mesh immersed in water in a Petri dish. After 10 min the grid, held by fine forceps, is gently moved around in the water and then removed and dried as above.[19]

Method 2

1. A sample of protists is placed in a 10-ml centrifuge tube and concentrated into a pellet in a bench centrifuge.

2. The supernatant liquid is discarded and the pellet resuspended in about 0.5 ml of liquid.

3. Three or four drops (about 10 μl) of 1% osmium tetroxide solution in 0.1 M cacodylate buffer at pH 7.0 are added and the suspension shaken.

4. After about 1 min the tube is filled with distilled water to wash cells and centrifuged to form a pellet.

5. This is repeated twice more.

6. After the final wash the pellet is resuspended in about 0.5 ml distilled water and small drops loaded on to Formvar/carbon coated grids and allowed to dry.

Shadowcasting Whole Amounts for TEM

Shadowcasting specimens with metal in a vacuum is a standard procedure in most EM laboratories and only brief mention of the technique will be made here.[20] Metals commonly used include gold/palladium, chromium, platinum/palladium, and platinum/carbon. Choice of metal depends on personal preference, although chromium and platinum are generally considered to produce a very fine grain deposit. Platinum/carbon shadowcasting is obtained by the simultaneous evaporation of platinum and carbon. Angles of shadowcasting vary from 15° for small objects with smooth surfaces to 45° for large objects. The optimal thickness of metal must be judged empirically.

Shadowcasting specimens with a metal has the effect of enhancing contrast and creating a three-dimensional effect. The technique has been used extensively with microscopical protists and is particularly suitable for enhancing surface features of flagella and cell coverings. Shadowcasting is also used in conjunction with freeze-fracture techniques.

Negative Staining

The negative staining technique involves surrounding or embedding a particle on a coated grid within an electron-dense material so that when viewed with a TEM the background appears electron dense while the particle is seen in relief.[21] Solutions of heavy metal salts such as uranyl acetate, ammonium molybdate, or sodium phosphotungstate provide the electron opaque deposit. This technique is most commonly used for observing surface features of cells such as scales, cellulose and carbohydrate fibrils, flagellar appendages, and trichocysts.

Problems commonly encountered with negative staining include (1) overstaining or understaining; judging the correct amount of stain can only be achieved with experience and (2) introduction of particles of dirt and debris to the grid. This may result from deposition of particles in the original cell sample or the stain may require filtering. Most stains will form colloidal precipitates if stored for more than a few days and should be membrane filtered prior to use. Stains should be stored in tightly stoppered bottles and kept in a refrigerator.

1. The method for the preparation of protists for negative staining is similar to that described for producing whole mounts with washed cells (see above). A small drop of cell suspension is placed on a grid and allowed to stand so that cells will settle onto the Formvar/carbon surface.

2. After 1 to 5 min, excess liquid is drawn off with a narrow sliver of filter paper (Whatman No. 1) but the surface of the grid must be kept moist.

3. A small drop of negative stain is then added with the capillary of a Pasteur pipette to the layer of cells on the grid and withdrawn with the pipette, and then a second drop is added and allowed to stand for 5 to 10 min.

4. After removal of the drop of stain with filter paper and drying at room temperature the grid can be viewed with a TEM.

Preparation for Embedding and Sectioning

There are many protocols available for fixation and embedding although most, in principle, consist of aldehyde fixation followed by postosmication, dehydration, and embedding. The protocol detailed here has been used by many authors.[22]

1. Samples of water containing protists, concentrated by filtration or centrifugation as described above, should be transferred into a 10-ml glass centrifuge tube or microcentrifuge tube for further concentration into a pellet of about 3 mm³. Pellets smaller than 3 mm³ will prove difficult to handle and larger size pellets may not process thoroughly.

2. Prior to fixation the pellet of cells should be resuspended in a small volume (0.5 ml) of sample liquid. All treatments should be carried out at 4°C on ice.

3. To the suspension of cells, about 4 ml of 5% glutaraldehyde in 0.1 M cacodylate buffer at pH 7.0 is added. For marine specimens the glutaraldehyde solution should also contain 0.25 M sucrose.

4. Fixation times vary but 1.5 h is usual. After this time the cell suspension is pelleted by centrifugation, the glutaraldehyde removed with a Pasteur pipette, and the pellet washed with three 10-ml rinses of an aqueous solution of 0.1 M cacodylate buffer at pH 7.0, 10 min for each wash. It is usual to resuspend the pellet before each wash and repellet cells by centrifugation prior to removing the buffer. If the pellet is unstable and likely to fragment, resuspension of cells should be avoided.

5. After removal of the third wash, 1 ml of a 2% solution of osmium tetroxide in 0.1 M cacodylate buffer is added and the pellet left for 1.5 h.

6. Three further washes with cacodylate buffer follow but at this stage the pellet must not be resuspended.

7. The pellet is next dehydrated in an ethanol series (10, 25, 50, 75, and 100%), 15 min in each solution.

8. When in absolute ethanol the tube is brought to room temperature and the pellet is dislodged from the bottom of the centrifuge tube by carefully cutting the pellet with the tip of a mounted needle.

9. The next steps must be accomplished with great care. A 7-ml glass specimen tube is filled with absolute alcohol. The fragments of pellet immersed in ethanol in the centrifuge tube

are very gently sucked up into a Pipette, consisting of a 10-cm length of 3-mm diameter glass tubing, fitted with a teat and quickly transferred to the specimen tube containing absolute alcohol.

10. The absolute alcohol is changed twice at 15-min intervals always leaving the pellets immersed in liquid and then the alcohol is exchanged for propylene oxide (care must be exercised when handling this reagent since it is toxic and rapidly vaporizes).

11. This is followed by three changes of propylene oxide for 15 min each.

12. The amount of propylene oxide is then reduced so that the pellet fragments are immersed in 2 to 3 ml and an equivalent volume of the embedding resin is added thereby giving a 1:1 mixture. Pellets are left in this mix for 12 h.

13. The mixture of resin and propylene oxide is removed from the specimen tube with a Pasteur pipette and replaced with resin.

14. The specimen tube is stoppered and placed in a rotating holder for 12 h.

15. Pellet fragments are then placed in embedding capsules, a numbered label added, and the capsule filled with resin.

16. The resin is hardened by placing in an oven at 60°C for 12 h.

17. Sections are cut by standard procedure on an ultramicrotome, picked up on Formvar/carbon coated grids, stained with saturated aqueous uranyl acetate for 30 min at 60°C, followed by Reynold's lead citrate at room temperature. Grids are washed with distilled water after treatment with each stain and viewed with a TEM.

Variations of this protocol sometimes encountered include (1) inclusion of cells into a pellet of agar prior to fixation. This necessitates adding molten agar at about 60°C to protists. The advantage of this technique is that cells are bound together but the temperature shock does not improve the state of cells. (2) 4% paraformaldehyde may be added to or substituted for glutaraldehyde. (3) Osmium tetroxide may be added to glutaraldehyde after 15-min fixation. (4) Uranyl acetate staining may be carried out *en bloc* either before or after dehydration. In the former case the stain is in aqueous solution and in the latter in ethanol solution.

An elaboration of this technique that is particularly useful for protists is flat embedding, whereby fixes and dehydrated cells are embedded in a thin layer of resin.[23] Individual protists can be viewed with light microscopy in the resin, ringed, cut out with a scalpel, and mounted onto a resin block for sectioning. In this way individually selected protistan cells can viewed with the TEM.

REFERENCES

1. Leadbeater, B. S. C., Identification, by means of electron microscopy, of flagellate nanoplankton from the coast of Norway, *Sarsia,* 49, 107, 1982.
2. Chrétiennot-Dinet, M. J., *Atlas du Phytoplancton Marin,* Editions du CNRS, Paris, 1990.
3. Dodge, J. D., *Atlas of Dinoflagellates,* Farrand Press, London, 1985.
4. Hallegraeff, G. M., *Plankton: A Microscopic World,* E. J. Brill, Leiden, 1988.
5. Ogden, C. G. and Hedley, R. H., *An Atlas of Freshwater Testate Amoebae;* Oxford University Press, Oxford, 1980.
6. Round, F. E., Crawford, R. M., and Mann, D. G., *The Diatoms; Biology and Morphology of Genera,* Cambridge University Press, Cambridge, 1990.

7. Lewis, J. and Dodge, J. D., The use of the SEM in dinoflagellate taxonomy, in *Scanning Electron Microscopy in Taxonomy and Functional Morphology,* Clanger, D., Ed., Clarendon Press, Oxford, 1990.
8. Manton, I. and Leadbeater, B. S. C., Fine-structural observations on six species of *Chrysochromulina* from wild Danish marine nanoplankton, including a description of *C. campanulifera* sp. nov. and a preliminary summary of the nanoplankton as a whole, *Det Kong. Dan. Vid. Selsk. Biol. Skrift.,* 20, 5, 1974.
9. Thomsen, H. A., A survey of the smallest eucaryotic organisms of the marine phytoplankton, in *Photosynthetic Picoplankton,* Platt, T. and Li, W. K. W., Eds., *Can. Bull. Fish. Aquat. Sci.,* 214, 121, 1986.
10. Booth, B. C., Lewin, J., and Norris, R. E., Nanoplankton species predominant in the subarctic Pacific in May and June 1978, *Deep-Sea Res.,* 29, 185, 1980.
11. Paerl, H. W. and Shimp, S. L., Preparation of filtered plankton and detritus for study with scanning electron microscopy, *Limnol. Oceanogr.,* 18, 802, 1973.
12. Johnson, P. W. and Sieburth, J. McN., *In situ* morphology and occurrence of eucaryotic phototrophs of bacterial size in picoplankton of estuarine and oceanic waters, *J. Phycol.,* 18, 318, 1982.
13. Silver, M. W., Gowing, M. M., and Daroll, P. J., The association of photosynthetic picoplankton and ultraplankton with pelagic detritus through the water column (0-2000m), in *Photosynthetic Picoplankton,* Platt, T. and Li, W. K. W., Eds., *Can. Bull. Fish. Aquat. Sci.,* 214, 311, 1986.
14. Kimball, J. F., Jr. and Wood, E. J. F., A simple centrifuge for phytoplankton studies, *Bull. Mar. Sci. Gulf Caribb.,* 14, 141, 1964.
15. Marchant, H. J., Choanoflagellates in the Antarctic marine food chain, in *Antarctic Nutrient Cycles and Food Webs,* Siegfried, W. R., Condy, P. R., and Laws, R. M., Eds., Springer-Verlag, Berlin, 1985, 271.
16. Fournier, R. O., Membrane filtering, in *Phytoplankton Manual,* Sournia, A., Ed., UNESCO, Paris, 1978, chap. 5.4.2.
17. Pickett-Heaps, J. D., Preparation of algae for scanning electron microscopy, in *Handbook of Phycologiocal Methods,* Vol. 3, Gantt, E., Ed., Cambridge University Press, Cambridge, 1980, chap. 29.
18. Marchant, H. J. and Thomas, D. P., Polylysine as an adhesive for the attachment of nanoplankton to substrates for electron microscopy, *J. Microsc. Soc.,* 131, 127, 1983.
19. Moestrup, Ø. and Thomsen, H. A., Preparation of shadowcast whole mounts, in *Handbook of Phycologiocal Methods,* Vol. 3, Gantt, E., Ed., Cambridge University Press, Cambridge, 1980, chap. 31.
20. Bradley, D. E., Replica and shadowing techniques, in *Techniques for Electron Microscopy,* 2nd ed., Kay, D., Ed., Blackwell Scientific, Oxford, 1967, chap. 5.
21. Gantt, E., Replica production and negative staining, in *Handbook of Phycologiocal Methods,* Vol. 3, Gantt, E., Ed., Cambridge University Press, Cambridge, 1980, chap. 30.
22. Reimann, B. E. F., Duke, E. L., and Floyd, G. L., Fixation, embedding, sectioning and staining of algae for electron microscopy, in *Handbook of Phycologiocal Methods,* Vol. 3, Gantt, E., Ed., Cambridge University Press, Cambridge, 1980, chap. 23.
23. Reymond, O. L. and Pickett-Heaps, J. D., A routine flat-embedding method for electron microscopy of microorganisms allowing selection and precisely oriented sectioning of single cells by light microscopy, *J. Microsc.,* 130, 79, 1983.

CHAPTER 29

A Rapid Technique for the Taxonomy of Methanogenic Bacteria: Comparison of the Methylreductase Subunits

Pierre E. Rouvière and Carla H. Kuhner

INTRODUCTION

Methanogenic bacteria are found in all aquatic environments where strict anaerobic conditions are met such as the digestive tract of animals, freshwater and marine sediments, decaying organic matter, and geothermal springs. They are responsible for the last step in the anaerobic degradation of organic matter by consuming reducing equivalents and producing CH_4.[1,2] Even though they form a very deep branching group phylogenetically,[3] their energy metabolism is highly conserved.[4] They are obligate methane producers and can only use a very limited number of simple carbon substrates (CO_2, formate, methanol, mono-, di-, and trimethylamines, dimethylsufide, acetate).[1,2,5] Their nutrition is therefore of limited help in assigning new isolates to a precise taxonomic position. Only one family, the Methanosarcineae, can be characterized by the carbon substrates it can specifically use (methanol, methylamines, and acetate). Morphology in a few instances is characteristic of a genus, i.e., *Methanobacterium, Methanosarcina, Methanospirillum,* or *Methanothrix*, but does not usually provide information on the relationship within a genus or a family. Other methods of comparison have been proposed for the taxonomy of methanogens (subunit composition of the RNA polymerase,[6] DNA-DNA hybridization, distribution and composition of polyamines,[7] or lipids,[8] antigenic cross-reactivity[9]). However, these techniques are often of limited use either because they do not allow comparison among all methanogen species, they are dependent on the growth conditions, or they are too complex. Also, other physiological characteristics such as growth condition optima, trophic requirements, or metabolic capabilities such as the utilization of alcohols as electron source or the ability to fix nitrogen are widespread among all methanogen species and cannot be used for taxonomic purposes. The detailed phylogenetic analysis of the methanogens has been established by comparing the sequences of 16S rRNA. It constitutes the framework for the taxonomy of methanogens based on evolutionary relationship.[2]

We describe here a simple taxonomic method based on the comparison of the apparent molecular weight of the three subunits of the methylreductase enzyme, also known as component C, the enzyme which catalyzes the last step of methane production.

This technique consists of a quick one-step chromatographic enrichment of the methylreductase protein, followed by its purification to homogeneity using nondenaturing polyacrylamide gel electrophoresis. The protein is then analyzed by denaturing sodium dodecylsulfate (SDS) polyacrylamide gel electrophoresis and the apparent molecular weights of its three subunits compared to those of proteins from known species.

0-87371-564-0/93/$0.00 + $.50
© 1993 by Lewis Publishers

Several characteristics make the methylreductase protein particularly useful to identify methanogenic bacteria and determine their phylogenetic positions and relationships:

1. This enzyme is present only in methanogenic bacteria and is conserved among all the species, implying that this technique can be used with all methanogenic species (unlike DNA-DNA hybridization which can only be used between closely related species).
2. The methylreductase is very easy to purify. It is extremely abundant, accounting for up to 10% of total cell protein in some species. In addition, it shows in all species the unusual behavior of not being retained on Phenyl-Sepharose hydrophobic interaction chromatography, even at very high salt concentration. This property is essential for its purification.
3. It contains a nickel tetrapyrrole cofactor, factor F_{430}, which gives the protein a characteristic yellow color, making it easily identifiable.
4. It is composed of three subunits from which three parameters (apparent molecular weights) can be derived. They allow a more detailed comparison between species than characteristics measured by only one parameter such as the $G + C$ mol% of the DNA. Most importantly, the information obtained by this analysis follows quite closely the evolutionary classification obtained by comparison of 16S rRNA.[10]

The technical advantages of the method are:

1. Its simplicity; the techniques involved do not require sophisticated material and are simpler than the sequencing of the 16S rRNA. It can be readily used by biologists who are not expert in molecular biology.
2. Its rapidity; results of the analysis are obtained in less than 2 d, most of the time being spent on the running of the electrophoresis gel.
3. Its relative accuracy, enabling the assignment of a new isolate to its phylogenetic group in the absence of any other characterization.

This technique is especially useful in the case of methanogens isolated from marine environments. Many species belonging to different families share a coccoid morphology (Table 1) which reflects a cell wall composed uniquely of proteins (S layer). This makes their classification and identification difficult. Several examples of utilization are described below.

MATERIALS REQUIRED

Equipment

- Centricon ultrafiltration microconcentrators (Amicon Corp., Beverly, MA)
- Electrophoresis apparatus

Supplies

- Glass beads 150 to 210 µm (Sigma Chemical Co., St. Louis, MO)
- A column of Phenyl-Sepharose CL-4B (Pharmacia LKB Biotechnology, Piscataway, NJ) (bed volume: 0.5 ml), equilibrated with $4 M$ NaCl in 50 mM sodium phosphate buffer, pH 7. Such a column is usually made by using a 1-ml glass syringe plugged with cotton and held upright by placing it into a small test tube.
- Molecular weight standard proteins (with their molecular weights in kilodaltons indicated in parentheses): soybean trypsin inhibitor (21.5), carbonic anhydrase (29), egg albumin (45), bovine serum albumin (66), and phosphorylase B (93) from Bio-Rad Laboratories (Richmond, CA); rabbit triose phosphate isomerase (26), rabbit lactate dehydrogenase phosphate kinase

Table 1. Taxonomy of Methanogens

Order
  Family
    Genus
	Contain species
	with coccoid morphology

Methanobacteriales	
Methanobacteriaceae	
Methanobacterium	
Methanobrevibacter	
Methanosphaera	x
Methanothermaceae	
Methanothermus	
Methanococcales	
Methanococcaceae	
Methanococcus	x
Methanomicrobiales	
Methanomicrobiaceae	
Methanoculleus (formerly *Methanogenium*)	x
Methanogenium	x
Methanolacinia (formerly *Methanomicrobium*)	x
Methanoplanus	x
Methanospirillum	
Methanocorpusculaceae	
Methanocorpusculum	x
Methanosarcinaceae	
Methanosarcina	x
Methanothrix	
Methanococcoides	x
Methanolobus	x
Methanohalophilus	x

Note: The classification of methanogenic bacteria is being reorganized and some of these names have not been formally accepted.[1]

(36.5), porcine heart fumarase (48.5), chicken pyruvate kinase (58), and rabbit fructose-6-phosphate kinase (84) from Sigma Chemical Co. (St. Louis, MO).

- Representative methanogen species (see below)

Solutions and Reagents

- 50 mM sodium phosphate buffer, pH 7
- 0.1% solution of DNAse I and ribonuclease A (Sigma Chemical Co., St. Louis, MO)
- 4 M NaCl in 50 mM sodium phosphate buffer, pH 7
- 2 M NaCl in 50 mM sodium phosphate buffer, pH 7
- 0.1% Coomassie Brilliant blue R250 in 10% acetic acid
- 10% acetic acid
- Denaturing solution: 0.2 M Tris-HCl, pH 6.8; 5% SDS; 0.01% bromophenol blue

PROCEDURES

Experimental Procedures

1. Growth of the cells — Cultures of methanogenic isolates are grown using strict anaerobic techniques under a pressurized gas atmosphere, as described in detail elsewhere;[2,11-13] 1-l cultures yielding between 0.3 and 1.0 g of cells should be sufficient. However, smaller cultures (100 ml) might be enough to produce material for one or two analyses. Cells are harvested by centrifugation and collected in a 1.7-ml eppendorf tube.

2. Disruption of the cells

   a. 100 μl of cell paste is resuspended in 200 μl of 50 mM sodium phosphate buffer, pH 7, with 100 μl of glass beads.

b. 1 μl of a 0.1% solution of DNAse I and ribonuclease A is added to digest nucleic acids during cell disruption (the supernatant should not be viscous).

c. The cell slurry is mixed vigorously using a bead beater or a vortex for 2 min.

d. The beads and cell debris are removed by centrifugation and the supernatant is collected.

e. The beads are washed once with 100 μl of buffer and after centrifugation this supernatant is added to the first one. This should yield between 2 and 5 mg of protein. If the supernatant is still turbid, it should be clarified by centrifugation. This disruption technique should be sufficient for most aquatic methanogenic species which have a relatively fragile protein cell wall. In fact, in many cases one freeze-thaw cycle promotes extensive lysis of the cells. The extent of cell disruption can be judged by the color of the extracts (cellular extracts of methanogenic bacteria are dark due to cofactors and metalloproteins), by direct microscopic observation, or by measure of the protein content of the lysate. If cell disruption is only partial, in particular for members of the *Methanobacteriales* which have a strong peptidoglycan cell wall, agitation with the glass beads for longer periods (10 min) can be achieved by taping the tube onto a vortex mixer.

3. After disruption, solid NaCl is added to the extract until saturation.

4. The extract is centrifuged for 3 min at 15,000 × g and the supernatant is collected.

5. The following steps pertain to partial purification of the methylreductase by Phenyl-Sepharose chromatography. The high salt extract is loaded on a small column of Phenyl-Sepharose CL-4B (bed volume: 0.5 ml) equilibrated with 4 M NaCl in 50 mM sodium phosphate buffer, pH 7. The extract is allowed to flow by gravity. At this point most of the colored material should bind the resin with the exception of the pink corrinoids particularly abundant in some species of the Methanosarcinaceae.

6. The column is then washed with 0.8 ml of the 4 M NaCl equilibrating buffer. Some colored material should elute, in particular factor F_{420}, a deazaflavin easily recognizable by its yellow-green color and its blue-green fluorescence under UV light.

7. The methylreductase still binds to the resin. The enzyme is eluted from the column by washing it with 0.8 ml of 2 M NaCl. The methylreductase can be identified by its yellow color due to factor F_{430}, its nickel tetrapyrrole prosthetic group. This fraction can be identified spectroscopically (absorption maximum at 425 nm with a shoulder at 455 nm).[14] The methylreductase constitutes between 30 and 90% of the protein of the 2-M NaCl fraction. Care should be taken to collect only the core of the yellow methylreductase peak to avoid dilution of the protein in the high salt buffer. When the methylreductase fraction is too dilute and no yellow color visible, it can be detected by measuring the protein concentration. The expected protein concentration should be about 1 mg/ml. If the methylreductase fraction is too dilute it can be concentrated using a Centricon ultrafiltration device equipped with a PM30 membrane.

8. The following steps pertain to purification by nondenaturing electrophoresis. 50 μg of protein from the 2-M NaCl eluate are loaded on an 8% acrylamide, 1 mm thick, discontinuous, nondenaturing polyacrylamide gel prepared according to Laemmli.[15]

9. Electrophoresis is performed at 70 V. The band corresponding to the methylreductase is identified as the most abundant protein after staining the gel with Coomassie Brilliant blue R250 and destaining with 10% acetic acid. It can also often be identified without staining the gel by its yellow color.

10. The band of methylreductase is excised from the gel and denatured for 30 min at room temperature in denaturing solution.

11. Denaturing electrophoresis — The gel slice after denaturation is inserted in the well of a thicker (1.5 or 2 mm thick) 10% acrylamide, SDS denaturing gel.[15] Care should be taken to insert the gel slice horizontally to ensure that the protein in the second electrophoresis

will migrate as a band. Other samples should be applied in adjacent wells to allow comparison with the subunits of other species. Molecular weight standards should be used every few lanes for estimation of the molecular weights as well as for internal standardization, to take into account variations of migration within the gel. It should be remembered that longer gels (15 to 20 cm) will have a better resolving power and might detect smaller mobility differences which is useful when comparing closely related organisms.

12. Determination of the molecular weights — Apparent molecular weights for the three subunits are estimated by comparison with protein molecular weight standards. The use of many standard proteins gives more precision to the correlation between apparent molecular weights and migration as seen on a semi-log plot.[16]

Data Analysis

The comparison of the apparent molecular weights of the subunits of the methylreductase can be used to derive different information.

Identifying the Taxonomic Group of a New Isolate

Methanogenic bacteria are divided into three orders and five families as presented in Table 1. In the case of morphologies indicative of a group (rods, spirilla, filaments, packets), the assignment of an isolate to a family or a genus is often facilitated and the technique presented here can be used as a confirmation. However, this technique is most helpful for isolates with a coccoid morphology (Table 1, 11 out of 16 genera) even before a thorough physiological characterization has been done. For example, utilization of this technique led to the taxonomic reassignment of the isolate *Methanococcus frisius* to the *Methanosarcina* genus.[10,17] It was also used to confirm that *Methanosphaera stadtmanae* belongs to the Methanobacterium family even though it has a spherical morphology and is capable of using methanol as a carbon source.[10] The apparent molecular weights can be used as shown in Table 2. For example, a relatively small α subunit (64,000 to 65,000) and a medium size γ subunit (35,000 to 37,000) are characteristic of a *Methanococcus* species. Similarly, small β and γ subunits are characteristic of the *Methanosarcina* genus. As an internal control, several representative species should be analyzed in parallel with the isolate studied, such as *Methanobacterium thermoautotrophicum, Methanococcus voltae, Methanogenium thermophilicum,* and *Methanosarcina barkeri.* The subunit patterns of the methylreductase of five representative species are shown in Figure 1.

Defining Close Relationship within a Family or a Genus

Once the family or the genus of the isolate has been identified, it is useful to determine its phylogenetic relationship with other species already characterized belonging to the same

Table 2. Apparent Molecular Weights ($\times$ 10^{-3}) of the Methylreductase Subunits from Representative Genera or Clusters of Genera

Genus/Genera cluster	Apparent molecular weights		
	α subunit	β subunit	γ subunit
Methanobacterium	66–67	48–49	38.5–39
Methanococcus	64–65	48–49	35–37
Methanogenium/Methanoplanus	64	45–47	29.5–30.5
Methanococcoides/Methanolobus	66–68	46–47	32
Methanosarcina	68–71	42–43	29–31

Note: So far the apparent molecular weights of the subunits from 20 species have been analyzed (References 10 and 18).

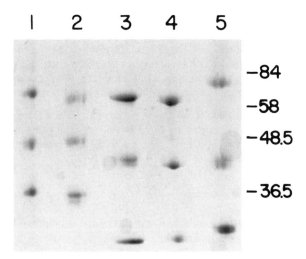

Figure 1. Subunit pattern of five representative methanogen species. Lane 1, *Methanobacterium thermoautotrophicum*; lane 2, *Methanococcus jannashii*; lane 3, *Methanogenium thermophilicum*; lane 4, *Methanothrix soehngenii*; lane 5, *Methanosarcina barkeri*. Numbers refer to molecular weights expressed in kilodaltons.

group or showing related properties. This was the case when trying to assign two coccoid marine isolates belonging to the Methanomicrobiales family to their genus. This family has been divided in several genera on the basis of physiological as well as subtle morphological characteristics. However, when studied at the molecular level by comparison of 16S rRNA sequences, it became obvious that the taxonomy did not reflect the evolutionary relationships. Figure 2 shows the application of this technique to elucidate the relationship of isolates TCl and CV. By comparing lanes 2 (isolate CV) and 3 (isolate TCl) with lanes 1 *(Methanococcus voltae)* and 4 *(Methanogenium marisnigri)* it is obvious that the two isolates are not part of the Methanococcaceae family but rather are related to the Methanomicrobiales order. Isolate TCl, now named *M. thermophilicum* was shown to be very closely related to *M. marisnigri* whereas isolate CV, now named *M. organophilum,* seemed slightly more distantly related. In fact, the measurement of the apparent molecular weights of the subunits showed that isolate CV was more closely related to the species *Methanoplanus limicola* than to *Methanogenium marisnigri*. The hierarchy of these close relationships was confirmed by comparison of the 16S rRNA sequences.[18]

Comparing a New Isolate with Characterized Species

It may be necessary to identify a new isolate or distinguish it from species already characterized. Along with the traditional characterization, comparison of the apparent molecular weights of the methylreductase can be helpful. Most of the species characterized so far can be obtained from the German Collection of Microorganisms (Deutsche Sammlung von Mikroorganismen, Mascheroder Weg 1b, D3300 Braunschweig, Germany) or from the OCG/CMA collection of Methanogenic Archaeae (Dr. D. Boone, curator, Oregon Graduate Center, Department of Environmental Sciences and Engineering, Beaverton, OR 97006-1999). Distinctions between closely related species are possible in some cases. For example, subtle but clear differences in migration patterns are seen between four *Methanosarcina* species which show only three percent sequence difference at the level of the 16S rRNA.[10]

NOTES AND COMMENTS

1. Sometimes, especially if very small quantities of cells are available, too little methylreductase is recovered after the Phenyl-Sepharose step. In this case the extract, without NaCl added, can be loaded on the nondenaturing gel directly. The relative mobility of the native methylreductase from different species varies: 0.2 to 0.3 for a 10% gel, 0.35 to 0.5 for an 8% gel, and 0.6 to 0.75 for a 6% gel. The methylreductase is almost always the most abundant protein. If there were an ambiguity between several major bands within this mobility range, the different candidates should be further soaked in the denaturing solution. A drawback of this more direct technique is that the methylreductase band can contain other proteins which might complicate the analysis after their second denaturing electrophoresis.

2. Presence of a second methylreductase — Recently it was shown that *Methanobacterium thermoautotrophicum* strains Marburg and ΔH possessed a second distinct methylreductase enzyme with apparently the same molecular weight for the α and β subunits but with a γ subunit 5000 Da smaller.[19] Although the presence of a second methylreductase has not been shown in other methanogen species so far, it is likely. This alternate methylreductase is synthesized only under early rapid growth conditions and not when the culture has entered the stationary phase. This is most likely why this second methylreductase was undetected during many years of investigation of the biochemistry of the methylreductase system or during our initial subunit comparison study. Growing the cultures to a maximum cell density should avoid this problem.

3. Species having unexpected methylreductase patterns — When studying the subunit pattern of *M. bryantii* strain MoH, it was found that the α subunit was noticeably smaller than that of the closely related species *M. formicicum*, whereas the γ subunit was larger. However, the sum of the molecular weight of the α and β subunits is unchanged.[10] This important variation most likely is not due to the analysis of an alternate methylreductase such as the one discussed above and is tentatively explained by a rearrangement between the contiguous genes of the α and β subunits. When analyzing the subunit pattern of a new isolate one should bear in mind that a similar situation might arise in which dramatic structural changes mask the small differences due to evolutionary drift.

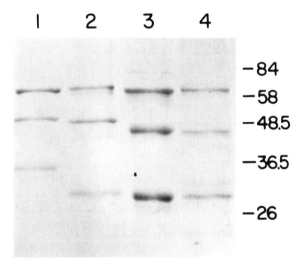

Figure 2. Utilization of the subunit comparison technique to compare closely related species. Lane 1, *Methanococcus voltae*; lane 2, *Methanogenium organophilum* CV; lane 3, *M. thermophilicum* TCl; lane 4, *M. marisnigri*. The γ subunit clearly separates *Methanococcus voltae* from the three other *Methanogenium* species. The β and γ subunits separate *M. organophilum* from the other two closely related species. The identical migration of *Methanoplanus limicola* and *Methanogenium organophilum* is not shown on the figure.

REFERENCES

1. Jones, W. J., Diversity and physiology of methanogens, in *Microbial Production and Consumption of Greenhouse Gases: Methane, Nitrogen Oxides, and Halomethanes,* Rogers, J. E. and Whitman, W. B., Eds., American Society for Microbiology, Washington, DC, 1991, chap. 3.

2. Whitman, W. B., Bowen, T. L., and Boone, D. R., The methanogenic bacteria, in *The Prokaryotes,* Balows, A., Truper, H. G., Dworkin, M., Harder, W., and Schleifer, K. H., Eds., Springer-Verlag, New York, 1991, chap. 33.

3. Woese, C. R., Bacterial evolution, *Microbiol. Rev.,* 51, 221, 1987.

4. Rouvière, P. E. and Wolfe, R. S., Novel biochemistry of methanogenesis, *J. Biol. Chem.,* 263, 7913, 1987.

5. Boone, D. R., Ecology of methanogenesis, in *Microbial Production and Consumption of Greenhouse Gases: Methane Nitrogen Oxides and Halomethanes,* Rogers, J. E. and Whitman, W. B., Eds., American Society for Microbiology, Washington, DC, 1991, chap. 4.

6. Berghofer, B., Krockel, L., Kotner, C., Truss, M., Schallenberg, J., and Klein, A., Relatedness of archaebacterial core subunits to their eukaryotic and eubacterial equivalents, *Nucleic Acids Res.,* 16, 8113, 1988.

7. Koga, Y., Ohga, M., Nishihara, M., and Morii, H., Distribution of diphytanyl ether analog of phosphatidylserine and an ethanolamine-containing tetraether lipid in methanogenic bacteria, *Syst. Appl. Microbiol.,* 9, 176, 1987.

8. Kneifel, H., Stetter, K. O., Andreseen, J. R., Wiegel, J., Konig, H., and Schobert, S. M., Distribution of polyamines in representative species of archaebacteria, *Syst. Appl. Microbiol.,* 7, 241, 1986.

9. Conway de Macario, E., Wolin, E. M. J., and Macario, J. L., Immunology of bacteria that produce methane gas, *Science,* 214, 74, 1981.

10. Rouvière, P. E. and Wolfe, R. S., Use of subunits of the methylreductase protein for the taxonomy of methanogenic bacteria, *Arch. Microbiol.,* 148, 253, 1987; erratum, 150, 208, 1988.

11. Balch, W. E., Fox, G. E., Magrum, L. L., Woese, C. R., and Wolfe, R. S., Methanogens: reevaluation of a unique biological group, *Microbiol. Rev.,* 43, 260, 1979.

12. Boone, D. R., OGC Oregon Collection of Methanogens, Catalog of Strains, Oregon Graduate Institute, Beaverton, OR, 1992.

13. Noll, K. M., Cultivation of strictly anaerobic archaea: Methanogens, in *Protocols for Archaebacterial Research,* Fleishmann, E. M., Place, A. R., Robb, F. P., and Schreier, H. J., Eds., 1991, 1.3.1., University of Maryland, Baltimore, MD.

14. Ellefson, W. B. and Wolfe, R. S., Component C of the methylreductase system of *Methanobacterium, J. Biol. Chem.,* 256, 4259, 1981.

15. Laemmli, U. K., Cleavage of structural protein during the assembly of the head of the bacteriophage T4, *Nature,* 227, 680, 1970.

16. Weber, K. and Osborne, M., The reliability of molecular weight determination by dodecyl sulfate polyacrylamide gel electrophoresis, *J. Biol. Chem.,* 244, 4406, 1969.

17. Blotevogel, K.-H. and Fischer, U., Transfer of *Methanococcus frisius* to the genus *Methanosarcina* as *Methanosarcina frisia, Int. J. Syst. Bacteriol.,* 39, 91, 1989.

18. Widdel, F., Rouvière, P. E., and Wolfe, R. S., Classification of a secondary alcohol-utilizing methanogens including a new thermophilic isolate, *Arch. Microbiol.,* 150, 477, 1988.

19. Rospert, S., Linder, D., Ellermann, J., and Thauer, R. K., Two genetically distinct methylcoenzyme M reductases in *Methanobacterium thermoautotrophicum* strain Marburg and ΔH, *Eur. J. Biochem.,* 194, 871, 1990.

CHAPTER 30

Extraction of DNA from Soils and Sediments

Ronald M. Atlas

INTRODUCTION

Two approaches have been developed with variations for the recovery of DNA from soils and sediments. The cell extraction approach involves the separation of bacterial cells from soil or sediment particles by differential centrifugation followed by lysis of the recovered cells, and recovery of the DNA.[1-8] The direct lysis method involves release of DNA from the cells by physical disruption, without separating the cells from the soil or sediment matrix, followed by extraction and purification of the DNA.[8-11] One variation of this method uses a bead beater to disrupt cells, alkaline extraction of the DNA in buffer, and purification of the extracted DNA by ethanol precipitation, cesium chloride density gradient centrifugation, and hydroxyapatite column chromatography.[9] Another variation uses freeze-thaw cycling to disrupt the cells followed by phenol-chloroform extraction, and column chromatographic purification of the DNA.[10,11] Yet other purification methods rely upon gel electrophoresis to separate recovered DNA from soil components contaminating the extracted DNA. Purified DNA recovered by these methods is of sufficient purity to perform gene probe detection by hybridization, to perform restriction enzyme and Southern blot analyses of the recovered DNA, and to amplify DNA by the polymerase chain reaction.

MATERIALS REQUIRED

Cell Extraction Method

Equipment

- Blender
- High-speed refrigerated centrifuge
- Ultracentrifuge with Ti70 motor
- Vortex mixer
- Spectrophotometer
- Heated water bath

0-87371-564-0/93/$0.00 + $.50

Solutions and Reagents

- 0.1 M sodium phosphate buffer (pH 4.5)
- Polyvinylpyrrolidone (PVPP)
- 20% sodium dodecyl sulfate (SDS)
- 0.1% sodium hexametaphosphate
- 0.1% sodium pyrophosphate
- Chrombach buffer (0.33 M Tris hydrochloride, 0.001 M EDTA, pH 8.0)
- Ammonium acetate
- Lysozyme
- TE buffer (10 mM Tris hydrochloride, 1 mM EDTA, pH 8.0)
- Ethanol
- Cesium chloride
- Ethidium bromide
- 1-butanol
- 0.01 M sodium phosphate buffer (pH 6.8)
- 0.014 M sodium phosphate (pH 6.8)
- 0.12 M sodium phosphate buffer (pH 6.8)
- SSC (1 × SSC is 0.15 M NaCl plus 0.015 M sodium citrate, pH 7.0)
- Urea phosphate buffer (8 M urea in 0.12 M sodium phosphate, pH 6.8)
- Hydroxyapatite, high-resolution DNA-grade

Direct Lysis/Alkaline Extraction Method

Equipment

- Bead beater cell disrupter (Bio Spec, Bartlesville, OK)
- Blender
- High-speed refrigerated centrifuge
- Ultracentrifuge
- Vortex mixer
- Millipore filter apparatus

Solutions and Reagents

- 10% SDS solution
- PVPP
- 0.12 M phosphate buffer (pH 8.0)
- Sodium chloride
- 50% polyethylene glycol (PEG) 8000
- TE-saturated phenol
- Phenol-chloroform-isoamyl alcohol (25:24:1)
- Chloroform-isoamyl alcohol (24:1)
- Potassium acetate
- Ethanol

Direct Extraction of DNA from Soil and Sediment Using Freeze-Thaw DNA Release

Equipment

- High-speed refrigerated centrifuge
- Heated water bath
- Vortex mixer

Supplies

- Prefilter (Schleicher & Schuell NA010/27, 0.45-μm pore size cellulose acetate)
- Elutip-d column (Schleicher & Schuell, Keene, NH)

Solutions and Reagents

- Sodium phosphate buffer (120 mM, pH 8.0)
- Lysis solution (0.15 M NaCl, 0.1 M Na$_2$EDTA, pH 8.0, containing 15 mg of lysozyme ml^{-1})
- 0.1 M NaCl-0.5 M Tris-HCl (pH 8.0), 10% sodium dodecyl sulfate
- 0.1 M Tris-HCl (pH 8.0) saturated phenol
- Phenol
- Chloroform mixture (chloroform-isoamyl alcohol ratio, 24:1)
- Isopropanol
- TE buffer (20 mM Tris-HCl, 1 mM EDTA, pH 8.0)
- Ethanol
- Dry ice
- Pancreatic RNase A (final concentration, 0.2 μg μl^{-1})

PROCEDURES

Cell Extraction Method

The cells are separated from the soil or sediment by differential centrifugation. Approximately 50% of the bacterial cells can be recovered by this method. The specific buffers used and the actual efficiency of recovery depends upon the specific properties of the soil or sediment. Clays and organics alter the extraction efficiencies. A polyvinylpolypyrolidone (PVPP) extraction step aids in the removal of humic substances that interfere with DNA analyses.

1. Suspend a 100 g (dry wt) amoung of sediment or soil in 300 ml of 0.1 M sodium phosphate buffer (pH 4.5).
2. Add 20 g of PVPP.
3. Homogenize with a blender at medium speed for three 1-min periods, with 1 min of cooling on ice between each blending cycle.
4. Add a 2 ml portion of 20% sodium dodecyl sulfate (SDS).
5. Blend for 5 s.
6. Place on ice for 5 min.
7. Transfer samples to 250-ml centrifuge bottles.
8. Shake bottles by hand for 1 min.
9. Centrifuge for 10 min at 1000 × g and 10°C.
10. Transfer supernatant to a flask and maintain on ice.
11. Wash the sediment or soil pellets back into the blender with 300 ml of 0.1 M sodium phosphate buffer (pH 4.5).
12. Blend and centrifuge as described above, but without further addition of SDS, for a total of three cycles.
13. Combine the supernatant with the earlier collected supernatant and maintain on ice.
14. Centrifuge the combined supernatants for 30 min at 10,000 × g and 10°C to collect the bacterial cells.

15. Resuspend pellet containing bacterial cells in 200 ml of a solution of 0.1% sodium hexametaphosphate-0.1% sodium pyrophosphate at 5°C.
16. Shake by hand for 1 min.
17. Centrifuge for 30 min at 10,000 × g and 10°C.
18. Discard supernatant and repeat steps three times.
19. As a final washing procedure, suspend the pellet in 150 ml of Chrombach buffer.
20. Centrifuge for 30 min at 10,000 × g and 10°C.
21. Transfer to a 50 ml centrifuge tube by washing with Chrombach buffer and bring final volume to 25 ml.
22. Mix the cell pellets with a vortex mixer.
23. Add lysozyme to a final concentration of 5 mg ml^{-1}.
24. Incubate for 2 h at 37°C.
25. Heat to 60°C.
26. Add SDS to a final concentration of 1.0% and incubate for 10 min.
27. Cool on ice for 2 h.
28. Centrifuge for 20 min at 12,000 × g at 5°C.
29. Transfer the supernatant (first lysate) to a sterile centrifuge tube.
30. Wash the pelleted material with 10 ml of Chrombach buffer.
31. Centrifuge for 30 min at 10,000 × g and 10°C.
32. Combine supernatant with the first lysate.
33. To purify the DNA in the combined cell lysate, add solid ammonium acetate to give a final concentration of 2.5 M.
34. Centrifuge immediately for 30 min at 12,000 × g and 5°C.
35. Add 2.5 volumes of ice-cold 95% ethanol and incubate at −20°C for 12 h to precipitate DNA in the supernatant fraction.
36. Centrifuge for 30 min at 12,000 × g and 5°C.
37. Decant the ethanol and dry under vacuum.
38. Suspend the DNA in 10 ml of TE buffer.
39. Purify the DNA. A variety of methods can be used. The degree of purification needed depends upon the analyses to be performed on the recovered DNA. Commonly ultracentrifugation, column chromatography, and/or electrophoresis are used to purify DNA.
40. Purification by cesium chloride-ethidium bromide density gradient centrifugation.

    a. Add 0.8 g of CsCl and 160 μg of ethidium bromide per milliliter to sample.
    b. Centrifuge in a Ti70 rotor at 50,000 rpm for 18 h at 15°C.
    c. Recover broad fluorescing regions from tubes.
    d. Remove ethidium bromide by extractions with 1-butanol.
    e. Dialyze against two changes of 0.12 M sodium phosphate buffer (pH 6.8).
    f. Precipitate DNA with cold ethanol.

41. Purification with hydroxyapatite column chromatography.

    a. Suspend hydroxyapatite in 0.01 M sodium phosphate buffer (pH 6.8).
    b. Boil for 10 min.
    c. Centrifuge for 2 min at 5000 × g.
    d. Remove supernatant by decanting.
    e. Mix hydroxyapatite with urea phosphate buffer.
    f. Centrifuge for 2 min at 5000 × g.
    g. Add samples to 2 g (dry weight) of hydroxyapatite.
    h. Mix with gentle intermittent swirling at 28°C for 1 h.
    i. Pour mixture into a column made of a 20-ml disposable syringe with siliconized glass wool as a support base and a 16-gauge needle as an outlet.
    j. Drain urea-containing buffer from the columns and pass additional urea-containing buffer through the columns to remove organic contaminants until the effluent absorbance between 320 and 220 nm reaches zero.

k. Pass a 50-ml (10-column volume) amount of 0.014 M sodium phosphate (pH 6.8) through the columns to remove the urea-containing buffer.

l. Elute DNA sequentially with 10-ml volumes of 0.20 and 0.25 M sodium phosphate (pH 6.8).

m. Collect 10-ml fractions.

n. Remove phosphate from the recovered fractions by dialysis for 12 h against $0.1\times$ SSC.

o. Recover DNA by ethanol precipitation and suspension in 100 μl of $0.1 \times$ SSC.

Direct Lysis Alkaline Extraction Method

DNA is recovered from soils and sediments by directly lysing the cells and then extracting the DNA with an alkaline extraction solution. This method is highly efficient at recovering the total DNA in a sample.

1. Add 100 ml 10% SDS solution to 100 g soil or sediment samples and incubate at 70°C for 1 h.
2. Physically disrupt cells using a bead beater.
3. Add 20 g of PVPP to the lysed cell mixture to remove humic material.
4. Recover DNA by repeated washings with 0.12 M phosphate buffer (pH 8.0).
5. Add sodium chloride to the DNA suspension to a final concentration of 0.5 M.
6. Precipitate the DNA by adding 0.5 volume of 50% polyethylene glycol (PEG) 8000 and incubating at 5°C for 12 h.
7. Recover the DNA by centrifugation at 5000 $\times$ g for 10 min at 4°C.
8. Remove supernatants from the loose pellets by aspiration.
9. Remove PEG from the DNA by one extraction with TE-saturated phenol, one extraction with phenol-chloroform-isoamyl alcohol (25:24:1), and two extractions with chloroform-isoamyl alcohol (24:1).
10. Add potassium acetate to achieve a final concentration of 0.5 M.
11. Incubate samples on ice for 2 h.
12. Add 2 g of acid-washed PVPP to each sample to further remove humic compounds and incubate the samples for 15 min at 20°C with intermittent swirling.
13. Filter samples through a 0.5- to 1.0-μm filter (Millipore 65) to remove PVPP and precipitated humic material.
14. Precipitate DNA by adding 2.5 volumes of 95% ethanol and incubating for 12 h at 20°C.
15. Dry resulting pellets and suspend in TE.
16. Purify DNA by cesium-chloride centrifugation and/or hydroxyapatite column chromatography as described in previous procedure.

Direct Extraction of DNA from Soil and Sediment Using Freeze-Thaw DNA Release

A rapid method for the direct extraction of DNA from soil and sediments involves lysing the soil or sediment microorganisms by using lysozyme and a freeze-thaw procedure. The lysate is extracted with sodium dodecyl sulfate and phenol-chloroform and the DNA is purified with an Elutip-d column. The procedure is far simpler than those previously described and can be performed on a small sample of soil or sediment.

1. Mix 1 g soil or sediment with 2 ml of 120 mM sodium phosphate buffer (pH 8.0) by shaking at 150 rpm for 15 min.
2. Pellet by centrifugation at 6000 $\times$ g for 10 min.
3. Wash the pellet with phosphate buffer.
4. Resuspend in 2 ml of lysis solution.

5. Incubate in a 37°C water bath for 2 h with agitation at 20- to 30-min intervals.
6. Add 2 ml of 0.1 *M* NaCl-0.5 *M* Tris-HCl (pH 8.0)-10% sodium dodecyl sulfate.
7. Freeze in a −70°C dry ice-ethanol bath.
8. Thaw in a 65°C water bath.
9. Repeat steps 7 and 8 two times.
10. Add 2 ml of 0.1 *M* Tris-HCl (pH 8.0)-saturated phenol and vortex briefly to form an emulsion.
11. Centrifuge at 6000 × g for 10 min.
12. Collect 3 ml from the top aqueous layer and mix with 1.5 ml of phenol and 1.5 ml of chloroform mixture.
13. Extract a 2.5-ml portion of the mixture with an equal volume of chloroform mixture.
14. Precipitate nucleic acids in 2 ml of the extracted aqueous phase with 2 ml of cold iso-propanol at −20°C for 1 to 12 h.
15. Centrifuge at 10,000 × g for 10 min.
16. Vacuum dry at 23°C.
17. Resuspend pellet in 100 ml of TE buffer.
18. Incubate with heat-treated pancreatic RNAse A for 2 h at 37°C.
19. Purify with an Elutip-d column attached to a prefilter.

REFERENCES

1. Faegri, A., Torsvik, V. L., and Goksoyr, J., Bacterial and fungal activities in soil: separation of bacteria by a rapid fractionated centrifugation technique, *Soil Biol. Biochem.*, 9, 105, 1977.
2. Holben, W. E., Jansson, J. K., Chelm, B. K., and Tiedje, J. M., DNA probe method for the detection of specific microorganisms in the soil bacterial community, *Appl. Environ. Microbiol.*, 54, 703, 1988.
3. Jansson, J. K., Holben, W. E., and Tiedje, J. M., Detection in soil of a deletion in an engineered DNA sequence by using DNA probes, *Appl. Environ. Microbiol.*, 55, 3022, 1989.
4. Pillai, S. D., Josephson, K. L., Bailey, R. L., Gerba, C. P., and Pepper, I. L., Rapid method for processing soil samples for polymerase chain reaction amplification of specific gene sequences, *Appl. Environ. Microbiol.*, 57, 2283, 1991.
5. Torsvik, V. L., Isolation of bacterial DNA from soil, *Soil Biol. Biochem.*, 12, 15, 1980.
6. Torsvik, V. L. and Goksoyr, J., Determination of bacterial DNA in soil, *Soil Biol. Biochem.*, 10, 7, 1978.
7. Torsvik, V., Goksoyr, J., and Daae, F. I., Diversity in DNA of soil bacteria, *Appl. Environ. Microbiol.*, 56, 782, 1990.
8. Steffan, R. J., Goksoyr, J., Bej, A. K., and Atlas, R. M., Recovery of DNA from soils and sediments, *Appl. Environ. Microbiol.*, 54, 2908, 1988.
9. Ogram, A., Sayler, G. S., and Barkay, T., DNA extraction and purification from sediments, *J. Microbiol. Methods*, 7, 57, 1988.
10. Tsai, Y.-L. and Olson, B. H., Method for direct extraction of DNA from soil and sediments, *Appl. Environ. Microbiol.*, 57, 1070, 1991.
11. Tsai, Y.-L. and Olson, B. H., Detection of low numbers of bacterial cells in soils and sediments by polymerase chain reaction, *Appl. Environ. Microbiol.*, 58, 754, 1992.

CHAPTER 31

Detecting Gene Sequences Using the Polymerase Chain Reaction

Ronald M. Atlas

INTRODUCTION

The polymerase chain reaction (PCR) is an *in vitro* method for amplification of DNA sequences.[1-8] PCR can be applied to DNA recovered from environmental samples.[8,9] A single copy of a DNA sequence in a sediment, soil, or water sample can be amplified to 1 million copies within a few hours by PCR. Detection of specific gene sequences can be used to diagnose the presence of specific microbial populations and the presence of specific functional genes. The target region for amplification is defined by two unique oligonucleotide primers flanking a DNA segment. The oligonucleotide primers are designed to hybridize to regions of DNA flanking a desired target gene sequence, annealing to opposite strands of the target sequence. The primers are then extended across the target sequence using a heat-stable DNA polymerase (frequently *taq* DNA polymerase) in the presence of free deoxynucleotide triphosphates. The product of each PCR cycle is complementary to and capable of binding primers, so that the amount of DNA synthesized is doubled in each successive cycle. The essential reagents for PCR are a thermostable DNA polymerase, oligonucleotide primers, deoxynucleotides (dNTPs), template (target) DNA, and magnesium ions. Generally the total reaction volume is 50 to 100 µl. An automated temperature cycler is used to cycle between the temperatures for DNA melting, primer annealing, and primer extension. The heat-stable DNA polymerase retains activity through sufficient PCR cycles to achieve better than 1 million-fold amplification of the target DNA sequence. The primer sequences and the primer annealing temperature determine the stringency (specificity) of the amplification.

Because of the amplification power of PCR, it is critical to avoid even a trace of contamination with DNA containing the target sequence. Precautions such as pre-aliquoting reagents, using pipettes exclusively for specific steps of the reaction, using positive displacement pipettes, or using tips with barriers to prevent contamination of the pipette barrel, and physical separation of the amplification reaction preparation from the area of amplified product analysis can minimize the possibility of such contamination. Addition of multiple negative control reactions without any target DNA added is essential as quality control with every set of PCR reaction to reveal and monitor possible contamination.

MATERIALS REQUIRED

Equipment

- Automated thermal cycler

Solutions and Reagents

- Target DNA, which may be obtained from a target organism or sample. The total amount of DNA typically used for PCR is 0.05 to 1.0 μg. Impurities (e.g., such as formalin, blood, humic acids, chelating agents, detergents, and heavy metals) must be eliminated or diluted so as not to inhibit DNA polymerization. DNA may be extracted and purified but extensive purification often is not required for successful DNA amplification. The target gene can be from <100 nucleotides to a few kilobases. There must be at least one intact copy of the target gene.
- 10× stock solution reaction buffer (500 mM KCl, 100 mM Tris-HCl, and 15 mM MgCl$_2$, pH 8.4). The recommended buffer for PCR is 10 to 50 mM Tris-HCl (pH 8.3 to 8.8). Changing the buffering capacity of the PCR reaction, for example, by increasing the concentrations of the Tris-HCl up to 50 mM (pH 8.9), sometimes increases the yield of the amplified DNA products. Up to 50 mM KCl can be included in the reaction mixture to facilitate primer annealing. KCl greater than 50 mM inhibits *taq* DNA polymerase activity. PCR requires divalent magnesium ions and the buffer should contain 0.5 to 2.5 mM magnesium over the total dNTP concentration. Concentrations of MgCl$_2$ in the range of 1.5 to 4 mM can be used to enhance specificity and to obtain a higher yield of the amplified products. The optimal Mg^{2+} ion concentration for each primer set should be determined by running replicate PCRs with Mg^{2+} concentrations varied by factors of two.
- 10× stock solution of dNTPs (2 mM dATP, 2mM dCTP, 2 mM dGTP, and 2 mM dTTP, pH 7.0). Free deoxynucleotide triphosphates (dNTPs) are required for DNA synthesis. The dNTP concentrations for PCR should be 20 to 200 μM to give optimal specificity and fidelity. The four dNTPs (dATP, dTTP, dGTP, dCTP) should be used at equivalent concentrations to minimize misincorporation errors. It is important to keep the four dNTP concentrations balanced for best base incorporation fidelity. A final concentration of greater than 50 mM total dNTP in the PCR reaction inhibits *taq* DNA polymerase activity. The lowest dNTP concentration appropriate for the length and composition of the target sequence should be used to minimize mispriming.
- 10× solution of primers (2.5 mM of primer 1 and 2.5 mM of primer 2 in TF buffer [10 mM Tris-HCl, 0.1 mM EDTA, pH 8.0]). Primers are oligonucleotides which are short single-stranded segments of DNA, complementary to the 5′ ends of the target DNA to be amplified. For primer annealing to occur specifically at the sites flanking the DNA region to be amplified, there must be nearly complete identity between the target DNA and the primer nucleotide sequence. The 5 to 6 nucleotides at the 3′ ends of the primers must exhibit precise base pairing with the target DNA. The terminal match at the 3′ end is critical. Complementarity between the paired primers at the 3′ ends should be avoided because they may produce primer dimers. All primers for a target amplification should have the same melting temperature (Tm) and an average G + C content of 40 to 60% with no stretches of polypurines or polypyrimidines. Significant secondary structures internal to the primers or immediately downstream (for left-hand side primer) or immediately upstream (for right-hand side primer) of the template should be avoided. Three or more C's or G's at the 3′ ends of primers may promote mispriming and should be avoided. Primer concentrations of 0.1 −0.5 μM are recommended. Higher primer concentrations may promote nonspecific product formation and may increase the generation of a primer dimer.
- Thermally stable DNA polymerase. The DNA polymerase from *Thermus aquaticus* (*taq* DNA polymerase) or that produced by a recombinant *Escherichia coli* (AmpliTaq) often are used, but other commercial thermally stable DNA polymerases may also be used. The

recommended concentration range of *taq* DNA polymerase for PCR is 1 to 2.5 units per 100 μl. If the enzyme concentration is too high, nonspecific background products often form. If the enzyme concentration is too low, and insufficient amount of desired product is made.

PROCEDURES

1. Prepare 2× reaction mixture by mixing 10 μl each of the 10× buffer, 10× dNTPs, and 10× primer solutions and 20 μl of distilled water.

2. Add 50 μl of 2× reaction mixture to 50 μl (0.5 to 1.0 μg) target DNA in a 0.5-ml microcentrifuge tube.

3. Add 2.5 units (0.5 μl) *taq* DNA polymerase. To improve specificity, addition of DNA polymerase can be delayed until after the reaction is heated to 94 to 95°C. This approach improves specificity and minimizes the formation of primer dimer.

4. Overlay with a drop of mineral oil. An overlay of 80 to 100 μl light mineral oil on top of the reaction mix often is used to prevent evaporation of the liquid. The overlap may maintain heat stability and limits evaporation so that salt concentrations are maintained. Newer thermal cyclers do not require addition of mineral oil overlay because of the specialized tubes used in the system.

5. Place the sample into the thermal cycler. As accurate temperatures are critical, it is important that thermal cyclers, which permit the automated cycling of temperatures according to a temperature program, have accurate temperature regulation. It is essential that a thermal cycler give uniform heating and cooling, consistently give the same results in all reaction wells, and produce a profile in the sample tube that is essentially the same as the cycle that had been programmed. The temperature in each reaction well must be very precise so as to provide the required stringency for primer annealing and to provide accurate conditions for denaturation and DNA extension. The temperature ramping between temperature settings must be rapid to permit accurate and short cycles. Since the emergence of the first DNA thermal cycler by Perkin-Elmer Cetus Corporation, a dozen more companies have begun to manufacture thermal cyclers with different designs. A thermal cycler with a metal block heated by a heating element pad or a cell cooled by a Peltier pump gives the best results.

6. Melt DNA by heating to 94°C for 3 min. To ensure complete denaturation of target DNA, a temperature of 94 to 95°C for 3 min typically is used prior to the first cycle. The half-life of *taq* DNA polymerase activity is 40 min at 95°C.

7. Anneal primers by cooling to 50°C 1 min for appropriate temperature based on Tm's of primers). The temperature for primer annealing depends upon the base composition, length, and concentration of the amplification primers. The ideal annealing temperature generally is 5°C below the true Tm of the primers. Annealing temperatures in the range of 55 to 72°C generally yield the best results. Increasing the annealing temperature enhances discrimination against incorrectly annealed primers and reduces addition of incorrect nucleotides at the 3′ end of the primers.

8. Raise temperature to 72°C for 1 min to extend primers. Extension time depends upon the length and concentration of the target sequence and upon the temperature used for the extension reaction. The activity of *taq* DNA polymerase varies by two orders of magnitude between 20 and 85°C. Primer extensions typically are performed at 72°C, which is optimal for *taq* DNA polymerase, but *taq* DNA polymerase also is active at lower temperatures. In two-step PCR, the primer extension step is set at the same temperature as the reannealing temperature.

9.   Raise temperature to 94°C for 1 min to remelt DNA. Typical denaturation conditions are 94 to 95°C for 30 to 60 s. The selection of temperature and denaturation time depend upon the length and G + C content of the target DNA sequences. Using too low a temperature may result in incomplete denaturation of the target template and/or the PCR product and, therefore, failure of the PCR. In contrast, DNA denaturation steps that are too high and/or too long cause enzyme denaturation and loss of essential enzyme activity.

10. Repeat steps 6 to 8 20 to 30 times. The number of cycles used in PCR depends upon the degree of amplification required and the need to amplify selectively the target DNA sequence. Too many cycles can increase the amount and complexity of nonspecific background products. Too few cycles give low product yield. The initial amount of target DNA and the sensitivity of the detection method are critical in determining the number of cycles; 20 cycles theoretically could produce 1 million copies of a target sequence, but because of reaction inefficiencies in most cases 30 to 40 cycles are used.

11. End last cycle at the primer extension step by maintaining 72°C for 3 min.

12. Cool to 5°C and remove tube for analysis. PCR-amplified DNA can be analyzed by various methods including gel electrophoresis and hybridization techniques. The method of detection selected depends upon the sensitivity required and whether additional information such as size of the amplified DNA is also to be determined in the analysis.

REFERENCES

1. Bej, A. K., Mahbubani, M. H., and Atlas, R. M., Amplification of nucleic acids by polymerase chain reaction (PCR) and other methods and their applications, *Crit. Rev. Mol. Biol.*, 26, 301, 1991.

2. Ehrlich, H. A., Ed., *PCR Technology: Principles and Applications for DNA Amplification*, Stockton Press, New York, 1989.

3. Ehrlich, H. A., Gelfand, D., and Sninsky, J. J., Recent advances in the polymerase chain reaction, *Science,* 252, 1643, 1991.

4. Gibbs, R. A., DNA amplification by the polymerase chain reaction, *Anal. Chem.,* 62, 1202, 1990.

5. Innis, M., Gelfand, D., Snisky, D., and White, T., Eds., *PCR Protocols: A Guide to Methods and Applications,* Academic Press, New York, 1990.

6. Mullis, K. B., The unusual origin of the polymerase chain reaction, *Sci. Am.,* 262, 56, 1990.

7. Saiki, R. K., Gelfand, D. H., Stoffel, S., Scharf, S. J., Higuchi, R., Horn, G. T., Mullis, K. B., and Ehrlich, H. A., Primer-directed enzymatic amplification of DNA with a thermostable DNA polymerase, *Science,* 239, 487, 1988.

8. Steffan, R. J. and Atlas, R. M., Polymerase chain reaction: applications in environmental microbiology, *Annu. Rev. Microbiol.,* 45, 137, 1991.

9. Bej, A. K. and Mahbubani, M. M., Applications of the polymerase chain reaction in environmental microbiology, *PCR Methods Appl.,* 1, 151, 1992.

CHAPTER 32

Quantitative Description of Microbial Communities Using Lipid Analysis

Robert H. Findlay and Fred C. Dobbs

INTRODUCTION

Determining the community structure of a microbial assemblage is one of the greatest challenges facing microbial ecologists. Unlike eukaryotes, the morphology of prokaryotes often yields little or no information concerning the phylogenetic affiliation or the ecological role of the organisms.[1] Hence, direct observations, while suitable for biomass determinations, do not allow the investigator to distinguish among many microbial populations. Classical approaches utilize enrichment and isolation of microorganisms. These approaches are suitable for biochemical, taxonomic, or autecological studies,[2] but it is clear that most microorganisms in the environment are viable but not culturable[3,4] and that these methods will not produce a quantitative description of microbial assemblages. Modern molecular techniques now being applied to environmental samples can produce detailed descriptions of the species (and even strains) of microorganisms present within a community.[5] Unfortunately, these techniques are labor intensive and a single investigator can realistically expect to completely describe the microbial community from only 10 to 20 samples per year. In addition, methods requiring amplification of the genetic material impose a potential selection on the sequences to be analyzed. Although this selection can be minimized, it is not yet possible to relate the relative recovery of different types of rRNA to the relative abundance of different organisms.[6] Lipid analysis offers an alternative method for the quantification of microbial community structure that does not rely upon the cultivation of microorganisms and that is free of potential selections of the biochemical compounds to be analyzed. In this chapter we review our methodology for the quantitative recovery of phospholipid fatty acids from environmental samples and illustrate the interpretation of fatty acid profiles and their usefulness in determining microbial community structure.

Advantages of the methods are (1) in addition to community structure, total microbial biomass can be assessed from the same sample; in fact, our protocol assumes that such is the case; (2) results integrate across the entire microbial community without the truncations of cultural studies; (3) there are none of the difficulties inherent in enumerative studies (e.g., dislodging microbes from sediment, subsampling, visualization); (4) these techniques can be used in sediments not amenable to analysis by microscopy (e.g., coarse sands); (5) relative to visual and molecular techniques, these biochemical methods are time and cost competitive and allow for experimental designs involving scores of analyses; (6) the techniques have high precision and as few as three to five replicate samples per treatment may be sufficient; and (7) if desired, further biochemical characterizations may be made using the samples extracted in these protocols.

Disadvantages of the methods are (1) results integrate across the entire microbial community; there is no discrimination at the level of individual cells; (2) these methods are relatively new, therefore not widespread in their use, and comparable data are rare; (3) although the fatty acid composition of many microorganisms is established, there remain some uncertainties in converting phospholipid fatty acid data to descriptions of microbial communities; (4) statistical methods for the comparison of complex data sets such as fatty acid profiles are still under development; and (5) there is a need for specialty equipment, such as a gas chromatograph, solvent-handling glassware, and a hood to protect laboratory personnel from exposure to solvent fumes.

MATERIALS REQUIRED

Sample Collection

Equipment

- None

Supplies

- Sediment corers
- 25 × 150 mm test tubes with PTFE-lined screw caps

Solutions and Reagents

- Solvents (methanol, chloroform: pesticide-residue grade or higher quality)
- 50 mM phosphate buffer, pH 7.4 (8.7 g K_2HPO_4 l^{-1} adjusted with 1 N HCl)[7]

Laboratory Procedures

Equipment

- Nitrogen evaporator
- Fume hood
- Centrifuge
- Vortex mixer
- Gas chromatograph (GC) — fitted with splitless injector, fused-silica capillary column, and flame-ionization detector

Supplies

- Disposable pipettes, Pasteur
- Disposable pipettes, large-volume
- 10-ml Griffin beakers
- Graduated cylinders
- Glass wool
- Nitrogen gas

- Silicic acid (e.g., Bio-Sil A, 100 to 200 mesh, Bio-Rad) or commercial columns (e.g., JT Baker #7086-3, 3-ml SPE columns)
- Octadecyl (C_{18}) bulk packing for flash chromatography (e.g., JT Baker #7025-00)
- 16 × 125 mm tubes with PTFE-lined screw caps
- Ring stand, cross-bar, and clamps

Solutions and Reagents

- Solvents (chloroform, methanol, acetone, toluene, hexane, acetonitrile: pesticide-residue grade or better)
- Deionized water
- Phosphate buffer (recipe above)
- 0.2 N methanolic KOH (make fresh daily, add 0.28 g of KOH to 25 ml methanol)
- 0.2 N acetic acid (1.15 ml glacial acetic acid diluted to 100 ml with DI water)
- Hexane:chloroform (95:5, make fresh daily, use chloroform preserved with 0.75% ethanol)
- Fatty acid methyl ester standards (Sigma)
- 20:0 and 20:4ω6 fatty acid ethyl ester (FAEE) standards (Sigma); these compounds are used as internal standards: prepare such that 50 μl of solvent contains 5 ng of each FAEE

PROCEDURES

The procedures used for sampling and for the extraction and purification of phospholipids are identical to those given in Chapter 40. We emphasize that the determination of microbial community structure does not require incubation with radiolabeled precursors. However, radiation-safety protocols must be adopted should radiolabeled phospholipid be used for determining community structure.

The procedures described in Chapter 40 separate the phospholipid fraction from the other lipids contained in the total lipid extract. What remains is to transmethylate the phospholipid fatty acids (PLFAs) and further purify the resulting fatty acid methyl esters (FAMEs) prior to separation and quantification using the gas chromatograph.

Formation of FAMEs

1. Dissolve the dry phospholipid in 0.5 ml of methanol:toluene (1:1).

2. Add 0.5 ml of 0.2 N KOH in methanol. Seal, vortex, and heat for 15 min at 37°C. This reaction transmethylates the PLFAs, forming FAMEs.

3. Cool test tubes to room temperature, add 0.5 ml of 0.2 N acetic acid, and vortex. Immediately add 2 ml of chloroform and 2 ml of deionized water and vortex for approximately 30 s.

4. To separate the aqueous and lipid phases, centrifuge the mixture at 1400 × g for 5 min. The lipid (bottom) phase now contains the FAMEs. Transfer the lipid phase to a clean test tube with a Pasteur pipette. Avoid transferal of any of the aqueous phase. This may necessitate leaving a small drop of chloroform in the reaction test tube.

5. Add 1 ml of chloroform to the reaction test tube, vortex, and repeat step 4.

6. Repeat step 5.

7. Add 1 ml of chloroform to the reaction test tube, but do not vortex or centrifuge. Transfer the chloroform to the clean test tube. The clean test tube should now contain a total of 5 ml of chloroform.

8. Add 5 ng each of 20:0 and 20:4ω6 ethyl ester internal standards. Add only the ethyl ester standards to three test tubes; these will serve to calibrate the GC detector response.

9. Remove the solvents using a nitrogen evaporator having a bath temperature of 37°C.

Purification of FAMEs

At this juncture, the FAMEs are ready to be purified in preparation for gas chromatographic separation and quantification. Although some types of samples yield FAME mixtures of sufficient purity that no further purification is necessary prior to gas chromatographic analysis, the risks of foregoing further purification are two. The first is that contaminants in the FAME mixture may corrupt the gas chromatographic column, degrading its performance. The second is that extraneous compounds may be interpreted to be FAMEs, confounding the determination of microbial community structure. Several strategies for the purification of FAMEs are available. Classically, FAMEs are purified using thin-layer chromatography.[8] While this method yields good results, it is tedious and time consuming. Alternately, the FAMEs may be passed down a second silicic acid column.[9] They are now neutral lipids and will elute in the chloroform fraction. Finally, one of us (Findlay) has developed a purification scheme using reverse-phase, SPE column chromatography that yields high-purity FAME preparations while maintaining high sample throughput. The procedures that follow are a synthesis of this third method.

1. Add 1 g of dry C_{18} packing to a 3-ml glass column fitted with a Teflon frit or to a large-volume disposable pipette into which a small amount of glass wool has been packed (tapered end).

2. Condition the column by passing a series of solvents through the packing. First hydrate the packing by adding 2 ml of deionized water to the column and initiate flow either by pulling a vacuum on the column or by pressurizing the headspace of the column. Continue the flow until the column appears dry. Except where noted, flow should be approximately 1 drop s^{-1} and columns should not be allowed to run dry.

3. Wash the packing with 2 ml of methanol. Stop flow as the solvent-air interface reaches the head of the column. Wash the packing with 1 ml of chloroform. Add 2 ml of chloroform and close the column such that solvent flow is inhibited. The column packing material will soon begin to effervesce and become translucent. Dislodge any air pockets and improve the column packing by first tapping the column and then pressurizing the headspace with a 20-ml syringe. After the C_{18} packing has become translucent allow the chloroform to drip from the column, but do not let the column run dry. Wash the packing with 2 ml of acetonitrile and 2 ml of acetonitrile:water (1:1) to complete the column conditioning. To prevent drying, leave approximately 1 ml of the acetonitrile:water wash over the packing and remove it just prior to the addition of the FAMEs to the column.

4. Add 250 μl of acetonitrile to the dry FAMEs. Vortex three times over the next 10 minutes. Then add 250 μl of deionized water, vortex, and transfer to the column and allow the solvent to drip through the column. To complete transfer of the FAMEs to the column, repeat this process three times, omitting the 10-min wait, for a total of four transfers.

5. Wash the column with 1 ml of acetonitrile:water (1:1) and allow the column to run dry. Wash the column with 200 μl of hexane and allow the column to run dry. Dry the column completely either by fitting a drying attachment and establishing a flow of nitrogen for 5 min ($= 1$ l min^{-1}) or by applying a vacuum to the bottom of the column for 15 min (15 psi).

6. Hydrate the packing by adding 2 ml of deionized water to the column and allow it to run dry.

7. Replace the waste container with a clean 16 × 125 mm test tube, close the column, and add 750 μl of hexane:chloroform (95:5). Allow the column to stand 2 min prior to initiating solvent flow.

8. Wash the column three times with 0.5 ml of hexane:chloroform (95:5) and collect the FAMEs.

9. Evaporate the solvent in a nitrogen evaporator having a bath temperature of 37°C.

Identification and Quantification of FAMEs (see Notes and Comments)

1. Dissolve FAMEs in the appropriate volume of hexane (usually 100 to 1000 μl) and inject them on a gas chromatograph.

2. Identify FAMEs by coelution with known standards.

3. Quantify FAMEs by developing appropriate integrator response factors.

Interpretation of Data

Approaches

The interpretive task is to relate the complex mixture of FAMES back to the organisms present in the sample, making use of phylogenetic relationships between organisms and their phospholipid fatty acids. These associations are so strong in some cases that "biomarker" fatty acids have been identified for particular microorganisms. On the other hand, the task is complicated in that all organisms contain a mixture of fatty acids — most often the common, infrequently the unique. The relationships between groups of microorganisms and specific fatty acids have been explored using pure culture studies, mixed enrichment cultures, and manipulative laboratory and field experiments. The intricacies of these relationships are beyond the scope of this paper and the reader is referred to References 10 to 13 and citations therein.

We employ a "functional-group" approach in our interpretation, i.e., we group together suites of microorganisms that share biochemical characteristics. For example, the two fundamental functional groups in sediments are eukaryotes and prokaryotes. Analysis of PLFA profiles can identify functional groups at a variety of levels, e.g., photosynthetic microeukaryotes, sulfate-reducing bacteria, the genus *Desulfobacter,* etc. Within this general approach, we have established the functional groups in two ways.

Dobbs and Guckert[11,12] used an *a priori* approach and defined eight functional groups based on PLFAs with known phylogenetic affinities. These groups were microeukaryotes (further divided into ω6 "animal" series, ω3 "plant" series and photoautotrophs) and prokaryotes (further divided into *Desulfobacter, Bacillus*-type Gram-positive bacteria, and bacteria utilizing an anaerobic desaturase pathway).

In an *a posteriori* approach, Findlay et al.[13] used multivariate analysis to determine clusters of fatty acids that showed similar patterns of change following disturbance. They then defined functional groups based on an integration of microbes' response to disturbance and information in the literature. Several fatty acids of broad phylogenetic distribution were included in the original analysis but are not considered "biomarkers" for their assigned functional group.[13] These fatty acids (14:0, 15:0, 16:0, 17:0, and 18:0) are present in a diversity of microorganisms and, although they showed patterns similar to fatty acids of restricted phylogenetic distribution, there is little reason to assume they will show similar covariance in other systems. Therefore, these PLFAs are not included in the community structure analysis illustrated below.

Table 1. Two Approaches for Assigning Phospholipid, Ester-Linked Fatty Acids to Functional Groups of Microorganisms

A. *A Priori* Approach

Functional Group	Fatty Acid(s)
Eukaryotes	16:4ω1,[a] 16:3,[a] 18:4ω3,[a] 18:3ω6, 18:2ω6, 18:3ω3, 20:4ω6, 20:5ω3, 22:5ω6,[a] 22:6ω3[a]
ω6 "animal" series	18:3ω6, 18:2ω6, 20:4ω6, 22:5ω6[a]
ω3 "plant" series	18:3ω3, 20:5ω3, 22:6ω3[a]
Eukaryotic photoautotrophs	16:1ω13t, 18:3ω3, 18:1ω9
Prokaryotes	i15:0, a15:0, 15:0, i17:0, a17:0, 17:0, 18:1ω7c, cy19:0(ω7,8), 10Me16:0, cy17:0(ω7,8)
Desulfobacter	10Me16:0, cy17:0(ω7,8)
Bacteria, anaerobic desaturase pathway	18:1ω7c
Bacillus-type Gram-positive bacteria	i15:0, a15:0, i17:0, a17:0

B. *A Posteriori* Approach

Functional Group	Fatty Acid(s)
Microeukaryotes	16:4ω1,[a] 16:3,[a] 18:4ω3,[a] 18:3ω3, 20:3ω6, 20:4ω6, 20:5ω3, 22:5ω6,[a] 22:6ω3[a]
Aerobic prokaryotes and eukaryotes	16:1ω5, 16:1ω7c, 17:1ω6, 17:1ω9, 18:1ω7c, 18:1ω9, 18:2ω6
Gram-positive prokaryotes and other anaerobic bacteria (except in the case of 16:1ω13t)	14:0,[b] a15:0, i15:0, 15:0,[b] i16:0, 16:1ω13t
Sulfate-reducing bacteria and other anaerobic prokaryotes	16:0,[b] 10Me16:0, a17:0, i17:0, cy17:0, 17:0,[b] 18:0,[b] cy19:0

Note: For expanded versions of these formats and references, refer to Reference 12 for approach A and Reference 22 for approach B, and to Reference 23.

[a]   Added to functional groups for this analysis; not listed in Reference 12.
[b]   Not included in analysis of this data set (see text for further comments).

Our functional-group assignments are summarized in Table 1. Note that our previous assignments are inadequate to assess completely the data set in this chapter. Five FAMEs (16:4ω1, 16:3, 18:4ω3, 22:5ω6, and 22:6ω3) present in Lowes Cove sediment that identified in the sediment used during the development of our functional-group approach. These FAMEs are clearly eukaryotic in origin, and most serve as markers for photoautotrophs. Indeed, 16:4ωl has been proposed as a biomarker for diatoms.[14] We have added these fatty acids to our eukaryotic functional groups.

Example

We illustrate the functional-group approach by comparing PLFA profiles from surface (0 to 1 cm) and subsurface (9 to 10 cm) sediments from Lowes Cove, Walpole, ME. We present the data for each fatty acid in two ways: (1) as absolute abundance (μg FAME g^{-1} dry weight of sediment) and (2) as relative abundance (i.e., the percentage each contributes to the total concentration of fatty acids).

We first note that 33 of the 34 fatty acids decreased in concentration with depth; the decrease was significant in 31 cases (Table 2). Thus, total microbial biomass decreased and the vast majority of organisms in surface sediments were less abundant at depth (but see Notes and Comments — 1).

One fatty acid, cyl7:0, exhibited a higher concentration at 9 to 10 cm, from which we tentatively consider that a functional group was more abundant at depth (but see Notes and Comments — 2). We expand on this thought when we consider cyl7:0 as one of the biomarkers for sulfate-reducing bacteria and other anaerobes.

We next determine whether any unusual FAMEs were present and if any expected ones were absent. In the surface samples 20:5ω3, 16:4ω1, 16:3, and 16:1ω13t were unusually

Table 2. Phospholipid Fatty Acid Profiles of Sediments Collected at Two Horizons from Lowes Cove, ME

Fatty acid	Absolute Abundance		Relative Abundance	
	0–1 cm	9–10 cm	0–1 cm	9–10 cm
i14:0	1.01 ± 0.49	0.08 ± 0.01*	0.57 ± 0.13	0.51 ± 0.05
14:0	3.64 ± 1.90	0.22 ± 0.00**	2.00 ± 0.52	1.41 ± 0.09
i15:0	2.57 ± 0.77	0.38 ± 0.03**	1.50 ± 0.14	2.45 ± 0.04*** ↑
a15:0	6.95 ± 1.31	1.42 ± 0.14***	4.22 ± 1.09	9.17 ± 0.25** ↑
15:0	2.72 ± 0.75	0.21 ± 0.02***	1.61 ± 0.27	1.35 ± 0.17
16:3	8.23 ± 4.73	0.12 ± 0.01**	4.43 ± 1.50	0.77 ± 0.09*** ↓
16:4ω1	1.94 ± 0.83	0.00 ± 0.00**	1.11 ± 0.21	0.00 ± 0.00** ↓
i16:0	1.24 ± 0.30	0.15 ± 0.01**	0.73 ± 0.10	0.97 ± 0.04* ↑
16:1ω9	1.94 ± 0.71	0.12 ± 0.02**	1.11 ± 0.03	0.79 ± 0.09**
16:1ω7	37.00 ± 11.58	3.08 ± 0.45***	21.46 ± 1.10	19.81 ± 1.47
16:1ω7t	1.08 ± 0.43	0.27 ± 0.04*	0.68 ± 0.32	1.77 ± 0.10* ↑
16:1ω5	2.64 ± 0.77	0.30 ± 0.05**	1.54 ± 0.12	1.91 ± 0.16* ↑
16:1ω13t	2.88 ± 1.38	0.02 ± 0.01**	1.59 ± 0.30	0.10 ± 0.07*** ↓
16:0	25.09 ± 9.03	2.19 ± 0.17***	14.33 ± 0.25	14.10 ± 0.11
br17:1	1.14 ± 0.28	0.16 ± 0.02**	0.68 ± 0.11	1.05 ± 0.04** ↑
10me16:0	2.19 ± 0.35	0.51 ± 0.06***	1.33 ± 0.33	3.29 ± 0.15** ↑
i17:0	0.84 ± 0.25	0.11 ± 0.00**	0.50 ± 0.11	0.73 ± 0.03* ↑
a17:0	3.50 ± 1.04	0.45 ± 0.02**	2.04 ± 0.16	2.90 ± 0.10** ↑
17:1ω6	2.33 ± 0.60	0.24 ± 0.04***	1.38 ± 0.21	1.54 ± 0.18
cy17:0	0.00 ± 0.00	0.11 ± 0.02** ↑	0.00 ± 0.00	0.70 ± 0.10*** ↑
17:0	2.14 ± 0.47	0.30 ± 0.02***	1.28 ± 0.22	1.94 ± 0.28* ↑
18:4ω3	2.67 ± 1.51	0.04 ± 0.01*	1.45 ± 0.47	0.24 ± 0.07** ↓
18:2ω6	0.25 ± 0.43	0.10 ± 0.04	0.24 ± 0.41	0.65 ± 0.28
18:1ω9c	4.44 ± 1.09	0.77 ± 0.07***	2.64 ± 0.42	4.97 ± 0.14** ↑
18:1ω7c	18.44 ± 4.49	2.44 ± 0.31***	10.94 ± 1.64	15.73 ± 0.92* ↑
18:0	2.64 ± 0.86	0.47 ± 0.08**	1.53 ± 0.05	3.03 ± 0.66* ↑
cy19:0	0.06 ± 0.10	0.03 ± 0.01	0.03 ± 0.05	0.20 ± 0.04* ↑
19:0	0.34 ± 0.17	0.03 ± 0.01*	0.19 ± 0.06	0.22 ± 0.05
20:4ω6	2.74 ± 0.91	0.42 ± 0.01**	1.58 ± 0.05	2.73 ± 0.18*** ↑
20:5ω3	24.86 ± 12.11	0.38 ± 0.12**	13.61 ± 2.77	2.51 ± 0.95** ↓
20:1ω9	0.45 ± 0.16	0.26 ± 0.13	0.26 ± 0.01	1.73 ± 0.94* ↑
20:0	1.61 ± 0.98	0.08 ± 0.00*	0.87 ± 0.32	0.50 ± 0.03
22:5ω6	0.33 ± 0.11	0.00 ± 0.00**	0.19 ± 0.02	0.00 ± 0.00*** ↓
22:6ω3	3.97 ± 2.25	0.02 ± 0.01**	2.12 ± 0.69	0.13 ± 0.04** ↓

Note: Absolute (μg FAME g^{-1} dry weight) and relative abundance (weight percent) of each fatty acid presented as a mean (n = 3) and standard deviation. For the percentage data, arrows indicate the direction of depth-related change. For absolute abundance data all significant changes are decreases, except where indicated. Asterisks denote the statistical significance of the depth-related differences (* = 0.05, ** = 0.01, *** = 0.001).

abundant. The generally common fatty acids, 18:2ω6 and 18:3ω3, were all but absent. These results suggest a relatively large biomass of a biochemically unique group; the fatty acid 16:1ω13t indicates photosynthetic organisms. (NOTE: The presence of 16:4ω1 suggests that diatoms are an important component of this group. Confirmation of diatoms' importance is provided by large amounts of 16:0, 16:3, and 20:5ω3.) On this basis, together with the abundance patterns of the above-mentioned fatty acids, we conclude that the biochemically unique group was primarily composed of diatoms. This conclusion can be (and was) confirmed microscopically.[15]

We now apply the functional groups defined by Dobbs and Guckert (Table 1, approach A).[11,12] Polyenoic fatty acids, indicative of microeukaryotes, accounted for 25% of all PLFAs in surface sediments, and only 7% at the 9- to 10-cm depth (Table 3). Their absolute abundance was 42 times lower at depth (Table 4). Omega-3 fatty acids ("plant" series) comprised 69% of the polyenoic fatty acids in surface sediments. When indicators of diatoms, 16:4ω1 and 16:3, are added to the ω3 series, the "plant" series fatty acids account for more than 90% of all polyenoic fatty acids in the surface sediments. At 9 to 10 cm, ω6 fatty acids (animal series) increased greatly in dominance relative to the surface and contributed 48%

Table 3. Changes in the Relative Abundance of Functional Groups (Approach A) of Sedimentary Microorganisms with Depth

	Relative Abundance	
Functional Group	0–1 cm	9–10 cm
Eukaryotes		
Total polyenoics/total PLFAs	24.75 ± 4.8	7.0 ± 1.2
ω3 "plant" series/total polyenoics	69.25 ± 3.7	40.3 ± 6.9
ω3 "plant" series + 16 polyenoics/total polyenoics	91.45 ± 3.9	51.4 ± 5.0
ω3 "animal" series/total polyenoics	8.65 ± 3.9	48.5 ± 5.0
Prokaryotes		
Total prokaryotes/total PLFAs	25.65 ± 3.9	41.2 ± 0.9
Desulfobacter/total PLFAs	1.35 ± 0.3	4.0 ± 0.0
Bacteria, anaerobic desaturase pathway/total PLFAs	10.92 ± 1.6	15.7 ± 0.9
Bacillus-type Gram-positive bacteria/total PLFAs	8.35 ± 1.4	15.3 ± 0.2

Note: Data are percentages, Mean ± SD, n = 3.

of the polyenoic fatty acids. The absolute abundance of the ω6 acids dropped much less with depth than that of ω3 acids, 6- vs. 72-fold. The fatty acids characteristic of prokaryotes, i.e., branched chain, odd-chain, and cyclopropyl derivatives, were 26% of all surface PLFAs, 41% at depth, and decreased in absolute abundance seven times with depth. The *Desulfobacter* functional group increased from approximately 1% of surface fatty acids to 4% at 9 to 10 cm, and decreased only 3.5-fold in absolute abundance. The biomarker for bacteria having the anaerobic-desaturase pathway, 18:1ω7c, comprised 11% of surface PLFAs, 16% at 9 to 10 cm, and decreased in abundance 7.5 times with depth. Fatty acids representative of *Bacillus*-type, Gram-positive bacteria increased from 8% at the surface to 15% at depth, and their abundance dropped sixfold.

Next, we apply the functional-group approach of Findlay et al. (Table 1, approach B).[13] In this approach, the patterns of change of the individual PLFAs are analyzed. Two factors are considered: (1) are patterns within the functional group consistent and (2) are there contrasts among the patterns for the various functional groups? All the PLFAs within the microeukaryote functional group, except 20:4ω6, decreased with depth in absolute abundance by 65-fold or greater (Table 5). Relative abundances were also significantly less. The PLFA 20:4ω6, in contrast, decreased in absolute abundance only 6.5-fold and increased in relative abundance. The PLFAs indicative of the aerobic prokaryotic and heterotrophic eukaryotes

Table 4. Changes in Absolute Abundance of the Functional Groups (Approach A) of Microorganisms with Depth

	Abundance		
Functional Groups	0–1 cm	9–10 cm	Change
Total microbial biomass			
Total PLFA	174.36 ± 60.55	15.50 ± 1.07	11.2
Phospholipid phosphate	402.83 ± 137.80	51.98 ± 5.04	7.8
Eukaryotes			
Total polyenoics	44.99 ± 21.58	1.08 ± 0.13	41.6
ω3 "plant" series	31.51 ± 15.73	0.44 ± 0.12	71.6
ω3 "plant" series + 16 polyenoics	41.68 ± 21.01	0.56 ± 0.11	74.6
ω6 "animal" series	3.32 ± 0.58	0.52 ± 0.05	6.4
Photoautotrophs	7.32 ± 2.33	0.79 ± 0.08	9.3
16:1ω13t	2.88 ± 1.38	0.02 ± 0.01	186.8
Prokaryotes			
Total prokaryotes	43.14 ± 10.15	6.39 ± 0.58	6.8
Desulfobacter	2.19 ± 0.35	0.62 ± 0.07	3.5
Bacteria, anaerobic desaturase pathway	18.48 ± 4.49	2.44 ± 0.32	7.5
Bacillus-type, Gram-positive bacteria	13.86 ± 3.14	2.36 ± 0.19	5.9

Note: Data are given as micrograms per gram dry weight of sediment, except phospholipid phosphate which is given as nanomoles per gram dry weight. Mean ± SD, n = 3.

Table 5. Changes in Absolute and Relative Abundance of the Functional Groups (Approach B) of Sedimentary Microorganisms with Depth

Functional Group	Abundance (μg g^{-1} dry wt)			Relative Abundance (%)		
	0–1 cm	9–10 cm	change	0–1 cm	9–10 cm	change
Eukaryotes						
16:4ω1	1.94	0.00	—	1.11	0.00	— ↓
16:3	8.23	0.12	68.6	4.43	0.77	5.8 ↓
18:4ω3	2.67	0.04	66.8	1.45	0.24	6.0 ↓
20:5	24.86	0.38	65.4	13.61	2.51	5.4 ↓
22:5ω6	0.33	0.00	—	0.19	nd	— ↓
22:6ω3	3.97	0.02	198.5	2.12	0.13	16.3 ↓
20:4ω6	2.74	0.42	6.5	1.58	2.73	1.7 ↑
Aerobic prokaryotes and eukaryotes						
16:1ω5	2.64	0.30	8.8	1.54	1.91	1.2 ↑
16:1ω7	37.00	3.08	12.0	21.46	19.81	=
17:1ω6	2.33	0.24	9.7	1.38	1.54	=
18:1ω7	18.44	2.44	7.5	10.94	15.73	1.4 ↑
18:1ω9	4.44	0.77	5.8	2.64	4.97	1.9 ↑
18:2ω6	0.25	0.10	=	0.24	0.65	=
Gram-positive prokaryotes and other anaerobic bacteria						
a15:0	6.95	1.42	4.9	4.22	9.17	2.2 ↑
i15:0	2.57	0.38	6.8	1.50	2.45	1.6 ↑
i16:0	1.24	0.15	8.3	0.73	0.97	1.3 ↑
Sulfate-reducing bacteria and other anaerobic bacteria						
10Me16:0	2.19	0.51	4.3	1.33	3.29	2.5 ↑
a17:0	3.50	0.45	7.7	2.04	2.90	1.4 ↑
i17:0	0.84	0.11	7.6	0.50	0.73	1.5 ↑
cy17:0	0.00	0.11	— ↑	0.00	0.70	— ↑
cy19:0	0.06	0.03	=	0.03	0.20	6.7 ↑

Note: — indicates that the calculation of change is made meaningless by zero data, = indicates that there was no statistically significant difference between mean values for the two depths; the arrows indicate the directions of significant changes, except for absolute abundance data in which all significant changes but one were decreases in abundance.

functional group were 6 to 12 times less abundant at depth; their relative abundance was unchanged or increased moderately. Fatty acids that define the anaerobic, Gram-positive functional group decreased in absolute abundance four- to eightfold with depth and their relative abundance increased. Of the fatty acids used to define the anaerobic, sulfate-reducing functional group, three of five decreased in absolute abundance with depth (four- to eightfold), one was unchanged, and cy17:0 increased. The relative abundance of all fatty acids in this group increased with depth.

To summarize, the distribution of total PLFAs indicates that total microbial biomass was about an order of magnitude greater in surface sediments than in sediments from the 9- to 10-cm horizon (see Notes and Comments — 1). Correspondingly, all previously defined functional groups followed this trend in absolute abundance, but the degree of decrease varied among them. In the 0- to 1-cm horizon, phototrophic microeukaryotes, predominantly diatoms, contributed most of the eukaryotic biomass and more than half of the total microbial biomass (see Notes and Comments — 1). There was a profound depth-related decrease in eukaryotic biomass because the strong contribution of diatoms in surface sediments was essentially absent at 9 to 10 cm. The biomass of heterotrophic eukaryotes declined only 6.5-fold and increased in relative abundance at depth, a trend similar to that observed for prokaryotic biomass. The prokaryotic community was a mix of aerobes and anaerobes and had a strong sulfate-reducing component. Among prokaryotic groups, sulfate-reducing and Gram-positive bacteria showed the largest relative increase with depth, three and two times, respectively.

NOTES AND COMMENTS

Extraction and Purification of PFLAs

1. Gas chromatographic analysis of environmental PLFAs usually requires nanomolar sensitivity. This necessitates good analytical technique and a high degree of cleanliness in glassware and sample handling equipment. Unfortunately, much of the available disposable plasticware (e.g., prepacked SPE column, pipette tips, etc.) yield significant contaminants when exposed to lipid solvents, hexane in particular. This problem is not intractable. For example, pipettors can be fitted with Pasteur pipettes instead of disposable plastic tips and columns can be packed in large-volume pipettes or reusable glass columns fitted with Teflon frits. We recommend investigators test all glassware and plasticware prior to first use as follows: wash with chloroform and hexane, collect and dry the solvents, redissolve the residue in a minimum of hexane, and inject this mixture into the GC. If peaks are noted, the glassware must undergo further cleaning (see Chapter 40) or the plasticware must be replaced with a glass equivalent. We also suggest that at least one "blank" extraction be conducted for each batch of samples analyzed. This blank should experience all analytical procedures and if it is "clean" it makes possible the conclusion that peaks observed with the GC originated in the environmental sample and not from the analytical reagents or procedures.

2. Light, molecular oxygen, and water are the most common causes of degradation of lipids after extraction. We recommend that lipid extracts be stored dark, frozen, dry, or in fresh solvent, and under a N_2 atmosphere.

3. Classically, chloroform has been stabilized using 0.75% ethanol. Recently, many high-purity grades of chloroform have been stabilized using other compounds (notably pentene). For the chromatographic procedures described here the change in polarity caused by the change in stabilization compound can be profound. If ethanol is not used for stabilization, it must be added to the chloroform for the procedures described above to yield quantitative results.

Formation of FAMEs

1. A simple column-height to packing-weight relationship can be established and used for setup of columns. Packing is provided "in excess" and need not be weighed after the initial setup.

2. There are a great variety of labor saving devices available to aid in the use of SPE columns. These devices are not necessary for successful completion of the above protocols but they do increase sample throughput, thereby decreasing costs.

3. The C_{18} packing must be hydrated prior to use. If not, recovery of polyenoic FAMEs may be nonquantitative.

Identification and Quantification of FAMEs

1. We assume an equal detector response for all FAMEs and FAEEs, that is, a 14-carbon, saturated, methyl ester is assumed to yield the same integrator counts per nanogram as does a 20-carbon, polyunsaturated, ethyl ester. Environmental FAME data have been presented in two ways — moles FAME g^{-1} dry wt substrate (typical range is nanomoles) or grams FAME g^{-1} dry wt substrate (typical range is micrograms). To proceed to grams g^{-1} dry wt of substrate calculate (counts FAME) $\times$ (counts standard^{-1}) $\times$ (grams standard) $\times$ (dilution^{-1}) $\times$ (percentage of lipid removed for other analyses^{-1}) $\times$ (g dry wt of substrate^{-1}).

2. One of us (Findlay) routinely conducts FAME analyses. The equipment and chromatographic conditions used are as follows. The gas chromatograph is a Hewlett-Packard 5890

series II fitted with a flame ionization detector and a Grobes splitless injector. The chromatographic column is a 60-m J & W Scientific DB-1. The film thickness (the liquid phase) is 0.25 μm (Catalog #122-1062). This is a nonpolar, SE 30 equivalent, bonded-phase, fused silica column. These columns give excellent overall separations, are quite durable and show good column to column reproducibility (the catalog claims they are made by elves!). Hydrogen is used as the carrier gas and the flow set at 50 cm s^{-1} at 80°C. Flow out of the split valve is set at 100 ml min^{-1} and the split valve is opened 0.5 min after the injection of the sample. (NOTE: Hydrogen, when mixed with oxygen, is quite explosive. Inexperienced users may opt for helium as a carrier gas, thereby eliminating the danger of explosion and fire, but sacrificing some resolution and speed of analysis.) The flame ionization detector is run with hydrogen as fuel (45 ml min^{-1}), breathing air as oxidant (350 ml min^{-1}) and nitrogen as a makeup gas (38 ml min^{-1}). The chromatogram is recorded and the peaks integrated using an HP 3396A integrator. The injector is heated to 250°C and the detector to 260°C. Samples are typically injected by hand using a hot, fast injection with the oven heated to 80°C. Immediately after injection the oven temperature is increased at a rate of 4°C min^{-1} to a final temperature of 250°. Figure 1 shows a "typical" chromatogram (for nonchromatographers — "one of our best") resulting from the use of the above protocol, equipment, and chromatographic conditions. The complexities of gas chromatography are many and this forum is not amenable to further discusssion.

3. An alternative to investing the effort and capital necessary to conduct gas chromatographic analysis is to extract and purify the phospholipids, then purchase FAME analysis. One of us (Dobbs) utilizes this option. Many workers who routinely analyze FAMEs run samples for other PI's on a subcontract basis; others have set up consulting firms to provide the service.

4. Identification of a fatty acid methyl ester is considered preliminary if the environmental FAME coelutes with an authentic standard. Identification is considered confirmed if the environmental FAME coelutes with an authentic standard using a second chromatographic column (composed of a different liquid phase). Typically, a nonpolar (e.g., DB-1) column and a polar (e.g., carbowax) column are used to confirm FAME identifications. For putative FAMEs for which no authentic standards can be obtained, mass spectral (MS) analysis is necessary to confirm identification. Identifications can be purchased commercially or in many cases can be obtained from a PI's chemistry department.

5. Fatty acids and their methyl esters are named based on the number of carbons in the molecule, the number of double bonds, and the position of the first double bond from the "omega" or aliphatic end of the molecule. For example, 16:0 refers to the fatty acid that is 16 carbons in length and has no double bonds: 20:5ω3 refers to the fatty acid that is 20 carbons in length, has 5 double bonds, and the first double bond occurs at the number 3 carbon from the aliphatic end of the molecule. All double bonds are presumed *cis* unless otherwise denoted as *trans* (with a suffix of "t"). The prefixes "a", "i", "br", and "cy" indicate antiiso-branched, iso-branched, branched (position undetermined), and cyclopropyl fatty acids, respectively.

Interpretation of Data

1. It is problematic to convert the concentration of PLFA from an environmental sample to the abundance of a particular functional group of microorganism. Empirical conversion factors have not been determined and a number of assumptions must be made. We provide an example using the data set. Polyenoic fatty acids, indicative of microeukaryotes, comprised 25% of the PLFAs from the surface sediment. Assume that 50% of the PLFAs of eukaryotic microorganisms are polyenoic (for examples, see References 16 and 17); then 50% of the PFLAs from the surface sediments originated in microeukaryotes. By difference, 50% of the PFLAs were from prokaryotes. These percentages remain the same when expressed for phospholipid, since there are two fatty acids on each phospholipid

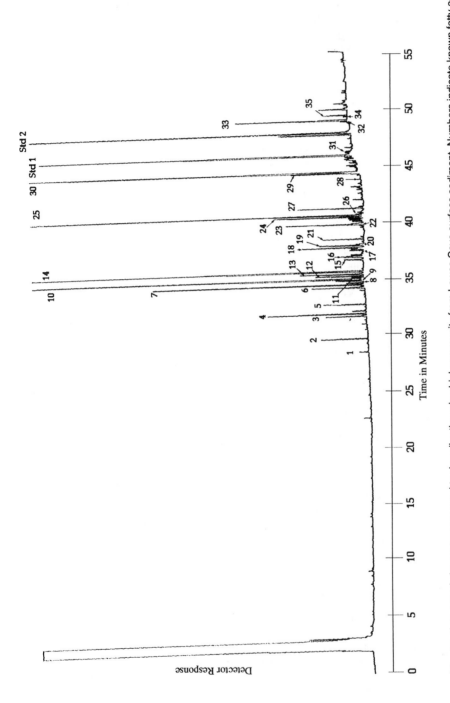

Figure 1. Scanned image of an actual chromatogram used to describe the microbial community from Lowes Cove surface sediment. Numbers indicate known fatty acid methyl esters. They are 1 — i14:0, 2 — 14:0, 3 — i15:0, 4 — a15:0, 5 — 15:0, 6 — 16:4ω1, 7 — 16:3, 8 — i16:0, 9 — 16:1ω9, 10 — 16:1ω7, 11 — 16:1ω7t, 12 — 16:1ω5, 13 — 16:ω13t, 14 — 16:0, 15 — br17:1, 16 — 10 me16:0, 17 — i17:0, 18 — a17:0, 19 — 17:1ω6, 20 — cy17:0, 21 — 17:0, 22 — 18:3ω6, 23 — 18:4ω3, 24 — 18:1ω9, 25 — 18:1ω7, 26 — 18:1ω7t, 27 — 18:0, 28 — 19:0, 29 — 20:4ω6, 30 — 20:5ω3, 31 — 20:0, 32 — 22:5ω6, 33 — 22:6ω3, 34 — 22:4ω6, 35 — 22:5ω3. Peaks labeled Std 1 and Std 2 indicate internal standards added prior to the C18 column cleanup procedure. They are the ethyl esters of the fatty acids 20:4ω6 and 20:0, respectively.

molecule. The relative biomass of microeukaryotes and the prokaryotes can be computed using conversion factors.[9] Assuming 50 μmol P g^{-1} C for eukaryotic cells and 100 μmol P g^{-1} C for prokaryotic cells, microeukaryotes comprised 75% and prokaryotes 25% of the total microbial biomass (as carbon). These calculations contain a number of reasonable assumptions; however, any or all could be in error for the system under study. The above predominance of microeukaryotic biomass is uncommon for sediments, in which it is generally perceived that prokaryotic biomass dominates.[18] Our contrary finding is not surprising, however, for sediments in this region during times of cold water and sediment temperature[19] (samples were taken in late November). It appears that macrofaunal activity (both feeding and bioturbation) is suppressed which allows the development of sedimentary diatom blooms.

2. While the phospholipid content of prokaryotic cells has been shown to be reasonably stable to culture and growth conditions, the composition of PLFA profiles are known to change. In particular, the proportion of cyclopropyl and *trans*-fatty acids can increase with changes in environmental or growth conditions, such as starvation, low oxygen tension, or increased temperature. Many Gram-negative organisms have been shown to increase their percentage of either 16:1ω7t or cy17:0 at the expense of 16:1ω7c or their percentage of 18:1ω7t or cy19:0 at the expense of 18:1ω7c.[20,21] In particular, we have utilized the ratio of these fatty acids (i.e., ratio of 16:1ω7t to 16:1ω7c) as a measure of stress in environmental samples.[22] In the present data set the ratio 16:1ω7t/16:1ω7c increased from 0.0312 ± 0.014 in surface sediment to 0.089 ± 0.002 in the 9- to 10-cm horizon. This change suggests that the microbes at depth experience environmental stress such as low oxygen tension (stressful to aerobic bacteria) or starvation. Such a conclusion, however, tempers the interpretation (made earlier) that increased levels of cy17:0 indicate an increase in biomass of a particular type of microorganism. An alternative (and regrettably, nonexclusive) interpretation is that the same suite of microbes occurs in both horizons, and have adjusted their PLFA profiles in response to stresses associated with life in subsurface sediment.

ACKNOWLEDGMENTS

This work was supported in part by NSF Grants OCE-9018599 (Dobbs), OCE-8315883 (Findlay, awarded to J. Fell, University of Miami), and OCE-8700358 (Findlay); and EPA grant R-817196-01-0 (Findlay). We gratefully acknowledge T. Sawyer who conducted the FAME analysis. This is the Darling Marine Center's Contribution Number 249 and the University of Hawaii's School of Ocean and Earth Sciences and Technology Contribution Number 3092.

REFERENCES

1. White, D. C., Analysis of microorganisms in terms of quantity and activity in natural environments, in *Microbes in their Natural Environments,* Slater, J. H., Whittenbury, R., and Wimpenny, J. W. T., Eds., *Society of General Microbiology Symposium,* 1983, 37.
2. Austin, B., *Methods in Aquatic Bacteriology,* John Wiley & Sons, Chichester, England, 1988.
3. Rollins, D. M. and Colwell, R. R., Viable but nonculturable stage of *Campylobacter jejuni* and its role in survival in the natural aquatic environment, *Appl. Environ. Microb.,* 52, 531, 1986.
4. Roszak, D. B. and Colwell, R. R., Survival strategies of bacteria in the natural environment, *Microb. Rev.,* 51, 365, 1987.
5. Pace, N. R., Stahl, D. A., Lane, D. J., and Olsen, G. J., The analysis of natural microbial populations by ribosomal RNA sequences, *Adv. Microb. Ecol.,* 9, 1, 1986.

6. Schmidt, T. M., DeLong, E. F., and Pace, N. R., Analysis of a marine picoplankton community using 16S rRNA gene cloning and sequencing, *J. Bacteriol.*, 173, 4371, 1991.

7. White, D. C., Davis, W. M., Nickels, J. S., King, J. D., and Bobbie, R. J., Determination of the sedimentary microbial biomass by extractible lipid phosphate, *Oecologia*, 40, 51, 1979.

8. Kates, M., *Techniques of Lipidology*, 2nd Ed., Elsevier, New York, 1986.

9. Dobbs, F. C. and Findlay, R. H., Analysis of microbial lipids to determine biomass and detect the response of sedimentary microorganisms to disturbance, in *Current Methods in Aquatic Microbial Ecology*, Kemp, P. F., Sherr, B. F., Sherr, E. B., and Cole, J. J., Eds., CRC Press, Boca Raton, FL, 1993, chap. 40.

10. Ratledge, C. and Wilkinson, S. G., *Microbial Lipids*, Vol. 1, Harcourt Brace Jovanovich, New York, 1988.

11. Dobbs, F. C. and Guckert, J. B., Microbial food resources of the macrofaunal-deposit feeder *Ptychodera bahamensis* (Hemichordata: Enteropneusta), *Mar. Ecol. Prog. Ser.*, 45, 127, 1988.

12. Dobbs, F. C. and Guckert, J. B., *Callianassa trilobata* (Crustacea: Thalassinidea) influences abundance of meiofauna and biomass, composition, and physiologic state of microbial communities within its burrow, *Mar. Ecol. Prog. Ser.*, 45, 69, 1988.

13. Findlay, R. H., Trexler, M. B., Guckert, J. B., and White, D. C., Laboratory study of disturbance in marine sediments: response of a microbial community, *Mar. Ecol. Prog. Ser.*, 62, 121, 1990.

14. Gillan, F. T. and Hogg, R. W., A method for the estimation of bacterial biomass and community structure in mangrove-associated sediments, *J. Microbiol. Methods*, 2, 275, 1984.

15. Watling, L., personal communication, 1992.

16. Findlay, R. H., Fell, J. W., Coleman, N. K., and Vestal, J. R., Biochemical indicators of the role of fungi and thraustochytrids in mangrove detrital systems, in *Biology of Marine Fungi*, Moss, S. T., Ed., Cambridge University Press, Cambridge, 1986, 91.

17. Wood, B. J. B., Lipids of algae and protozoa, in *Microbial Lipids*, Ratledge, C. and Wilkinson, S. G., Eds., Harcourt Brace Jovanovich, New York, 1988, 807.

18. Federle, T. W., Livingston, R. J., Wolfe, L. E., and White, D. C., A quantitative comparison of microbial community structure of estuarine sediments from microcosms and the field, *Can. J. Microbiol.*, 32, 319, 1986.

19. Findlay, R. H., unpublished data, 1992.

20. Guckert, J. B., Hood, M. A., and White, D. C., Phospholipid ester-linked fatty acid profile changes during nutrient deprivation of *Vibrio cholerae*: increases in the trans/cis ratio and proportions of cyclopropyl fatty acids, *Appl. Environ. Microbiol.*, 46, 930, 1986.

21. Wilkinson, S. G., Gram-negative bacteria, in *Microbial Lipids*, Ratledge, C. and Wilkinson, S. G., Eds., Harcourt Brace Jovanovich, New York, 1988, 299.

22. Findlay, R. H., Trexler, M. B., and White, D. C., Response of a benthic microbial community to biotic disturbance, *Mar. Ecol. Prog. Ser.*, 62, 135, 1990.

23. Vestal, J. R. and White, D. C., Lipid analysis in microbial ecology, *Bioscience*, 39, 535, 1989.

CHAPTER 33

Single-Cell Identification Using Fluorescently Labeled, Ribosomal RNA-Specific Probes

Edward F. DeLong

INTRODUCTION

In recent years fluorescent DNA stains, used in conjunction with epifluorescence microscopy, have become standard tools for obtaining quantitative information on microbial communities. Most studies to date have utilized DNA- or RNA-binding fluorochrome stains, such as acridine orange or DAPI,[1,2] which indiscriminately bind to the nucleic acids of all bacterial cells in a given sample. Field studies using these fluorochromes have provided a wealth of information on total bacterial numbers in a variety of environments. Though extremely useful, particularly for studying aquatic microbial populations, these general fluorochrome stains provide little information on the identity of specific bacterial population constituents. For detailed analyses of microbial population structure and dynamics, methods for the identification and enumeration of specific microbial taxa are necessary. Recent methodological advances, including the application of fluorescently labeled immuno-chemical[3-5] or nucleic acid probes,[6,7] provide a sensitive means for taxonomically identifying individual cells present in mixed populations. Though these techniques are relatively new, they have already facilitated rapid identification of culturable and "unculturable" microbial taxa, and the localization of bacterial symbionts in host tissues.[6-11]

The procedure described here employs oligonucleotide probes complementary to ribosomal RNA (rRNA) sequences. The probes can be targeted toward either small subunit rRNA (ssu rRNA = 16S-like rRNA) or large subunit rRNA (lsu rRNA = 23S-like rRNA). The large amount of phylogenetic information contained within rRNA sequences[12] facilitates the design of oligonucleotide probes of varying specificities. Blocks of contiguous rRNA sequence can be identified which are unique to and diagnostic for very closely related taxa. More conserved regions in the molecule can be used to specifically identify higher level taxa (e.g., kingdoms, genera).[12-16] Probe specificity is achieved empirically by adjusting hybridization conditions such that stable probe-target duplexes form only when there is near perfect Watson-Crick base pair complementarity. With proper attention to hybridization conditions, discrimination at the level of one nucleotide mismatch is possible.[7,17]

Due to the cellular abundance of ribosomes, ribosomal RNA-targeted hybridization probes can be used to identify individual cells. Detection sensitivity will depend in part on the physiological state of the target cells. Faster growing or more physiologically active cells tend to have more ribosomes, and hence bind proportionately more probe molecules.[6] The

0-87371-564-0/93/$0.00 + $.50

basic technique described here employs fluorescently labeled, rRNA-targeted oligonucleotide probes which bind to specific rRNA target sequences in the ribosomes of fixed, intact cells. Cells are detected via standard epifluorescence microscopy. Scanning laser confocal microscopy[6] and flow cytometry[18] have also been used to detect cells which have bound the rRNA-targeted probes.

One advantage of using single, fluor-labeled oligonucleotides is that labeling and detection is simple and rapid. However, low cellular ribosome content or high fluorescent background may limit detection sensitivity, and in these situations alternative isotopic or nonisotopic labels may be preferable. Increasing the signal strength and reducing background are major goals for the future.[6,8,19] Attempts to label oligonucleotides with multiple fluors have resulted in poor or nonspecific binding, probably due to the hydrophobicity of the fluorescent moieties. Potential labeling alternatives include indirect labeling with biotin/avidin systems[20] (also Eee Lin Lim, David Caron, and Ed DeLong, unpublished data) or the digoxigenin system of Boeringer Mannheim.[19] These indirect labeling methods can be coupled to a variety of detection systems including fluorescence detection, enzyme-linked colorimetric or chemiluminescent detection, or immunogold detection at the level of electron microscopy.

There is a good deal of room for improvement of these techniques for practical field application. Adaptation of the methods for use with field samples concentrated on membrane filters, or for in solution hybridization and detection via flow cytometry,[18] represent a few of the possibilities. Future refinement and improvement of the basic technique should facilitate rapid phylogenetic identification of single cells in field samples, and promises to increase our understanding of the structure and dynamics of natural microbial assemblages.

MATERIALS REQUIRED

Probe Labeling

Equipment

- Speed Vac or a similar evaporator/concentrator with vacuum pump
- Vertical gel electrophoresis apparatus
- Electrophoresis power supply
- Glass plates, tape
- Vacuum aspirator
- Hand held UV illuminator (long wave)

Supplies

- 500-ml vacuum sidearm flask
- Plastic disposable 10-ml pipettes
- Glass wool
- 1.5-ml microfuge tubes
- 8-ml plastic scintillation vials

Solutions

- Amino modified oligonucleotide (1 mg ml^{-1} solution in H_2O)
- Reactive dye (FITC or Texas Red, freshly dissolved to 10 mg ml^{-1} in dimethylformamide)
- 500 mM sodium carbonate buffer, pH 9.2 (mix 10 ml of 500 mM Na_2CO_3 with 90 ml 500 mM HaHCO$_3$; Measure pH and adjust to 9.2. Store frozen in small aliquots.)

- 50% nondenaturing acrylamide stock (50% acrylamide/2.5% bisacrylamide in H_2O 0.22-μm filter. Store in the dark at 4°C)
- 10× NNB (Dissolve 162 g Tris base, 27.5 g boric acid, 9.3 g Na_2EDTA in 870 ml H_2O)
- N,N,N',N'-tetramethylethylenediamine (TEMED)
- 40% ammonium persulfate (dissolve 0.4 g ammonium persulfate in 1 ml H_2O. Prepare fresh.)
- 10% sucrose (w/v)
- 0.1% bromphenol blue (w/v)
- 0.1% xylene cyanol
- Sephadex G25 — (equilibrate overnight in 200 ml of 10 mM Tris-HCl, pH 8.0. Use 2 g for each 8-ml column to be poured.)

Cell Fixation/Immobilization

Equipment

- Slide racks and staining jars
- Teflon coated slides with wells (Cell Line Assoc., Newfield, NJ)
- 4°C incubator or refrigerator

Supplies/Solutions

- Gelatin-subbed slides: soak clean slides for 2 min in a filtered 65°C solution of 0.1% gelatin/ 0.01% $CrK(SO_4)_2$(chrom alum) for 2 min. Air dry and protect from dust.
- 10× phosphate buffered saline (10× PBS), pH 7.6. Mix 12.63 g Na_2PO_4, 1.8 g NaH_2PO_4 and 85 g NaCl in 1 l H_2O.
- Sterile filtered seawater
- 8% paraformaldehyde/1× PBS (freshly prepared): dissolve 4 g paraformaldehyde in 5 ml 10× PBS + 45 ml H_2O at 65°C. Cool and store at 4°C.
- Methanol:formaldehyde post-fixative: mix 30 ml Formalin (37% formaldehyde solution) with 270 ml methanol.

Hybridization/Detection

Equipment

- Epifluorescence microscope fitted with appropriate filter sets
- Dry-type bacteriological incubator or oven
- Water baths

Supplies

- Coplin jars (or the equivalent for washing slides)
- 50-ml polyethylene tubes (Falcon #2098 tubes)
- Immersion oil (Zeiss type 518C)
- Coverslips (No. 1, 24 × 60 mm)

Solutions

- Fluorescent oligonucleotides (50 ng μl^{-1} in H_2O)
- 25× Saline-Tris-EDTA (25× SET): for 1 l, dissolve 219.2 g NaCl, 50 ml of 0.5 M EDTA, and 60.5 g Tris base in 800 ml H_2O. Adjust pH to 7.8 with concentrated HCl ($\approx$32 ml).

Bring volume to 1 1, check, and adjust the pH after the solution has cooled, and store at 4°C.
- 10% sodium dodecyl sulfate (SDS): dissolve 10 g SDS in 100 ml H_2O.
- Citifluor mountant (Citifluor, Ltd., London)

PROCEDURES

The methods described below have been optimized for fixed cells cultivated in the laboratory. They should be viewed as a starting point, to be modified for the application to specific cell types, environments, or samples.

Probe Design, Synthesis, and Testing

The first step in designing an rRNA probe specific for a given species (or other phylogenetic group) is to obtain rRNA sequences from a collection of the specie(s) of interest. Current protocols for the direct sequencing of ribosomal RNA templates, or their cloned or amplified genes,[21-23] have recently facilitated rapid accumulation of SSU rRNA sequences. In addition, development of techniques for clonally isolating rRNA genes from mixed microbial assemblages has greatly expanded our ability to characterize the phylogenetic diversity of microbial populations.[24-28] Hence, it is now possible to design probes even for "unculturable" microbes.[7,24-26] Analysis of a set of aligned rRNA sequences from as many organisms as possible is the starting point for identifying sequence regions which are diagnostic for particular taxa. Highly conserved sequence regions, as well as the conserved secondary structure of SSU rRNA, aids in the alignment of homologous nucleotides.[29] Publicly available databases now contain a large number of rRNA and rRNA gene sequences. A large database of aligned rRNA sequences is currently available through the NSF-funded Ribosomal RNA Database Project,[30] which can be accessed via e-mail or Internet.

Computer-assisted analyses of aligned rRNA sequences is employed to identify a contiguous region of nucleotide sequence that is unique to a given species (or other taxon). Ideally, this 15- to 30-base oligonucleotide sequence should have at least 2 to 3 mismatches with the homologous sequence region from closely related species. The base composition and placement of mismatches will substantially influence probe-target duplex stability. In general, oligonucleotides with a G + C content of around 50% or greater provide adequate duplex stability. Mismatches in the interior of the oligonucleotide sequence will have a greater destabilizing effect than those at the 3' or 5' ends, and so interior mismatches afford greater specificity. Care should be taken during the design stage that the oligonucleotide does not contain substantial self-complementarity. Once an appropriate target region is identified, the corresponding oligonucleotide probe is synthesized *in vitro* with an automated DNA synthesizer.

The specificity and fidelity of probe binding may be confirmed by hybridization of the probe with a collection of rRNAs from a variety of different species. Typically, the probe is radiolabeled and hybridized with rRNA immobilized on a nylon support. Stahl and Amman[15] provide an excellent methodological summary of isotopic and nonisotopic probe labeling for further reference.

A rough estimate for the dissociation temperature (Td, temperature at which half the bound oligonucleotide is dissociated from the target) of short oligonucleotide probes in 1 M Na^+ is Td $\approx 4°$ × (number of G + C residues) + $2°$ × (number of A + T residues).[17] In general, hybridization and wash temperatures about 5 to 10°C below the Td of a perfectly matched duplex yield strong hybridization signals with good specificity. A well-designed probe should give a positive hybridization signal with rRNA from the targeted organism, but should not bind to rRNA of heterologous species. As a first test, hybridization of isotopically or nonisotopically labeled probes with bulk nucleic acids fixed to solid supports is convenient for optimizing hybridization and wash temperatures, and screening a large variety of rRNAs. Though the binding of probes to extracted, immobilized rRNA is a good indicator of probe specificity, binding properties may vary for *in situ* hybridizations. Hence, it is important to characterize the hybridization properties of the fluorochrome-labeled probes with fixed whole cells as well.

Fluorescent Labeling of Oligonucleotide Probes

1. Mix the following in a 1.5-ml tube

    100 μl of 1 mg ml^{-1} amino-oligonucleotide in H_2O (100 μg total)
    50 μl of 500 mM carbonate buffer, pH 9.2
    60 μl H_2O

2. Remove 1 μl, and check the pH with pH paper. Adjust the pH to >9.0 with 0.2 M NaOH, if necessary.

3. Add 40 μl of 10 mg ml^{-1} reactive dye solution.

4. Incubate in the dark at room temperature overnight.

5. Pour one Sephadex G-25 column for each reaction. Plastic disposable 10-ml pipettes, plugged with glass wool, work well. A bed volume of 5 to 7 ml provides good separation of the oligo from unincorporated dye. Pour the preswollen sephadex (in 10 mM Tris-HCl) into the column (plugged at the end with parafilm) and allow it to gently settle. Bring the meniscus of the column buffer just to the top of the sephadex bed by allowing the column to drain — do not allow the column bed to dry out!

6. Plug the end of the column again and carefully load the 250 μl of the dye-oligonucleotide reaction mixture onto the column bed. Start the column dripping, allowing the reaction mix to fully enter the sephadex bed.

7. Top off the column with 10 mM Tris-HCl, pH 8.0, and start collecting the column eluate. Collect fractions ($\approx$8 drops each) in a microtiter plate or in 1.5-ml tubes. The labeled oligonucleotide is visible as a lightly colored band, and should elute in the void volume, well before the intensely fluorescent, trailing band which is the unincorporated fluor.

8. Examine the collected fractions under a long-wave UV lamp. The first few fractions should be colorless, followed by faintly fluorescent fractions (usually fractions #4 to 8). Pool these fractions containing labeled oligonucleotide probe.

9. Lyophilize the pooled product in a Speed-vac or the equivalent. Resuspend the dried product in 25 μl of 10% sucrose, to increase the density for gel loading.

10. While the oligonucleotide is lyophilizing, pour a 20 $\times$ 20 cm preparative gel with a spacer thickness of 2 to 3 mm. In a 500-ml sidearm flask, mix

    40 ml of 50% acrylamide stock
    5 ml 10$\times$ NNB
    55 ml H_2O

    Attach an aspirator to the sidearm, place a rubber stopper on the top of the flask, and apply a weak vacuum for about 3 min. To the above degassed solution, add and mix

50 µl TEMED
150 µl 40% ammonium persulfate

Pour the gel, and allow to polymerize at least 2 h.

11. Load the fluor-labeled oligonucleotide in 10% sucrose (step c) into appropriate wells in the gel. Load an extra lane with gel-loading buffer containing 10% sucrose, 0.1% bromphenol blue (BPB), 0.1% xylene cyanol (XC). This serves as a convenient marker for the progress of the electrophoresis. Oligonucleotides of 15 to 20 nucleotides in length should run approximately midway between the BPB and XC, and should be visible as faint, fluorescent bands. Electrophorese at 20 W constant power, in $1 \times$ NNB for 3 to 4 h.

12. Remove the gel from the apparatus. Place on saran wrap, and set the saran on parafilm (parafilm fluoresces under UV excitation). Visualize the nucleic acids by illuminating the gel with a hand-held UV lamp. Unlabeled oligonucleotide should cast a visible shadow on the parafilm. Above this shadow band on the gel, a fluorescent band containing the fluor-labeled oligonucleotide should be visible. Excise the portion of gel containing this fluorescent band, and place in 1.0 to 1.5 ml sterile distilled H_2O in a plastic scintillation vial. Elute the fluorescent oligonucleotide overnight, with mixing at room temperature.

13. The resulting eluate can be passed through a 0.22-µm filter to remove gel fragments, and used without further purification. (If desired the oligonucleotide can be further purified on commercially available columns, such as Nensorb 20.[15])

14. The fluor-labeled oligonucleotide is quantified by measuring the absorbance at 260 nm (assuming 1 A 260 $\approx$20 µg/ml of pure oligonucleotide). A rough estimate of purity can be obtained by measuring the absorbance at the absorbance maxima of the dye (fluorescein $\approx$495 nm; Texas Red $\approx$595 nm). The ratio of $A_{260}/A_{max\ dye}$ should be about 3.0 for pure preparations of fluor-labeled oligonucleotides.

15. Lyophilize the purified fluorescent oligonucleotide, and resuspend in sterile H_2O to a final concentration of 50 ng $µl^{-1}$. Store in small aliquots in the dark at $-20°C$.

Cell Fixation

1. Cell samples may be fixed *in situ* by adding an equal volume of cold 8% paraformaldehyde/$1 \times$ PBS (final concentration of 4% paraformaldehyde/$0.5 \times$ PBS). Alternatively, cells can be centrifuged, washed, and resuspended in cold $1 \times$ PBS or sterile filtered seawater. (A good concentration for cell suspensions is about 1×10^7 cells ml^{-1}). Washed, resuspended cells are then fixed by adding an equal volume of freshly prepared 8% paraformaldehyde/$1 \times$ PBS.

2. Samples should be incubated in fixative for 2 to 24 h at 4°C. Cells can be stored in fixative longer, but prolonged storage (over several weeks to months) may result in lower hybridization signal. It is preferable to apply the cells to slides, and store them dry as indicated in steps 3 to 5, below. (Fixation by addition of 1/10 volume 37% formaldehyde at 4°C is a good alternative which works well for most cell types).

3. Post fixation: Spread 10 to 30 µl of the fixed cell suspension directly onto the gelatin subbed slide. (It is preferable, but not absolutely necessary, to remove the fixative first by resuspending cells in $1 \times$ PBS or sterile filtered seawater, and then applying them to the slide). Use the backside of a 200-µl pipette tip to spread the cells on the slide. Allow to air dry.

4. Place slides in a slide rack. Set the rack in a staining jar containing a mixture of methanol:formaldehyde post-fixative.

5. Incubate at room temperature 20 min. Remove slides and dip briefly in H_2O to rinse. Air dry. At this stage the samples may be stored dry at room temperature for prolonged periods.

An alternative post-fixation treatment which yields equally good results involves dipping the slides serially in 50% ethanol, 80% ethanol, and 95% ethanol, 3 min in each solution.[7] This is followed by air drying.

Hybridization Reactions and Detection

1. Make fresh hybridization buffer as follows:

Stock reagent	Volume to add
Distilled H_2O	3.16 ml
25× SET	800 μl
10% SDS	40 μl

Mix the hybridization buffer and preequilibrate to the appropriate hybridization temperature in the dry incubator.

2. Add 10 to 20 μl of the hybridization buffer onto each of the smears, directly on top of the cell smear.

3. Add 1 to 2 μl of the 50 ng μl^{-1} fluorescent oligonucleotide stock to the hybridization buffer on the smear.

4. Next, carefully place slides in an airtight chamber containing a tissue soaked with hybridization buffer, to prevent drying of the probe-hybridization mixture. For single slides, 50-ml Falcon tubes make convenient hybridization chambers. Place the airtight chamber containing the slides in an incubating oven, and hybridize 2 h to overnight at the appropriate temperature, usually between 37 and 55°C. (The hybridization and wash temperatures are determined empirically for each different probe.)

5. Place a Coplin jar or the equivalent in a temperature-equilibrated waterbath and fill it with preequilibrated 0.2× SET (the wash temperature will depend on the particular probe being used; usually between 45 and 55°C). Place the hybridization slides in the 0.2× SET at the appropriate wash temperature. Note the time.

6. After 10 min, pour off the 0.2× SET and replace with fresh 0.2× SET. Wash for 10 more minutes, and again replace with fresh 0.2× SET. After 10 more minutes, remove the slides and allow them to air dry in subdued light.

7. Hybridized samples are mounted in Citifluor or glycerol:phosphate buffered saline (9:1 v/v). Commercial mountants such as Citifluor contain reagents which help retard oxidation and photobleaching. When slides are dry, place a small drop of mountant directly on the smear, and cover with a coverslip.

8. Samples are viewed in the epifluorescence microscope, using excitation and emission filters appropriate for the specific fluorochrome. Negative and positive controls are viewed in parallel. For photomicroscopy, black and white (Kodak TriX Pan 400) or color (Kodak Ektachrome 400) films are appropriate. Exposure times may range from 15 s to several minutes.

NOTES AND COMMENTS

1. Fluorescent labeling of oligonucleotides is achieved by incorporating an amino group either internally or on the 5' or 3' end of the oligonucleotide probe. Most probes used to date have employed 5' end labeling. Several commercial reagents (AminolinkII, Applied Biosystems, Foster City, CA; Amino-Modifier-dT, Glen Research, Sterling, VA; Amine-ON, Clonetech, Palo Alto, CA) are available for incorporating amino groups into oligonucleotides during automated DNA synthesis. The free amino group serves as a linker for covalent coupling of the fluorescent dye to the oligonucleotide. A variety of dyes linked to nucleophilc reactive groups, such as fluorescein isothiocyanate (FITC, isomer I) or Texas Red sulfonyl chloride (Molecular Probes, Eugene, OR) are available. Amino-modified oligonucleotides are first mixed with the reactive, nucleophilic dye, to produce the fluor-labeled oligonucleotide. It is critical that the pH be greater than 9.0 for FITC labeling. After the reaction is complete, oligonucleotides are separated from unincorporated dye by gel filtration chromatography. A final, polyacrylamide gel purification step is employed to separate labeled from unlabeled oligonucleotide.

2. Several types of fixation protocols appear to work well with rRNA-targeted *in situ* hybridization protocols. The fixative should adequately preserve the morphological structure of the cells and allow free access of the oligonucleotide probe to the cell's interior. In addition, the fixative should not contribute to cellular autofluorescence. Glutaraldehyde fixation can result in intense autofluorescence, and should be avoided for fluorescence applications. Fixation in either 3.7% formaldehyde or 4% paraformaldehyde generally yields good results. It is important that cells be fixed in the cold (4°C). Subsequent treatment with ethanol or methanol often improves the hybridization signal. The specific fixation protocols will depend in part on the particular cell type tissues or environmental sample being examined. In general, cells may be fixed in solution prior to immobilization on gelatin-coated slides. Terrestrial microorganisms may be washed and resuspended in 0.22-μm filtered phosphate-buffered saline prior to fixation. Marine bacteria can be resuspended sterile filtered seawater. For any particular cell or tissue type a variety of fixation procedures should be tested and compared for optimal results. Several surveys of appropriate fixatives for *in situ* hybridization protocols have been published, and may be consulted for further information.[20,31]

3. In practice, the hybridizations are simple to perform. Prefixed smears or sections are covered with 10 to 20 μl of a hybridization buffer containing 5 ng μl^{-1} of fluor-labeled, oligonucleotide probe. The slides are then incubated in an airtight chamber to prevent evaporative losses of the hybridization mixture from the slide. After hybridization the slides are immersed in wash buffer preequilibrated to the appropriate wash temperature. After three washes of 10-min duration, at temperatures and salt concentrations which afford the appropriate probe specificity, the slides are air dried. The samples may then be stored dry in the dark, or viewed immediately. Relevant parameters to consider in design of *in situ* hybridization experiments include hybridization time, temperature, and probe concentration. In general, hybridization and wash temperatures around 5 to 10°C below the Td of a perfectly matched duplex yield strong hybridization signals with good specificity. The kinetics of *in situ* hybridization with appropriate probes is quite rapid, with maximal binding generally occurring within 2 h. Hybridization times of 2 to 16 h generally yield comparable results. The concentration of nucleic acid probe also has a pronounced effect on the extent of hybridization. Saturating conditions are achieved at probe concentrations of about 2 to 5 ng μl^{-1} (approximately 350 nM for fluorescent probes and formalin-fixed whole cells; DeLong, unpublished data), in good agreement with results obtained with radioactively labeled mRNA-specific *in situ* hybridization probes.[32,33]

REFERENCES

1. Hobbie, J. E., Daley, R. J., and Jasper, S., Use of nuclepore filters for counting bacteria by fluorescence microscopy, *Appl. Environ. Microbiol.*, 33, 1225, 1977.
2. Porter, K. G. and Feig, Y. S., The use of DAPI for identifying and counting aquatic microflora, *Limnol. Oceanogr.*, 25, 943, 1980.
3. Ward, B. B. and Carlucci, A. F., Marine ammonia and nitrite-oxidizing bacteria serological diversity determined by immunofluorescence in culture and in the environment, *Appl. Environ. Microbiol.*, 50, 194, 1985.
4. Campbell, L. and Iturriaga, R., Identification of *Synechococcus* spp. in the Sargasso Sea by immunofluorescence and fluorescence excitation spectroscopy performed on individual cells, *Limnol. Oceanogr.*, 33, 1196, 1988.
5. Currin, C. A., Paerl, H. W., Suba, G. K., and Alberte, R. S., Immunofluorescence detection and characterization of N_2-fixing microorganisms from aquatic environments, *Limnol. Oceanogr.*, 35, 59, 1990.
6. DeLong, E. F., Wickham, G. S., and Pace, N. R., Phylogenetic stains: ribosomal RNA-based probes for the identification of single cells, *Science*, 243, 1360, 1989.
7. Amann, R. I., Krumholz, L., and Stahl, D. A., Fluorescent-oligodeoxynucleotide probing of whole cells for determinative, phylogenetic and environmental studies in microbiology, *J. Bacteriol.*, 172, 762, 1990.
8. DeLong, E. F. and Shah, J., Fluorescent ribosomal RNA probes for clinical application: a research review, *Diagnos. Clin. Test.*, 28, 41, 1990.
9. Tsien, H. C., Bratina, B. J., Tsuji, K., and Hanson, R. S., Use of oligodeoxynucleotide signature probes for identification of physiological groups of methylotrophic bacteria, *Appl. Environ. Microbiol.*, 56, 2858, 1990.
10. Amann, R. I., Springer, N., Ludwig, L., Gortz, H., and Schleifer, K. H., Identification *in situ* and phylogeny of uncultured bacterial endosymbionts, *Nature*, 351, 161, 1991.
11. Distel, D. L., DeLong, E. F., and Waterbury, J. B., Phylogenetic characterization and *in situ* localization of the bacterial symbiont of shipworms (Teredinidae: Bivalvia) using 16S rRNA sequence analysis and oligodeoxynucleotide probe hybridization, *Appl. Environ. Microbiol.*, 57, 2376, 1991.
12. Woese, C. R., Bacterial evolution, *Microbiol. Rev.*, 51, 221, 1987.
13. Giovannoni, S. J., DeLong, E. F., Olsen, G. J., and Pace, N. R., Phylogenetic group-specific oligodeoxynucleotide probes for identification of single microbial cells, *J. Bacteriol.*, 170, 720, 1988.
14. Woese, C. R., Kandler, O., and Wheelis, M. L., Towards a system of organisms; proposal for the domains Archaea, Bacteria aned Eucarya, *Proc. Natl. Acad. Sci. U.S.A.*, 87, 4576, 1990.
15. Stahl, D. A. and Amann, R. I., Development and application of nucleic acid probes, in *Nucleic Acid Techniques in Bacterial Systematics*, Stackebrandt, E. and Goodfellow, M., Eds., John Wiley & Sons, New York, 1991, 205.
16. Winker, S. and Woese, C. R., A definition of the domains Archaea, Bacteria and Eucarya in terms of small subunit ribosomal RNA characteristics, *Syst. Appl. Microbiol.*, 14, 305, 1991.
17. Wallace, R. B. and Miyada, C. G., Oligonucleotide probes for the screening of recombinant DNA libraries, *Meth. Enzymol.*, 152, 432, 1987.
18. Amann, R. I., Binder, B. B., Olson, R. J., Chisholm, S. W., Devereux, R., and Stahl, D. A., Combination of 16S rRNA-targeted oligonucleotide probes with flow cytometry for analyzing mixed microbial populations, *Appl. Environ. Microbiol.*, 56, 1919, 1990.
19. Zarda, B., Amann, R., Wallner, G., and Schleifer, K. H., Identification of single bacterial cells using digoxigenin-labelled, rRNA-targeted oligonucleotides, *J. Gen. Microbiol.*, 137, 2823, 1991.
20. Singer, R. H., Lawrence, J. B., and Villnave, C., Optimization of *in situ* hybridization using isotopic and non-isotopic detection methods, *Biotechniques*, 4, 230, 1986.

21. Lane, D. J., Pace, B., Olsen, G. J., Stahl, D. A., Sogin, M. L., and Pace, N. R., Rapid determination of 16S ribosomal RNA sequences for phylogenetic analyses, *Proc. Natl. Acad. Sci. U.S.A.,* 82, 6955, 1985.

22. Sakai, R. K., Gelfand, D. H., Stoffel, S., Scharf, S. J., Higuchi, R., Horn, G. T., Mullis, K. B., and Erlich, H. A., Primer-directed enzymatic amplification of DNA with a thermostable DNA polymerase, *Science,* 239, 487, 1988.

23. Medlin, L., Ellwood, H. J., Stickel, S., and Sogin, M. L., The characterization of enzymatically amplified eukaryotic 16S-like rRNA encoding regions, *Gene,* 71, 491, 1988.

24. Lane, D. J., Giovannoni, S. J., Pace, N. R., and Stahl, D. A., Microbial ecology and evolution: a ribosomal RNA approach, *Annu. Rev. Microbiol.,* 40, 337, 1986.

25. Pace, N. R., Stahl, D. A., Lane, D. J., and Olsen, G. J., The analysis of natural microbial populations by ribosomal RNA sequences, in *Advances in Microbial Ecology,* Vol. 9, Marshall, K. C., Ed., Plenum Press, New York, 1986, 1.

26. Giovanonni, S. J., Britschgi, T. B., Moyer, C. L., and Field, K. G., Genetic diversity of Sargasso Sea bacterioplankton, *Nature,* 345, 60, 1990.

27. Ward, D. M., Weller, R., and Bateson, M. M., 16S rRNA sequences reveal numerous uncultured microorganisms in a natural community, *Nature,* 344, 63, 1990.

28. Schmidt, T. M., DeLong, E. F., and Pace, N. R., Analysis of a marine picoplankton community by 16S rRNA gene cloning and sequencing, *J. Bacteriol.,* 173, 4371, 1991.

29. Olsen, G. J., Phylogenetic analysis using ribosomal RNA sequences, *Meth. Enzymol.,* 164, 793, 1988.

30. Olsen, G. J., Overbeek, R., Larsen, N., and Woese, C. R., The ribosomal RNA database project: an updated version, *Nucleic Acids Res.,* 19, 4817, 1991.

31. Bresser, J. and Evinger-Hodges, M. J., Comparison of *in situ* hybridization procedures yielding rapid, sensitive mRNA detections, *Gene Anal. Techn.,* 4, 89, 1987.

32. Cox, K. H., DeLeon, D., Angerer, L. M., and Angerer, R. C., Detection of mRNAs in sea urchin embryos by *in situ* hybridization using assymetric RNA probes, *Dev. Biol.,* 101, 485, 1984.

33. Tanjea, K. and Singer, R. H., Use of oligodeoxynucleotide probes for quantitative *in situ* hybridization to actin mRNA, *Anal. Biochem.,* 166, 389, 1987.

CHAPTER 34

Immunofluorescence Method for the Detection and Characterization of Marine Microbes

Lisa Campbell

INTRODUCTION

Immunological techniques take advantage of the specificity and sensitivity of the antigen-antibody reaction of mammalian systems. Antigens are substances which elicit an immune response in a host animal. Because antibodies can be produced to almost any organic compound that is considered "foreign" to the host, the possibilities for generating specific probes are great. Whole cells or large proteins, generally, are more antigenic than small, low molecular weight molecules. The antigen-antibody reaction can be visualized by a number of different techniques (e.g., agglutination, ELISA, radioimmunoassay, immuno-fluorescence, or immunoblot), but fluorescent tags are useful for both visualization and quantification using epifluorescent microscopy primarily and, more recently, flow cytometry. Immunofluorescence assays can either be a direct, one-step procedure in which the specific antibody is conjugated with the fluorochrome or an indirect, two-step procedure in which a sample is first labeled with antibodies directed against the cell of interest and second with a fluorescently labeled antibody directed against the first antibody. For example, if an antibody directed against a bacterial strain is produced in a rabbit host, the secondary antibody would be produced in another host (e.g., swine or sheep) and directed against rabbit antibodies.

Antibodies can be generated in either a polyclonal or monoclonal system. Polyclonal antisera, by definition, contain a mixture of antibodies specific for the various antigenic determinants of the immunizing organism. Antigens are injected into the host, and antibodies are isolated from the serum. For monoclonal antibodies, spleen cells harvested from an immunized mouse are fused with cells from a myeloma cell line in the presence of polyethylene glycol, which promotes the creation of hybrid cells. Fused cells are grown on medium which selects for hybrid growth, but repeated cloning is performed to verify only one cell fusion type is present. Monoclonal antibodies generally will react only with a single determinant (epitope).

Applications of immunofluorescence methods in aquatic systems include marine bacteria,[1-3] cyanobacteria,[4-6] eukaryotic ultraplankton,[7] and freshwater systems.[8,9]

The principal advantage of the immunofluorescence approach in microbial ecology is that identification and enumeration of a single cell type within a natural sample is possible. The specificity of the antibodies may be such that recognition of individual clones, or species, is possible; or, with other antibodies, specificity may be at the family level. Immunochemical

0-87371-564-0/93/$0.00 + $.50

investigations can also be of taxonomic importance because cell surface antigens can be important species-specific markers.[10-12] Furthermore, for photosynthetic picoplankton, which are also recognized by their characteristic autofluorescence, the diversity within a group of organisms in the field can be investigated.[5,7] With monoclonal antibodies, which recognize a single epitope, one can obtain various levels of specificity, and, theoretically, have an infinite supply; whereas polyclonals, although easier to obtain (produced easily and inexpensively) and perhaps more useful in some applications, are limited to the supply obtained from one animal.

Chief among the disadvantages is that one needs to have in culture the organism to be used as the antigen. If the organism of interest is amenable to culture, this may not present a problem; however, if the organism is difficult to isolate, or to grow in sufficient quantities for antigen preparation, this may be a limitation. The specificity of each antibody preparation must also be characterized by screening for all potential cross-reacting strains. Potential nonspecific staining due to physical properties and from unrelated nonspecific antibodies present must be controlled by determining proper primary and secondary antibody dilutions and by the use of blocking buffers.[13,14] Finally, if one has opted to produce monoclonal antibodies, these can be more costly to produce and require special facilities and may have weaker affinity for the antigen than polyclonal antibodies, although this is not always the case.[15,16]

MATERIALS REQUIRED

Antigen Preparation

Equipment

- High-speed centrifuge
- Autoclave

Supplies

- Polycarbonate centrifuge bottles

Solutions, Media, and Cultures

- 0.02 M phosphate buffered saline (PBS) or filtered seawater (FSW) diluted 1:1 FSW:PBS, autoclaved and filter sterilized. Make a 0.2-M stock: first dissolve 47.17g $Na_2HPO_4 \cdot 7H_2O$; then 3.75 g $NaH_2PO_4 \cdot 2H_2O$ and 87.5 g NaCl and adjust final volume to 1 l with ddH_2O. Dilute 1:10 for working dilution and adjust pH to 7.3.
- Culture of organism: Approximately 10 ml of culture concentrated to 10^9 ml^{-1} is required for immunization protocol. Thus, in general several liters of culture may be necessary for algae which do not grow to high cell densities, but only 1 l, or less, may be necessary for bacteria which can attain much higher cell densities. The culture is concentrated by centrifugation (at 4°C) down to approximately 100 ml. Cells are washed in sterile filtered (0.2 μm seawater, FSW), then washed three more times in a small volume (25 ml) of PBS. Most bacteria can be fixed with paraformaldehyde (1% final concentration), but some algae require 0.6% glutaraldehyde.[17] Cells are resuspended in the preservative and shaken gently overnight at 4°C. Fixed cells are concentrated by centrifugation, washed two times with sterile PBS, resuspended in 0.2% paraformaldehyde and stored at 4°C. (Note: For delicate microalgal clones, PBS cannot be used because it causes lysis of cells. Instead, a 1:1 FSW:PBS is used and the temperature is decreased gradually at each of the three initial washing steps [10, 8, and 6°C] before fixing at 4°C).

Epifluorescence Filter Assay

Equipment

- Filtration unit for 25-mm diameter polycarbonate filters; a multiplace unit is desirable (e.g., Hoefer 10-place stainless steel unit with individual valves for each filter is convenient).
- Vacuum pump
- Microscope equipped for epifluorescence microscopy; blue (488 nm) filters for FITC conjugates; UV for AMCA conjugates; green (546 nm) for phycoerythrin.

Supplies

- Micropipettes
- 25-mm diameter polycarbonate filters (0.2 or 0.4 μm)

Solutions and Media

- Primary antibody (polyclonals produced in rabbits; MAb in mice)
- Secondary, fluorochrome-conjugated anti-rabbit IgG or anti-mouse IgG or IgM (can be obtained from a commercial supplier)
- Normal serum of primary host (e.g., pre-immunization rabbit serum)
- Normal serum of secondary antibody host (e.g., pre-immunization swine) (both of the above can be obtained from a commercial supplier)
- 0.02 M phosphate-buffered saline (PBS) or filtered seawater (FSW)
- Blocking buffer: final concentrations in PBS: 0.1% azide, 0.05% Tween-20, 5% normal serum of host for secondary antibody (e.g., Normal Swine serum)
- Buffered mounting medium (9:1 glycerol: 0.05 M carbonate-bicarbonate buffer): Make 0.5 M carbonate buffer (to 8 ml of 0.5 M NaHCO$_3$ add 0.5 M Na$_2$CO$_3$ until pH is 9 and adjust final volume to 10 ml with 0.5 M NaHCO$_3$) add 1 ml of this buffer to 9 ml glycerol.
- 10% paraformaldehyde: To 900 ml boiling water, add 100 g paraformaldehyde. Add 1 N NaOH dropwise to clear solution while stirring. Cool to room temperature and add 100 ml of FSW or PBS, adjust pH to 7.4, and filter through GF/F filter.

PROCEDURES

Antigen Preparation

Antibody production can be performed in-house or by a commercial laboratory. For monoclonal antibody production, a collaboration with a laboratory regularly producing such antibodies or a commercial firm is recommended. For polyclonal production, there are numerous references.[18,19] Included here is a brief overview.

1. Wash antigen twice with sterile PBS before injection and adjust final cell concentration to 10^9 cells ml^{-1}. After obtaining a sample of pre-immune serum (for use in experimental controls), a typical immunization schedule might be: day 1, 0.5 ml; day 4, 0.75 ml; day 6, 1.0 ml; day 7 to 13 rest; day 14 test bleed and booster; boosters at 14-d intervals thereafter until sufficient titer is obtained (see Considerations section below). At the end of the immunization protocol to obtain a large quantity of serum, the animal is exsanguinated by cardiac puncture.

2. Test bleeds are allowed to clot at room temperature for 2 h, then refrigerated overnight at 4°C.

3. Serum is separated from clot and centrifuged to remove any red blood cells.

4. For most applications, the serum is decomplemented by heating to 56°C for 30 min, then aliquoted into small volumes (to avoid repeated freezing and thawing) and stored frozen (−20°C in a non-self-defrosting freezer).

Epifluorescence Filter Assay

1. Prepare filtered (0.2 μm) seawater (FSW) or PBS (FPBS).

2. Filter sterilize blocking buffer.

3. Prepare antibody dilution in blocking buffer.

4. Set up filters in filtration unit: Millipore HA-type filter is used as backing to disperse vacuum; 0.2- or 0.4-μm polycarbonate filters. It is not necessary to stain, or use black filters.

5. Collect fixed sample onto filters using <100 mmHg vacuum.

6. Wash (10 ml rinse volume) once with FSW and twice with PBS.

7. Incubate with blocking buffer: 0.6 ml (or volume to cover filter) for 20 min (all antibody incubation times must be verified for each application).

8. Incubate 1 h with primary antibody. A duplicate sample incubated with pre-immune or control serum should also be run to verify absence of nonspecific staining and false-positive staining.

9. Wash five times with PBS.

10. Incubate 1 h with secondary, fluorochrome-conjugated antibody.

11. Wash five times with PBS.

12. Mount on slide affixing coverslip with carbonate-buffered mounting medium.

13. Examine using epifluorescence microscope with appropriate filter sets.

14. Samples should be stored frozen if not examined immediately.

15. Calculation: Slides are counted and cell concentration determined as for standard epifluorescence microscopy cell counts: (cell/grid area counted)/(grid area/filter)/volume filtered.

In some applications, FSW can be used in place of PBS, where PBS may cause cells to lyse. This must be determined with each antibody preparation; it may not be appropriate in some systems (e.g., monoclonals).

NOTES AND COMMENTS

1. Before an IF probe can be used to identify populations in field samples, both the strength of the antibody preparation (titer) and its specificity must be characterized by testing for cross reactions with other isolates. The appropriate dilutions of all antibodies to be used must be determined. For examining test bleeds, the presence of antibody can be determined by various methods (e.g., immunodiffusion, agglutination, immunoblotting), but it is best to make these tests in the same assay system in which the antibodies are to be used. Using the immunofluorescence filter staining technique, the titer of the antisera can be determined by making a twofold dilution series of the antibody (e.g., 1:10 to 1:2560) and testing each against the antigen. Typically, the lowest dilution producing no false positives is used. Following this determination, cross reaction tests are performed with all available strains for potential labeling.

2. In all such tests, controls are essential. A dilution series of the fluorochrome-conjugated secondary antibody must be performed with both positive and negative controls. Positive controls, consisting of the antigen (cell) and its corresponding antibody, and negative controls, consisting of the antigen and normal serum (or an unrelated antiserum), must be run simultaneously to determine that there are no false positives (i.e., labeling due to nonspecific staining). Typically, dilutions can be 1:100 to 1:2000, but must be verified for each batch of secondary antibody with each primary antibody.

3. Unwanted background staining can be a problem, and is more common with polyclonal antisera because these may contain a variety of unrelated antibodies than with monoclonal antibodies. Therefore, it is imperative to determine the proper dilutions of both primary and secondary antibodies to reduce nonspecific staining. If necessary, the antisera can be purified.[19-21] Autofluorescence due to glutaraldehyde fixation can produce a weak green fluorescence. This artifact can be difficult to distinguish from weak labeling with FITC. Thus, Campbell et al.[14] found paraformaldehyde superior to glutaraldehyde fixation because autofluorescence was reduced. In addition, for the ultraphytoplankton, glutaraldehyde caused considerable nonspecific staining.

4. Another consideration is that antibodies may label nonviable cells and, therefore, in some cases, yield inaccurate cell counts. This question can be addressed by combining IF with other methods, such as autoradiography.[22]

5. A final consideration is the stability of the antigen. Because antigens are phenotypic characteristics, they may be influenced by environmental changes. Previous studies have found the major characterizing epitopes to be stable.[23,24]

REFERENCES

1. Ward, B. B. and Perry, M. J., Immunofluorescent assay for the marine ammonium-oxidizing bacterium *Nitrosococcus oceanus, Appl. Environ. Microbiol.,* 39, 913, 1980.

2. Ward, B. B. and Carlucci, A. F., Marine ammonia- and nitrite-oxidizing bacteria: serological diversity determined by immunofluorescence in culture and in the environment, *Appl. Environ. Microbiol.,* 50, 194, 1985.

3. Dahle, A. B. and Laake, M., Diversity dynamics of marine bacteria studied by immunofluorescent staining on membrane filters, *Appl. Environ. Microbiol.,* 43, 169, 1982.

4. Campbell, L., Carpenter, E. J., and Iacono, V. J., Identification and enumeration of marine chroococcoid cyanobacteria by immunofluorescence, *Appl. Environ. Microbiol.,* 46, 553, 1983.

5. Campbell, L. and Carpenter, E. J., Characterization of phycoerythrin-containing *Synechococcus* spp. populations by immunofluorescence, *J. Plankton Res.,* 9, 1167, 1987.

6. Currin, C. A., Paerl, H. W., Suba, G. K., and Alberte, R. S., Immunofluorscence detection and characterization of N$_2$-fixing microorganisms from aquatic environments, *Limnol. Oceanogr.,* 35, 59, 1990.

7. Shapiro, L. P., Campbell, L., and Haugen, E. M., Immunochemical recognition of phytoplankton species, *Mar. Ecol. Prog. Ser.,* 57, 219, 1989.

8. Fliermans, C. B., Bohlool, B. B., and Schmidt, E. L., Autecological study of the chemoautotroph *Nitrobacter* by immunofluorescence, *Appl. Environ. Microbiol.,* 27, 124, 1974.

9. Reed, W. M. and Dugan, P. R., Distribution of *Methylomonas methanica* and *Methylosinus trichosporium* in Cleveland Harbor as determined by an indirect fluorescent antibody-membrane filter technique, *Appl. Environ. Microbiol.,* 35, 422, 1978.

10. Fuhrmann, B., Roquebert, M. F., Van-Hoegaerden, M., and Strosberg, A. D., Immunological differentiation of *Penicillium* spp., *Can. J. Microbiol.,* 35, 1043, 1989.

11. Conway de Macario, E., Wolin, M. J., and Macario, A. J. L., Antibody analysis of relationships among methanogenic bacteria, *J. Bacteriol.,* 149, 316, 1982.

12. Conway de Macario, E., Konig, H., and Macario, A. J. L., Antigenic determinants distinctive of *Methanospirillum hungatei* and *Methanogenium cariaci* identified by monoclonal antibodies, *Arch. Microbiol.,* 144, 20, 1986.

13. Bohlool, B. B. and Schmidt, E. L., Nonspecific staining: its control in immunofluorescence examination of soil, *Science,* 162, 1012, 1968.

14. Campbell, L., Shapiro, L. P., Haugen, E. M., and Morris, L., Immunochemical approaches to the identification of the ultraplankton: assets and limitations, in *Novel Phytoplankton Blooms: Causes and Impacts of Recurrent Brown Tides and other Unusual Blooms,* Cosper, E. M., Bricelj, V. M., and Carpenter, E. J., Eds., Springer-Verlag, Berlin, 1989, 39.

15. Mason, D. Y., Cordell, J. L., and Pulford, K. A. F., Production of monoclonal antibodies for immunocytochemical use, in *Techniques in Cytochemistry,* Vol. 2, Bullock, G. R. and Petrusz, P., Eds., Academic Press, Orlando, FL, 1983, 175.

16. Campbell, L., unpublished data, 1991.

17. Murphy. L. S. and Haugen, E. M., The distribution and abundance of phototrophic ultraplankton in the North Atlantic, *Limnol. Oceanogr.,* 30, 47, 1985.

18. Campbell, D. H., Garvey, J. S., Cremer, N. E., and Sussdorf, D. H., *Methods of Immunology,* W. A. Benjamin, New York, 1964, 545.

19. Hudson, L. and Hay, F. C., *Practical Immunology,* 3rd ed., Blackwell Scientific, Oxford, 1989, 507.

20. Heyman, U., Heyman, B., and Ward, B. B., Cell affinity chromatography for a marine nitrifying bacterium, in *Immunochemical Approaches to Coastal, Estuarine and Oceanographic Questions,* Yentsch, C. M., Mague, F. C., and Horan, P. K., Eds., Springer-Verlag, New York, 1988, 100.

21. Fliermans, C. B. and Schmidt, E. L., Immunofluorescence for autecological study of a cellular bluegreen alga, *J. Phycol.,* 13, 364, 1977.

22. Ward, B. B., Combined autoradiography and immunofluorescence for estimation of single cell activity by ammonium-oxidizing bacteria, *Limnol. Oceanogr.,* 29, 402, 1984.

23. Bohlool, B. B. and Schmidt, E. L., The immunofluorescence approach in microbial ecology, *Adv. Microb. Ecol.,* 4, 203, 1980.

24. Campbell, L., Identification of chroococcoid cyanobacteria by immunofluorescence, in *Immunochemical Approaches to Coastal, Estuarine and Oceanographic Questions,* Yentsch, C. M., Mague, F. C., and Horan, P. K., Eds., Springer-Verlag, New York, 1988, 208.

Section III
Biomass

CHAPTER 35

The Relationship Between Biomass and Volume of Bacteria

Svein Norland

INTRODUCTION

The attention to the relationship between biovolume and biomass stems mainly from the need for a conversion between the two. Though the objective of most studies is the assessment of microbial biomass, usually as carbon,[1,2] biovolume is more often determined, as this parameter is more easily obtained from dimensions of individual organisms observed, e.g., under the light microscope. Next, biomass is estimated, usually assuming that a constant ratio exists between carbon and biovolume,[2] by applying an experimentally obtained conversion factor.

A large number of values for conversion factors has been estimated using different measuring methods, preparation techniques, organisms, and growth conditions.[2-6] As these constants vary considerably, it has stirred a discussion on their true size and general applicability, methodological differences and errors,[2] as well as their natural variation.[6] Less focus has been placed on the more basic nature of the relationship between biomass and biovolume: can it be described by a constant ratio or does it exhibit a size dependence?

This problem has been addressed during the last few years.[6-8] I will here review the models and methods involved, some central results, and a discussion of their practical implications.[9] The term biomass will be used to mean any unspecified measure of main cell material (e.g., dry weight, carbon content, or protein) when the exact nature of the biomass is inconsequential for the discussion.

METHODS FOR DETERMINING THE VOLUME-TO-BIOMASS CONVERSION FACTOR

Approaches Used to Estimate Conversion Factors

1. Pure bulk methods are based on determination of carbon-wet weight ratio and bacterial buoyant density.[3,10] Even though this was the first method to be applied for this purpose, it has provided no data on the size dependency of the factor, so it will not be discussed further in this context.
2. An alternative is based on single cell estimation of volume accompanied by an estimate of the total number of cells in the bulk sample and bulk analysis of some representative fraction of the cells material (typically carbon or dry weight).[2]

3. Recently a new single cell technique has emerged where biovolume and dry weight are
    determined simultaneously on individual cells. It is based on X-ray microanalysis in
    transmission electron microscope to determine single cell dry weight accompanied by
    simultaneous measurement of bacterial dimensions.[11] Norland et al.[6] used this method to
    study the relationship between dry weight and volume in bacteria. The method is reviewed
    in Chapter 44.

The Models

Three different models for the relationship between volume and biomass have been formulated or assumed. They differ in their experimental support as well as in their implications.

Constant Ratio Model

The idea that a general conversion factor from biovolume to biomass exists implies a
constant ratio between the two.[2] Users of conversion factors therefore at least implicitly
accept that these factors are size independent. There has been a discussion of how general
conversion factors are; however, this discussion has focused on topics like definition of
biovolume and biomass, species differences, the effect of growth conditions, and techniques
for preparation and measurement.[2,4,10] Obviously a considerable part of the variation of
conversion factors may be attributed to these factors. Simultaneously, evidence has accumulated for a correlation between conversion factor and size.

Constant Biomass Model

The most extreme opposite to the constant ratio model is phrased by Lee and Fuhrman.[7]
Based on measurement of natural bacterioplankton assemblages, they stated that per-cell
carbon was rather constant (mean ~20 fg cell^{-1}). Thus, they support the view that small
bacteria tend to have a relatively higher dry weight than larger ones. Another implication
is that they find a cell number to biomass conversion factor more appropriate than a factor
converting biovolume into biomass, at least for cells in the range reported (0.036 to 0.073
μm^3). It should be noted that Lee and Fuhrman[7] include extracellular particulate organic
material as a part of the bacterial cell.

Allometric Model

This model assumes that the dry weight-to-volume ratio is linearly dependent on volume,
so that smaller organisms tend to have a higher dry weight-to-volume ratio than larger ones.
Such relationships have been reported between carbon content and volume in algae.[12,13] For
bacteria Norland et al.[6] have shown an allometric relationship between dry weight measured
on individual bacteria using X-ray microanalysis and volume from the transmission electron
microscope. Simon and Azam[8] have reported that protein content and volume of size-
fractionated bacterial assemblages from seawater show an allometric relationship. An allometric relationship is described by a power function:

$$m = CV^a \tag{1}$$

where m is biomass, V is volume, C is conversion factor between biomass and volume for
unity volume ($V = 1$), and a is a scaling factor. The two previous models are special cases
of the allometric model: for the constant ratio model $a = 1$ and for the constant biomass
model $a = 0$. All values reported so far are less than unity, suggesting that smaller cells

tend to have a higher biomass-to-volume ratio than larger ones. Simon and Azam[8] reported a scaling factor of 0.59, a value lower than both the other value for bacteria, 0.91 ± 0.02 (Reference 6) and for algae, 0.892 (Reference 13). Simon and Azam[8] found that protein made up a constant fraction of dry weight independent of bacterial size. Their scaling factor thus also applies to other measures of biomass. The conversion factor C is subject to essentially the same sources of errors and variability as the conversion factors for the constant ratio model.

Statistical Considerations

Norland et al.[6] have shown that there is a considerable variation (approximately 20% from standard error of estimate for the regression) in the volume to dry weight ratio for individual bacteria. A similar-sized relative variation of the protein-to-volume ratio can be estimated from data given by Simon and Azam.[8] The precision with which the scaling factor can be determined will be limited by this variation, but it also will depend on the range of bacterial sizes that the determination is based on. Simon and Azam[8] base their conclusions on a size range slightly higher than one order of magnitude, whereas Norland et al.[6] measured individual bacteria over a size range of three orders of magnitude. In this context it should be mentioned that Lee and Fuhrman[7] base their "constant biomass model" on only a twofold volume range.

The estimated scaling factor also will depend on the way the regression is done. Simon and Azam[8] use a predictive regression with volume as the independent axis. This is not quite appropriate as none of the axes are independent (type II problem), and they should be treated in a symmetric way. Norland et al.[6] used functional regression.[14] The functional regression coefficient (scaling factor) can be determined by dividing the predictive coefficient by the correlation coefficient. As the correlation coefficient tends to decrease with wider size ranges, the difference between the two methods will be largest for narrow ranges. The recalculated scaling factor from Simon and Azam[8] is 0.72, reducing the difference between the two estimates of scaling factor for bacteria considerably.

Implications and Conclusion

Bacterial volumes range over three orders of magnitude. Our perception of the composition of small bacteria is for large part based on extrapolation from cultured bacteria with a volume one to two orders of magnitude larger.[8] Even a small-size dependence of the volume-to-biomass ratio may give pronounced effects when extrapolated. Psenner[9] applied three sets of model parameters to estimate carbon distribution from his volume distribution of 1082 bacteria determined by image analysis of electron micrographs. The results are shown in Table 1. The constant ratio model is not included as it will produce percent carbon distribution identical with that of biovolume. The differences are pronounced and are also reflected in estimated mean cell carbon content per cell. The more the scaling factor deviates from unity, the more the estimated carbon distribution differs from the biovolume distribution. Actually the carbon distribution of (d) in Table 1 is best compared with cell number distribution.

The carbon-to-volume ratios reflect variation in both calibration and scaling, but it is apparent that choice of scale factor is of consequence for estimation of biomass. The effect becomes even clearer from the computed carbon-to-volume ratio for size classes. Formula (d) has a range of more than 500 over the size classes, whereas (c) has a range of less than two. A large fraction of the values is so extreme that they must be refused as physiologically unreasonable. Lee and Fuhrman[7] warn against extrapolations of their results to bacteria that are much smaller or larger. Extrapolation of the power function also will magnify any error and produce unrealistic results.

Table 1. Percent Distribution of Carbon (Upper Numbers), Biovolume, and Cell Number in Different Size Classes of Pelagic Bacteria from Mondsee, August 9 and 10, 1988.

Formula	Size Classes				Mean C content, fg C cell^{-1}
	0.0005–0.001	0.001–0.01	0.01–0.1	0.1–0.5	
(a) $C = 0.09V^{0.6}$	0.3	25.7	33.5	40.4	8.4
	1.77	0.96	0.37	0.17	
(b) $C = 0.12V^{0.7}$	0.2	15.2	38.1	46.7	6.0
	0.88	0.58	0.30	0.18	
(c) $C = 0.09V^{0.9}$	0.1	12.3	26.5	61.2	3.7
	0.17	0.15	0.12	0.10	
(d) $C = 0.014V^{-0.15}$	3.0	65.9	25.5	5.6	28.7
	5.95	1.07	0.08	0.01	
Biovolume	0.1	9.2	23.6	67.2	
Number	2.1	59.2	29.3	9.1	

Note: The conversion factors in formulas have units of pg C μm^{-3}. The size class carbon-to-volume ratios (pg C μm^{-3}) are computed (lower number). Mean C content is also listed. Model parameters are from (a) Simon and Azam,[8] (b) as in (a) but recalculated using functional regression, (c) Norland et al.,[6] (d) Lee and Fuhrman,[7] as calculated by Psenner.[9] Data modified from Psenner.[9]

As it is generally agreed that carbon comprises approximately 50% of the dry weight, carbon-to-volume ratio above 0.5 pg μm^{-3} must be considered physiologically dubious. High ratios that have also been estimated experimentally, e.g., 0.56 pg μm^{-3} (Reference 2) and 0.6 pg μm^{-3} (Reference 7), probably can be explained by underestimation of volume due to use of fixatives.

Finally, it should be stressed that there is no biological interpretation of the constants of the power function; it is an empirical function fitted to the data. Other functions may be chosen that when extrapolated to smaller cells will give a different result. Still it is biologically reasonable that smaller bacteria have a higher biomass per volume than larger ones. Simon and Azam[8] have made a rough macromolecular inventory for bacterial cells in size classes from 0.026 to 0.4 μm^{-3}. Protein and DNA was measured; other components were extrapolated. They show that there is a progressive increase in the "concentration" of protein, DNA, cell walls, and cell membranes. DNA makes up 13% of dry weight in the smallest bacteria, but only 5% in the largest. The overall effect is that small cells tend to be very "dry" (only 46% v/v water) and they suggest that reducing the water content may be a mechanism to reduce energetic costs. Size reduction has also been suggested as a way to escape predation.[15]

REFERENCES

1. Fuhrman, J. A. and Azam, F., Bacterioplankton secondary production estimates for coastal waters of British Columbia, Antarctica and California, *Appl. Environ. Microbiol.*, 39, 1085, 1980.
2. Bratbak, G., Bacterial biovolume and biomass estimations, *Appl. Environ. Microbiol.*, 49, 1488, 1985.
3. Bakken, L. R. and Olsen, R. A., Buoyant densities and drymatter contents of microorganisms: conversion of a measured biovolume into biomass, *Appl. Environ. Microbiol.*, 45, 1188, 1983.
4. van Veen, J. A. and Paul, E. A., Conversion of biovolume measurement of soil organisms, grown under various moisture tensions, to biomass and their nutrient content, *Appl. Environ. Microbiol.*, 37, 686, 1979.
5. Luria, S. E., The bacterial protoplasm: composition and organization, in *The Bacteria,* Vol. 1, Gunsales, I. C. and Stanier, R. Y., Eds., Academic Press, New York, 1960, 1.
6. Norland, S., Heldal, M., and Tumyr, O., On the relation between dry matter and volume of bacteria, *Microb. Ecol.*, 13, 95, 1987.

7. Lee, S. and Fuhrman, J. A., Relationship between biovolume and biomass of natural derived marine bacterioplankton, *Appl. Environ. Microbiol.,* 53, 1298, 1987.
8. Simon, M. and Azam, F., Protein content and protein synthesis rates of planktonic marine bacteria, *Mar. Ecol. Prog. Ser.,* 51, 201, 1989.
9. Psenner, R., From image analysis to chemical analysis of bacteria: a long-term study?, *Limnol. Oceanogr.,* 35, 234, 1990.
10. Bratbak, G. and Dundas, I., Bacterial Dry matter content and biomass estimation, *Appl. Environ. Microbiol.,* 48, 755, 1984.
11. Heldal, M., Norland, S., and Tumyr, O., X-ray microanalytic method for measurement of dry weight and elemental content of individual bacteria, *Appl. Environ. Microbiol.,* 50, 1251, 1985.
12. Mullin, M. M., Sloan, P. R., and Eppley, R. W., Relationship between carbon content, cell volume and area in phytoplankton, *Limnol. Oceanogr.,* 11, 307, 1966.
13. Strathmann, R. R., Estimating the organic carbon content of phytoplankton from cell volume and plasma volume, *Limnol. Oceanogr.,* 12, 411, 1967.
14. Ricker, W. W., Linear regression in fishery research, *J. Fish. Res. Board. Can.,* 30, 409, 1973.
15. Fenchel, T., Relation between particle size selection and clearance in suspension-feeding ciliates, *Limnol. Oceanogr.,* 25, 733, 1980.

CHAPTER 36

Microscope Methods for Measuring Bacterial Biovolume: Epifluorescence Microscopy, Scanning Electron Microscopy, and Transmission Electron Microscopy

Gunnar Bratbak

INTRODUCTION

The biomass of a population is perhaps the most basic parameter that may be considered in microbial ecology. Knowing the biomass we may evaluate the population's activity potential in the environment and its potential as a food source for higher trophic levels.

The biomass of a bacterial population may be defined as the mass of living bacteria in a given habitat. The biomass, *Sensu stricto,* is in practice very difficult to determine and most methods have therefore aimed at measuring a parameter that is assumed to correlate with biomass. These parameters include total cell counts as well as several different chemical constituents such as lipopolysaccharide, muramic acid, bacteriochlorophyll, specific lipids, etc. One of the most popular methods, however, has been to estimate bacterial biomass from cell volume in combination with total cell counts.

While fluorescent staining and counting under epifluorescence microscopy has become a standard method for enumerating bacteria, no single method for measuring bacterial cell volume has been adopted as a standard. The methods used for measuring bacterial volume in aquatic environments include different kinds of microscopical techniques, electronic sizing, and flow cytometry. Microscope methods most widely used for measuring bacterial cell size include epifluorescence microscopy, scanning electron microscopy (SEM), and transmission electron microscopy (TEM).

The number of applicable methods for size measurement of bacteria in natural waters is limited by the relatively low concentration of bacteria found in most environments and by the small sizes of individual cells. When working with cultures or with bacteria concentrated from natural water samples by hollow-fiber filtration, etc., several other methods may be used. These include various light microscopy techniques in combination with agar slide preparation of bacteria and electron microscopy in combination with agar-filtration of bacteria.[1]

The general advantages of microscope methods are that they are direct and specific for bacteria and that the basic underlying assumptions for estimating biomass from biovolume are few and relatively simple.

The general disadvantages of the microscope methods are that they do not distinguish between living and dead bacteria and that some of them are labor intensive and depend on expensive equipment (SEM, TEM). The limited resolution of light microscopes makes precise and accurate size measurements of small natural bacteria difficult.

MATERIAL REQUIRED

Epifluorescence Microscopy

The list below includes only the equipment, solutions, and supplies specifically needed for sizing of AO- or DAPI-stained bacteria. The protocols of staining and counting bacteria are described in detail in other chapters in this volume.

Equipment

- Epifluorescence microscope equipped with a camera system or an eyepiece graticule such as New Porton G12 (Graticules, Ltd., Morley Road, Tonbridge, Kent, TN9 1RN England)
- Stage micrometer
- Slide projector (photomicrography only)

Supplies

- Fluorescent latex beads, two or three different sizes: 0.25, 0.5, and/or 1.0 μm diameter (Polysciences, Inc., 400 Valley Road, Warrington, PA 18976-2590)
- Film (Kodak Ektachrome 100 ASA, or a similar color dispositive, fine-grained film)
- Slide frames (may be reused)

Scanning Electron Microscopy

Equipment

- SEM microscope
- Critical point drying apparatus
- Sputtercoater
- Filter holder for 25-mm filters
- Pipettes

Supplies

- Small vial or beaker
- Polycarbonate membranes, 25 mm, 0.2-μm pore size
- Preparate stubs
- Double sided tape
- Latex beads, e.g., 0.25- and 0.5-μm diameter (Polysciences, Inc., 400 Valley Road, Warrington, PA 18976-2590)

Solutions

- Ethanol (concentration: 50, 75, 95, and 100%) or 2,2-dimethoxypropane (DMP) (Sigma D-8761)
- Fixative (glutaraldehyde)
- 0.2-μm filtered distilled water

Transmission Electron Microscopy

Equipment

- TEM microscope
- Centrifuge with swing-out rotor
- Centrifuge tubes with flat bottom. The flat tube bottoms may be molded in epoxy glue (e.g., Araldit, Ciba-Geigy)
- Balance

Supplies

- Forceps
- Scalpel
- Grid storage box
- EM-grids (nickel or copper grids, 100 to 400 mesh) with carbon-coated formvar or colloidone film. Glow-discharged grids are recommended, but not necessary. (see Chapter 44).
- Filters for grid support, e.g., Sartorius cellulose nitrate filters (or any other kind with good wetting properties)
- Double-sided tape
- Filter paper
- Latex beads, e.g., 0.25- and 0.5-μm diameter (Polysciences, Inc., 400 Valley Road, Warrington, PA 18976-2590)

Solutions

- 2% Uranyl acetate (or another electron microscope stain).
- Distilled water

PROCEDURES

Epifluorescence Microscopy

Counting of acridine orange (AO)[2]- or 4′,6-diamidino-2-phenylindole (DAPI)[3,4]-stained cells using epifluorescence microscopy has become a standard method for counting bacteria in aquatic environments. The microscope preparations made for counting may also be used for sizing of bacteria. The sizing may be done directly using an eyepiece graticule or indirectly by taking photomicrographs and using these as the basis for sizing. A third possibility is to use a TV camera and an image analysis system (see Chapters 38 and 39). It is imperative to calibrate the size measurement procedures against fluorescent latex beads with known diameter. The latex beads are very bright and will therefore appear larger than they really are. The fluorescence intensity of the beads should be reduced to a level more like the bacteria by inserting a neutral gray filter in the light path. Preparations of latex beads for calibration of size measurements are made in the same way as the bacterial preparations, but staining is omitted. The bead stock solutions should be diluted 10^4- to 10^5-fold in distilled water before use. Mix the solutions to make a preparation containing all bead sizes.

Using an Eyepiece Graticule

1. Filter, stain, and mount bacterial samples as described for counting of bacteria.

2.   The New Porton G12 graticule has an array of 11 globes and circles with different diameters. The diameter increases with square root of 2 progression. At a microscope magnification of 1000 × the smallest (no. 0) and largest (no. 10) graticule circles have diameters equivalent to about 0.2 and 7.6 μm, respectively. The accurate size of the different circles may be determined by measuring the distance between the graticule lines denoted "Z" (zero) and "14" and dividing progressively by the square root of two to obtain the diameter of the circles numbered 0 to 10. Select bacterial cells at random and make a record of the graticule circle number (0 to 10) corresponding to the length (L) and the width (W) of each cell.

3.   Use the mixed bead preparation described above and compare the different beads to the graticule circles with corresponding size and determine how to measure the beads to obtain the correct results. The bead preparate should be checked every time you start a sizing sequence to assure repeatable results.

By Photomicrography

Epifluorescence preparations with bright but rapidly fading bacteria are difficult objects to photograph. Correct exposure time is critical for size measurements of bacteria as the cells will appear larger the more overexposed they are. The optimal exposure time may be determined by varying the exposure time or, if an automatic exposure system is used, by varying the exposure compensation factor. Once the optimal exposure time has been determined one should attempt to stain the preparations to the same intensity and produce pictures as identical as possible with respect to background darkness, fluorescent intensity, color, etc.

1.   Filter, stain, and mount bacterial samples as described for counting of bacteria.

2.   Select microscope fields at random and take a number of pictures that will allow you to measure the size of an adequate number of cells. Include one picture of the stage micrometer (use transmitted light for this).

3.   Adjust the illumination so that the beads fluoresce with an intensity similar to that of the bacteria, and take photographs of the bead preparations described above. When using the preparation with mixed bead sizes take several pictures and make sure to focus both on the larger and on the smaller beads.

4.   After the film has been developed and the pictures mounted in frames they are projected onto a white wall at a distance that gives a final magnification of 10,000 to 20,000 × (i.e., 1 μm appears as 1 to 2 cm). The final magnification may be measured from the slide of the stage micrometer.

5.   Establish the criteria for excluding the dim halo around the fluorescent images by comparing the diameter of the bead images to the known diameter of the beads (determined from the final magnification). Use the same criteria when measuring bacteria.

6.   Hold a white sheet of paper against the wall and draw with a sharp pencil around the entire outline of the bacteria but exclude the dim fluorescent halo around the cell images. With some practice this procedure can be done quickly and provides a permanent record of the measured cells. Abandon cells that are out of focus.

7.   Measure the length and width of the drawn cell images to the closest 2 mm (i.e., an accuracy of 0.2 to 0.4 μm) with a ruler or a digitizing tablet.

Scanning Electron Microscopy

The size of natural bacteria are often close to the resolution of light microscopes, i.e., about 0.2 μm. Electron microscopes have a much higher resolution and they are therefore, in this respect, better suited for sizing of bacteria than epifluorescence microscopes. The sizing may be done directly on the microscope CRT-screen or on photomicrographs or videoprints using a ruler or a digitizing tablet.

1. Make sure that the top side of the filters and the identity samples can be recognized by cutting small, differently-sized, and asymmetrically placed triangles in the edge of the polycarbonate membranes.

2. Filter water samples preserved with 3% v/v glutaraldehyde onto the marked filters. The volume to be filtered depends on the concentration of bacteria, but a volume three to five times the volume that gives good preparates for counting in the epifluorescence microscope is usually enough.

3. Rinse the filters three times with 3 ml prefiltered distilled water.

4. Dehydration: The preparations may be dehydrated in an ethanol series or by using 2,2-dimethoxypropane (DMP).[5] DMP reacts chemically with water to form acetone and methanol.

   a. Ethanol dehydration: Pass the filter through a series of ethanol solutions with increasing concentration by adding 2 to 3 ml ethanol solution to the filter funnel and draining through the filter (apply suction if necessary). Leave for 10 min at each concentration. Repeat the procedure once for the 100% solution.

   b. DMP dehydration: Add 2 to 3 ml acidified DMP (2 drops of concentrated HCl to 100 ml DMP) to the filter and leave for 5 min. Drain through the filter and add 3 ml pure DMP.

5. The dehydrated filters may be air dried or critical point (CP) dried. Air drying will usually work well when the bacterial cells are small such as in natural water samples. Cultured bacteria and other bacterial populations with large cells may require critical point drying to avoid distortion. For critical point drying, transfer the filter from the filter holder to a small vial or beaker and keep it submersed in 100% ethanol or DMP, respectively, until CP drying can take place. The CP drying should be carried out according to the operation manual of the instrument. After the bulk of the ethanol or DMP has been replaced with liquid CO_2, it should be sufficient to leave the preparations for 10 to 20 min in liquid CO_2 to ensure complete exchange of ethanol and DMP.

6. Mount pieces of the dried filters on EM stubs using double-sided tape. To ensure good electric contact, make sure that the filters do not cover the stubs completely. The stubs are then placed in a sputtercoater and coated with gold or gold-palladium. The sputtercoater should be operated according to the operation manual of the instrument.

7. The size (i.e., the length and the width) of the bacteria may be measured directly with a ruler on the microscope's CRT screen. It may be preferable, however, to obtain a permanent record of bacteria by taking photographs or making videoprints of the preparations at a magnification of 5000 to $10,000 \times$. The size of the bacteria may then later be measured with a ruler or on a digitizing tablet.

8. The following steps are required for calibration.

9. Dilute the bead stock solutions 100 to 1000 times in distilled water and filter a few drops on polycarbonate membranes.

10. Air dry the filters, mount on EM stubs, and sputtercoat as described above.

11. Use the same magnification as when measuring the bacteria and take photographs or make videoprints of areas where linear aggregation of three or more beads are found (if bead aggregates are few, make a new preparate with higher bead concentration).

12. Measure the diameter of single beads and the length of linear bead aggregates.

13. Plot single bead diameter and aggregate length against number of beads and estimate the thickness of the Au/AuPd coating by extrapolating to zero bead number ($= 2\times$ coating thickness). This value should be used to correct the bacterial size measurements.

Transmission Electron Microscopy

The transmission electron microscope (TEM) offers the same high resolution as the scanning electron microscope but the sample preparation is more simple and there is no need for metal coating of the preparates. The sizing may be done directly on the microscope fluorescent screen, or on photomicrographs using a ruler or a digitizing tablet.

1. Cut the filters to fit on the flat bottom of the centrifuge tubes and attach a small piece (2 × 5 mm) of double-sided tape to the center of the filter.

2. Tape the EM grids to the filter but make sure that it is only the edge of the grids that touches the tape.

3. Fill the centrifuge tubes with sample water, put the filter with attached grids into the tubes, and let it sink to the bottom. Avoid trapping air bubbles under the grids. Balance the tubes in pairs.

4. Centrifuge the samples at 5000 g for 20 min.

5. Remove the supernatant with a pipette and pick out the filters. Loosen the grids from the tape with a scalpel and place them in a grid storage box.

6. Stain the grids with 2% uranyl acetate (UAc) for 30 s.

7. Drain off the UAc by touching the grid with a piece of filter paper and rinse once by dipping the grid in distilled water.

8. Drain off the water and air dry.

9. Assess the length and the width of the bacteria at a magnification of 10,000 to 20,000 × by the aid of a straight line of known length or another kind of indication on the TEM's fluorescent screen. If a permanent record of the bacteria is preferred, take photographs of the preparates at a magnification of 5000 to 10,000 ×. With a lower magnification more cells may be included on each picture, but the pictures must be enlarged before cell size measurement can take place. The size of the bacteria may then later be measured with a ruler or on a digitizing tablet.

10. The following steps are required for calibration.

11. Dilute the bead stock solutions 100 to 1000 times in distilled water, mix them, and place a drop of the mixture on a grid.

12. Drain off after a few seconds by touching the grid with a filter paper and air dry.

13. Check the size of the beads as measured in the microscope against the manufacturers information.

Data Interpretation

1. The volume of the bacteria is calculated using the formula

$$V = (\pi/4)\, W^2(L - W/3) \tag{1}$$

where V is volume, L is cell length, and W is cell width. Although this formula is based on the assumption that the bacteria are straight rods with hemispherical ends, it works equally well for cocci.

2. Cell carbon content may be calculated from volume by the use of a conversion factor. Each operator should determine his/her own conversion factor for each the volume measuring methods and procedures used. Alternatively, one may rely on data from the literature. The average of experimentally determined conversion factors reported in the literature is about 350 fg C μm^{-3}.[6-8] This conversion factor may be used as a general value (see Chapter 37). If a more conservative factor is called for, a value of 200 fg C μm^{-3} may be used.[9,10]

NOTES AND COMMENTS

Statistics

For natural populations with many small and few large cells it may be necessary to measure as many as 200 cells to ensure that a representative number of large cells are included. For populations with a relatively narrow cell size range, it may be adequate to measure 50 to 100 cells.

The microscope methods for measuring bacterial cell size are not necessarily precise or accurate. The reasons for this are

1. The cells may shrink (or swell) due to fixation, dehydration, drying, etc. during sample preparation.[1,11-13]
2. The fluorescent dye may not stain the whole cell. AO-stained cells will for example appear larger than DAPI-stained cells because DAPI is very DNA specific and only stains the central nucleoplasm of the cells, while AO is less specific and also stains the peripheral parts of the cell body.
3. Brightly stained cells may appear larger than weaker stained cells.[8] Differences in staining intensity may be caused by a number of factors: (a) the amount of dye bound by a specific cell may depend on the type of cell, cell content and composition, physiological condition, etc.; (b) variations in staining procedure or staining condition; (c) variations in sample properties such as pH, salt concentration, cell number, detritus, DOM etc.; (d) fluorescence intensity will depend on the stain used. DAPI may produce more intensely stained cells than AO and this may in part compensate for the difference between DAPI and AO with respect to partial staining of the cells (see point 2).
4. The AuPd coating used on SEM preparates will add to the size of the cell.[6]
5. The size measurement depends on the skills and training of the operator.[14] Problems related to precision of size estimates may be minimized by strictly and always following the same procedure when preparing the samples and during size measurement, and by thorough calibration of the sizing procedure. Any systematic errors (i.e., accuracy) in the carbon estimates may be corrected for by choosing an appropriate volume-to-carbon conversion factor.

Epifluorescence Microscopy

1. The above size measurement procedures rely heavily on the subjective judgment of the operator. It is therefore recommended that people within each laboratory together work up an internal agreement on the size measurement procedure to ensure repeatable results. Blind testing using several different bead sizes is another possibility. Each sample should also be measured independently by at least two different operators.
2. It is possible to estimate the halo effect by an alternative method. Use a single-size bead preparation and measure the diameter of single beads and the length of linear bead aggregates (if bead aggregates are few, make a new preparation with higher bead concentration). Plot the diameter/length measurements against the number of beads and estimate the halo effect by extrapolating to zero bead number ($= 2 \times$ halo effect).

SEM

The most frequent sources of difficulty are

1. Low cell number which makes it difficult to obtain a reasonable high cell concentration on the filters. One possibility is to concentrate the bacteria from larger volumes of waters using hollow fiber filtration or similar techniques.
2. A high concentration of detrital and inorganic particles which obscures the bacteria on the filter. This problem is difficult, but prefiltration on a larger filter (e.g., 1 μm) may be attempted.
3. Salt precipitation on the filter. Wash more carefully with distilled water and avoid drying of the filter during filtration, washing, and dehydration.

REFERENCES

1. Woldringh, C. L., de Jong, M. A., van den Berg, W., and Koppes, L., Morphological analysis of the division cycle of two *Escherichia coli* substrains during slow growth, *J. Bacteriol.*, 131, 270, 1977.
2. Hobbie, J. E., Daley, R. J., and Jasper, S., Use of Nuclepore filters for counting bacteria by fluorescence microscopy, *Appl. Environ. Microbiol.*, 33, 1225, 1977.
3. Porter, K. and Feig, Y. S., The use of DAPI for identifying and counting aquatic microflora, *Limnol. Oceanogr.*, 25, 943, 1980.
4. Paul, J. H., The use of Hoechst dyes 33258 and 33342 for enumeration of attached and planktonic bacteria, *Appl. Environ. Microbiol.*, 43, 939, 1982.
5. Muller, L. L. and Jacks, T. J., Rapid chemical dehydration of samples for electron microscopy examination, *J. Histochem. Cytochem.*, 23, 107, 1975.
6. Bratbak, G., Bacterial biovolume and biomass estimations, *Appl. Environ. Microbiol.*, 49, 1488, 1985.
7. Bjørnsen, P. K., Automatic determination of bacterioplankton biomass by image analysis, *Appl. Environ. Microbiol.*, 51, 1199, 1986.
8. Lee, S. and Fuhrman, J. A., Relationships between biovolume and biomass of naturally derived marine bacterioplankton, *Appl. Environ. Microbiol.*, 53, 1298, 1987.
9. Bratbak, G. and Dundas, I., Bacterial dry matter content and biomass estimations, *Appl. Environ. Microbiol.*, 48, 755, 1984.
10. Kogure, K. and Koike, I., Particle counter determination of bacterial biomass in seawater, *Appl. Environ. Microbiol.*, 53, 274, 1987.
11. Trueba, F. J. and Woldringh, C. L., Changes in cell diameter during division cycle of *Escherichia coli*, *J. Bacteriol.*, 142, 869, 1980.
12. Fuhrman, J. A., Influence of method on the apparent size distribution of bacterioplankton cells: epifluorescence microscopy compared to scanning electron microscopy, *Mar. Ecol. Prog. Ser.*, 5, 103, 1981.
13. Montesinos, E., Esteve, I., and Guerrero, R., Comparison between direct methods for determination of microbial cell volume: electron microscopy and electronic particle sizing, *Appl. Environ. Microbiol.*, 45, 1651, 1983.
14. Allan, T., *Particle Size Measurements,* 3rd ed., Chapman and Hall, London, 1981, 187.

Additional References

Edwards, R. T., Sestonic bacteria as a food source for filtering invertebrates in two southeastern blackwater rivers, *Limnol. Oceanogr.*, 32, 221. 1987.

Fry, J. C., Determination of biomass, in *Methods in Aquatic Bacteriology,* Austin, B., Ed., John Wiley & Sons, New York, 1988, chap 2.

Fry, J. C. and Davis, A. R., An assessment of methods for measuring volumes of planktonic bacteria, with particular reference to television image analysis, *J. Appl. Bacteriol.*, 58, 105, 1985.

Nagata, T., Carbon and nitrogen content of natural planktonic bacteria, *Appl. Environ. Microbiol.*, 52, 28, 1986.

Norland, S., Heldal, M., and Tumyr, O., On the relation between dry matter and volume of bacteria, *Microb. Ecol.*, 13, 95, 1987.

Scavia, D. and Laird, G. A., Bacterioplankton in Lake Michigan: dynamics, controls and significance to carbon flux, *Limnol. Oceanogr.*, 32, 1017, 1987.

Simon, M., Specific uptake rates of amino acids by attached and free-living bacteria in a mesotrophic lake, *Appl. Environ. Microbiol.*, 49, 1254, 1985.

Watson, S. W., Novisky, T. J., Quinby, H. L., and Valois, F. W., Determination of bacterial number and biomass in the marine environment, *Appl. Environ. Microbiol.*, 33, 940, 1977.

CHAPTER 37

Measurement of Carbon and Nitrogen Biomass and Biovolume from Naturally Derived Marine Bacterioplankton

SangHoon Lee

INTRODUCTION

Bacterial production or cell concentration often needs to be presented in carbon (C) or nitrogen (N) biomass. Conversions to such universal units are particularly necessary to interpret the data in terms of energy or material flow in relation to other trophic groups of organisms. For field applications, bacterial biomass is routinely estimated from cell concentrations or cell volumes (biovolume) with appropriate conversion factors.[1-3] There are several biomass conversion factors reported to date (Table 1) from direct determinations, using a variety of methods and organisms.[3-10]

For the direct determination of a conversion factor, in principle, three independent parameters are measured: number of cells, biovolume of the cells, and biomass. Current methods for all three are subject to potential errors. This is particularly true for the biovolume estimation, since errors in measuring linear dimension amplify to the power of 3 when converted to biovolume.

Described here is a method that combines the acridine orange (AO) direct count (DC) method for cell enumeration, AO epifluorescence photomicrography for the biovolume measurement, and CHN analysis for C and N biomass.[9] Compared to other similar studies, this method has several improvements. For example, (1) bacterial cells are derived from natural seawater and grown in unsupplemented, filter-sterilized, particle-free natural seawater; (2) corrections are made for the bacterial cells passing through glass fiber GF/F filters; and (3) biovolume measurement is calibrated with fluorescent microspheres of known diameters.

MATERIALS REQUIRED

Equipment

- Epifluorescence microscope and photomicrography system
- CHN analyzer and related material
- Slide projector
- Vacuum pump

Table 1. Conversion Factors Directly Determined from Biovolume and Biomass Measurements

| | Method for | | | | Conversion factor | | |
| | | | | Cell volume (μm^3) | (pg C μm^{-3}) | (pg N μm^{-3}) | C/N |
Source[a]	Culture[b]	Biovolume[c]	Biomass[d]	mean (range)	mean (range)	mean (range)	mean (range)	Ref.
Pure/SW	ENR	SEM,OG,EFPM	CHN	0.376 (0.11–0.71)	567 (238–964)	104 (45–202)	5.5 (4.6–6.7)	5
FW/EST	ENR/UN	EFM,IA	CIRA	0.195 (0.08–0.34)	351 (178–728)	ND[e]	ND[e]	4
FW	UN	EFM,OG	CHN	0.163 (0.09–0.26)	106 (39–188)	25 (NS[f]–41)	4.8 (3.3–6.8)	7
Pure	ENR	PC	CHN	1.253 (0.52–2.02)	209 (155–292)	62 (32–94)	3.5 (2.8–5.9)	6
SW	UN	PC	CHN	0.290 (0.17–0.53)	186 (84–353)	30 (11–50)	6.3 (3.4–12.2)	6
SW	UN	EFPM	CHN	0.056 (0.04–0.07)	380 (230–600)	110 (60–220)	3.7 (2.5–4.3)	9
FW	ENR/UN	EFPM	CHN	0.545 (0.17–1.75)	136 (59–252)	27 (7–42)	5.5 (3.4–10.5)	8

a Pure, pure culture; SW, seawater; FW, fresh water; EST, estuary brackish water.
b ENR, enriched medium; UN, unenriched natural (in situ) medium.
c SEM, scanning electron microscopy; OG, ocular grid; EFPM, epifluorescence photomicrography; EFM, epifluorescence microscopy; IA, image analysis; PC, particle counter.
d CHN, CHN analysis; CIRA, combustion-infrared absorption method.
e ND = not determined.
f NS = no significant numbers.

Supplies

- Nuclepore polycarbonate filters (0.6 μm pore size, 47-mm diameter)
- Millipore filters (type GS; 0.22-μm pore size, 47-mm diameter)
- Combusted (at 450°C for >4 h) material:
   Glass fiber GF/F filters (24-mm diameter or smaller)
   Forceps, aluminum foil
- Acid-rinsed opaque plastic bottles (1 l)
- Acid-rinsed or combusted filtration units for the 47-mm diameter type GS filters and for the precombusted glass fiber GF/F filters
- Nuclepore filters (0.2-μm pore size, 25-mm diameter)
- Microscope slides, cover slip, immersion oil
- Kodak Ektachrome 400

Solutions

- Irgalan black for staining filters (prestained Nuclepore filters are also available)
- Acridine orange
- Culture medium and inoculum, prepared from freshly collected natural sample by filtration

PROCEDURES

Inoculation and Culturing

1. Soak 47-mm diameter-type GS (0.22-μm pore size) Millipore filters (Millipore Corp., Bedford, MA) in deionized water before use. This is to screen out defective filter discs and to leach out foreign chemicals, often detergent, from filter materials. Use organic-free (acid-rinsed or combusted) labware, and rinse the containers thoroughly several times with the sample (or filtered) seawater before use.

2. Culture medium: Using a large (47-mm diameter) filtration unit (e.g., Millipore Sterifil unit), filter approximately 1 l of fresh natural seawater through the GS filters with vacuum of <20 cm Hg; the filtrate is the culture medium. Prepare another 1 l of GS filtrate to estimate organic material that is contributed by particles shed from the GS filter (see Notes and Comments).

3. Inoculum: Use the filtrate of 0.6-μm pore size Nuclepore (Costar Corp., Cambridge, MA) filters. These filters remove most of nonbacterial particles as well as large bacterial cells and particle-associated cells (see Notes and Comments). Using the Sterifil unit, prepare the 0.6-μm filtrate from the same seawater sample with vacuum of <20 cm Hg.

4. Start the culture by mixing about 10 to 20 ml of inoculum to 1000 ml of media in a dark bottle. The use of dark bottle is to suppress the growth of autotrophic cells possibly present in the filtrate. Monitor the bacterial cell concentration periodically (e.g., every 12 h) by the AODC method (see below). Harvest the cells at the end of the logarithmic growth, normally a few to several days after the inoculation.

CHN Samples

Sample Collection (Cell Harvesting)

1. Collect subsample for AODC from the culture before cell harvesting, and preserve with formalin (1 to 5% by vol). Using organic-free labware, collect the cultured cells on

precombusted glass fiber GF/F filters (Whatman) by filtration with vacuum of <20 cm Hg. Always handle the CHN samples with combusted forceps on a sheet of combusted aluminum foil to avoid contamination. Glass fiber filters of a smaller diameter (e.g., 1.3 cm) will be better, if a CHN analyzer requires the samples to be wrapped in tiny tin boats. Save the filtrate.

2. Measure and note the volume of the filtrate, i.e., the culture filtered through the GF/F filter. Collect subsample from the filtrate for AODC samples and preserve as described above. A significant fraction of the bacterial cells pass through glass fiber filters (see Notes and Comments).

3. After the cell collection, wrap the GF/F filters with tin boats (or with combusted aluminum foil), and dry thoroughly at approximately 60°C for >6 h to prevent the degradation of organic materials. Store the samples in a dessicator until CHN analysis.

Controls, Blanks, and Standards

1. Filter shedding control: Filter the GS filtrate (i.e., culture media, see above) through a precombusted GF/F filter. It will be better if the volume of the GS filtrate that is filtered through the GF/F filter is comparable to the volume of the culture.

2. Blank: The blank may be the absolute blank, i.e., the precombusted GF/F, or the precombusted GF/F wetted with the culture media.

3. Internal standard: Apply small known amount of high-purity amino acids (e.g., alanine has a C:N ratio similar to that of bacteria) to a precombusted GF/F filter. Thoroughly dry and store the filters as described above. See Notes and Comments for the use of the controls and blanks.

Cell Enumeration and Size Measurements

Cell Enumeration by Epifluorescence Microscopy

Hobbie et al.[11] described a standard protocol with detailed background information. Methods for direct counts are presented elsewhere in this book. Measure the cell concentrations of the unfiltered culture and the GF/F filtrate. Combining the culture volume and the cell concentrations will give the total number of cells collected on the GF/F filter.

Cell Size Measurements by Photomicrography

1. Photographic slides for the cell size measurements are prepared from the same microscope slide used for the cell enumeration. Use Ektachrome 400 (Kodak). The optimum exposure settings must be empirically determined for the particular photomicrography system being used. Normally, short exposure time gives a better contrast between the dark background and bright cells. Note that the pictures should be taken from microscope fields chosen "randomly", the same as the AODC is done.

2. Fluorescent microspheres (Covalent Technology Corp., Ann Arbor, MI; Polyscience, Inc., Warrington, PA) of known diameter may help to check your size measurement criteria. Use microspheres with similar size, color, and intensity of fluorescence to those of the AO-stained bacterial cells. Note that apparent size of the fluorescent microsphere is affected by the fluorescence intensity and the photographic setting.[3,8,9] Therefore, an exposure setting and optical filters different from those used for the AO-stained bacterial cells may need to be tried out, in order to produce photographs similar to those of the AO-stained

bacterial cells with respect to the fluorescence intensities, contrast, and overall darkness.[3,9] The microsphere suspensions usually need to be sonicated and diluted before use. Make microscope slides and take pictures of the microspheres as described above.

3. Take pictures of a stage micrometer (10-μm spacing). It is to be used as a standard scale when projecting the photographic slides onto a screen for the size measurements (see step 4).

4. Set up the slide projector in such a way that 1 μm in the real scale is enlarged to 1 cm. Measure and record the lengths and widths of the projected cell images to the nearest 1-mm unit. Note that no cells from a slide should be excluded. However, you may measure cell images from a part of the projection, when there are too many cells in a slide. When measuring the size, it is important to use a consistent criteria for where-to-cut (see Notes and Comments). Strict cutting off of the fuzzy "halo" of the cell images undermeasures by approximately 10% in linear scale; therefore, individual operators should establish and check the cell-boundary criteria.[9] Large sample sizes are required for differentiation between two populations (see Notes and Comments).

5. There are a few different ways to calculate cell volumes from the linear measurements. Normally, cells of the same length and width are assumed to be "spheres" (Equation 1, see below) and the other elongated forms are treated as "rods". However, in practice, it is not that simple since there are occasionally cells of some irregular shapes, e.g., crescent, curvy filament, etc. Consideration should also be given to whether the model rod is a cylinder with two spherical ends (Equation 2) or an oblate spheroid of a cylindrical body with tapered ends (Equation 3). Following are the formulas for

$$\text{Volume of sphere (formula 1)} = (4/3) \times \text{pi} \times R^3 \qquad (1)$$

$$\text{Volume of cylindrical rod (formula 2)} = \text{pi} \times (3L - W)/3 \times (W/2)^2 \qquad (2)$$

$$\text{Volume of oblate spheroid (formula 3)} = (4/3) \times \text{pi} \times (L/2) \times (W/2)^2 \qquad (3)$$

where R = radius, L = length, W = width, and L > W.

6. From the size measurements, cell size-frequency distribution of the cells is obtained. From the cell enumeration (cell number ml^{-1}) and the culture volume (ml), total cell numbers in the unfiltered culture and in the GF/F filtrate are obtained. Combining the total cell number and the size-frequency distributions will give the total biovolume before and after the GF/F filtration, and the difference is the biovolume that is collected on the GF/F filter.

7. The biovolume conversion factor (biovolume-to-biomass conversion) is calculated as the carbon and nitrogen biomass from the CHN analysis, divided by the biovolume of the cells collected on the GF/F filter. The biomass conversion factor for cell number (cell number-to-biomass conversion), i.e., "per cell biomass", is calculated as the biomass divided by the number of cells collected on the GF/F filter.

NOTES AND COMMENTS

1. See Inoculation and Culturing — The type GS filter is made of mixed esters of nitrocellulose, and particles shed from the filter contain significant amount of carbon and nitrogen. Controls for the particle shedding are made from the separate preparation of GS filtrate (see below for CHN controls and blanks). Alternatively, boiling for 15 min three times in distilled water was recommended to minimize the shedding.[8] Inoculum may be prepared

with larger pore-size Nuclepore filters, e.g., 0.8 or 1.0 μm to include larger bacterial cells. However, filtrate of the pore size >0.8 μm often contains substantial amount of cells other than heterotrophic natural bacteria.

2. See Sample Collection (Cell Harvesting) — Glass fiber filters by no means collect all the particles of bacteria size. The retention efficiency varies among individual filters, filter types, and brands. There is no clear-cut nominal size (diameter) for the retention. The GF/F filtrate contains bacterial cells of almost all size classes present in the unfiltered culture, but the size-frequency distribution is skewed to smaller cells.[9] The cells collected on the GF/F filter is the fraction (from approximately 2/3 to 1/2) of the total cells in the culture. To correct for the cells passing through the GF/F, it is important to measure cell concentrations and size-frequency distributions from the GF/F filtrate as well as from the unfiltered culture (see Cell Size Measurements by Photomicrography).

3. See Controls, Blanks, and Standards — Blanks and filter shedding controls are subtracted from samples. Internal standards are for an independent check of the CHN analysis. Since the standard contains both carbon and nitrogen, CHN analysis of the standard will produce a calibration factor (i.e., expected amount vs. measured amount) for carbon and nitrogen, separately. The two calibration factors should be reasonably close to 1, as well as close to each other. The raw CHN data are multiplied by the calibration factor. However, there may be more than one way to incorporate those control and blank CHN samples into the biomass calculation.[9]

4. See Cell Enumeration by Epifluorescence Microscopy — Because the samples have various cell concentrations, the sample volumes to be filtered for AO slides should be adjusted to yield optimum cell densities on a Nuclepore filter (normally a few 10^6 cells on a filter). The sample size for AODC is very small compared to the population size; therefore, it is important to reduce the sampling error. Optimum number and the level (e.g., subsample, slide, or fields) of replicates in the direct counting method are discussed in Reference 12 and Chapter 14).

5. See Cell Size Measurements by Photomicrography — Since images of bacterial cells have a "halo" when projected onto a screen, certain criteria must be applied consistently for the cell size measurements. To aid the consistency, picture qualities (sharpness, contrast, overall darkness, etc.) should be reasonably same. For this reason, photographic settings should not be changed throughout an experiment. Size-frequency distributions typically have a widespread range of size classes.[13] A certain minimum sample size is required for a resolution of statistical parameters between two populations. For a better estimation and comparison of the population statistics, a relatively large size of samples is recommended (see Reference 14 for related discussion).

6. Concluding remarks — An accurate derivation of a biomass conversion factor is primarily based on the accurate measurements of the cell size. Different sizing methods have different resolution power.[5-7] Apparent cell size varies with the sample preparation and sizing method used.[5,13,15] It is not a simple matter to isolate true variations due to those artifacts. Among the current methods for biovolume measurement, image analysis and photomicrography are generally known to yield more realistic measurements.[3-5,9,13,15,16] However, the measuring criteria and the consistency seem equally or even more important than the method used.

Table 1 lists directly measured conversion factors published to date (conversion factors from partial measurements employing indirect estimations are not listed). There are variations between studies as well as within a study. The variation may be a true variation because of different organisms (different species compositions) and different physiological state of the cells. However, a considerable fraction of the variation may be attributable to the large potential errors in size measurements. Unfortunately, no absolute calibration is currently available for biovolume measurements. Use of per-cell biomass conversion factors, being free of the bias due to biovolume measurements, would be more practical for natural planktonic bacteria. Cellular C and N contents ratios (C/N) are relatively less variable regardless of the methods used; mostly in the range from approximately 3.5 to 6.5 except for a few cases. Those high C/N ratios may be due to interference from detritus or nonbacterial organisms.

A trend of higher conversion factors for smaller cells was reported by some, whereas no such trend was found by others (see Figure 2 of Reference 8). It has been argued whether high conversion factors are physiologically possible.[17] Since no intercalibration method is currently available between studies where different methodologies and different source organisms were employed, it is not easy to compare the conversion factors obtained in diverse studies.[8,16] A conversion factor is most useful when applied to a set of data obtained using similar procedures, making the conversion factor internally consistent as well as convenient.

REFERENCES

1. Watson, S. W., Novitsky, T. J., Quinby, H. L., and Valois, F. W., Determination of bacterial number and biomass in the marine environment, *Appl. Environ. Microbiol.,* 33, 940, 1977.
2. Cho, B. C. and Azam, F., Major role of bacteria in biogeochemical fluxes in the ocean's interior, *Nature,* 332, 441, 1988.
3. Scavia, D. and Laird, G. A., Bacterioplankton in Lake Michigan: dynamics, controls, and significance to carbon flux, *Limnol. Oceanogr.,* 32, 1017, 1987.
4. Bjornsen, P. K., Automatic determination of bacterioplankton biomass by image analysis, *Appl. Environ. Microbiol.,* 51, 1199, 1986.
5. Bratbak, G., Bacterial biovolume and biomass estimation, *Appl. Environ. Microbiol.,* 49, 1488, 1985.
6. Kogure, K. and Koike, I., Particle counter determination of bacterial biomass in seawater, *Appl. Environ. Microbiol.,* 53, 274, 1987.
7. Nagata, T., Carbon and nitrogen content of natural planktonic bacteria, *Appl. Environ. Microbiol.,* 52, 28, 1986.
8. Nagata, T. and Watanabe, Y., Carbon- and nitrogen-to-volume ratios of bacterioplankton grown under different nutritional conditions, *Appl. Environ. Microbiol.,* 56, 1303, 1990.
9. Lee, S. and Fuhrman, J. A., Relationships between biovolume and biomass of naturally derived marine bacterioplankton, *Appl. Environ. Microbiol.,* 53, 1298, 1987.
10. Heldal, M., Norland, S., and Tumyr, O., X-ray microanalytic method for measurement of dry matter and elemental content of individual bacteria, *Appl. Environ. Microbiol.,* 50, 1251, 1985.
11. Hobbie, J. E., Daley, R. J., and Jasper, S., Use of Nuclepore filters for counting bacteria by fluorescence microscopy, *Appl. Environ. Microbiol.,* 33, 1225, 1977.
12. Kirchman, D., Sigda, J., Kapuscinski, R., and Mitchell, R., Statistical analysis of the direct count method for enumerating bacteria, *Appl. Environ. Microbiol.,* 44, 376, 1982.
13. Fuhrman, J. A., Influence of method on the apparent size distribution of bacterioplankton cell: epifluorescence microscopy compared to scanning electron microscopy, *Mar. Ecol. Prog. Ser.,* 5, 103, 1981.
14. Sokal, R. R. and Rohlf, F. J., *Biometry,* W. H. Freeman, San Francisco, 1969.
15. Montesinos, E., Esteve, I., and Guerrero, R., Comparison between direct methods for determination of microbial cell volume: electron microscopy and electronic particle sizing, *Appl. Environ. Microbiol.,* 45, 1651, 1983.
16. Psenner, R., From image analysis to chemical analysis of bacteria: a long term study?, *Limnol. Oceanogr.,* 35, 234, 1990.
17. Joint, I. R. and Pomroy, A. J., Activity of heterotrophic bacteria in the euphotic zone of the Celtic Sea, *Mar. Ecol. Prog. Ser.,* 41, 155, 1987.

CHAPTER 38

Use of Color Image Analysis and Epifluorescence Microscopy to Measure Plankton Biomass

Peter G. Verity and Michael E. Sieracki

INTRODUCTION

Fundamental size- and biomass-dependent relationships are apparent within and among trophic groups of plankton. These include allometric dependence of growth and respiration in autotrophs and heterotrophs;[1-4] size dependence of light absorption,[5] nutrient uptake,[6] photosynthesis,[7] and sinking rates[8] of phytoplankton; size ratios between predator and prey;[9] and inverse logarithmic relationships between cell size and population abundance across size classes.[10] Allometric considerations may also constrain plankton community composition, species diversity, and perhaps even cell shape.[9,11,12]

Functional relationships between prey concentration and ingestion by (or growth of) planktonic predators/grazers are now well known. Thus, a first-order quantitative description of material fluxes among trophic compartments may be derived from biomass estimates of predator and prey populations. Their accuracy can be enhanced when coupled with estimates of rate parameters specific to each trophic group or compartment. Determination of population or community biomass requires (1) accurate and precise measurements of individual cells and (2) a sampling scheme in which temporal scales of population assessment converge with relevant physical processes and size-dependent metabolic rates.

It is now generally accepted that pico-, nano-, and microplankton are central components of pelagic food webs. These cells, however, have characteristics not conducive to simple traditional methods of biomass estimation: they are very small and numerous; they can change rapidly in abundance; and cells of similar size may include autotrophic or heterotrophic taxa. Thus, the optimal technique of biomass estimation must be able to accurately measure numerous small cells in large numbers of samples and assess their trophic status.

Color image-analyzed fluorescence microscopy provides such a tool. It has been successfully applied to biomass measurements of coastal and oceanic plankton ranging in size from approximately 1 to 100 μm.[13-16] The combination of fluorochrome-induced fluorescence and autofluorescence of plant pigments provides a powerful way of characterizing plankton communities using true-color image analysis (Figure 1). Using DAPI/proflavine dual stain procedures (described below), detrital particles are discriminated from living cells by the presence of blue fluorescence of DAPI-stained DNA under UV excitation. Phototrophic and heterotrophic cells are discriminated under blue excitation by the presence and absence, respectively, of red or orange autofluorescence. Phycoerythrin-containing cells can be

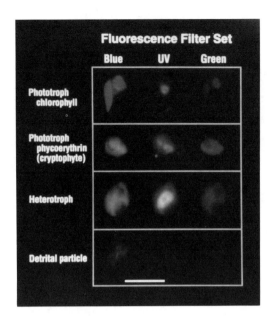

Figure 1. Example of using object color to classify cell types and detritus. Three cell types (phototrophic flagellate, cryptomonad, heterotrophic flagellate) and a detrital aggregate are shown under blue (BP 450-490 nm, Dichroic 510 nm, LP 520 nm), ultraviolet (BP 365 nm, Dichroic 395 nm, LP 397 nm), and green (BP 510-560 nm, Dichroic 580 nm, LP 590 nm) excitation. Scale bar is 5 μm.

discriminated from eukaryotic cells by their characteristic orange fluorescence. Mixotrophic cells (plastidic taxa which also ingest prey) can potentially be recognized by short-term incubations with fluorescent beads, although this method will only provide a minimum estimate of total mixotrophic cells.

Additional applications of generic image analysis systems in plankton studies, other than biomass estimation, include morphometrics and pattern recognition,[17-20] *in situ* particle size analysis,[21] and feeding studies.[22] In this chapter, we describe the system components and measurement protocols necessary to rapidly and accurately recognize and quantify the abundance and biomass of different trophic components within natural plankton communities. Thus, we emphasize true-color image-analyzed fluorescence microscopy. The reader will recognize that specific individual applications may require only a subset of these procedures.

GENERIC COMPONENTS AND PROTOCOLS

In the simplest sense, image analysis of plankton comprises three basic steps: image rendering and capture, image enhancement, and morphometry. The hardware required to complete these tasks (Figure 2) includes an epifluorescence microscope to visualize the sample; one or more cameras (often with associated control units) to optimize and transmit an electronic image to the processor; a video digitizer, or "frame grabber", to convert the analog image to digital form; a host computer to house the frame grabber and run image processing and data analysis software; and one or two color monitors for image display and data operations. The best contemporary systems can display live or "grabbed" images in a window of the computer monitor, obviating the need for a second monitor. The accuracy of an image analysis system is limited by microscope optics, camera, and digitizer, whereas the power of an image analysis system is determined by the computer processor(s) and

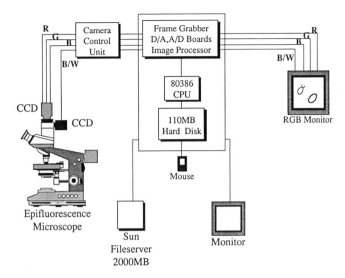

Figure 2. Schematic of a typical desktop imaging system. See text for details. (Reprinted from *Deep Sea Res.*, 40(1/2), Verity, P. G. et al., 227, Copyright 1993, with permission from Pergamon Press, Ltd., Headington Hill Hall, Oxford OX3 0BW, UK.

quality and flexibility of the imaging software. The sequential steps and principles of operation of image analysis applied to plankton studies, in terms of hardware and software, are described below. Subsequently, we discuss details of sample preparation and processing.

Several types of image sensors are available. The most commonly used are tube-based TV cameras (Vidicon, Plumbicon, Newvicon, Chalnicon) and solid-state cameras (charge-coupled devices, or CCDs). The most important parameters which define them are light sensitivity, resolution, signal:noise ratio, and spectral response. In general, the tube-style cameras have greater resolution than CCDs; for low light levels, image-intensifying devices are useful, although signal:noise ratios decrease significantly. CCDs offer the advantages of very low distortion due to the fixed geometric array of the silicon-based photodetectors, broad dynamic range, high quantum efficiency, and small size. They are now available with sensitivities of 10^{-2} to 10^{-3} lux (single chip, B&W CCDs) to 10^1 lux (three-chip, color CCDs), which are sufficiently sensitive for epifluorescence plankton microscopy at $1000\times$. CCD cameras are available with read-out rates synchronized to the standard NTSC video rate of 1/30 s. This makes them compatible with standard video equipment (monitors, frame grabbers, tape recorders, etc.).

Newer, cooled CCDs have even greater signal:noise ratios, quantum efficiencies, and dynamic range.[23,24] The real advantages of CCD technology for scientific imaging is realized when they are cooled to reduce dark current noise, and are used as a digital camera, with exposure time controlled by a shutter. The CCD array accumulates charge over the time it is exposed, analogous to photographic film. The resulting digital image is then read out and transferred to a host computer directly, without the need for a frame grabber. These cameras can only be used with fixed, nonmoving images, such as fixed plankton cells on a filter. They have proven especially useful in imaging the smaller picoplankton cells and submicrometer particles.[25,26] Multispectral imaging cytometry can be done with these cameras using automatic filter changers on the microscope.[27]

The electronic signal of a video camera is digitized by the frame grabber into one or more two-dimensional arrays of picture elements, or "pixels", which can be displayed on a raster screen monitor. The standard video picture has an aspect ratio of 1.25:1. Thus, the frame grabber will produce rectangular pixels unless specifically designed to produce square pixels. For quantitative image analysis, square pixels are highly desirable. Each pixel contains two types of information: x and y coordinates, and intensity (or gray scale) level. Black and white (B&W) images are recorded as a single gray scale, while color images contain gray-scale information of the three primary colors (red, green, blue, or RGB). In a typical application with an image resolution of 512 pixels per line and 512 horizontal lines, and an intensity scale range of 0 to 255, a B&W image requires at least 262K of computer memory while a true-color (RGB) image requires at least 786K. Once in memory, image processing software is used to enhance image contrast. Since each operation must be completed 786 $\times$ 10^3 times for a full color image, image analysis places tremendous demands on computational speed and memory. Currently used host computers include 386/486 AT-class machines, Macintosh II's, and a variety of workstations. When using a color video camera we have found that rapid ("real-time") image averaging is necessary for noise reduction. This results from the use of a high gain amplification setting on the camera to adequately detect faintly fluorescing cells. To average images at normal frame rates of 1/30 s requires a pipeline processor, or similar unit, integrated with the frame grabber.

Numerous generic image processing routines are available in most commercial systems, including background subtraction, image averaging, digital filtering, and histogram equalization. Detailed technical descriptions of these operations are available elsewhere.[28,29] For determination of plankton biomass, the most significant software features are those which enhance detection of cell boundaries and those which interpolate cell depth to calculate biovolume. Methods of automating edge detection and biovolume calculation are discussed below. It should be noted that the authors of this chapter represent different approaches toward application of image-processing software to determination of plankton biomass. One of us (MES), after early work[13] with a dedicated system, has developed software which has been formatted for different operating systems.[30,31] The other (PGV) uses commercial software developed by Analytical Imaging Concepts, Inc. and Belvoir Consulting, Inc., who were willing to work with the senior author to modify routines which perform contrast enhancement, image averaging, and background subtraction. Because it is commercially available, a description of the general features and capabilities of the AIC/Belvoir Consulting software is given in the section Example of a Commercially Available True-Color Image Analysis System. Additional specific software routines for automatic thresholding and biovolume calculation, developed by Sieracki et al.[30,31] and described below, have been ported over to the AIC package so that the two perform similarly.[15,16] The measurement procedures outlined later are followed by both systems.

Epifluorescent images are generally composed of bright fluorescing cells against a dark background, allowing a computer to detect cells in the image using a brightness threshold. Pixels exceeding that threshold represent part of a cell and those below it are background. In a typical two-color plankton image (containing red and green fluorescing cells), the two color frames are averaged to measure cell size. Image detectors cannot render a perfect image, because the edge of a cell is not distinct but appears as a gradient between the dark background and the bright cell. Thus, varying thresholds produce different apparent cell sizes, making accurate threshold selection necessary to precisely measure cell size. An objective, automated method for determining a threshold based on the image characteristics was developed which finds the threshold as an extrema of the second derivative of the cell brightness profile.[30] Although derived from analyzing images of standard fluorescent microspheres of different sizes and brightnesses, it is accurate and reliable for measuring a variety of pico- and nanoplankton cells.

Once a cell (or all the cells) in an image have been thresholded, a variety of cell measurements are made. Gray-scale intensity is measured in each of the colors, including average and maximum brightness. Red brightness should relate to chlorophyll content per cell, though this has not yet been verified. Direct cell size measurements include area, perimeter, longest and shortest dimensions, and biovolume. Cell biovolume is calculated from each cell's perimeter array and orientation using a recently developed algorithm.[31] The distance from the cell perimeter pixel to a center line on the longest axis of each cell is mathematically "rotated" through 180° and these half-cylinder volumes, each 1 pixel wide, are summed. This measuring algorithm is more precise and accurate than currently used visual or automated methods based on geometrical calculations using length and width, or area and perimeter.

NOTES ON SAMPLE PREPARATION AND PROCEDURES

1. Water samples are preserved in glutaraldehyde to a final concentration of 0.3% and, if not processed immediately, refrigerated at 4°C temporarily (<1 d). Each preserved sample is stained with 4′6-diamidino-2-phenylindole (DAPI) for 4 min (5 μg ml^{-1} final concentration). Samples are then filtered onto 0.2- to 0.8-μm black Nuclepore filters, depending upon the smallest dimensions of the plankton of interest. To achieve an even distribution for counting and measurement purposes, black filters are placed on top of prewetted 0.45-μm Millipore backing filters. Samples are then stained with 3-6-diaminoacridine hemi-sulfate (proflavin) for 1 min (5 μg ml^{-1} final concentration), filtered under low vacuum, rinsed with distilled water, and placed on a glass slide. A drop of low-fluorescence immersion oil is placed on top of the filter, which is covered with a cover slip, and the slide is stored frozen at −20°C. Samples should be image analyzed as soon as possible to avoid loss of fluorescence, especially chlorophyll autofluorescence of small cells.

2. Calibration of pixels to real linear units (e.g., micrometers) must be done once for each objective and magnifier combination used with the system. This is usually done by digitizing an image of a stage micrometer and counting the number of pixels between each line on the micrometer. If a system has square pixels, the vertical and horizontal pixels per micrometer will be equal. This can be easily determined by rotating the camera 90° on the microscope, digitizing another image of the stage micrometer, and calibrating it vertically.

3. Under the microscope, a field of view is located, either randomly or on a transect, and visually focused on the video monitor (Figure 3). It is then digitized using frame-rate image averaging to reduce noise. Digitization with 16 frames averaged takes less than 2 s so fluorescence fading is minimized. At this point, the image is edited, if necessary, to remove detrital particles or to separate adjoining cells. Editing is typically required in only about 10% of the images from estuarine samples, and less from oceanic samples which contain relatively little detritus. Once the image is judged to be adequate either the whole image is segmented (cells are thresholded as distinct from background) or individual cell subimages are saved for later analysis.

4. The volume of sample filtered per slide should be adjusted according to the populations encountered to optimize the cell density on the slide for image analysis. It is generally preferable to have more cells per field for image analysis than visual enumeration. Since each field image is completely scanned by the computer regardless of the number of cells it contains, it is more efficient for each image to contain as many cells as possible without excessive cell overlap. Optimally, microscopes should be equipped with a wide range of objectives (e.g., 16, 40, 63, and 100×) and additional magnifier settings (e.g., 1.25, 1.6, and 2.0×). This permits considerable flexibility in optimizing the analysis to different plankton communities. Two approaches are used depending upon the density of the particular cell population being analyzed.

5. For cells which are numerous per field and relatively uniform in brightness in a given sample (e.g., cyanobacteria and perhaps flagellates), randomly chosen whole fields are analyzed. In this case all the cells in a field are segmented and measured automatically.

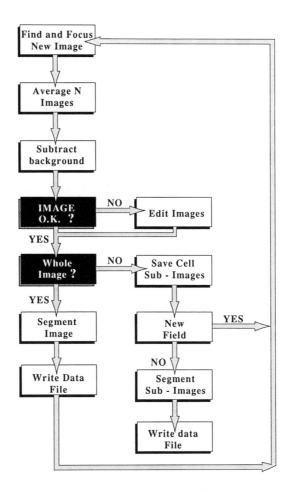

Figure 3. Flow diagram of image analysis measurement procedures. See text for details.

This approach requires that a single threshold be used for all cells in the image, but is faster than automatically finding individual cell thresholds. Acquisition and analysis of an image with 40 cells takes about 4 s.

6. For more rare cells (e.g., ciliates and phototrophic eukaryotes), transects of the slide are scanned and individual cells are isolated and identified by the operator interactively. If the system has a motorized stage and trackball control, the operator can easily scan across transects of the slide while the computer keeps account of the area of the slide examined. In this way densities of rare cells are calculated per volume of sample. Subimages containing the individual cells are temporarily stored and analysis done overnight. The analysis procedure for these subimages should involve automatic threshold determination, segmentation, and cell measurement. A combination of these two approaches is used to optimize sample analysis for the particular populations. Cell measurement data and example images are archived on tape or floppy disk for later analysis.

7. Individual cell biovolume measurements may be converted to carbon biomass for intercomparison of trophic and size classes of the plankton community. This conversion is based primarily on literature values of carbon density (pg C μm^{-3}). Mathematical formulations of carbon content, image-analyzed cell volume, and stereometric formula volumes for photosynthetic nanoplankton have been recently described.[32]

METHODOLOGICAL CONSIDERATIONS

Procedures

Traditional descriptive biological oceanography, such as surveys of population abundance, is now predominantly a thing of the past. The value of abundance data is increased immeasurably when coupled with information on biovolume, carbon or nitrogen content, and expressed as population biomass. However, counting cells and sizing cells are very different tasks, the latter requiring more sophisticated technology and software, and more objective definition of cell perimeter. Whereas the accuracy of size measurements is limited by initial image quality, optical capture, and digitization (e.g., pixel density), the precision of size measurements is determined by the method of thresholding.

Highly accurate and precise measurements of individual cells are of little use if the cells measured are not representative of the entire population. Therefore, proper sampling and statistical considerations are paramount, e.g., how many cells of each type must be counted and measured, and how best to do it? Within a given trophic group which may be taxonomically diverse, e.g., photosynthetic nanoplankton or heterotrophic dinoflagellates, the variance around estimates of mean cell biovolume is often 50 to 100%.[15,16] As a minimum, 200 cells of each type should be measured.[33] For cell types in which entire fields of view can be thresholded (because they contain only the cells of interest, e.g., cyanobacteria), more measurements improve the accuracy of population biomass estimates.

There is always a trade-off between accuracy and efficiency when enumerating and sizing plankton, whether determined by machine or human vision. Experience quickly teaches the optimal density of cells of each type on the filter. Unfortunately, the area of the filter which must be scanned to encounter the required number of cells varies with cell type and among samples, and it is often unnecessary (indeed, undesirable) to scan an entire filter. In such cases, a procedure must be instituted to determine the fraction of cells collected on a filter which have been measured. One approach is to record the number of fields which have been examined, or the distance which has been scanned, and to compare this area to the known collecting surface area of the filter. This manual approach is inexpensive but tedious. A more satisfactory but costlier solution is to interface a motorized microscope stage and associated x-y controller with the imaging software or macro shell operation, to automatically record the area which has been scanned. An intermediate option (both in cost and benefit) is to add an electronic x-y encoder with a manual microscope stage, which simply records the total distance traveled in x or y directions.

Imaging Systems

There has been a rapid expansion in breadth of applications of image analysis in science and industry. While expansion has fostered development of new software and more sophisticated sensors, and made entry-level desktop imaging systems affordable ($25 to $75,000), a general lack of industry standards is prevalent. Interestingly, the sheer diversity of sizes and shapes of plankton is typically much greater than that of objects in other scientific and industrial applications. Plankton science, however, does not offer to the imaging industry the investment potential needed to leverage the development of focused software routines specific to determining plankton biomass. To the plankton scientist, this means that commercially available, off-the-shelf imaging systems often do not provide an ideal solution. Many of the basic capabilities are shared among systems, such as contrast enhancement, digital filters, image averaging, and backround subtraction. However, important details such as thresholding, the user interface, count and size data tabulation, and degree of automation vary greatly among systems.

Thus, in shopping for a new imaging system, upgrading an existing one, or designing novel source codes, it is essential to specify and prioritize scientific objectives in advance of purchase. Am I counting, sizing, or both? Do I need true color, or will B&W be acceptable? Does the required level of resolution dictate a three-chip color CCD, or will a less expensive single-chip provide the information required? Am I interested in only one trophic group which can be uniquely identified (e.g., stained), or must my software be able to address varying sizes, shapes, abundances, and colors of cells? To answer these questions, the typical plankton scientist who is not image-literate will find texts such as *Video Microscopy*[34] and magazines such as *Advanced Imaging* to be invaluable references. Without adequate background knowledge, proper comparison of needs with existing solutions will be difficult.

The following comments may aid in evaluating existing imaging systems. (1) Experiment with as many systems as you can, using your own samples. If you intend to use a system to size fluorescent bacteria, do not let a salesperson show you how it thresholds heart muscle tissue under phase contrast illumination. (2) Do not be overly impressed by software options of little use to your application; conversely, a vendor may offer a novel approach which previously you had not considered. (3) Match specifications of system components: a high-resolution CCD is only as good as the resolution of the frame grabber and monitor. (4) Flexibility and breadth of software may be advantageous in meeting future research needs or in attaching keystroke macros or shell operations. In this regard, small vendors may be more willing than large vendors to implement software changes for you. (5) Consider system portability if you anticipate transporting it to other sites or for use aboard ship. (6) The host computer should be as fast and reliable as funds permit. Estimate your storage needs and purchase memory capacity twice that estimate. (7) Imaging technology is evolving on time scales shorter than publication times, and the ideal imaging system will be able to incorporate such progress. Does the vendor have an upgrade path or policy? (8) You will likely be surprised at the length of the learning curve and the initial investment in time required to fully utilize image analysis. Product support and training are highly valuable attributes of vendors.

EXAMPLE OF A COMMERCIALLY AVAILABLE TRUE-COLOR IMAGE ANALYSIS SYSTEM

Hardware

The presently configured system utilizes an Olympus BH-2 epifluorescence microscope equipped with a mercury burner and appropriate exciter/barrier filter sets for UV, blue, and green excitation. Magnifications include $125\times$, $250\times$, $500\times$, $750\times$ (oil), and $1250\times$ (oil). A Hitachi DK-7000 microsurgical three-chip color CCD senses the image with a minimum illumination of 11 Lux and a resolution exceeding 600 horizontal lines. A Dage 72 single-chip CCD is used for applications where color is relatively unimportant, e.g., DAPI-stained bacteria. It has superior sensitivity at the low illuminations associated with epifluorescence microscopy. Both CCDs have camera control units which permit gain, color balance, and automatic black- and white-level adjustments. The separate RGB (red, green blue) signals from the color CCD are digitized in 1/30 s and stored in 12 frame buffers using a customized Matrox board. The digitized image can be resolved with 512×512 pixels using 8 bits of memory for each color plane. Interimage operations can be performed to 16-bit precision with no loss of image resolution. Object segmentation by gray level utilizes the maximum 256 gray levels since none are sacrificed for overlays or menus. The image processing board, which contains a real-time ALU, is housed within an 80386 Compaq computer operating at 20 mHz. Digitized images or reduced data can be stored on an internal 110-MB hard disk or ported via Ethernet to a Sun 3/280 fileserver. Stored or live images are displayed on a Sony PVM-1342Q 13'' RGB monitor with a resolution exceeding 600 lines at center. Data manipulation and DOS shell operations are viewed on a Compaq color monitor. Hard copy is available using an Epson dot matrix printer. A Microsoft mouse provides sensitive interactions between the operator and the image analysis system.

Software

The software was developed by Analytical Imaging Concepts and Belvoir Consulting, with refinements suggested by the senior author. Images may be analyzed live or after storage to disk. Live, focused images are digitized as many times as desired and subsequently averaged, reducing noise. A background image is subtracted to eliminate variability associated with nonuniformity in illumination or electronics. On-screen peak level indicators optimize illumination, gain, and background levels which do not result in loss of usable gray-scale information (''image saturation''). Image contrast may be enhanced by adjusting input to output lookup tables which can be stored and applied to live or frozen images. Cells to be measured can be thresholded by gray values, and the thresholded image can be superimposed on the original image. This permits assessment of the accuracy of the threshold settings. Detection of cell edges and highlights of details can be enhanced using digital filters run as 3 × 3, 5 × 5, or 7 × 7 kernels. These include sharpen, smooth, erode, dilate, laplacian, and sobel filters. A line may be drawn across a cell, which generates a histogram of gray-scale intensities that can then be used to enhance edge detection. All of these settings may be stored and applied to any or all future images.

Once image contrast and enhancement are optimized, cells may be measured using several modes. The single object mode automatically traces the boundary and measures a segmented cell. Parameters which can be measured include area, perimeter, width, true length, height, shape factor, hole area, maximum horizontal and vertical chords, average gray value, calibrated density, percent area, center of gravity, and angle of orientation. Upper and lower values can be set to select or exclude cells from consideration. Logical equations can be defined by the user between any combination of parameters, and cells not meeting criteria are highlighted. A region of interest mode permits measurements of cells contained within regular- or irregular-shaped regions. An automeasure mode allows the user to define a measuring frame and automatically measure all cells or only selected objects within that frame. All settings can be stored and recalled for subsequent images or subimages. Images can also be interactively edited, including separation of touching objects and adding text to images.

Sorted images are archived and retrieved using pulldown menus, each of which contains on-screen selection of path and storage device. Images can be moved or deleted without exiting to the DOS shell. Data from single objects, regions of interest, and entire fields can be stored in ASCII format and loaded into most major statistical packages. Threshold and biovolume algorithms calculate cell biovolume directly from each cell's perimeter array and orientation.[30,31]

CONCLUDING REMARKS

More than ever, aquatic biology is riding a technological wave toward the future. Imaging cytometry is just one approach whereby electronic sensors and automated binomial calculations can address old questions and provide increasingly accurate answers to problems over greater temporal and larger spatial scales. Presently, color image analysis provides aquatic science with an improved tool to determine the abundance and biomass of plankton. Significant progress in the future will likely occur in at least three directions. Morphometric software will be coupled to pattern recognition schemes to permit automated enumeration, size determination, and taxonomic categorization of planktonic organisms. Another area of expanding interest is the application of biomedical techniques to aquatic biology, which will provide more sophisticated fluorescent cellular tags and tracers which can be imaged. Finally, optical sensing and presentation of microscope images is continually improving.[35,36] Laser confocal microscopy already permits measurement and display of microscope images in true 3-D, albeit at a price tag of $100,000 to $200,000. Digital confocal imaging systems are being developed which permit volume rendering of objects visualized using traditional epifluorescence or transmitted illumination. All of these developments bode well for the future of imaging applications to plankton studies.

ACKNOWLEDGMENTS

Contributions from numerous people and organizations made this paper possible. The Skidaway Institute of Oceanography and the University of Georgia provided the funds to purchase the image analysis system in the senior author's laboratory. The research to develop and implement software routines and measurement protocols was supported by National Science Foundation grants OCE 88-17399 and OCE 88-13356. P. Belvoir, M. Goldberg, and K. Kutz are commended for learning sufficient plankton science to improve a commercial imaging package for use in measuring plankton biomass. C. Viles and C. Tronzo interfaced the thresholding and biovolume algorithms discussed in this chapter with that software. D. Peterson prepared the manuscript.

REFERENCES

1. Fenchel, T., Intrinsic rate of natural increase: the relationship with body size, *Oecologia*, 14, 317, 1974.
2. Banse, K., Rates of growth, respiration, and photosynthesis of unicellular algae as related to cell size — a review, *J. Phycol.*, 12, 135, 1976.
3. Banse, K., Cell volumes, maximal growth rates of unicellular algae and ciliates, the role of ciliates in the marine pelagial, *Limnol. Oceanogr.*, 27, 1059, 1982a.
4. Banse, K., Mass-scaled rates of respiration and intrinsic growth in small invertebrates, *Mar. Ecol. Prog. Ser.*, 9, 281, 1982b.
5. Agusti, S., Allometric scaling of light absorption and scattering by phytoplankton cells, *Can. J. Fish. Aquatic Sci.*, 48, 763, 1991.
6. Munk, W. H. and Riley, G. A., Absorption of nutrients by aquatic plants, *J. Mar. Res.*, 11, 215, 1952.
7. Geider, R. J., Platt, T., and Raven, J. A., Size dependence of growth and photosynthesis in diatoms: a synthesis, *Mar. Ecol. Prog. Ser.*, 30, 93, 1986.
8. Smayda, T. J., The suspension and sinking of phytoplankton in the sea, *Oceanogr. Mar. Biol. Annu. Rev.*, 8, 353, 1970.

9. Fenchel, T., The role of protozoa in nature in terms of functional properties related to size, *Zool. Sci.,* 7(Suppl.), 51, 1990.

10. Sheldon, R. W., Prakash, A., and Sutcliffe, W. H., Jr., The size distribution of particles in the ocean, *Limnol. Oceanogr.,* 17, 327, 1972.

11. Lewis, W. M., Jr., Surface/volume ratio: implications for phytoplankton morphology, *Science,* 192, 885, 1976.

12. Agusti, S. and Kalff, J., The influence of growth conditions on the size dependence of maximal algal density and biomass, *Limnol. Oceanogr.,* 34, 1104, 1989.

13. Sieracki, M. E., Johnson, P. W., and Sieburth, J. McN., Detection, enumeration, and sizing of planktonic bacteria by image-analyzed epifluorescence microscopy, *Appl. Environ. Microb.,* 49, 799, 1985.

14. Estep, K. W. and MacIntyre, F., Counting, sizing, and identification of algae using image analysis, *Sarsia,* 74, 261, 1989.

15. Sieracki, M. E., Verity, P. G., and Stoecker, D. K., Plankton community response to sequential silicate and nitrate depletion during the 1989 North Atlantic spring bloom, *Deep-Sea Res.,* 40, 213, 1993.

16. Verity, P. G., Stoecker, D. K., Sieracki, M. E., Burkill, P. H., Edwards, E. S., and Tronzo, C. R., Abundance, biomass, and distribution of heterotrophic dinoflagellates during the North Atlantic spring bloom, *Deep-Sea Res.,* 40, 227, 1993.

17. Gorsky, G., Guilbert, P., and Valenta, E., The Autonomous Image Analyzer — enumeration, measurement, and identification of marine phytoplankton, *Mar. Ecol. Prog. Ser.,* 58, 133, 1989.

18. Sheath, R. G., Applications of image analysis and multivariate morphometrics for algal systematics, *J. Phycol.,* 25, 3, 1989.

19. Berman, M. S., Application of image analysis in demographic studies of marine zooplankton in large marine ecosystems, in *Large Marine Ecosystems,* Sherman, K., Alexander, L. M., and Gold, B. D., Eds., American Association for the Advancement of Science, Washington, DC, 1990, chap. 10.

20. Tsuji, T. and Nishikawa, T., Automated identification of red tide phytoplankton *Prorocentrum triestinum* in coastal areas by image analysis, *J. Oceanogr. Soc. Jpn.,* 40, 425, 1984.

21. Eisma, D., Schumacher, T., Bockel, H., van Heerwaarden, J., Franken, H., Laan, M., Vaars, A., Eijgenraam, F., and Kalf, J., A camera and image analysis system for *in situ* observation of flocs in natural waters, *Neth. J. Sea Res.,* 27, 43, 1990.

22. Noji, T. T., Estep, K. W., MacIntyre, F., and Norrbin, F., Image analysis of faecal material grazed upon by three species of copepods: evidence for coprorhexy, coprophagy, and coprochaly, *J. Mar. Biol. Assoc. U.K.,* 71, 465, 1991.

23. Hiraoka, Y., Sedat, J. W., and Agard, D. A., The use of a charge-coupled device for quantitative optical microscopy of biological structures, *Science,* 238, 36, 1987.

24. Sieracki, M. E. and Viles, C. L., Color image-analyzed fluorescence microscopy: a new tool for marine microbial ecology, *Oceanography,* 3, 30, 1990.

25. Sieracki, M. E. and Viles, C. L., Distributions and fluorochrome staining characteristics of bacteria and sub-micrometer particles in the North Atlantic ocean, *Deep-Sea Res.,* 39, 1919, 1992.

26. Viles, C. L. and Sieracki, M. E., Measurement of marine picoplankton size and biomass using a cooled, charge coupled device (CCD) camera with image analyzed fluorescence microscopy, *Appl. Environ. Microb.,* 58, 584, 1992.

27. Galbraith, W., Wagner, M. C. E., Chao, J., Abaza, M., Ernst, L. A., Nederlof, M. A., Hartsock, R. J., Taylor, D. L., and Waggoner, A. S., Imaging cytometry by multiparameter fluorescence, *Cytometry,* 12, 579, 1991.

28. Walter, R. J., Jr. and Berns, M. W., Digital image processing and analysis, in *Video Microscopy,* Inoue, S., Ed., Plenum Press, New York, 1986, chap. 10.

29. Jackman, P. H., Image analysis, in *Computers in Microbiology,* Bryant, T. N. and Wimpenny, J. W. T., Eds., Oxford University Press, New York, 1989, chap 2.

30. Sieracki, M. E., Reichenbach, S. E., and Webb, K. L., Evaluation of automated threshold selection methods for accurately sizing microscopic fluorescent cells by image analysis, *Appl. Environ. Microbiol.*, 55, 2762, 1989a.
31. Sieracki, M. E., Viles, C. L., and Webb, K. L., Algorithm to estimate cell biovolume using image analyzed microscopy, *Cytometry*, 10, 551, 1989b.
32. Verity, P. G., Robertson, C. Y., Tronzo, C. R., Andrews, M. G., Nelson, J. R., and Sieracki, M. E., Relationships between cell volume and the carbon and nitrogen content of marine photosynthetic nanoplankton, *Limnol. Oceanogr.*, 37, 1434, 1992.
33. Venrick, E., How many cells to count?, in *Phytoplankton Manual*, Sournia, A., Ed., UNESCO, Paris, 1978, sect. 7.1.2.
34. Inoue, S., *Video Microscopy*, Plenum Press, New York, 1986.
35. Boatman, E. S., Berns, M. W., Walter, R. J., and Foster, J. S., Today's microscopy, *BioScience*, 37, 384, 1987.
36. Taylor, D. L. and Wang, Y., Fluorescence microscopy of living cells in culture. Part B. Quantitative Fluorescence Microscopy — Imaging and Spectroscopy, in *Methods in Cell Biology*, Vol. 30, Academic Press, San Diego, 1989.

CHAPTER 39

Determination of Size and Morphology of Aquatic Bacteria by Automated Image Analysis

Roland Psenner

INTRODUCTION

Over the past decade, numerous studies have shown that bacteria are a major link for the flux of organic matter in freshwater and marine ecosystems.[1] Although bacterial biomass generally represents only a small amount of particulate organic carbon,[2] it is of paramount importance, especially with regard to its role in microbial food webs. In the euphotic zone of Sargasso Sea, for example, bacteria contain >70 and >80% of the microbial C and N, respectively, and >90% of the biological surface area.[3] Bacteria may, therefore, function as a major reservoir for particulate carbon, nitrogen, and phosphorus[4] which can be utilized by mixotrophic algae and Protozoa.

Although much effort has been spent in order to acquire more reliable data on bacterial growth rates and some progress has been made in this field,[5,6] the estimation of bacterial biomass is still a weak point in the study of microbial food webs.[7] This is very astonishing, since biomass seems to be a better indicator of the trophic state of marine and freshwater ecosystems than growth rates, as stated by Billen et al.[8] Cytoplasm synthesis is the driving force regulating the cell cycle of bacteria, and a shift up in substrate concentrations leads to an increase in the specific mass per cell, i.e., the mean cell volume.[9] Therefore, the argument which proposes a constant carbon value for bacteria regardless of their size should be put at rest because it has no biochemical or physiological basis, and also ecological studies suggest that cell size is a significant indicator of the nutrient status and activity of bacteria.[10,11]

What we really need now is a reliable method to measure bacterial biomass. The determination of size and some basic morphometric parameters has made one large step forward by applying image analysis techniques to epifluorescence microscopy[12,13] because quantification of size and shape of objects is performed much better by analytical systems than by the human eye. Pattern recognition, however, is still a problem requiring time and effort.[14-16]

Biovolume, shape factors such as elongation, and information on phototrophy of picoplankton can be obtained by using image analysis techniques in combination with the epifluorescence microscope. The analysis starts with a digital image of specifically stained or autofluorescent cells generated by a sensitive video camera and composed of approximately 300,000 pixels each containing, for example, 8-bit information (256 gray levels). Three

0-87371-564-0/93/$0.00 + $.50

gray-level images (red-green-blue) might be combined to a color image. Image analysis techniques are used to (1) improve the quality of the picture by enhancing differences between objects and background and (2) quantify number, size, and shape of the detected objects. Especially this part of the job is performed much better and quicker by computers than by human observers, whereas detection of objects is still a problem even for sophisticated image analysis systems.

After different steps of transformation[17] a binary image is generated, consisting of only two gray levels. On this image, which might be further processed, the number of objects and morphometric parameters such as length, width, perimeter, area, and roughness are measured and stored. This information is used to calculate volume, length:width ratio, and other values indicative of size and shape of the cells. Biomass calculations are performed by multiplying the measured biovolume ($\mu m^3 L^{-1}$) by empirical factors ranging between 0.12 and 0.35 g cm^{-3} or by applying special scale factors assuming a nonlinear relationship between cell volume and cell mass or protein content.

Among the various attempts to estimate bacterial biomass, the microscopic image analysis of specifically stained cells has many advantages over other methods. It is (1) sensitive: single bacterial cells corresponding to carbon amounts in the range of 10^{-15} g are detectable; (2) specific: bacteria are analyzed in the microscope, thus interference with other organisms or detritus particles is avoided; notwithstanding its better resolution, such crucial qualitative information (presence of DNA or chlorophyll) is not achieved by scanning electron microscopy which suffers also from shrinking artifacts and tedious preparation procedures; (3) informative: not only an average biomass value is registered, but numbers and individual morphometric parameters.

Principal problems of the method are (1) the resolution of the light microscope is in the range of the smallest bacteria, although fluorescent particles of less than 0.1-μm diameter are detectable; (2) until now, only a relative calibration with fluorescent microspheres is possible; (3) biomass conversion factors (carbon per volume) are still in dispute, differing by a factor of three; (4) systems are becoming cheaper and better, but the method is still expensive: approximately \$5,000 to \$10,000 per frame grabber and \$10,000 to \$20,000 per videocamera (epifluorescence microscope, monitor, and computer not included); (5) although done semiautomatically, it still takes approximately 30 min to size about 500 bacteria per sample (preparation time included); (6) sizing bacteria on detritus particles is difficult and time consuming.

MATERIALS REQUIRED

Equipment

- Epifluorescence microscope (high light intensity) with UV filter set. DAPI staining and UV epifluorescence (excitation/beam splitter/barrier = 365/395/397 nm) provides high light energy and slow fading. The microscope should be of good quality and high light intensity (short light paths): this allows the use of less-sensitive video cameras which generally have better resolution and lower price. Object detection must be done semiautomatically; in the presence of bacteria-sized detritus which normally gives a bright yellow fluorescence and of autofluorescent picoplankton, this can be achieved by using an interference filter (450- to 490-nm transmission) placed between UV filter set and video camera; in this case, however, the light energy is strongly reduced requiring more sensitive video cameras (see next item).
- Video camera with high sensitivity times resolution factor and shading correction. The video camera (Newvicon or Chalnicon tubes, Silicon Intensified Target tube, CCD camera with

image intensifier) must have an appropriate combination of sensitivity and resolution in order to work on magnifications of approximately 2000: one pixel should correspond to <0.1 μm; automatic shading correction is of great help in the further processing.

- Frame grabber (gray or red/green/blue) with software. Frame grabber and software should allow basic segmentation techniques, e.g., subtraction of images and pixel averaging. (Segmentation = all steps necessary to separate regions of interest from the background).
- PC (high speed) with a second monitor (color, high resolution) and digitizer tablet (optional, some programs may work with a mouse). Computers used should be fast because some algorithms are time consuming; an image screen (color, high resolution) is recommended.

Supplies

- Filtration apparatus
- Black or nonfluorescent membrane filters (Nuclepore, Poretics, Anotec). Membrane filters (Nuclepore, Poretics, Anotec) should be completely plane with low and uniform backscatter of incident light (black or gold sputtered); so all small objects are in focus and contrast with an inert background.
- Fluorescent microspheres for calibration in the size range of 0.2 to 1 μm (Polysciences)

Solutions

- DAPI
- Formalin

PROCEDURES

Preparation of Samples

1. Water samples are fixed with formalin (final concentration 2% v/v) and stored at 4°C in the dark. Chlorophyll autofluorescence will fade after several months, but heterotrophic bacteria can be sized also after 2 years of storage.

2. A suitable amount of sample, normally between 5 and 20 ml, is filtered onto black or gold-sputtered membrane filters of 0.2-μm pore size after staining with DAPI (4',6-diamidino-2-phenylindole; final concentration 1 μg ml^{-1}) which binds to DNA.[18] Bacteria and other plankton organisms as well as detritus should not be too dense making image analysis unreliable or even impossible; on the other hand, numbers should be high enough to allow the analysis of several tens or hundreds of bacteria per image captured in order to speed up operation times. Therefore, you cannot enlarge the original microscopic image by more than approximately 2000 to 2500 ×.

Image Capture

Sites for image capture are chosen by chance, avoiding, however, areas with detritus flocks or large plankton organisms because bacteria in such aggregates cannot be sized correctly. The first and main problem is inhomogeneity of the background which may derive from the filter material, the light source (xenon or mercury lamp), the video camera, or fluorescent matter (detritus, plankton) inside or outside the field of vision. Such changes in the background can be greater than the difference between the local background and very small bacteria. Several algorithms, e.g., tophat correction (LUCIA, Laboratory Imaging, Prague, CSFR), or combinations of treatments (dilation/erosion and background averaging) have been developed to smooth the background. A very simple though effective measure consists of storing the original image, then to defocus by approximately 0.5 μm in the direction of the filter surface and to store this second image. It is subtracted from the first one thereby removing background inhomogeneities almost completely. By this procedure, called background subtraction, the background becomes a uniform gray level and differences in brightness of the objects (depending on size and DNA content) are averaged out. As demonstrated for fluorescent microspheres differing by one order of magnitude in volume, the correct volume is represented much better if background subtraction is applied.[16]

Segmentation

After background subtraction, a constant threshold for all objects (defining the border value between object and background) can be set automatically, thus speeding up operation times. Other methods for automatic thresholding have been proposed by Sieracki et al.[14] and Schröder et al.[19] which, however, are not implemented in standard software.

Before setting a threshold, however, some other segmentation routines[17] may be used to improve contrasts, to define edges, to remove noise, to detach adhering cells, etc. I recommend

1. To be very cautious with all segmentation steps and to test their plausibility and applicability with a mixture of fluorescent spheres differing in volume by at least one order of magnitude. One has to keep in mind that removing or adding a shell of only 1 pixel to small objects, typically having areas of approximately 10 pixel, can change the calculated volume by a factor of 5 or so.

2. To compare the effects implied by segmentation with the original digital image and the microscopic image.

3. To keep in mind that also the most sophisticated algorithms are of little help if the original image is of low quality.

As a minimal treatment before thresholding, averaging may help to remove some noise without changing size and aspect ratio of small objects. Averaging might be done pixel selectively[20] or by a 3×3, 4×4, or 5×5 mask.

The binary (black and white) image resulting from thresholding can be further processed, for example, by eroding or dilating objects. Also on this stage, one must be very cautious with small objects. Interactive correction of the binary image by the use of a digitizer tablet or a mouse might be necessary for difficult images. This procedure, however, is time consuming and in conflict with the attempt to run image analysis automatically. In such cases, one should try to improve the quality of the image on previous stages beginning with sample preparation.

Measurement and Calculation of Size and Shape

The measurement is what quantitative image analysis is all about. It may include determination of the "roughness" of objects by consecutive erosions and measurements with masks of different size, shape, and loading factor. For bacteria having no branches or specific appendices, in most cases area, perimeter, x, and y (length and width of a rectangular box surrounding the object) are sufficient to calculate cell size. x and y might be measured at different angles but the real length and width can be approximated only for objects of relatively simple shape (cocci or rods). The volume of vibrios or spirillum-like forms must be inferred from two-dimensional data as mentioned above. All calculations are based on the assumption that bacteria are spheres or rods with rounded ends. The real length:width ratio of the object (F_{object}) can be infered from the aspect ratio of the Feret-box (F_{Feret}) and from the ratio of the maximum area an object surrounded by the Feret-box can assume (A_{Feret}) to the measured area of that object (A_{object}). It can be approximated by the formula

$$F_{object}/F_{Feret} = 0.1 * 10^{(A_{object}/A_{feret})}$$ (1)

In this way more realistic values for the real length and width of the objects can be calculated. Both parameters can be controlled by comparing the perimeter resulting from this calculation with the perimeter measured independently. One additional correction of the data should be considered, especially for very small cells: the perimeter consists of many small squares (pixels) which do not represent the real shape of bacteria; bacterial cells, therefore, look more rough than they are and a smoothing factor of 0.9 or so should be used.

Biomass Calculation

Finally we may boil down all data to a single number such as biovolume, mean cell volume, or mean elongation factor. This procedure may be acceptable when comparing large amounts of samples. It must be considered, however, that aquatic bacteria differ in volume by several orders of magnitude. This results in statistical problems: one large bacterium may have the same volume as 1000 small ones which implies that the number of cells measured must be large, depending on the variability of cell size. Measuring approximately 500 objects per sample could be a compromise between statistical requirements and time and effort spent per sample. Thus, data are better presented in the form of frequency diagrams or as tables with deciles, means, maxima, and minima.

The calculation of biomass is a weak point although it does not primarily involve image analysis, but rather chemical analysis of bacteria.[7] Carbon measurements of aquatic bacterial cells yield a ratio of 120 to 350 fg C μm^{-3}. If we assume a "mean" carbon content of 0.25 g cm^{-3} and a "mean" cell size of 0.1 μm^3 we obtain 20 or 25 fg C $cell^{-1}$ (see Reference 21) which was often considered as an ecological constant. Cells of 0.001 μm^3 reported in image analysis studies must be regarded as methodological artifacts or DNA-containing particles (e.g., virus) because they would be too small to contain any considerable amount of DNA. Many bacteria, however, are very small, around 0.2 μm in diameter, yielding a cell volume of approximately 0.004 μm^3 which in turn yields not more than 1 fg C $cell^{-1}$. These results can hardly be reconciled with a theoretical DNA content of approximately 2 fg $cell^{-1}$. This assumption is based on a genome size of 1 to 3 $\times$ 10^9 Da resulting in a DNA content from 1.7 to 5 fg $cell^{-1}$. Since there is no biochemical or physiological reason to assume a constant carbon content per cell (see Reference 9) I recommend using a conversion factor of 0.25 g C cm^{-3} or the formula given by Simon and Azam,[6] i.e., carbon = 90 $\times$ $volume^{0.9}$.

NOTES AND COMMENTS

Progress in video camera technology has shifted the standards: cooled CCD cameras available on the market (e.g., Hamamatsu Photonics, 812 Joko-cho, Hamamtsu City, 431-32 Japan, Fax: 81-534-35-1574) have a high dynamic range (10 to 14 bit), ultra-high sensitivity (through long exposure, i.e., for nonmoving objects: up to 10^{-8} lux $= 10^4$ photons cm^{-2} s^{-1}) in the spectrum of 200 to 1000 nm, low image distortion, and high resolution (1000 $\times$ 1000 pixel). These cameras require special frame grabbers and software as well as very fast computers and high-resolution monitors.

Recently, microbiologists started to count viruses with the epifluorescence microscope (EFM) with DAPI staining using sensitive video cameras,[22] but one can hardly say more about those particles other than "containing DNA".

Another development regards the use of color image analysis for samples of low-light intensity[23] which allows to distinguish between fluorescently labeled bacteria and autofluorescent algae within the food vacuole of ciliates. Color video cameras are not as sensitive as the most sensitive black and white cameras, but color image analysis can be done also by superimposing three gray images taken up by the same camera with different filter sets. Problems in this case can arise from minimal shifts of the objects during the change of filter sets but this problem can be overcome by isolating the filter holder from the microscope or by using acousto-optic tunable filters which can spectrally modulate light. Problems with larger objects such as ciliates with ingested prey (see Chapter 26) originate from different focus planes. In such cases, confocal laser scanning microscopes with three-dimensional reconstruction would be a good although very expensive solution.

Recent advances in the development of a digital confocal microscopy [VayTec Inc., 305 W. Lowe, Suite 109, Fairfield, IA 52556, Fax: (515)472-8131] have shown that the reconstruction of three-dimensional objects is possible by using the components as described above: epifluorescence microscope, video camera, and frame grabber. Additional equipment (approximately $20,000) consists of an array processor for quick deconvolutions of the digital image by using the images taken above and below the image being processed, and a stage controller (optional). Digital images can be sent to VayTek for processing.

Obviously much information on the shape of bacteria is lost during image processing. Storage of the original images is possible, although requiring much memory, but this information is not accessible to rapid or quantitative comparison of samples. Mathematical morphology might by applied to separate cells of different shape into distinct classes. Methods and algorithms for doing a straightforward digital taxonomy are implemented in the LUCIA system (Laboratory Imaging, Prague 10 — Strasnice, Pretlucka 41, CSFR, Tel/Fax: +42-2-7818025) and should be further developed in the future.

Still an open question is the transformation of cell volumes into carbon. Here more progress should be expected on the field of chemical analysis of single cells[24] or the chemical analysis of cells after selective fractionation by especially designed filters.

REFERENCES

1. Azam, F., Fenchel, T., Field, J. G., Gray, J. S., Meyer-Reil, L.-A., and Thingstad, F., The ecological role of water-column microbes in the sea, *Mar. Ecol. Prog. Ser.,* 10, 257, 1983.
2. Simon, M., Biomass and production of small and large free-living and attached bacteria in Lake Constance, *Limnol. Oceanogr.,* 32, 591, 1987.
3. Fuhrman, J. A., Sleeter, T. L., Carlson, C. A., and Proctor, L. M., Dominance of bacterial biomass in the Sargasso Sea and its ecological implications, *Mar. Ecol. Prog. Ser.,* 57, 207, 1989.
4. Güde, H., The role of grazing on bacteria in plankton succession, in *Plankton Ecology,* Sommer, U., Ed., Springer-Verlag, Berlin, 1990, 337.
5. Bell, T. R., An explanation for the variability in the conversion factor deriving bacterial cell production from incorporation of [³H]thymidine, *Limnol. Oceanogr.,* 35, 910, 1990.

6. Simon, M. and Azam, F., Protein content and protein synthesis rates of planktonic marine bacteria, *Mar. Ecol. Progr. Ser.,* 51, 201, 1989.

7. Psenner, R., From image analysis to chemical analysis of bacteria: a long-term study?, *Limnol. Oceanogr.,* 35, 234, 1990.

8. Billen, G., Servais, P., and Fontigny, A., Growth and mortality in bacterial population dynamics of aquatic environments, *Arch. Hydrobiol. Beih.,* 31, 173, 1988.

9. Cooper, S., *Bacterial Growth and Division,* Academic Press, New York, 1991.

10. Psenner, R. and Sommaruga, R., Control of Bacterial Biomass in Aquatic Systems: Grazing and Nutrient Supply, presentation at Fifth International Workshop on the Measurement of Microbial Activities in the Carbon Cycle in Aquatic Environments, Helsingør, Denmark, August 18–23, 1991.

11. Servais, P., Bacterial Production Measured by [³H]Thymidine and [³H]Leucine Incorporation in Various Aquatic Ecosystems, presentation at Fifth International Workshop on the Measurement of Microbial Activities in the Carbon Cycle in Aquatic Environments, Helsingør, Denmark, August 18–23, 1991.

12. Bjørnsen, P. K., Automatic determination of bacterioplankton by image analysis, *Appl. Environ. Microbiol.,* 51, 1199, 1986.

13. Sieracki, M. E., Johnson, P. W., and Sieburth, J. McN., Detection, enumeration, and sizing of planktonic bacteria by image-analyzed epifluorescence microscopy, *Appl. Environ. Microbiol.,* 49, 799, 1985.

14. Sieracki, M. E., Reichenbach, E., and Webb, K. L., Evaluation of automated threshold selection methods for accurately sizing microscopic fluorescent cells by image analysis, *Appl. Environ. Microbiol.,* 49, 2762, 1989.

15. Krambeck, C., Krambeck, H. J., Schröder, D., and Newell, S. Y., Sizing bacterioplankton: a juxtaposition of bias due to shrinking, halos, subjectivity in image interpretation and asymmetric distributions, *Binary,* 2, 5, 1990.

16. Psenner, R., Detection and sizing of aquatic bacteria by means of epifluorescence microscopy and image analysis, *Microsc. Anal.,* 26, 13, 1991 (special issue entitled *Image Enhancement and Analysis*).

17. Loebel, J., *Image Analysis. Principle and Practices.* Short Run Press, Exeter, England, 1985.

18. Porter, K. G. and Feig, Y. S., The use of DAPI for identifying and counting aquatic microflora, *Limnol. Oceanogr.,* 25, 943, 1980.

19. Schröder, D., Krambeck, C., and Krambeck, H. J., Einflüsse digitaler Bildsegmentierungstechniken auf die quantitative Analyse fluoreszierenden mikrobiellen Planktons, in *Informatik für den Umweltschutz,* Pillmann, W. and Jaeschke, A., Eds., Springer-Verlag, Berlin, 1990, 827.

20. Nagao, M. and Matsuyama, T., Edge preserving smoothing, *CGIP,* 9, 394, 1979.

21. Azam, F. and Cho, B. C., Bacterial utilization of organic matter in the sea, in *Ecology of Microbial Communities,* Fletcher, M., Gray, T. R. G., and Jones, J. G., Eds., Cambridge University Press, Cambridge, 1987, 262.

22. Fuhrman, J. A., The Role of Viruses in Aquatic Environments, Fifth International Workshop on the Measurement of Microbial Activities in the Carbon Cycle in Aquatic Environments, Helsingør, Denmark, August 18–23, 1991.

23. Sieracki, M. E. and Webb, K. L., The application of image analysed fluorescence microscopy for characterising planktonic bacteria and protists, in *Protozoa and their Role in Marine Processes,* Reid, P. C., Turley, C. M., and Burkill, P. H., (NATO ASI Series Vol. 25), Springer-Verlag, Berlin, 1991, 77.

24. Norland, S., Heldal, M., and Tumyr, O., On the relation between dry matter and volume of bacteria, *Microb. Ecol.,* 13, 95, 1987.

CHAPTER 40

Analysis of Microbial Lipids to Determine Biomass and Detect the Response of Sedimentary Microorganisms to Disturbance

Fred C. Dobbs and Robert H. Findlay

INTRODUCTION

While ecologists uniformly agree that disturbance is a factor important in structuring natural communities, they likely would not agree on its definition. We begin this chapter, therefore, by defining disturbance as any process or punctuated event that removes or reduces microbial biomass.[1] In the context of sedimentary ecosystems, a "disturbance" may be one of many phenomena that translocate sediment or alter chemical gradients.[2] Thus, disturbance can result from biotic and abiotic environmental factors. Here we review our methodology, biochemical in character and with specific focus on lipids, for detecting disturbance in sedimentary microbial communities. We have used these techniques to detect disturbance from predation,[3] bioturbation,[3] and physical disruption of sediments.[2,4] Although these techniques have been used almost exclusively in marine sediments, they are transferable, with some modifications, to other habitats.

Our quantitative indicators of disturbance are two. The first involves tracking the dynamics of microbial biomass following the disturbance. The magnitude of the disturbance is reflected in the loss of biomass and its pattern of recovery over time. We refer not only to temporal change in total biomass, but also to change in biomass-specific growth rates. In this chapter, however, we shall restrict ourselves to determinations of biomass and leave to the reader the task of assimilating the controversial subject of growth-rate determinations.

Our second indicator of disturbance reflects the change in growth state of the microbial community as the cells respond to the disturbance. With the caveat that we refer to the community and not to specific populations, sedimentary microbes begin to grow or increase their rate of growth following disturbance; this growth is often at the expense of their lipid stores.[2,3] In so doing, they increase the ratio of carbon incorporated into cell membrane to that incorporated into cell-storage product. This ratio may be quantified through use of ^{14}C-labeled carbon compounds.

The dominant component of sedimentary microbial communities is bacteria. Their endogenous storage lipid, poly-β-hydroxyalkanoate (PHA), accumulates under conditions of unbalanced growth when cell division does not occur, whereas phospholipid ester-linked fatty acids (PLFA) accumulate as the cellular membrane expands during balanced growth.[5-7] The ratio of incorporation of ^{14}C into PHA to that in PLFA has been shown to be a sensitive measure of the metabolic status of bacteria *in situ*[8] and has been used to detect disturbance in sedimentary microbial communities[2-4] and biofilms.[9] In a functionally similar manner, neutral lipid (e.g., triglyceride) is the storage lipid of eukaryotic organisms. Thus, the ratio in which carbon is incorporated into neutral lipid and PLFA indicates the metabolic status of eukaryotes.[10]

Advantages of the methods are (1) total microbial biomass and growth state of the microbial community can be assessed from the same sample; in fact, our protocol assumes that such is the case; (2) results integrate across the entire microbial community without the truncations of cultural studies; (3) there are none of the difficulties inherent in enumerative studies (e.g., dislodging microbes from sediment, subsampling, visualization); (4) these techniques can be used in sediments not amenable to analysis by microscopy (e.g., coarse sands); (5) these biochemical methods are time and cost competitive and allow for experimental designs involving scores of analyses; (6) the techniques have high precision and as few as three to five replicate samples per treatment may be sufficient; (7) if desired, further biochemical characterizations may be made using the samples extracted in these protocols.

Disadvantages of the methods are (1) results integrate across the entire microbial community; there is no discrimination at the level of individual cells or populations of cells (NOTE: there can be discrimination at the level of populations if PLFA analysis is extended to identification and quantification of individual fatty acids — see Chapter 32); (2) these methods are relatively new and comparable data are correspondingly rare; (3) although factors for converting phospholipid data to carbon are established, there remain some uncertainties in converting phospholipid data to cell numbers or biovolume; (4) there is a need for specialty equipment, such as solvent-handling glassware and a hood to protect laboratory personnel from exposure to solvent fumes.

MATERIALS REQUIRED

Sample Collection

Supplies

- Sediment corers (with silicone-covered ports if horizontal injections are to be made)
- Microliter syringes
- Glass tubes (25 × 150 mm) with PTFE-lined screw caps

Solutions and Reagents

- Solvents (methanol, chloroform; pesticide-residue grade or higher quality)
- ^{14}C-acetate (or another lipid precursor)
- Chloroform:methanol:buffer (1:2:0.8)[11]
- 50 mM phosphate buffer, pH 7.4 (8.7 g K_2HPO_4 l^{-1} adjusted with 1 N HCl)[12]

Laboratory Procedures

Equipment

- Nitrogen evaporator
- Hood
- Liquid scintillation counter

Supplies

- Nitrogen gas
- Disposable pipets, Pasteur
- Disposable pipets, large-bore
- 10-ml Griffin beakers
- Graduated cylinders
- Glass wool
- Silicic acid (e.g., Bio-Sil A, 100 to 200 mesh, Bio-Rad) or commercial prepacked columns (e.g., SPE, Baker)
- Glass tubes (16 × 125 mm) with PTFE-lined screw caps
- Ring stand, cross-bar, and clamps
- 2-ml glass ampules
- Whatman chromatography paper, grade no. 3-mm chromatography, 2 cm × 2 cm squares
- Hair dryer
- Scintillation vials

Solutions and Reagents

- Solvents (chloroform, methanol, acetone, diethyl ether, absolute alcohol; pesticide-residue grade or higher quality)
- Phosphate buffer (recipe above)
- Scintillation fluor
- Lipid standards, reagent (e.g., glycerol phosphate) (Sigma)
- Polyvinyl alcohol, 100% hydrolyzed (Janssan Chemica)
- Saturated potassium persulfate (5 g in 100 ml of 0.36 N H_2SO_4; solution is unstable in light; store in refrigerator; use as long as crystals are visible)
- Ammonium molybdate (2.5% $(NH_4)_6Mo_7O_{24}.4H_2O$ in 5.72 N H_2SO_4)
- Malachite green (Sigma) in polyvinyl alcohol (dissolve 0.111% polyvinyl alcohol in water at 80°C, cool to room temperature, then add 0.011% malachite green)
- NaOH, 0.02 N

PROCEDURES

Sample Collection

1. Collect sediment cores.

2. Inject the cores with ^{14}C-acetate.

3. Incubate the injected cores at appropriate temperature and light levels.

4. Terminate incubations by transferring sediment to the chloroform:methanol:buffer solution and vortex or shake thoroughly. Use sufficient chloroform such that there is at least seven times the weight of chloroform relative to the wet weight of the sediment. A single-phase system should result; if two phases are present, add methanol in 1-ml increments, mixing thoroughly after each addition, until a single-phase system is obtained.

5. Prepare "extraction" controls by adding the radiolabeled lipid precursor to unamended sediment already in the extraction solution.

6. Extract in the dark for 24 h, with occasional agitation.

Laboratory Procedures

Extraction of Total Lipid

1. After extracting the sediment for 24 h in chloroform:methanol:buffer, add more buffer and chloroform to break phase and partition the lipid-containing chloroform from the other solvents. Add volumes of chloroform and buffer equivalent to the volume of chloroform in the first step of the extraction. The final ratio of chloroform:methanol:buffer should be 1:1:0.9.

2. Add 0.7 ml of 0.02 N NaOH (final concentration 0.001 M) to the aqueous phase (do not shake) to reduce the concentration of ^{14}C-acetate (as free acid) in the lipid phase.

3. Allow the extraction mixture to separate overnight into lipid (lower) and aqueous (upper) phases. Separation may be enhanced using centrifugation, which may in fact be necessary in fine-grained sediments. However, centrifuge only after allowing the extraction mixture to stand overnight; failure to wait increases the likelihood of contamination of the lipid by non-phospholipid phosphate.[17]

4. Pipet away most of the aqueous fraction, being careful not to disturb the interface between the two phases.

5. Transfer a subsample of the organic phase (e.g., 10 ml of the 15-ml total) to a glass tube (16 × 125 mm), and evaporate the chloroform in a nitrogen evaporator having a bath temperature of 37°C. The nonvolatile material is considered to be the total extractible lipid. Keep the lipid in a nitrogen atmosphere and frozen (-70°C if possible) until it is to be fractionated. Some researchers consider that storage in a small volume of chloroform (e.g., 1 ml) further reduces oxidation artifacts.

6. Dry the extracted sediment and determine its weight. It is convenient to know the relationship of sediment dry weight to sediment wet volume; the weight of a volumetric sample can then easily be calculated from a regression.

Fractionation of Total Lipid

1. If commercial silicic-acid columns are used, flush them with methanol to dehydrate the column, then with chloroform in preparation for loading the lipid. Skip to step #7.

2. Dehydrate and activate the silicic acid by heating in an oven (120°C) for at least 2 h.

3. Weigh out into a 10-ml beaker an amount of silicic acid appropriate to the amount of lipid to be fractionated and the pipet in which the procedure will take place; see notes below.

4. Rinse the insides of the pipets first with methanol, then chloroform. Place 100-ml beakers underneath the pipets to collect these and other washes.

5. Place a small amount of glass wool into the large opening of the pipet and pack it into the taper using a pipet. Wash the pipet and glass wool with methanol, then chloroform.

6. Add about 3 ml of chloroform to the beaker containing the silicic acid. Remove all air bubbles by stirring with a pipet. Then quickly pour the mixture into the large opening of the pipet. Repeat as necessary, using smaller volumes of chloroform, to transfer all of the silicic acid to the pipet. Dislodge any air pockets and improve the column packing by tapping the sides of the pipet. If the top-most part of the silicic acid is not level, make it so by tapping the sides of the pipet.

7. Add very small amounts of chloroform (about 0.1 ml) to the extracted, dried lipid to dissolve it. After each addition of chloroform, transfer the lipid to the silicic-acid column. Be careful not to overload the column with too much lipid (see note #1 below) or to add so much chloroform that the neutral lipid begins to elute.

8. Elute three lipid fractions using 10 ml of each of a series of increasingly polar solvents. Elute the neutral-lipid fraction with chloroform, the glycolipid fraction with acetone, and the phospholipid fraction with methanol. The front between chloroform and acetone is distinctly visible; that between acetone and methanol is scarcely discernible and requires experience to recognize. Collect fractions to be analyzed in glass tubes (16 × 125 mm) having PTFE-lined screw caps.

9. From each fraction, evaporate the solvent in a nitrogen evaporator having a bath temperature of 37°C.

Determination of Biomass

1. Use material from the total lipid extract or from the phospholipid fraction (See Notes and Comments, Laboratory Procedures, Extraction of Total Lipid).

2. Dilute the lipid samples in chloroform (2 to 3 ml) such that a 100-μl subsample contains between 1 and 20 nmol of phosphate.

3. Remove two 100-μl subsamples, place each in a separate 2-ml glass ampule, and remove the solvent under a stream of nitrogen.

4. Add 0.5 ml of a saturated solution of potassium persulfate, seal the ampules using a flame, and incubate overnight or longer at 95°C.

5. Remove ampules from oven and let them cool. Break open the vials. Begin the assay for inorganic phosphorus released by digestion by adding 0.1 ml of a solution of ammonium molybdate to each vial. Allow to stand for 10 min at room temperature.

6. Add 0.45 ml of the solution containing malachite green and polyvinyl alcohol and allow to stand at room temperature. Color development should be complete in 20 min and is stable for 24 h.[23]

7. Read the absorbance at 610 nm. Set zero absorbance using deionized water.

8. Calculate the concentrations of phosphate using the regression line from a standard curve prepared by digesting glycerol phosphate.

9. Convert phosphorus concentrations to microbial biomass values and normalize for sediment dry weight or carbon content. Compare the results with either contemporaneous or historical control values.

Purification of PHA

1. Use material from the total-lipid extract or from the glycolipid fraction.

2. Transfer the lipid to a square piece of Whatman 3-mm chromatography paper. Slowly pipet the lipid solution onto the paper while removing the solvent in a stream of warm air from a hair dryer.

3. Fix the PHA to the paper by heating in an 80°C oven for 20 minutes.

4. Remove lipids other than PHA by washing the paper in absolute ethanol and diethyl ether.

Radioassays and Calculations

1. Quantitatively transfer to scintillation vials samples of the lipid fractions to be radioassayed. In the case of PHA, place the filter paper in the scintillation vial.

2. If any color is present, e.g., from photosynthetic pigments, bleach the lipid fraction with sunlight or a bright light. An alternative is to bleach with 1.0 N benzoylperoxide in chloroform at 60°C for 2 h.[8]

3. Evaporate the solvent, add fluor, and radioassay.

4. Normalize the results of the radioassay (dpm's) for dry weight or carbon content of the extracted sediment.

5. Calculate the ratio of radioactivity in the cell-membrane fraction (i.e., phospholipid) to that incorporated into the cell-storage products (i.e., PHA or neutral lipid). Follow the ratio over time if the experiment is so designed. Compare the results with either contemporaneous or historical control values.

NOTES AND COMMENTS

Sample Collection

1. In taking samples to assess disturbance, it is ironic that the investigator causes a disturbance! It is possible, however, to minimize artifacts, e.g., by carefully collecting hand-held cores. Of critical importance is the introduction of the radiolabeled lipid precursor. Injection techniques disturb sedimentary microbes to a lesser degree than do slurries. We strongly feel that core injections are preferable, for reasons presented elsewhere.[4,8,13] Regardless of the method of introducing the lipid precursor into the sediments, extraction and fractionation procedures remain the same.

2. Depending upon the question being addressed, inject cores vertically or horizontally through small, silicone-covered ports in the plastic corer. Introduce acetate at a concentration no greater than 5 to 10% of the porewater's concentration. Typical values range from 10 μM for sandy beaches in northern Florida and 11 μM for mudflats in Maine to 100 μM for deep-sea sediments.[14]

3. Timing is everything when attempting to quantify the effects of disturbance. The ratio of ^{14}C-incorporation into PLFA vs. PHA can change within minutes, possibly seconds, of the disturbance[15] and remain elevated for hours.[2,8] Biochemically determined measures of microbial biomass may or may not decrease instantaneously. Predation, a disturbance that eliminates organisms, may result in an immediate decrease.[16] Other disturbances may kill or injure organisms but not remove them, and the response may not be detectable until several hours later.[2] Biomass can return to predisturbance levels within hours to days,[2,3] but sometimes there is a secondary, less severe, loss of biomass 2 to 5 d post-disturbance.[2,17] Given the complexity of the temporal response, we suggest that investigators consider time, at the levels of hours and days, as components of their experimental designs.

4. The optimal time for incubation must be determined empirically and will vary with the environment and the question being asked. We have used incubations as short as 10 min[15] and as long as 5 h.[4] Short incubations are more appropriate to investigations of prokaryotic processes, longer ones when incorporation of carbon into neutral lipids of eukaryotes is important. Keep in mind, however, that the longer the incubation, the greater the chance of recycling the radioisotope.

5. Glassware to be used in extraction and fractionation procedures should at minimum be washed with phosphate-free detergent, rinsed with tap water (5×), then deionized water (10×), dried, and rinsed with solvent just prior to use. It is preferable to bake all glassware in a furnace or clean with acid to remove organic material completely prior to washing with solvent. (Although the furnace and acid cleaning is not required in a strict sense for radioisotopic work, it is nonetheless a good general technique.)

6. It is convenient and efficient[18] to perform the extractions in glass tubes (25 × 150 mm) having PTFE liners in their screw-on caps. These tubes have sufficient volume for a sediment sample of about 1 to 2 ml and the first step of the extraction begins with 7.5 ml of chloroform, 15 ml of methanol, and 6 ml of buffer. The rest of this protocol assumes such an extraction. If larger sediment samples are to be extracted, separatory funnels are more appropriate; a direct comparison of these extraction methodologies is found in Reference 18.

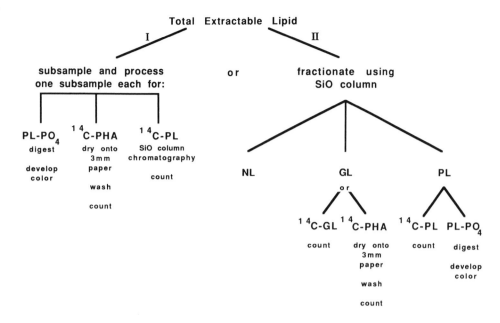

Figure 1. Analytical options for purification and analysis of microbial lipids to determine biomass and detect response to disturbance. Option I has fewer steps and is faster, but does not yield the better-defined, higher-purity lipids obtained with option II. NL = neutral lipid; GL = glycolipid; PL = phospholipid; PO$_4$ = phosphate; PHA = poly-β-hydroxyalkanoate; SiO = silicic acid; a prefix of ^{14}C indicates the incorporation of radioactivity from ^{14}C-acetate into a lipid class. See text for additional details.

7. Sediments having a high water content, such as the centimeter immediately below the sediment-water interface, are problematic. The amount of methanol necessary to create a single-phase system greatly changes the chloroform:methanol ratio of 1:2. Alternatives include reducing the amount of the buffer in the first step of the extraction or lyophilizing the sediment prior to its extraction.
8. As an alternative to immediate extraction in the field, the sample may be quickly frozen using liquid nitrogen, then maintained at dry-ice temperatures (approximately −70°C) or colder until extraction. It is important to chill samples quickly,[19] to keep them cold,[20] and to place them into the extraction mixture while still frozen.[17,21]

Laboratory Procedures

Extraction of Total Lipid

Assays can be performed on either the total-lipid extract or an appropriate lipid fraction (Figure 1). For example, values for phospholipid phosphate may be determined using the total-lipid extract or the phospholipid fraction. The simplest analytical scheme is to partition the total lipid into three parts and use one part each to measure phospholipid phosphate, incorporation of ^{14}C into PHA, and incorporation of ^{14}C into phospholipid. Alternatively, an analytical scheme yielding products of higher purity starts with the fractionation of the total lipid into neutral-lipid, glycolipid, and phospholipid fractions. Using subsamples from the phospholipid fraction, separately analyze for phospholipid phosphate and incorporation of ^{14}C into phospholipid. Further purify the glycolipid fraction, then estimate incorporation of ^{14}C into PHA. The choice of analytical scheme depends upon a balance of cost, convenience, complexity of the sample matrix, sources of possible contamination, and is particular to an investigator's needs. The procedures detailed above represent a synthesis of methods currently used in our laboratories.

Fractionation of Total Lipid

1. As rules of thumb, prepare columns with the approximate ratio of 50:1 stationary phase:lipid (dry weight). The ratio of the stationary phase's bed height to its cross-sectional area should be about 5:1[22] (an odd ratio in that it has a unit of centimeters), although we consider that minor deviations will have little effect; 1 g of silicic acid should be used for no more than 30 μmol of phosphate.[22]

2. We use 1 g of silicic acid in wide-bore disposable pipets, an arrangement that yields a ratio of bed height to cross-sectional area of about 9:1. We clamp the pipets, small end pointing down, to a bar held horizontal by ring stands.

3. Chloroform traditionally has been stabilized using 0.75% ethanol. Recently, many high-purity grades of chloroform have been stabilized using other compounds, notably pentene. For the chromatographic procedures described here, the change in polarity caused by the change in stabilization compound can be profound. If ethanol is not used for stabilization, it must be added to the chloroform for the procedures described above to yield quantitative results.

4. Warming of solvents (approximately 40°C) aids in quantitative transfer of PHA, a complex substance of high but variable molecular weight. Warming also helps in transferring samples having large amounts of total lipid.

Determination of Biomass

1. This protocol is the one detailed by Findlay et al.[18] Briefly, the phosphorus-containing, cell-membrane lipids are oxidized and the phosphate is released. An inorganic-phosphate determination follows, and phosphate concentrations are related back to phospholipid phosphate, then to biomass. Lower limits of detection are about 0.1 nmol phosphate. A modification of this technique has been described.[24]

2. Using material from the phospholipid fraction decreases the possibility of contamination from non-phospholipid, phosphorus-containing components that may be extracted from sediments.[17,25,26]

3. See Table 1 for a list of representative phospholipid-phosphate values from sedimentary environments.

4. Conversion of phosphate measurements to number of cells is problematic at best. Reported values for bacteria range from 3.4×10^7 to 2×10^9 cells nmol^{-1} of phosphate, e.g., References 18, 24, and references therein. The difficulty lies in applying to natural systems a conversion factor (usually) generated from monocultures in the laboratory. More complications arise if eukaryotes account for a significant portion of the microbial biomass; see note #7. Unless a factor is specifically determined for a site, e.g., by regressing direct counts on phosphate concentrations, we care little for cell counts inferred from conversion factors.

5. See Table 2 for a compilation of factors used to convert phospholipid-phosphate values to carbon biomass. These factors have been presented in several formats: micromoles phospholipid phosphate per gram bacterial carbon,[24] micromoles phospholipid phosphate per gram bacterial biomass,[27] and micrograms C per 100 nanomoles phospholipid phosphate.[18] In Table 2, we have converted all values to the first format. A recent review asserts that these factors generally are insensitive to cellular growth state and are relatively constant as a function of substrate concentration.[24] We suggest, however, establishing a conversion factor specific for the study area.

6. Phosphate data have been normalized in two ways: per gram dry weight of sediment[2-4] or per gram carbon content of the sample.[24] The first method is relatively insensitive and

Table 1. Representative Values of Phospholipid Phosphate from Various Sedimentary Environments

Location	Value	Ref.
Marine intertidal		
Florida sand	15–36	2–4,[a] 16, 31
Maine mud 1–2 cm	125*	18[b]
Shallow subtidal		
Florida		
0–1 cm	44	12
12–13 cm	27	12
Maine		
Pristine	26–167*	17
Organically enriched	159–464*	17
Deep sea		
HEBBLE site		
0–1 cm	16	32
9–10 cm	4	32
Clipperton-Clarion fracture zone		
0–0.5 cm	7–8*	33
9–10 cm	1–2*	33
Patton escarpment (area of phytodetritus fall)		
0–0.5 cm	22–32*	33
9–10 cm	1–5*	33
Freshwater streams	48–217	34[c]

Note: Units are nanomoles of phosphate per gram dry weight of sediment. Data are from surface sediment, unless indicated otherwise as depth (in centimeters) below the sediment-water interface. Concentrations marked with an asterisk were obtained using the method of Findlay et al.;[18] all others were determined using the method of White et al.[12]

[a] Erroneously printed as μmoles g^{-1} in References 2 and 3.
[b] Converted from value for sediment wet weight in their Table 4.
[c] Back calculated from their Table 3.

Table 2. Representative Values for Converting Phospholipid-Phosphate Values to Carbon Biomass for a Variety of Microbial Organisms

Organism	μmol P g^{-1} C	Ref.
Bacteria		
Pure cultures		
Aerobic	63–168	24
Anaerobic	118–253	24
Various	100	27
Enrichment cultures		
Aerobic	520	18
Aerobic	215, 190[a]	24
Anaerobic	100, 100[a]	24
Cyanobacteria	14	24
Yeasts	29–39[b]	35
Diatoms	53–109	24
	30–46[b]	36
Chlorophyta	54–116[b]	37

Note: Values are expressed as micromoles of phospholipid phosphate per gram carbon.

[a] The first value is from their cultures; the second value represents their integration of their results with the literature.
[b] Calculated assuming dry cell mass is 50% carbon.

does not allow rigorous intersite comparisons, as volume-specific surface area is more relevant to bacterial colonization than is volume-specific mass. The second method may well be the better metric, although difficult to apply to environmental samples; the small-scale, spatially heterogeneous distribution of carbon in sediments will require many carbon analyses to determine a useful mean value.

7. We emphasize that the phospholipid-phosphate technique, in a manner similar to ATP determinations, yields a total microbial biomass. This characteristic of the assay poses two potential difficulties. The first is that factors for converting phospholipid-phosphate values to cell numbers or carbon biomass are not easily applied when microorganisms other than bacteria are significant components of the microbial community. For example, in surface horizons from an intertidal area, eukaryotic microalgae can be abundant. Since the phosphate:carbon ratio in most eukaryotes is lower than in most bacteria (Table 2), there exists the likelihood of underestimating carbon biomass in such environments. The second potential difficulty is the contribution of nonmicrobial components to the putative microbial biomass. For example, Yingst[28] reported that meiofauna at times accounted totally for sedimentary ATP concentrations. While Yingst's results are at variance with current paradigms (e.g., data in Reference 29), very few investigators have emulated her painstaking methods.

Purification of PHA

1. Using material from the glycolipid fraction decreases the possibility of contamination from non-PHA, non-glycolipid components into which acetate may incorporate.[30,31]

2. It has been shown of one system that the incorporation of ^{14}C from acetate into glycolipid is equivalent to incorporation into PHA. We suggest that such be proved, not assumed, for the area to be studied. If such is the case, PHA need not be purified and the glycolipid fraction can be radioassayed.

3. Warming of solvents (approximately 40°C) aids in quantitative transfer of PHA.[30]

Radioassays and Calculations

1. There are few comparative data from this technique and we list them all here. In a laboratory study, Findlay et al.[2] created a disturbance and measured over time the ratio of radioactivity in phospholipid ("PLFA" in their paper) to that in PHA. They observed an immediate increase from a control value of 0.37 ± 0.04 to 0.61 ± 0.10 after 2 h, then a rapid decline to control levels after 24 h. Findlay now considers the slow decline to control levels after 24 h to be a "container" effect.[17] In a field study of biological disturbances, Findlay et al.[3] found a similar trend, i.e., an immediate increase in the ratio of ^{14}C-activity followed by a decrease. Values for control samples ranged from 0.36 to 0.42, those for sediments immediately after disturbance from 0.63 to 0.75, and after 4 h from 0.44 to 0.49 (from their Table 1; standard deviations all ≤0.10). Finally, Dobbs et al.[4] compared three methods of administering ^{14}C-acetate to intertidal sediments and determined the following ratios of radioactivity in phospholipid to that in glycolipid: injection, 0.70 ± 0.04 (mean $\pm$ standard deviation); slurry, 0.81 ± 0.02; porewater replacement, 0.81 ± 0.01. From these results, they concluded that the injection method disturbed the sediments to a lesser extent than the other two methods. Findlay et al.[2,3] used 10-min incubations, Dobbs et al.[4] 5-h incubations.

2. We strongly recommend that all workers applying the techniques outlined here first study the pertinent literature, then proceed with a logical course of experiments to test recovery, accuracy, and precision. Only when a worker is confident that s/he can routinely reproduce results should a technique be applied in an ecological context.

ACKNOWLEDGMENTS

We thank P. Kemp and K. Carman for comments on earlier drafts of this chapter. Support for this work was provided in part by NSF grants OCE 87-00358 (Findlay), OCE 90-18599 (Dobbs), and EPA grant R-817196-01-0 (Findlay). This publication is The University of Maine's Darling Marine Center Contribution Number 247 and The University of Hawaii's School of Ocean and Earth Science and Technology Contribution Number 3093.

REFERENCES

1. Grime, J. P., Evidence for the existence of three primary strategies in plants and its relevance to ecological and evolutionary theory, *Am. Nat.,* 111, 1169, 1977.
2. Findlay, R. H., Trexler, M. B., Guckert, J. B., and White, D. C., Laboratory study of disturbance in marine sediments: response of a microbial community, *Mar. Ecol. Prog. Ser.,* 62, 121, 1990.
3. Findlay, R. H., Trexler, M. B., and White, D. C., Response of a benthic microbial community to biotic disturbance, *Mar. Ecol. Prog. Ser.,* 62, 135, 1990.
4. Dobbs, F. C., Guckert, J. B., and Carman, K. R., Comparison of three techniques for administering radiolabeled substrates to sediments for trophic studies: incorporation by microbes, *Microb. Ecol.,* 17, 237, 1989
5. Herron, J. S., King, J. D., and White, D. C., Recovery of poly-β-hydroxybutyrate from estuarine microflora, *Appl. Environ. Microbiol.,* 35, 251, 1978.
6. Nickels, J. S., King, J. D., and White, D. C., Poly-β-hydroxybutyrate accumulation as a measure of unbalanced growth of estuarine detrital microbiota, *Appl. Environ. Microbiol.,* 37, 459, 1979.
7. Dawes, E. A. and Senior, P. J., The role and regulation of energy reserve polymers in microorganisms, *Adv. Microb. Physiol.,* 10, 135, 1973.
8. Findlay, R. H., Pollard, P. C., Moriarty, D. J. W., and White, D. C., Quantitative determination of microbial activity and community nutritional status in estuarine sediments: evidence for a disturbance artifact, *Can. J. Microbiol.,* 31, 493, 1985.
9. White, D. C. and Findlay, R. H., Measurement of predation effects on the biomass, community structure, nutritional status, and metabolic activity of microbial biofilms, *Hydrobiologica,* 159, 119, 1988.
10. Gehron, M. J. and White, D. C., Quantitative determination of the nutritional status of detrital microbiota and the grazing fauna by triglyceride glycerol analysis, *J. Exp. Mar. Biol. Ecol.,* 64, 145, 1982.
11. Bligh, E. G. and Dyer, W. J., A rapid method of total lipid extraction and purification, *Can. J. Biochem. Physiol.,* 31, 911, 1959.
12. White, D. C., Davis, W. M., Nickels, J. S., King, J. D., and Bobbie, R. J., Determination of the sedimentary microbial biomass by extractable lipid phosphate, *Oecologia,* 40, 51, 1979.
13. Carman, K. R., Radioactive labeling of a natural assemblage of marine sedimentary bacteria and microalgae for trophic studies: an autoradiographic study, *Microb. Ecol.,* 19, 279, 1990.
14. King, G. M., Measurement of acetate concentration in marine pore waters using an enzymatic approach, *Appl. Environ. Microbiol.,* 57, 3476, 1991.
15. Findlay, R. H. and White, D. C., *In situ* determination of metabolic activity in aquatic environments, *Microbiol. Sci.,* 1, 90, 1984.
16. Dobbs, F. C. and Guckert, J. B., Microbial food resources of the macrofaunal deposit feeder *Ptychodera bahamensis* (Hemichordata: Enteropneusta), *Mar. Ecol. Prog. Ser.,* 45, 127, 1988.
17. Findlay, R. H., unpublished data.
18. Findlay, R. H., King, G. M., and Watling, L., Efficacy of phospholipid analysis in determining microbial biomass in sediments, *Appl. Environ. Microbiol.,* 55, 2888, 1989.
19. Dobbs, F. C. and LaRock, P. A., The freezing of marine sediment for subsequent extraction of adenosine triphosphate, *J. Microbiol. Methods,* 10, 113, 1989.

20. Sasaki, G. C. and Capuzzo, J. M., Degradation of *Artemia* lipids under storage, *Comp. Biochem. Physiol.*, 78B, 525, 1984.

21. Federle, T. W. and White, D. C., Preservation of estuarine sediments for lipid analysis of biomass and community structure of microbiota, *Appl. Environ. Microbiol.*, 44, 1166, 1982.

22. Kates, M., *Techniques of Lipidology*, 2nd revised ed., Elsevier, New York, 1986, chap. 5.

23. Van Veldhoven, P. P. and Mannaerts, G. P., Inorganic and organic phosphate measurements in the nanomolar range, *Anal. Biochem.*, 161, 45, 1987.

24. Brinch-Iversen, J. and King, G. M., Effects of substrate concentration, growth state, and oxygen availability on relationships among bacterial carbon, nitrogen and phospholipid phosphorus content, *FEMS Microbiol. Ecol.*, 74, 345, 1990.

25. Harvey, H. R., Richardson, M. D., and Patton, J. S., Lipid composition and vertical distribution of bacteria in aerobic sediments of the Venezuela Basin, *Deep-Sea Res.*, 31, 403, 1984.

26. Guckert, J. B., Personal communication.

27. White, D. C., Analysis of microorganisms in terms of quantity and activity in natural environments, *Soc. Gen. Microbiol. Symp.*, 34, 37, 1983.

28. Yingst, J. Y., Patterns of micro- and meiofaunal abundance in marine sediments, measured with the adenosine triphosphate assay, *Mar. Biol.*, 47, 41, 1978.

29. Rowe, G., Sibuet, M., Deming, J., Khripounoff, A., Tietjen, J., Macko, S., and Theroux, R., 'Total' sediment biomass and preliminary estimates of organic carbon residence time in deep-sea benthos, *Mar. Ecol. Prog. Ser.*, 79, 99, 1991.

30. Findlay, R. H. and White, D. C., A simplified method for bacterial nutritional status based on the simultaneous determination of phospholipid and endogenous storage lipid poly-β-hydroxyalkanoate, *J. Microbiol. Methods*, 6, 113, 1987.

31. Findlay, R. H. and White, D. C., The effects of feeding by the sand dollar *Mellita quinquiesperforata* (Leske) on the benthic microbial community, *J. Exp. Mar. Biol. Ecol.*, 72, 25, 1983.

32. Baird, B. H., Nivens, D. E., Parker, J. H., and White, D. C., The biomass, community structure, and spatial distribution of the sedimentary microbiota from a high-energy area of the deep sea, *Deep-Sea Res.*, 32, 1089, 1985.

33. Dobbs, F. C., unpublished data.

34. Bott, T. L. and Kaplan, L. A., Bacterial biomass, metabolic state, and activity in stream sediments: relation to environmental variables and multiple assay comparisons, *Appl. Environ. Microbiol.*, 50, 508, 1985.

35. Tunblad-Johansson, I. and Adler, L., Effects of sodium chloride concentration on phospholipid fatty acid composition of yeasts differing in osmotolerance, *FEMS Microbiol. Lett.*, 43, 275, 1987.

36. Cooksey, K. E., Guckert, J. B., Williams, S. A., and Callis, P. R., Fluorometric determination of the neutral lipid content of microalgal cells using Nile Red, *J. Microbiol. Methods*, 6, 333, 1987.

37. Guckert, J. B. and Cooksey, K. E., Triglyceride accumulation and fatty acid profile changes in *Chlorella* (Chlorophyta) during high pH-induced cell cycle inhibition, *J. Phycol.*, 26, 72, 1990.

CHAPTER 41

Total Microbial Biomass Estimation Derived from the Measurement of Particulate Adenosine-5'-Triphosphate

David M. Karl

INTRODUCTION

Microbial biomass, defined here as the total amount of living cellular material and generally expressed as g C m^{-3} for water samples or g C m^{-2} for sediments, is a fundamental state variable in quantitative microbial ecology. In trophic studies of aquatic environments it is necessary to know the total standing stock or biomass of microorganisms (e.g., all organisms less than or equal to approximately 200 μm) present in a given habitat or size-fractionated subsample (i.e., protozoan, algal, or bacterial cell-enriched fraction). The biomass of viable microorganisms, when used in conjunction with estimates of metabolic activity, productivity, or rates of cell division (see Notes and Comments) can be used to place limitations on material and energy fluxes through aquatic microbial food webs.

Conventional laboratory-derived methods including measurement of dry weight or carbon to estimate biomass or cell enumeration by direct microscopy are generally unreliable for estimating microbial biomass in natural aquatic environments. Causes for the relative ineffectiveness of these methods include the presence of a relatively large and variable proportion of nonliving particulate matter and the difficulty in distinguishing viable from nonviable cells. Furthermore, the heterogeneous distribution of cell sizes and morphologies and variable cell number-to-biovolume-to-biomass extrapolation factors complicate methods involving direct microscopy to determine total microbial biomass.

ATP has several unique characteristics which make it a reliable indicator of microbial biomass in aquatic environments. It is ubiquitous in all living cells, has a relatively short half-life following cell death and autolysis, and is present at a fairly constant intracellular concentration regardless of nutritional mode (e.g., photoautotroph, chemoheterotroph, chemolithoautotroph, phagotroph) and growth rate. Furthermore, particulate ATP (P-ATP) can be rapidly and efficiently extracted from cells and stabilized in solution using boiling buffers, cold mineral or organic acids, or a variety of organic solvents. The preferred method of ATP quantification is by the firefly bioluminescence reaction; however, a variety of analytical techniques is available for either discrete sample or continuous flow analyses. Data on the P-ATP content of a water or sediment sample can be extrapolated to total microbial biomass using C:ATP relationships derived from either laboratory or field studies.

The advantages of using P-ATP to measure total microbial biomass are (1) the relative ease of P-ATP preconcentration (if necessary), extraction, and analysis including the simplicity of the required equipment, (2) the low detection limit (less than 10^{-12} mol) and acceptable level of precision for field replicates (typically 1 to 10%, depending upon ATP concentration and operator), (3) the high degree of objectivity compared with methods requiring operator recognition of "live" cells and estimation of cell dimensions, and (4) the potential for near "real time" analyses and continuous flow applications. Furthermore, the P-ATP biomass assay is founded on fundamental first principles of cell physiology; both the obligate association of P-ATP with viable organisms and the constancy of intracellular ATP pool concentrations are well established. Compared to most other biomass assays, there is also an extensive laboratory and field database for comparisons, conclusions, and ecological interpretations.

The P-ATP method has been used, and abused, but for the most part has withstood the critical test of time since the first field application to oceanic waters off Southern California.[1] Often, the P-ATP biomass method is criticized for its failure to differentiate among individual groups of microorganisms (except by size or specific density partitioning). Although this clearly is a limitation of the technique, one should hasten to add that the unique value of P-ATP measurements is for the determination of total microbial biomass. Other limitations include the presence of potentially interfering substances or adsorption losses in certain sample materials (especially marine sediments) which require careful internal standardization, variations in cellular C:ATP ratio as a result of nutrient starvation and possibly other forms of environmental stress, and the absence of commercially available certified reference standards for estimating the accuracy of field determinations.

MATERIALS REQUIRED

Equipment

- Sampling gear (water column): Niskin bottles (or equivalent), kevlar line or equivalent, subsampling bottles, 202-μm Nitex mesh
- Sampling gear (sediment column): corer device, core tubes, syringe samplers, glove box (optional)
- Filtration gear: PVC or stainless steel manifold (3 or 6 place), equipped with glass filter bases with stainless steel screens for 25-mm diameter filters and large volume funnels (100- to 200-ml capacity)
- Vacuum pump with gauge (compressed N_2 or air can be used as an alternative to vacuum filtration)
- Block heater (Sybron Thermolyne Type #16500 Dri-bath, or equivalent): capable of heating extraction menstruum in test tubes to $100 \pm 1°C$ (1 atm) and maintaining boiling temperatures throughout the extraction period
- Storage freezer ($-20°C$)
- pH meter and reference buffers
- Magnetic stirrer and stir bars
- Light-detection instrument: Any one of a number of commercially available general instruments, such as fluorometers, spectrophotometers, or liquid scintillation counters, are suitable for measuring light emission. However, in order to obtain efficient and reliable data, with maximum sensitivity, a specially designed ATP photometer is required. Instruments specifically designed for ATP analyses are marketed by several corporations including, but not limited to, E. I. DuPont, LKB Instruments, Biospherical Instruments Corp., Turner Designs, and United Technologies. Whatever light-measuring device is selected, it is imperative to have either a strip chart recorder, integrator, or suitable computer-assisted data station to quantify light emission.

Supplies

- Filters (24/25 mm diameter Whatman GF/F, Millipore HA/GS, or equivalent)
- Filter forceps
- Extraction and assay glassware: test tubes for ATP extraction (approximately 15 mm diameter × 125- to 160-mm length); assay vials (18-mm diameter × 40-mm height). Size requirements may vary depending upon equipment/instruments used.
- Adjustable automatic pipettes (0.1- to 1-ml and 1- to 5-ml capacities)

Solutions and Reagents

(Unless otherwise indicated, catalog numbers or information refers to items sold by Sigma Chemical Company.)

- Firefly lantern extract (FLE-50)
- Potassium arsenate buffer, 100 mM, pH 7.4
- Tris(hydroxy)aminomethane-HCl buffer, 20 mM, pH 7.75
- ATP, ADP, and AMP, sodium salt, 2 μM solutions in Tris buffer, in 1- to 5-ml aliquots, and stored frozen
- Magnesium sulfate solution, 40 mM
- Potassium phosphate buffer, 60 mM, pH 7.4
- D-Luciferin, optional (L-9504)
- H_3PO_4 (1 to 1.5 M)
- NaOH (1 and 0.1 M)

PROCEDURES

Sampling, Subsampling, Extraction, and Sample Storage

Water Column

Typically, oceanic samples are collected at predetermined depths throughout the water column using standard, commercially available polyvinylchloride (PVC) bottles (e.g., General Oceanics Niskin or Go-Flo bottles, or equivalent), mounted on a CTD-rosette sampler. In addition to obtaining complementary information on physical and chemical characteristics of the water column, this sampling protocol allows interactive, directed sampling at specific regions of interest (e.g., particle, fluorescence or oxygen maxima or minima, density discontinuities, etc.). Alternatively, water can be obtained using submersible pumps, manually operated evacuated bottles, syringe samplers, or any other effective means. Selected habitats, such as the sea-surface microlayer or high-temperature hydrothermal vents require the use of specialized samplers.

1. Prior to use, the samplers are cleaned with dilute HCl (0.5 to 1 M), or ethyl alcohol (95%) and rinsed thoroughly with distilled water; sterilization of the samplers is neither required nor practical for most field studies.

2. To measure microbial ATP, it is necessary to remove metazoans and other nonmicrobial ATP prior to sample extraction. This is done most conveniently by passing the water sample through a 202-μm Nitex mesh as part of the subsampling procedures (e.g., during subsampling from Niskin bottles). It is conceivable that, in certain aquatic environments, this procedure may also remove large detrital particles to which microorganisms are attached. In these extreme cases, metazoans can be hand sorted from the respective water samples prior to analyses.

3. Following collection and subsampling, the particulate matter is concentrated and extracted as soon as possible. Although numerous methods have been described for the extraction of ATP from microorganisms,[2] the most commonly used method for aqueous samples involves P-ATP concentration by vacuum or pressure filtration and extraction into boiling TRIS (0.02 M, pH 7.4 to 7.7) or boiling phosphate (60 mM, pH 7.4) buffers. The latter is recommended for samples suspected of containing alkaline phosphatase.[3] After concentration of microbes onto a filter, the filter should be immersed into the boiling buffer as quickly as possible after the final portion of liquid passes through the filter. If left on the filtration manifold for an extended period of time, measured in seconds, the cells dessicate causing a loss of viability. It is imperative that the extraction buffer is at its boiling point at the time of filter insertion. At temperatures below approximately 90°C, inefficient extraction occurs.

4. After the filter is placed into the boiling buffer, the sample is heated for an additional 5 min during which time the test tubes are partially covered to minimize evaporation and resultant volume changes.

5. Following extraction, the samples are removed from the heating block or temperature-controlled bath, cooled, then stored frozen (-20°C) until assayed. At this point the sample extracts are extremely stable with ATP losses of less than 1% year^{-1} in properly buffered solutions.

Sediment Column

Many aquatic sediments are well stratified and characterized by steep depth gradients in microbial biomass. Consequently, it is imperative that the sampling and subsampling methods used to collect sediment for microbial ATP analysis preserve the unique depth distribution. For intertidal or shallow subtidal habitats, sediment cores can be collected manually by inserting PVC or acrylic tubes (10- to 15-cm diameter, 30- to 50-cm length) into the sediment and placing stoppers on both ends during retrieval. Deeper samples require the use of a spade box corer (or equivalent device) which is designed to minimize both sediment disruption during sampling and winnowing during sample recovery. Unfortunately, most gravity corers, which are easier to operate than box corers, create an unacceptable bow wave prior to penetration into the sediment. This results in a disruption of the microbial biomass gradients that are of greatest interest to the microbial ecologist. As for water column samples, the greatest variability is expected to occur at the level of sample (i.e., sediment core) replication. This variability is also of greatest relevance to the microbial ecologist.

Once collected, the core samples should be processed as quickly as possible and, as discussed above, with care taken to minimize changes in environmental conditions. For most anoxic sediments, this requires the use of a N_2-filled glove box to prevent sample oxidation and subsequent transitions in intracellular ATP pools or potential loss of obligate anaerobe viability.

A variety of procedures has been described for core subsampling, including techniques for the collection of the water-sediment interface where microbial biomass and activity are expected to be the highest,[4] and for the preservation of millimeter-scale habitat variability.[5] If required, the sediment samples can be screened to remove ATP-containing metazoa, macroalgae, or higher plant rhizomes prior to analysis. The replicated subsamples from the different depth strata are then mixed thoroughly using a spatula in preparation for final subsampling for ATP extraction.

1. Triplicate subsamples (1 to 2 cm³) of each sediment fraction are collected using a 3-cc plastic syringe barrel (luer lock end removed). Additional replicates are also taken for the determination of wet volume-to-dry weight conversion and for other bulk chemical parameters (e.g., total carbon).

2. The plugs of sediment are immediately discharged into test tubes containing 10 ml of cold H_3PO_4 (0.5 M, 4°C), capped, and thoroughly mixed. Additional subsamples should be prepared with a known amount of ATP added as an internal standard in order to assess and correct for ATP losses by the combined effects of adsorption, hydrolysis, and various potential sources of chemical interference. Representative sediment porewaters should also be collected (by centrifugation or pressure filtration through a 0.2-μm filter) to assess the potential interference due to dissolved ATP (D-ATP). If the sediment samples contain $CaCO_3$, occasional venting to release accumulated CO_2 gas may be required.

3. After an extraction period of 15 to 20 min at 4°C, the extracted nucleotides are separated from the solid phase by centrifugation or vacuum filtration.

4. The pH of the acid extracts is adjusted to 7.4 by titration with NaOH (1.0 and 0.1 M). At this point, the ATP is stable in samples stored at −20°C. If the ATP concentration in the acid extract is >100 nM, the sample can be diluted with 60 mM PO_4 buffer as an alternative to base titration.

5. If the ATP concentration is ≤1 nM, the acid extract must be concentrated by either the activated charcoal procedure[6] or brucite coprecipitation[7] prior to analysis. At this point, the ATP is relatively stable and the samples may either be stored at 4°C for up to 2 to 3 weeks or processed immediately.

Detection of ATP by Firefly Bioluminescence

Although ATP (and other adenine and non-adenine nucleotides) can be measured using any one of a variety of analytical detection systems, the firefly bioluminescence assay and high-performance liquid chromatography (HPLC) are most commonly used in ecological studies. A major advantage of HPLC is the ability to separate complex nucleotide mixtures (i.e., cell extracts) into individual components (ATP, ADP, AMP, etc.) that can be quantified during a single sample run. This additional information on the concentrations of non-ATP nucleotides can provide useful data on the metabolic states and *in situ* growth rates of microbial communities in nature.[2,8] The more commonly used bioluminescence assay, however, has a much lower ATP detection limit, is straightforward and inexpensive to perform, has a high level of precision, and requires less-specialized instrumentation. Furthermore, non-ATP nucleotides (e.g., ADP, AMP, GTP, etc.) can also be measured by the firefly bioluminescence reaction following stoichiometric generation of ATP from other nucleotide triphosphates via specific transphosphorylation reactions.[9]

Several reviews have been published concerning the specificity, kinetics, and mechanism of the firefly bioluminescence reaction. The postulated steps are[10]

$$LH_2 \text{ (luciferin)} + ATP \xrightarrow{\text{Mg}^{2+}\text{–luciferase}} \text{E-}LH_2\text{-AMP} + PP_i \qquad (1)$$

$$\text{E-}LH_2\text{-AMP} + O_2 \xrightarrow{\text{neutral pH}} \text{oxyluciferin} + E + CO_2 + \text{light} \qquad (2)$$

When all necessary reactants are present in excess, the *in vitro* light emission is directly proportional to the concentration of ATP in solution. Reaction kinetics and specificity are dependent upon the purity of the enzyme preparation; sensitivity is controlled by luciferin concentration.[11] The measurement of the initial rise of the luminescence curve (0 to 2 s), the peak height of luminescence (0 to 5 s), or a predetermined integrated portion (e.g., 15 to 75 s) of the light emission decay curve can be used to relate ATP concentrations in reference standards to those in the unknown sample extracts. The major advantages of the integrated mode are increased sensitivity, ease and reliability of mixing, and nonreliance on the peak-height response, which is difficult or impossible to measure using certain instruments. However, a major disadvantage of the integrated mode is the nonspecificity of light emission with certain crude enzyme preparations.[12] Reliability of peak-height analyses depends upon a very rapid and complete mixing of all reactants. This is best accomplished using an automatic injection system, which ensures consistent mixing velocities for all samples. The peak-height mode of analysis offers the advantages of speed of assay and minimum interference from other enzymes or substrates (e.g., non-ATP nucleotides) that may affect the rate of the luciferase-catalyzed reaction.

1. Firefly lantern extract (catalog FLE-50 or FLE-250, Sigma Chemical Co., St. Louis, MO) is stored with dessicant at −20°C. To activate lyophilized FLE, a 50-mg vial is hydrated in 5 ml of distilled water. This enzyme preparation is allowed to ''age'' at room temperature for a minimum of 4 to 6 h to a maximum of 24 h, depending upon the desired sensitivity. During the aging process, endogenous ATP initially present in the crude extract is consumed, thereby decreasing the background light emission. Due to variations among individual enzyme preparations, it is imperative that only a single batch of enzyme be used for the analysis of a given set of samples and ATP standards.

2. Next, the hydrated, aged FLE is further diluted using equal volumes (generally 10 ml each for a single 50-mg vial of FLE) of $MgSO_4$ (0.04 M) and $KHAsO_4$ buffer (0.1 M, pH 7.4) and incubated at room temperature for 1 h.

3. If desired, the FLE preparation can be diluted to greater final volume depending upon the required sensitivity of the assay. However, if a single vial of FLE-50 is diluted to a working volume of >50 ml, the bioluminescence reaction can become limiting for luciferin which must be added back in order to maintain first-order reaction kinetics with respect to ATP.[11]

4. Immediately prior to use, the insoluble residue is removed by vacuum filtration (Whatman GF/F filter) or centrifugation (1500 × g, 5 min).

ATP Standards

For each enzyme preparation, an ATP standard curve is prepared and analyzed with the sample extracts.

1. A stock ATP solution containing 1 to 2 μM ATP is prepared in TRIS buffer (0.02 M, pH 7.4 to 7.7) and stored at −20°C in 1 ml-aliquots until needed. Under these conditions, ATP hydrolysis is <1% $year^{-1}$. The exact concentration of the stock solution is determined by absorption spectrophotometry at 259 nm, using the relationship

$$A = Cle \qquad (3)$$

where C = ATP (M); 1 = absorption pathlength (cm); and e = 15.4 × 10^3 (ATP molar extinction coefficient at pH 7.4).

2. A working ATP standard solution is prepared on the day of the assay by diluting the stock ATP preparation with the appropriate extraction menstruum (TRIS or phosphate buffer depending upon sample type, and preferably the same batch as used for sample extraction).

3. Between seven and eight ATP standards (including a buffer blank) covering the expected range of the sample concentrations (approximately 1 to 100 nM ATP) are prepared. If ≥100 samples are analyzed or if the analysis time exceeds approximately 1 h, it is desirable to measure a set of standards at the beginning and at the end of the analysis to monitor temporal changes in the response of the FLE preparation. Otherwise, a single standard curve measured midway through the experiment is sufficient. Generally, there is no significant hydrolysis of the diluted ATP standard solutions during a typical 3- to 4-h working period.

4.   Following a single use, all thawed ATP stock standards and serial dilutions thereof should be discarded.

Data Reduction and Extrapolations of ATP to Biomass

The relationship between ATP and light emission is linear provided substrate (O_2 and luciferin) limitation does not occur. Peak heights or integrated areas are regressed on ATP standard concentrations. From a model-1 linear regression analysis of these data, the ATP concentrations in sample extracts can be calculated. By correcting for the proportion of the sample actually assayed and the volume of medium originally extracted, the ATP L^{-1} of the original water sample can be determined.

The C:ATP ratio in microorganisms varies considerably, although somewhat predictably, among taxa and even for a given species when grown under different nutritional constraints (data summarized by Karl[2]). Among the most conspicuous differences in C:ATP ratios are those observed between unicellular microorganisms (i.e., bacteria and microalgae: C:ATP $\approx$ 200 to 350, by weight) and micrometazoa (C:ATP = 50 to 150), and the large increases in the C:ATP ratio when microorganisms are starved for phosphorus.[2,8] However, under most conditions found in nature, the C:ATP ratio of the microbial community is about 250:1. Although originally developed to estimate total microbial "biomass," ATP concentrations are, in theory, more closely coupled to "protoplasm" biomass and, more specifically, total biovolume. Because of the obligate role of ATP in cellular bioenergetics, intracellular ATP levels are carefully maintained at a concentration of approximately 1 to 2 mM.[13] Unfortunately it is difficult to use biovolume estimates directly, however accurate, in most studies of microbial ecology. Consequently, we rely upon empirically determined C:ATP ratios to extrapolate P-ATP determinations to estimates of total microbial biomass. In so doing, we decrease both the level of precision and the accuracy of the initial ATP determination. Furthermore, in habitats where copious amounts of capsular materials, extracellular secretions, or slimes occur,[14] ATP-based values of total biomass probably provide only minimum estimates.

NOTES AND COMMENTS

Sampling

The investigator must be aware of three basic areas where variability can be introduced into field measurements: replication at the level of sampling (i.e., multiple water bottles at a given depth), replication at the level of subsampling (multiple samples from a single collection), and analytical replication (i.e., multiple analyses of a single sample extract). Because of the heterogeneous distribution of microbial communities in nature and problems inherent to the collection of particulate matter from aquatic environments, variance between sampling bottles is generally the largest. Ideally, the entire contents of multiple water samples collected at each depth should be analyzed for P-ATP.

Exposure of viable microorganisms to environmental conditions that are substantially different from the collection site should be avoided to minimize short-term transitions in intracellular ATP pools. However, this becomes nearly impossible during the collection of many samples, for example, abyssal water samples from the equatorial ocean. Decreases in pressure and increases in temperature, even during the time required for the samples to reach the ocean's surface, are almost certain to alter microbial ATP concentrations and perhaps even cell viability. However, at the present time these effects have not been systematically evaluated. In the future, a technique which provides for the *in situ* extraction of ATP from microorganisms needs to be developed and compared to conventional sampling procedures in order to provide quantitative constraints on the potential changes in P-ATP during routine sample recovery.

Extraction

The concentration step, which is necessary for most low biomass environments (i.e., <1 g C m^{-3}), must be performed with extreme care and predetermined judgment regarding total sample volume. It has been shown that the recovery of microbial ATP from certain high particulate load samples is volume dependent.[15] Initially, the observed P-ATP losses were thought to be the result of cell lysis during prolonged filtration. However, it is now known to be caused by a filtration-induced metabolic stress that results in the hydrolysis of ATP to ADP and AMP.[16] Consequently, if the total adenine nucleotide pool (the sum of ATP + ADP + AMP) rather than ATP is measured, filtration volume becomes less critical.

In certain eutrophic habitats where total microbial biomass exceeds 1 g C m^{-3}, water samples can be extracted directly, thereby eliminating the preextraction concentration step.[17] It is imperative, however, that the sample volume injected does not exceed 5% (by volume) of the extractant volumes; otherwise temperature changes affect the efficiency of ATP extraction. If desired, multiple injections can be made allowing time for the extraction menstruum to return to its boiling temperature. Furthermore, if direct injection is employed for P-ATP extraction it is necessary to measure, and correct for, D-ATP (operationally defined as passing through a 0.2-μm filter) which is also present in most marine[18,19] and freshwater[20] ecosystems.

Sensitivity, Precision, and Accuracy

The sensitivity of the ATP assay is determined by the instrumentation used to detect light emission and by the purity of the luciferase preparation. Using the crude FLE-50 luciferase preparation prepared as described in this chapter and a commercially available ATP photometer (Biospherical Instruments Corp.), the lower limit of ATP detection (i.e., twice the background light emission) is about 0.2 nM ATP. For greater sensitivity, exogenous luciferin is added to the enzyme preparations enabling the detection of 1 pM ATP.[11] The precision of the peak height assay procedure as routinely performed in our laboratory is ±1 to 2% of the mean (n = 8) throughout the entire range of ATP standards. Accuracy is estimated by analyzing diluted ATP standards and treating them as unknown samples. At the present time, there are no commercially available certified reference materials available for independent determination of accuracy.

Analytical Interferences and Use of Internal Standards

In addition to the potential problems discussed in the previous sections, several sources of analytical interference are possible. These include (1) the presence of inorganic and organic ions in the sample extracts, resulting in loss of ATP in solution (i.e., through chelation) or in decreased luciferase activity, (2) the presence of humic acid-like substances in the sample extracts that impart a yellow color to the solution, thereby resulting in attenuation of the emitted light, (3) turbidity of the final extracts resulting in light scattering and absorption, (4) the presence of a high concentration of inorganic particulate material in the final extracts resulting in loss of ATP through adsorption, and (5) the presence of contaminating enzymes, in either the sample extracts or the luciferase preparation, that compete with luciferase for the ATP in solution (e.g., ATPase or adenylate kinase) or that result in the production of ATP through transphosphorylase reactions (e.g., nucleoside diphosphate kinase or pyruvate kinase).

Most of the above sources of error are detected, and corrected for, through the use of an ATP internal standard, as discussed by Strehler.[21] The internal standard may be added in the form of an ATP salt solution, as live or lyophilized bacterial cells, or as radiolabeled ATP. To minimize the effects of ionic interference, it is imperative that the standard ATP solutions be prepared in an ionic medium identical to that of the samples. Peak-height measurements significantly decrease the analytical interference due to the presence of non-adenine nucleotide triphosphates and therefore, are strongly recommended. If available, a strip-chart recorder or oscilloscope should be interfaced with the photometer in order to detect deviations from the standard ATP-dependent light-emission reaction kinetics.

REFERENCES

1. Holm-Hansen, O. and Booth, C. R., The measurement of adenosine triphosphate in the ocean and its ecological significance, *Limnol. Oceanogr.*, 11, 510, 1966.
2. Karl, D. M., Cellular nucleotide measurements and applications in microbial ecology, *Microbiol. Rev.*, 44, 739, 1980.
3. Karl, D. M. and Craven, D. B., Effects of alkaline phosphatase activity on nucleotide measurements in aquatic microbial communities, *Appl. Environ. Microbiol.*, 40, 549, 1980.
4. Novitsky, J. A., Heterotrophic activity throughout a vertical profile of seawater and sediment in Halifax Harbor, Canada, *Appl. Environ. Microbiol.*, 45, 1761, 1983.
5. Craven, D. B., Jahnke, R. A., and Carlucci, A. F., Fine-scale vertical distributions of microbial biomass and activity in California Borderland sediments, *Mar. Biol.*, 83, 129, 1986.
6. Hodson, R. E., Holm-Hansen, O., and Azam, F., Improved methodology for ATP determination in marine environments, *Mar. Biol.*, 34, 143, 1976.
7. Karl, D. M. and Tien, G., MAGIC: a sensitive and precise method for measuring dissolved phosphorus in aquatic environments, *Limnol. Oceanogr.*, 37, 105, 1992.
8. Karl, D. M., Determination of *in situ* microbial biomass, viability, metabolism, and growth, in *Bacteria in Nature*, Vol. 2, Poindexter, J. S. and Leadbetter, E. R., Eds., Plenum Press, New York, 1986, 85.
9. Karl, D. M., A rapid sensitive method for the measurement of guanine ribonucleotides in bacterial and environmental extracts, *Anal. Biochem.*, 89, 581, 1978.
10. DeLuca, M. and McElroy, W. D., Purification and properties of firefly luciferase, in *Methods in Enzymology*, Volume 57, DeLuca, M. A., Ed., Academic Press, New York, 1978, 3.
11. Karl, D. M. and Holm-Hansen, O., Effects of luciferin concentration on the quantitative assay of ATP using crude luciferase preparations, *Anal. Biochem.*, 75, 100, 1976.
12. Karl, D. M., Occurrence and ecological significance of GTP in the ocean and in microbial cells, *Appl. Environ. Microbiol.*, 36, 349, 1978.
13. Chapman, A. G. and Atkinson, D. E., Adenine nucleotide concentrations and turnover rates. Their correlation with biological activity in bacteria and yeast, *Adv. Microb. Physiol.*, 15, 253, 1977.
14. Costerton, J. W., Irvin, R. T., and Cheng, K. J., The bacterial glycocalyx in nature and disease, *Annu. Rev. Microbiol.*, 35, 299, 1991.
15. Sutcliffe, W. H., Jr., Orr, E. A., and Holm-Hansen, O., Difficulties with ATP measurements in inshore waters, *Limnol. Oceanogr.*, 21, 145, 1976.
16. Karl, D. M. and Holm-Hansen, O., Methodology and measurement of adenylate energy charge ratios in environmental samples, *Mar. Biol.*, 48, 185, 1978.
17. Jones, J. G. and Simon, B. M., Increased sensitivity in the measurement of ATP in freshwater samples with a comment on the adverse effect of membrane filtration, *Freshwater Biol.*, 7, 253, 1977.
18. Azam, F. and Hodson, R. E., Dissolved ATP in the sea and its utilization by marine bacteria, *Nature*, 267, 696, 1977.
19. Nawrocki, M. P. and Karl, D. M., Dissolved ATP turnover in the Bransfield Strait, Antarctica during a spring bloom, *Mar. Ecol. Prog. Ser.*, 57, 35, 1989.

20. Maki, J. S., Sierszen, M. E., and Remsen, C. C., Measurements of dissolved adenosine triphosphate in Lake Michigan, *Can. J. Fish. Aquat. Sci.,* 40, 542, 1983.
21. Strehler, B. L., Bioluminescence assay: principles and practice, *Methods Biochem. Anal.,* 16, 99, 1968.

CHAPTER 42

Microphytobenthic Biomass Measurement Using HPLC and Conventional Pigment Analysis

Catherine Riaux-Gobin and Bert Klein

INTRODUCTION

The microphytobenthos is far less investigated than phytoplankton; therefore, methods in current use, and especially for standing crop measurement, have been adapted from those used in plankton research. Concentration of photosynthetic pigments (generally chlorophyll *a,* Chl *a*) in sediments is widely used as an indicator of microphytic biomass, in spite of the remarks of some authors concerning the variability of the C:Chl *a* ratio in benthic diatoms[1] and the sometimes poor agreement when comparing cell numbers with pigment concentrations.[2] Different techniques have been developed and modified during the last 3 decades which can be separated in two groups: the "conventional" methods (spectrophotometry and fluorometry) and chromatography (separation of pigments prior to detection of their absorbance and/or fluorescence).

There is a huge literature concerning the so-called "conventional" techniques. Worth mentioning are the fluorometric method[3-5] and the spectrophotometric methods (trichromatic[6] and acidification[7] methods). The more recent spectrofluorometric method[8] is more discriminating and sensitive than the conventional spectrophotometric one, but has not yet received much attention. Many reviews give exhaustive bibliographies on these methods, their history and evolution, their specificities and adaptation to sediments, and their inter-comparisons.[9-12] The most popular of the conventional techniques used in oceanography is fluorometry. In spite of numerous criticisms concerning its inadequacy when working on samples containing high concentrations of degraded pigments,[13] interference with Chl *b* and Chl *c*[13,14] or problems related to acidification,[15,16] this technique is sensitive, rapid, relatively cheap, and easy to use (see description below).

There are, however, several problems with the conventional methods. In the trichromatic method the postulations for the validity of the Lambert-Beer's law (on which spectrometry depends) are not fulfilled.[17] Spectrophotometric and fluorometric methods both determine the concentration of a mixture of pigments because chlorophyllide *a,* isomer, and allomer Chl *a* all have similar absorption and fluorescence properties as Chl *a*. Therefore, the conventional techniques overestimate the Chl *a* concentrations. Moreover, sediments are generally enriched with Chl *a* degradation products or phaeopigments.[18-20] The term "phaeopigments", often called phaeophytin *a* in the older literature, includes many different Chl *a* degradation products (several phaeophytins and phaeophorbides) that can be present in great numbers and high concentrations.[21,22] Since phaeophytins and phaeophorbides have

0-87371-564-0/93/$0.00 + $.50
© 1993 by Lewis Publishers

**Table 1. Selected References on Pigment Determination in Sediments
with Chromatographic Methods**

Chloropigments	Carotenoids	Chloropigments and carotenoids
Daley et al.[50]	*Watts and Maxwell[51]	Jeffrey[52]
*Brown et al.[39]	Züllig[53]	*Hajibrahim et al.[54]
*Liebezeit and Bartel[55]	*Repeta and Gagosian[56]	*Riaux-Gobin et al.[21]
*Daemen[9]		*Abele[57,58]
*Keely and Brereton[59]		*Klein and Riaux-Gobin[22]

Note: * = HPLC method.

different molecular weights it is impossible to determine the concentration (weight per liter) of the phaeopigments in extracts containing a mixture of these compounds.[23,24] Accurate analysis of these derivatives can only be obtained with chromatographic methods.

Various methods of chromatography have been used extensively for the analysis of algal pigments in phytoplankton and microphytobenthos research. These include column, paper, thin layer (TLC), high-performance thin layer chromatography (HPTLC), and high-performance liquid chromatography (HPLC). The great advantage of HPLC over TLC is the direct detection of the pigment concentrations. Several authors[25-27] have published valuable technical papers on HPLC techniques which they developed for the analysis of algal pigments, whereas others have only published the applications.[28-31] Roy[32] reviewed the development of the separation of chloropigments and Rowan[33] reviewed both chloropigments and carotenoids. It is generally acknowledged that the best results are obtained with reversed-phase HPLC with gradient elution. However, at the moment there is no generally accepted standard method for pigment analysis because there are still new pigments being discovered (e.g., divinyl chlorophyll *a*). These "new" compounds generally coelute with known pigments (causing overestimation of the concentrations of these pigments) and therefore new gradients, columns, and/or solvents mixtures are used. Recently Wright et al.[34] described a new ternary gradient reversed-phase HPLC system, which gives an enhanced resolution of many pigments, but it is not yet perfect and will not be further dealt with here. Depending on the detector, chloropigments (fluorometer or spectrophotometer 666 nm) or chloropigments and carotenoids (spectrophotometer 436 or 440 nm eventually combined with a second detector for chloropigments) can be detected (see Table 1). Simultaneous detection of chloropigments and carotenoids can indicate which algal classes are present in the sample since microphyte groups contain specific pigments or pigment combinations.[35-37] Furthermore, specific carotenoids may be useful in describing changes in biomass composition.[38] The few studies on sediments employing chromatographic techniques have given most attention to technical questions or to geological processes. We present below the method we used for the determination of both chloropigments and carotenoids.

Some authors[26] used a more rapid HPLC method for the accurate determination of Chl *a* concentrations. However, with these rapid methods complete separation of all chlorophyll derivatives remains problematic.

Intercomparisons have been undertaken with conventional and chromatographic methods[9,25,29,40-43] but results are somewhat contradictory.

In conclusion, rapid HPLC technique[26] may be a good alternative to the conventional methods but the full HPLC method, giving much more information, especially on phaeopigments, is recommended when these phaeopigments are expected in high concentration and the number of samples to analyze is low. When only the standing crop is wanted, the fluorometric method may be preferable in view of its easy use, rapidity, and lower cost.

MATERIALS REQUIRED

Equipment

- Precision balance
- Refrigerator
- Oven (60°C)
- Multitube corer: Hern Wuttke, Industriestrasse 3, Postfach 1326, 2359 Henstedt Ulzburt 3, Germany
- Refrigerated centrifuge with swing-out rotor
- Turner Design Fluorometer
- Scanning spectrophotometer
- For HPLC method: a dual pump system with gradient controller. High-pressure mixing systems often show better reproducibility than low-pressure mixing systems (Klein, unpublished). Various manufacturers offer different mixing systems, which show differences in volumes. Therefore, even when gradient programs are identical, the solvent gradient passing through the column shows some differences which cause differences in retention times when comparing them with literature values. Also required are injector with sample loop (large loops, e.g., 500 μl, can eventually be used in order to decrease the detection limits) and syringes; two detectors in series, i.e., spectrophotometer (436 or 440 nm) or preferably a diode array spectrophotometer (400 to 700 nm), and a fluorometer (excitation: 400 to 440 nm; emission: 600 to 700 nm); and data acquisition system or integrator (eventually a two-channel recorder). The detector with the smallest flow cell is used as the first detector, because the flow cell is a mixing chamber, which causes, for example, peak distortion.

Supplies

- Borosilicate tubes (Corning pyrex culture tubes)
- Pyrex tubes with teflon stoppers
- Micropipette
- For HPLC: C18 reverse phase column (e.g., Lichrosorb 7 μm, 25 cm, and 4 mm i.d.). Smaller particles and shorter columns can give similar resolutions as larger particles in long columns. Furthermore, solvent consumption is lower with these columns.
- Millex-SR filter units (0.5 μm; Millipore Corporation) and syringes

Solutions and Reagents

- 1 N HCl
- Acetone (p.a.)
- For HPLC: solvent A consists of 70% methanol (analytical or HPLC grade), 20% double-distilled or deionized (e.g., Milli-Q) water, and 10% PIC (ion-pairing agent: 1.5 g tetra-butylammonium acetate (Fluka) and 7.7 g ammonium acetate (Merck) in 100 ml water, pH 7.1[25]). The PIC concentration can often be reduced to 2% and the water content increased to 28% (Klein, unpublished). Solvent B consists of 80% methanol and 20% ethyl acetate.

PROCEDURES

Sediment Sampling and Preparation of Extracts

1. The best way to collect sediment samples for quantitative analysis of biomass is to use corers. Caution must be taken to obtain undisturbed interfaces. Intertidal and shallow areas can be sampled by hand (e.g., by divers; plexiglass cores, 5.4 cm internal diameter 20 to 25 cm long, rubber stoppers) whereas for deeper sediments an automatic Multitube Corer is recommended (see address above).

2. For delayed analysis cores are deep frozen ($-20°C$), immediately after sampling (storage for several months does not apparently affect the pigment concentrations[22,28]). Some authors recommended freeze drying prior to extraction[44] but Lenz and Fritsche[45] and Riaux-Gobin et al.[21] observed a consistant reduction of pigment concentrations due to this procedure.

3. For all following manipulations it is recommended to work at dim light, low temperature, and to avoid exposition to acid vapors. The procedure of pigment extraction is similar for the fluorometric and HPLC methods discussed here.

4. During thawing, while the sediment is still partially frozen, the cores are cut into several layers (use piston and steel spatula) that are collected in glass vials. The top layer, that is generally microphyte enriched, must be 5 mm or less. Deeper layers are also interesting for analysis in order to evaluate the biomass gradient with depth.

5. Each layer is manually homogenized.

6. A subsample (1 g $\pm$ 0.0001 g wet weight) is placed on a preweighed aluminum foil for moisture determination by drying at $60°C$ for 48 h.

7. Subsamples for pigment analysis (1 g $\pm$ 0.0001 g wet weight) are extracted in 10 ml 90% acetone p.a. in teflon stoppered pyrex tubes and kept cool ($4°C$) and dark for 24 h. Use of $MgCO_3$ is not recommended (see contradictory references in Reference 33). If the sediment is extremely liquid (high water content), adjust the acetone concentration in order to obtain a final concentration close to 90% (below 80% degradation of Chl a appears). In order to avoid too many manipulations after adding acetone, grinding samples is not recommended; however, for very coarse sediments tests must be done (sonication may be a good alternative, but can produce silt particles too small for elimination by centrifugation).

8. During extraction shake twice (manually or with a vortex), one time just after the weighing and another a few hours before analysis.

Pigment Analysis

Fluorometric Method[4,5]

Advantages and disadvantages of the method include rapidity, sensibility, reproducibility, easy use at shipboard or in precarious conditions, low cost, and permits comparisons with literature; however, phaeopigment concentrations may be overestimated.[23,24]

1. Immediately before analysis decant the extracts (6 ml) in the tubes that will be used for the fluorometric analysis (Corning culture tubes, parafilm stoppered); centrifuge (3000 rpm, 2 min); wait before analysis until thermic equilibrium with the room.

2. The fluorescence of the extract (6 ml) is measured on a Turner Design Fluorometer, before (F_0) and after (F_a) acidification with 80 μl 1 N HCl.

3. The equations used for calculations are adapted from Reference 5, as follows:

$$\text{Chlorophyll } a \ (\mu g \ g^{-1} \ DW) = (F_0 - F_a) \cdot \frac{(F_0/F_a)_{max}}{(F_0/F_a)_{max} - 1} \cdot \frac{v}{x \cdot 1000} \cdot K_x \qquad (1)$$

Phaeopigments a (μg g^{-1} DW, Chl a eq.) =

$$[(F_0/F_a)_{max} \cdot F_a - F_0] \cdot \frac{(F_0/F_a)_{max}}{(F_0/F_a)_{max} - 1} \cdot \frac{v}{x \cdot 1000} \cdot K_x \qquad (2)$$

where $(F_0/F_a)_{max}$ = acidification ratio of pure Chl a (specific for the used fluorometer); F_0 = sample fluorescence before acidification; F_a = sample fluorescence after acidification; v = volume of acetone used for extraction (ml); x = sediment sample dry weight (g); and K_x = calibration constant (specific for each fluorometer).

$(F_0/F_a)_{max}$ and K_x are recalculated every 3 or 4 months by calibrating the Turner Design fluorometer: use pure Chl *a* from spinach (e.g., Sigma); concentration determined by spectrophotometry [with the extinction coefficient of pure Chl *a* and Lambert-Beer's law $(1000 \cdot OD/\alpha = mg \ l^{-1}; \alpha = 87.67;$ see Reference 6)].

4. Results may be expressed per dry weight unit, after calculating the percentage of moisture content of each sample, or per wet surface unit, after calculation of the conversion factor obtained by weighing (before and after drying for 48 h at 60°C) a sample corresponding to a known surface.

Chromatographic Method

The HPLC method discussed here (ion-pairing chromatography) is based on, and resembles, the techniques developed by Mantoura and Llewellyn[25] and Gieskes and Kraay.[29] It has been used for the determination of pigment composition in phytoplankton and algal cultures[46,47] and sediments.[21,22] This technique is time consuming (up to 35 min per analysis), but permits the separation, in one run, of a great number of chloropigments and carotenoids. However, for the separation of the carotenoids zeaxanthin and lutein and Chl *a* and divinyl Chl *a,* the methods described by Wright and Shearer[27] and Gieskes and Kraay,[28] respectively, should be used.

1. Gradient program: The gradient (flow rate 1.5 ml min^{-1}) shows a linear increase of solvent B from 20 to 60% in 7 min, a hold at 60% for 5 min, a linear increase from 60 to 100% solvent B from 12 to 20 min, and a hold for 10 min at 100%. After this gradient the solvent mixture is brought back to initial conditions in 1 min and kept unchanged until next analysis. The hold at 60% of solvent B improves the separation of the pigments in the middle of the chromatogram considerably.

2. Standards and calibration: Only Chl *a* and Chl *b* and β-carotene can be obtained commercially (Sigma). Chlorophyllide *a* can be prepared by extraction of the pigments in 50% acetone.[48] Phaeophytin *a* and *b* and phaeophorbide *a* can be produced by acidification of Chl *a,* Chl *b,* and chlorophyllide *a,* respectively. These acidified solutions should be neutralized with NaOH before injection into HPLC system. All other pigments have to be isolated from algal cultures and/or samples. They can be identified after evaporation of the solvents containing the eluted pigment and redissolution in the appropriate solvents (e.g., ethanol, hexane, chloroform, acetone). An absorption spectrum can be obtained and maxima compared with literature data[33,49] for identification of the pigments. However, complete identification of the pigments requires techniques like NMR and mass spectrometry. Foppen[49] and Rowan[33] give specific absorption coefficients, with which the concentrations of the individual compounds can be calculated which can subsequently be used for calibration of the apparatus. This is done by injecting known amounts of the pigments (x μl of y mg l^{-1}) into the system. From the resulting peak area, the pigment-specific calibration coefficient can be calculated.

ACKNOWLEDGMENTS

Thanks are due to Anthony Grehan for improving the english text.

REFERENCES

1. de Jonge, V. N., Fluctuations in the organic carbon to chlorophyll-a ratios for estuarine benthic diatom populations, *Mar. Ecol. Prog. Ser.*, 2, 345, 1980.
2. Taasen, J. P. and Hoisaeter, T., The shallow-water soft-bottom in Lindaspollene, Western Norway. IV. Benthic marine diatoms, seasonal density fluctuations, *Sarsia*, 66, 293, 1981.
3. Yentsch, C. S. and Menzel, D. W., A method for determination of phytoplankton chlorophyll and phaeophytin by fluorescence, *Deep-Sea Res.*, 10, 1221, 1963.
4. Holm-Hansen, O., Lorenzen, C. J., Holmes, R. W., and Strickland, J. D. H., Fluorometric determination of chlorophyll, *J. Cons. Perm. Int. Explor. Mer.*, 30, 3, 1965.
5. Lorenzen, C. J., A method for the continuous measurement of *in vivo* chlorophyll concentration, *Deep-Sea Res.*, 13, 223, 1966.
6. Jeffrey, S. W. and Humphrey, G. F., New spectroscopic equations for determining chlorophylls a, b, c_1 and c_2 in higher plants, algae and natural phytoplankton, *Biochem. Physiol. Pflanzen.*, 167, 191, 1975.
7. Lorenzen, C. J., Determination of chlorophyll and phaeophytin: spectrophotometric equations, *Limnol. Oceanogr.*, 12, 343, 1967.
8. Neveux, J. and Panouse, M., Spectrofluorometric determination of chlorophylls and pheophytins, *Arch. Hydrobiol.*, 109, 567, 1987.
9. Daemen, E. A. M. J., Comparison of methods for the determination of chlorophyll in estuarine sediments, *Neth. J. Sea Res.*, 20, 21, 1986.
10. Neveux, J., Delmas, D., Romano, J. C., Algarra, P., Ignatiades, L., Herbland, A., Morand, P., Neori, A., Bonin, D., Barbe, J., Sukenik, A., and Berman, T., Comparison of chlorophyll and pheopigment determinations by spectrophotometric, fluorimetric, spectrofluorimetric and HPLC methods, *Mar. Microb. Food Webs*, 4, 217, 1990.
11. Plante-Cuny, M.-R., Pigments photosynthétiques et production primaire des fonds meubles néritiques d'une région tropicale (Nosy-Bé, Madagascar). Thèse Doc. Sci. Nat. Aix-Marseille. Tome I. 206 p. *Trav. Docum. ORSTOM*, 96, 1978.
12. Rai, H. and Marker, A. F. H., Eds., The measurement of photosynthetic pigments in freshwaters and standardization of methods, *Arch. Hydrobiol. Beih. Ergebn. Limnol.*, 16, 1–130, 1982.
13. Lorenzen, C. J. and Jeffrey, S. W., Determination of chlorophyll in seawater. Report of intercalibration tests, *UNESCO Tech. Papers Mar. Sci.*, 35, 1–20, 1980.
14. Gibbs, F., Chlorophyll *b* interference in the fluorometric determination of chlorophyll *a* and phaeopigments, *Aust. J. Mar. Freshwater Res.*, 30, 596, 1979.
15. Usacheva, M. N., Nature of the effect of addition of acids on the spectra and photonics of the molecules of pheophytin, *Biophysics*, 16, 1021, 1971.
16. Moed, J. R. and Hallegraeff, G., Some problems in the estimation of chlorophyll-*a* and phaeopigments from pre- and post-acidification spectrophotometric measurements, *Int. Rev. Ges. Hydrobiol.*, 63, 787, 1978.
17. Rott, E., Spectrophotometric and chromatographic chlorophyll analysis: comparison of results and discussion of the trichrometric method, *Arch. Hydrobiol. Beilh. Ergebn. Limnol.*, 14, 37, 1980.
18. Riaux, C., La chlorophylle *a* dans un sédiment estuarien de Bretagne Nord, *Ann. Inst. Océanogr.*, 58, 185, 1982.
19. Plante-Cuny, M.-R., Evaluation par spectrophotométrie des teneurs en chlorophylle *a* fonctionnelle et en phéopigments des substrats marins, *Doc. Sci. Mission ORSTOM Nosy-Bé*, 45, 1–76, 1974.
20. Orr, W. L., Emery, K. O., and Grady, J. R., Preservations of chlorophyll derivatives in sediments of Southern California, *Bull. Am. Assoc. Petrol. Geol.*, 42, 925, 1958.
21. Riaux-Gobin, C., Llewellyn, C. A., and Klein, B., Microphytobenthos from two subtidal sediments from North Brittany. II. Variations of pigment compositions and concentrations determined by HPLC and conventional techniques, *Mar. Ecol. Prog. Ser.*, 40, 275, 1987.
22. Klein, B. and Riaux-Gobin, C., Algal pigment diversity in coastal sediments from kerguelen (sub-Antarctic Islands) reflecting local dominance of green algae, euglenoids and diatoms, *Polar Biol.*, 11, 439, 1991.

23. Helling, G. R. and Baars, M. A., Changes of the concentrations of chlorophyll and phaeopigment in grazing experiments, *Hydrobiol. Bull.,* 19, 41, 1985.

24. Conover, R. Y., Durvasala, R., Roy, S., and Wang, R., Probable loss of chlorophyll-derived pigments during passage through the gut of zooplankton, and some of the consequences, *Limnol. Oceanogr.,* 31, 878, 1986.

25. Mantoura, R. F. C. and Llewellyn, C. A., The rapid determination of algal chlorophyll and carotenoid pigments and their breakdown products in natural waters by reverse-phase high performance liquid chromatography, *Anal. Chim. Acta,* 151, 297, 1983.

26. Murray, A. P., Gibbs, C. F., and Longmore, A. R., Determination of chlorophyll in marine waters: intercomparison of a rapid HPLC method with full HPLC, spectrophotometric and fluorometric methods, *Mar. Chem.,* 19, 211, 1986.

27. Wright, S. W. and Shearer, J. D., Rapid extraction and high performance liquid chromatography of chlorophylls and carotenoids from marine phytoplankton, *J. Chromatogr.,* 294, 281, 1984.

28. Gieskes, W. W. C. and Kraay, G. W., Unknown chlorophyll *a* derivatives in the North Sea and the tropical Atlantic Ocean revealed by HPLC analysis, *Limnol. Oceanogr.,* 28, 757, 1983.

29. Gieskes, W. W. C. and Kraay, G. W., Phytoplankton, its pigments, and primary production at a central North Sea station in May, July and September 1981, *Neth. J. Sea Res.,* 18, 51, 1984.

30. Gieskes, W. W. C. and Kraay, G. W., Analysis of phytoplankton pigments by HPLC before, during and after mass occurrence of the microflagellate *Corymbellus aureus* during the spring bloom in the open northern North Sea in 1983, *Mar. Biol.,* 92, 45, 1986.

31. Gieskes, W. W. C., Kraay, G. W., Nontji, A., Setiapermana, D., and Sutomo, Monsoonal alternation of a mixed and a layered structure in the phytoplankton of the euphotic zone of the Banda Sea (Indonesia): a mathematical analysis of algal fingerprints, *Neth. J. Sea Res.,* 22, 123, 1988.

32. Roy, S., High-performance liquid chromatographic analysis of chloropigments, *J. Chromatogr.,* 391, 19, 1987.

33. Rowan, K. S., *Photosynthetic Pigments of Algae,* Cambridge University Press, Cambridge, 1989.

34. Wright, S. W., Jeffrey, S. W., Mantoura, R. F. C., Llewellyn, C. A., Bjornland, T., Repeta, D., and Welschmeyer, N., Improved HPLC method for the analysis of chlorophylls and carotenoids from marine phytoplankton, *Mar. Ecol. Prog. Ser.,* 77, 183, 1992.

35. Liaaen-Jensen, S., Carotenoids—a chemosystematic approach, *P. Appl. Chem.,* 51, 661, 1979.

36. van den Hoek, C., *Algen Einfuhrung in der Phykologie,* Thieme-Verlag, Stuttgart, 1978.

37. Parsons, T. R., Takahashi, M., and Hargrave, B., *Biological Oceanographic Processes,* 3rd ed., Pergamon Press, Oxford, 1984.

38. Lehman, P. W., Comparison of chlorophyll *a* and carotenoid pigments as predictors of phytoplankton biomass, *Mar. Biol.,* 65, 237, 1981.

39. Brown, L. M., Hargrave, B. T., and MacKinnon, M. D., Analysis of chlorophyll *a* in sediments by high-pressure liquid chromatography, *Can. J. Fish. Aquat. Sci.,* 38, 205, 1981.

40. Jacobsen, T. R., Comparison of chlorophyll *a* measurements by fluorometric, spectrophotometric and high pressure liquid chromatographic methods in aquatic environments, *Arch. Hydrobiol. Beih. Ergebn. Limnol.,* 16, 35, 1982.

41. Hafsaoui, M., Fertilisation d'un système eutrophe à forte variabilité saisonnière et annuelle. Facteurs limitant de la production phytoplanctonique, assimilation simultanée des différentes formes d'azote inorganique et organique, Thèse 3 ème Cycle, UBO, Brest, 1984.

42. Sartory, D. P., The determination of algal chlorophyllous pigments by high performance liquid chromatography and spectrophotometry, *Water Res.,* 19, 605, 1985.

43. Trees, C. C., Kennicut, M. C. J. M., II, and Brooks, J. M., Errors associated with standard fluorimetric determination of chlorophylls and phaeopigments, *Mar. Chem.,* 17, 1, 1985.

44. Hansson, L. A., Chlorophyll *a* determination of periphyton on sediments: identification of problems and recommendation of method, *Freshwater Biol.,* 20, 347, 1988.

45. Lenz, J. and Fritsche, P., The estimation of chlorophyll *a* in water samples: a comparative study on retention in glass-fibre and membrane filter and on the reliability of two storage methods, *Arch. Hydrobiol. Beih. Ergebn. Limnol.,* 14, 46, 1980.

46. Klein, B., Variations in pigment content in two benthic diatoms during growth in batch cultures, *J. Exp. Mar. Biol. Ecol.,* 115, 237, 1988.

47. Klein, B. and Sournia, A., A daily study of the diatom springbloom at Roscoff (France) in 1985. II. Phytoplankton pigment composition studied by HPLC analysis, *Mar. Ecol. Prog. Ser.,* 37, 265, 1987.

48. Hallegraeff, G. M. and Jeffrey, S. W., Description of new chlorophyll *a* alteration products in marine phytoplankton, *Deep-Sea Res.,* 32, 697, 1985.

49. Foppen, F. H., Tables for the identification of carotenoid pigments, *Chromatogr. Rev.,* 14, 133, 1971.

50. Daley, R. J., Brown, S. R., and McNeely, R. N., Chromatographic and SCDP measurements of fossil phorbins and postglacial history of Little Round Lake, Ontario, *Limnol. Oceanogr.,* 22, 349, 1977.

51. Watts, C. D. and Maxwell, J. R., Carotenoid diagenesis in a marine sediment, *Geochim. Cosmochim. Acta,* 41, 493, 1977.

52. Jeffrey, S. W., Paper chromatographic separation of chlorophylls and carotenoids from marine algae, *Biochem. J.,* 80, 336, 1961.

53. Züllig, H., Untersuchungen über die Stratigraphie von Carotenoiden im geschichteten Sediment von 10 Schweizer Seen zur Erkundung früherer Phytoplankton-Entfaltungen, *Schweiz. Z. Hydrol.,* 44, 1, 1982.

54. Hajibrahim, S. K., Tibbetts, P. J. C., Watts, C. D., Maxwell, J. R., Eglinton, G., Colin, H., and Guiochon, G., Analysis of carotenoid and porphyrin pigments of geochemical interest by high-performance liquid chromatography, *Anal. Chem.,* 50, 549, 1978.

55. Liebezeit, G. and Bartel, J., Reverse phase HPLC separation of fossil chlorins, *J. High Res. Chrom. Chrom. Comm.,* 6, 573, 1983.

56. Repeta, D. J. and Gagosian, R. B., Carotenoid diagenesis in recent marine sediments. I. The Peru continental shelf (15°S, 75°W), *Geochim. Cosmochim. Acta,* 51, 1001, 1987.

57. Abele, D., Carotinoide als biogene Marker für benthische Makroalgen im Sediment der Kieler Bucht, *Berichte Inst. f. Meereskunde* (Universität Kiel, Germany), 183, 1, 1988.

58. Abele, D., Potential of some carotenoids in two recent sediments of kiel Bight as biogenic indicators of phytodetritus, *Mar. Ecol. Prog. Ser.,* 70, 83, 1991.

59. Keely, B. J. and Brereton, R. G., Early chlorin diagenesis in a recent aquatic sediment, *Org. Geochem.,* 10, 975, 1986.

Additional References

For principles of HPLC techniques and knowledge of the equipment, see the following handbooks:

Snyder, L. R., Glajch, J. L., and Kirkland, J. J., *Practical HPLC Method Development,* John Wiley & Sons, New York, 1988.

Snyder, L. R. and Kirkland, J. J., *Introduction to Modern Liquid Chromatography,* John Wiley & Sons, New York, 1979.

Ahuja, S., *Selectivity and Detectability Optimizations in HPLC,* John Wiley & Sons, New York, 1989.

CHAPTER 43

Microphotometric Analysis of the Spectral Absorption and Fluorescence of Individual Phytoplankton Cells and Detrital Matter

Rodolfo Iturriaga and Susan L. Bower

INTRODUCTION

The application of microphotometry to the study of optical properties of individual phytoplankton and detrital particulates has been facilitated by the availability of computer-interfaced microscope photometer systems. These systems are capable of analyzing spectral absorption properties of individual particles as small as 0.5 μm.[1] Most of the information about spectral absorption and fluorescence properties of marine particulates has been obtained by concentrating samples on glass fiber filters.[2-4] Such procedures are easy to perform, but only provide information on the properties of the bulk particulates; discrimination between the phytoplankton and detrital components is only possible by indirect analysis.[5,6] Microphotometric analysis of individual phytoplankton cells or detrital matter provides direct determination of taxonomic and particle morphometry, as well as spectral absorption and fluorescence analysis.

Advantages of the method are (1) allows for simultaneous direct determinations of taxonomic, morphometric, spectral absorption, and spectral fluorescence (excitation and emission) properties of individual particles; (2) once embedded onto a gelatin-coated slide and maintained in cold storage (− 20°C), the optical properties of the particles remain unchanged; (3) data on individual particulates may be collected and compared to that of bulk specimens from the same water sample; (4) allows for collection of a large number of field samples due to the ability to preserve the specimens for analysis at a later time.

Disadvantages of the method are (1) although the particle transfer efficiency, from filter to gelatin-coated slide, approaches 100% for small cells, such as coccoid cyanobacteria, some selection due to particle size does occur for larger particles, such as diatoms (efficiency approaches 75%) and (2) analysis is time consuming.

MATERIALS REQUIRED

Equipment

- Filtration manifold with pump
- Filter funnel assembly glass (25-mm diameter)

- Water bath/stirrer set at 50°C
- Magnetic stirring bar
- Drying oven (50°C)
- Universal microscope (e.g., Carl Zeiss, Germany)
- Photometer, type 03
- Tungsten-halogen light source (150 W, Xenophot, Osram, Germany)
- Scanning monochromators, type H10 (Instruments SA, New Jersey) — located at excitation or emission sites
- Personal computer (to read continuous spectra, 400 to 750 nm) with data acquisition board (Zeiss, Germany). An electronic robotic controller unit is necessary to interface the monochromators' stepper motors to the data acquisition system.
- Darkfield Ultracondenser (Zeiss, Germany)
- Band pass filter, yellow (575 nm) for phycobiliproteins
- Band pass filter, red (683 nm) for chlorophyll *a*
- Objective lens fitted with an iris diaphragm
- Calibration system (45-W quartz-halogen tungsten standard lamp and precision current source Model 65, Optronic Laboratories, Inc., Silver Spring, MD)

Supplies

- 25-mm diameter Nuclepore filters (0.4-μm pore size)
- Forceps (to handle filters)
- Microscope slides, precleaned
- Cover slips (22-mm circles)
- Cotton swabs
- Microscope slide boxes, plastic (airtight)
- Erlenmeyer flask, glass (250 ml)
- Aluminum foil
- 50-ml syringe
- Millipore (HA) filters (0.45-μm pore size)
- 20 ml vials, glass (5)
- Laboratory labeling tape
- Dropper
- Stainless steel microscope slide with 10-mm diameter hole in center
- Amber glass microvial
- Electrical tape
- Immersion oil, low fluorescence
- Lens cleaning paper
- Fluorescent microbeads

Solutions and Reagents

- Glycerol, ACS reagent grade, 30% v/v (anhydrous/water) with 0.2% sodium azide. NOTE: this solution should be filtered through a 0.2-μm pore size filter and stored in a dropper bottle for immersion of transferred sample.
- Gelatin powder, USP or NF reagent grade (5 g)
- Rhodamine B (Kodak, reagent grade), 125 mg in 250 μl propylene glycol (1,2-propanediol, 99.5% pure)

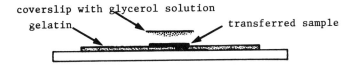

Figure 1.   Gelatin-transfer prepared microscope slide.

PROCEDURES

Preparation of Samples

Concentration of Particulates

The phytoplankton cells and detrital particulates found in natural systems are usually too dilute for direct analysis, so it is often necessary to concentrate the particles. Once the particulates have been concentrated, they are immobilized in embedding medium which allows for preservation and determination of the optical properties.[7]

1.   Water samples are concentrated onto 25-mm diameter Nuclepore filters (0.4-μm pore size) by filtration at low vacuum pressure (not exceeding 5 mmHg). A sample volume of 100 ml is usually appropriate; however, this volume may vary depending upon the particle concentration present within the sample.

2.   To minimize cell rupture, filter the last 1 ml of sample by gravity alone.

3.   The particles collected on the upper surface of the filter are then transferred to a gelatin-coated microscope slide by inverting the filter and placing it in contact with the gelatin coating; the filter is then gently swept with a damp cotton swab. The filter is removed and discarded. See Figure 1.

4.   A drop of 30% glycerol solution is placed on the transferred sample, followed by a coverslip. The prepared slide should be placed in a slide box and kept frozen ($-20°C$) until measurements are performed.

Preparation of Coated Microscope Slides

The embedding medium is a gelatin mixture with a refractive index which nearly matches that of the particles to be embedded.[7] The following will allow for the preparation of at least 500 coated slides. Care should be taken to maintain a clean, dust-free preparation area; avoid contamination of gelatin mixture and slides with foreign particles from the environment which may interfere with future sample analysis. Avoid exposure of the gelatin mixture to temperatures exceeding 60°C which may result in denaturation of the gelatin.

1.   Pour 30 ml of glycerol into a 100-ml graduated cylinder. Add enough distilled water to the cylinder to obtain a final volume of 100 ml. Invert the cylinder to mix, and pour the 30% glycerol solution into a 250-ml Erlenmeyer flask; add stirring bar.

2.   Place the flask in a 50°C water bath with stirrer. Weigh 5 g of gelatin powder and add slowly to the 30% glycerol solution, allowing powder to dissolve.

3.   Weigh 0.2 mg of sodium azide and add to the gelatin mixture.

4.   Cover the flask with foil to reduce moisture loss due to evaporation, and allow mixture to dissolve (approximately 20 min). During this time, heat a 50-ml syringe in a drying

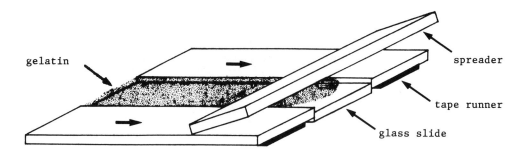

Figure 2. Preparation of gelatin-coated microscope slide.

oven (50°C) to facilitate filtration of the gelatin. Once completely dissolved, the mixture is filtered through a Millipore (HA) filter (0.45-μm pore size) using the syringe, and stored in 20-ml glass vials until used.

5.   Laboratory labeling tape is then applied across the length of the lower surface of two microscope slides. The thickness of the tape will determine the thickness of applied gelatin. A precleaned microscope slide is then positioned between the two runner slides, and coated by adding three drops of warm gelatin mixture to one end of the pre-cleaned slide with a dropper. The gelatin is spread to the other end of the slide by sweeping the edge of another microscope slide across the upper surface of the precleaned slide (Figure 2).

6.   The prepared slides are cold stored in microscope slide boxes (sealed with a strip of electrical tape to prevent dehydration of the gelatin) at 4°C.

Preparation of Quantum Counter

For the calculation of quantum yields and overlap integrals, it is necessary that the fluorescence spectra be corrected to account for distortions by the wavelength-dependent intensity of the exciting light. Although several methods have been developed, we have found that a concentrated solution of rhodamine B, which absorbs virtually all incident light from 220 to 600 nm, is a convenient quantum counter.[8] Due to the configuration of the microphotometric system, we have designed a 2-mm thick stainless steel microscope slide with a 10-mm diameter hole in its center. The quantum counter slide is prepared as follows:

1.   Add 250 μl of propylene glycol to 125 mg of rhodamine B in an amber glass microvial, and allow to dissolve. Filter solution through GF/F filter.

2.   Attach a glass coverslip to the lower surface of the steel microscope slide with silicone sealant to create a well. Fill the well with rhodamine B solution, and seal the solution within the well by attaching a glass coverslip to the top of the well with sealant. Care must be taken to avoid trapping any air bubbles between the rhodamine B solution and the upper coverslip.

3.   Once prepared, the quantum counter slide should be stored at 4°C in darkness.

Microphotometric Determinations

Spectral Absorption

The determination of the spectral transmittance of individual particles is obtained by first focusing on a selected particle. The appropriate optical slit is determined by the size of this target particle, allowing for the sampling of the cross-sectional area of the particle. Due to the irregular shape of some particles, it may be necessary to obtain several determinations on a single particle. A mean value is then obtained; thus, a more representative absorption efficiency factor may be obtained for such individual particles. By selecting a particle-free area adjacent to the target particle as the blank, it is possible to determine the spectral radiant flux for the sample $[I_s(\lambda)]$, as well as the blank $[I_b(\lambda)]$. With these flux determinations, the absorption efficiency factor $[Q_a(\lambda)]$ of the particle is obtained. $Q_a(\lambda)$ is defined as the ratio of energy absorbed by the particle to the energy impinging upon its cross-sectional area,[9,10] as follows:

$$Q_a(\lambda) = 1 - [I_s(\lambda)/I_b(\lambda)] \qquad (1)$$

To compensate for the small energy loss by particle scattering, the value of Q_a at 750 nm is subtracted from the measurements taken at all the other wavelengths.[11]

Spectral Fluorescence

To obtain fluorescence measurements (either excitation or emission spectra), the system is altered by replacing the bright-field condenser with a special dark-field condenser (Ultracondenser, Zeiss). The dark-field condenser forms a hollow cone of excitation light whose apex falls in the plane of the specimen, so light only enters the objective lens if a specimen is present to refract, or scatter, the incident light. Thus, the only light to pass through the objective is fluoresced or scattered light. To insure that only the fluoresced light contributes to the fluorescence excitation measurement, a bandpass filter for the desired emission wavelength is positioned prior to the photomultiplier tube.

If a fluorescence emission spectrum is being run, a scanning monochromator must be positioned in front of the photomultiplier. The use of the dark-field condenser with monochromatic illumination has the advantage that no optical filters interfere with the action spectra, as occurs in epifluorescence microscopy. Only the emission is set up at a desired wavelength with a second monochromator or bandpass filter, as stated above.

Monochromatic dark-field illumination has been useful in the discrimination of different fluorescent pigments among individual microalgae in the 1- to 5-μm diameter range, such as coccoid cyanobacteria.[12] Since dark-field condensers will accept objectives of a lower numerical aperture, the high-power objectives for dark-field microscopy are usually fitted with an iris diaphragm to allow for the continuous reduction of the objective numerical aperture.[13] Further, careful focusing and exact centering of the condenser are necessary to create reproducible illumination conditions, which are required for a consistent fluorescence signal. We have found fluorescent microbeads, embedded in gelatin as described above, to aid in maintaining consistent illumination intensities.

Correction of Fluorescence Spectra

Several techniques can be found in the literature regarding correction for the wavelength-dependent effects of excitation and emission spectra. However, none are completely satisfactory, though they are needed in order to account for the distortion of the wavelength-dependent intensity of the incident light of excitation spectra, or wavelength-dependent efficiency of the detection system present in emission spectra. Additional information can be found in Lakowicz (1983).

Corrected Excitation Spectra. In order to correct for the intensity of the light source, we used a quantum counter, Rhodamine B, in ethylene glycol (mg/ml). This concentrated solution absorbs virtually all light from 220 to 600 nm. For our purposes, excitation spectra were within the 400 to 600 nm spectral range. For microphotometric corrections, the Rhodamine solution was placed on a custom-made stainless steel microscope slide (5 mm thickness) with a hole (5 mm diameter) in the center. The solution was contained within two coverslips glued with silicon adhesive. The slide was then placed in the microscope in the same fashion used to perform action spectra measurements. The action spectra of the Rhodamine solution accounts for the wavelength-dependency of the light source and is used as a correction factor for the samples.

Corrected Emission Spectra. Correction factors can be obtained by measuring the wavelength-dependent output from a calibrated light source. In our system an Optronic Calibration system was used. This consisted of a precision current source Model 65, equipped with a 45-W Tungsten-Halogen standard lamp. The intensity of the standard lamp versus wavelength $I(\lambda)$ is measured using the detection system of the microphotometer. Then the sensitivity of the microscope photometer $S(\lambda)$ is calculated using:

$$S(\lambda) = I(\lambda)/L(\lambda)$$

where $L(\lambda)$ is the spectral output data provided by the manufacturer with the calibration lamp. The correction factor $S(\lambda)$ is then used to correct the emission spectra of the samples. By obtaining the ratio of a specimen's emission spectrum to the sensitivity factor $[S(\lambda)]$, the fluorescence emission may be corrected.[8]

Transferred Specimens and Associated Spectra

The condition of microalgal specimens after being transferred to a gelatin-coated microscope slide is demonstrated for *Dinophysis* and *Bacteriastrum*. The *Bacteriastrum* absorption spectrum is provided (Figure 3). Also shown are the absorption and fluorescence action spectra for the *Dinophysis* specimen (Figure 4). The spectra were normalized to the chlorophyll *a* peak.

ACKNOWLEDGMENTS

The development of the microphotometric technique was supported by the Ocean Optics Program of the Office of Naval Research under Contract N00014-89-J-1047 (R.I.).

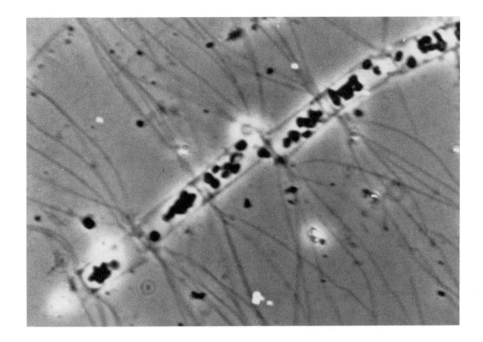

A

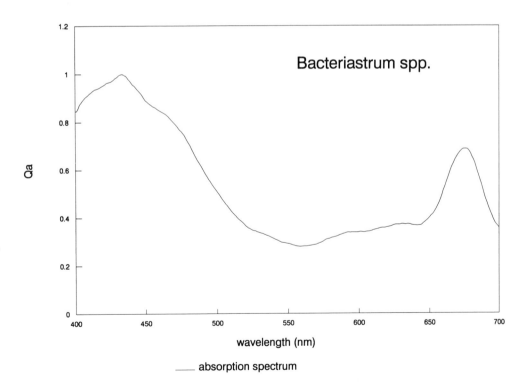

_____ absorption spectrum

B

Figure 3. The *Bacteriastrum* absorption spectrum.

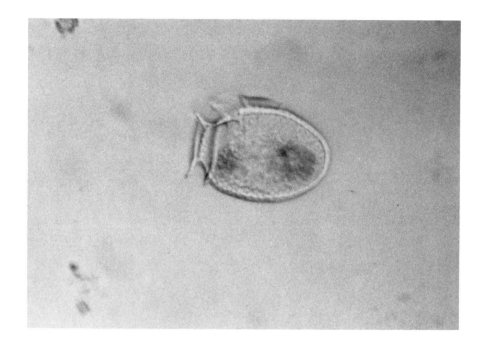

A

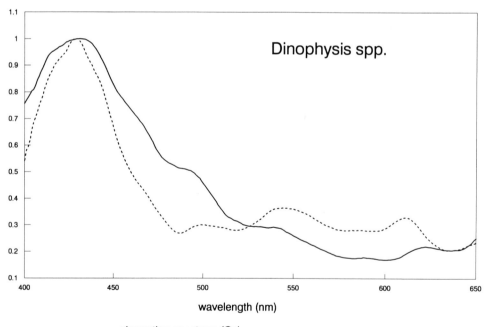

wavelength (nm)

_____ absorption spectrum (Qa)

..... fluorescence excitation spectrum (intensity)

*normalized to maximum value.

B

Figure 4.   The absorption and fluorescence action spectra for the *Dinophysis* specimen.

REFERENCES

1. Piller, H., *Microscope Photometry,* Springer-Verlag, New York, 1977.
2. Yentsch, C. S., Measurement of visible light absorption by particulate matter in the ocean, *Limnol. Oceanogr.,* 7, 207, 1962.
3. Kiefer, D. A. and SooHoo, J. B., Spectral absorption by marine particles of coastal waters of Baja California, *Limnol. Oceanogr.,* 27, 492, 1982.
4. Mitchell, B. G. and Kiefer, D. A., Determination of absorption and fluorescence excitation spectra for phytoplankton, in *Marine Phytoplankton and Productivity,* Holm-Hansen, O., et al., Eds., Springer-Verlag, Berlin, 1984, 157.
5. Bidigare, R. R., Ondrusek, M. E., Morrow, J. H., and Kiefer, D. A., *In vivo* absorption properties of algal pigments, *SPIE-Ocean Optics X,* 1302, 290, 1990.
6. Kishino, M., Booth, C. R., and Okami, N., Underwater radiant energy absorbed by phytoplankton, detritus, dissolved organic matter and pure water, *Limnol. Oceanogr.,* 29, 340, 1984.
7. Iturriaga, R., Mitchell, B. G., and Kiefer, D. A., Microphoto-metric analysis of individual particle absorption spectra, *Limnol. Oceanogr.,* 33, 128, 1988.
8. Lakowicz, J. R., *Principles of Fluorescence Spectroscopy,* Plenum Press, New York, 1983.
9. Bricaud, A., Morel, A., and Prieur, L., Optical efficiency factors of some phytoplankters, *Limnol. Oceanogr.,* 28, 816, 1983.
10. Morel, A. and Bricaud, A., Theoretical results concerning light absorption in a discrete medium, and application to the specific absorption of phytoplankton, *Deep-Sea Res.,* 28, 1375, 1981.
11. Iturriaga, R. and Siegel, D. A., Microphotometric characterization of phytoplankton and detrital absorption properties in the Sargasso Sea, *Limnol. Oceanogr.,* 34(8), 1706, 1989.
12. Campbell, L. and Iturriaga, R., Identification of *Synechococcus* spp. In the Sargasso Sea by immunofluorescence and fluorescence excitation spectroscopy performed on individual cells, *Limnol. Oceanogr.,* 35, 1196, 1988.
13. Pluta, M., *Advanced Light Microscopy,* Vol. 2, Elsevier, New York, 1989.

CHAPTER 44

Measurement of Elemental Content and Dry Weight of Single Cells: X-Ray Microanalysis

Mikal Heldal

INTRODUCTION

Microbial communities from various environments usually consist of diverse groups of phytoplankton, protists, bacteria, virus, and other particles of various sizes and origins. Only a small fraction of such communities can be analyzed through experimental work on isolated species or strains: generally the viable counts are about 1% of the total count of bacteria in natural waters.[1] According to these findings, most of the aquatic bacteria are not accessible for experimental work using pure cultures. This is also true for other microorganisms in mixed microbial communities. We may conclude that for the vast majority of microorganisms in natural waters, we have a very scarce knowledge of how various species are related to both biotic and physiochemical environments. For estimation of microbial activity in terms of energy flow, respiratory rates or carbon flow etc., the carbon content of the various fractions of microorganisms is required. The common methodology is (1) to count the total numbers of cells by fluorescence microscopy,[2,3] (2) estimate cell sizes and calculate biovolumes,[4] and (3) convert from biovolume to carbon.[5,6]

By X-ray microanalysis with the transmission electron microscope (TEM), more direct measurements may add to and improve estimates based on the above-mentioned methods.

Advantages of the method are: (1) direct determination of dry weight or carbon in single cells without any need for biovolume measurements;[7-9] (2) good estimates of size distribution of cells and of frequency of dividing cells; (3) direct measurements of elemental composition of single cells like C:N:P ratios and the content of K, Cl, and Na related to osmotic and physiological conditions of cells;[10] and (4) characterization of certain structures in cells, e.g., polyphosphate granules.[11,12] It is also an advantage of the method that the preparation is easy and can be combined with preparation for total viral counts.[13] Total counts of bacteria can be determined with good accuracy. For discrimination between different groups of particles, e.g., between large viruses and small bacteria, TEM is by far better than fluorescence microscopy.

The disadvantages of the method are: (1) an analytical equipped TEM is not generally available and may be prohibitively expensive for extensive use in ecology studies and (2) X-ray analysis is time consuming: 100 to 200 s for each particle (analysis of particle and background).

0-87371-564-0/93/$0.00 + $.50

MATERIALS REQUIRED

Equipment

Centrifuge with Swinging-Bucket Rotor

X-Ray Detector Systems. Energy dispersive detectors are produced by several companies. The following will be related to the Tracor® Si(Li) (Beryllium window detector) and Z-MAX 30® (Norvar) thin window detectors. The difference between these two detectors is the opportunity to analyze light elements with the latter. For technical specifications and performance of various detectors, further information is given by the producers and through general texbooks.

Supplies

Centrifuge Tubes

  Parafilm

- Grids: Various types of grids may be used: nylon grids, Cu grids, (100 to 150 mesh) and Al grids (150 mesh) (Agar Scientific, Herts, England). The nylon grids are woven mesh grids and accordingly, the support film will not be as flat as for other grids. The nylon grids are mainly used for reduction of background counts, and part from light elements, they contain only trace amounts of Ti (Kα 5.51 keV). The choice of Cu or Al grids should be related to the purpose of the analysis; the energy peaks (Kα) are 8.04 for Cu and 1.49 for Al, thus different parts of the energy spectra will be influenced by these different grids.
- Grid support films: The grid support film may be of various types but collodion and formvar are both applicable and generally used.

Solutions

- Formvar: 0.25 to 0.5% in chloroform
- Collodion: 3 to 4% in anhydrous *n*-amyl acetate

PROCEDURES

Preparation of Support Films

Collodion Support Films

1. Fill a glass dish with distilled water. Add 1 to 2 drops of collodion solution from 1 to 2 cm above water surface (Pasteur pipette). The interference color of a thin film is seen. Remove the film with a clean glass rod to clean the surface.

2. Make a fresh film by adding one drop of collodion to the water surface. This film should be seen as a silvery film. If the film shows gold or higher interference color, the film is too thick.

3. Place grids on the collodion film for Cu or Ni grids with the shining side up.

4. From above, attach parafilm to the collodion film with the grids and remove the collodion coated grids carefully.

5. Air dry the grids overnight in a dust-free environment.

6. Stabilize the grids by a light carbon coating (pale gray coating seen on an exposed part of a clean white paper).

7. Store the grids on the parafilm protected from dust, preferably in a vacuum desiccator. From our experience, the grids can be stored for several months. The collodion solution should be kept water-free and stored in a closed jar containing silica gel. Water can be removed from *n*-amyl acetate by treatment with anhydrous magnesium sulfate.

Formvar Support Films

1. Prepare dust-free microscope slides by use of lens paper.

2. Place the slide in a funnel which can be filled and drained with formvar solution with as stable flow as possible.

3. Coat a slide and after a short period of drying, cut along the edges of the slide with a razor blade.

4. Push the slide slowly into the water (dust-free surface). The film on the slide (uppermost) should detach from the glass surface and float on the water.

5. Use formvar films which show silver-gray interference colors.

6. Place the grids on the film and pick them up as described above. For stabilizing and storage see above. From our experience, the formvar support film is more stable than the collodion film, but the formvar film has a relatively high Si content compared to collodion films.

Preparation of Cells

Harvesting by Centrifugation

For analysis of discrete particles from aquatic environments, it is convenient to harvest the particles directly onto the grids by centrifugation. Use a centrifuge equipped with a swing-out rotor. Mold a plastic (e.g., Araldite), flat bottom in the centrifuge tubes. For ultraspeed centrifugation, we generally use Polyallomer® centrifuge tubes (Beckman).

1. Attach a small piece of double-sided tape at the center of a membrane filter (e.g., cellulose nitrate filters, Sartorius) which fits into the centrifuge tube.

2. Place the grids along the edge of the tape by attaching the rim of the grids at the edge of the tape.

3. Fill sample into the centrifuge tube and slide the filter with the grid(s) into the tube at an angle of about 80°C, to avoid trapping air bubbles underneath the grid. Balance and place the tubes in the swing-out rotor and run the centrifuge at the conditions preferred. For harvesting bacteria and particles of similar size and density, $4000 \times g$ for 20 min is appropriate for quantitative harvesting from a 60-mm water column. For quantitative harvesting of particles down to the size of viruses, for both X-ray microanalysis and total counts of virus, we apply $100,000 \times g$ for 2.5 h for a 60-mm water column.

This preparation method allows quantitative counting of various groups of particles, since all particles from a defined part of the sample will be sedimented on the grid.[13] Harvesting can be performed without use of glow-discharged grids and both microorganisms and other particles are distributed as discrete particles on the grid. Samples containing high particle densities may either be diluted in particle-free water or the height of the water column in the centrifuge tube may be reduced. The latter is preferred since it is considered to give less perturbation of the sample. (Warning: if an ultraspeed centrifuge is used at high speed, the tubes must be shortened if short water heights are harvested).

Preparation by Adding Drops Directly on Grids

For preparation of microorganisms from dense cultures or samples, it may be convenient just to add a drop of culture to a grid, wait for 1 to 2 min, remove excess water by blotting with a filter paper, and air dry the preparations. The grids can be checked by light microscopy to ensure that the grids are appropriate for analysis. To enhance adhesion of negatively charged particles, the carbon-coated grids should be made hydrophilic or positively charged. For this purpose, glow-discharging of the grids (glow-discharge at 0.1 torr for several minutes) is a common method. The glow-discharged grids should be used within 1 to 2 h.

In Situ Attachment of Cells

This preparation method is mainly suitable for sampling of bacterioneuston, e.g., from the air-water interface. Grids are placed with the carbon-coated side on the water surface for 30 s, removed, and air dried.[14] Longer exposure times (2 to 4 d) can be used for microbial growth and colonization on the grid surface (unpublished results).

Calibration

Calibration for quantitative analysis of single particles is obtained by (1) measurements of particles with known masses (i.e., monodisperse beads), (2) measurement of specific elements in microdroplets of known mass, and (3) determination of the detector efficiency for elements of interest.

1. Mixed samples of monodisperse latex beads with diameters of 0.25, 0.50, 1.04, 2.0, 5.0, and 10 μm should be prepared to give equal frequency of the various beads. The beads can be sprayed onto grids in microdroplets or added directly in a drop to a glow-dischraged grid. Remove excess water and air dry the sample.

2. For calibration measurements of specific elements, microdroplets of calcium glycero-phosphate (0.1% wt/vol, Merck, analytical grade) in distilled water is sprayed onto grids which are then air dried.

3. The detector efficiency is determined from analysis of microdroplets of a standard solution with equimolar amounts of Na, Mg, P, S, K, and Co.[15]

Quantitative Analysis of Particles

The general principles considered in the following will be related to biological thin specimens and thus, analytical conditions where X-ray absorption and secondary X-ray fluorescence within the specimen can be ignored (Be window detectors).

Briefly, the two main sources of X-ray signals are:

1. One source is the interaction between incident electrons and nucleic field of atoms in the specimen. The incident electrons will decelerate and emit the lost energy as photons in an energy range from zero to that of the incident electrons. This will, accordingly, give a continuous energy spectrum: continuum or white radiation (WR) without any specific peaks.

2. Another source is the interaction between incident electrons and orbital electrons within atoms of the specimen. This will give ionization and X-rays with energy levels characteristic for the energies between various orbitals. These energies will be element-specific and they will be seen as peaks superimposed on the continuous energy spectrum described above.

The measurements include: (1) energy spectrum (0 to 10.24 keV) of particle, (2) background spectrum of grid support film close to the particle, (3) length and width of the analyzed area, (4) length and width of the particle, and (5) the orientation angle α = length axis of the particle related to the tilt axis on the screen.

Determination of the total mass (dry weight) is based on net WR intensity in the energy window of 4.5 to 6 keV, where peaks of specific elements do not interfere:

$$W_{sp} = W - W_{bk} \tag{1}$$

and

$$m_{sp} = [(W - W_{bk}) \cdot a]/C_{sp} \tag{2}$$

where: W_{sp} = the WR intensity from the particle;
W = the intensity from particle and background;
W_{bk} = the background intensity;
m_{sp} = the total mass of the particle;
a = analyzed area; and
C_{sp} = calibration constant for WR.

Scanned (analyzed) area, a, is included in Equation 2, since in scanning mode of the intensity from the specimen particle (W_{sp}) is inversely proportional to the scanned area circumscribing the particle. The WR intensity is dependent on the elemental composition of the specimen by:

$$G = \text{average } (Z^2/A) \tag{3}$$

where Z is the atomic number and A is the atomic weight. The average value of Z^2/A is calculated for all elements present related to their mass fraction. G may be estimated if the specimen composition is known; G has been estimated to 3.28 for soft tissue[16] and 3.6 for bacteria.[7] For calibration by monodisperse latex beads with known mass and G value of 2.7, the bacterial dry weight is estimated by calculation of C_{bac}:

$$C_{bac} = C_{lat} \cdot (G_{bac}/G_{lat}) = C_{lat} \cdot 3.6/2.7 \tag{4}$$

Quantitative Determination of Specific Elements

From analysis of microdroplets of calcium glycerophosphate (g) with known G (G_g = 5.34), the total mass of each microdroplet is estimated from:

$$C_g = C_{lat}(G_g/G_{lat}) \tag{5}$$

As the mass fraction of phosphorus is known, the calibration factor for total amount of phosphorus from the intensity of P peaks is established:

$$C_p = (I_p - I_{bk} - WR_p) \cdot a/m_p \qquad (6)$$

where:
- m_p = total mass of phosphorus;
- I_p = intensity of the P peak;
- I_{bk} = intensity of the film background in the energy window;
- WR_p = the white radiation contribution in the energy window;
- a = analyzed area; and
- C_p = calibration constant for the element.

The calibration constants for other elements are calculated from the calibration of phosphorus, or similar measurements for other elements, related to the detector efficiency for the various elements.[15] Microcrystals have also been used as primary standard combined with analysis of metalloproteins as secondary standards.[17]

Particle Size Estimation

For biological particles, or particles of similar density, the WR intensity (4.5 to 6.0 keV) has been shown to be proportional to the particle mass in the range from less than 5×10^{-3} to 550 pg.[18] This gives the method a dynamic range for quantifying cellular mass of small bacteria to medium-sized microalgae with a diameter of about 15 μm. Length and width of particles are measured directly on the CRT screen. Since the grids are tilted 38° toward the detector, corrections have to be made for the particle orientation by:

$$BL = [(BL_d \cdot \cos(\alpha))^2 + ((BL_d \cdot \sin(\alpha))/\cos(t))^2]^{1/2} \qquad (7)$$

and

$$BW = [(BW_d \cdot \sin(\alpha))^2 + ((BW_d \cdot \cos(\alpha))/\cos(t))^2]^{1/2} \qquad (8)$$

where:
- BL and BW = the length and width of the particles after correction;
- BL_d and BW_d = length and width measured from the CRT screen;
- t = the tilt angle (38°); and
- α = the angle for the orientation of the particle (length axis of particle − horizontal CRT axis), measured on the screen.

The cell volume V is calculated from:

$$V = 3.14 \cdot BW^2 \cdot [BL - (BW/3)]/4 \qquad (9)$$

Measurement of cell sizes from air-dried preparations may give a 20 to 30% underestimation of calculated cell volume compared to light microscopy measurements of living cells (*E. coli*, unpublished results). Similar measurements have shown that cells which are fixed (2% formaldehyde) and then air dried have a 50% volume reduction compared to nonfixed, air-dried cells. For size-distribution measurements in TEM we may conclude that air drying of unfixed cells is preferred.

NOTES AND COMMENTS

1. The operating conditions generally used for analysis are: accelerating voltage = 80 kV; filament current = 85 mA; and objective aperture = none or 1, depending on particle size. For analysis of a broad range of particle sizes, objective aperture 1 is used, but then with low count rates and signal-to-noise ratios for the smallest bacteria. Condenser aperture = none and tilt angle = 38°. For a beam current some instruments allow continuous measurements, but special equipment (i.e., Faraday cup) should be used for calibration and as control during a long series of analyses. A stable beam current is important for the analyses.

2. Fixation and staining — for X-ray microanalysis, fixation and staining should be avoided. Even the smallest bacteria are easily seen in TEM without staining due to their electron density. The contrast is sufficient for image analysis but some care should be taken to balance brightness, contrast, and intensity since the exact outline of unstained bacteria can be difficult to see on the cathode ray tube. Fixation of cells will give an unpredictable loss of elements from the cells. Elements like Na, K, and Cl are mainly dissolved in the cells and may be lost extensively by disruption of cell membranes through fixation. Phosphorus may also get lost from the cells by fixation, but this effect seems to vary among different samples and may be related to the physiological conditions of the cells (unpublished observation).

3. Light element detection — low energy X-rays (<1 keV) will be absorbed in conventional Be windows and the K-lines for elements lighter than sodium are undetectable. During recent years, a range of different detectors has been introduced to allow detection of light elements: windowless detectors, ultrathin window detectors (UTW), and detectors with windows of materials which transmit low energy X-rays. The choice of detector will, to some extent, depend on the microscope performance (e.g., vacuum) and the purpose of analysis. Calibration and operation will be as for Be window detectors, although a more direct estimation of total dry weight is obtained since all major elements except hydrogen are included in the analysis.

ACKNOWLEDGMENT

I am indebted to Cand. real Svein Norland for his indispensable contributions through many years of cooperation and for critically reading this manuscript.

REFERENCES

1. Fry, J. C. and Zia, T., A method for estimating viability of aquatic bacteria by slide culture, *J. Appl. Bacteriol.*, 53, 189, 1982.

2. Zimmermann, R. and Meyer-Reil, L. A., A new method for fluorescence staining of bacterial populations on membrane filters, *Kieler Meeresforsch.*, 30, 24, 1974.

3. Hobbie, J. E., Daley, R. J., and Jasper, S., Use of Nuclepore filters for counting bacteria by fluorescence microscopy, *Appl. Environ. Microbiol.*, 33, 1225, 1977.

4. Bratbak, G., Microscope methods for measuring bacterial biovolume: epifluorescence microscopy, scanning electron microscopy and transmission microscopy, in *Current Methods in Aquatic Microbial Ecology*, CRC Press, Boca Raton, FL, 1993, chap. 36.

5. Norland, S., The relationship between biomass and volume of bacteria, in *Current Methods in Aquatic Microbial Ecology*, CRC Press, Boca Raton, FL, 1993, chap. 35.

6. Lee, S., Measurement of carbon and nitrogen biomass and biovolume from naturally delivered marine bacterioplankton, *Current Methods in Aquatic Microbial Ecology*, CRC Press, Boca Raton, FL, 1993, chap. 37.

7. Heldal, M., Norland, S., and Tumyr, O., X-ray microanalytic method for measurement of dry matter and elemental content of invididual bacteria, *Appl. Environ. Microbiol.*, 50, 1251, 1985.

8. Norland, S., Heldal, M., and Tumyr, O., On the relation between dry matter and volume of bacteria, *Microb. Ecol.*, 13, 95, 1987.

9. Heldal, M. and Tumyr, O., Morphology and content of dry matter and some elements in cells and stalks of *Nevskia* from an eutrophic lake, *Can. J. Microbiol.*, 32, 89, 1986.

10. Fagerbakke, K. M., Heldal, M., and Norland, S., Variation in elemental content among and within trichomes in *Nostoc calcicola* 79WA01 measured by x-ray micro-analysis, *FEMS Microbiol. Lett.*, 81, 227, 1991.

11. Kjeldstad, B., Heldal, M., Nissen, H., Bergan, A. S., and Evjen, K., Changes in polyphosphate composition and localization in *Propionibacterium acnes* after near-ultraviolet irradiation, *Can. J. Microbiol.*, 37, 562, 1991.

12. Jensen, T. E., Rachlin, J. W., Vandana, J., and Warkentine, B. E., Heavy metal uptake in relation to phosphorus nutrition in *Anabaena variabilis* (Cyanophyceae), *Environ. Pollut. (Ser. A)*, 42, 261, 1986.

13. Børsheim, K. Y., Bratbak, G., and Heldal, M., Enumeration and biomass estimation of planktonic bacteria and viruses by transmission electron microscopy, *Appl. Environ. Microbiol.*, 56, 352, 1990.

14. Fuerst, J. A., McGregor, A., and Dickson, M. R., Negative staining of freshwater bacterioneuston sampled directly with electron microscope specimen support grids, *Microb. Ecol.*, 13, 219, 1987.

15. Davis, T. W. and Morgan, A. J., The application of X-ray analysis in the electron analytical microscope (T.E.A.M.) to the quantitative bulk analysis of biological microsamples, *J. Microsc.*, 107, 47–54, 1976.

16. Roomans, G. M., Quantitative X-ray microanalysis of thin sections, in *X-ray Microanalysis in Biology,* Hayat, M. A., Ed., Macmillan Publishers, London, 1981.

17. Booth, K. N., Sigee, D. C., and Bellinger, E., Studies on the occurrence and elemental composition of bacteria in fresh water plankton, *Scanning Micros.*, 1, 2033, 1987.

18. Bergh, Ø., Røntgenmikroanalyse av enkeltceller i mikrobielle populasjoner (X-ray microanalysis of single cells in microbial populations) Cand. Scient. thesis (in norwegian), University of Bergen, 1988.

Section IV
Activity, Respiration, and Growth

CHAPTER 45

Microautoradiographic Detection of Microbial Activity

Kevin R. Carman

INTRODUCTION

Brock and Brock[1] introduced microautoradiography (MA) as a tool for the study of activities by individual aquatic microorganisms from natural samples. The technique has subsequently been used to address a variety of ecological questions regarding bacterial activities in natural aquatic systems.[2-5] In particular, MA can be used in radiotracer studies to determine the proportion of microorganisms in a sample that is metabolizing a given radiolabeled compound. MA is also useful for determining labeling specificity, i.e., whether targeted microorganisms take up labeled substrates at the exclusion of nontarget microorganisms.[6,7] MA has also been used to check assumptions of radiotracer-grazing techniques, i.e., whether label is taken up by grazers via consumption of labeled microbes or as a consequence of activity by microorganisms associated with the grazers.[8]

The MA methods described here are based primarily on the "MARGE-E" procedure described by Tabor and Neihof,[9] and secondarily on suggestions from Rogers[10] and Meyer-Reil.[3] These procedures are designed for use in radiotracer studies that involve the use of compounds labeled with ^3H or ^{14}C. With relatively minor modifications (e.g., the use of different emulsions), the basic methods described here should be applicable for studies involving other β-emitting isotopes (e.g., ^{32}P, ^{33}P, and ^{35}S). Readers who wish to further investigate practical and theoretical aspects of autoradiography are referred to Rogers'[10] excellent book.

MATERIALS REQUIRED

Sample Preparation

Equipment

- Centrifuge
- 0.2-μm filter-sterilization device
- Waring blender: semi-micro blender container (Eberbach, Ann Arbor, MI)

0-87371-564-0/93/$0.00 + $.50

Supplies

- 50-ml centrifuge tubes
- 5-ml pipettes

Solutions

- 0.01% sodium pyrophosphate solution, filter sterilized (0.2 μm) (in deionized water or appropriate saline solution)
- 3% glutaraldehyde solution in 0.1 M cacodylate buffer
- Sterile, 15-ml, screw-cap test tube or comparable sample containers
- 25 mm, 0.2-μm membrane filters, prestained with Irgalan black (Poretics Corporation, Livermore, CA or Nuclepore Corporation, Pleasanton, CA)

Microautoradiography

Equipment

- Refrigerator set at 4°C
- Safelight with 30-W bulb and Kodak #2 filter
- Foot switch (to control safelight)
- Water bath (to maintain temperature at 43°C)
- Epifluorescence microscope
- Filtration apparatus for 25-mm filters

Supplies

- 10-μl Hamilton syringe
- 25-mm and 0.2-μm membrane filters
- Membrane-filter forceps
- Microscope slides with frosted end
- Autoradiography slide boxes (RPI, Mount Prospect, IL; standard plastic slide boxes may also be used)
- 30-ml beaker (used as a dipping container for emulsion)
- 100-ml beaker (in which 30-ml dipping container is placed)
- Kimwipes (or equivalent)
- Cold metal plate (chilled in freezer compartment of standard refrigerator)
- Paper towels
- Silica-gel desiccant
- Desiccator (large enough to hold slide boxes)
- 60-ml, plastic, screw-cap bottles
- Aluminum foil
- Coplin jars

Solutions

- Chromic acid (dissolve 100 g $K_2Cr_2O_7$ in 850 ml deionized water, add 100 ml H_2SO_4)
- Deionized water
- Glycerin
- Autoradiographic emulsion (e.g., NTB-2, Kodak, Rochester, NY)
- Dektol developer (Kodak; 1:2, stock solution:deionized water)
- Kodak fixer solution (NOT Rapid fixer), make up in deionized water
- Glycerin solution, 1% in deionized water

- Low-fluorescence immersion oil
- Acridine orange, 0.04% in pH 6.6 citrate buffer
- Citrate buffer, 0.004 M, pH 6.6, pH 5.0 and pH 4.0

Citrate buffer recipe:
1. Make 0.1 M stock solutions of citric acid and sodium citrate in 25% methanol (MeOH; prevents microbial growth in stock solution):
    citric acid: 2.10 g/100 ml 25% MeOH
    sodium citrate: 2.94 g/100 ml 25% MeOH
2. Mix 20 ml of stock solutions in the following ratios to achieve the appropriate pH:

pH	ml citric acid	ml sodium citrate
4.0	13	7
5.0	8	12
6.6	1.5	18.5

3. For working solution, dilute the 20 ml of mixed stock solution with 480 ml deionized water. Final pH can be adjusted by adding small amounts of citric acid or sodium citrate.

PROCEDURES

Sample Preparation

If the study involves the use of plankton samples, microorganisms can be concentrated onto membrane filters of appropriate pore size (e.g., 0.2 μm for bacteria, 1 to 5 μm for eukaryotic algae) using standard filtration procedures. For studies involving sedimentary microorganisms, it is recommended that microorganisms be separated from sediments before they are concentrated on filters. The procedure described below for separating microbes from sediments is derived from Balkwill et al.[11] It is suitable for a 1-cm³ sediment sample. If other sediment volumes are used, appropriate adjustments to solution volumes should be employed:

1. Place 1-cm³ sediment in sterile test tube, fix in 6 ml of 3% glutaraldehyde solution. Store at 4°C until use.

2. Transfer entire sample from test tube to a clean, sterile, semi-micro blender container.

3. Add 24 ml of 0.2-μm-filtered (FS), 0.01% sodium pyrophosphate solution to the blender container.

4. Blend sample for 1 minute on, 30 s off, 30 s on, 30 s off, and 30 s on.

5. Transfer blender contents to a clean, sterile 50-ml centrifuge tube.

6. Pelletize sediment and cells by centrifugation at 2350 × g for 10 min.

7. Taking care not to disturb pellet, remove and discard 25 ml of supernatant using pipette.

8. Add 5 ml of FS 0.01% sodium pyrophosphate solution to centrifuge tube, vortex tube for 30 s.

9. Centrifuge at 650 × g for 5 min. This pelletizes sediment but leaves cells in suspension.

10. Decant supernatant into a separate clean, sterile 50-ml centrifuge tube.

11. Add 5 ml 0.01% sodium pyrophosphate solution to tube containing sediment pellet, vortex 30 s.

12. Centrifuge at 650 $\times$ g for 5 min.

13. Repeat steps 10 through 12 until 25 ml of supernatant has been collected.

14. Stain an aliquot of the supernatant with acridine orange (or DAPI) and filter onto a 0.2-μm membrane filter prestained with Irgalan black. Determine supernatant volume required to obtain 50 to 100 bacteria per field of view when viewed with a 100$\times$ oil-immersion objective. If MA is to be performed on microalgae, microalgal abundance may be determined using membrane filters of a larger pore size (1 to 5 μm).

Microautoradiography

Microautoradiography is performed on microbes that have been concentrated on membrane filters. Briefly, slides are coated with emulsion and the filter is placed face down on the liquid emulsion. The emulsion is then hardened by cooling and drying. After exposure, development, staining, and redrying, filters are peeled from the emulsion, leaving the cells embedded in the emulsion. Cells and developed silver grains can then be detected with fluorescence and bright-field microscopy. Experience in our lab has been that slides are most conveniently processed in batches of 10. Several batches may be processed in a single day.

Advance Preparations

1. Soak microscope slides and 30-ml beaker in chromic acid overnight, then rinse thoroughly with FS deionized water. Store acid-cleaned slides and glassware so that they will not be exposed to dust.

2. Label (sample #) frosted end of microscope slide with indelible ink.

3. Using a 10-μl syringe, place a 1-μl drop of glycerin on a corner of the frosted end of the slide.

4. Filter an aliquot of the sample supernatant onto a 25-mm, 0.2-μm membrane filter. Rinse twice with deionized water to remove unincorporated radioactivity.

5. Attach filter (filtered side up) to corner of slide by placing an edge of the filter on the glycerin drop. This minimizes the chance that filters will become disassociated from labeled slides (leading to confusion over sample i.d.s), and greatly facilitates handling of slides and filters in the darkroom.

Sample Processing

Darkroom preliminaries:

1. Place emulsion in 43°C water bath. Emulsion should liquify in approximately 1 h. This can be performed in the light if lid of the emulsion container is not removed. Each of the following steps should be performed in total darkness.

2. Dilute emulsion with distilled water, two parts emulsion to one part water (this dilution yields an emulsion layer 1 to 2 μm thick).[10]

3. Dispense 20-ml aliquots of diluted emulsion into 60-ml screw-cap plastic bottles. Place bottles in a film-development tank and store at 4°C for future use.

Continue processing:

4. Place approximately 1 cm of water in a 100-ml beaker (so that it will not float) and place in water bath.

5. Place 30-ml beaker inside 100-ml beaker.

6. Liquify a 20-ml aliquot of diluted emulsion (43°C water bath).

7. Pour diluted emulsion into a 30-ml beaker.

8. Be certain that no air bubbles are in the emulsion. To determine if air bubbles are in the emulsion, dip a slide into the emulsion, turn on the safelight (safelight should be approximately 2 m from area where slides are being processed) and examine the slide (walk over to safelight and observe the slide while holding it next to the light). If bubbles can be seen in the emulsion, repeat test every 2 min until no bubbles can be seen on slide. When a smooth coat of emulsion is achieved, you can proceed with sample processing.

9. Turn on safelight with foot switch.

10. Detach filter from the slight using forceps; hold filter in one hand, slide in the other.

11. Turn off safelight with foot switch.

12. Dip slide into a 30-ml beaker containing the emulsion; remove slide and allow to drain in a vertical position for 20 s.

13. Hold slide horizontally (frosted edge up) and turn on safelight. Position slide so that it is in your line of vision to the safelight (you should be able to see the silhouette of the slide).

14. Position filter behind slide (between slide and safelight) and align with emulsion-coated end of slide.

15. Draw filter toward the slide until the lower edge of the filter touches the edge of the slides.

16. Release filter allowing it to lie face down on the emulsion. N.B. Leave at least one edge of the filter hanging from the edge of the slide to facilitate removal of the filter after development and staining.

17. Wipe the back side of the slide clean with a Kimwipe.

18. Place slide (emulsion side up) on a cold metal plate. Allow approximately 30 min on cold plate (emulsion gels).

19. Remove slides from cold plate and place on a paper towel to dry (allow approximately 1 h).

20. Place slides in autoradiography slide box; place slide box and lid (do not place lid on slide box) in a desiccator containing silica gel desiccant. Leave overnight in light-tight space, such as sealed cabinet.

21. Open desiccator and place lid on slide box. (Silica gel desiccant should be in slide box. When using RPI slide boxes, desiccant can be placed in a special chamber in the lid, otherwise, a small packet of desiccant should be placed in the slide box). Wrap slide box in two layers of aluminum foil. Label wrapped box (e.g., sample numbers — this also serves as a reminder of which side of the slide box should remain ''up'').

22. Place slide box in a refrigerator at 4°C.

23. Expose slides for a suitable period of time (this may range from 3 days to 3 weeks depending on the experimental treatment[4]).

Development of Autoradiographs

1. Remove slides from refrigerator and allow to warm to room temperature (about 1 h).

2. Dispense 250 ml each of deionized water, Dektol, and fixer solutions into separate 500-ml beakers. Adjust temperature of these solutions to 17°C.

The following steps are carried out in total darkness:

3. Open slide box and remove rack of slides.

4. Fill one slide box with developer solution, another with deionized water, and a third with fixer solution. Place slide rack into box containing developer. Develop slides for 2 min.

5. Transfer slide rack to box containing deionized water. Soak for 30 s.

6. Transfer slide rack to box containing fixer solution. Fix for 5 min.

7. Remove slides from fixer solution and place in a flat pan. GENTLY rinse slides 15 min with tap water.

At this point, lights can be turned on.

8. Place slides (face up) on paper towels to dry.

9. Dry slides further by placing in a desiccator overnight.

Staining

After the autoradiographs have been developed and dried, slides can be prepared for microscopy by staining with acridine orange. I have found that attempts to prestain microbial cells (prior to development of emulsion) result in very high densities of background silver grains.

1. Soak slides for 3 min in pH 6.6 citrate buffer solution.

2. Stain slides for 7 min in a 0.04% solution of acridine orange in citrate buffer (pH 6.6).

3. Destain for 10 min in each of the following series of citrate buffer solutions: pH 6.6, pH 5.0, pH 5.0, pH 4.0, and pH 4.0.

4. Soak slides for 2 min in 1% glycerin solution.

5. Dry slides on towel then place in a desiccator overnight.

6. Trace an outline of the filter on the back of the slide with indelible ink.

7. Use membrane forceps to peel the filter away from the emulsion.

Microscopy

Autoradiographs are examined using a combination of epifluorescence and bright-field illumination. Bright-field illumination can be used to scan autoradiographs for clusters of silver grains, while epifluorescence illumination is used to verify that the cluster of silver grains is associated with a microbial cell. The filtered area of the slide is centered on the microscope stage, immersion oil placed directly on the autoradiograph, and randomly selected fields of view are scanned for active cells.

In samples incubated with ^{3}H substrates, cells may be designated as "active" if three or more silver grains are associated with a cell.[5] In samples incubated with ^{14}C bicarbonate, cells may be designated as active if 10 or more silver grains are associated with a cell.[7] In practice, these criteria are generally not necessary, as cells that are labeled usually have a large number of silver grains associated with them.

NOTES AND COMMENTS

1. Autoradiographic emulsion is extremely sensitive to light. For this reason I (and other authors) recommend that the safelight should be on only during periods when it is absolutely necessary, e.g., when picking up the filter with forceps and when placing the filter on the emulsion-coated slide. The use of a foot switch for the safelight allows you to accomplish this goal with minimal effort.
2. The use of frosted-end slides makes handling slides in total darkness much less difficult, i.e., the frosted surface can easily be detected by touch, and indicates which side of the slide the sample is on and which end is coated with emulsion.
3. Slides should not be exposed in a refrigerator in which radioactive materials are normally stored.
4. The same basic procedure can be used to perform autoradiography on embedded sections of animals or plants. In this case, sections are fixed to the microscope slide then dipped in emulsion and processed as described above. Development of exposed autoradiographs is the same.
5. A handy substitute for the 100-ml beaker is the 4 oz. cup in which Kodak emulsion is shipped. After emulsion has been dispensed into individual containers, the cup can be cleaned and used as a holding vessel for the 30-ml dipping beaker. If the lights must be turned on at some point, the plastic lid can be placed on the cup, creating a light-tight container for the emulsion.
6. The volume of developer, water, and fixer should be sufficient to cover slides completely. If using containers other than RPI slide boxes for development, the appropriate volume should be determined empirically.
7. Previous authors, including myself, have used sodium thiosulfate solution ($Na_2S_2O_3 \cdot 5H_2O$; 30 g/100 ml deionized water) as a fixer. When using sodium thiosulfate, however, I have frequently had problems with emulsion detaching from the slide during the development process. Kodak fixer contains a hardener that eliminates the problem of emulsion detachment.
8. Soaking in 1% glycerin facilitates removal of the filter from the slide.
9. Nonradioactive (no labeled substrate added) and poisoned controls should be processed along with experimental samples.
10. If only a small percentrage of bacteria have taken up labeled substrates, the concentration of cells used in autoradiographs may be increased to reduce the number of fields required to count a statistically meaningful number of active cells. The higher density of cells, however, makes direct counts of bacterial abundance unfeasible. In this case, bacterial abundance can be determined independently from direct counts on filters.

REFERENCES

1. Brock, T. D. and Brock, M. L., Autoradiography as a tool in microbial ecology, *Nature,* 209, 734, 1966.
2. Brock, T. D., Bacterial growth rate in the sea: direct analysis by thymidine autoradiography, *Science,* 155, 81, 1967.
3. Meyer-Reil, L.-A., Autoradiography and epifluorescence microscopy combined for the determination of number and spectrum of actively metabolizing bacteria in natural waters, *Appl. Environ. Microb.,* 36, 506, 1978.
4. Novitsky, J. A., Heterotrophic activity throughout a vertical profile of seawater and sediment in Halifax Harbor, Canada, *Appl. Environ. Microb.,* 45, 1735, 1983.
5. Douglas, D. J., Novitsky, J. A., and Fournier, R. O., Microautoradiography-based enumeration of bacteria with estimates of thymidine-specific growth and production rates, *Mar. Ecol. Prog. Ser.,* 36, 91, 1987.

6. Munro, A. L. S. and Brock, T. D., Distinction between bacterial and algal utilization of soluble substances in the sea, *J. Gen. Microbiol.*, 51, 35, 1968.

7. Carman, K. R., Radioactive labeling of a natural assemblage of marine sedimentary bacteria and microalgae for trophic studies: an autoradiographic study, *Microb. Ecol.*, 19, 279, 1990a.

8. Carman, K. R., Mechanisms of uptake of radioactive labels by meiobenthic copepods during grazing experiments, *Mar. Ecol. Prog. Ser.*, 68, 71, 1990b.

9. Tabor, P. S. and Neihof, R. A., Improved microautoradiographic method to determine individual microorganisms active in substrate uptake in natural waters, *Appl. Environ. Microb.*, 44, 945, 1982.

10. Rogers, A. W., *Techniques of Autoradiography,* Elsevier/North-Holland Biomedical, New York, 1977.

11. Balkwill, D. L., Labeda, D. P., and Casida, L. E., Simplified procedures for releasing and concentrating microorganisms from soil for transmission electron microscopy viewing as thin-sectioned and frozen-etched preparations, *Can. J. Microb.*, 21, 251, 1975.

CHAPTER 46

¹⁴C Tracer Method for Measuring Microbial Activity in Deep-Sea Sediments

Jody W. Deming

INTRODUCTION

Rates of consumption of dissolved organic compounds (DOC) by heterotrophic bacteria in deep-sea sediments may be estimated by evaluating the time-course fate (incorporation into macromolecules and release as respired CO_2) of tracer amounts of ¹⁴C-labeled DOC in cold, repressurized sediment slurries. With the advent of affordable and easy-to-use pressure equipment, the recreation of deep-sea hydrostatic pressure during sample incubation shipboard is becoming standard practice for evaluating not only DOC consumption,[1-3] but also a variety of other deep-sea microbial activities.[4-10] The concepts that heterotrophic bacteria in deep-sea sediments are barophilic (achieving optimal rates under elevated pressures) and dominant over other size classes of benthic organisms in the cycling of organic carbon at the abyssal seafloor have been confirmed using the ¹⁴C-incubation approach.[2,11-14] In the future, this method may prove useful to assessments of carbon flux and burial rates on an oceanic and global scale.[3,15-16]

The ¹⁴C-tracer method for deep-sea sediments borrows from protocols developed by Wright and Hobbie,[17] Williams and Askew,[18] and Hobbie and Crawford[19] for measuring heterotrophic microbial activity in aquatic waters, and from later modifications by Paul and Morita,[20] Schwarz and Colwell,[21] and Tabor et al.[22] for pressure studies. A typical experiment begins with rapid preparation of sediment slurries at atmospheric pressure and *in situ* temperature (-1 to $4°C$, depending on latitude), as soon as a sediment core is recovered shipboard. Replicate samples amended with the desired ¹⁴C-substrate(s) are then returned to *in situ* T/P conditions in multiple pressure vessels in a temperature-controlled laboratory or water bath. One vessel, containing duplicate subsamples of each sediment sample under study, is sacrificed (decompressed) at time zero and at each of four subsequent time intervals until the experiment is terminated at 48 h. Each decompressed sample is transferred quickly to a serum bottle and processed further (acidified, incubated with agitation, filtered) for analysis of radioisotope present in both the solid phase (incorporated into macromolecules, after corrections for sediment adsorption and quench) and gaseous phase (respired as CO_2). Early work with deep-sea sediments employed excessively long incubation periods (weeks to months) in anticipation of very slow rates with the result that significant activity early in the experiment was missed.

Variations on the ¹⁴C-tracer method for repressurized sediment slurries include repressurizing whole (unslurried) subcores (J. W. Deming, unpublished observations) that have been injected with the labeled substrate through side ports in the subcore wall[23,24] and injecting substrate directly into the seafloor via free vehicle[25] or submersible.[14] The former approach appears promising, since it avoids diluting and mixing the sediments, treatments that can stimulate bacterial activity. It remains problematic, since ideal controls have not been devised to account for constraints on substrate-microbe interactions imposed by diffusion of the ^{14}C substrate away from the point of injection. The latter *in situ* methods appear ideal, since sample decompression does not occur until after the measured activity period. However, *in situ* approaches suffer not only from the same diffusion-related problems of repressurized cores, but also from limitations inherent to remote seafloor operations. These include the inability to ensure reaction termination prior to and during ascent and gear retrieval, expense of the specialized gear, and restrictions on number of deployments (rate measurements) per cruise.

Alternatives to the ^{14}C-tracer method for evaluating microbial utilization of DOC in deep-sea sediments include: (1) use of multiple concentrations of ^{14}C substrate in endpoint incubations of repressurized sediments (kinetic approach); (2) use of tritiated substrates *in situ* or with recovered samples; (3) inference from experimentally determined microbial growth rates; (4) inference from vertical sediment profiles of naturally occurring organic compounds or from porewater profiles of dissolved or gaseous constituents (DOC, CO_2, O_2); and (5) inference from dissolved compounds or gases fluxing from the seabed into overlying waters, measured *in situ* via remote-controlled incubation chambers. The ^{14}C kinetic approach requires prior knowledge of DOC consumption rates in order to determine an appropriate single incubation period.[11] It assumes that all rates are linear within the incubation period (regardless of sample type or substrate concentration) and requires mathematical extrapolation from a Lineweaver Burke plot to derive an activity rate at natural (unsupplemented) substrate concentrations.[19,26,27] Tritium is frequently the radiolabel of choice in oligotrophic environments, because the high specific activities that can be achieved ensure substrate addition at tracer level. However, ^{3}H substrates do not allow a direct measure of carbon loss via respiration,[28] which appears to constitute a major pathway of carbon flow in deep-sea sediments.[2,11] Inferring DOC utilization rates from bacterial growth rates, whether measured by tritiated adenine,[29] thymidine,[7] or epifluorescence microscopy,[4,30] suffers not only from the need to assume a respiratory quotient, but also from specific problems inherent to each growth-rate method. Because most bacteria in deep-sea (and shallow) sediments are metabolically active in the cycling of DOC but "nongrowing",[4,30] extrapolations of total carbon consumption from a minimal growth rate will likely involve unacceptable magnification of experimental errors. The last two alternative approaches practiced by geochemists (chemical profiles and flux chambers) require numerous assumptions, conversion factors, and modeling exercises and do not discriminate bacterial activities from the benthic community at large.

Advantages of the ^{14}C-tracer method are that: (1) microbial activity is measured in short (by deep-sea standards), time-course measurements prior to development of "bottle effects"; (2) both substrate incorporation into macromolecules and loss via respiration are tracked directly; (3) measured activity reflects consumption of natural levels of substrate, since DOC levels in deep-sea porewaters are sufficiently elevated (micromolar range) to allow addition of ^{14}C substrates at <10% of that present naturally (thus, meeting the theoretical requirements of the tracer approach, according to Wright[26] and Wright and Burnison;[27] and (4) rates can be obtained in replicate with appropriate controls on multiple sample types, the number of which is limited only by space in the available pressure vessels and hands available for postincubation sample processing.

Some disadvantages of the method are that: (1) the samples undergo decompression during recovery; (2) they are disrupted by dilution prior to incubation; (3) inhabitants other than bacteria (e.g., protozoa and invertebrate larvae) may consume some of the added substrate (and some of the active bacteria); (4) the standard protocol does not account for intracellular pools of substrate, not yet incorporated into macromolecules, or volatile fatty acids that may be released from organisms, especially upon sample acidification;[31] and (5) the method requires pressure equipment not typically available (though now quite affordable) in microbiology or oceanography laboratories. The problems of decompression, which conceivably eliminates some members of the microbial community otherwise active *in situ*, and loss of intracellular substrate pools should each lead to underestimates of true *in situ* rates. However, there is still no evidence for the existence of decompression-sensitive bacteria in the deep-sea[32] (even obligately barophilic bacteria tolerate multiple decompression periods, as shown by Yayanos et al.[33] and Deming et al.[34]) and the fraction of substrate pooled intracellularly is likely to be minor. The problems of sediment dilution and presence of nonbacterial DOC consumers (and bacterivores) should each result in overestimates of *in situ* microbial rates. However, dilution should eliminate the nonmicrobial DOC consumers (and bacterivores), given their typically low densities in deep-sea sediments. On balance, the ^{14}C-tracer method as applied to repressurized sediments has yielded substrate incorporation rates that are conservative or equivalent to *in situ* rates measured via free vehicle on the seafloor.[11]

MATERIALS REQUIRED

Sampling

Equipment

- Deep-sea box corer or multi-corer
- Refrigerated laboratory (or assorted freezers, refrigerators, and ice and temperature-controlled water baths for chilling solutions, tools, samples, etc.)
- Ice machine (or other source of ice)
- Oven at 60°C

Supplies

- Inert tubing to siphon water from core surface
- Modified plastic syringes as subcoring devices
- Sterile spatulas for slicing desired sediment strata
- Sterile 100-ml containers for diluted samples
- Preweighed glass fiber filters for dry weight analyses
- Sterile plastic 3-cc syringes for subsamples
- Sterile 23-gauge needles and small rubber plugs
- 3-mm glass beads

Solutions and Reagents

- 0.2-μm filtered bottom seawater for sediment dilutions
- Stock solution of ^{14}C substrate (prepared in sterile artificial seawater; 20 to 100 ng substrate/ml, depending on specific activity of the substrate used)

Incubations

Equipment

- Stainless steel pressure vessels and one quick-coupling device (the models that we use are available commercially from the Tem-Pres Division of Leco Corporation, Blanchard, PA; ask for the 1983 Deming/Johns Hopkins University prototypes)
- Hydraulic hand pump and high-pressure gauge (Enerpac, Butler, WI)
- High-pressure capillary tubing, fittings, valves, and O-rings for vessels and pump (High Pressure Equipment Company, Erie, PA)
- Portable vise
- Temperature-controlled ice or water bath
- Ice machine (or other source of ice)

Solutions and Reagents

- Distilled water as hydraulic fluid

CO$_2$ Recovery and Macromolecular Analyses

Equipment

- Shaker table with dark incubation box
- Vacuum pump and tubing
- Scintillation counter

Supplies

- Filtration glassware for 25-mm filters
- 50-ml (or larger) serum bottles with rubber caps
- Plastic center wells (Kontes, Hayward, CA)
- Fluted 2 × 5 cm pieces of Whatman No. 1 filter paper
- 1-cc syringes and 2-inch fine-gauge needles
- 0.2-μm, 25-mm nitrocellulose filters
- Supplies for work with radioactivity (gloves, absorbent towels and benchcoat; paper, solid, and liquid waste disposal containers, shipping declaration forms, etc.)
- Scintillation vials
- Rubber cement
- Forceps

Solutions and Reagents

- 4 N H$_2$SO$_4$
- Phenethylamine
- Sterile artificial seawater
- Scintillation cocktail

PROCEDURES

Coring Devices

A range of sediment coring devices can be used to recover sediments from the deep ocean floor. The most widely used for biological purposes has been the large US-NEL (U.S. Naval Electronics Laboratory) spade or box corer that samples a 0.25- m² area of the seafloor.[35] A more recently developed multiple corer or "multicorer"[36] that collects circular 7-cm² areas of sediment in plastic tube corers improves upon the box corer by recovering the sediment-water interface virtually undisturbed. The multicorer may not be the best choice for microbial activity studies in temperate or tropical oceans, however, because the mass of mud recovered in a single core tube can be insufficient to protect against warming during the recovery process. The GOMEX box corer, recently described by Boland and Rowe,[37] is less cumbersome and more efficient than the larger US-NEL corer, while still collecting sufficient mass (625 cm² area; penetration to >20 cm) to protect against warming (J. W. Deming, unpublished observations).

Like the tube corers of the multicorer, the GOMEX box can also be removed immediately to a cold room for temporary storage and subsampling. All subsequent steps in preparation for microbial activity measurements are best conducted in a cold room maintained at deep-sea temperature. However, if a cold room is not available shipboard, refrigerators, freezers, ice, and temperature-controlled water baths can also be used to good advantage to prechill solutions and tools and to keep samples at the proper temperature during the course of an experiment.

Sample Preparation and Incubation

1. As soon as overlying seawater has been siphoned carefully from the surface of the sediment, samples are removed. A (prechilled) calibrated, sterile plastic syringe with the needle end precut and beveled like a corer makes a convenient subcoring device for collecting both surface and subsurface samples.

2. A calibrated portion of the desired sediment stratum is placed in a (prechilled) sterile plastic container and diluted 1:10 (v:v) with cold sterile (0.2-μm prefiltered) near-bottom seawater. The latter is typically collected in a Niskin bottle deployed earlier at the same station as closely as possible to the seafloor. As an alternative, sterile artificial seawater can be used (never use seawater of elevated DOC content from shallower depths).

3. The sample is then homogenized by vortex mixer, taking care not to allow warming during the process, and distributed into pressure-sensitive sample containers, a pair for each desired time point on the activity and killed-control curves. Any sterile container that has a flexible diaphragm or closure point and can be sealed without leaving an air space inside is suitable for incubation under elevated hydrostatic pressure. The most common container has been a sterile plastic syringe. Air bubbles can be expelled from the needle end prior to plunging the needle into a rubber plug, thus, sealing the syringe. The syringe barrel responds easily to pressure changes, preventing collapse of the syringe. Note that the rubber fittings on some syringe barrels contain hidden pockets of air. The air should be displaced with sterile water prior to using the syringe. Also, placing a sterile glass bead into the syringe prior to use will assist in later sample-substrate mixing.

4.   Additional subsamples are reserved in duplicate for dry weight determinations (and for any other desired analyses, such as epifluorescence microscopy), according to standard practice. Filter a known volume of the diluted sediment onto a preweighed glass fiber filter, dry at 60°C for about 18 h, and weigh again to determine gram dry weight sediment per volume of experimental sample.

5.   Immediately prior to pressurizing the cold, sample-filled syringes, draw into the syringe sufficient volume of a stock solution of the ^{14}C-labeled substrate to achieve the desired final concentration. The latter is usually a balance between achieving absolute tracer-level additions of the substrate and spiking the sample with sufficient radioactivity so that the amount consumed microbially during the experiment is detectable above background (the time-zero values) but not greater than 20% of the total added. In practice, the final added concentration of most ^{14}C-labeled substrates falls in the range of nanograms per milliliter. It is convenient to prepare a solution of 10-fold greater concentration in sterile artificial seawater and draw 0.2 ml of it into calibrated 3-ml syringes preloaded with 1.8 ml of the diluted sediment sample.

6.   Each pair of syringes prepared for a given incubation period should be pressurized immediately after addition of the ^{14}C substrate and prior to spiking the next pair of syringes. By starting the process with syringes destined for the longest incubation period and working backward to time zero, variability in the resulting uptake curves due to the time required for pressurization is minimized.

7.   Each pressure vessel and its required volume of hydraulic fluid must be chilled to deep-sea temperature prior to loading and pressurizing the syringes. Use a separate vessel for each incubation period.

8.   Secure the vessel upright in a vise and fill it with hydraulic fluid (we use distilled water). Adding food coloring to the fluid allows a later visual check of the integrity of the syringe samples.

9.   Carefully load the syringes, needle end down, so that air spaces in the barrel end are displaced with water.

10. Cap the vessel, allowing water to spill freely and displace air in the cap, and bring the vessel to *in situ* pressure using a portable hydraulic hand pump with an in-line high-pressure gauge. For every 10 m of sample depth, pressurize 1 atm (or 14.7 psi on a standard U.S.-made hydraulic gauge).

11. Close the pressure valve and disconnect the vessel from the pump. A unique device that couples the pressure vessel to the pumping system, greatly facilitating quick pressurization and release, was designed and described in detail by Yayanos and Van Boxtel.[38] The whole process, using steady strokes of the pump, usually requires less than a minute.

12. Return the vessel to an ice or water bath at deep-sea temperature, if a refrigerated laboratory is not available shipboard. A water-filled pressure vessel will remain ice-cold for many minutes if the vessels must be pressurized outside of a cold room.

13. At the end of each incubation period, secure the pressure vessel in the vise, connect it to the pump, confirm that the vessel is still at desired pressure, and release internal pressure steadily through the pump valve over a period of 20 to 30 s. It is possible that more rapid or jerky motions may compromise the integrity of the syringes as they decompress or effect the resulting activity curve; however, I have not observed evidence of either problem.

14. Immediately uncap the vessel and remove the syringes to an ice bath, processing them as quickly as possible.

Sample Processing After Incubation

1. Each syringe sample must be transferred to a separate gas-tight container to recover respired $^{14}CO_2$. Clean (acid-rinsed), 50- or 100-ml glass serum bottles capped with rubber septa are convenient for this purpose. Prepare the cap by piercing it with a needle and inserting the arm of a plastic "center well" so that the well will hang suspended above the sample in the bottle. Place a fluted, 2×5 cm piece of Whatman No. 1 filter paper into the center well. Prepare all serum bottles and caps in advance of decompressing a vessel so that samples can be transferred into bottles immediately after decompression.

2. As soon as the syringe contents are emptied into a bottle, cap the bottle, pierce the cap with a 2-inch, fine-gauge needle and inject 0.4 ml of $4 N$ H_2SO_4 below the center well into the sample (do not get acid on the fluted paper).

3. Swirl the bottle to mix well, but avoid sample contact with the suspended center well. The objective is to acidify the sample to a pH of about 1.0, which simultaneously ends microbial activity and releases respired $^{14}CO_2$. Postdecompression substrate utilization will be negligible if the time interval between sample decompression and acidification is kept brief (less than 2 min) and the sample, ice-cold.

4. Collect $^{14}CO_2$ by piercing the cap with another 2-inch, fine-gauge needle and injecting 0.2 ml of the CO_2-adsorber, phenethylamine (PEA), into the fluted filter paper (do not get PEA in the sample). A quick brushing of the serum cap with rubber cement prevents minor leaks of gas through the needle pierce holes.

5. Gently shake the serum bottles in the dark (PEA is light-sensitive) on a shaker table for at least 1 h to maximize recovery of CO_2 on the fluted filter paper.

6. Use clean forceps to remove the fluted paper to an appropriately labeled scintillation vial.

7. Cap the vial tightly until scintillation cocktail is added prior to counting.

8. Vacuum-filter the entire volume of acidified sample onto a 0.2-μm nitrocellulose filter.

9. Chase the sample with a 5-ml rinse of the serum bottle and another 5-ml rinse of the filtration tower, using sterile artificial seawater (the minimum effective rinse volume is twice the sample volume).

10. Use forceps to transfer the filter to a scintillation vial.

11. Allow the filter to air dry in the vial, then cap tightly until scintillation cocktail is added prior to scintillation counting.

12. Prior to counting the samples, fill each scintillation vial with 10 ml of scintillation cocktail. We use an all-purpose, biodegradable, nontoxic cocktail called Ecolume. If the vials are allowed to equilibrate with the cocktail for about 24 h at 2 to 4°C prior to counting, chemiluminescence and quenching in the sediment-containing samples will be minimized.

Poisoned and Quench Controls

1. Poisoned controls to demonstrate the biological basis of the resulting activity measurements can be run in parallel with the experimental samples. A common and effective poison is formaldehyde at a final concentration of about 2%. Duplicate poisoned control samples for each time interval of the overall incubation period would be ideal; however, time-zero and endpoint incubations are usually sufficient to demonstrate negligible abiotic substrate conversion.

2.   Quench controls for each filter and sediment type are essential.

- Prepare and process to completion replicate "mock" (no ^{14}C-substrate added) time-zero samples of each distinctive sediment type.

- To the final PEA-saturated and sediment-loaded filters in scintillation vials, add a known amount of the labeled substrate.

- Determine quench factors by comparing counts of the latter with counts from vials containing the same amount of substrate but no filter-sample. PEA quench factors are usually minor, but the quenching effects of sediment particles, as well as substrate adsorption onto the particles, can be more significant. The latter effects should be determined rigorously for each sediment sample type.

Reporting and Interpreting Results

Separate uptake curves can be generated for substrate respiration, substrate incorporation, and "total" utilization (the sum of the two). Rates are calculated from the linear portion of each curve. If no portion of the curve is linear, assumptions inherent to the method (ultimately based on Michaelis-Menton kinetics) are not met and deriving a meaningful rate calculation from the data will be problematic. Even so, nonlinear curves resulting from different treatments of the same sample (for example, different incubation temperatures or pressures) can be compared effectively to determine treatment effects.

When the objective is a test of treatment effects, rates are usually reported as dpm or nanogram substrate per hour per gram dry weight sediment.[2,22] When attempting to reach broader conclusions about benthic bacterial carbon cycling, some investigators have applied conversion factors that reduce rates further to total protein or carbon consumed per day per seafloor area.[11,28] Given caveats of the method itself and the potential to magnify experimental and theoretical errors when extrapolating to a grander scale, caution should be exercised when interpreting these rates. In particular, direct comparisons among activity rates from a broad geographic spectrum of sample types should be avoided, unless it is clear that the basic assumption of the tracer method has been met in each case; i.e., that the amount of substrate added has not exceeded 10% of that present naturally. Organic analyses of sediment porewaters from replicate cores can provide confidence in the validity of the method and such comparisons.[11]

NOTES AND COMMENTS

1.   Strict maintenance of a deep-sea temperature during sample collection, preparation, and subsequent incubation is essential to the measurement of deep-sea microbial activities.[2,39]

2.   If a scintillation counter is not available shipboard, tightly capped plastic scintillation vials containing the processed samples (but not cocktail) can be shipped home.

3.   Diluting sediments 10-fold represents a compromise between: (1) the need to eliminate nonbacterial inhabitants, filter the sample, and minimize quenching effects and (2) the desire to minimize sample disturbance (no dilution or mixing). Maintaining anaerobic conditions during dilution and incubation can reduce disturbance effects when studying oxygen-limited subsurface sediments.[40]

4.   The ^{14}C-tracer method has been used with single substrates like glutamic acid, glucose, and acetate but also with more naturally derived organics like phytoplankton extracts of mixed amino acids. The former have been useful in tracing specific metabolic pathways in sediments or documenting treatment effects, while the latter hold promise for estimating overall heterotrophic microbial activity in sediments.

5. A total "budget" of added radioisotope can be obtained by sampling and counting a known volume of the filtrate from each sample. The sum of ^{14}C respired (on the fluted filter paper), incorporated into macromolecules (on the nitrocellulose filter), and not incorporated (in the filtrate) should approach the initial amount added (some volatile fatty acids may be lost), after accounting for quenching effects of the filters and sediment. However, the extra glassware and labor involved in collecting separate filtrates is not justified by the return, in my experience; spot budget checks are sufficient to develop confidence in the overall procedure.

REFERENCES

1. Deming, J. W. and Colwell, R. R., Barophilic bacteria associated with the digestive tracts of abyssal holothurians, *Appl. Environ. Microbiol.,* 44, 1222, 1982.
2. Deming, J. W. and Colwell, R. R., Observations of barophilic microbial activity in samples of sediments and intercepted particulates from the Demerara Abyssal Plain, *Appl. Environ. Microbiol.,* 50, 1002, 1985.
3. Deming, J. W. and Yager, P. L., Natural bacterial assemblages in deep-sea sediments: towards a global view, in *Deep-Sea Food Chains and the Global Carbon Cycle,* Rowe, G. T. and Pariente, V., Eds., Kluwer Academic Publishers, The Netherlands, 1992, p. 11.
4. Deming, J. W., Bacterial growth in deep-sea sediment trap and boxcore samples, *Mar. Ecol. Prog. Ser.,* 25, 305, 1985.
5. Deming, J. W., Ecological strategies of barophilic bacteria in the deep ocean, *Microbiol. Sci.,* 3, 205, 1986.
6. Deming, J. W. and Baross, J. A., The early diagenesis of organic matter: bacterial activity, in *Organic Geochemistry,* Vol. 6, Topics in Geobiology, Engel, M. and Macko, S., Eds., Plenum Press, New York, 1993, in press.
7. Alongi, D., Bacterial growth rates, production and estimates of detrital carbon utilization in deep-sea sediments, *Deep-Sea Res.,* 37, 731, 1990.
8. Lochte, K. and Turley, C. M., Bacteria and cyanobacteria associated with phytodetritus in the deep sea, *Nature,* 333, 67, 1988.
9. Cowen, J. P., Positive pressure effect on manganese binding by bacteria in deep-sea hydrothermal plumes, *Appl. Environ. Microbiol.,* 55, 764, 1989.
10. deAngelis, M. A., Baross, J. A., and Lilley, M. D., Enhanced microbial methane oxidation in water from a deep-sea hydrothermal vent field at simulated in situ hydrostatic pressures, *Limnol. Oceanogr.,* 36, 565, 1991.
11. Rowe, G. T. and Deming, J. W., The role of bacteria in the turnover of organic carbon in deep-sea sediments, *J. Mar. Res.,* 43, 925, 1985.
12. Rowe, G. T., Sibuet, M., Deming, J. W., Khripounoff, A., and Tietjen, J., Organic carbon residence time in the deep-sea benthos, *Prog. Oceanogr.,* 24, 141, 1990.
13. Rowe, G. T., Sibuet, M., Deming, J. W., Khripounoff, A., Tietjen, J., Macko, S., and Theroux, R., "Total" sediment biomass and preliminary estimates of organic carbon residence time in deep-sea benthos, *Mar. Ecol. Prog. Ser.,* 79, 99, 1991.
14. Cahet, G. and Sibuet, M., Activité biologique en domaine profond: transformations biochimiques in situ de composes organiques marques au carbone-14 a l'interface eau-sediment par 2000 m de profondeur dans le golfe de Gascogne, *Mar. Biol.,* 90, 307, 1986.
15. Cole, J. J., Honjo, S., and Erez, J., Benthic decomposition of organic matter at a deep-water site in the Panama Basin, *Nature,* 327, 703, 1987.
16. Jahnke, R. A., Reimers, C. E., and Craven, D. B., Intensification of recycling of organic matter at the sea floor near ocean margins, *Nature,* 348, 50, 1990.
17. Wright, R. T. and Hobbie, J. E., Use of glucose and acetate by bacteria and algae in aquatic ecosystems, *Ecology,* 47, 447, 1966.
18. Williams, P. J. LeB. and Askew, C., A method of measuring the mineralization by microorganisms of organic compounds in seawater, *Deep-Sea Res.,* 15, 365, 1968.

19. Hobbie, J. E. and Crawford, C. C., Respiration corrections for bacterial uptake of dissolved organic compounds in natural waters, *Limnol. Oceanogr.,* 14, 528, 1969.

20. Paul, K. L. and Morita, R. Y., Effects of hydrostatic pressure and temperature on the uptake and respiration of amino acids by a facultatively psychrophilic marine bacterium, *J. Bacteriol.,* 108, 835, 1971.

21. Schwarz, J. R. and Colwell, R. R., Heterotrophic activity of deep-sea sediment bacteria, *Appl. Microbiol.,* 30, 639, 1975.

22. Tabor, P. S., Deming, J. W., Ohwada, K., and Colwell, R. R., Activity and growth of microbial populations in pressurized deep-sea sediment and animal gut samples, *Appl. Environ. Microbiol.,* 44, 413, 1982.

23. Meyer-Reil, L.-A., Measurement of hydrolytic activity and incorporation of dissolved organic substrates by microorganisms in marine sediments, *Mar. Ecol. Prog. Ser.,* 31, 143, 1986.

24. Meyer-Reil, L.-A., Seasonal and spatial distribution of extracellular enzymatic activities and microbial incorporation of dissolved organic substrates in marine sediments, *Appl. Environ. Microbiol.,* 53, 1748, 1987.

25. Jannasch, H. W. and Wirsen, C. O., Studies on the microbial turnover of organic substrates in deep-sea sediments, in *Biogeochemie de la Materiere Organique a l'Interface Eau-Sediment Marin,* Colloques Internationaux du C.N.R.S., Marseilles, France, 1981, p. 285.

26. Wright, R. T., Mineralization of organic solutes by heterotrophic bacteria, in *Effect of the Ocean Environment on Microbial Activities,* Colwell, R. R. and Morita, R. Y., Eds., University Park Press, Baltimore, MD, 1974, p. 546.

27. Wright, R. T. and Burnison, B. K., Heterotrophic activity measured with radiolabeled organic substrates, in *Native Aquatic Bacteria: Enumeration, Activity, and Ecology,* Costerton, J. W. and Colwell, R. R., Eds., ASTM Spec. Tech. Publ. 695, Philadelphia, 1979, p. 140.

28. Smith, K. L., Jr., Carlucci, A. F., Jahnke, R. A., and Craven, D. B., Organic carbon mineralization in the Santa Catalina Basin: Benthic boundary layer metabolism, *Deep-Sea Res.,* 34, 185, 1987.

29. Craven, D. B. and Karl, D. M., Microbial RNA and DNA synthesis in marine sediments, *Mar. Biol.,* 83, 129, 1984.

30. Novitsky, J. A., Microbial growth rates and biomass production in a marine sediment: evidence for a very active but mostly nongrowing community, *Appl. Environ. Microbiol.,* 53, 2368, 1987.

31. Baross, J. A., Hanus, F. J., Griffiths, R. P., and Morita, R. Y., Nature of incorporated ^{14}C-labeled material retained by sulfuric acid fixed bacteria in pure cultures and in natural aquatic populations, *J. Fish. Res. Board Canada,* 32, 1976, 1975.

32. Jannasch, H. W., Wirsen, C. O., and Taylor, C. D., Deep-sea bacteria: Isolation in the absence of decompression, *Science,* 216, 1315, 1982.

33. Yayanos, A. A., Dietz, A. S., and Van Boxtel, R., Obligately barophilic bacterium from the Mariana Trench, *Proc. Natl. Acad. Sci. U.S.A.,* 78, 5212, 1981.

34. Deming, J. W., Somers, K. L., Straube, W. L., Swartz, D. G., and MacDonell, M. T., Isolation of an obligately barophilic bacterium and description of a new genus, Colwellia gen. nov., *System. Appl. Microbiol.,* 10, 152, 1988.

35. Hessler, R. R. and Jumars, P. A., Abyssal community analyses from replicate box cores in the central North Pacific, *Deep-Sea Res.,* 21, 185, 1974.

36. Barnett, P. R. O., Watson, J., and Connelly, D., A multiple corer for taking virtually undisturbed samples from shelf, bathyal and abyssal sediments, *Oceanologica Acta,* 7, 399, 1984.

37. Boland, G. S. and Rowe, G. T., Deep-sea benthic sampling with the GOMEX box corer, *Limnol. Oceanogr.,* 36, 1015, 1991.

38. Yayanos, A. A. and Van Boxtel, R., Coupling device for quick high-pressure connections to 100 MPa, *Rev. Sci. Instrum.,* 53(5), 704, 1982.

39. Yayanos, A. A. and Dietz, A. S., Thermal inactivation of a deep-sea barophilic bacterium, isolate CNPT-3, *Appl. Environ. Microbiol.,* 43, 1481, 1982.

40. Burdige, D. J., The effects of sediment slurrying on microbial processes, and the role of amino acids as substrates for sulfate reduction in anoxic marine sediments, *Biogeochemistry,* 8, 1, 1989.

CHAPTER 47

Evaluating Bacterial Activity from Cell-Specific Ribosomal RNA Content Measured with Oligonucleotide Probes

P. F. Kemp, S. Lee, and J. LaRoche

INTRODUCTION

Few methods are available to evaluate the distribution of activity among marine bacterial cells. Two approaches are the INT-formazan method[1,2] (see Chapter 53) and microautoradiography (see Chapter 45). Both methods are laborious, and consequently, neither is in common use. These methods are qualitative, i.e., cells are scored as positive (metabolically active) or negative (no evidence of metabolic activity). Both methods require incubation of bacteria with specific substrates. Sample manipulations associated with such incubations have been shown to perturb the composition, activity, and growth rate of bacteria.[3,4] There is a pressing need for a quantitative approach to measuring cell-specific activity, preferably without requiring incubations.

In this chapter, we describe a procedure for measuring the cell-specific quantity of ribosomal RNA (rRNA) and DNA in order to evaluate the frequency distribution of activity among cells. The procedure is inherently quantitative, does not require sample incubation, and can potentially be taxon-specific. Fluorescently labeled oligonucleotide probes are hybridized to the complementary 16S rRNA sequences in preserved, intact cells[5] (see Chapter 33). The resulting cell fluorescence is proportional to cellular rRNA content and can be measured with a microscope-mounted photometer system, or by other instrument systems (e.g., image analysis, flow cytometry). Similarly, DNA content is measured as fluorescence of cells stained with the DNA-specific fluorochrome DAPI. These are either prepared as separate samples for purposes of enumeration and DNA measurements, or are dual-labeled cells which are also hybridized with oligonucleotide probes. In the latter case, we can measure an rRNA:DNA ratio for individual cells.

A strong correlation between RNA content, or the RNA:DNA ratio, and growth rate has been demonstrated repeatedly in studies of copiotrophic "model" bacteria such as *Escherichia coli, Salmonella typhimurium,* and *Aerobacter aerogenes,* grown at rates ranging from 0.2 to 20 h^{-1}.[6-10] This correlation exists because the specific growth rate μ of bacteria is dependent on the cellular quantities and synthesis rates of rRNA, RNA polymerase, and protein.[6,7] Nutrient limitation reduces rRNA synthesis rates, leading to a reduction in cellular rRNA content and consequently, in protein synthesis required for growth.

We have found that rRNA cell^{-1} and the RNA:DNA ratio are also highly correlated to specific growth rate for marine bacteria.[11] rRNA was measured as probe fluorescence in

0-87371-564-0/93/$0.00 + $.50
© 1993 by Lewis Publishers

individual cells, unlike the previous studies where measurements were based on extracted nucleic acids. Marine bacteria were grown in batch or chemostat culture at much slower specific growth rates (0.01 to 0.23 h^{-1}) which were comparable to published *in situ* rates. Our data suggest that the RNA:DNA ratio is a more robust predictor of growth rate than rRNA cell^{-1} alone, with a greater proportional response to changes in growth rate. Correlation coefficients between the RNA:DNA ratio and specific growth rate for four marine bacterial isolates were $r^2 = 0.88, 0.98, 0.98$, and 0.99 ($n = 4$ or 5 growth rates, $p < 0.01$). As yet, we are not able to convert a measured cellular rRNA content (or RNA:DNA ratio) to the equivalent specific growth rate in field samples: the relationship between RNA:DNA and growth rate appears to be species specific, and we lack data on the effects of environmental variables such as temperature. However, it is clear that rRNA content and the RNA:DNA ratio are robust measures of activity.

Oligonucleotide probes are inherently taxon-specific at some level, which depends on the degree to which the probe sequence is evolutionarily conserved, and on the experimental conditions used to control hybridization stringency. In principle, taxon-specific oligonucleotide probes can be used to measure rRNA content in selected taxa within a mixed microbial community (see Chapter 33). To date, we have used mixtures of universal-level and eubacterial probes to measure the distribution of rRNA content for cultured marine bacteria, and for marine bacteria in coastal-water samples.

MATERIALS AND METHODS

Equipment

- Microphotometer (we use an Optical Technology Devices, Inc. MSA scanning microphotometer), or other instrument capable of measuring fluorescence intensity of individual bacterial cells
- Epifluorescence microscope with Hg lamp
- Vacuum pump and filtration manifold (e.g., Hoefer manifold)
- Incubation chamber, 37°C
- High-speed benchtop centrifuge

Supplies

- Glass slides, cover slips
- Glass beakers
- 0.2-μm Acrodisc (Gelman)
- Calibration-standard microbeads (Flow Cytometry Standards Company)

Solutions and Reagents

- Formalin
- DAPI fluorochrome, 1 mg ml^{-1} in distilled water
- Glycerol
- 100 and 95% ethanol
- Gelatin (Bio-Rad, Richmond, CA)
- Sodium dodecyl sulfate (SDS), 1%
- Phenol
- Chloroform
- Isoamyl alcohol

- $CrK(SO_4)_2$ (Aldrich Chemical, Milwaukee, WI)
- PBS: $1\times$ PBS = 145 mM NaCl, 100 mM sodium phosphate, pH 7.5
- Nonidet P-40 in distilled water (Sigma), 1 and 0.1%
- TE buffer: 10 mM Tris, 1 mM EDTA, pH 8.0
- SET: $5\times$ and $0.2\times$ concentrations
  $1\times$ SET = 0.15 M NaCl, 20 mM Tris, 1 mM EDTA, pH 7.8
- STE buffer: 0.1 M NaCl, 10 mM Tris, 1 mM EDTA, pH 7.6
- Hybridization solution: contains the following in $5\times$ SET: probe(s) (50 ng each/30 μl solution); dextran sulfate (10%); bovine serum albumin (0.2%); polyadenylic acid (0.01%); and SDS (0.1%)
- DNase-free RNase (Boehringer Mannheim, Indianapolis, IN)
- Calf thymus DNA standard (Sigma)
- *E. coli* rRNA standard (Sigma)

PROCEDURES

Procedures described here are modified from DeLong et al.[5] and Giovannoni et al.[12] Modifications introduced to the original methods include cell permeabilization, composition of the hybridization solution, and dual-labeling of cells with DAPI in order to measure DNA as well as rRNA per cell.

Sample Preparation

1. Cells are collected on a 0.2-μm Nuclepore filter and preserved immediately by dipping the filter in an ice-cold mixture of PBS and formalin (1:9 volumes of formalin:PBS, filtered through a GS filter [Millipore, Bedford MA]). Samples are then stored at 4°C until processed.

2. Add 1 volume of 1% Nonidet P-40 to 9 volumes of the cell suspension. Incubate overnight at 4°C, in order to increase cell permeability.

3. Slide preparation:

   - Dissolve gelatin (final concentration 0.1%) in a boiling water bath.

   - After the gelatin cools to room temperature, add $CrK(SO_4)_2$ (final concentration 0.01%).

   - Filter the gelatin solution through a 0.2-μm Acrodisc.

   - Prepare clean glass microscope slides by dipping in 95% ethanol for about 2 min. Wipe and dry the slides.

   - Three to four drops of the gelatin solution are squirted onto a slide, spread with a clean applicator (e.g., cotton-tipped swab), and dried vertically in a rack overnight.

4. Add DAPI (final concentration 10 μg ml^{-1}) to the Nonidet P-40-treated cell suspension. Allow to stain for 10 min. Collect cells by centrifugation (10,000 $\times$ g, 15 min), and resuspend in 0.1% Nonidet P-40. For convenience, the cell suspension should be adjusted at this time to about 5×10^7 cells ml^{-1}.

5. Smear about 10 to 15 μl of the cell suspension on the gelatin coated slide (about 1-cm diameter circle). Dry the smear at room temperature (normally $>$1 h).

6. Dehydrate the smear by sequential ethanol baths for 3 min each in 50, 80, and 100% ethanol, and dry again at room temperature.

Hybridization

1. We have oligonucleotides custom-synthesized with a 5′ aminolink modifier (our oligo-nucleotides are synthesized by Oligos Etc., Inc., Wilsonville, OR). Oligonucleotides are conjugated with the fluorochrome Texas Red sulfonyl chloride as described by DeLong (Chapter 33). Briefly, one volume of Texas Red sulfonyl chloride (Molecular Probes, Eugene, OR) dissolved in dimethyl formamide (10 μg μl^{-1}) is mixed with 5 volumes of the oligonucleotide dissolved in 50 mM borate buffer (pH 9.2), and incubated overnight at 22°C. Fluorochrome-conjugated oligonucleotides (fluorescent probes) are purified from unconjugated fluorochromes and oligonucleotides without conjugated fluorochrome. Seph-adex G-25 (Pharmacia, Uppsala, Sweden) column chromatography and polyacrylamide gel electrophoresis are used for the purification of the fluorescent probes (see Chapter 33).

2. Then 30 μl of the hybridization solution is gently pipetted onto the cell smear, and a siliconized cover slip is placed on the droplet.

3. Slides are incubated overnight (15 to 18 h) at 37°C on a raised bed with a small volume of 5× SET at the bottom in an air-tight container.

4. After the incubation, the cover slips are gently removed by dipping in 5× SET, and the slides are washed 3 times in 0.2× SET bath at 37°C for 10 min each time.

5. After the wash, the slides are dried at room temperature. Care is taken to handle the fluorochrome under subdued light at all times.

6. Three to four drops of mountant (9 volumes of glycerol mixed with 1 volume of 1× PBS) are dropped on the cell smear and covered with a cover slip. Excess mountant is removed by lightly pressing the slide against absorbant paper tissue.

7. Slides are stored in the dark, either at 4°C for several days or at −70°C for long-term storage.

Measurement of rRNA and DNA by Microphotometry

Instrument

We measure fluorescence of individual cells with a scanning microphotometer (Optical Technology Devices, Inc.) mounted on a Nikon Optiphot microscope. Photometer output is passed to a Metrabyte analog/digital converter and recorded on an IBM-compatible computer. The microphotometer incorporates a target-spotting aperture system which precisely defines the area over which fluorescence intensity is measured. Detector apertures of 0.09 to 0.35 are used to restrict the target area to a minimum of 1 μm diameter at 1250× total magni-fication. The microphotometer has been modified so that its photomultiplier detector is parfocal with the binocular eyepieces. Thus, the operator can simply center a cell or cell-free background area at the target location (for example, using an ocular grid) while viewing the field normally. Automatic scanning and recording of excitation wavelength and fluo-rescence intensity are triggered by appropriate single-key commands at the computer key-board. Background fluorescence associated with fluorescence of the mounting medium and glass is measured and automatically subtracted from cell fluorescence.

Instrument Calibration and Measurement

Calibration-standard fluorescent microbeads (e.g., from Flow Cytometry Standards Com-pany) are used to correct for day-to-day fluctuations in instrument readings, which result from variations in line voltage, lamp output, and gain and sensitivity settings. Daily stand-ardization is absolutely essential, because day-to-day variation in photometer output can be substantial.

1. Establish a reference calibration curve by measuring mean bead fluorescence at full ex-citation intensity, then at lower intensities using combinations of neutral-density filters to achieve $1/2$, $1/4$, and $1/8$ intensity levels. Mean bead fluorescence should be linearly related to intensity with a zero-intercept. This reference curve should be repeated periodically.

2. Daily calibration can consist of a single-point measurement of the mean fluorescence of standard beads (n $\approx$ 100) at full excitation intensity.

3. Fluorescence of cells in samples is measured at emission wavelengths of 455 nm (for DNA by DAPI fluorescence) or 600 nm (for RNA by Texas Red fluorescence). About 150 to 180 cells are measured for each subsample.

4. Daily measurements of cell fluorescence are expressed as fractions or multiples of the mean fluorescence of standard beads on that day. Measurements expressed as "bead units" are reproducible and comparable among days and samples.

Calibration of Probe and DAPI Fluorescence

We obtain independent estimates of RNA cell^{-1} using ethidium bromide (EtBr) fluoro-metry, and construct the regression of EtBr-measured RNA cell^{-1} on probe fluorescence. This regression is used therefore to convert fluorescence measurements to rRNA cell^{-1}. Similarly, the regression of EtBr-measured DNA cell^{-1} on DAPI fluorescence is used to convert fluorescence to DNA cell^{-1}. Note that although the RNA extraction procedure described below isolates both rRNA and other forms of RNA, nearly all (approximately 90%) of cellular RNA is ribosomal. Therefore, calibration of probe fluorescence (measures rRNA) against EtBr-RNA (measures total RNA) introduces very little bias. Procedures similar to the following can be used to calibrate the user's particular instrument.

1. Cultures of marine bacterial isolates are grown at various rates, using batch or chemostat culture methods. Samples are collected on 0.2-μm Nuclepore filters.

2. The cells are resuspended in a small volume of ice cold STE buffer, then a subsample is taken to estimate (by Acridine orange or DAPI direct counts) the total number of cells collected. Samples are kept at <6°C during handling.

3. The cells are pelleted (10,000 $\times$ g, 15 min), and stored for <2 weeks at -70°C until nucleic acid extraction.

4. Cell pellets are resuspended in TE buffer and cells are lysed with 1% SDS.

5. The lysate is extracted once with an equal volume of phenol-chloroform-isoamyl alcohol mixture (25:24:1, equilibrated with STE), and once with an equal volume of chloroform. In each extraction, the aqueous phase is transferred to the next step while the phenol phase is discarded. Material at the interface between the aqueous phase and the phenol mixture is carried to the chloroform extraction in order to minimize the possible loss of nucleic acid during the extraction.

6. The final aqueous phase is precipitated overnight with two volumes of ethanol and am-monium acetate (final concentration 2.3 M) at -20°C.

7. The precipitate is collected by centrifugation at 10,000 $\times$ g for 15 min. The pellet is briefly rinsed with 70% ethanol, and a small volume of TE buffer is added.

8. The solution is stored at 4°C for 1 day. The pellet is thoroughly dissolved by vortexing occasionally.

9. After dissolution, subsamples of the nucleic acid extract are digested with DNase-free RNase (Boehringer Mannheim, Indianapolis, IN) for 2 h at 37°C (DNA sample). The untreated extract (containing both DNA and RNA) and the RNase-digested samples (containing DNA only) are stored at -70°C until fluorimetric measurement.

10. Then, 3 to 12 µl of the DNA or DNA + RNA samples are added to 3 ml of TE with 1 µg of EtBr.

11. Fluorescence is measured in duplicate or triplicate samples using a Perkin-Elmer spectrophotometer (excitation 300 nm, emission 600 nm).

12. Standard curves for DNA or RNA are generated with known amounts of calf thymus DNA (Sigma) or *E. coli* rRNA (Sigma), over the concentration range of 0.1 to 0.8 µg DNA (or RNA) ml^{-1}. Standard curves generated with RNase-digested calf thymus DNA showed no difference from the curves generated with untreated calf thymus DNA.

13. The standard curves are used to convert fluorescence of DNA or DNA + RNA samples to the equivalent weight of DNA or RNA. The amount of DNA in a sample is calculated directly from the fluorescence of the RNase-digested DNA sample. The amount of RNA is calculated by difference using the disappearance of fluorescence after RNase digestion: i.e., the difference between fluorescence in the undigested DNA + RNA sample and in the matching, RNase-digested DNA sample. DNA or RNA content per cell is then calculated by dividing the total amount of DNA or RNA by the total number of cells in the extract.

14. Extraction efficiency is checked by determining the total recovery of known quantities of DNA and RNA standards added at concentrations equivalent to the nucleic acid concentrations in samples and processed identically to normal samples. We have found that extraction efficiency is relatively high (about 75 to 100%) using this procedure.

Example of rRNA Frequency Distribution Data

Figure 1 shows the distributions of rRNA cell^{-1} at four growth rates from a chemostat culture of a bacterium isolated from the Sargasso Sea water column (50 m). The mean rRNA cell^{-1} for this isolate was highly correlated to specific growth rate ($\mu = 0.01$ to 0.24 h^{-1}, $r^2 = 0.996$).

NOTES AND COMMENTS

1. The primary obstacle to measuring rRNA cell^{-1} in natural bacterial assemblages is the low rRNA content of many slow-growing cells. We have found that with a single probe, as few as 33% of bacteria in coastal water samples are visibly labeled.[13] The fluorescence of the dimmest visibly labeled cells falls below the detection limit of the microphotometer. However, it is possible to design and utilize multiple probes with comparable levels of specificity.[13] Since each probe acts independently of the others, the fluorescence per rRNA increases in additive fashion with an increasing number of probes. Using this approach, we have found that the percent of natural bacterial cells (Long Island coastal water, in spring) which contain a measurable quantity of rRNA increases asymptotically with an increasing number of probes (up to 5) to approximately 75%. Even in these relatively eutrophic coastal waters, the remaining 25% of cells containing extremely little rRNA, i.e., less than our current detection limit of 0.24 fg rRNA cell^{-1}. The majority of cells which remained unmeasurable were coccoidal and relatively small.

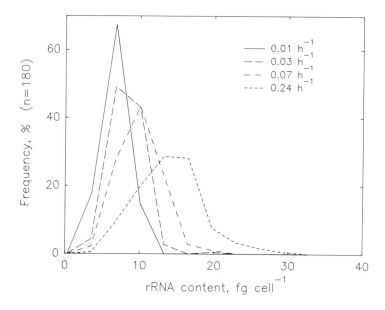

Figure 1. Distribution of rRNA content among cells at four growth rates.

2. The procedures described here should be readily applicable to other instruments such as image analysis systems (see Chapter 38) and flow cytometers (see Chapters 21 and 22). Although the photometry system we use is designed to have excellent sensitivity to low-fluorescence signals, alternative instruments may provide greater sensitivity or faster sample processing times.

3. Calibration of probe fluorescence — we used the regression line obtained from culture studies to estimate RNA cell^{-1} from probe fluorescence in a natural bacterial community sample. When compared to independent measurements of RNA cell^{-1} using EtBr fluorometry, the two estimates consistently agree to within 15%.[13] We conclude that our culture-based calibration of probe fluorescence is robust and applicable to natural samples.

4. The fluorochrome DAPI intercalates between adenosine and thymine base pairs in DNA. Because DAPI is AT-specific, it can be used to measure the relative DNA content per cell if the exact GC:AT ratio is unknown. If the GC:AT ratio is known, measured, or assumed (e.g., to be 1:1), then the absolute DNA content per cell can be estimated. It may be possible to use AT-specific and GC-specific fluorochromes simultaneously to measure the GC:AT ratio directly in natural bacterial cells.

5. Nucleic acid samples prepared for EtBr fluorometry by the method described above were clean enough for RNase and DNase digestion. When checked by agarose gel electrophoresis, characteristic bands of DNA or RNA disappeared completely after 2 h digestion with DNase or RNase, respectively.

REFERENCES

1. Zimmerman, R., Iturriaga, R., and Becker-Birck, J., Simultaneous determination of the total number of aquatic bacteria and the number of thereof involved in respiration, *Appl. Environ. Microbiol.*, 36, 926, 1978.

2. Newell, S. Y., Modification of the gelatin-matrix method for enumeration of respiring bacterial cells for use with salt-march water samples, *Appl. Environ. Microbiol.*, 47, 873, 1984.

3. LaRock, P. A., Schwartz, J. R., and Hofer, K. G., Pulse labelling: a method for measuring microbial growth rates in the ocean, *J. Microbiol. Method*, 8, 281, 1988.

4. Carman, K. R., Dobbs, F. C., and Guckert, J. B., Consequences of thymidine catabolism for estimates of bacterial production: An example from a coastal marine sediment, *Limnol. Oceanogr.*, 33, 1595, 1988.

5. DeLong, E. F., Wickman, G. S., and Pace, N. R., Phylogenetic stains: ribosomal RNA-base probes for the identification of single cells, *Science*, 243, 1360, 1989.

6. Bremer, H., Parameters affecting the rate of synthesis of ribosomes and RNA polymerase in bacteria, *J. Theor. Biol.*, 53, 115, 1975.

7. Ryals, J., Little, R., and Bremer, H., Control of rRNA and tRNA syntheses in *Escherischia coli* by guanosine tetraphosphate, *J. Bacteriol.*, 151, 1261, 1982.

8. Bremer, H. and Dennis, P. P., Modulation of chemical composition and other parameters of the cell by growth rate, in *Echerischia coli and Salmonella typhimurium Cellular and Molecular Biology*, Neidhardt, F. C., et al., Eds., American Society of Microbiology, 1987, 1527.

9. Schaechter, E., Maaloe, O., and Kjeldgaard, N. O., Dependence on medium and temperature of cell size and chemical composition during balanced growth of *Salmonella typhimurium*, *J. Gen. Microbiol.*, 19, 592, 1958.

10. Dortsch, Q., Roberts, T. L., Clayton, J. R., Jr., and Ahmed, S. I., RNA/DNA ratios and DNA concentrations as indicators of growth rate and biomass in planktonic marine organisms, *Mar. Ecol. Prog. Ser.*, 13, 61, 1983.

11. Kemp, P. F., Lee, S., and LaRoche, J., Estimating population growth rates from fluorescence of single cells hybridized with 16S rRNA-targeted oligonucleotide probes, *Appl. Environ. Microbiol.*, in preparation.

12. Giovannoni, S. J., DeLong, E. F., Olsen, G. J., and Pace, N. R., Phylogenetic group-specific oligodeoxynucleotide probes for identification of single microbial cells, *J. Bacteriol.*, 170, 720, 1988.

13. Lee, S., Malone, C., and Kemp, P. F., Use of multiple 16S rRNA-targeted fluorescent probes to increase signal strength and measure cellular RNA from natural planktonic bacteria, *Mar. Ecol. Prog. Ser.*, submitted.

CHAPTER 48

Use of Fluorogenic Model Substrates for Extracellular Enzyme Activity (EEA) Measurement of Bacteria

Hans-Georg Hoppe

INTRODUCTION

The bulk of organic substances in aquatic environments is macromolecular and not ready for incorporation into the bacterial cell. These materials have to be preconditioned by extracellular enzymes, so that they can be available for bacterial growth and nutrient cycles. The activity of these enzymes is, in many cases, a limiting factor for substrate decomposition and bacterial growth. As demonstrated by size fractionation experiments,[1] extracellular enzymes are mainly produced by bacteria. To a minor extent, they may also originate from autolytic processes and from other organisms. According to the definition given by Priest,[2] extracellular enzymes are generally located outside the cytoplasmic membrane. In order to distinguish between enzymes which are still associated with their producers and those which occur dissolved in the water or adsorbed to particles, Chróst[3] suggested calling the former "ectoenzymes" and the latter "extracellular" enzymes.

Substrates for hydrolysis are generally proteins, carbohydrates, fats and organic P- or S-compounds. Mechanisms of decomposition of individual compounds within these groups may be studied by *in vitro* experiments. However, aquatic microbial ecologists require, in many cases, a more general measurement of the *in situ* hydrolytic capacity of the prevailing bacterial community. This has led to the adaptation of biochemical methods for determination of overall bacterial extracellular enzyme activities (peptidases, α- and β-glucosidases, chitinases, etc.) in natural waters (Table 1). These methods enable us to study the impact of extracellular enzyme activity (EEA) on bacterial substrate uptake, bacterial growth, and water chemistry. The quantitative estimates of total bacterial extracellular enzyme activity are completed by rapid and sensitive tests for the detection of enzymatic properties of bacterial isolates. The methods used for these purposes are based on the application of fluorogenic model substrates. These substrates have some characteristics in common: (1) they contain an artificial fluorescent molecule and one or more natural molecules (e.g., glucose, amino acids), linked by a specific binding (e.g., peptide binding, ester binding); fluorescence is observed after enzymatic splitting of the complex molecule (Figure 1); (2) the hydrolysis of model substrates is competitively inhibited by a variety of natural compounds with the same structural characteristics; (3) hydrolysis of model substrates follows first order enzyme kinetics; and (4) application of those model substrates allows enzyme activity measurements under natural *(in situ)* conditions within short incubation periods. The latter is highly important for microbial ecologists, because the process of enzymatic hydrolysis is fully

0-87371-564-0/93/$0.00 + $.50

Table 1. Catalog of Fluorogenic Analog (Model) Substrates used for the Study of Extracellular Enzyme Activity in Aquatic Habitats and of Natural or Hygienically Important Bacteria

Substrate	Extracellular enzyme	Application	Ref.
MUF-N-acetyl-glucosaminide	N-acetyl-glucosaminidase (chitinase)	Brackish water	11
L-Leucyl-β-naphthylamide	Aminopeptidase	Natural waters	17
		North Sea	5
L-Leucine-4-methylcoumarinyl-7-amide (Leu-MCA)	Aminopeptidase	Lakes	4
		Marine sediment	18
		Brackish water	19
		Coral reef	19
MUF-β-D-galactopyranoside	β-D-galactosidase	Natural waters	20
MUF-α-D-glucopyranoside	α-D-glucosidase	Brackish water	11
		Eutrophic water	21
MUF-β-D-glucopyranoside	β-D-glucosidase	Brackish water	11
		Eutrophic water	21
		Marine sediment	22
		Lake water	23
MUF-β-D-glucuronide	β-D-glucuronidase	Fecal coliform test	15,26
MUF-butyrate-heptanoate-palmitate	Lipase	Brackish water bacterial colonies	13
MUF-phosphate	Phosphatase (acid, alcaline)	Lake water	9
		Lake water bacteria	24
Methylfluorescein-phosphate	Phosphatase	North Pacific	25
MUF-sulfate	Sulfatase	River epilithon	8

MUF = 4-methylumbelliferone; MCA = 4-methylcoumarinyl-7-amide; the fluorochrome included in MCA-substrates is AMC = 7-amino-4-methylcoumarin.

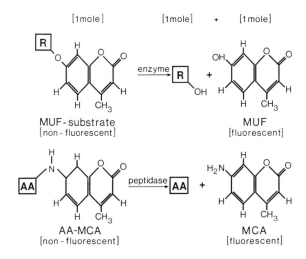

Figure 1.   Molecular structure and enzymatic hydrolysis products of 4-methylumbelliferyl (MUF) substrates and 4-methylcoumarinyl-7-amide (MCA) substrates.

integrated into the dynamics of the substrate pool in question (e.g., input by primary production, output by bacterial substrate uptake), which should not be changed or interrupted. Short incubation times avoid or minimize changes of the microbial community structure and the substrate regime (which are due to the bottle effect) during the experiment. Fluorogenic model substrates involving methylumbelliferone (MUF) or 7-amino-4-methylcoumarin (AMC) as a fluorescing agent are not toxic and they are supplied to natural water in low quantities for measurement of their hydrolysis rate (HR), which is dependent on the concentration of their natural competitors, and in high or increasing quantities for the measurement of the maximum velocity of hydrolysis (V_{max}).

The method of determining bacterial EEA is valid for all types of natural waters and also, for waste water and drinking water. Using special application techniques, it can also be applied to undisturbed sediment cores.[4]

MATERIALS REQUIRED

Equipment

- Epifluorescence microscope (used with low magnification objective lens or without objective for specific UV excitation of bacterial colonies)
- Fluorometer (type: filter, spectral, cuvette, or microtiter plate)
- Temperature-controlled incubator
- Deep-freezer (storage of stock solutions and incubation devices inoculated with model substrate for later use)

Supplies

- Micropipettes
- Incubation devices (optional scintillation vials, plastic cuvettes, or microtiter plates)

Solutions

See Table 1 for list of substrates. Concentrations of model substrate solutions and specifications of buffers are provided in the section Preparation and Application of Substrate Solutions.

PROCEDURES

Quantitative Determination of EEA in Water

Sample Preparation

In general, water samples are taken by sterile bacteriological samplers and subsamples are distributed to sterile flasks for incubation with the model substrate. Alternatively, water samples may also be taken from oceanographic samplers and distributed in clean scintillation vials, plastic cuvettes, or microtiter plates for incubation. The risk of contamination is, of course, greater in the latter case.

The measurement of EEA is independent of water volume, because the fluorescent end-product of hydrolysis is not bound to particles or incorporated by them, but exists dissolved in the water. The organic compound linked to the fluorescent tracer may be incorporated by bacteria after hydrolysis of the combined molecule; however, this does not influence the measurement.

Sediment cores are treated with injections of the fluorogenic model substrate in order to establish an even distribution of substrate in the core and not to disturb the sediment structure. For details see Meyer-Reil.[4] Special sample preparations and methical details for the determination of EEAs of bacteria attached to surfaces (plant, detritus, stones, artificial surfaces) submerged in natural water are supplied by Rego et al.,[5] Goulder,[6] Hoppe,[7] and Chappell and Goulder.[8]

Model Substrates

A variety of fluorogenic model substrates is in use for the most common types of natural macromolecules. Occasionally, two or more types of model substrate are appropriate for one and the same natural substrate; comparative studies, however, are very rare.[9] A model substrate is usually composed of a fluorescent tracer molecule and an organic molecule which are linked by a specific binding. Despite their dimeric character, these model substrates are not only competitively inhibited by natural dimers but also by oligomers and polymers.[3] The most commonly used model substrates, together with some of their specifications, are listed in Table 1. Model substrates are supplied by companies Serva, Sigma, Fluca, and others.

Preparation and Application of Substrate Solutions

In order to measure enzyme kinetics, model substrate concentrations should cover a sufficient range, depending on the trophic status of the investigated water. The number of concentrations used for a study depends on its specific aim. For instance, the application of a single, high concentration is recommended, only if measurements of relative activity are desired. The following description refers to 4-methylumbelliferyl (MUF)-substrates and 4-methylcoumarinyl-7-amide (MCA)-substrates, the model substrates which are most commonly used in water laboratories.

Model substrate solutions are prepared in sterile water, as long as they dissolve completely in water at the desired concentration. Water solubility differs with type of substrate. For dissolving higher concentrations of substrate, parts of the water may be substituted by an organic solvent, such as Methylcellosolve (ethylene glycol monomethyl ether). Low concentrations of Methylcellosolve do not influence the result of the measurement. Stirring supports the process of model substrate dissolution.

For our measurements in coastal waters of the Baltic Sea, we prepare a set of six concentrations whose molarities are 2, 20, 100, 200, 2500, and 5000 μM. Aliquots of these (50 μl) are pipetted into $1/2$ microcuvettes prior to the addition of 1 ml sample water, resulting in final concentrations of 0.1, 1, 5, 10, 125, and 250 μM. We have chosen relatively high volumes of model substrate in order to avoid mistakes due to inaccurate pipetting. $1/2$ Microcuvettes give higher fluorescence readings than other types. Positioning of their small side toward the emission collecting lens of the fluorometer is recommended. If incubation is not made in cuvettes but in greater or smaller volumes of water (e.g., in scintillation vials or microtiter plates), model substrate additions have to be adjusted to these volumes, or substrate solution concentrations have to be changed. Deep-frozen substrate solutions can be stored in sterile bottles for several weeks and have always to be kept in the dark.

Incubation

All incubations have to be made in the dark. Shaking of the water enhances enzyme activity in most cases, but in order to obtain representative results, it should be adjusted to natural water turbulence. Temperature of incubation depends on the aim of the study. The period of incubation depends on several preconditions: (1) the nature of water samples or sediment samples with respect to the EEA; (2) the mode of fluorescence reading — continuously or start-endpoint determination (see below); and (3) the sensitivity of the fluorometer. Incubation of 1 to 3 h is recommended for eutrophic limnetic waters and coastal areas. In the open sea and clean inland waters 3 to 8 h are appropriate and up to 48 h for deep water samples. Continuous fluorescence reading, of course, reduces required incubation time considerably.[10]

Reading Fluorescence

Any filter fluorometer or spectral fluorometer may be used for fluorescence measurement. Excitation and emission characteristics of the fluorochrome have to be installed in the instrument (364 and 445 nm, respectively for MUF; and 380 and 440 nm for AMC, which is the fluorochrome of MCA substrates).

Start-Endpoint Measurement

Initial fluorescence is read shortly after mixing the model substrate with the sample. A second reading is made at the end of the incubation time. Time between the two fluorescence readings must be recorded for later calculations. Because the intensity of fluorescence is influenced by pH, a decision about pH adjustment has to be made before measurement. Maximum fluorescence of MUF is obtained at pH 10.3. Thus, 200 μl of pH 10.3-Tris/HCl buffer are placed in the cuvette prior to the addition of 1 ml of the water to measure. When the cuvette itself serves as an incubation vessel, of course, measurements have to be made at the natural pH of the water. This can only be done when no pH changes are expected during incubation, e.g., with naturally buffered sea water. Depending on temperature and type of substrate, fluorescence increases are linear for several hours.[11] Otherwise, time series experiments have to be made before starting routine work to check for the period of linear fluorescence increase.

The initial fluorescence reading detects some fluorescence, which is related to model substrate concentrations. This is due to impurities of the model substrates and does not influence the enzymatic reaction. This background fluorescence increases during prolonged storage of model substrate stocks. Heavy turbidity of the probe causes quenching. Centrifugation rather than filtration is recommended to avoid this effect. The endpoint measurement of fluorescence should be made with the same fluorometer-settings as used for the initial measurement, otherwise, factors for correction have to be considered.

Time Series, Continuous Fluorescence Reading

In the case of time-series measurements, subsamples are taken in short intervals from the incubation vessels for fluorescence detection. If incubation takes place in cuvettes, these are repeatedly checked for fluorescence development. Most suitable for this type of measurement, however, is an automatic microtiter plate fluorometer, e.g., the Fluoroskan II fluorometer supplied by Flow Laboratories Ltd. Incubations are made in the holes of the microtiter plates, which allows running many parallel samples. Microtiter plates are scanned automatically for fluorescence at any desired time interval. Temperature control of the microtiter plate chamber is an option of necessity. Microtiter plate fluorometers are not as sensitive as conventional types of fluorometers, and thus, their application is restricted to waters with high enzyme activity.

Calibration

Fluorochromes included in the model substrate are produced by the companies: Sigma, Serva, and Fluca. Molecular weights are 175.2 for AMC and 176.2 for MUF. Fluorochrome standard solutions in water are prepared according to the fluorescence readings of the enzyme assay, because their concentrations should be in the same range as fluorochrome concentrations produced by model substrate hydrolysis. Suitable concentrations for fluorochrome standard solutions would be 0.1, 1, 5, 10, and 50 μM. Employing the method, standard addition is the best way of calibration. In this case, 20 μl aliquots of a suitable fluorochrome standard solution are carefully mixed with the contents of the cuvette after the last fluorescence reading of the enzyme assay and then measurement is repeated. By standard addition, all problems which may arise including matrix effects, water turbidity, aging of the lamp, etc., are avoided. Alternatively a 100% setting of an appropriate fluorochrome standard solution can also be used for calibration.

Blanks

It is not necessary to correct fluorescence readings for the autofluorescence of the water sample unless it has been proven that autofluorescence changes during incubation. Theoretically, measured fluorescence changes during incubation may also be attributed to reactions other than enzymatic hydrolysis. This can only be tested by boiling the sample water prior to incubation with model substrates. Boiled water samples, in most cases, do not show any changes in fluorescence after incubation, and thus, nonenzymatic hydrolysis appears to be negligible.

Calculations and Interpretation

Fluorescence readings of the replicates (four recommended) at any reading time are averaged. Corrections for autofluorescence or "boiled water" blank are made, if necessary. Fluorescence increases per hour are calculated by regression. Added standard fluorescence increases are also averaged and the velocity of enzymatic hydrolysis is calculated by the equation:

$$V = (A * C * D)/B \qquad (1)$$

where V = velocity of hydrolysis (μg C L^{-1} h^{-1}), in terms of the C content of the organic component hydrolyzed from the model substrate; A = fluorescence increase per hour; B = added fluorochrome standard fluorescence; C = added fluorochrome standard concentration (μM L^{-1}); D = carbon content of the organic component of the model substrate which is equivalent to 1 μM of fluorochrome. D is 72 μg C μM^{-1} in cases where the organic component is glucose or a 6-C amino acid.

Depending on the design of the study, this calculation is done for one, two, or more model substrate concentrations used in the enzyme assay. If a series of increasing model substrate concentrations has been applied, individual V-calculations may be used for enzyme kinetic calculations of the Lineweaver-Burk or Eadie-Hofstee type on the basis of the Michaelis-Menten equation.[12] This provides data of Vm (maximum velocity of hydrolysis) and Ht (hydrolysis time of the model substrate). In addition, the Michaelis constant (Km) can be calculated from experiments where natural competitors of the model substrates were excluded.

The sensitivity of fluorescence detection of the applied fluorochromes is very high. Concentrations and increases of concentrations of these fluorochromes in the nanomolar range can be reliably measured by their fluorescence, if the power supply of the instrument is sufficiently stabilized. The standard error of the method depends on water quality and fluorescence intensity; in most cases it is <5% of the mean value.

Qualitative Determination of Extracellular Enzymatic Properties of Bacterial Isolates

Sample Preparation

The basis for this variation of the method are undefined bacteria colonies or pure culture colonies grown on agar plates. Tracing bacteria colonies for a variety of extracellular enzymatic properties with conventional "specific media" techniques is a laborious and time-consuming task. Using fluorogenic model substrates, a complete analysis can be made within minutes.[13] Bacterial colonies may be grown on any kind of agar medium. Spread plate techniques are more suitable than pore plate techniques. Colonies should not be grown too big and they should not overlap.

We use this technique for the analysis of saprophytic bacteria colonies growing on conventional ZoBell-agar plates (prepared with fresh water or sea water). In the case of pure culture analysis, an agar plate is inoculated with 16 to 20 bacteria in a defined order, which are subsequently processed simultaneously. The analysis results in percentages of total saprophytes, which are able to hydrolyze given compounds, or in the detection of individual pure-culture enzymatic properties.

Model Substrates

Model substrates are again those which are listed in Table 1. For the quick identification of *E. coli*, there are some fluorogenic model substrates in use which refer to its specific metabolism. These are 4-methylumbelliferyl-β-D-galactoside and 4-methylumbelliferyl-β-D-glucuronide.[14,15] Commercial media containing the latter substrate (e.g., Brila-MUG-broth, Merck) are already extensively used for *E. coli* tests and they are on their way to becoming international standards.

Results obtained with model substrates should be comparable with those obtained by conventional agar techniques (e.g., growth characteristics on starch, protein, cellulose, and chitin). Comparative studies, however, are rare. It has been demonstrated, that chitin-agar and MUF-*N*-acetyl-glucosaminide medium give exactly the same results for all species tested.[16] In the case of other enzyme identifications (e.g., peptidases and esterases), results obtained by the two techniques were the same with the majority of test species.[13]

Preparation and Application of Substrate Solutions

Model substrate stock solutions (5 mM L^{-1}) are prepared in Methylcellosolve (ethyleneglycolmonomethylether, $C_3H_8O_2$) and stored at $-25°C$ in the dark. Working solutions of 0.1 mM L^{-1} are made before the experiment by the dilution of stock solutions with specific buffers, which create the optimal pH for the desired enzyme reaction. Optimal buffer systems are: Tris/HCl, pH 7.4 for MUF-α-D-glycopyranoside, MUF-β-D-glucopyranoside, Leu-MCA, and MUF-butyrate and Tris/HCl, pH 8.3 for MUF-phosphate and Phosphate/Citrate, pH 4.95 for MUF-*N*-acetyl-glucosaminide.

Chromatography filter disks of the same size as the Petri dishes used are soaked with model substrate working solutions. The disks are then carefully laid on the bacteria colonies growing on the agar surface. Inclusion of air bubbles has to be avoided. After approximately 3 min of exposure at room temperature the disks are removed from the Petri dishes and placed in a clean and empty Petri dish. For the establishment of optimal pH conditions for fluorescence (pH 10.3) they may be exposed to NH_3 vapors for about 1 min.

Detection of Enzyme Activity

Pretreated filter disks are illuminated with excitation light (MUF-substrates, 364 nm; MCA-substrates, 380 nm) e.g., from an HBO epifluorescence microscope lamp. A conventional UV lamp cannot be used for this purpose; but it can be used for MPN (most probable number) *E. coli* fluorescence detection, e.g., in Brila-MUG-broth.

The whole filter shows a blue fluorescence, due to background fluorescence of model substrates. Sites of enzyme activity, however, can be clearly recognized by their bright fluorescence. Bright spots can easily be coordinated with colonies on the plates and thus, the percentage of total colony counts, which exhibits special extracellular activities, can be calculated. If the plate was inoculated with known species colonies, individual spectra of their enzymatic properties can be estimated. Enzyme spectra of total colonies show characteristic patterns in different habitats. They reflect the prevailing nutrient situation and other environmental conditions.

REFERENCES

1. Hollibaugh, J. T. and Azam, F., Microbial degradation of dissolved proteins in seawater, *Limnol. Oceanogr.*, 28, 1104, 1983.
2. Priest, F. G., *Extracellular Enzymes,* Van Nostrand Reinhold (UK), Wokingham, 1984, 79.
3. Chróst, R. J., Environmental control of the synthesis and activity of aquatic microbial ectoenzymes, in *Microbial Enzymes in Aquatic Environments,* Springer-Verlag, New York, 1991, 29.
4. Meyer-Reil, L.-A., Seasonal and spatial distribution of extracellular enzymatic activities and microbial incorporation of dissolved organic substrates in marine sediments, *Appl. Environ. Microbiol.,* 53, 1748, 1987.
5. Rego, J. V., Billen, G., Fontigny, A., and Somville, M., Free and attached proteolytic activity in water environments, *Mar. Ecol. Prog. Ser.,* 21, 245, 1985.
6. Goulder, R., Extracellular enzyme activity associated with epiphytic microbiota on submerged stems of the reed Phragmites australis, *FEMS Microbiol. Ecol.,* 73, 323, 1990.
7. Hoppe, H.-G., Microbial extracellular enzyme activity: a new key parameter in aquatic ecology, in *Microbial Enzymes in Aquatic Environments,* Chróst, R. J., Ed., Springer-Verlag, New York, 1991, 60.
8. Chappell, K. R. and Goulder, R., Epilithic extracellular enzyme activity in acid and calcareous headstreams, *Arch. Hydrobiol.,* 125, 129, 1992.
9. Pettersson, K. and Jansson, M., Determination of phosphatase activity in lake water — a study of methods, *Verh. Int. Verein. Theor. Angew. Limnol.,* 20, 1226, 1978.
10. Jacobsen, T. R. and Rai, H., Determination of aminopeptidase activity in lakewater by a short term kinetic assay and its application in two lakes of differing eutrophication, *Arch. Hydrobiol.,* 113, 359, 1988.
11. Hoppe, H.-G., Significance of exoenzymatic activities in the ecology of brackish water: measurements by means of methylumbelliferyl-substrates, *Mar. Biol. Prog. Ser.,* 11, 299, 1983.
12. Armstrong, F. B., *Biochemistry,* 2nd edition, Oxford University Press, New York, 1983, 653.
13. Kim, S. J. and Hoppe, H.-G., Microbial extracellular enzyme detection on agar plates by means of methylumbelliferyl-substrates, in GERBAM — Deuxienne Colloque International de Bactériologie Marine, Actes de Colloque, 3, IFREMER, Brest, 1986, 175.
14. Berg, J. D. and Fiksdal, L., Rapid detection of total and fecal coliforms in water by enzymatic hydrolysis of 4-methylumbelliferyl-β-D-galactoside, *Appl. Environ. Microbiol.,* 54, 2118, 1988.
15. Hofmann, G. and Schwien, U., Eds., *Fluoreszenzmikroskopischer Nachweis von E. coli, Grundlagen und Praxis,* GIT-Verlag, Darmstadt, 1989, 74.
16. O'Brien, M. and Colwell, R. R., A rapid test for chitinase activity that uses 4-methylumbelliferyl-N-acetyl-β-D-glucosaminide, *Appl. Environ. Microbiol.,* 53, 1718, 1987.

17. Somville, M. annd Billen, G., A method for determining exoproteolytic activity in natural water, *Limnol. Oceanogr.*, 28, 190, 1983.
18. Hoppe, H.-G., Kim, S. J., and Gocke, K., Microbial decomposition in aquatic environments: combined process of extracellular enzyme activity and substrate uptake, *Appl. Environ. Microbiol.*, 54, 784, 1988.
19. Hoppe, H.-G., Schramm, W., and Bacolod, P., Spatial and temporal distribution of pelagic microorganisms and their proteolytic activity over a partly destroyed coral reef, *Mar. Ecol. Prog. Ser.*, 44, 95, 1988.
20. Chróst, R. J. and Krambeck, H. J., Fluorescence correction for measurements of enzyme activity in natural waters using methylumbelliferyl-substrates, *Arch. Hydrobiol.*, 106, 79, 1986.
21. Somville, M., Measurement and study of substrate specificity of exoglucosidase activity in eutrophic water, *Appl. Environ. Microbiol.*, 48, 1181, 1984.
22. King, G. M., Characterization of β-glucosidase activity in intertidal marine sediments, *Appl. Environ. Microbiol.*, 51, 373, 1986.
23. Chróst, R. J., Characterization and significance of β-glucosidase activity in lake water, *Limnol. Oceanogr.*, 34, 660, 1989.
24. Chróst, R. J. and Overbeck, J., Kinetics of alkaline phosphatase and phosphorus availability for phytoplankton and bacterioplankton in lake Plußsee (north German eutrophic lake), *Microbiol. Ecol.*, 13, 229, 1987.
25. Perry, M. J., Alkaline phosphatase activity in subtropical Central North Pacific waters using a sensitive fluorometric method, *Mar. Biol.*, 15, 113, 1972.
26. Shadix, L. C. and Rice, E. W., Evaluation of β-glucuronidase assay for the detection of *Escherichia coli* from environmental waters, *Can. J. Microbiol.*, 37, 908, 1991.

CHAPTER 49

Photoassimilation of Acetate by Algae

Russell L. Cuhel

INTRODUCTION

Acetic acid, found as the acetate anion (CH_3COO^-) in neutral solution, is a principal intermediate in a number of intracellular reactions of biosynthesis and energy metabolism. In particular, much of the energy of aerobes is generated through the tricarboxylic acid cycle starting with acetyl-CoA, and this same molecule is the starting material for fatty acid (lipid) biosynthesis. Sources of acetate in aquatic habitats include excretion by some photoautotrophs and, more importantly, release from anaerobic zones where fermentation is active. Acetate concentrations in organic-rich anaerobic surface sediments may reach 300 μM (see Reference 1) with associated high flux into the water column, and fall within the range of reported half-saturation constants for acetate uptake by algae (about 1 mM; cf. Reference 2). While CO_2 is quantitatively the most important carbon source for most photoautotrophs, several groups of algae (notably the diatoms and cyanobacteria) are capable of assimilating acetate at a low concentration. Hence, higher affinity transport systems may exist, based on uptake at nanomolar concentration,[3] but have not been investigated. This may permit these algae to grow mixotrophically on the sediment surface or in low CO_2 environments such as extremely softwater lakes.[4]

In the presence of light, much of the acetate carbon is used specifically for lipid synthesis in algae. This attribute lends itself as a diagnostic feature for estimating the role of phototrophs in the biogeochemical cycling of acetate in aquatic systems. Through comparison of light and dark total acetate uptake and light stimulation of acetate incorporation into the lipid fraction of microplankton, several aspects of microbial acetate cycling may be investigated.

Advantages of the method are: (1) the subcellular distribution of acetate carbon is little influenced by substrate concentration, permitting the use of a single level of addition in light/dark experiments; (2) the proportion of acetate respired to CO_2 by bacteria is independent of light, whereas much more of the label is respired by algae in the dark than in the light; (3) the light-induced change in acetate carbon distribution among subcellular fractions of algae is pronounced (5- to 10-fold enhancement of incorporation into lipids in the light) and provides a clear signal for photoassimilation measurements, moreover, it is independent of biomass and total uptake, permitting measurements in highly heterogeneous environments such as microbial mats where reproducible subsampling is difficult or impossible; (4) the subcellular distribution of acetate carbon in bacteria is independent of light, hence the degree of light effect is qualitatively related to the contribution of algae to acetate cycling in mixed natural populations; and (5) specifically labeled [2-¹⁴C]-acetate (i.e., the methyl carbon) is available and imparts added specificity for fatty acid biosynthesis investigations.

0-87371-564-0/93/$0.00 + $.50

Disadvantages of the method are: (1) the acetate concentration in the natural environment is rarely known and hence, quantitative cycling of acetate is difficult to assess; (2) during condensation of two-carbon fragments in fatty acid biosynthesis, one of the methyl hydrogens is lost to water, rendering high specific activity tritiated acetate unsuitable for use in distributional analysis; (3) the relatively low specific activity of ^{14}C acetates (up to 60 mCi $mmol^{-1}$) leads to the addition of approximately 70 nM acetate at 10,000 dpm, often overwhelming likely natural substrate concentrations; (4) fatty acid synthesis occurs in the mitochondrion of eukaryotes, and is hence susceptible to many of the same inhibitors as might be used to prevent bacterial assimilation; and (5) small (< 1 μm) cyanobacteria such as *Synechococcus* spp. actively photoassimilate acetate, defeating the possible use of size fractionation to separate bacterial and algal activities.

MATERIALS REQUIRED

Equipment

- Filtration assembly for 25-mm filters and 100- to 250-ml samples
- Light source (deck box or artificial light incubator)
- Shaker table for land-based studies if respiration is measured
- Liquid scintillation counter

Supplies

- Automatic microliter pipettes
- Incubation vessels (glass, polycarbonate, polystyrene)
- Liquid scintillation vials with plastic cone-style caps (two per sample)
- Glass fiber filters, 25 mm
- 1- and 10-cc syringes
- 18 and 22° needles
- 25-ml Erlenmeyer flasks or 30-ml serum bottles (one per sample) with serum stoppers
- Respiration cups (#K-882320-0000, Kontes Corporation, Vineland, NJ 08360), one per sample
- Filter paper for respiration wicks (Whatman 3MM, cut into 1'' squares)

Solutions and Reagents

- Inert gas (N_2 or Ar) supply and regulator
- CO_2 trapping agent (ethanolamine, phenethylamine, Protosol, etc.)
- 1 N HCl
- 2N H_2SO_4
- 1 M Tris buffer, pH 6.5
- [2-^{14}C]-acetic acid, sodium salt, >50 mCi $mmol^{-1}$
- Distilled water
- Liquid scintillation fluid (fluor) capable of high water content (Aquasol, Universol, or similar)

PROCEDURES

Incubation of Natural Plankton Assemblages

Preparation of [¹⁴C]-Acetate Stock Solutions

It is desirable to add the smallest amount of labeled acetate possible, consistent with obtaining sufficient uptake to measure the subcellular distribution of incorporated label (>5000 dpm total). This reduces perturbation of the natural acetate concentration (usually unknown but probably in the low nanomolar range) by added [¹⁴C]-acetate to a minimum. Ideally, a range-finding experiment using a low addition (about 10,000 dpm ml⁻¹) and several time points should be undertaken if a liquid scintillation counter is readily available. In its absence, effective measurements can often be made using 100- to 250-ml filtered samples after 2 to 4 h of incubation at 10,000 dpm ml⁻¹.

In preparation of the stock solution, the often ignored problem of contamination with other organic compounds must be taken into consideration. High specific activity [2-¹⁴C]-acetate is usually shipped in ethanol as a preservative and free-radical scavenger. Even a 1:10,000 dilution of the ethanolic stock (17 M ETOH) adds more than 1 mM final concentration of metabolizable dissolved organic carbon and thus may result in stimulation of bacterial growth. In addtion, it has been reported that [¹⁴C]-acetate stocks are often contaminated with [¹⁴C]-carbonate,[5] which can be assimilated by algae in the light.

To simultaneously eliminate both of these problems:

1. Add one drop (about 50 μl) of 1 N HCl to the stock (or aliquot thereof, depending on amount purchased).

2. Insert an 18° needle attached to a clean source of N₂ or Ar gas and a 22° vent needle into the septum of the vial.

3. Evaporate to dryness with a slow stream of the gas (the smaller vent needle keeps slight positive pressure in the vial, preventing loss of aerosol).

4. Reconstitute the acetate in sterile distilled water at 50 μCi ml⁻¹ (gives 10,000 dpm ml⁻¹ at 1:10,000 dilution) for the working stock solution and filter sterilize (0.2 μm syringe filter) into a sterile vial.

This procedure should result in a volatile loss of much less than 5% of the [¹⁴C]-acetate stock. The stock may be kept refrigerated during regular use and should be frozen during the interim.

Sampling and Incubation

To minimize the effects of containment-induced bacterial growth, acetate photoassimilation studies should be kept to as short an incubation as possible (1 to 4 h). Since the method is largely qualitative, simulation of *in situ* conditions (e.g., light intensity and chromatic quality, day/night cycles, etc.) is not really necessary, although experiments should be conducted during the normal light period if possible. Initial sample handling should be undertaken in subdued light, but the extreme precautions required for carbon dioxide fixation measurements are unnecessary. Sufficient sample should be obtained for zero time, dark, the intended number of light incubations, and ancillary analyses (e.g., chlorophyll, species composition, bacterial counts, etc.).

Subsamples may be incubated conveniently in 100- to 250-ml polycarbonate flasks or bottles, or in disposable tissue culture flasks. A relatively constant light field, such as an artificial light incubator or deck box during mid-day is preferable. The minimum number of treatments includes a dark and light level which saturates but does not photoinhibit CO_2 fixation (i.e., approximately 25% clear noonday sunlight or 500 μEinsteins m^{-2} s^{-1}). Ideally three or four subsaturating light intensities should also be used. While the level of addition of [^{14}C]-acetate and length of incubation are best determined on site or by experience, additions of 1:10,000 (10 to 25 μl depending on sample volume) and incubations of 2 h will almost always provide useful results. Additions are best made with a repeating micropipette to individual containers, but it is sometimes convenient to add the [^{14}C]-acetate to a bulk sample and then quickly dispense this into the incubation vessels. Zero-time blanks for total uptake/subcellular distribution and respiration should be taken at this time.

After incubation, remove the samples to the dimly lit filtration area. Prior to processing, one total added activity sample (1 ml) should be removed from each incubation bottle (or several from the bulk sample if the latter method of addition is used) and added directly to a liquid scintillation vial containing 10 ml of fluor. Then proceed with the respiration analysis and filtering below.

Respiration of [^{14}C]-Acetate

In pure cultures of algae, the proportion of [^{14}C]-acetate respired to $^{14}CO_2$ can be much greater in the dark than in the light,[5,6] whereas this difference is expected to be negligible for bacterial metabolism. In addition, the turnover time for acetate must include consideration of respiratory as well as assimilatory transformation. Hence analysis of conversion of [^{14}C]-acetate to $^{14}CO_2$, a relatively simple procedure, is recommended. However, the procedure is optional since the subcellular distribution of assimilated acetate (below) will identify the presence or absence of the phenomenon.

While samples are incubating, prepare respiration flasks as follows (be sure to include a zero-time control as well):

1. Fold a 1'' (25 mm) square piece of filter paper into a pleated fan that will fit into the respiration cup (about half will stick out the top of the cup). Insert the stem through a hole in the serum stopper (the slit from a large-bore needle puncture works fine) so that the bottom of the cup will be about halfway down the small flask or serum bottle when inserted tightly.

2. Moisten the wick with 0.25 ml of the CO_2 trapping agent delivered from a 1-cc syringe or repeating pipette. Assemble the stopper/cup combination tightly to the small flask or bottle.

3. When samples are harvested, withdraw 10 ml into a syringe fitted with an 18° needle. Punch a second 18° needle through the stopper to serve as a vent. Carefully insert the sample needle and inject the 10 ml into the bottom of the small flask, taking care not to moisten any part of the cup assembly with the sample (aim the needle toward the side of the container and eject the sample slowly). Withdraw the sample syringe and remove the vent needle.

4. Using a 1-cc syringe and 22° needle, carefully dispense 0.25 ml of 2 N H_2SO_4 into each small flask, again letting it run down the side of the container and avoiding the wick assembly.

5. Incubate the samples for at least 2 h with mild agitation. For shipboard experiments, the inevitable motion of the ship will be sufficient, but for land-based studies a shaker table set to just swirl the liquid is necessary. Meanwhile, proceed with filtration of the remaining sample.

6. Remove the stopper and cup, taking care not to touch the cup to the side of the small flask. Holding the cup in a liquid scintillation vial, use scissors to cut the stem just below the stopper. Add 0.25 ml 1 M Tris buffer pH 6.5 (neutralizes the base of the trapping agent, reducing chemiluminescence) and 10 ml fluor. After capping, shaking vigorously to dislodge the wick from the cup. Allow at least an hour (preferably in the dark) before counting on a liquid scintillation counter.

Filtration of Assimilated Radioactivity

The remainder may then be filtered through fine mesh glass fiber filters (25 mm Whatman GF/F is recommended), rinsed lightly with filtered natural water, and filters frozen at $-20°C$ until processing. The volume filtered must be known, either through measurement or by averaged actual volume of completely filled incubation vessels (less the volume analyzed for respiration if included). Although sample processing time should be short relative to the duration of light exposure, the actual incubation time will be the period from isotope addition to the end of filtration for each sample.

Subcellular Fractionation of Labeled Plankton

The various ^{14}C-labeled components of microbial biomass are separated by a serial extraction scheme based on the work of Roberts et al.[7] (see Figure 1) with minor modifications for filtered samples. The procedure is described in Chapter 72 and should be followed with the omission of the barium sulfate precipitation step for the cold and hot acid soluble fractions. This method may be simplified for acetate photoassimilation measurements by drawing supernatant fractions for "low molecular weight" (LMW) and "hot trichloroacetic acid (TCA) soluble" (HTCA) directly into liquid scintillation vials and counting the entire combined extracts without subsampling.

Up to 40 samples are conveniently processed at one time (12 to 15 h total) if facilities permit. Calculations (see below) are amenable to spreadsheet format, and to maximize simplicity it is convenient to count all (n) samples from each fraction in order, i.e., total activities $(1 - n)$, total LMW $(1 - n)$, alcohol, lipid, total HTCA, protein, and respiration wicks (if done). The statistical terminator of 20 min or 1% error (whichever comes first) is recommended.

Calculations

The subcellular fractionation procedure produces five components and a total uptake which is the sum of the fractions. For ^{14}C]-acetate, these components are:

1. Low molecular weight: cold acid-soluble carbon compounds including (but not restricted to) untransformed acetate; amino acids; sugars; nucleotides; metabolic intermediates; and vitamins

2. Alcohol-soluble protein (ALC): membrane-localized proteins having a relatively hydrophobic nature

3. Lipids (LIP): ether-soluble nonpolar components including mono-, di-, and triglycerides; phospholipids; sterols; and most chlorophylls

4. Hot TCA-soluble (HTCA): hydrolyzed polysaccharides and nucleic acids

5. Protein (PROT): the majority of cellular proteins

6. Total uptake (TUPTAKE): sum of 1. through 5. above

PELLET   TREATMENT   SUPERNATANT   FRACTION

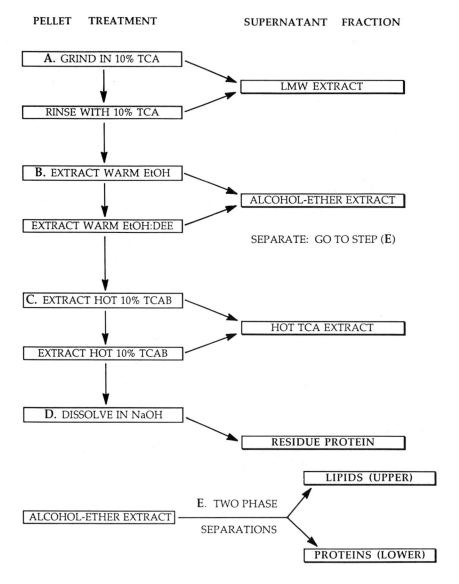

Figure 1.   Various ^{14}C-labeled components of microbial biomass, separated by a serial extraction scheme with minor modifications for filtered samples. (Based on the work of Roberts, R. B. et al.[7])

In addition, if respiration (RESP) to $^{14}CO_2$ is measured, then

7.   Total metabolism (TMETAB): total transformation of acetate to metabolic products is the sum of (TUPTAKE + RESP) and is the term appropriate for use in determining the turnover time (T_t) of acetate by the population at the concentration of measurement.

To simplify calculations to integer format it is convenient to work at the dpm L^{-1} level, since less than 1% of the label is often taken up under the incubation conditions described, and that amount is further divided among the subcellular fractions. Minor variations in the added total activity (TACT) rarely cause concentration effects on uptake but complicate comparative calculations. It is, therefore, useful to normalize all sample values (and blanks) to a constant activity, applying minor sample-to-sample adjustments through a correction factor (CF). In this example, 10,000 dpm ml^{-1} will be used as the normalizing activity. For each incubated sample and the zero-time blank (BLK), the data are adjusted by the CF:

$$CF = (10,000 \text{ dpm ml}^{-1}) (TACT)^{-1} \tag{1}$$

The BLK for each fraction should be subtracted prior to incubation time correction, hence for each fraction:

$$dpm \ L^{-1} = [(\text{aliquot dpm}) (\text{ml extract/ml counted}) (CF)/(\text{L filtered})] - BLK \tag{2}$$

and for the total radioactivity assimilated into biomass:

$$TUPTAKE = (LMW) + (ALC) + (LIP) + (HTCA) + (PROT) \tag{3}$$

Likewise, for respiration determination:

$$dpm \ L^{-1} = \{[(\text{wick dpm}) (CF)/(\text{ml analyzed})] - BLK\}(1000 \text{ ml } L^{-1}) \tag{4}$$

The most readily obtained diagnostic for photoassimilation is the percentage of total assimilated label in the lipid fraction, i.e.:

$$\% \ LIPID = 100(LIP)/TUPTAKE \tag{5}$$

Interpretation

In the absence of the respiration measurement, there are four kinds of light response which I have observed:

No light response — This most likely represents bacterial dominance in acetate uptake.

Little or no light effect on total uptake, but light stimulation of the proportion of ^{14}C incorporated into lipid — This is the most common positive result,[6,8] and is probably linked to light-dependent biomass increase (hence, biovolume and cell surface area) due to photoautotrophic CO_2 fixation in the light.

Total uptake proportional to light, but no light effect on lipid incorporation — While rare, this could reflect extreme energy limitation restricting active transport by algae (and would be associated with high acetate respiration) and has been reported for the cyanobacterium *Anacystis nidulans*.[9]

Total uptake inversely proportional to light with no light effect on subcellular distribution — This would be expected if excretion of photosynthetically produced acetate by algae were to cause dilution of the label for subsequent uptake by bacteria.

The respiration measurement adds another series of possibilities which need not be iterated here (but see Note 3 below). In culture studies dark incubation always leads to enhanced respiration of acetate in algae, and is directly interpretable in that context. Since bacterial acetate metabolism can be dominated by respiration, however, light/dark changes in algal acetate respiration may not be detectable in heterotroph-dominated systems.

NOTES AND COMMENTS

1. Much of the literature on acetate photoassimilation concerns pure culture physiological measurements using extremely high concentrations of acetate (up to 5 mM). Little has been published regarding this process with natural populations at low substrate concentrations.
2. With higher added radioactivity it may be possible to identify active organisms by autoradiographic techniques on incubation subsamples.
3. The photosystem II inhibitor DCMU may provide a useful means of delineating photoassimilation of acetate because it specifically increases the production of $^{14}CO_2$ from acetate in algae.[6] Thus, a DCMU-enhanced proportion of [^{14}C]-acetate respired would be clearly indicative of algal acetate metabolism.

REFERENCES

1. Sansone, F. J. and Martens, C. S., Volatile fatty acid cycling in organic-rich marine sediments, *Geochim. Cosmochim. Acta,* 46, 1575, 1982.
2. Hellebust, J. A. and Lewin, J., Heterotrophic nutrition, in *The Biology of Diatoms,* Werner, D., Ed., University of California Press, Berkeley, 1977, chap. 6.
3. Cuhel, R. L. and Waterbury, J. B., Biochemical composition and short-term nutrient incorporation patterns in a unicellular marine cyanobacterium, *Synechococcus* (WH8703), *Limnol. Oceanogr.,* 29, 370, 1984.
4. Provasoli, L., Organic regulation of phytoplankton fertility, in *The Sea,* Hill, M. N., Ed., Interscience Publishers, New York, 1963, 165.
5. Cooksey, K. E., Acetate metabolism by whole cells of *Phaeodactylum tricornutum* Bohlin, *J. Phycol.,* 10, 253, 1974.
6. Cooksey, K. E., The metabolism of organic acids by a marine pennate diatom, *Plant Physiol.,* 50, 1, 1972.
7. Roberts, R. B., Abelson, P. H., Cowie, D. B., Bolton, E. T., and Britten, R. J., Studies of biosynthesis in *Escherichia coli, Carnegie Inst. Wash. Publ.,* 607, 1–521, 1963.
8. Kannangara, C. G., Henningsen, K. W., Stumpf, P. K., Appelqvist, L.-A., and von Wettstein, D., Lipid biosynthesis by isolated barley chloroplasts in relation to plastid development, *Plant. Physiol.,* 48, 526, 1971.
9. Hoare, D. S., Hoare, S. L., and Moore, R. B., The photoassimilation of organic compounds by autotrophic blue-green algae, *J. Gen. Microbiol.,* 49, 351, 1967.

ADDITIONAL REFERENCES

Cuhel, R. L., Jannasch, H. W., Taylor, C. D., and Lean, D. R. S., Microbial growth and macromolecular synthesis in the northwestern Atlantic Ocean, *Limnol. Oceanogr.,* 28, 1, 1983.

Ihlenfeldt, M. J. and Gibson, J., Acetate uptake by the unicellular cyanobacteria *Synechococcus* and *Aphanocapsa, Arch. Microbiol.,* 113, 231, 1977.

Thomas, G. and Mercer, E. I., Biosynthesis of the chlorosulpholipids of *Ochromonas danica, Phytochemistry,* 13, 797, 974.

CHAPTER 50

Starvation-Survival Strategies in Bacteria

Richard Y. Morita

INTRODUCTION

Most ecosystems are oligotrophic. Even in eutrophic ecosystems there will be physiological types of bacteria that will not have the proper substrates for growth and reproduction. In nature, the heterotrophic bacteria rely mainly on syntrophy for the energy source, because most bacteria cannot use cellulose (the main constituent of plant debris). The lack of bioavailable energy sources is attributed to all of the various physiological types of bacteria utilizing the various types of organic matter, leaving only a small amount of organic matter in the environment, most of it recalcitrant. The lack and bioavailability of energy is discussed in detail by Morita.[1,2] In this chapter I outline a method for producing and investigating starved cells and the background and rationale of this approach is explained.

All organisms are influenced by their chemical and physical environments. The most important environmental factor is the availability of energy since energy is necessary to respond when any of the other chemical or physical environmental factors become favorable. Furthermore, energy, within limits, aids the microorganisms to ward off the unfavorable physical and chemical environmental factors.

Most of the bacteria in any given ecosystem are not active. The amount of organic material in soil, which has much more organic matter than aquatic environments, does not fulfill the requirements for the energy of maintenance. Thus, we recognize that most of the bacteria in any oligotrophic environments must be in a starvation mode. When cells are deprived of energy for a sufficient length of time, they enter into the starvation-surgical mode. Starvation-survival is defined as a physiological state resulting from an insufficient amount of nutrients, mainly energy, for growth and reproduction of the microbes.[3]

When energy is supplied to starved cells, it is taken up immediately and the longer the starvation period, the longer the lag period for the organisms.[4,5] Thus, when energy, especially a labile substrate such as glucose or an amino acid, is added to an oligotrophic environmental sample, one must take into account the priming effect of the added substrate. The amount of substrate, in addition to the amount of energy in the sample, may just be enough to permit the cells to function. In addition, perturbation effects must also be taken into consideration. Unfortunately, researchers do not often take into account either the priming effect or the perturbation effect on the activity of the indigenous microflora in the environmental sample. In essence, what is being observed is the response of the starved cells to the added substrates. Yet, readers generally overlook this fact.

0-87371-564-0/93/$0.00 + $.50

When cells are starved, there are four patterns observed in the starvation-survival process. One of the results of starvation of the cells is the formation of ultramicrocells. The most studied pattern of starvation-survival is one where the cells increase in number after the onset of starvation and then decrease in number, but the cell size decreases drastically.[4] The other patterns are where (1) cells immediately decrease in number with the advent of starvation, (2) cells neither increase nor decrease in number with starvation time, and (3) cells increase in number on the onset of starvation and reach a plateau that remains higher than the original number of cells that were starved.[6] Thus, it is recognized that starvation-survival is actually the survival of the species in the face of the lack of energy. If we take into consideration the residence time of water masses in the ocean, then we realize that the period of survival of microbes without energy for over a thousand years. In long-term survival of the cell, it enters into a state of metabolic arrest, a situation analogous to a bacterial spore. Although in most patterns the number of viable cells is reduced drastically, by no means does the complete clone expire. It only takes one cell of any species per environment for continuation of the species.

In environmental samples, it is well recognized that ultramicrocells dominate. With the advent of epifluorescent microscopy, ultramicrocells have been observed in most aquatic environments. Viable bacteria, belonging to the genera *Vibrio, Aeromonas, Pseudomonas,* and *Alcaligenes,* were recovered from the estuarine waters passed through a 0.2-μm Nuclepore filter.[7] By image analysis, Maeda and Taga[8] demonstrated that they fell into the bacterial cell size range between 0.4 and 0.8 μm range (73% in the open ocean; in Tokyo Bay, 90%; Sagami Bay, 58%, and Ohtsuchi Bay, 48.3%). Kogure and Koike,[9] after filtering seawater through a 1.0 μm Nuclepore filter, found most of the particles were less than 0.6 μm, employing an Elzone particle counter. Li and Dickie[10] were able to demonstrate bacteria in the Sargasso Sea that were able to pass through a 0.2-μ Nuclepore filter. In the deep sea, bacteria capable of passing through a 0.45-μm membrane filter, were ubiquitous.[11] Even in soil, many bacteria are less than 1.0 μm. What percentage of the ultramicrocells are the result of starvation is not known. However, most bacteria will form ultramicrocells when starved. The presence of ultramicrocells is a good indication that most bacteria in various ecosystems are in some stage of starvation.

When cells grown in the laboratory are subjected to a starvation menstruum, an initial intense metabolic activity occurs, as measured by oxygen uptake.[12] This should be expected because in most patterns of starvation-survival the cell numbers are increasing. At this time all the endogenous energy reserves are being used. Lipids, poly-β-hydroxybutyric acid, and carbohydrates disappear from the cell early in the starvation process.[13-15] During the starvation process, the amount of protein, DNA, and RNA decline.[13,16] The survival process was divided into three stages with the AODC counts remaining high throughout the stages.[17,18] During the first stage (0 to 14 d) large fluctuations in viable cell numbers occur along with fluctuations in the protein, DNA, and RNA. In stage 2 (14 to 70 d), an overall decrease in viability occurs, which is dependent upon the growth rate of the cell population, and the DNA per cell drops to between 4.2 and 8.3% of the original amount. RNA and protein also fluctuate during stage 2, except when the cells are grown at a very low dilution rate in a chemostat. In stage 3, the protein, DNA, and RNA stabilize. In stage 3 (70 to 98 d), viable cell counts stabilized at 0.3% of the AODC counts. Long-term starvation is the prolongation of stage 3 and the cells enter a state of metabolic arrest — a situation analogous to a bacterial spore. Cells that have a slow growth are better adapted for starvation-survival.

One of the more fascinating features of the starvation process is the synthesis of new proteins reported by Amy and Morita[16] and later found in other cells undergoing starvation-survival.[19,20] These were then termed starvation proteins. However, starvation proteins are probably in bacteria in the various ecosystems. Thus, we are actually look at the loss of the starvation proteins when cells are grown under more optimal conditions. In actuality, when labile substrates are added to ecological samples, what we are observing is the response of the starved cells to the added substrate.

In the environmnent, it should be realized that the various bacteria present are in some degree of starvation-survival, mainly because only a fraction of the cells present are active. Thus, we can expect to see cells that have just recently used substrate(s) to those that have not seen substrates for an extensive period of time. Starved cells still retain their ability to use substrate.

MATERIALS REQUIRED

Only a small fraction of the bacteria in aquatic systems can be cultivated on artificial media. As a result, only the starvation-survival of those bacteria that can be grown on artificial media can be addressed.

Equipment

- Incubator, preferably set to the organism's optimal growth temperature
- High-speed centrifuge
- Rotary incubator shaker
- Magnetic stirrer

Supplies

- Pipettes (sterile) or automatic pipette with sterile tips
- Spreaders (hockey sticks)
- Centrifuge tubes with caps (sterile)
- Starvation bottle (cotton-plugged; 1-l aspirator bottle with silicone tubing attached to a large pyrex filling attachment fitted with a cotton plug with a large silicone stirring bar)
- Bunsen burner

Solutions and Media

- Sterile starvation menstruum (artificial seawater or normal saline [sterile], depending if a marine or freshwater bacterium is employed; 500 ml should be in the starvation bottle
- Sterile dilution blanks (10 ml) containing the starvation menstruum
- Sterile medium (the formulation depends on the organism in question; 20 ml in a 500-ml flask)
- Agar plates (same formulation as medium but solidified with 1.2% agar; however, 1/10 the concentration sometime works better)

PROCEDURES

Aseptic techniques should be used throughout the procedures.

1. The known culture is inoculated into the flask containing the appropriate medium and incubated in a rotary incubator shaker at the proper temperature and aerated by shaking (about 200 rpm).

2. The culture is permitted to grow to the mid to late log phase.

3. The cells are harvested by centrifugation in sterile screw-capped centrifuge tubes at 22,000 × g for 12 to 20 min. The supernatant fluid is decanted off and the cells washed with sterile starvation menstruum. This is repeated three times.

4. The final suspension of the washed cells are then added to the starvation bottle, but first adjusted so that the starvation bottle will have about 10^7 bacteria per ml.

5. Spread plates are made from the starvation bottle at various intervals to determine the number of viable cells ml^{-1}. Prior to sampling the starvation bottle, it should be placed on a magnetic stirrer to insure even distribution of the cells. Small portions of the starvation menstruum can be withdrawn through the filling attachment fitted to the starvation bottle. Proper dilutions of the starved cell suspension are made before the spread plates are undertaken.

6. Spread plates should be incubated at the proper temperature and incubated for 6 to 7 d before the viable cell counts are made. (Be prepared to find more bacteria per ml than at zero time in the first several days of the starvation process. The counts should not go down to zero with prolonged incubation.)

7. Optional: AODC counts can also be made during the starvation process. The decrease in cell size is another indication of the starvation-process. Protocols in this volume may be used to estimate cell size.

REFERENCES

1. Morita, R. Y., Bioavailability of energy and its relationship to growth and starvation survival in nature, *Can. J. Microbiol.*, 34, 436, 1988.
2. Morita, R. Y., The starvation-survival state of microorganisms in nature and its relationship to the bioavailable energy, *Experientia*, 46, 813, 1990.
3. Morita, R. Y., Starvation-survival of heterotrophs in the marine environment, *Adv. Microb. Ecol.*, 6, 171, 1982.
4. Novitsky, J. A. and Morita, R. Y., Morphological characterization of small cells resulting from nutrient starvation of a psychrophilic marine vibrio, *Appl. Environ. Microbiol.*, 32, 616, 1976.
5. Amy, P. S. and Morita, R. Y., Protein patterns of growing and starved cells of a marine *Vibrio* sp., *Appl. Environ. Microbiol.*, 43, 1748, 1982.
6. Morita, R. Y., Starvation and miniaturisation of heterotrophs, with special emphasis on maintenance of the starved viable stage, in *Bacteria in the Natural Environments: The Effect of Nutrient Conditions,* Fletcher, M. and Floodgate, G., Ed., Academic Press, NY, 1985, 111.
7. MacDonnell, M. T. and Hood, M. A., Isolation and characterization of ultramicrobacteria from a Gulf Goast Estuary, *Appl. Environ. Microbiol.*, 43, 566, 1982.
8. Maeda, M. and Taga, N., Comparisons of cell size of bacteria in four marine localities, *La Mer (Tokyo)*, 21, 207, 1983.
9. Kogure, K. and Koike, I., Particle counter determination of bacterial biomass in seawater, *Appl. Environ. Microbiol.*, 53, 274, 1987.
10. Li, W. K. W. and Dickie, P. M., Growth of bacteria in seawater filtered through 0.2 μm Nuclepore membranes: implication for dilution experiments, *Mar. Ecol. Prog. Ser.*, 26, 245, 1985.
11. Tabor, P. S., Ohwada, K., and Colwell, R. R., Filterable marine bacteria found in the deep sea: distribution, taxonomy, and response to starvation, *Microb. Ecol.*, 7, 67, 1981.
12. Kjelleberg, S., Humphrey, B. A., and Marshall, K. C., Initial phases of starvation and activity of bacteria at surfaces, *Appl. Environ. Microbiol.*, 46, 978, 1983.
13. Hood, M. A., Guckert, J. B., White, D. C., and Deck, F., Effect of nutrient deprication on lipid, carbohydrate, DNA, RNA, and protein levels in *Vibrio cholerae*, *Appl. Environ. Microbiol.*, 52, 788, 1986.
14. Oliver, J. D. and Stringer, W. F., Lipid composition of a psychrophilic marine *Vibrio* during starvation-induced morphogenesis, *Appl. Environ. Microbiol.*, 47, 461, 1984.

15. Malmcrona-Friberg, K., Tunlid, A., Mården, P., Kjelleberg, S., and Odham, G., Chemical changes in cell envelope and poly-β-hydroxybutyrate during short term starvation of a marine bacterial isolate, *Arch. Microbiol.*, 144, 340, 1986.
16. Amy, P. S., Pauling, C., and Morita, R. Y., Recovery from nutrient starvation by a marine *Vibrio* sp., *Appl. Environ. Microbiol.*, 45, 1685, 1983.
17. Moyer, C. L. and Morita, R. Y., Effect of growth rate and starvation-survival on the viability and stability of a psychrophilic marine bacterium, *Appl. Environ. Microbiol.*, 55, 1122, 1989.
18. Moyer, C. L. and Morita, R. Y., Effect of growth rate and starvation-survival on cellular DNA, RNA and protein of a psychrophilic marine bacterium, *Appl. Environ. Microbiol.*, 55, 2710, 1989.
19. Jouper-Jaan, Å., Dahllöf, B., and Kjelleberg, S., Changes in the protein composition of three bacterial isolates from marine waters during short term energy and nutrient deprivation, *Appl. Environ. Microbiol.*, 52, 1419, 1986.
20. Reeve, C. A., Amy, P. S., and Matin, A., Role of protein synthesis in the survival of carbon-starved *Echerichia coli* K-12, *J. Bacteriol.*, 160, 1041, 1984.

CHAPTER 51

Community Respiration Measurements Using a Pulsed O$_2$ Electrode

Christopher Langdon

INTRODUCTION

The biological cycle in the ocean at its very simplest can be viewed as the production and destruction of organic matter. The process of production of organic matter by photosynthesis is limited to a shallow near surface layer of the ocean during the daylight hours. The destruction of organic matter by the process of respiration is carried out throughout the water column at all times. The relative constancy of our atmosphere indicates that on a global scale the processes of photosynthesis and respiration are basically in balance. However, in most environments the processes are not tightly coupled. Organic matter produced in the euphotic zone supports a complex food chain which extends 4000 m to the ocean floor. Ocean currents may transport organic matter formed in the shelf seas long distances before it is ultimately metabolized. Understanding the balance between production and destruction of organic matter is important to the general issue of biogenic element cycling in the ocean.

The various methods of measuring respiration were reviewed comprehensively in Williams,[1] who discusses the merits of: (1) *in situ* O$_2$, ΣCO$_2$, pH, or pCO$_2$ changes; (2) *in vitro* O$_2$ or ΣCO$_2$ changes; (3) ^{14}CO$_2$ release after uptake of labeled organic compounds; and (4) measurement of electron transport system activity. At that time, the most widely used technique was the *in vitro* O$_2$ change method. This was due to improvements in the precision of the Winkler titration[2-4] which had raised the sensitivity of the method to the point where it was useable in most oceanic environments. The precision of ΣCO$_2$ analyses have also improved with the advent of coulometers.[5] However, the precision is still significantly lower than that achievable with O$_2$ analyses, i.e., ± 1 for ΣCO$_2$ vs. ± 0.1 μM for O$_2$.

Within the last six years there have been improvements in the precision of oxygen sensors, which means that they can now be used to make respiration measurements with the same sensitivity as the Winkler technique and with considerably more ease. Oxygen consumption, and the flow-rate sensitivity which occurs as a result of it, turns out to be the major cause of poor stability and therefore, low precision. Different approaches were taken to reduce oxygen consumption. The sensor used by Griffith[6] (available from L. G. Nestor Company) employs a third electrode which regenerates oxygen at the same rate as it is consumed at the cathode. Langdon[7] reduced the consumption of a conventional Yellow Springs Instrument electrode by applying the polarizing voltage as a very short duty cycle pulse. The theory of operation of oxygen sensors during the first few seconds after turn on is quite different than for the same sensor operated with a continuous polarizing voltage.[8,9] Sensor current just after turn on is much less sensitive to temperature than the steady-state current as well as being insensitive to flow rate. For this reason, I feel the pulsed sensor offers some advantages over the Nestor sensor.

The most important advantage of using an O_2 sensor over the Winkler titration method is that it yields a time course of respiration. This is important because rates may vary over time due to changes in substrate availability, population size, or activity level of the organisms. The second advantage over the titration method is that it is nondestructive. Subsamples may be obtained during or after the experiment for microscopic, gravimetric, or chemical analysis. Substates could also be added to alter the respiration of portions of the microbial community. Unlike other less sensitive sensor methods, preconcentration of natural samples is not necessary in most environments. Finally, because of the very low oxygen consumption no stirring of the sample is required. The disadvantages of the technique are that: (1) it does not measure anaerobic respiration which may be important in some specialized environments; (2) enclosing sample in a bottle may alter rates of some components of the community;[10] and (3) the method is limited to environments where rates exceed 0.6 mmol O_2 m^{-3} d^{-1}.

MATERIALS REQUIRED

An automated *in situ* respirometer can be assembled from an Endeco 1133 Pulsed Dissolved Oxygen Sensor Controller and an Onset Tattletale Model 5 data logger (Figure 1). The choice of material for the incubation chamber is important. I have found that Plexiglass reacts with dissolved oxygen and may release toxic compounds. Polycarbonate is nontoxic and impermeable to oxygen. The main body of the incubation chamber is made from a polycarbonate vacuum jar. These PC jars are available in a range of sizes. I use a 5-L jar to ensure that a representative sample of the plankton community is obtained. In near-shore environments such a large volume may be overkill. A valve also made of polycarbonate is mounted on the open end of the jar. The valve is opened and closed by a small motor; I use a motor scavenged from an inexpensive cordless screwdriver. The Tattletale data logger closes one of two relays which energizes the motor for clockwise or counter clockwise operation. The motor turns a threaded rod. The value is connected to the rod by a nut. Depending on the direction of rotation of the motor, the valve moves in or out on the rod. The Tattletale monitors the motor current and cuts power when it senses a stall at either end of the valve's travel. The Tattletale logs the serial data from the Endeco 1133 at its auxiliary serial input. Eight analog channels (0 to 5 V) are available to log additional sensors.

PROCEDURES

Field Calibration of Sensors

General Procedure for Comparisons

I have found that despite taking great care there is usually a difference between laboratory and field calibrations. Generally, the sensors read 10 to 20 μM lower in the field. This may have something to do with how the electronics handle the ubiquitous 60 Hz signal present in the lab. For this reason calibration is best done in the field. The simplest way is to collect water samples alongside the respirometer with a Niskin bottle during times when the incubator is flushing and analyze for oxygen concentration by the Winkler method.[11,12] A sample for oxygen determination should also be obtained from the chamber at the end of the experiment.

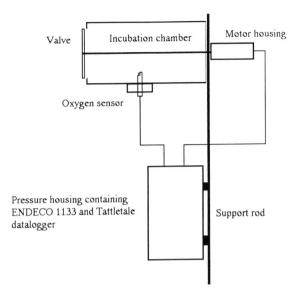

Figure 1.   Schematic diagram of respirometer.

Calculations and Equations

1.   Titrate the bottles and compute the oxygen concentration in units of μmol L^{-1}.

2.   Compute the partial pressure of oxygen (pO_2) in units of kilo Pascals (kPa) by dividing the oxygen concentration by the solubility coefficient S_s. S_s (μmol L^{-1} kPa^{-1}) is defined as:

$$S_s = (\alpha * 10^6)/(\text{molar volume} * K) \qquad (1)$$

where α is the Bunsen adsorption coefficient in dm^3 dm^{-3} atm^{-1}, the molar volume is 22.393 dm^3 mol^{-1} and K is a conversion factor depending on the unit of pressure, here, 101.325 kPa.[13] α may be computed as a function of temperature and salinity according to Equation 2:[14]

$$\alpha = \exp(A_1 + A_2(100/T) + A_3\ln(T/100) + S[B_1 \qquad (2)$$
$$+ B_2(T/100) + B_3(T/100)^2])$$

where $A_1 = -58.3877$, $A_2 = 85.8079$, $A_3 = 23.8439$, $B_1 = -0.034892$, $B_2 = 0.015568$, and $B_3 = -0.0019387$.

3.   Do a linear regression of sensor reading (0-8192) vs. pO_2 to obtain an equation of the form:

$$\text{Count} = a + b(pO_2) \qquad (3)$$

4.   The oxygen concentration at any temperature and salinity is given by:

$$[O_2] = S_s(T,S)(\text{Count} - a)/b \qquad (4)$$

where $S_s(T,S)$ is the solubility coefficient as a function of temperature and salinity in units and the oxygen concentration is in units of μmol L^{-1}.

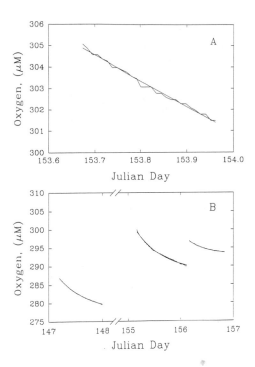

Figure 2. Community respiration measurements made in 1989 south of Iceland (59°N 20°W) at: (A) 5-m and (B) 30-m depths. Regression lines (linear in A and decaying exponential in B) have been drawn through the data.

Interpretation of Field Data: Specific Examples of Field Applications

Northeastern North Atlantic

Figure 2 shows the results of some community respiration measurements made in the northeastern North Atlantic (59°N 20°W) at 5- and 30-m depth. Daily rates of community respiration were 12.2 mmol O_2 m^{-3} d^{-1} at 5 m on June 2 and 7.8, 9.4, and 3.4 mmol O_2 m^{-3} d^{-1} for May 27 (JD 147), June 4 and June 5, respectively, at 30 m. The data set is not complete because of electrical problems with the sampler.

It is difficult to compare these rates to the few other measurements of oceanic community respiration because the rates are highly variable in time. The rates seem high. If they are normalized to chlorophyll a (chl a), the average rate is 0.23 mmol O_2 (mg chl a)$^{-1}$ h^{-1}. Algal rates of respiration range from 0.02 to 0.08 mmol O_2 (mg chl a)$^{-1}$ h^{-1} for diatoms to 0.24 for some species of dinoflagellates.[15] Chlorophyll-specific community respiration would obviously be expected to be higher than the pure algal rate because of the respiratory contribution of the bacteria, protozoa, micro- and macrozooplankton. Chlorophyll-specific community respiration ranges from 0.24 to 0.44 in productive coastal waters[16] and 0.08 mmol O_2 (mg chl a)$^{-1}$ h^{-1} in oligotrophic waters.[17] In light of this analysis, the measurements reported with the respirometer are reasonable.

Note that the rate of respiration (i.e., the slope of the curves in Figures 2A and 2B) is constant at 5 m and decreases exponentially at 30 m. One could argue that the decrease is due to die-off of organisms in the 30-m respirometer. However, this seems unlikely in view of the fact that the organisms evidently did not die off in the 5-m respirometer. I think the shape of the time course may be telling us something very interesting about the energy reserves of the community. During the period of these measurements, the water column was

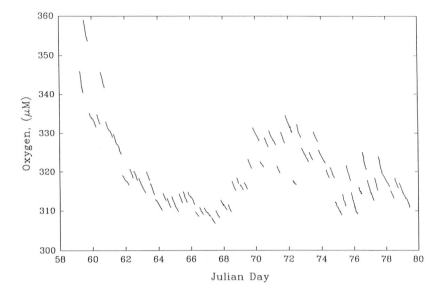

Figure 3. Morning, afternoon, and night measurements of community respiration rate at 2-m depth in Narragansett Bay, RI, from February 27 to March 21, 1991.

strongly stratified with a thermocline at approximately 15 m. The base of the euphotic zone based on *in situ* [14]C incubations was 30 to 40 m (G. Savidge and P. Boyd, unpublished data). If one assumes that respiration is proportional to the availability of respiratory substrates, then simple first order kinetics applies. In that case, the decline in respiration rate over the course of the incubation indicates the size of the populations' energy reserves relative to the metabolic demand. If the rate is constant then the pool of respiratory substrates is large relative to the metabolic demands of the community. However, if the rate of respiration declines several-fold over the course of a day then we know that energy reserves are low relative to metabolic demand. Based on this analysis, the 5-m community situated high up in the euphotic zone has plentiful energy reserves to support growth while the 30-m community located at the base of the euphotic has limited energy reserves and cannot sustain its level of metabolism without daily resupply of energy in some form. It may be that the degree of decline in respiration over some standard time interval could become a useful index of a population's energy reserves that might reflect light, nutrient, or food limitation.

Narragansett Bay

A more successful test of the respirometer was conducted in Narragansett Bay, RI. To examine how respiration rates vary over the course of a day, the respirometer is programmed to sample at 0500, 1100, and 1800 h daily. Thus morning (0600 to 1100), afternoon (1200 to 1800) and night (1900 to 0500) respiration rates were obtained. The respirometer ran continuously for three weeks. The data are shown in Figure 3.

The average of all 62 measurements was 10.8 ± 4 (1σ) with a minimum of 4.8 and a maximum of 29.2 mmol O_2 m^{-3} d^{-1}. The average chlorophyll-specific community respiration rate was 0.07 mmol O_2 (mg chl a)$^{-1}$ h^{-1} suggesting that the community was an almost pure diatom culture. The ratio of afternoon:morning rate was 1.3 ± 0.2 (95% CI) and the ratio of day:succeeding night rate was 1.5 ± 0.3 (95% CI). Both of these ratios were significantly different than the one at the 95% CI. Elevated rates of mitochondrial respiration in the light have been demonstrated for several species.[18-20] Turpin et al.[19] showed how mitochondrial respiration of nitrogen-limited microalgae increased in the light in response to the addition of NH_4. They hypothesized that the algae were activating the tricarboxylic acid cycle to produce C skeletons for amino acid synthesis. Weger et al.[20] showed that addition of dissolved inorganic carbon to algae held at the CO_2 compensation point in the light resulted in a rapid increase in both photosynthetic O_2 evolution and mitochondrial respiration. They argued that the higher mitochondrial respiration rate in the light was probably due to an increase in substrate supply from photosynthesis.

In addition to the relationship between substrate availability and respiration, there is evidence in the Narragansett Bay data of an endogenous rhythm. Morning rates were significantly greater than the previous nights, i.e., 1.3 ± 0.2 (95% CI), despite the fact that the incubation was started before any significant photosynthesis could have replenished the energy reserves of the cell. It may be that microalgae possess an endogenous rhythm wherein mitochondrial respiration accelerates in the morning to support increased biosynthesis with the onset of photosynthesis.

I have illustrated just a few of the new avenues of research in respiration made possible by the new instrumentation. Respiration is not the insignificant loss term many phytoplankton ecologists seem to think. Respiration is tightly coupled to biosynthesis and hence, growth of all organisms.

NOTES AND COMMENTS

Detection Limits

I have not had the opportunity to use the respirometer in oligotrophic waters. Given the noise level of the instrument (0.2 mmol m^{-3}) and the excellent stability of the sensor, I would estimate the lowest rate which could be measured with a confidence of $\pm 30\%$ as 0.6 mmol m^{-3} d^{-1}.

Troubleshooting

If sensor output is erratic or drifts excessively, the problem is often due to a hole in the membrane. To confirm this press gently on the center of the membrane with a blunt object. A spray of electrolyte or a small drop will form if a hole is present. If there is a hole replace the membrane and electrolyte using the supplies provided with the sensors. If the user wants to add any additional sensors to the respirometer, such as a light or conductivity sensor, he must ensure that their pressure cases are floating with respect to the electrical ground of the respirometer. Failure to do this will result in the oxygen sensor producing a very low reading as a result of a portion of the oxygen sensor's current following the easier path to ground.

ACKNOWLEDGMENTS

Comments and suggestions made by John Marra and Roger O. Anderson were helpful and are appreciated. This work was supported by Office of Naval Research contract N00014-89-J-1160, contribution No. 4872 of the Lamont-Doherty Geological Observatory, and contribution 3 of ML-ML.

REFERENCES

1. Williams, P. J. LeB., A review of measurements of respiration rates of marine plankton populations, in *Heterotrophic Activity in the Sea,* Hobbie, J. and Williams, P. J. LeB., Eds., Plenum, New York, 1984, 357.

2. Bryan, J. R., Riley, J. P., and Williams, P. J. LeB., A Winkler procedure for making precise measurements of oxygen concentration for productivity and related studies, *J. Exp. Mar. Biol. Ecol.,* 21, 191, 1976.

3. Culberson, C. H. and Huang, S., Automated amperometric oxygen titration, *Deep-Sea Res.,* 34, 875, 1987.

4. Williams, P. J. LeB. and Jenkinson, N. W., A transportable microprocessor controlled precise Winkler titration suitable for field station and shipboard use, *Limnol. Oceanogr.,* 27, 576, 1982.

5. Johnson, K. M., King, A. E., and Sieburth, J. McN., Coulometric ΣCO_2 analysis for marine studies: an introduction, *Mar. Chem.,* 16, 61, 1985.

6. Griffith, P. C., A high-precision respirometer for measuring small rates of change in oxygen concentration of natural waters, *Limnol. Oceanogr.,* 33, 632, 1988.

7. Langdon, C., Dissolved oxygen monitoring system using a pulsed electrode: design, performance, and evaluation, *Deep-Sea Res.,* 31, 1357, 1984.

8. Short, D. L. and Shell, G. S. G., Fundamentals of Clark membrane configuration oxygen sensors: some confusion clarified, *J. Phys. E: Sci. Instrum.,* 17, 1085, 1984.

9. Short, D. L. and Shell, G. S. G., Pulsing amperometric oxygen sensors: earlier techniques evaluated and a technique implemented to cancel capacitive charge, *J. Phys. E: Sci. Instrum.,* 18, 79, 1985.

10. Venrick, E. L., Beers, J. R., and Heinbokel, J. F., Possible consequences of containing microplankton for physiological rate measurements, *J. Exp. Mar. Biol. Ecol.,* 26, 55, 1977.

11. Strickland, J. D. H. and Parsons, T. R., *The Practical Handbook of Seawater Analysis,* Bull. 167, Fish. Res. Board Canada, 1972, 310.

12. Knapp, G. P., Stalcup, M. C., and Stanley, R. J., Iodine losses during Winkler titrations, *Deep-Sea Res.,* 38, 121, 1991.

13. Forstner, H. and Gnaiger, E., Calculation of equilibrium oxygen concentration, in *Polarographic Oxygen Sensors,* Gnaiger, E. and Forstner, H., Ed., Springer-Verlag, New York, 1983, 321.

14. Weiss, R. F., The solubility of nitrogen, oxygen and argon in water and seawater, *Deep-Sea Res.,* 17, 721, 1970.

15. Langdon, C., On the causes of interspecific differences in the growth irradiance relationship for phytoplankton. I. A comparative study of the growth irradiance relationship of three marine phytoplankton species: *Skeletonema costatum, Olisthodiscus luteus* and *Gonyaulax tamarensis, J. Plank. Res.,* 9, 459, 1987.

16. Packard, T. T. and Williams, P. J. LeB., Respiration and respiratory electron transport activity in sea surface seawater from the northeast Atlantic, *Oceanol. Acta,* 4, 351, 1981.

17. Williams, P. J. LeB., Heinemann, K. R., Marra, J., and Purdie, D. A., Comparison of ^{14}C and oxygen measurements of phytoplankton production in oligotrophic waters, *Nature,* 305, 49, 1983.

18. Grande, K. D., Marra, J., Langdon, C., Heinemann, K., and Bender, M., Rates of respiration in the light measured in marine phytoplankton using [18]O isotope labelling technique, *J. Exp. Mar. Biol. Ecol.,* 129, 95, 1989.

19. Turpin, D. H., Elrifi, I. R., Birch, D. G., Weger, H. G., and Holmes, J. J., Interactions between photosynthesis, respiration and nitrogen assimilation in microalgae, *Can. J. Bot.,* 66, 2083, 1988.

20. Weger, H. G., Herzig, R., Falkowski, P. G., and Turpin, D. H., Respiratory losses in the light in a marine diatom: measurements by short-term mass spectrometry, *Limnol. Oceanogr.,* 34, 1153, 1989.

CHAPTER 52

Sediment Community Production and Respiration Measurements: The Use of Microelectrodes and Bell Jars

P. A. G. Hofman and S. A. de Jong

INTRODUCTION

In estuaries a major part of primary production is performed by benthic diatoms.[1-4] In relatively turbid estuaries the highest benthic photosynthesis rates are reached during periods of low tide when sediment-inhabiting microalgae receive maximal irradiance.[1,5-7] Mutual enhancement of autotrophic and heterotrophic activity in the aerobic layer of the sediment not only allows benthic photosynthesis and aerobic mineralization to be maintained at high levels, but also forms the basis for other forms of life present in or on the sediment.[6,8,9]

Sediment oxygen production rates and actual oxygen concentrations can be measured directly in the sediment with oxygen microelectrodes.[6,9-13] In bell jars, sediment oxygen production or consumption rates can be measured indirectly as the increase or decrease of oxygen concentration occurs in the enclosed overlying water column.[9,14,15]

Oxygen concentrations in sediments and in overlying water can be measured very accurately with oxygen microelectrodes. In sediments these measurements show oxygen gradients. Oxygen gradients can be used to calculate oxygen fluxes into or out of the sediment when the porosity and the diffusion coefficient are known. Primary oxygen production rates can be measured directly in the sediment by the light/dark shift method.[11,12] For total production fluxes, the production at each depth interval has to be totalled. Oxygen consumption rates can be expressed as oxygen fluxes, using Fick's first law of diffusion:

$$F = - D * \delta c / \delta x \qquad (1)$$

in which F = oxygen flux, D = the diffusion coefficient, and $\delta c / \delta x$ = the slope of the oxygen concentration profile. In sediments the oxygen flux is influenced by the apparent sediment diffusion coefficient, which might be several times higher than the molecular diffusion coefficient.[9,14] Porosity also plays a role.

In bell jars or chambers, net oxygen fluxes can be derived by transforming the change of oxygen concentration in the bell jar into a flux through the sediment-water interface of the bell jar (the oxygen production or consumption in the included water has to be substracted). The basic assumption in oxygen production or consumption measurements by incubations in chambers is that the rate of oxygen release or uptake is independent of the oxygen concentration in the water overlying the sediment.

0-87371-564-0/93/$0.00 + $.50

These methods can be utilized to calculate the apparent sediment diffusion coefficient in two approaches.

1.   The bell jar/microelectrode method,[10,14] in which the apparent sediment diffusion coefficients can be obtained by measuring oxygen fluxes using a chamber, and oxygen gradients measured with microelectrodes.

2.   The microelectrode oxygen gradient/production method,[9] in which the apparent sediment diffusion coefficient is calculated from oxygen gradient measurements and independently measured downward-directed oxygen production fluxes, both measured in the light with microelectrodes.

Oxygen consumption fluxes can be calculated when the apparent sediment diffusion coefficient, the dark oxygen gradient, and the porosity are known, using Fick's first law of diffusion (see below).

Advantages of microelectrode method include: (1) microelectrodes are commercially available today; (2) oxygen production rates and oxygen concentrations can be measured in 100 μm steps or less; (3) measurements reflect concentrations and processes as they are or take place in the sediment, biofilms, or boundary layer; (4) microelectrodes respond very fast, so five-fold measurements can be done in a few minutes; and (5) although lots of measurements are performed in cores, direct *in situ* measurements can also be done, either on emerged or submerged sediments.[14]

Disadvantages of the microelectrode method include: (1) for production measurements microelectrodes with a short 90% response time are needed, such microelectrodes are very thin and may break easily; (2) oxygen consumption rates cannot be acquired directly so the oxygen gradients measured have to be converted to fluxes using the appropriate apparent sediment diffusion coefficient;[9] (3) microelectrode measurements are limited to a small area determined by their tip size and the number of profiles measured and for ecosystems covering larger areas such as estuaries with a variety of intertidal and subtidal flats, measurements have to include all representative areas; (4) measuring needs special precautions such as stable electrical power and stable equipment; and (5) for sediments, consumption fluxes calculated from microelectrode measurements include only consumption of microorganisms, micro- and meiofauna.

Advantages of the bell jar method include: (1) in chambers whole community metabolism or net production rates can be measured continuously over longer periods; (2) chamber measurements represent the production or consumption of larger areas; and (3) measurements can be performed under controlled circumstances.

Disadvantages of the bell jar method include: (1) bell jars cannot be used for emerged sediments, nor in water with strong currents as strong currents cannot be simulated within the bell jar, or cause suspending of sediment; (2) larger enclosed animals can influence measurements dramatically by suspending sediment; (3) in bell jars, net oxygen fluxes over the sediment-water interface are measured and there is no distinction between gross production, consumption, and transport of oxygen during incubation in the light; (4) benthic chambers eliminate the supply of organic matter to the sediment surface and disturb the natural flow regime of the water; and (5) oxygen concentration increases or decreases within the chambers during incubation, and may therefore, interfere with natural circumstances.

MATERIALS REQUIRED

Bell Jar Measurements

Equipment

- Bell jar (see Figure 1), e.g., made of translucent Plexiglass, removable lid, inner diameter 45 cm, height 33 cm, when large macrofauna or macroflora is present larger bell jars must be used

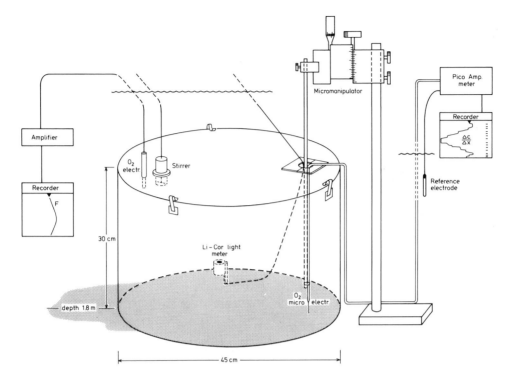

Figure 1. Bell jar equipped with a normal oxygen electrode and an oxygen microelectrode. With the bell jar oxygen fluxes can be obtained, whereas with the microelectrode oxygen profiles can be measured. (From Lindeboom, H. J. et al., *Limnol. Oceanogr.*, 30, 693, 1985.)

- Oxygen electrode (Model no. 5739, Yellow Springs Instruments, Inc.), oxygen monitor, and (multichannel) strip chart recorder
- Thermistor
- Stirring device, e.g., a motor-driven magnetic bar (length 4 cm) or impeller
- Electrode, thermistor, and stirring device are watertight and mounted in the removable lid
- For continuous light measurements a Li-Cor light meter can be used

Supplies

- Weighted black plastic sheets to darken the chambers

Microelectrode Measurements

Equipment

- Sturdy rack for micromanipulator and to mount sediment corer (Figure 1, sediment corer not shown).
- Oxygen microelectrodes — for production measurements fast responding microelectrodes with a tip diameter <10 μm are needed. The response time can be measured by quickly inserting the electrode from air into deoxygenated water. Ninety percent of the difference in oxygen concentration must be measured within 1 s. Suitable electrodes are insensitive for stirring (stirring effect $<5\%$), and their drift is $<0.5\%$ min^{-1}. The current output is linear with oxygen, 2 to 100 pA at a range of 0 to 100% oxygen saturation. They are insensitive for light-dark shifts (e.g., Eschweiler, Kiel, FRG, or Diamond Electro-Tech). For gradient measurements more sturdy types of microelectrodes can be used (tip size 700 μm[16]).

- Ag/AgCl single junction reference electrode (Radiometer, Copenhagen, Denmark).
- Micromanipulator, hand-driven (Figure 1), or electrically driven (Uhl/Aßlar, FRG, or Plato B.V., Diemen, The Netherlands).
- Picoammeter (Kiethley 480, equipped with an IC-stabilized polarization unit, the adjustable polarization voltage set at -0.75 V[16]).
- Strip chart recorder.

Supplies

- Large box to darken the core during production measurements.
- Plexiglass sediment corers with rubber stoppers (e.g., inner diameter 6 cm, length 15 cm, 30 cm for underwater sediment).

PROCEDURES

Bell Jar Measurements

In flux chambers both net oxygen production rates and oxygen consumption rates of inundated sediments can be measured as the change in oxygen concentration in the overlying water column.

1.   Calibrate oxygen electrode with ambient water prior to measurement.

2.   Collect water and mix thoroughly.

3.   Measure oxygen concentration with electrode.

4.   Take five 100-ml samples for Winkler titration analysis.[17]

5.   Place the bell jar (without lid) on the sediment without disturbing the sediment surface. Insert the bell jar approximately 3 cm into the sediment to ensure an adequate seal.

6.   Measure the exact height of the water column for inner volume calculation.

7.   Check oxygen electrode and stirring device.

8.   Close the bell jar after 5 to 10 min for adaptation and start oxygen monitoring. For oxygen consumption measurement, darken the bell jar with black plastic sheet.

9.   Take three sediment samples with sediment corers from ambient sediment outside the bell jar, and measure oxygen gradients in the upper bottom layers with microelectrodes within 10 min (see below).

10.   Stop oxygen measurement when 150% oversaturation (in the light) or 50% undersaturation (in the dark) is reached.

11.   Calculate oxygen production flux, respectively, consumption flux when oxygen signal is stable and increase or decrease is linear.

12.   Remove the lid from the bell jar when oxygen concentration is too low or two high. Ventilate bell jar without disturbing the sediment surface and measurements can be restarted again.

13.   Repeat measurements 5 to 10 times, if possible with several bell jars.

14.   Remove the bell jar lid at the end of the measurements. Take three undisturbed sediment cores inside the bell jar and measure oxygen gradients within 10 min (see below).

15.   Sediment oxygen gradients inside and outside the bell jar should have comparable slopes, if not, the stirring device has to be adjusted.

16.   Recalibrate the oxygen electrode as in step 1.

Microelectrode Measurements

Oxygen concentrations and gross oxygen production can be measured directly in the sediment with oxygen microelectrodes. Oxygen consumption rates can be calculated from the O_2 gradient, oxygen diffusion coefficient, and porosity according to Fick's first law of diffusion (see below).

1. Take undisturbed sediment sample with Plexiglass corer. (For underwater sediment, fill corer with about 10 cm water above the sediment.)

2. Calibrate oxygen microelectrode with ambient water prior to measurement. Collect water and mix thoroughly. Measure oxygen concentration with electrode. Take five 100-ml samples for Winkler titration analysis.[17]

3. Start microelectrode measurements as soon as possible (e.g., within 10 min at 20°C) to prevent aberration from *in situ* situation.

4. Mount corer tight into the frame, and place reference electrode in the upper layer of the sediment (in case of emerged sediment) or into the overlying water (remove overlying water with siphon till 5 cm above the sediment).

5. Lower microelectrode near the sediment and fix the electrode to the micromanipulator (electrode can be lengthened with Plexiglass bar).

6. Lower microelectrode carefully toward the sediment. The sediment surface can be defined at the electrical current read upon the first contact between the microelectrode tip and the sediment when the sediment is exposed to air (e.g., intertidal flats). For inundated sediment the absence of a time-lag in the electrode response upon darkening (see step 8) is used as sediment surface (see production measurements), or the electrode reaching the surface can be visually detected with a dissection microscope.

7. Wait until stable signal and measure oxygen concentration and oxygen production by darkening the core.

The following steps pertain to production measurements.

8. Measure the rate of photosynthesis (gross oxygen production) by determining the rate of decrease in oxygen just after a light-dark shift.[6,11] At steady-state conditions in the light (stable oxygen concentration), darkening results in an immediate linear decrease in oxygen concentration for 5 to 30 s. This decline in oxygen concentration after darkening is due to photorespiration, oxygen respiration, and diffusion, and is defined as gross oxygen production. Oxygen production can be measured at depth intervals of 100 μm (e.g., Revsbech and Jørgensen[11,12]) or even 50 μm (e.g., De Jong et al.[6]).

9. After measuring production at a certain depth, oxygen concentration must return to >95% of its initial value when exposed to *in situ* light again.

10. Lower electrode to next depth and repeat measurement. At the end of the photic zone there will be some delay in oxygen decrease after darkening. Gross oxygen production is actually measured as the decrease in oxygen concentration per volume of interstitial water. For flux calculations, gross oxygen production at any depth interval in the photic zone has to be summed up and transferred to sediment surface area.

$$\text{Flux} = \left(\sum_{i=1}^{n} P_i \right) \cdot \Phi \qquad (2)$$

where P_i = oxygen production at interval i (production per volume has to be converted to production per area), n = total number of intervals, and Φ = porosity.

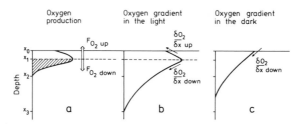

Figure 2.   Scheme of microelectrode oxygen gradient/production method: (a) oxygen production (mM h^{-1}) measured with microelectrode, and oxygen fluxes (mmol m^{-2} h^{-1}); (b) oxygen gradients (mmol m^{-4}) measured with microelectrode in the light; and (c) oxygen gradient measured in the dark. (From Hofman, P. A. G. et al., *Mar. Ecol. Prog. Ser.*, 69, 261, 1991. With permission.)

11.   Oxygen profiles, belonging to the oxygen production profiles, are measured simultaneously with production measurements.

The following steps pertain to gradient measurements.

12.   Measure oxygen profiles for oxygen consumption rate calculations in the dark, just after the oxygen peak has vanished (this peak in the O_2 profile is caused by photosynthesis). Oxygen decrease in the upper part of the profile is almost linear (Figure 2c). The slope of the gradient can be calculated with linear regression using at least 5 points. Oxygen concentrations at the sediment surface must be higher than 100 μM.[9,15]

13.   Recalibrate oxygen microelectrode immediately after measuring each core by reading the electrode signal in well-mixed water (see step 2). The zero current of the microelectrode is recorded in the anaerobic layer of the sediment.

Assumptions

For oxygen consumption rate and diffusion coefficient calculations the following assumptions have to be made.

1.   The oxygen flux into the sediment can be described by Fick's first law of diffusion:

$$F_{O_2} = - \, \Phi \, D_a \, (\delta O_2/\delta x)_{x=0} \tag{3}$$

where F_{O2} = flux of oxygen (mmol O_2 m^{-2} h^{-1}) and
Φ = mean porosity for aerobic sediment layer. Porosity is defined as the volume of interconnected porewater relative to the volume of total sediment, provided that it is saturated with water.[18]
D_a = apparent sediment diffusion coefficient (m^2 h^{-1}) $\delta O_2/\delta x$ = oxygen concentration gradient over depth interval x (mmol O_2 m^{-4})

2.   The apparent sediment diffusion coefficient is constant over the oxic sediment layer.

Calculation of the Apparent Sediment Diffusion Coefficient

Bell Jar/Microelectrode Method[10,14]

The net oxygen flux ($F_{O2\ bell\ jar}$) across the sediment-water interface, is calculated from the rate of change in oxygen concentration in the overlying water. The oxygen gradient

$(\delta O_2/\delta x)$ in the sediment is measured with microelectrodes (Figure 2). The diffusion coefficient is calculated with:

$$D_a = - F_{O_2 \text{ bell jar}}/(\Phi \; \delta O_2/\delta x) \qquad (4)$$

Apparent sediment diffusion coefficients, measured during a dark period with downward oxygen fluxes (Figure 2c) or during a light period with upward oxygen fluxes (Fig. 2b), have shown to be of the same order of magnitude.[9,14]

Microelectrode Oxygen Gradient/Production Method[9]

Oxygen production, oxygen consumption, and diffusion of oxygen over depth interval $x_0 - x_1$ result in an upward-directed oxygen gradient ($\delta O_2/\delta x_{up}$; Figure 2b) and an upward-directed oxygen flux ($F_{O_2 \text{ up}}$; Figure 2a). Oxygen production at depth interval $x_1 - x_2$, oxygen consumption, and diffusion of oxygen over depth interval $x_1 - x_3$ together result in a downward-directed oxygen gradient ($\delta O_2/\delta x_{down}$; Figure 2b).

The oxygen production at depth interval $x_1 - x_2$ is the only source of oxygen over interval $x_1 - x_3$ and can be calculated as a downward-directed oxygen flux ($F_{O_2 \text{ down}}$; Figure 2a).

The apparent diffusion coefficient can now be calculated from:

$$D_a = - F_{O_2 \text{ down}}/(\Phi \; \delta O_2/\delta x_{down}) \qquad (5)$$

Oxygen consumption rates during the light period on emerged flats or light receiving underwater sediments can now be calculated from the dark profiles (Figure 2c) by combining the value of D_a, calculated according to Equation 5, with the O_2 gradient measured in the dark. For this latter calculation $\delta O_2/\delta x_{dark}$ should be constant in time.

NOTES AND COMMENTS

1. In bell jars the water column has to be stirred continuously at such a rate that the water column is homogeneous and water velocity rates correspond with *in situ* rates. The latter can be checked with microelectrodes measuring the boundary layer[19] or the oxygen gradients in the sediment inside and outside the chamber.[9]

2. Oxygen increase or decrease in bell jars has to be linear and measurements have to be stopped before oxygen concentrations reach 150% oversaturation or 50% undersaturation in the overlying water.[6,9,15]

3. Chambers for benthic community respiration must not be made of stainless steel or other conducting material. Cramer[20] found oxygen losses due to reduction of oxygen by the stainless steel bell jar, which acted as a cathode when brought in contact with the sediment.

4. The size of the chamber in itself is not an important parameter in chamber design (although it has to be appropriate with respect to the size of macrofauna, see next note), but the stirring should be scaled with chamber size.[15]

5. Coastal and estuarine sediments with abundant macrofauna contain burrows that may be truncated or disturbed during the coring process and by pushing the chambers into the sediment. Such disturbances may seriously affect the observed fluxes.[21]

6. In a closed system, the rate of oxygen uptake can be measured at the same time with the flux of carbon dioxide, silicate, phosphate, and other nutrients from the sediment.[22]

7. Sediment oxygen uptake in chambers does not always reflect *in situ* oxygen consumption rates. Reimers,[13] for instance, found that microelectrode measurements in box-cored, deep-sea sediments never precisely duplicated the *in situ* oxygen distributions, especially near the sediment-water interface.

8.   Estimates of benthic oxygen uptake rates from oxygen gradients measured with micro-electrodes (which assume oxygen transport by molecular diffusion) may underestimate oxygen demand in sediments where turbulence or irrigation by benthic organisms influence transport of oxygen across the sediment-water interface.[15]

9.   Some microelectrodes show a velocity sensitivity and must be checked for this phenom-enon,[23,24] especially in the diffusive boundary layer this might result in important errors. The diffusive boundary layer is the thin layer of water adjacent to the sediment surface through which molecular diffusion is the dominant transport mechanism for dissolved material.[25]

10. Microelectrodes must be checked for linearity in response and light independency.[11]

11. New types of oxygen electrodes and combined oxygen-N_2O electrodes are developed by Revsbech.[24,26]

12. In coastal environments, a considerable part of the benthic metabolism proceeds through pathways of nitrate, manganese, iron, and sulfate respiration. Part of this anaerobic res-piration is included in the benthic oxygen uptake through oxidation of the reduced com-pounds. Anderson et al.[27] concluded from this that the oxygen consumption rate is not an ideal measure of benthic mineralization in coastal sediments. This is only the case when losses of reduced compounds such as N_2 (which diffuses to overlying water or air) and FeS_2 (which is buried permanently) are considerable.[27,28]

13. Apparent sediment diffusion coefficients calculated with the microelectrode oxygen gra-dient/production method are not influenced by manipulation with overlying water and approach actual values.[9]

REFERENCES

1. Cadée, G. C. and Hegeman, J., Primary production of the benthic microflora on tidal flats in the Dutch Wadden Sea, *Neth. J. Sea Res.*, 8, 260, 1974.

2. Cadée, G. C. and Hegeman, J., Distribution and primary production of benthic microflora and accumulation of organic matter on a tidal flat area, Balgzand, Dutch Wadden Sea, *Neth. J. Sea Res.*, 11, 24, 1977.

3. Admiraal, W., Peletier, H., and Zomer, H., Observations and experiments on the population dynamics of epipelic diatoms from estuarine mudflats, *Est. Coast. Shelf Sci.*, 14, 471, 1982.

4. Admiraal, W., The ecology of estuarine sediment-inhabiting diatoms, in *Progress in Phycol-ogical Research*, Round, F. E. and Chapman, D. J., Eds., Biopress, Bristol, England, 1984, Vol. 3, 269.

5. Colijn, F. and De Jonge, V. N., Primary production of microphytobenthos in the Ems-Dollard estuary, *Mar. Ecol. Prog. Ser.*, 14, 185, 1984.

6. De Jong, S. A., Hofman, P. A. G., Sandee, A. J. J., and Jansen, H. A. P. M., An inorganic carbon budget for benthic microalgal photosynthesis on intertidal flats in the Oostertschelde estuary, The Netherlands, submitted.

7. Rasmussen, M. B., Henriksen, K., and Jensen, A., Possible causes of temporal fluctuations in primary production of the microphytobenthos in the Danish Wadden Sea, *Mar. Biol.*, 73, 109, 1983.

8. Scholten, H., Klepper, O., Nienhuis, P. H., and Knoester, M., Oosterschelde estuary (S.W. Netherlands); a self-sustaining ecosystem? *Hydrobiologia*, 195, 201, 1990.

9. Hofman, P. A. G., de Jong, S. A., Wagenvoort, E. J., and Sandee, A. J. J., Apparent sediment diffusion coefficients for oxygen and oxygen consumption rates measured with microelectrodes and bell jars: applications to oxygen budgets in estuarine intertidal sediments (Oosterschelde, S.W. Netherlands), *Mar. Ecol. Prog. Ser.*, 69, 261, 1991.

10. Baillie, P. W., Oxygenation of intertidal estuarine sediments by benthic microalgal photosyn-thesis, *Estuar. Coast. Shelf Sci.*, 22, 143, 1986.

11. Revsbech, N. P. and Jørgensen, B. B., Photosynthesis of benthic microflora measured with high spacial resolution by the oxygen microprofile method: capabilities and limitations of the method, *Limnol. Oceanogr.*, 28, 749, 1983.

12. Revsbech, N. P. and Jørgensen, B. B., Microelectrodes: their use in microbial ecology, *Adv. Microb. Ecol.*, 9, 293, 1986.

13. Reimers, C. E., An in situ microprofiling instrument for measuring interfacial pore water gradients: methods and oxygen profiles from the North Pacific Ocean, *Deep-Sea Res.*, 34, 2019, 1987.

14. Lindeboom, H. J., Sandee, A. J. J., De Klerk, V. D., and Driessche, H. A. J., A new bell jar/microelectrode method to measure changing oxygen fluxes in illuminated sediments with a microalgal cover, *Limnol. Oceanogr.*, 30, 693, 1985.

15. Hall, P. O. J., Anderson, L. G., Rutgers van der Loeff, M. M., Sundby, B., and Westerlund, S. F. G., Oxygen uptake kinetics in the benthic boundary layer, *Limnol. Oceanogr.*, 34, 734, 1989.

16. Helder, W. and Bakker, J. F., Shipboard comparison of micro- and minielectrodes for measuring oxygen distribution in marine sediments, *Limnol. Oceanogr.*, 30, 1106, 1985.

17. Bryan, J. R., Riley, J. P., and Williams, P. J. LeB., A Winkler procedure for making precise measurements of oxygen concentrations for productivity and related studies, *J. Exp. Mar. Biol. Ecol.*, 21, 191, 1976.

18. Ullman, W. J. and Aller, R. C., Diffusion coefficients in nearshore marine sediments, *Limnol. Oceanogr.*, 27, 552, 1982.

19. Sweerts, J.-P. R., Louis, V. ST., and Cappenberg, T. E., Oxygen concentration profiles and exchange in sediment cores with circulated overlying water, *Freshwater Biol.*, 21, 401, 1989.

20. Cramer, A., A common artefact in estimates of benthic community respiration caused by the use of stainless steel, *Neth. J. Sea Res.*, 23, 1, 1989.

21. Hammond, D. E., Fuller, C., Harmon, D., Hartman, B., Korosec, M., Miller, L. G., Rea, R., Warren, S., Berelson, W., and Hager, S. W., Benthic fluxes in San Francisco Bay, *Hydrobiologia*, 129, 69, 1985.

22. Blackburn, T. H., Nitrogen cycle in marine sediments, *Ophelia*, 26, 65, 1986.

23. Gust, G., Booij, K., Helder, W., and Sundby, B., On the velocity sensitivity (stirring effect) of polarographic oxygen microelectrodes, *Neth. J. Sea Res.*, 21, 255, 1987.

24. Revsbech, N. P., An oxygen microsensor with a guard cathode, *Limnol. Oceanogr.*, 34, 474, 1989.

25. Jørgensen, B. B. and Des Marais, D. J., The diffusive boundary layer of sediments: oxygen microgradients over a microbial mat, *Limnol. Oceanogr.*, 35, 1343, 1990.

26. Revsbech, N. P., Nielsen, L. P., Christensen, P. B., and Sørenson, J., Combined oxygen and nitrous oxide microsensor for denitrification studies, *Appl. Environ. Microbiol.*, 54, 2245, 1988.

27. Anderson, L. G., Hall, P. O. J., Iverfeldt, Å., Rutgers van der Loeff, M. M., Sundby, B., and Westerlund, S. F. G., Benthic respiration measured by total carbonate production, *Limnol. Oceanogr.*, 31, 319, 1986.

28. Jørgensen, B. B. and Revsbech, N. P., Oxygen uptake, bacterial distribution, and carbon-nitrogen-sulfur cycling in sediments from the Balthic Sea-North Sea transition, *Ophelia*, 31, 29, 1989.

CHAPTER 53

Distinguishing Bacterial from Nonbacterial Decomposition of *Spartina alterniflora* by Respirometry

David E. Padgett

INTRODUCTION

The importance of salt marsh cordgrass, *Spartina alterniflora* Loisel, in the estuarine habitat is abundantly documented. Because most of its organic carbon makes its way into the estuarine and marine food webs via the detrital pathway, the role of decomposers to cordgrass cycling is of considerable importance. This fact has prompted a good deal of research ranging from determining the time course of its disappearance, to listings of microorganisms *(sensu lato)* capable of degrading its moribund tissues. An outgrowth of these research efforts has been disagreement concerning the relative importance of various categories of microorganisms (primarily bacteria and filamentous fungi) to this decomposition process. Attempts to resolve this question have centered principally on estimations of biomass of decomposer groups, but no attempts had been made prior to 1985 to determine the level of metabolic activity of suspected decomposers.

There is recent evidence that aerobic filamentous fungi are active in the decomposition of cordgrass roots and rhizomes at considerable depths within anoxic salt marsh soils (Padgett et al.[1,2] and Padgett and Celio[3]). Discovery of active fungal biomass in this hostile environment where sulfate-reducing bacteria have been considered the preeminent decomposers reemphasizes the importance of resolving the question discussed above.

The procedure to be outlined herein (developed by Padgett et al.[4]) was designed to quantify aerobic respiration of decomposers established on cordgrass following suppression of the bacterial component by antibiotics. The utility of this procedure toward determining the relative importance of competing decomposer groups lies mainly in the fact that it provides a technique for quantifying the extent of suppression imposed by the antibiotic treatment. Such quantification, in turn, facilitates determining which group of decomposers is respiring most rapidly.

0-87371-564-0/93/$0.00 + $.50

MATERIALS REQUIRED

Equipment

- Epifluorescent microscope equipped with phase contrast condenser and phase optics (1000×
  total magnification needed); we use an Olympus BH-B epifluorescent scope with a 100 W
  high pressure mercury vapor lamp, a BG-12 excitation filter, and an O-530 barrier filter
- Waring blender for tissue maceration
- Differential respirometer; we employed a Gilson IGRP-20 differential respirometer operated
  at 20°C according to manufacturer's guidelines for manometric trials, but standard Warburg
  respirometers will do just as well (Umbreit et al.[5])
- Drying oven (105°C heat needed) and muffle furnace for determining ash-free dry weights

Supplies

- 0.2-μm sterile disposable membrane filters and syringes for sterilizing antibiotics
- 100-μm mesh Nitex screen for filtering tissue grindate (Tetco Inc., Lancaster, NY)
- Sterile Petri dishes (if plate count verification of results is desired)

Solutions and Media

- Natural or artificial seawater
- Agar for plate counts (if desired); the medium selected should contain appropriate nutrients
  to facilitate growth and enumeration of the bacteria likely to be contained in the plant tissue
  being analyzed. We recommend a mixture containing (g L^{-1}): 7.5 g soluble starch; 1.0 g
  yeast extract; 1.0 g Bacto peptone; 15.0 g agar-agar made up in dilute seawater of desired
  salinity.
- Antibacterial antibiotic(s): we recommend chloramphenicol (60 μg ml^{-1} final concentration)
  for use in saline solutions. Other potentially useful antibiotics or combinations (with concentrations) are listed in Padgett et al.[4]
- Iodonitrotetrazolium formazan (INT) is available from Sigma Chemical Co., St. Louis, MO.
  Sigma stock number 17375. Make up a fresh 0.21% aqueous working solution of this every
  24 h and store refrigerated in a dark bottle.
- Acridine Orange hemi-zinc chloride salt (also available from Sigma, stock number A-6014).
  Prepare a fresh 0.1% aqueous working solution daily. Note that this stain is carcinogenic
  and should be handled with care.
- Isoamyl alcohol
- Tween 20 surfactant
- Formaldehyde

PROCEDURES

Experimental Approach and Assumptions

The extent to which a particular antibiotic treatment suppresses the bacteria on and within
dead cordgrass tissue can be assessed by staining representative bacterial cells with INT-
formazan vital stain (Zimmerman et al.[6]) and then with acridine orange following treatment.
Calculations (described below) which compare staining properties of treated vs. control
samples permit determination of the percentage of the bacterial community that is inactivated
by the particular antibiotic. Armed with an assessment of bacterial suppression, antibiotic
treated and untreated cordgrass samples can be subjected to respirometric analysis. If, for
example, a cordgrass sample, 90% of whose bacterial flora is inactivated, utilizes only 30%

less oxygen than an untreated sample, it is justifiable to conclude that the majority of metabolic activity resulting in decomposition is due to nonbacterial decomposers.

It should be understood that the accuracy of this research approach is based on the following assumptions:

1. That tissue maceration to liberate contained bacteria into solution yields a representative sample of active decomposers. If, for some reason, an important species is not easily recovered by tissue grinding, subsequent calculations will be biased.

2. That there is a direct correlation between the aerobic respiration rate and production of degradative enzymes by a decomposer or decomposer assemblage.

3. That all unmacerated samples used for respirometric analysis permit the same degree of antibiotic penetration as did the sample used to determine antibiotic effectiveness.

Antibiotic Screening

1. Antibiotic screening is initiated by macerating a sample of decomposing plant tissue in sterile, dilute seawater (isotonic). A Waring blender works well for this purpose, but avoid overdoing it lest excessive tissue shearing cause heating of the solution. Our preliminary experiments revealed that addition of Tween 20 surfactant (0.1% final concentration) facilitates dislodging embedded bacteria without altering antibiotic effectiveness or staining affinities, but that its use causes foaming during the maceration unless one drop of isoamyl alcohol also is added per 50 ml of solution.

2. Filter the slurry through sterile, 100-μm mesh Nitex screen and retain the filtrate in a sterile flask.

3. Pipette 10-ml aliquots of filtrate into separate, sterile flasks and add enough filter-sterilized antibiotic to achieve the desired final concentration. We usually use three sample aliquots for antibiotic treatment and retain three untreated as controls.

4. Incubate treated and control flasks on a shaker at room temperature for 3 h and 20 min.

5. Initiate vital staining of bacteria toward the end of the antibiotic treatment period (i.e., at 3 h and 20 min) by aseptically adding 1 ml of aqueous 0.21% INT to each flask and incubating in the dark for 40 min.

6. Add 0.1 ml of 37% formaldehyde solution to fix the sample and refrigerate until acridine orange staining can be performed.

7. Treated solutions will contain many small particles that are difficult to distinguish from bacteria without the use of acridine orange stain. This staining is best accomplished by transferring 2.0 ml subsamples to small test tubes before adding 0.2 ml of the fresh 0.1% aqueous acridine orange stain. Let these tubes sit for five minutes at room temperature.

8. Prepare a careful wet mount of the stained subsample using a very clean microscope slide and cover glass.

9. Observe each microscope field first using epifluorescent illumination and count all fluorescing (viable) bacteria. The small size of most bacterial cells (and granules) will necessitate $1000\times$ magnification for counting. We recommend using a $100\times$ phase contrast objective.

10. Without moving the stage, switch to phase contrast and count all bacteria in the same field that contain INT-formazan granules. A little practice will reveal that actively respiring aerobic bacteria will deposit the INT-formazan vital stain as discrete, dark red granules (usually one per cell but several is not uncommon) the total size of which is proportional to the rate of aerobic respiration. Bacterial endospores can cause some confusion, but these do not stain red.

11. Continue counting randomly selected fields until at least 200 bacteria in each subsample have been tallied. Combine results of all replicates and calculate the percentage of respiring bacteria by dividing the total INT-positive by the total number of fluorescing cells.

Calculations and Interpretations

Determining the effectiveness of the test antibiotic merely involves comparing numbers derived from counts of treated vs. untreated samples taken from the same original slurry. Before doing final calculations (as in the example below) be aware that the population in the control samples may well have grown during the 4-h period that treated samples were undergoing antibiotic exposure. Not accounting for this (by standard plate count procedures or by comparing "time zero" with "time + 4 h" AODC counts) will result in overestimating antibiotic effectiveness.

Consider a sample calculation (from Padgett et al.[4]).

"A 4-h treatment with 60 μg ml^{-1} chloramphenicol resulted in an average of 72.9% fewer fluorescing bacteria in treated than in adjusted control samples. This represents the percentage of bacterial cells that died (subsequent to chloramphenicol-induced protein synthesis shutdown) and underwent nucleic acid degradation. Of the remaining viable bacteria (27.1% of the original population), the INT staining revealed that treated samples contained an average of 53.3% fewer respiring cells than in controls. Thus, if 72.9% were killed and an additional (0.271) (53.9%) = 14.4% were induced to dormancy, we can say that 72.9% + 14.4% = 87.3% of the original population was killed or inactivated by the treatment."

Respirometry

Once the antibiotic (or combination) of choice has been identified, comparative respirometry can be done on representative control and treated samples. Note that tissue maceration likely will have an adverse effect on metabolic activity of filamentous fungi. For this reason, it is important to do respirometric measurements on unmacerated samples. Be sure, however, to select samples that are at the same stage of decomposition as those used for antibiotic screening. This is essential to ensure that antibiotic penetration will be as expected. Our preliminary trials verified that unmacerated cordgrass tissue was subject to the same degree of antibiotic penetration as was comparable macerated tissue as long as it was decomposed enough for any waxy cuticle to have been lost.

1. Four hours prior to respirometric analysis, initiate antibiotic treatment of test samples.

2. 15-ml reaction vessels accommodate small samples (1 to 2 g fresh weight) of treated or untreated tissue. We recommend doing at least six replicates of both test and control samples. Operate according to manufacturer's instructions. We conducted experiments at 20°C.

3. After gathering suffcient manometric data, carefully remove all plant tissue from each flask and (separately) determine the ash-free dry weights. These figures will permit computing the oxygen utilization per hour per gram ash-free dry weight for each replicate. Comparing the average respiration rate of control and treated samples (in light of previously quantified antibiotic effectiveness) will permit conclusions to be drawn concerning the relative contribution of bacterial vs. nonbacterial decomposers of cordgrass.

NOTES AND COMMENTS

1. Note that the use of polycarbonate filters cleared with immersion oil (a commonly used procedure for acridine orange direct bacterial counts) should be avoided because immersion oil will extract the INT-formazan granules that indicate respiring bacteria.
2. Antibiotics should be made fresh and filter-sterilized (not autoclaved) just before use. A variety of antibiotic types with appropriate concentrations for each are listed in Padgett et al.[4] It should be kept in mind that the most effective antibiotic for a given application is influenced not only by the bacterial population present in the sample, but also by the chemical makeup of the solution to which it is added. For example, Connamacher[7] demonstrated that divalent magnesium ions (an important constituent of seawater) inhibits tetracycline. More recently, Samuelson[8] found that oxytetracycline degrades rapidly in seawater as a function of temperature and light. It is recommended that preliminary screening be done using several antibiotics to determine the most appropriate one for each experimental protocol.
3. Avoid drawing the conclusion that fungi alone are responsible for all oxygen consumed in treated flasks. Unless other steps are taken to determine the relative composition of the eukaryotic decomposer community (e.g., ciliates vs. flagellates vs. fungi), inferences about a particular component will be tentative.

REFERENCES

1. Padgett, D. E., Hackney, C. T., and de la Cruz, A. A., Growth of filamentous fungi into balsa wood panels buried in North Carolina salt marsh sediments, *Trans. Br. Mycol. Soc.,* 87, 155, 1986.
2. Padgett, D. E., Celio, D. A., Hearth, J. H., and Hackney, C. T., Growth of filamentous fungi in a surface-sealed woody substratum buried in salt-marsh sediments, *Estuaries,* 12, 142, 1989.
3. Padgett, D. E. and Celio, D. A., A newly discovered role for aerobic fungi in anaerobic salt marsh soils, *Mycologia,* 82, 791, 1990.
4. Padgett, D. E., Hackney, C. T., and Sizemore, R. K., A technique for distinguishing between bacterial and non-bacterial respiration in decomposing *Spartina alterniflora, Hydrobiologia,* 122, 113, 1985.
5. Umbreit, W. W., Burris, R. H., and Stauffer, J. F., *Manometric Techniques: A Manual Describing Methods Applicable to the Study of Tissue Metabolism,* 4th ed., Burgess Publishing, Minneapolis, MN, 1964.
6. Zimmerman, R., Iturriaga, R., and Becker-Birck, J., Simultaneous determination of the total number of aquatic bacteria and the number thereof involved in respiration, *Appl. Environ. Microbiol.,* 36, 926, 1987.
7. Connamacher, R., Nongenetic adaptation of *Bacillus cereus* 569H to tetracycline in *Bacterial Plasmids and Antibiotic Resistance,* Krcmery, V., Rosival, L., and Watanabe, T., Eds., Springer-Verlag, New York, 1972, 425.
8. Samuelson, O. B., Degradation of oxytetracycline in seawater at two different temperatures and light intensities, and the persistence of oxytetracycline in the sediment from a fish farm, *Aquaculture,* 83, 7, 1989.

CHAPTER 54

Microbial RNA and DNA Synthesis Derived from the Assimilation of [2,³H]-Adenine

David M. Karl

INTRODUCTION

The measurement of microbial growth in nature has presented, and continues to present, a challenge to experimental microbial ecologists. In microorganisms, nucleic acids are vital cellular constituents with well-known metabolic functions.[1] The observed correlations between rates of stable RNA and DNA synthesis and cell growth[2-4] are so universally applicable among unicellular microorganisms that they appear to be well suited for the analysis of complex communities such as one finds in nature. Consequently, measurements of total microbial RNA and DNA production should provide useful information regarding microbial growth and metabolism in nature.

Several different experimental methods have been described for measuring rates of nucleic acid synthesis of either the entire microbial community or isolated portions thereof. The various methods described rely upon the incorporation of either thymidine, adenine, or uridine as presumed measures of polynucleotide (DNA and RNA) biosynthesis. Although there is continuing debate on this issue, it can be stated without reservation that the mere incorporation of a given nucleic acid precursor into cold acid-insoluble cellular materials cannot be accepted as a measure of the *in situ* rates of RNA or DNA synthesis without a careful and comprehensive consideration of: (1) the community potential for assimilating the exogenous precursor; (2) the presence of intracellular and extracellular pools of structurally related compounds which dilute the specific radioactivity of the incorporated molecules; (3) a detailed analysis of the pathways and regulation of precursor metabolism including the balance between *de novo* synthesis and salvage pathways (i.e., utilization of exogenous pools); and (4) the specificity and extent of macromolecular labeling. These are fundamental concerns for the uptake and assimilation of all radiotracers and this situation is certainly not unique for ³H-adenine assimilation.

A unique characteristic of the ³H-adenine method is the ability to measure the specific radioactivity (nCi pmol^{-1}) of the immediate precursor pool of the adenine incorporated into nucleic acids (i.e., ATP for RNA synthesis and dATP [deoxyadenosine triphosphate] for DNA synthesis[5,6]). This measurement permits correction for isotope dilution when the added radiotracer mixes with exogenous pools of structurally related compounds before uptake and with endogenous pools after transport into the cells. If the specific activity of the precursor pool is not measured, then it is impossible to extrapolate incorporation measurements to estimates of nucleic acid synthesis with any degree of accuracy. Even comparisons of "relative" rates are probably not justified since differences in the measured rates of RNA and DNA labeling may result either from variations in the specific radioactivity of the precursor pool or from actual differences in the rates of nucleic acid synthesis.

An additional advantage of the ecological use of ^{3}H-adenine is the fact that (unlike commonly used nucleosides) it is incorporated into both RNA and DNA from intracellular precursor pools which are in isotopic equilibrium.[5,7] Consequently, it is possible to simultaneously measure the rates of both RNA and DNA synthesis (and the rate ratio) from a single incubation procedure. Under these conditions, habitats supporting biosynthesis or macromolecular turnover (RNA synthesis) in the absence of cell division (DNA synthesis) and other forms of uncoupled or transient growth can be readily identified.[8]

This chapter will focus primarily on the protocols used to measure RNA and DNA synthesis in naturally occurring communities of microorganisms. Detailed presentations of the biochemical and physiological principles, evaluation of the underlying assay assumptions, and numerous ecological applications of this technique are presented elsewhere[6,7,9,10] and will only be mentioned here as they apply to the conduct of the field protocol. The reader is encouraged to refer to the primary literature for additional details on the rationale, unique advantages, and limitations of the ^{3}H-adenine technique.

Most microorganisms possess the ability to utilize exogenous supplies of adenine, and certain other nucleic acid bases, as a supplement to *de novo* synthesis. The mechanisms and implications of these so-called "salvage pathways" have been reviewed elsewhere.[11] When adenine is transported into a cell, it is rapidly incorporated into a number of derivatives including ATP, dATP, ADP, and AMP, and eventually into the metabolically stable nucleic acids. It has been demonstrated that adenine uptake is directly coupled to nucleic acid synthesis so that the rate of entry of adenine into the cell (k_1, Figure 1) does not exceed the biosynthetic requirements for ATP and dATP (sum of $k_2 + k_3$). However, the actual quantity of adenine taken up by microbial cells is much less than the value $k_2 + k_3$ due to variable but substantial contributions of adenine from *de novo* synthesis and internal recycling of nucleic acid bases (RNA turnover).

When ^{3}H-adenine is added to a culture or seawater sample, it is diluted to an unknown extent by existing pools of adenine and adenine-containing compounds (e.g., ATP, ADP, AMP, and adenosine). This process decreases the specific radioactivity of the introduced radiotracer prior to transport into the cells. Since the salvage pathways generally cannot supply the total amount of adenine required by the cells, this process must be supplemented by *de novo* synthesis of adenine. This further decreases the specific activity of the introduced radiotracer after transport into the cells but before incorporation into nucleic acids. Consequently, before an absolute rate of RNA or DNA synthesis can be obtained, it is necessary to determine the specific radioactivity of the immediate nucleotide triphosphate precursor, as well as the total amount of radioactivity incorporated into the individual nucleic acid pools. Without knowledge of both precursor-specific activity (ATP in the case of ^{3}H-adenine assimilation) and the extent of endproduct labeling, rates of nucleic acid synthesis in seawater samples cannot be measured. This concern, of course, is not unique to ^{3}H-adenine experiments, but it is true for the use of all stable or radioisotopic tracers.

The uptake and incorporation of ^{3}H-adenine as a measure of total microbial (bacteria and unicellular algae) nucleic acid synthesis was first used in oceanographic field experiments by Karl.[12] Since that time, the method has been substantially improved[5,13-15] and adapted for use in marine sediments.[16,17] There are several testable assumptions implicit in the application of the ^{3}H-adenine method to the analysis of natural microbial assemblages, including: (1) all (or in practice, most) microorganisms assimilate exogenous adenine by a

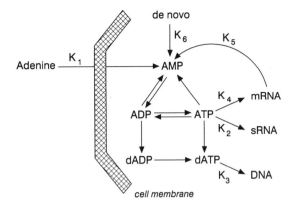

Figure 1. Schematic diagram depicting the flux of adenine through the intracellular adenine nucleotide pool (AMP, ADP, ATP, dADP, and dATP) and into the metabolically stable nucleic acid fractions (sRNA and DNA). The rate of removal of adenine for nucleic acid synthesis ($k_2 + k_3$) does not exceed the combined rate of salvage (k_1) and *de novo* synthesis (k_6). mRNA is an unstable intermediate and does not represent a net sink. The two alternative reductive pathways indicated for the formation of the DNA precursor pool (dATP) assure that radioisotopic equilibrium between ATP and dATP is maintained. (Adapted from Karl[5] and Winn and Karl.[6])

common pathway and exhibit a similar salvage response under *in situ* conditions; (2) addition of radioactive adenine does not affect the ATP cell quota, ATP turnover rate, or the rate of microbial RNA and DNA synthesis; and (3) there is no intracellular compartmentalization of microbial ATP pools or, if there is, such compartmentalization does not affect the accuracy of the measured rates. A thorough discussion of the validity of each of these assumptions has been presented elsewhere,[7] and will not be repeated here. The ^{3}H-adenine method has also been calibrated using chemostat and batch cultures of representative marine microorganisms[6] and with natural populations under field conditions.[18,19] The results show excellent agreement between the ^{3}H-adenine-derived rates of nucleic acid synthesis and the rates derived from independent measurements.

MATERIALS REQUIRED

Equipment

- Water sampling gear (Niskin bottles, or equivalent)
- Polycarbonate incubation bottles (0.1 to 4 l, depending upon design of labeling experiments)
- On-deck temperature- and light-controlled incubator or *in situ* array
- Filtration gear: PVC or stainless steel manifold (3- or 6-place), equipped with glass filter bases with stainless steel screens for 25-mm diameter filters and large volume funnels (100 to 200 ml capacity), and an adjustable vacuum pump with gauge (compressed N_2 or air can be used as an alternative to vacuum filtration), filters (24/25 mm diameter Whatman GF/F, Millipore HA/GS, or equivalent), filter forceps
- Block heater (Sybron Thermolyne Type #16500 Dri-bath, or equivalent) for ATP extraction, capable of heating extraction menstruum in test tubes to $100 \pm 1°C$ (1 atm) and maintaining boiling temperatures throughout the extraction period
- Adjustable automatic pipettes (0.01 to 0.1 ml, 0.1 to 1 ml, and 1 to 5 ml capacities)
- Storage freezer ($-20°C$)
- Light-detection instrument: any one of a number of commercially available general instruments, such as fluorometers, spectrophotometers or liquid scintillation counters, are suitable

for measuring light emission; however, in order to obtain efficient and reliable data, with maximum sensitivity, a specially designed ATP photometer is required. Instruments specifically designed for ATP analyses are marketed by several corporations including, but not limited to, E. I. DuPont, Turner Designs, LKB Instruments, Biospherical Instruments Corporation, Turner Designs and United Technologies. Whatever light-measuring device is selected, it is imperative to have either a strip chart recorder, integrator, or suitable computer-assisted data station to quantify light emission.

- Thin-layer chromatography (TLC) development tank (with glass cover)
- High intensity ultraviolet (UV) lamp
- Liquid scintillation counter
- Gyrotatory shaker table
- Refrigerated centrifuge with swinging-bucket rotor
- Vacuum evaporator (Savant Speed-vac concentrator or equivalent)
- Tissue grinder with Teflon pestle
- Water bath incubator (for 37°C and 95 to 100°C)

Supplies

- TLC plates, 20 × 20 cm, polyethyleneimine (PEI) impregnated cellulose (MN #300 PEI, Brinkmann Instrument Company, Westbury, NY)
- ATP extraction and assay glassware: test tubes for ATP extraction (approximately 15 mm diameter × 125 to 160 mm length); ATP assay vials (18 mm diameter × 40 mm height), size requirements may vary depending upon equipment/instruments used
- Disposable glass culture tubes (12 × 75 mm) and glass liquid scintillation counting (LSC) vials
- Pasteur pipettes

Solutions and Reagents

- Firefly lantern extract (Sigma Chemical Company, cat. #FLE-50)
- [2,^{3}H]-adenine radioisotopic tracer (specific radioactivity approximately 15 Ci mmol^{-1}; DuPont cat. #NET-350, or equivalent)
- Liquid scintillation counting fluid
- Sodium arsenate buffer (100 mM, pH 7.4)
- Tris(hydroxy)aminomethane-HCl buffer (20 mM, pH 7.75)
- ATP (5 × 10^{-4} M) sodium salt solutions in Tris buffer (20 mM, pH 7.7)
- Magnesium sulfate solution (40 mM) and magnesium chloride solution (0.7 M)
- Potassium phosphate solution (0.85 M, pH 3.4)
- RNA (baker's yeast; Sigma Chemical Company cat. #R-7125) and DNA (calf thymus; Sigma Chemical Company cat. #D-1501) solutions (5 mg ml^{-1})
- HCl (2 M), NaOH (1 M) and trichloroacetic acid (TCA; 100%, wt/vol)
- Ethanol (95%), reagent grade

PROCEDURES

Summary

A freshly collected water sample is incubated with ^{3}H-adenine for a predetermined period under *in situ* environmental conditions. A time course is preferred to single endpoint measurements, although endpoint measurements can also be used to estimate rates of nucleic acid synthesis. Following the incubation, subsamples are removed and extracted for ATP and nucleic acids. The ATP is purified by one-dimensional TLC and radioassayed. Total ATP is measured by firefly bioluminescence. The mean ATP pool-specific radioactivity (nCi ^{3}H-ATP mol^{-1} total ATP) is then calculated from the chemical and radiochemical ATP data. RNA and DNA are also separated and purified and their total radioactivities measured. From the incorporation and precursor specific radioactivity data, the rates of RNA and DNA synthesis can be calculated and expressed as mass of adenine (or deoxyadenine) incorporated into RNA (or DNA) per unit volume per unit time. Biomass-specific rates of RNA and DNA synthesis can also be derived by dividing the measured rates by total microbial biomass (ATP $\times$ 250; see Karl[20]). Alternatively, relative RNA-to-DNA synthesis rate ratios can be determined directly from the incorporation data, thereby eliminating the necessity for ATP pool-specific radioactivity measurements. If time-course data are available, an independent calculation of the rates of RNA and DNA synthesis can be performed using the total radioactivity accumulated in RNA and DNA and the integral of the ATP-specific activity over the length of the time-course experiment[6,21] in order to obtain the most accurate estimate (see section on data from time-course experiments). The protocol presented here is focused on water column measurements. For sediment applications, refer to Craven and Karl[16] and Novitsky and Karl.[17]

Sample Collection, Incubation, and ATP Pool Extraction

1. Collect water sample for analysis using clean, noncontaminating PVC bottles, or equivalent. Try to minimize the time between sample collection and the beginning of the experiment if results are to be extrapolated to *in situ* rate processes. Keep water samples in ambient conditions (especially with regard to light and temperature regimes) during intermediate storage of samples in order to avoid carbon and energy flux transitions or perturbations caused by change in the environmental conditions. Remove subsamples of each water type for complementary chemical and microbiological measurements (e.g., total biomass, primary and secondary production, bacterial cell numbers, dissolved nutrients and gases, particulate matter composition, etc.).

2. Subsamples are placed into acid-cleaned, polycarbonate incubation bottles. The required volume is dependent upon microbial biomass. Typically for surface ocean samples, between 100 to 250 ml are used for each determination. Because time-course information on labeling kinetics is desirable, multiple (between 4 and 8) bottles may be required for each water sample. Alternatively, one large polycarbonate bottle (2 to 4 L) that can be conveniently subsampled with minimum disruption to the microbial communities (light or temperature shock, turbulence, etc.) can also be employed. Depending upon the design of the experiment, size-fractionated water samples may also be prepared.

3. The timed incubations begin with the addition of [2,^{3}H]-adenine to each sample to achieve a final activity of between 0.05 to 0.2 μCi ml^{-1}, depending upon the desired sensitivity. The incubation should extend for a period of greater than approximately 10% of the anticipated mean doubling time of the substrate-responsive population.[6] This approach will reduce potential problems caused by isotopic disequilibrium and kinetic compartmentalization which are especially important if endpoint measurements are made.[6,7] However, prolonged incubations (e.g., >3 to 4 times the mean population doubling time) should be avoided.

4. At each predetermined sampling time (or at the end of the incubation period for endpoint measurements), an aliquot is removed and filtered through a 25-mm diameter, microfine glass fiber filter (Whatman GF/F type) for the measurement of ^{3}H-RNA and ^{3}H-DNA. This filter is placed into a glass test tube and usually stored frozen for subsequent nucleic acid purification (see section on nucleic acid purification). A second aliquot is filtered

(also through a GF/F filter) and the filter is immediately immersed into boiling Tris (20 mM, pH 7.7) buffer for ATP extraction. The sample is extracted at 100°C for 5 min, and is then stored frozen for subsequent ATP purification (see section on ATP purification). For additional details on the ATP extraction technique refer to Chapter 41.

Nucleic Acid Purification

In this procedure, RNA and DNA (and protein) are separated by their differential susceptibility to alkaline hydrolysis using a modified Schmidt-Thannhauser procedure.[14,22]

1. Remove test tubes containing GF/F filters (see section on sample collection, step 4) from −20°C storage. Add 5 ml 5% (wt/vol) ice-cold (4°C) TCA and 1 mg each of reagent RNA and DNA to each tube. The latter are added to help catalyze the precipitation and subsequent quantitative isolation of the separate macromolecular fractions. Mix thoroughly and allow to extract, at 4°C, for at least 1 to 2 h.

2. Pulverize filters in a tissue grinder, or equivalent. Rinse the pestle and grinding vessel carefully with fresh 5% TCA. Combine sample and rinse volumes. Centrifuge (2500 × g, 5 min) at 4°C, and discard supernatant. (Note: a Pasteur pipette attached to a vacuum line is convenient for removal of supernatant).

3. "Wash" (i.e., add solvent, vortex vigorously, centrifuge at 2500 × g for 5 min and discard supernatant) the GF/F filter pellets three times with 5 to 8 ml each of 5% TCA and three times with 95% ethanol (ETOH), all at 4°C. Be careful not to lose any portion of the pellet. Evaporate the residual ETOH *in vacuo* or by gentle heating of the samples. RNA and DNA remain insoluble during these procedures.

4. Resuspend the pellets in exactly 2 ml 1 M NaOH. Cap the tubes and place into water bath at 37°C for 1 h. This will solubilize both RNA and DNA, but hydrolyze only the RNA.

5. The samples are transferred to an ice water bath (4°C) and allowed to chill for at least 15 min before acidification to a pH<1 using a mixture of HCl and TCA to yield a final TCA concentration of 5% (Note: we typically add 0.5 ml of a 2:1:1 mixture, by volume, of 9 M HCl:100% TCA:distilled water). Allow samples to stand at 4°C for an additional 15 min prior to centrifugation.

6. A quantitative portion (usually 0.5 ml) of the supernatant containing ³H-RNA is transferred to a clean vial; the remainder of the supernatant is discarded. An appropriate scintillation cocktail is added to each vial and the samples are radioassayed for ³H-RNA.

7. The pellet, containing ³H-DNA, is washed as in step 3, but using only two rinses each of 5% TCA and 95% ETOH, all at 4°C, and residual liquid is evaporated. Finally, 2 ml of 5% TCA is added to each dried pellet and the samples are incubated at 95 to 100°C for 30 min. This procedure hydrolyzes DNA, but not protein.

8. Cool, add 2 ml of 1 M NaOH, mix thoroughly, and centrifuge.

9. Remove 0.5 ml of the supernatant and place into an LSC vial containing 0.5 ml of 2 M HCl to acidify the sample (Note: basic solutions result in high chemiluminescence background when in contact with certain LSC fluids). Add cocktail and radioassay for ³H-DNA.

10. If desired (primarily for radiochemical mass balance purposes[13,14]) the ³H-protein remaining in the pellet is hydrolyzed at 37°C for 18 h in 1 M NaOH, and counted as in the previous step.

ATP Purification

1. The aqueous ATP extracts are thawed at room temperature, mixed thoroughly, and centrifuged ($1500 \times g$, 5 min) to remove particulate materials. Then 4 ml are removed, placed into a clean glass vial, and evaporated to dryness *in vacuo*.[14] Distilled water is added (100 to 200 μl) and the residue is redissolved. This procedure effects a 20- to 40-fold concentration of the ATP extract in preparation for TLC purification. Total ATP concentration in a diluted portion of the concentrated extract is determined by firefly bioluminescence (see Chapter 41).

2. Radiolabeled ATP (i.e., ^{3}H-ATP) is isolated by one-dimensional TLC using plastic backed (20×20 cm) sheets coated with PEI impregnated cellulose. Prior to use, the commercially available plates are soaked in a bath of 10% NaCl, air dried at room temperature and subjected to an ascending development with distilled water. These pretreated TLC plates are dried and stored at 4°C until needed.

3. In a clean 12×75 mm disposable test tube, mix 10 μl of concentrated ATP extract and 10 μl of a nonradioactive solution of ATP (5×10^{-4} M) and spot 10 μl of this mixture on to a predetermined location approximately 1 cm from the bottom edge of a PEI plate (i.e., "origin"). Approximately 8 to 10 lanes can be prepared per plate.

4. The plate is air dried, then washed twice (5 min each) by consecutive immersion into two 1-L distilled-water baths. This procedure separates unused precursor (^{3}H-adenine) from labeled adenine nucleotides.[14]

5. The plate is then air dried (alternatively, drying can be accelerated using an electric hair dryer), and developed in one dimension with potassium phosphate (0.85 M, pH 3.4) in a closed glass TLC tank.[23] When the solvent front has ascended to within 2 cm of the top edge, the plate is removed from the tank and air dried. Total separation time is approximately 2 to 3 h.

6. ATP (and ^{3}H-ATP) is visualized by irradiating the PEI plate with short-wavelength UV light and the spots are localized by circling lightly with a sharpened lead pencil. The spots are cut out with a pair of scissors, placed into a liquid scintillation counting vial containing 1 ml of $MgCl_2$ (0.7 M) and the ATP is eluted by shaking the samples for 1 h. This elution step is necessary to maximize liquid scintillation counting efficiency.

7. Liquid scintillation cocktail is added directly to each vial in preparation for ^{3}H counting. The counts per min^{-1} (cpm) are converted to disintegrations per min^{-1} (dpm) using standard techniques for quench correction. ATP pool-specific radioactivity (nCi $pmol^{-1}$) is calculated from the dpm in the isolated ATP spot (nCi) and ATP (pmol) in the concentrated extract after appropriate volume corrections have been applied. These data are used for the calculation of RNA and DNA synthesis rates (see section on data analysis).

Data Analysis

Data from Time-Course Experiments

The ^{3}H-adenine assimilation model (Figure 1) assumes that adenine flows through a single, well-mixed precursor pool and then into nucleic acids. If this representation is accurate, then the relationship between nucleic acid labeling and the specific activity of its direct precursor can be defined precisely as:[6]

$$d*NA/dt = K_s (*ATP/ATP) - K_b (*NA/NA) \qquad (1)$$

where *NA and NA are the radiolabel and chemical concentration of nucleic acid, respectively, K_s is the rate of synthesis of nucleic acid, K_b is the rate of breakdown of nucleic acid due to cell death, and * ATP and ATP are the radiochemical and chemical concentrations of the direct precursor (i.e., ATP). During short-term labeling experiments (i.e., less than the doubling time of the population), the specific activity of the endproducts (i.e., *NA/NA) will be much less than the specific activity of the precursor (.e., *ATP/ATP) so that the second term in the equation can be ignored. Equation 1 can then be simplified to:

$$d\text{*NA}/dt = K_s (\text{*ATP}/\text{ATP}), \text{ or} \qquad (2)$$

$$\text{*NA} = K_s \int_0^t (\text{*ATP}/\text{ATP}) \qquad (3)$$

and can be rearranged to solve for K_s, the rate of RNA or DNA synthesis:

$$K_s = \text{*NA}/ \int_0^t (\text{*ATP}/\text{ATP}) \qquad (4)$$

Consequently, the rate of either RNA (K_2; Figure 1) or DNA (K_3; Figure 1) synthesis can be calculated by dividing the total radioactivity accumulated into the respective nucleic acid pool (*RNA or *DNA) by the time-dependent integral of the ATP pool-specific radioactivity. This procedure has been shown to provide an accurate estimate of RNA and DNA synthesis rate for a variety of prokaryotic and eukaryotic microorganisms.[6]

Data from Endpoint Experiments

If time-course data are not available, the rates of RNA and DNA synthesis can be estimated by dividing the total radioactivity incorporated into the respectively RNA or DNA pool by the measured endpoint ATP pool specific radioactivity. This expression is precisely equal to the true rate of nucleic acid synthesis only when the specific activity of the precursor pool is constant with time and incorporation into macromolecules is linear with time. Because the ATP pool-specific activity is zero at the start of the experiment, this is never the case. However, because the intracellular adenine nucleotide pool turns over approximately 40 to 50 times per generation[24,25] (also see Chapter 55), the specific radioactivity of the ATP pool changes rapidly during the initial phases of the incubation period and achieves isotopic equilibrium. Consequently, the difference between the rate estimated by time-course and endpoint determinations decreases with time.[6]

Synthesis Rates and Extrapolations

RNA and DNA Synthesis. From data on ^{3}H-RNA, ^{3}H-DNA, and specific activity of the ^{3}H-ATP pool, rates of RNA and DNA synthesis can be calculated by dividing the total radioactivity incorporated into RNA or DNA (nCi L^{-1} of water sample per unit time) by the integral of the ATP pool specific radioactivity (nCi pmol^{-1}) during the appropriate incubation period (or if only endpoint data are available, divide by the endpoint ATP pool-specific radioactivity). This will provide nucleic acid synthesis rates in units of pmol A (or dA) incorporated into RNA (or DNA) L^{-1} h^{-1}.

Extrapolations. The majority of investigations concerned with the production, flux, and turnover of organic matter in natural ecosystems have relied upon carbon as the basic unit of cell mass and growth. If the assumptions inherent in the application of nucleic acid precursors to ecological studies were determined to be valid under *in situ* conditions, the ³H-adenine method as described would yield reliable estimates of the total rates of microbial RNA and DNA synthesis under natural environmental conditions. While one might argue that nucleic acid synthesis rate data, by themselves, provide useful information regarding the growth of microorganisms in nature, it is often necessary to extrapolate the DNA synthesis data to estimates of total microbial production (μg C produced per unit volume per unit time) for comparison with other ecological data. Of course, the precision and accuracy of the estimates are adversely affected by the common practice of extrapolation, since the "conversion factors" are, for the most part, highly variable.

Total microbial carbon production can be extrapolated from rates of DNA synthesis by converting pmol dA to pg DNA (by assuming dA is 25 mol% of total DNA) then to carbon production (by assuming DNA is 2% of total cell carbon). The rationale for these extrapolation factors is presented elsewhere.[5-7,26]

Specific growth rate can be determined from the above-referenced carbon production estimate and independent estimates of total microbial cell carbon based on chemical (particulate carbon, ATP) or direct microscopic methods and assuming: μ = (doubling day^{-1}) $\times$ 0.693 (Karl[5]). A more precise and accurate refinement of this general approach employs the measured DNA synthesis rates and total particulate DNA in the environment, corrected for the "nonreplicating" (e.g., detrital) fraction.[26]

NOTES AND COMMENTS

Any technique which requires the uptake and assimilation of a radioactive precursor, as for the method described in this chapter, is limited in that it can only detect the "substrate responsive" portion of the total microbial community. For ³H-adenine uptake, this is known to include bacteria and unicellular algae, even when the substrate is added at nanomolar concentrations.[6,7,26] Nevertheless, the presence of nonresponsive organisms, whether bacteria, algae, protozoa or metazoa, could affect the accuracy of RNA and DNA synthesis measurements and extrapolations thereof. This is because the specific radioactivity of the ATP pool will be influenced by the presence of viable (ATP-containing) nonresponsive biomass. Fuhrman et al.[27] have challenged the ³H-adenine technique on this basis; however, a detailed response placed their criticism into a proper quantitative perspective.[26]

The nucleic acid synthesis rates derived from the uptake and assimilation of ³H-adenine represent mass-weighted mean estimates for the growing population. A method now exists to determine the standard deviation of the average ATP pool turnover rate.[28] With this additional information, it is possible to assess population heterogeneity. If there is a large variance in community growth rates, then pre- or postincubation size fractionation procedures can be employed in order to help ascertain the source of this population variability. Future applications using flow cytometry and cell sorting capabilities might also provide an opportunity to obtain population-specific estimates of nucleic acid synthesis.

It should be emphasized that extrapolation of nucleic acid synthesis data to estimates of carbon production assumes that marine microorganisms in nature exhibit "balanced growth", with all cellular components increasing at the same rate. As Eppley[29] points out, the concept of balanced growth is fundamental (but generally not explicitly stated) to the use of nutrient assimilation rates as measures of rate constants for growth. Karl[5] has already demonstrated that cell growth (as measured by RNA synthesis) and cell division (as measured by DNA synthesis) may be uncoupled in mixed assemblages of asynchronously growing marine microorganisms over incubation periods as short as 1 h. By measuring both RNA and DNA synthesis, we may be in a better position to assess discontinuous or unbalanced growth[8] than with a less complete evaluation of growth processes.

Accurate measurements of microbial biomass, metabolic activity, productivity, and specific growth rate are essential for testing hypotheses in microbial ecology. Progress in our discipline is limited by methods. The use of ^{3}H-adenine to estimate rates of microbial nucleic acid synthesis in aquatic ecosystems is a relatively new and potentially useful approach for the study of marine microbial ecology. Methods employing other nucleic acid precursors, such as ^{3}H-thymidine, are discussed elsewhere in this book. In my view, nucleic acid synthesis measurements employing ^{3}H-adenine and ^{3}H-thymidine are not mutually exclusive methods in microbial ecology since each approach yields different information regarding microbial growth in nature. Unless evidence is presented to suggest otherwise, they should be considered complementary techniques. Finally, the potential significance of ecological information to be derived from estimates of *in situ* rates of RNA and DNA synthesis more than justifies the level of research effort that is required to ensure a proper interpretation and accurate extrapolation to estimates of microbial growth and production.

REFERENCES

1. Ingraham, J. L., Maaloe, O., and Neidhardt, F. C., *Growth of the Bacterial Cell,* Sinauer Associates, Sunderland, MA, 1983.
2. Kjeldgaard, N. O., Regulation of nucleic acid and protein formulation in bacteria, *Adv. Microbiol. Physiol.,* 1, 39, 1967.
3. Maaloe, O. and Kjeldgaard, N. O., *Control of Macromolecular Synthesis,* W. W. Benjamin, New York, 1966.
4. Nierlich, D. P., Regulation of bacterial growth, RNA, and protein synthesis, *Annu. Rev. Microbiol.,* 32, 393, 1978.
5. Karl, D. M., Simultaneous rates of ribonucleic acid and deoxyribonucleic acid syntheses for estimating growth and cell division of aquatic microbial communities, *Appl. Environ. Microbiol.,* 42, 802, 1981.
6. Winn, C. D. and Karl, D. M., Laboratory calibrations of the [^{3}H]adenine technique for measuring rates of RNA and DNA synthesis in marine microorganisms, *Appl. Environ. Microbiol.,* 47, 835, 1984.
7. Karl, D. M. and Winn, C. D., Adenine metabolism and nucleic acid synthesis: applications to microbiological oceanography, in *Heterotrophic Activity in the Sea,* Hobbie, J. E. and Williams, P. J. LeB., Eds., Plenum Publishing, 1984, 197.
8. Karl, D. M. and Novitsky, J. A., Dynamics of microbial growth in surface layers of a coastal marine sediment ecosystem, *Mar. Ecol. Prog. Ser.,* 50, 169, 1988.
9. Karl, D. M., Determination of in situ microbial biomass, viability, metabolism, and growth, in *Bacteria in Nature, Vol. 2,* Poindexter, J. S. and Leadbetter, E. R., Eds., Plenum Publishing, 1986, 85.
10. Karl, D. M. and Winn, C. D., Does adenine incorporation into nucleic acids measure total microbial production?: a response to comments by Fuhrman et al., *Limnol. Oceanogr.,* 31, 1384, 1986.
11. Hochstadt, J., The role of the membrane in the utilization of nucleic acid precursors, *CRC Crit. Rev. Biochem.,* 2, 259, 1974.
12. Karl, D. M., Measurement of microbial activity and growth in the ocean by rates of stable ribonucleic acid synthesis, *Appl. Environ. Microbiol.,* 38, 850, 1979.
13. Karl, D. M., Selected nucleic acid precursors in studies of aquatic microbial ecology, *Appl. Environ. Microbiol.,* 44, 891, 1982.

14. Karl, D. M., Winn, C. D., and Wong, D. C. L., RNA synthesis as a measure of microbial growth in aquatic environments. I. Evaluation, verification and organization of methods, *Mar. Biol.,* 64, 1, 1981.

15. Karl, D. M., Winn, C. D., and Wong, D. C. L., RNA synthesis as a measure of microbial growth in aquatic environments. II. Field applications, *Mar. Biol.,* 64, 13, 1981.

16. Craven, D. B. and Karl, D. M., Microbial RNA and DNA synthesis in marine sediments, *Mar. Biol.,* 83, 129, 1984.

17. Novitsky, J. A. and Karl, D. M., Characterization of microbial activity in the surface layers of a coastal sub-tropical sediment, *Mar. Ecol. Prog. Ser.,* 28, 49, 1986.

18. Karl, D. M. and Bossard, P., Measurement of microbial nucleic acid synthesis and specific growth rate by $^{32}PO_4$ and ^{3}H-adenine: a field comparison, *Appl. Environ. Microbiol.,* 50, 706, 1985.

19. Laws, E. A., Redalje, D. G., Haas, L. W., Bienfang, P. K., Eppley, R. W., Harrison, W. G., Karl, D. M., and Marra, J., High phytoplankton growth and production rates in oligotrophic Hawaiian coastal waters, *Limnol. Oceanogr.,* 29, 1161, 1984.

20. Karl, D. M., Cellular nucleotide measurements and applications in microbial ecology, *Microbiol. Rev.,* 44, 739, 1980.

21. Yamazaki, H. and Leung, K., Determination of the total rates of synthesis and degradation of RNA in bacterial cultures, *Can. J. Microbiol.,* 27, 168, 1981.

22. Munro, H. N. and Fleck, A., The determination of nucleic acids, in *Methods of Biochemical Analysis, Vol. 14,* Glick, D., Ed., Interscience Publishers, New York, 1966, 113.

23. Cashel, M., Lazzarini, R. A., and Kalbacher, B., An improved method for thin-layer chromatography of nucleotide mixtures containing ^{32}P-labeled orthophosphate, *J. Chromatogr.,* 40, 103, 1969.

24. Chapman, A. G. and Atkinson, D. E., Adenine nucleotide concentrations and turnover rates. Their correlation with biological activity in bacteria and yeast, *Adv. Microbiol. Physiol.,* 15, 253, 1977.

25. Karl, D. M., Jones, D. R., Novitsky, J. A., Winn, C. D., and Bossard, P., Specific growth rates of natural microbial communities measured by adenine nucleotide pool turnover, *J. Microbiol. Methods,* 6, 221, 1987.

26. Winn, C. D. and Karl, D. M., Diel nucleic acid synthesis and particulate DNA concentrations: Conflicts with division rate estimates by DNA accumulation, *Limnol. Oceanogr.,* 31, 637, 1986.

27. Fuhrman, J. A., Ducklow, H. W., Kirchman, D. L., Hudak, J., McManus, G. B., and Kramer, J., Does adenine incorporation into nucleic acids measure total microbial production?, *Limnol. Oceanogr.,* 31, 627, 1986.

28. Laws, E. A., Jones, D., and Karl, D. M., Method for assessing heterogeneity in turnover rates within microbial communities, *Appl. Environ. Microbiol.,* 52, 866, 1986.

29. Eppley, R. W., Relations between nutrient assimilation and growth in phytoplankton with a brief review of estimates of growth rate in the ocean, in Physiological bases of phytoplankton ecology, Platt, T., Ed., *Can. Bull. Fish. Aquat. Sci.,* 210, 1981, 251.

CHAPTER 55

Adenosine Triphosphate (ATP) and Total Adenine Nucleotide (TAN) Pool Turnover Rates as Measures of Energy Flux and Specific Growth Rate in Natural Populations of Microorganisms

David M. Karl

INTRODUCTION

Two fundamental goals in aquatic microbial ecology are to elucidate the pathways and to quantify the fluxes of carbon and energy through microbial communities in nature. In recent years, much attention has been focused on the measurement of microbial metabolism and productivity in natural aquatic habitats and a variety of experimental protocols is now available.[1] Among the most direct and meaningful assessments, however, is the measurement of microbial community specific growth rate (μ; or alternatively, generation time or doubling time) which can be compared to the extensive database derived from carefully controlled laboratory studies and to theoretical models relating specific growth rates to growth temperature or nutrient conditions.

Two general approaches have been used to estimate μ in natural populations of microorganisms: (1) direct measurements of the increase in cell numbers or biomass during timed bottle incubations, usually after removing the influence of grazer populations by size fractionation, chemical inhibition, or sample dilution and (2) direct and simultaneous measurements of standing stock (biomass) and production (rate of biomass increase) in order to calculate the mass-specific rate of biomass increase. A major limitation, however, with most experimentally derived estimates of specific growth rate, is the inability to ascertain heterogeneity in specific growth rate that is generally acknowledged or presumed to occur in natural populations or microorganisms.

In contrast to the availability of the above-referenced experimental methods providing either direct or derived estimates of carbon (or mass) flux, relatively few techniques exist for the measurement of microbiological energy flux. The generally assumed direct coupling between carbon and energy fluxes in nature require balanced cell growth and constant growth efficiencies. However, in nature, microbial cells and even entire populations of cells are known to exhibit unbalanced or transient growth dynamics. This not only invalidates the assumptions inherent in the application of certain productivity protocols, but also serves to uncouple the flux of energy from the production or turnover of cell biomass.

In a series of recent papers, Karl and colleagues[2-5] have described novel approaches for the independent estimation of energy flux and carbon flux by direct measurements of the turnover of the cellular adenosine triphosphate (ATP) and total adenine nucleotide (TAN) pools, respectively. In addition, a unique data analysis method has been devised for determining both the average turnover rate and the standard deviation thereof as a direct assessment of the heterogeneity of ATP or TAN pool turnover rates within microbial communities in nature.[6]

THEORETICAL PRINCIPLES

ATP Pool Turnover

The central role of ATP in the stoichiometric coupling of energy-yielding and energy-requiring metabolic reactions has been known since the pioneering research of Lipmann.[7] The steady-state intracellular concentration of ATP (referred to as the "ATP pool") in viable microorganisms (prokaryotes and eukaryotes) appears to be well-regulated at a value of 2 to 6 nmol ATP mg^{-1} dry weight ($\sim$1 to 3 mM) regardless of growth rate, culture condition, or mode of nutrition.[8] Previous studies of ATP pools in microorganisms and in environmental samples have emphasized the futility of extrapolating these static measurements to estimates of metabolic energy flux. Consequently, it is the turnover rate of the ATP pool, rather than the steady-state concentration of that pool, which varies in proportion to cellular metabolic energy requirements. It follows then, that direct measurements of ATP pool turnover rates, when coupled with independent estimates of ATP pool size, should provide useful information on biological energy flux in cells, populations, or natural microbial assemblages.

Cellular ATP pool turnover results from the hydrolysis of one or both of the anhydride-bound PO$_4$ groups (the β-P and γ-P or "high-energy" phosphate bonds) via orthophosphate or pyrophosphate cleavage, followed by subsequent regeneration of ATP by either substrate-level, oxidative, or photophosphorylation processes (Figure 1A and B). Because ATP pool turnover processes define a cycle, this essential set of metabolic processes neither results in net removal of adenine nucleotide molecules from the intracellular pool nor requires coupled biosynthesis thereof. This will become important in the analytical procedures employed to distinguish between the turnover cycles of ATP and TAN pools.

In theory, ATP pool turnover rates could be measured under steady-state metabolic conditions (i.e., d[ATP]/dt = 0) by estimating either the rate of ATP formation or rate of ATP utilization, if either process could be temporarily suspended without affecting the other. Such ATP pool transitions have been observed in microorganisms following the rapid removal of oxygen (for aerobic heterotrophs) or light (for phototrophs), but it is uncertain whether the results derived from these harsh experimental perturbations yield reliable estimates of ATP pool dynamics. Recently, Karl and Bossard[3] devised a novel method for ATP pool turnover in cell cultures or natural populations of microorganisms. The procedure relies upon the use of ^{32}PO$_4$ (or ^{33}PO$_4$) as a tracer for P-flux through the acid anhydride-bound P (β-P and γ-P; Figure 1A and B) groups of cellular ATP. The uptake of ^{32}PO$_4$ also results in the labeling of the α-P group of ATP, if the population in question is actively growing (Figure 1C). Consequently, in order to uniquely assess the labeling of ATP derived from energy flux, one must be able to separate β-P and γ-P labeling from α-P.[3] On the other hand, α-P labeling defines TAN pool turnover and therefore, ^{32}PO$_4$ labeling can be used to estimate both ATP pool and TAN pool turnover rates.

TAN Pool Turnover

The uptake, or salvage, of exogenously added nucleic acid precursors (e.g., adenine, thymidine, uridine) in preference to *de novo* synthesis is a well-documented characteristic

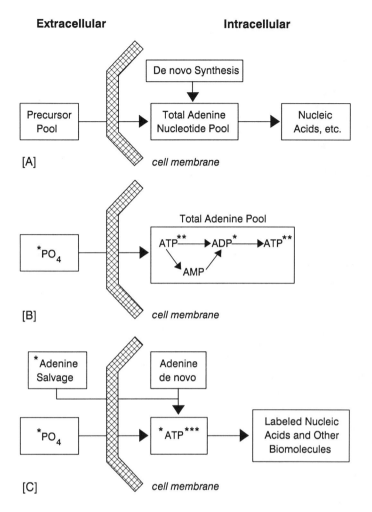

Figure 1. Schematic models for $^{32}PO_4$ ($^{33}PO_4$) labeling of the α-, β-, and γ-P positions of ATP and ^{3}H-adenine assimilation during microbial metabolism and biosynthesis. (A) Generalized model for the uptake of exogenously added precursor, incorporation into intermediate nucleotide pool, and into cellular macromolecules. The turnover rate of the adenine nucleotide pool is controlled by net removal as required for cellular biosynthesis. (B) Specific model for ATP pool turnover (energy flux) in absence of biosynthesis. Under these conditions, there is no net removal of nucleotide molecules and, therefore, no incorporation of radiolabeled phosphate (*P) into the α-P (i.e., AMP) position of ATP. Because of rapid interconversion between ATP, ADP, and AMP, the kinetic constants for labeling the β-P and γ-P positions of ATP are indistinguishable. Since there is no net removal of adenine for biosynthesis, ^{3}H-adenine would not be expected to label the ATP pool under these restrictive conditions. (C) During net macromolecular biosynthesis, $^{32}PO_4$ also labels the α-P of the ATP pool (although with a kinetic constant uncoupled from the labeling of β-P and γ-P positions of ATP) and ^{3}H-adenine is incorporated into cellular ATP pool. Net removal of adenine nucleotides for biosynthesis is balanced by the sum of salvage and *de novo* synthesis processes. In theory, the kinetic constant for labeling ATP by ^{3}H-adenine is identical to the $^{32}PO_4$ labeling constant of the α-P of ATP. (Adapted from Karl and Bossard.[3])

of aquatic microbial communities (References 9 and 10; see Figure 1C). This phenomenon comprises the theoretical basis for the use of [^{3}H]adenine, [^{3}H]thymidine, and [^{3}H]uridine in ecological studies of nucleic acid synthesis. An important aspect of the incorporation of nucleic acid precursors is the observation that the total flux of precursor into the nucleotide triphosphate pools (i.e., ATP, TTP, UTP, etc.), which is the result of the combined effects of salvage and *de novo* synthesis, is in equilibrium with the removal of the triphosphate precursors required for cellular biosynthesis. Furthermore, it is well-documented that the TAN pool (i.e., steady-state intracellular concentration of [ATP + ADP + AMP]) in microorganisms does not vary with changes in the specific growth rate.[8,11] Because cellular biosynthesis is directly related to growth rate, a positive correlation is expected to exist between cellular adenine nucleotide flux and μ. It follows then that the turnover rate of the TAN pool (and most likely, of all ribonucleotide and deoxyribonucleotide pools in general) must vary in direct proportion to the rates of nucleic acid synthesis and hence, net growth.

The TAN pool turnover time is defined as the average residence time of a molecule in the intracellular pool before it is removed for macromolecular biosynthesis (i.e., the steady-state pool concentration divided by the steady-state rate of synthesis or removal). If a radioactive precursor such as [³H]adenine is added to a growing culture or natural population of microorganisms, the TAN pool turnover time can be calculated by monitoring the change in the specific activity (SA) of the ATP pool with incubation time. TAN pool turnover rate can also be measured by monitoring the labeling kinetics of the α-P position of ATP following the addition of $^{32}PO_4$ (or $^{33}PO_4$) to a sample[3] (also see the section on ATP pool turnover). Exactly one turnover cycle has been completed when the ATP pool has achieved an SA that is equal to 50% of the value at isotopic equilibrium. Chapman and Atkinson[11] have estimated TAN pool turnover rates as a function of growth rate for *Escherichia coli* and *Salmonella typhimurium*. They conclude that the TAN pool "turns over" (i.e., is completely utilized for biosynthesis and is replenished through salvage and *de novo* synthesis) 30 to 50 times per generation regardless of generation time. This prediction has been tested using a diverse variety of marine microorganisms.[5] The results indicate that the TAN pool turnover time is positively correlated with generation time and averages 2.2% of the generation time (i.e., TAN pool turns over 45 times per generation).

MATERIALS REQUIRED

Equipment

- Water sampling gear (Niskin bottles, or equivalent)
- Polycarbonate incubation bottles (0.1 to 4 L, depending upon design of labeling experiments)
- On-deck temperature- and light-controlled incubator or *in situ* array
- Filtration gear: PVC or stainless steel manifold (3- or 6-place), equipped with glass filter bases with stainless steel screens for 25 mm diameter filters and large volume funnels (100 to 200 ml capacity), an adjustable vacuum pump with gauge (compressed N_2 or air can be used as an alternative to vacuum filtration), filters (24/25 mm diameter Whatman GF/F, Millipore HA/GS, or equivalent), filter forceps
- Block heater (Sybron Thermolyne Type #16500 Dri-bath, or equivalent) for ATP extraction, capable of heating extraction menstruum in test tubes to $100 \pm 1°C$ (1 atm) and maintaining boiling temperatures throughout the extraction period
- Adjustable automatic pipettes (10 to 100 μl, 100 to 1000 μl, and 1 to 5 ml capacities)
- Storage freezer ($-20°C$)
- Light-detection instrument: any one of a number of commercially available general instruments, such as fluorometers, spectrophotometers or liquid scintillation counters, are suitable for measuring light emission. However, in order to obtain efficient and reliable data, with maximum sensitivity, a specially designed ATP photometer is required. Instruments specifically designed for ATP analyses are marketed by several corporations including, but not limited to, E. I. Dupont, Turner Designs, LKB Instruments, Biospherical Instruments Corporation, Turner Designs and United Technologies. Whatever light-measuring device is selected, it is imperative to have either a strip chart recorder, integrator, or suitable computer-assisted data station to quantify light emission.

- Thin-layer chromatography (TLC) development tank (with glass cover)
- High-intensity ultraviolet (UV) lamp
- Liquid scintillation counter
- Gyrotatory shaker table
- Microcentrifuge
- Vacuum evaporator (Savant Speed-vac concentrator or equivalent)

Supplies

- TLC plates, 20 × 20 cm, polyethyleneimine (PEI) impregnated cellulose (MN #300 PEI, Brinkmann Instrument Company, Westbury, NY)
- ATP extraction and assay glassware: test tubes for ATP extraction (approximately 15 mm diameter × 125 to 160 mm length); ATP assay vials (18 mm diameter × 40 mm height), Note: size requirements may vary depending upon equipment/instruments used
- Disposable glass culture tubes (12 × 75 mm) and glass liquid scintillation counting (LSC) vials
- $^{32}PO_4$ ($^{33}PO_4$) or [2,^{3}H]-adenine (DuPont #NET-350, or equivalent) radioisotopic tracer
- LSC cocktail
- Microcentrifuge tubes, 1.5 ml capacity

Solutions and Reagents

- Firefly lantern extract (Sigma Chemical Company, cat. #FLE-50)
- ATPase, porcine cerebral cortex (EC 3.6.1.3; Sigma Chemical Company, cat. #A-7510) prepared in distilled water to a concentration of 1 unit ml^{-1} and stored frozen until needed
- Sodium arsenate buffer (100 mM, pH 7.4)
- Tris(hydroxy)aminomethane-HCl buffer (20 mM, pH 7.75)
- ATP (5×10^{-4} M) and AMP (3 mM) sodium salt solutions in Tris buffer
- Magnesium sulfate solution (40 mM)
- Potassium phosphate buffer (60 mM, pH 7.4) and potassium phosphate solution (0.85 M, pH 3.4)
- Magnesium chloride (0.7 M and 0.35 M), sodium chloride (5 M) and potassium chloride (1 M) solutions
- Activated charcoal (Sigma Chemical Company, cat. #C-4386)
- Phosphoric acid (concentrated)
- Formic acid (0.2 M)

PROCEDURES

Brief Summary

A freshly collected seawater sample is incubated under defined temperature and light conditions (e.g., *in situ*, simulated *in situ*, dark) in the presence of an appropriate radiolabeled precursor. After predetermined incubation periods, subsamples are removed and particulate ATP pools are extracted and purified. The isolated radiolabeled ATP is either directly counted (e.g., ^{3}H-adenine based determinations of TAN pool turnover rate) or subjected to enzymatic hydrolysis designed to selectively hydrolyze the γ-P position of ATP (e.g., $^{32}PO_4$ or $^{33}PO_4$ based determinations of ATP and TAN pool turnover rates). From time-course information on the change in ATP and TAN pool specific radioactivity, the turnover rates can be quantified and the results extrapolated to estimates of energy flux and specific growth rate.

Sample Collection, Incubation, and ATP Pool Extraction

1. Collect water sample for analysis using clean, noncontaminating PVC bottles, or equivalent. Try to minimize the time between sample collection and the beginning of the experiment if results are to be extrapolated to *in situ* rate processes. Keep water samples in ambient conditions (especially with regard to light and temperature regimes) during intermediate storage of samples in order to avoid carbon and energy flux transitions or perturbations caused by change in the environmental conditions. Remove subsamples of each water type for complementary chemical and microbiological measurements (e.g., total biomass, primary and secondary production, bacterial cell numbers, dissolved nutrients and gases, particulate matter composition, etc.)

2. Subsamples are placed into acid-cleaned, polycarbonate incubation bottles. The required volume is dependent upon microbial biomass. Typically for surface ocean samples, between 100 to 250 ml are used for each determination. Because time-course information on labeling kinetics is required, multiple (between 4 to 8) bottles are required for each water sample. Alternatively, one large polycarbonate bottle (2 to 4 L) which can be conveniently subsampled with minimum disruption to the microbial communities (light or temperature shock, turbulence, etc.) can also be employed. Depending upon the design of the experiment, light vs. dark bottles or size-fractionated water samples may also be prepared.

3. The timed incubations begin with the addition of either $^{32}PO_4$ (or $^{33}PO_4$) or ^{3}H-adenine, or both, to each sample to a final activity of between 0.1 to 0.5 μCi ml^{-1} ($^{32}PO_4$) or 0.05 to 0.2 μCi ml^{-1} (^{3}H-adenine) depending upon the desired sensitivity and expected ambient pool concentrations of nonlabeled substrates. The use of $^{32}PO_4$ enables one to measure both ATP and TAN pool turnover rates; ^{3}H-adenine labeling can only be used for growth rate estimation by TAN pool turnover rate measurements (also see the section on assay limitations and future applications).

4. At predetermined times, subsamples are filtered through 25 mm diameter, microfine glass fiber filters (Whatman GF/F type), and the filters containing the concentrated particulate matter are immersed into boiling phosphate (60 mM, pH 7.4) or Tris (20 mM, pH 7.7) buffer for ATP extraction. Alternatively, cellular ATP can be extracted by acidification of the entire water with H_3PO_4 (1 M final concentration). For additional details on ATP extraction techniques refer to Chapter 41.

ATP Purification

1. The aqueous ATP extracts are concentrated 10- to 20-fold by evaporation *in vacuo*.[12] The acid extracts are concentrated by the charcoal adsorption technique.[13] ATP concentration in an aliquot of the concentrated extract is determined using firefly bioluminescence (see Chapter 41).

2. Radiolabeled ATP is isolated by one-dimensional TLC using plastic backed (20 × 20 cm) sheets coated with PEI impregnated cellulose. Prior to use, the commercially available plates are soaked in a bath of 10% NaCl, air dried at room temperature, and subjected to an ascending development with distilled water. These pretreated TLC plates are dried and stored at 4°C until needed. Because ^{3}H-adenine and $^{32}PO_4$ ($^{33}PO_4$) label different suites of intracellular molecules, slightly different TLC procedures are required for ATP purification.

^{3}H-ATP Isolation

1. In a clean 12 × 75 mm disposable glass test tube, mix 10 μl of concentrated ATP extract and 10 μl of a nonradioactive solution of ATP (5 × 10^{-4} M) and spot 10 μl of this

mixture onto a predetermined location approximately 1 cm from the bottom edge of a PEI plate (i.e., "origin"). Approximately 8 to 10 lanes can be prepared per plate.

2. The plate is air dried, then washed twice (5 min each) by consecutive immersion into two 1-L distilled water baths. This procedure separates unused precursor (^{3}H-adenine) from labeled adenine nucleotides.[12]

3. The plate is then air dried (alternatively, drying can be accelerated using an electric hair dryer), and developed in one dimension with potassium phosphate (0.85 M, pH 3.4) in a closed glass TLC tank.[14]

4. When the solvent front has ascended to within 2 cm of the top edge, the plate is removed from the tank and air dried. Total separation time is approximately 2 to 3 h.

5. ATP (and ^{3}H-ATP) is visualized by irradiating the PEI plate with short-wavelength UV light and the spots are localized by circling lightly with a sharpened lead pencil.

6. The spots are cut out with a pair of scissors, placed into an LSC vial containing 1 ml of $MgCl_2$ (0.7 M) and the ATP is eluted by shaking the samples for 1 h. This elution step is necessary for ^{3}H-adenine labeled samples due to self-absorption on solid supports.

7. Liquid scintillation cocktail is added directly to the vials in preparation for ^{3}H counting.

8. The counts per min^{-1} are converted to disintegrations per min^{-1} (dpm) using standard techniques for quench correction. ATP pool SA (nCi pmol^{-1}) is calculated from the dpm in the isolated ATP spot (nCi) and ATP (pmol) in the concentrated extract after appropriate volume corrections have been applied. These data are used directly for the calculation of TAN pool turnover, as described in the section on data analysis.

^{32}P (or ^{33}P)-ATP Isolation

Because $^{32}PO_4$ labels many small molecular weight intracellular compounds including a variety of different nucleotides, the chromatographic purification procedure requires a two-step, ascending one-dimensional separation on PEI impregnated cellulose plates.

1. After sample application, distilled water washing and drying (as in the previous ^{3}H-ATP isolation section), the TLC plate is developed in formic acid (0.2 M) to the top of the plate. Under these conditions, nucleotide triphosphates (including ATP) remain at the origin while most other ^{32}P (^{33}P)-labeled molecules move with the solvent front.

2. The plate is removed and immediately washed in a distilled water bath to remove the formic acid and air dried before final development, in the same direction, with potassium phosphate (0.85 M, pH 3.4) in a closed TLC tank.

3. When the solvent front is within 2 cm of the top edge, the plate is removed, washed in two successive distilled water baths and air dried. The final wash is required to remove residual nonradioactive PO_4 (the developing solvent) which, if present, will inhibit the subsequent ATPase reaction (see section on selective ATP hydrolysis by ATPase). Total separation time is approximately 4 to 5 h.

4. The ATP spots are localized by UV light, circled with a pencil line and excised, as described previously.

5. The spots are eluted into 1 ml of $MgCl_2$ (0.35 M) by shaking for 1 h at room temperature.

6. A 10 μl aliquot is diluted into 1 ml of Tris buffer (pH 7.4, 0.02 M) and assayed for total ATP by the firefly bioluminescence reaction (see Chapter 41). This value is compared to a measurement of the total ATP applied to the plate to determine the overall recovery of ATP during the purification procedure, which is generally >90%.[3]

7. The remainder of the purified ATP (or a portion thereof) is transferred to a clean vial in preparation for selective enzymatic hydrolysis by ATPase, described in the following section.

Selective ATP Hydrolysis by ATPase

1. Stock solutions of ATPase are prepared in distilled water to a concentration of 1 unit ml^{-1} and stored frozen ($-20°C$) until needed. Under these conditions the ATPase reagent is stable for at least 1 month without appreciable loss of activity. For short-term storage (<2 to 3 d) the stock ATPase reagent should be stored at 4°C, rather than $-20°C$, in order to maximize stability.

2. To perform the ATPase hydrolysis reaction, 750 μl of the "TLC-purified" ATP solution (in 0.35 M MgCl$_2$) is placed into a 12 × 75 mm culture tube containing 250 μl distilled water, 10 μl of a 1 M KCl solution, 20 μl of a 5 M NaCl solution, and 25 μl of a 3 mM AMP solution, followed by the addition of 50 μl of the stock ATPase reagent. The AMP is added to repress the ADP dismutase reaction (2ADP → AMP + ATP) which is also catalyzed by ATPase. The contents of the tube are mixed thoroughly and a 10 μl subsample is immediately removed, diluted in 1 ml of 0.02 M Tris (pH 7.4) buffer to quench the reaction and assayed for determination of "time-zero" ATP concentration using the firefly bioluminescence reaction (Chapter 41).

3. The ATP assay procedure is repeated after approximately 10 to 15 min with one or two representative samples in order to establish the extent of ATP hydrolysis. Once the initial rate of hydrolysis is known one can estimate the time required for complete hydrolysis of ATP to ADP, which in our experience, ranges from 20 to 40 min. Since all samples contain identical concentrations of carrier ATP (from the TLC procedure) and equivalent amounts of enzyme, the rates of hydrolysis are expected to be similar. However, complete ATP hydrolysis should be confirmed for each individual sample before proceeding with the separation and purification of products.

4. Following ATP hydrolysis, a 250 μl aliquot of the sample is removed for measurement of "total radioactivity" (i.e., α-^{32}P + β-^{32}P + γ-^{32}P). A second 500 μl aliquot is added to a small plastic microfuge tube containing 500 μl of an activated charcoal slurry (25 mg charcoal per ml of 0.1 M H$_3$PO$_4$), mixed thoroughly and centrifuged for 10 min at 13,000 × g. A portion of the supernatant (usually 750 μl) is removed for LSC for determination of "^{32}PO$_4$ radioactivity" (i.e., γ-^{32}P). The phosphoric acid-charcoal treatment removes >99.5% of the ADP from solution and catalyzes a <1% nonspecific binding of ^{32}PO$_4$.[3] Since γ-^{32}P- and β-^{32}P-specific activities are equal over all time periods (experimental evidence for this assumption is presented by Karl and Bossard[3] and Bossard and Karl[2]), the α-^{32}P activity can be calculated by difference (i.e., α-^{32}P = [total-^{32}P] − [2 × γ-^{32}P]). The γ- and β-P data are used for ATP pool turnover rate estimation and the α-P data are used for TAN pool turnover rate estimation (see following section).

Data Analysis and Derived Estimates of Energy Flux and Growth Rate

As predicted by radiotracer theory (Figure 2), the change in the ATP or TAN pool-specific radioactivity (SA; nCi pmol^{-1}) following the addition of an appropriate radioactive precursor is an exponential function of incubation time (t). The initial SA (at t = 0) is zero and the relative SA (at isotopic equilibrium with precursor pool) is 1.000. Under these conditions, the decimal equivalent of SA at any time (SA$_t$) can be described by the equation:

$$SA_t = 1 - (2^{-N}) \tag{1}$$

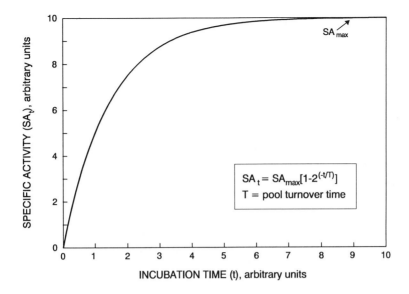

Figure 2. Change in ATP or TAN pool specific radioactivity (SA) vs. incubation time (t) following the addition of a suitable radiotracer. The general equation for this relationship is: $SA_t = SA_{max}[1 - 2^{-t/T}]$, which can be used to estimate SA_{max} and T from data derived from incubation experiments. When SA_t is expressed as a decimal equivalent of SA_{max} (e.g., $SA_{max} = 100$), the general relationship can also be expressed as: $SA_t = 1 - (2^{-N})$ where N is the number of turnover cycles.

where N is equal to the number of turnover cycles (N is dimensionless) observed during the incubation period. ATP or TAN pool turnover time (T) can then be calculated from the expression:

$$T = t/N \qquad (2)$$

This relationship is applicable for all incubation periods (t) less than or equal to 5N, at which time the pools are expected to be near isotopic equilibrium and, in theory, would not change until the exogenous radioactive precursor pool is exhausted (Figure 2).

Karl et al.[5] have compared two independent statistical curve-fitting methods for the analysis of the experimental kinetic data: (1) a computer-assisted nonlinear least squares regression program (PROC NLIN method, DUD; Statistics for the Applied Sciences [SAS] Institute Inc.) to obtain a best fit of the experimental data to the model equation (Equation 1), and (2) the Mangelsdorf[15] linear transformation method in which the ATP or TAN pool SA is determined at regular intervals of incubation time (Δt) and then analyzed by a linear regression analysis of a plot of SA at time (t) as the independent variable vs. the value at time ($t + \Delta t$) as the dependent variable. For the samples tested, both methods resulted in similar estimates of SA_{max} and turnover time (T) of the TAN pool,[5] but for reasons discussed by Karl et al.,[5] the authors recommended the use of the PROC-NLIN method of data analysis.

Once the turnover time (T) of the ATP or TAN pool has been determined, it is straightforward to extrapolate these data to the more meaningful ecological parameters of energy flux and specific growth rate. If we assign a value of -11 (± 1) kcal mol^{-1} to the free energy of ATP hydrolysis for either the β-P and γ-P positions under *in vivo* conditions,[16] then the total microbiological energy flux (EF = expressed in units of kcal L^{-1} of sample h^{-1}) can be calculated as:

$$EF \ (kcal \ L^{-1} \ hr^{-1}) \ = \ - \ 22 \ ([ATP]/T_{ATP}) \qquad (3)$$

where [ATP] is equal to the total particulate ATP pool in (mol L^{-1} of sample) T_{ATP} is ATP pool turnover time (h). This value for total available free energy can be directly compared to the solar energy flux, to the energy stored by photosynthesis, or to respiration. In this way, carbon and energy fluxes can be directly compared.

Extrapolation of the measured TAN turnover time (T_{TAN}) to specific growth rate (μ) is based on the theoretical predictions[11] and empirical observations[5] that the T_{TAN} is equivalent to a value that is 2 to 3% of the generation time (i.e., TAN pool turns over, on average, 40 times per generation). Consequently, the doubling time (T_d, expressed in h) of the substrate responsive population is:

$$T_d \ = \ T_{TAN} \ \times \ 40 \qquad (4)$$

and specific growth rate (μ, expressed in units of h^{-1}) is:

$$\mu \ = \ 1/T_d \ \times \ \ln(2) \qquad (5)$$

These kinetic model formulations will provide "ideal" results when all of the radiotracer responsive microorganisms in a given habitat are growing at identical rates. A novel, statistical treatment of these data, described in detail by Laws et al.,[6] provides a method for determining the coefficient of variation among the individual estimates of T_{ATP} and T_{TAN}. Consequently, one can quantify the variability among individual subcomponents of the microbial community. Although applied here for the turnover of ATP and TAN pools, this mathematical formulation can be used to assess heterogeneity in the turnover rate of any intracellular pool for which a suitable radioactive or stable isotopic precursor exists.

COMMENTS: ASSAY LIMITATIONS AND FUTURE APPLICATIONS

Any technique which requires the uptake and assimilation of a radioactive precursor, as for the methods described in this chapter, are limited in that they can only detect the "substrate responsive" portion of the total microbial community. For $^{32}PO_4$ ($^{33}PO_4$), however, this is likely to be a larger subset (if not all viable cells) than for the uptake of more complex or exotic precursors. Since both algae and bacteria are able to assimilate PO_4, the methods are designed to measure total microbial energy flux and specific growth rate. If one desires information on the magnitude of the photophosphorylation energy flux, a light vs. dark incubation experiment could be performed.

With regard to ^{3}H-adenine assimilation, we are also confident that both bacteria and unicellular algae are substrate responsive even at nanomolar concentrations.[17-19] Consequently, TAN pool turnover rate determination by measuring the labeling kinetics of either the α-P position of the ATP pool or ^{3}H-ATP (from ^{3}H-adenine incubations) should provide complementary, redundant information on the specific growth rates of microorganisms in nature.[4] We should, however, emphasize the fact that the presence of nonresponsive microorganisms or metazoans will not adversely influence the growth rate estimated for the substrate responsive subpopulations.[5]

The growth rate derived from our analysis of TAN pool turnover rate is considered to represent a mass-weighted mean estimate for the growing population. As mentioned previously, a method now exists to determine the standard deviation of the average ATP pool turnover rate.[6] With this additional information, it is possible to assess population heterogeneity. If there is a large variance in community growth rates, then the use of pre- or postincubation size fractionation procedures in order to obtain phytoplankton- and bacterial-enriched fractions for the measurement of size-fractionated TAN pool turnover rates may be used to help ascertain the source of this population variability. Future applications using flow cytometry and cell-sorting capabilities might also provide an opportunity to obtain population-specific estimates of growth using the TAN turnover method.

The theoretical basis of the method described in this chapter for total microbial community specific growth rate estimation could, in principle, be applied for the uptake of other radiolabeled molecules to provide complementary information on specific subcomponents of the total population. For example, the use of [^{3}H]thymidine, to the extent that it predominantly labels only heterotrophic bacteria,[20] is one such example. As expected from known metabolic models and experimental data, exponential labeling of the cellular TTP pool is similar to that described herein for ATP.[21] An exciting possibility is that the measurement of the thymidine nucleotide pool turnover in mixed assemblages of microorganisms may provide a direct measure of heterotrophic bacterial specific growth rate. Of course, the field application of this analysis would first require a laboratory calibration of the quantitative relationship between thymidine nucleotide pool turnover and generation time; however, from the data presented by Werner,[21] it appears that the quantitative relationship between TN pool turnover rate and generation time may be similar to that described herein for the TAN pool.

Another potential variation of the TAN pool turnover method might be to assess photoautotrophic (or chemolithoautotrophic) population growth rates using $^{14}CO_2$ labeling techniques. In theory, $^{14}CO_2$ should be incorporated into the ATP pool of microorganisms which use CO_2 as their principal carbon source (i.e., autotrophs). This method could be used to assess the specific growth rates of phytoplankton populations in mixed microbial assemblages without independent estimates of biomass or production. When used in conjunction with [^{3}H]adenine (bacteria plus algae), $^{32}PO_4$ (bacteria plus algae), and [^{3}H]thymidine (heterotrophic bacteria) the general approach of nucleotide pool turnover could eventually become a powerful method for assessing the growth rates of all microbial populations in nature.

REFERENCES

1. Karl, D. M., Determination of *in situ* microbial biomass, viability, metabolism and growth, in Poindexter, J. S. and Leadbetter, E. R., Eds., *Bacteria in Nature, Vol. 2,* Plenum Publishing, 1986, 85.
2. Bossard, P. and Karl, D. M., The direct measurement of ATP and adenine nucleotide pool turnover in microorganisms: a new method for environmental assessment of metabolism, energy flux and phosphorus dynamics, *J. Plankton Res.,* 8, 1, 1986.
3. Karl, D. M. and Bossard, P., Measurement and significance of ATP and adenine nucleotide pool turnover in microbial cells and environmental samples, *J. Microbiol. Methods,* 3, 125, 1985.
4. Karl, D. M. and Bossard, P., Measurement of microbial nucleic acid synthesis and specific growth rate by $^{32}PO_4$ and [^{3}H]-adenine: a field comparison, *Appl. Environ. Microbiol.,* 50, 706, 1985.
5. Karl, D. M., Jones, D. R., Novitsky, J. A., Winn, C. D., and Bossard, P., Specific growth rates of natural microbial communities measured by adenine nucleotide pool turnover, *J. Microbiol. Methods,* 6, 221, 1987.
6. Laws, E. A., Jones, D., and Karl, D. M., Method for assessing heterogeneity in turnover rates within microbial communities, *Appl. Environ. Microbiol.,* 52, 866, 1986.

7. Lipmann, F., Metabolic generation and utilization of phosphate bond energy, *Adv. Enzymol.*, 1, 99, 1941.

8. Karl, D. M., Cellular nucleotide measurements and applications in microbial ecology, *Microbiol. Rev.*, 44, 739, 1980.

9. Karl, D. M., Measurement of microbial activity and growth in the ocean by rates of stable ribonucleic acid synthesis, *Appl. Environ. Microbiol.*, 38, 850, 1979.

10. Fuhrman, J. A. and Azam, F., Bacterioplankton secondary production estimates for coastal waters of British Columbia, Antarctica, and California, *Appl. Environ. Microbiol.*, 39, 1085, 1980.

11. Chapman, A. G. and Atkinson, D. E., Adenine nucleotide concentrations and turnover rates. Their correlation with biological activity in bacteria and yeast, *Adv. Microbiol. Physiol.*, 15, 253, 1977.

12. Karl, D. M., Winn, C. D., and Wong, D. C. L., RNA synthesis as a measure of microbial growth in aquatic environments. I. Evaluation, verification and optimization of methods, *Mar. Biol.*, 64, 1, 1981.

13. Hodson, R. E., Holm-Hansen, O., and Azam, F., Improved methodology for ATP determination in marine environments, *Mar. Biol.*, 34, 143, 1976.

14. Cashel, M., Lazzarini, R. A., and Kalbacher, B., An improved method for thin-layer chromatography of nucleotide mixtures containing ^{32}P-labeled orthophosphate, *J. Chromatogr.*, 40, 103, 1969.

15. Mangelsdorf, P. C., Jr., Convenient plot for exponential functions with unknown asymptotes, *J. Appl. Physiol.*, 20, 442, 1959.

16. Wilson, D. F., Stubbs, M., Veech, R. L., Erecinska, M., and Krebs, H. A., Equilibrium relations between the oxidation-reduction reactions and the adenosine triphosphate synthesis in suspensions of isolated liver cells, *Biochem. J.*, 140, 57, 1974.

17. Karl, D. M. and Winn, C. D., Adenine metabolism and nucleic acid synthesis: applications to microbiological oceanography, in *Heterotrophic Activity in the Sea*, Hobie, J. E. and Williams, P. J. LeB., Eds., Plenum Publishing, 1984, 197.

18. Karl, D. M. and Winn, C. D., Does adenine incorporation into nucleic acids measure total microbial production?: a response to comments by Fuhrman et al., *Limnol. Oceanogr.*, 31, 1384, 1986.

19. Winn, C. D. and Karl, D. M., Laboratory calibrations of the [^{3}H]-adenine technique for measuring rates of RNA and DNA synthesis in marine microorganisms, *Appl. Environ. Microbiol.*, 47, 835, 1984.

20. Moriarty, D. J. W., Measurement of bacterial growth rates in aquatic systems using rates of nucleic acid synthesis, *Adv. Microbiol. Ecol.*, 9, 245, 1986.

21. Werner, R., Nature of DNA precursors, *Nature New Biol.*, 233, 99, 1971.

CHAPTER 56

Estimating Production of Heterotrophic Bacterioplankton via Incorporation of Tritiated Thymidine

Russell T. Bell

INTRODUCTION

Rates of planktonic bacterial growth may be estimated by measuring the incorporation of various radiolabeled precursors into macromolecules. For example, protein synthesis is assessed by the incorporation of [³H]leucine into protein, whereas DNA synthesis is measured from the incorporation of [³H]thymidine. To calculate bacterial growth from thymidine incorporation requires knowledge of various biochemical parameters that may be difficult to measure, or alternatively, an empirical calibration using bacterioplankton cultures. If bacterioplankton are in balanced growth, rates of incorporation into different macromolecular fractions (RNA, DNA, protein) will give equivalent estimates of growth rate. In natural waters, balanced growth of bacteria may be an exception, thus, it is advantageous to occasionally use several methods to estimate bacterial growth.

The use of thymidine as a radiolabeled exogenous precursor of DNA synthesis has a long history of use in microbiology and biochemistry. As estimator of bacterial growth *in situ*, this approach has several advantages: (1) DNA synthesis is stringently regulated and is related to cell division and thymidine is rapidly taken up by cells, is stable during uptake and is converted rapidly into nucleotides;[1] variation in the rates of synthesis of RNA and protein may not reflect rates of cell division; (2) a typical experiment in natural waters requires only the use of nanomolar concentrations during incubations of less than one hour and at these nM concentrations and short incubations, thymidine is not incorporated into the DNA of eukaryotic cells or cyanobacteria[2] either because they lack thymidine kinase or lack the transport mechanism for assimilating thymidine; thus in aerobic, pelagic waters thymidine is specifically incorporated into heterotrophic bacteria; and (3) the environment can be studied with fine resolution both spatially and temporally. Incubations are short, samples may be fixed with formaldehyde in the field, and the basic extraction and filtration procedures are relatively easy.

The method has several uncertainties: (1) macromolecules other than DNA may become labeled; (2) the exogenously added thymidine may be diluted by *de novo* sources of deoxythymidine monophosphate (dTMP) from dUMP via the action of thymidylate synthetase; (3) the ability of bacteria with more fastidious nutritional requirements to incorporate thymidine is not well studied, consequently, thymidine incorporation is not a perfect method for assessing total bacterial activity: in any anaerobic environment, the method should be used with caution, likewise, not all heterotrophic bacteria may have the ability to incorporate thymidine into DNA; and (4) a number of conversion factors are required before the incorporation of [³H]thymidine can be expressed as units of carbon production.

MATERIALS REQUIRED

Equipment

- Micropipette (e.g., Gilson or Finnpipette) with appropriate tips (20 to 200 μl)
- 2 Eppendorf multipipettes (e.g., model 4870): 1 with 50-ml combitips for quickly dispensing water samples, and the other (using e.g., the 5-ml volume combitips) for dispensing formaldehyde
- A multiple filtration manifold unit with stainless steel funnels to permit simultaneous filter of 10 to 12 samples
- Dispensers for 5% trichloroacetic acid (TCA) and cold ethanol solutions
- Pump

Supplies

- Sealable plastic bag for discarded pipette tips
- 20-ml sterile glass scintillation vials (can be used for incubations as well as scintillation counting, or approximately 30-ml glass-stoppered flasks for incubations only)
- A Plexiglass tube for *in situ* incubations
- Petri dish
- Forceps
- Scintillation solution
- Disposable vinyl gloves
- For cold-TCA extraction of macromolecules: 0.45-μm pore size, 25-mm diameter, cellulose acetate membrane filters (e.g., Schleicher & Schuell OE 67; Note: not to be used in incorporation into DNA procedure)
- For chloroform-phenol extraction (incorporation into DNA procedure): 0.45-μm pore size, 25-mm diameter, cellulose nitrate membrane filters (e.g., Schleicher & Schuell OE 85)
- Plastic scintillation vials

Solutions and Reagents

- Formaldehyde (37%)
- 1 mM nonradioactive thymidine (Sigma)
- 100% TCA (make by adding just a few drops of water to ≈100 g TCA)
- 50% TCA (50.0 g TCA in 100 ml distilled water; *(Caution: Wear gloves as TCA is caustic, adheres to skin and is very corrosive)*
- 5% TCA (5.0 g TCA in 100 ml distilled water)
- 5 N NaOH
- For chloroform:phenol extraction: 50% (w/v) phenol-chloroform solution (50 g phenol in 100 ml chloroform)
- 80% ethanol
- Stock and working [³H]-thymidine solutions: obtain high specific activity [*methyl*-³H] thymidine (e.g., Amersham TRI 418; 40 to 60 Ci mmol^{-1} in 2% ethanol); store at 2°C. There should be negligible self-decomposition for at least two months.[3] For longer periods, store in 10% ethanol. Before use the thymidine must be diluted with sterile water so that a 20 nM addition is approximately 50 to 200 μl of the diluted thymidine solution. Always remove

the [³H]thymidine through the septum using a sterile syringe. Dispense into sterile vials (e.g., glass scintillation vials) and add an appropriate amount of sterilized water. If the working solution is stored for >1 week, add ethanol to 2%. It is preferable to resterilize the stock [³H]thymidine solution by filtering through a sterile 0.2-μm pore-sized filter unit. Autoclaving *may* produce a small amount of thymine or tritiated water by exchange. We find that there is no need to remove this small amount of ethanol before use (it will eventually be diluted about 1:100). If stored in 10% ethanol, however, the solution should be freeze dried or evaporated to dryness with a stream of filtered air then reconstituted in sterilized water.

- Prepare a working solution equal to about 10% more than required. Transport to the field in a cool thermos but do not let the isotope freeze. Place several ice cubes in a sealable bag (e.g., Whirl-pak) and place in the bottom of the thermos. Place some insulating material between the ice and the vial containing thymidine.

PROCEDURES

In Situ Incubations

1. Add water samples (5 to 20 ml) to vials using the Eppendorf pipette. Keep out of direct sunlight. Always wear gloves during the entire procedure.

2. Prepare blank by adding formaldehyde to give a final concentration of 1 to 2% (the concentration most effective for a given environment should be determined from preliminary experiments).

3. Add [³H]thymidine to three replicate experimental vials and to one or two blanks to give the final desired concentration ($\approx$20 nM; see Notes and Comments 4). Shake well. Note the time to the nearest minute. Always change the pipette tip after adding thymidine to the blank or if contamination is suspected.

4. Place the vials in the Plexiglass tube and lower to appropriate depth.

5. After the appropriate incubation time (usually 30 to 60 min) retrieve the samples and wearing plastic gloves add formaldehyde to a final concentration of 1 to 2%. Place in an ice chest until filtration.

Thymidine Incorporation into Total Macromolecules (Cold TCA Insoluble Fraction)

There are several alternative treatments that will allow an estimation of incorporation into a total macromolecular fraction ($\approx$DNA + protein + RNA) or DNA specifically. The two procedures presented here are widely used. It is best to become familiar with the basic macromolecular fractionation before attempting more complicated procedures. Although the theoretical basis for using thymidine is based on DNA synthesis, this does not mean that better results are guaranteed by determining the incorporation into DNA for every sample (see section on conversion factors). It has repeatedly been shown in bacterioplankton cultures that there is an excellent correlation between bacterial cell production and the incorporation into total macromolecules (probably in most situations at least 70% of the radioactivity in the total macromolecular fraction is in DNA). The specificity of DNA labeling must be frequently checked, however, for natural samples. This is especially true when studies encompass a wide variation in productivity. Thymidine is readily degraded within cells, initially by the inducible enzyme thymidine phosphorylase. Nonspecific labeling is minimized by keeping incubation times short. For example, after 20 min of incubation, approximately 80% of the labeled macromolecules is DNA, but after 60 min, DNA may constitute <50% of the labeled macromolecules. In such cases, there has probably been an increase in [³H] label appearing in protein.

Note: TCA is caustic, adheres to the skin, and is corrosive. Wear gloves and work in a well ventilated area.

1. Keep the samples chilled at all times. Use an ice bath or keep in refrigerator (<5°C).

2. Place the required number of filters in distilled water [containing ≈1 mM of nonradioactive thymidine] in a Petri dish. Presoaking in thymidine gives lower and more uniform blanks.

3. If the entire incubated sample will not be filtered then remove an appropriate amount of sample (e.g., 5 ml for eutrophic or 10 to 20 ml from oligotrophic systems) into new glass scintillation vials.

4. Add 50% TCA to the samples to give a final TCA concentration of ≈5%, then shake well. Keep on ice for ≈15 min, but no longer than 1 h, shake again.

5. Place the filters on the manifold. Give pressure briefly. Carefully set the cooled stainless steel funnels in place. Keep the funnels cold (but not <0°C: sample may freeze to the sides during filtration).

6. Carefully pour the samples into the funnel. Filter at a pressure ≤100 mmHg. Rinse each vial 1× with 2 ml ice-cold 5% TCA and add the rinse to the funnel.

7. Rinse 3× with 1 ml portions of ice-cold 5% TCA. It is practical to use a glass dispenser with a long thin tube to allow careful rinsing of the sides of the funnels. Carefully remove the funnels.

8. Rinse the filters 5× with 1 ml portions of ice-cold 80% ethanol to remove thymidine taken into the cells but not incorporated into DNA (e.g., adsorbed to cell-wall lipids, etc.[4]). Some wells may filter more rapidly than others. When dry, close the stopcock.

9. Carefully grab the filters by sliding the forceps into the groove at the base of the filtering well and place in plastic scintillation vials. The material precipitated on the filters is the macromolecular fraction.

10. A common procedure is to add 1 ml ethyl acetate to dissolve the filters. Shake well. After about 30 min add 10 ml of scintillation solution (e.g., Aquasol).

11. Adjust counting time to give a cpm counting error of ≤1%. Make the appropriate quench curve to allow conversion of cpm to dpm following instructions given by the manufacturer.

Incorporation into DNA (Chloroform-Phenol Extraction)[5]

Note: Procedures must be performed in a fume hood. Wear gloves and long-armed clothing. If the solution touches your skin, place the exposed area immediately in ice-cold water.

1. Add 5 N NaOH to give a final concentration of ≈0.25 N. Keep at room temperature for ≈60 min or heat at 60°C for 30 min. This step removes RNA. Cool the samples.

2. Acidify the samples to pH ≈1 with 100% TCA. The procedure of Wicks and Robarts[5] suggests 1.4 ml of TCA for 5 ml of sample. In our experience, 1 ml of 100% TCA for 10 ml of sample is often sufficient. Shake well and keep on ice for ≈15 min.

3. Filter the samples and rinse the vials as above (Steps 6 to 7).

4. Rinse 1× with 5 ml of chloroform-phenol. This removes protein (Note: cellulose nitrate filters must be used as cellulose acetate filters will "dissolve" when the chloroform-phenol solution is added).

5. Remove the funnels and rinse the filters 5× with 1 ml portions of ice-cold 80% ethanol.

6. Continue as in thymidine incorporation procedure above, starting with step 9.

Bacterial Productivity Calculations and Conversion Factors

The rate of thymidine incorporation must always be converted to moles of [³H]thymidine per unit volume and time to allow comparisons between studies. The conversion of moles incorporated to cells produced per unit time requires a conversion factor (TCF = thymidine conversion factor). Dividing cells produced by the bacterial abundance will give an index of the specific growth rate. If the percentage of cells actively incorporating [³H]thymidine is determined by microautoradiography, a more accurate specific growth rate can be calculated. A subsequent calculation from cells to productivity in units of carbon (CCF = carbon conversion factor) is the most uncertain step, since knowledge of the average bacterial size is required, and the factors for estimating carbon per cell vary by a factor of five.[6]

$$\text{moles thymidine } L^{-1} \, h^{-1} = \left(\frac{[\text{dpm}_{\text{sample}} - \text{dpm}_{\text{blank}}] * (4.5 \times 10^{-13})}{SA * t * v} \right) * (10^{-3}) * (1.03) \quad (1)$$

where 4.5×10^{-3} is the number of curies per dpm; SA is the specific activity of the [³H]thymidine solution in curies per mmol; t is the incubation time in h; v is the filtered volume in l; 10^{-3} is mmol per mole; and 1.03 is a correction factor for volume of formaldehyde added (assuming 1% final volume; not needed if the entire incubated volume is filtered).

$$\mu g \, C \, L^{-1} \, h^{-1} = (\text{moles } 1^{-1} \, h^{-1}) * (\text{cells mole}^{-1}) * (\text{Carbon cell}^{-1}) \quad (2)$$

It is clear that the choice of the conversion factors and their combination can have a great impact on the levels of bacterial productivity calculated for a given environment. This underscores the value in using a combination of methods. Nonetheless, progress has been made in understanding the reasons for variability in conversion factors, and we can now place constraints on the range of plausible conversion factors.

On a theoretical basis, based on the best known estimates of bacterial DNA content, assuming no isotope dilution (i.e., no *de novo* synthesis: 100% of the thymine in DNA synthesis was from the exogenously added [³H]thymidine), and that all heterotrophic bacteria are capable of incorporating thymidine, a theoretical TCF is $\approx 0.5 \times 10^{18}$ cells produced per mole thymidine incorporated into DNA. In actuality, although *de novo* synthesis can be minimized to a relatively constant level, it is probably rarely turned off completely,[7] and some heterotrophic bacterial strains may not incorporate thymidine,[8] thus, the theoretical factor given above is certainly a conservative one.

The most widely used empirical TCF is 2×10^{18} cells mol^{-1}. This factor is the median of 97 marine studies,[9] and is commonly derived in freshwater as well.[10,11] Considering all sources of error and the heterogenous poorly known as "bacterioplankton", the agreement between theoretical and empirical factors is actually quite good. Simon and Azam[12] found that this TCF gave results comparable to bacterial productivity based on the leucine incorporation method (where all sources of leucine isotope dilution could be measured via HPLC).

The CCF varies with bacterial size: smaller cells have more carbon per unit biovolume,[12] thus, use of a general CCF for different environments is not recommended. Use of 10 fg C cell^{-1} for oligotrophic marine systems and 25 fg C cell^{-1} in eutrophic waters gives an approximation, but ultimately bacterial cell volumes must be determined. For a thorough treatment of empirical conversion factors and the CCF consult see Chapter 59.

NOTES AND COMMENTS

In Situ Incubations

1. The incubations of water samples with the radioisotope should be made at *in situ* temperatures. It is often practical for limnologists to incubate *in situ* from either a boat or a platform. This also eliminates light intensity as an unknown factor (via changes in algal metabolism). In practice, however, there is no difference between light and dark treatments of samples incubated for minutes to hours shortly after sampling. This is because DNA synthesis does not respond immediately to changes in environmental conditions.[13] If incubations will be made in the laboratory within several hours, then it is best not to chill the water samples during transport, but keep them out of direct sunlight. Try to avoid storing water overnight.

2. A practical arrangement is to incubate water samples with the radioisotope in 20-ml glass scintillation vials. These can be hung in the lake at the desired depth by placing the samples in a plexiglass tube that can be closed at the ends. The subsequent extraction and filtration procedure is most efficient, however, if the volume incubated is equivalent to the volume to be filtered (if ≥ 5 ml), unless several different extractions will be performed from the same incubated sample. Alternatively, samples can be incubated in 30- to 60-ml ground glass-stoppered bottles in a manner similar to the traditional radiocarbon method for primary productivity (e.g., Wetzel and Likens[14]). Such large volumes require more of the isotope, thus increasing the cost of the assay.

3. A combination of filtered sample and time of incubation that gives $\approx 10^4$ counts is sufficient. In general, 5 ml incubated for 30 min is sufficient in eutrophic environments, whereas in oligotrophic situations, 20 ml incubated for one hour may be necessary. Always aim to keep the incubation time ≤ 1 h (even < 30 min if possible), thus, increase the filtered volume if higher counts are required. Although the procedures described here use 25-mm diameter filters, it may be necessary in extremely oligotrophic environments to use 47-mm diameter filters and filter larger volumes (≥ 20 ml) to guarantee sufficient counts. Generally, rate of thymidine incorporation are strongly correlated with temperature. When temperatures are extremely low ($< 5°C$), longer incubations may be unavoidable.

4. 20 nM of [^{3}H]thymidine is a recommended concentration for most environments. In some cases, a lower concentration may be sufficient. In environments with high particle loading, or where many bacteria may be attached, a higher concentration may be required. Thus, always do preliminary tests (range of about 2 to 50 nM). After a concentration is chosen, it is recommended to still do concentration tests at least occasionally. The concentration of thymidine at which no further radioactivity is incorporated should be used. If the concentration of [^{3}H]thymidine is too low, isotope dilution may be variable, too much [^{3}H]thymidine may stimulate degradation of thymidine or uptake by phytoplankton.

Thymidine Incorporation into Total Macromolecules (Cold Insoluble Fraction)

5. Optional steps during the assay: some workers add a carrier solution of DNA + protein to aid precipitation of macromolecules (e.g., Wetzel and Likens[14]). In my experience this

      mainly increases filtration time, but this, like all steps in this assay, should be tested by the investigator.

6.   The original fractionation scheme[15] included cold TCA insoluble treatment for total macromolecules; warm base (NaOH) for DNA + protein; and hot-TCA insoluble material for protein. Doing this entire scheme is not worth the effort. Other extraction procedures include: (1) a moderately complicated perchloric acid treatment that gives a soluble DNA + RNA fraction;[14] (2) NaOH extraction combined with centrifugation;[16] and (3) DNase treatment.[17]

7.   On theoretical grounds, little [3H]thymidine should appear in RNA during the *short* incubations (30 to 60 min) used by aquatic microbial ecologists. It is noteworthy that substantial incorporation of [3H]thymidine into RNA is usually reported when the acid-base hydrolysis scheme was used.[11] RNA is only determined by subtraction in the latter procedure. Labeling of RNA from thymidine has been negligible when checked with enzymatic approaches.[5,18,19]

Incorporation into DNA (Chloroform-Phenol Extraction)[5]

8.   Wicks and Robarts[5] showed, via DNase digestion, that DNA was the only labeled macromolecule present after phenol-chloroform and ethanol washes. Torreton and Bouvy,[19] however, found poor agreement in some warm-water environments among the acid-base hydrolysis, phenol-chloroform extraction, and enzymatic hydrolysis approaches in determining DNA labeling from thymidine incorporation. On the other hand, Riemann[18] obtained good agreement between enzymatic digestion and acid-base hydrolysis of thymidine-labeled molecules, and Chrost et al.[20] found good agreement between the acid-base hydrolysis method and an inhibitor method. Bell and Riemann,[21] using [3H]adenine, found close agreement between acid-base hydrolysis and phenol-chloroform extraction. Wicks and Robarts[5] also corrected for loss of DNA (75% recovery) during the phenol-chloroform procedure. A similar recovery was found by Findlay et al.[16] using another approach for isolating DNA. Few other workers have made similar tests (this is done more routinely in sediments), although it is easily done using ^{14}C-DNA labeled *E. coli* (Amersham) or ^{3}H-labeled *E. coli* (New England Nuclear). Consequently, evaluating discrepancies among extraction techniques is difficult.

Bacterial Productivity Calculations and Conversion Factors

9.   The isotope dilution approach (see Chapter 57) used routinely in sediments is also used by some workers for water column samples (e.g., Chrzanowski and Hubbard[22]). Although Moriarty[23] states that dilution in the water column is usually minimal, there may be significant dilution in rivers[16,24] suggesting a relationship with high particulate levels. In theory, the isotope dilution factor multiplied by the "theoretical conversion factor" should give values similar to the empirically derived factors. There have been few direct checks but in a detailed study, Chrzanowski and Hubbard[22] derived TCF of 0.55 to 1.6×10^{18} cells per mole using the isotope dilution approach. When used for water samples, at least five levels of unlabeled thymidine (triplicates of each) should be used. I prefer to add unlabeled thymidine concentrations ranging from ≈ 0.3 to ≤ 5 times the [3H]thymidine concentration. Some workers dilute the [3H]thymidine up to >20 times with unlabeled thymidine. Because 1/dpm is plotted, this gives extremely low counts that, in turn, give high values that, in my experience, may unjustly influence the slope of the dilution plot.

REFERENCES

1. Kornberg, A. and Baker, T. A., *DNA Replication*, 2nd ed., Freeman, 1991.
2. Bern, L., Autoradiographic studies of [*methyl*-³H]thymidine incorporation into a cyanobacterium *(Microcystis wesenbergii)* — bacterium association and in selected algae and bacteria, *Appl. Environ. Microbiol.*, 49, 232, 1985.
3. Evans, E. A., *Self-Decomposition of Radiochemicals: Principles, Control, Observations, and Effects*, Review 16, Amersham International, England, 1976.
4. Robarts, R. D. and Wicks, R. J., [*Methyl*-³H]thymidine macromolecular incorporation and lipid labelling: their significance to DNA labelling during measurements of aquatic bacteria growth rate, *Limnol. Oceanogr.*, 34, 213, 1989.
5. Wicks, R. J. and Robarts, R. D., The extraction and purification of DNA labelled with [*methyl*-³H]thymidine in aquatic bacterial production studies, *J. Plankton Res.*, 9, 1159, 1987.
6. Riemann, B. and Bell, R. T., Advances in estimating bacterial biomass and growth in aquatic systems, *Arch. Hydrobiol.*, 118, 485, 1990.
7. Bell, R. T., An explanation for the variability in the conversion factor deriving bacterial cell production from incorporation of [³H]thymidine, *Limnol. Oceanogr.*, 35, 910, 1990.
8. Pollard, P. C. and Moriarty, D. J. W., Validity of the tritiated thymidine method for estimating bacterial growth rates: measurement of isotope dilution during DNA synthesis, *Appl. Environ. Microbiol.*, 48, 1076, 1984.
9. Ducklow, H. W. and Carlson, C. A., Oceanic bacterial production, *Adv. Microb. Ecol.*, 12, in press.
10. Bell, R. T., Ahlgren, G. A., and Ahlgren, I., Estimating bacterioplankton production by measuring [³H]thymidine incorporation in a eutrophic Swedish lake, *Appl. Environ. Microbiol.*, 45, 1709, 1983.
11. Smits, J. D. and Riemann, B., Calculation of cell production from [³H]thymidine incorporation with freshwater bacteria, *Appl. Environ. Microbiol.*, 54, 2213, 1988.
12. Simon, M. and Azam, F., Protein content and protein synthesis rates of planktonic marine bacteria, *Mar. Ecol. Prog. Ser.*, 51, 201, 1989.
13. Findlay, R. H., Pollard, P. C., Moriarty, D. J. W., and White, D. C., Quantitative determination of microbial activity and community nutritional status in estuarine sediments: evidence for a disturbance artifact, *Can. J. Microbiol.*, 31, 493, 1985.
14. Wetzel, R. G. and Likens, G. E., Bacterial growth and productivity, in *Limnological Analyses*, Wetzel, R. G. and Likens, G. E., Springer-Verlag, 1991, 255.
15. Furhman, J. and Azam, F., Thymidine incorporation as a measure of heterotrophic bacterioplankton production in marine surface waters: evaluation and field results, *Mar. Biol.*, 66, 109, 1982.
16. Findlay, S., Meyer, J. L., and Edwards, R. T., Measuring bacterial production via rate of incorporation of [³H]thymidine into DNA, *J. Microbiol. Methods*, 2, 57, 1984.
17. Servais, P., Martinez, J., Billen, G., and Vives-Rego, J., Determining [³H]thymidine incorporation into bacterioplankton DNA: improvement of the method by DNase treatment, *Appl. Environ. Microbiol.*, 53, 1977, 1987.
18. Riemann, B., Determining growth rates of natural bacteria by means of [³H]thymidine incorporation into DNA: comments on methodology, *Ergeb. Limnol.*, 19, 67, 1984.
19. Torreton, J. P. and Bouvy, M., Estimating bacterial DNA synthesis from [³H]thymidine incorporation: discrepancies among macromolecular extraction procedures, *Limnol. Oceanogr.*, 36, 299, 1991.
20. Chrost, R. J., Overbeck, J., and Wcislo, R., Evaluation of the [³H]thymidine method for estimating bacterial growth rates and production in freshwaters: re-examination and methodological comments, *Acta Microbiol. Pol.*, 37, 95, 1988.
21. Bell, R. T. and Riemann, B., Adenine incorporation into DNA as a measure of microbial production in freshwaters, *Limnol. Oceanogr.*, 34, 435, 1989.
22. Chrzanowski, T. and Hubbard, J. G., Primary and bacterial secondary production in a Southwestern Reservoir, *Appl. Environ. Microbiol.*, 54, 661, 1988.

23. Moriarty, D. J. W., Measurement of bacterial growth rates in aquatic systems from rates of nucleic acid synthesis, *Adv. Microb. Ecol.*, 9, 245, 1986.

24. Findlay, S., Pace, M. L., Lints, D., Cole, J. J., Caraco, N. F., and Peierls, B., Weak coupling of bacterial and algal production in a heterotrophic ecosystem: the Hudson River estuary, *Limnol. Oceanogr.*, 36, 268, 1991.

CHAPTER 57

Thymidine Incorporation into DNA as an Estimate of Sediment Bacterial Production

Stuart Findlay

INTRODUCTION

Measurement of the rate of incorporation of tritiated thymidine (^{3}H-TdR) into bacterial DNA has become the mostly commonly used method for measuring bacterial production in both water column and sediments. The basic rationale for the technique is that DNA synthesis is tightly coupled to cell division (carbon production) and that most bacteria actually incorporate exogenous TdR into newly synthesized DNA. Advantages of the method include: (1) fairly short-term incubations yielding instantaneous rates of production; (2) high sensitivity in the sense that even slow turnover times yield measurable rates of TdR incorporation; and (3) relatively simple sample incubation and preparation techniques (an individual can run about 50 samples in a day). There seem to be more methodological problems associated with sediments than water column samples. The major added difficulties when applying the method to sediments are: (1) uncertainties in extraction of labeled DNA from sediment; (2) potentially high isotope dilution; and (3) the need to perturb the sediment to introduce label. Each of these problems and a few general considerations are discussed after the general protocol is described.

MATERIALS REQUIRED

Equipment

- Tabletop centrifuge and ~10 ml plastic disposable centrifuge tubes
- Vacuum pump and aspirator with Pasteur pipette for washing sediment
- High-speed refrigerated centrifuge capable of ~20,000 × g; rotor to hold 10 to 15 ml polypropylene tubes
- Water bath at 100°C
- Scintillation counter

Supplies

- Scintillation vials

Solutions

- Low specific activity tritiated thymidine (methyl-^{3}H) (New England Nuclear high specific activity stock [NET027Z] diluted to ~20 Ci mmol^{-1} with unlabeled TdR from Sigma [#T9250]), need 20 to 50 μCi for each sample depending on whether high or low rates anticipated
- 5% formalin for washing sediment; ~15 ml per sample
- Scintillation cocktail (Fisher Scintiverse E will hold roughly 1 ml of final TCA hydrolysate)
- Alkaline extractant (0.3 N NaOH + 0.1% SDS + 25 mM EDTA)
- 5 and 50% (w/w) TCA
- 3 N HCl
- Nonradiolabeled ("cold") DNA made up 10 mg ml^{-1} in alkaline extractant; make fresh each day

PROCEDURES

Incubations

Time Course Determination. Sediment slurries (1 to 2 ml) are incubated with ~20 μCi ^{3}H-TdR for various times (10 min to 2 h) to determine maximum time of linear incorporation. Zero-time controls are run by killing samples with 2 ml 5% formalin immediately after isotope is added to sediment. Incubation of living samples is terminated by addition of formalin. After time course of incorporation is estimated for a certain set of conditions, a standard length of incubation can be used which is short enough to give measurable incorporation of radioactivity.

Washing. Sediment is washed to remove unincorporated ^{3}H-TdR by resuspending sediment in 5 ml of formalin, centrifuging at about 1000 rpm for 5 min, then aspirating and discarding supernatant. Washing should be repeated at least twice, more if radioactivity appearing in the "DNA fraction" of killed controls is more than 10% of live samples. After sediment has been washed, samples may be frozen for subsequent DNA analysis.

Extraction. (1) DNA is extracted from sediment for 12 h at 25°C in 5 ml of alkaline extractant. Tubes should be thoroughly mixed at the beginning and end of extraction period, more often if possible. (2) After extraction, a brief centrifugation (2000 × g, 5 min) is used to pellet the sediment, and the supernatant containing labeled DNA, protein, and RNA is transferred to a high-speed centrifuge tube. (3) The tubes are placed on ice and the solution neutralized with 0.5 ml of 3 N HCl and acidified with 1.0 ml of 50% TCA. Carrier DNA (1 mg per sample) is added and should immediately appear as a stringy substance indicating insolubility of DNA under these conditions.

Cleanup. (1) After 30 min tubes are centrifuged at 20,000 × g for 15 min at 5°C. After centrifugation, the pellet contains DNA and protein while the supernatant contains RNA, any residual unincorporated ^{3}H-TdR, and ^{3}H$_2$O. (2) The supernatant is aspirated and discarded after a small amount (100 μl) is saved for radioassaying. (3) The pellet is resuspended in cold 5% TCA, centrifuged, and again, a small amount of supernatant is radioassayed. If the second supernatant is more than about 1000 dpm ml^{-1}, the pellet should be washed again. High amounts of radioactivity in the washes indicate inefficient removal of unincorporated ^{3}H-TdR from the sediment and will contribute to high blanks.

DNA Hydrolysis. After washing, the pellet is resuspended in 3 ml of 5% TCA and the DNA hydrolyzed at 100°C for 30 min. After a final centrifugation (5000 × g, 10 min) to remove any particulates, a 1-ml aliquot is radioassayed.

Conversion Factor

The rate of incorporation of thymidine is calculated from:

$$I = dpm/SA \qquad (1)$$

where I = incorporation rate in nmol (ml sediment)$^{-1}$ h^{-1}; dpm = dpm (ml sediment)$^{-1}$ h^{-1}; and SA = specific activity of added thymidine, in dpm nmol^{-1} TdR. Note that 1 µCi = 2.2×10^6 dpm.

Actual cell or carbon production is calculated from the rate of TdR incorporation using conversion factors for cells produced per nmol TdR incorporated (commonly 1 to 2×10^9 cells nmol^{-1})[1] (see Chapter 59) and for carbon per cell volume (see Chapter 37). This calculation assumes negligible isotope dilution by other exogenous sources of thymidine or *de novo* synthesis. In sediments, practically all workers find significant dilution of added TdR[2-5] unless very large amounts of TdR are added to the sediment.[6] Isotope dilution is determined by incubating samples with different specific activities of labeled TdR.[3,7] In general, one uses a constant amount of labeled TdR and adds differing amounts of unlabeled TdR. Unfortunately, the appropriate range of specific activities for constructing an isotope dilution curve is largely a function of the size of the (unknown) diluting pool. As a first approximation, one should use a series of specific activities ranging from 20 Ci mmol^{-1} to 2 Ci mmol^{-1}. The isotope dilution curve is plotted as the inverse of dpm vs. the amount of added unlabeled TdR (see details and Figure 4 in Findlay et al.[3]). The line should have a positive slope (i.e., the dpms decrease as lower specific activities are used). The X-intercept of the line represents the amount of added, labeled TdR plus diluting TdR in the sample. This value is the total amount of TdR available to the bacteria for DNA synthesis and should be used as the denominator when calculating the specific activity to arrive at the rate of TdR incorporation.

NOTES AND COMMENTS

As stated in the introduction, three major problems with applying TdR incorporation to sediments are extraction, isotope dilution, and disturbance.

1. There have been relatively few detailed studies of the best extraction procedure for recovering labeled DNA from sediments, but there is almost universal agreement that DNA must be separated from other macromolecules.[3,8-10] In two direct comparisons of techniques, it was found that a 25°C extraction with a NaOH/SDS/EDTA mixture gave higher recoveries than harsher conditions.[3,10] Very brief (15 min) autoclaving of sediment in an NaOH/EDTA/SDS mixture also gives high recovery of labeled macromolecules.[4]
2. Measurement of isotope dilution is problematic because it requires running 12 to 18 samples simply for determination of the degree of isotope dilution. Moreover, the appropriate range of unlabeled TdR additions is hard to ascertain *a priori*. If added cold TdR levels are too high, then insufficient radioactivity is recovered in DNA. If cold TdR levels are too low, then the specific activity of the total TdR pool is not markedly changed, and the isotope dilution plot is essentially a flat line.
3. Most work on sediment bacterial production has used slurries of surface sediments for reasons of convenience. Dobbs et al.[11] found that different techniques for introducing labeled compounds into sediments (injection into intact cores, slurries, porewater replacement) did not affect TdR incorporation, but did significantly alter lipid synthesis and other processes.
4. There is clear evidence that certain bacteria, particularly fastidious anaerobes[9] and some chemolithotrophs,[12] are not capable of using exogenous TdR for DNA synthesis. Therefore,

low rates of TdR incorporation from anaerobic sediments[13] may be underestimating total bacterial production.

5. Catabolism of ^{3}H-TdR has been reported for both water-column and sediment samples.[8,14] The fact that catabolism occurs is not at all surprising but does confound the analysis if (1) recycling of degradation products results in added label incorporation into TdR and/ or (2) TdR added to the sediment serves as a growth substrate for the bacteria and artificially increases their growth rate. Nonlinear time courses are the simplest way to look for changes in either growth rates or changes in the specific activities in TdR pools.

REFERENCES

1. Moriarty, D., Accurate conversion factors for calculating bacterial growth rates from thymidine incorporation into DNA: elusive or illusive? *Arch. Hydrobiol. Ergeb. Limnol.*, 31, 211, 1988.
2. Bell, R. T. and Ahlgren, I., Thymidine incorporation and microbial respiration in the surface sediment of a hypereutrophic lake, *Limnol. Oceanogr.*, 32, 476, 1987.
3. Findlay, S. E. G., Meyer, J. L., and Edwards, R. T., Measuring bacterial production via rate of incorporation of [^{3}H]thymidine into DNA, *J. Microb. Methods*, 2, 57, 1984.
4. Fallon, R. D. and Boylen, C. W., Bacterial production in freshwater sediments: cell specific versus system measurements, *Microb. Ecol.*, 19, 53, 1990.
5. van Duyl, F. C. and Kop, A. J., Seasonal patterns of bacterial production and biomass in intertidal sediments of the western Dutch Wadden Sea, *Mar. Ecol. Prog. Ser.*, 59, 249, 1990.
6. Pollard, P. C. and Moriarty, D., Validity of the tritiated thymidine method for estimating bacterial growth rates: measurement of isotope dilution during DNA synthesis, *Appl. Environ. Microbiol.*, 48, 1076, 1984.
7. Moriarty, D. and Pollard, P., DNA synthesis as a measure of bacterial productivity in seagrass sediments, *Mar. Ecol. Prog. Ser.*, 5, 151, 1981.
8. Carman, K. R., Dobbs, F. C., and Guckert, J. B., Consequences of thymidine catabolism for estimates of bacterial production: an example from a coastal marine sediment, *Limnol. Oceanogr.*, 33, 1595, 1988.
9. Moriarty, D., Measurement of bacterial growth rates in aquatic systems from rates of nucleic acid synthesis, in *Advances in Microbial Ecology*, Vol. 9, Marshall, K. C., Ed., Plenum Publishing, New York, 1986, 245.
10. Thorn, P. M. and Ventullo, R. M., Measurement of bacterial growth rates in subsurface sediments using the incorporation of tritiated thymidine into DNA, *Microb. Ecol.*, 16, 3, 1988.
11. Dobbs, F. C., Guckert, J. B., and Carman, K. R., Comparison of three techniques for administering radiolabeled substrates to sediments for trophic studies: incorporation by microbes, *Microb. Ecol.*, 17, 237, 1989.
12. Johnstone, B. H. and Jones, R. D., A study on the lack of [methyl-^{3}H]thymidine uptake and incorporation by chemolithotrophic bacteria, *Microb. Ecol.*, 18, 73, 1989.
13. Austin, H. K. and Findlay, S., Benthic bacterial biomass and production in the Hudson River Estuary, *Microb. Ecol.*, 18, 105, 1989.
14. Brittain, A. M. and Karl, D. M., Catabolism of tritiated thymidine by aquatic microbial communities and incorporation of tritium into RNA and protein, *Appl. Environ. Microbiol.*, 56, 1245, 1990.

CHAPTER 58

Leucine Incorporation as a Measure of Biomass Production by Heterotrophic Bacteria

David L. Kirchman

INTRODUCTION

The leucine (Leu) method for estimating bacterial production consists of measuring the incorporation of radiolabeled leucine into bacterial protein over time. The physiological basis of the leucine method is protein synthesis. Biomass production can be calculated from rates of protein synthesis because protein comprises a large, fairly constant fraction (approximately 60%) of bacterial biomass. By knowing the ratio of protein to total biomass, rates of protein synthesis can be converted to total biomass production. It is possible to calculate biomass production without information about the cell sizes of bacterial assemblages.

Leucine incorporation into protein is measured by following the appearance of radioactivity into material that is insoluble in hot trichloroacetic acid (TCA). This precipitate is mainly protein and radioactive Leu is essentially associated with only protein,[1] although other macromolecules are also insoluble in hot TCA. In addition, Leu is not transformed to other amino acids, which would also be incorporated into protein and would lead to overestimates of the production rate. Finally, Leu comprises a fairly constant fraction of bacterial protein,[1,2] which implies that changes in Leu incorporation are not due to changes in the Leu/protein ratio.

Two processes can contribute to variations in Leu incorporation that are independent of net biomass production and possibly may lead to errors in estimating bacterial production. First, Leu can be synthesized from other compounds, which leads to isotope dilution of the added radiolabeled Leu. The problem is minimized by adding Leu to concentrations high enough (e.g., 10 nM for marine waters and oligotrophic lakes) to "swamp" unlabeled Leu and to repress *de novo* synthesis of intracellular Leu. Isotope dilution experiments can help in selecting the proper concentration,[3] although this approach apparently does not guarantee that isotope dilution will be zero.[4,5] Simon and Azam[2] directly measured intracellular isotope dilution using OPA-HPLC and found that it was about two-fold when 10 nM Leu was added to coastal waters of southern California. Estimates of intracellular isotope dilution can be very useful, but the methodology is difficult and depends on a reasonable separation of phytoplankton from bacteria.[2] Addition of high Leu concentrations should be avoided because some of the radiolabel may diffuse into or may be taken up by microorganisms other than bacteria, e.g., phytoplankton.

The other potential problem with the Leu method is protein turnover. Microbial cells can synthesize and degrade some proteins, i.e., protein turnover, independent of net growth.

0-87371-564-0/93/$0.00 + $.50
© 1993 by Lewis Publishers

Kirchman et al.[4] found that protein turnover was not important in the only published experiments with natural waters, but protein turnover cannot be ignored, especially when bacterial growth rates are low. If protein turnover is important, Leu incorporation would tend to overestimate biomass production because the radiolabel would be incorporated into new proteins while little radioactivity would be lost as old proteins are degraded. Kirchman et al.[4] argued that it may be useful to measure protein turnover if organic matter is mineralized during protein turnover. Even so, it complicates interpretation of Leu incorporation.

MATERIALS REQUIRED

Equipment

- Scintillation counter
- Vacuum pump
- Filter manifold
- Heating block or water bath, 80°C

Supplies

- Scintillation vials, 7 ml
- 0.45 μm × 25 mm filters, cellulose nitrate or mixed esters of cellulose filters (Millipore HAWP 025 00)

Solutions and Reagents

- [4,5-³H]-leucine, 40 to 60 Ci mmol⁻¹ (NEN NET-135H)
- Nonradioactive L-leucine (Sigma L 8000)
- Scintillation cocktail (Packard Ultima-gold)
- TCA; 50% w/v
- Ethyl acetate
- Ethanol 80% v/v

PROCEDURES

Experimental Procedures

1. Sample water using "clean techniques".[6] Use plastic gloves to avoid contact with sample. Handling can add amino acids. Acid-rinse sample containers before use. Start incubations as soon as possible (within minutes) after water is sampled.

2. Place sample into appropriate incubation containers (two to three replicates) and add ³H-Leu (final concentration 10 nM). Set up killed control by adding TCA (5% final concentration) to a sample. The sample volume will depend on the environment. For eutrophic environments, 5 or 10 ml will be sufficient. For oligotrophic environments or other environments with low rates, 25 ml may be necessary.

3. Incubate from 10 min to 10 h, depending on sample.

4. After incubation, add enough 50% TCA to obtain 5% TCA, final concentration. This kills the incubation and starts the extraction.

5. Heat sample to 80°C for 15 min.

6. After it has cooled, filter sample through 0.45-μm cellulose filters (e.g., Millipore or Gelman). The vacuum is not critical but should not exceed 150 mmHg.

7. Rinse filters twice (3 ml) with cold 5% TCA. Rinse twice (2 ml) with cold 80% ethanol.[7] Remove filter towers and gently rinse (1 ml) with 80% cold ethanol.

8. Place filters in scintillation vials. When dry, add 0.5 ml of ethyl acetate to dissolve filter. Filter must be completely at the bottom so that this volume of ethyl acetate is effective. After the filter is dissolved, add appropriate scintillation cocktail and radioassay.

Calculating Biomass Production, Theoretical Approach

This approach is called "theoretical" because it is based on literature values of the various parameters needed to relate Leu incorporation to biomass production. Some of these parameters have been measured for samples from natural equatic environments.[1,2] The equation for relating Leu incorporation to biomass production gC L^{-1} h^{-1} is:

$$\text{Production} = \text{Leu} * 131.2 * (\% \text{ Leu})^{-1} * (\text{C/Protein}) * \text{ID} \qquad (1)$$

where Leu is the rate of Leu incorporation (mol/L/h). The other parameters are as follows, with the best, current estimates provided by Simon and Azam:[2]

Parameter	Interpretation	Best Estimate
131.2	Formula weight of Leu	
% Leu	Fraction of Leu in protein	0.073
C/Protein	Ratio of cellular carbon to protein	0.86
ID	Isotope Dilution	2

When these best estimates are used, the resulting conversion factor is 3.1 kgC mol^{-1} which is multiplied times the Leu incorporation rate to obtain rates of bacterial biomass production.

Calculating Biomass Production, Empirical Approach

The other approach to relate Leu incorporation to bacterial production is the empirical approach which is described in Chapter 59. This procedure is used to estimate a conversion factor (cells or gC per mol of Leu incorporation) that converts Leu incorporation into biomass production. This empirical factor, in theory, includes all possible relationships between Leu incorporation and biomass production, and thus, should not be "corrected" further by other factors.

NOTES AND COMMENTS

1. The goal of the Leu method is not to obtain turnover rates of amino acids at *in situ* concentrations. The added concentration of Leu is purposely much higher than the *in situ* concentration (usually <1 n*M*). Also, organic contamination (unless extremely severe) will not change short-term rates.[8] Contamination by amino acids and other compounds is potentially a serious problem and obviously should be avoided.

2. Formalin can also be used for killed controls as abiotic adsorption of radiolabeled Leu in formalin-killed controls is the same as that with TCA. The problem with formalin is that any surface in contact with it should not be used in incubations with live samples. The fumes from formalin are also noxious and could affect live samples.

3.   The added concentration of ^{3}H-Leu should be tested in separate experiments.[3] For many environments 10 nM of added Leu has proven to be adequate, although much higher concentrations may be necessary in some eutrophic lakes (R. Bell, personal communication). If 10 nM is used, it is not necessary that the entire added Leu be radioactive. Leu incorporation is usually high enough so that rate can be measured with a mixture of 0.5 to 1.0 nM ^{3}H-Leu plus 9 to 9.5 nM nonradioactive Leu. This mixture is also quite inexpensive. Rates using this mixture should be corrected for the addition of nonradioactive Leu with the following equation:

$$\text{rate} = \text{rate in mixture} * [\text{nonradioactive} + {}^3\text{H-Leu}]/[{}^3\text{H-Leu}] \qquad (2)$$

Note only ^{3}H-Leu, without any nonradioactive Leu, should be used in environments where rates are expected to be low, e.g., deep oceans and highly oligotrophic lakes.

4.   Hot TCA extractions of large volumes (>10 ml) is inconvenient. Alternatively, one can extract the material collected on filters after killing the incubation with a low TCA concentration (0.5%). That is, after filtering the killed sample, the filter is then placed in 5-ml 5% TCA and heated. After extraction and cooling, the 5% TCA is filtered and rinsed. Both filters are radioassayed.

5.   Because nearly all Leu assimilated is incorporated directly into protein,[1] a simpler TCA extraction is often possible.[9] Instead of the hot extraction, the sample is killed with TCA and then filtered (cellulose acetate filters) without the 80°C extraction. The filter is then rinsed as described above.

REFERENCES

1. Kirchman, D. L., K'nees, E., and Hodson, R. E., Leucine incorporation and its potential as a measure of protein synthesis by bacteria in natural aquatic systems, *Appl. Environ. Microbiol.*, 49, 599, 1985.

2. Simon, M. and Azam, F., Protein content and protein synthesis rates of planktonic marine bacteria, *Mar. Ecol. Prog. Ser.*, 51, 201, 1989.

3. Moriarty, D. J. W. and Pollard, P. C., DNA synthesis as a measure of bacterial productivity in seagrass sediments, *Mar. Ecol. Prog. Ser.*, 5, 151, 1981.

4. Kirchman, D. L., Newell, S. Y., and Hodson, R. E., Incorporation versus biosynthesis of leucine: implications for measuring rates of protein synthesis and biomass production by bacteria in marine systems, *Mar. Ecol. Prog. Ser.*, 32, 47, 1989.

5. Ellenbroek, F. M. and Cappenberg, T. E., DNA synthesis and tritiated thymidine incorporation by heterotrophic freshwater bacteria in continuous culture, *Appl. Environ. Microbiol.*, 57, 1675, 1991.

6. Fuhrman, J. A. and Bell, T. M., Biological considerations in the measurement of dissolved free amino acids in seawater and implications for chemical and microbiological studies, *Mar. Ecol. Prog. Ser.*, 25, 13, 1985.

7. Wicks, R. J. and Roberts, R. D., Ethanol extraction requirements for purification of protein labeled with [^{3}H]leucine in aquatic bacterial production studies, *Appl. Environ. Microbiol.*, 54, 3191, 1988.

8. Kirchman, D. L., Limitation of bacterial growth by dissolved organic matter in the subarctic Pacific, *Mar. Ecol. Prog. Ser.*, 62, 47, 1990.

9. Chin-Leo, G. and Kirchman, D. L., Estimating bacterial production in marine waters from the simultaneous incorporation of thymidine and leucine, *Appl. Environ. Microbiol.*, 54, 1934, 1988.

CHAPTER 59

Estimating Conversion Factors for the Thymidine and Leucine Methods for Measuring Bacterial Production

David L. Kirchman and Hugh W. Ducklow

INTRODUCTION

Two common approaches for estimating bacterial production, the thymidine (TdR) and leucine (Leu) methods, share the same problem in relating incorporation rates to accurate rates of bacterial production, i.e., cells (or cellular C or N) produced per unit volume or area per unit time. Since both methods are based on measuring some aspect of macromolecular synthesis (DNA for TdR, protein for Leu), it is possible to convert these incorporation rates to rates of macromolecular synthesis and, in turn, to biomass production. But this conversion depends on detailed information about several cellular components, e.g., the amount of DNA and protein per cell, the ratio of thymine to total DNA, and perhaps the most troublesome, the intracellular isotope dilution of the incorporated radiolabeled compound. Some of these cellular components are difficult, if not impossible, to measure routinely for a natural bacterial assemblage, and literature values, even from pure bacterial cultures, are often used. Hence, the resulting factors for relating incorporation rates to bacterial production are often referred to as "theoretical conversion factors".

An alternative approach is to estimate these conversion factors with experiments using natural bacterial assemblages taken directly from the aquatic systems being examined. These "empirical conversion factors" are calculated by directly comparing incorporation of radiolabeled TdR or Leu with the increase in bacterial biomass over time, which is an unambiguous measure of bacterial biomass production. This is analogous to standardizing a chemical assay with a known compound over a range of concentrations. The conceptual disadvantage of using empirical conversion factors is that all other information about the physiology and biochemistry of macromolecular synthesis is ignored and not used.

The advantages of using empirical conversion factors include: (1) they are calculated with natural bacterial assemblages; (2) the factors are measured for the particular system being studied, and literature values are not assumed; and (3) many processes that would cause problems with theoretical conversion factors are "corrected" by using empirical conversion factors. For example, TdR incorporation rates into protein, if ignored, lead to overestimating bacterial production when theoretical factors are used. In contrast, although TdR incorporation rates are higher when protein is included, the empirical factor would be lower and thus, the net result would be the correct estimate of bacterial production when the rates are multiplied by the empirical conversion factor.

Normally, bacterial growth is matched by loss processes (e.g., grazing) and thus, bacterial abundance is relatively constant over time (hours to days). Since it is necessary to obtain an absolute measure of bacterial growth (the gross production of bacterial biomass), the key is to eliminate grazing and other processes that lead to loss of bacterial biomass. This has been attempted with two different approaches: (1) grazers such as microflagellates can be removed by filtration (0.6- to 1.0-μ filter); (2) loss due to both grazers and probably viruses can be substantially reduced by dilution, usually 1 part of unfiltered water to 9 parts of water filtered through 0.2-μm filters.[1] The latter procedure works because grazing decreases with a decrease in prey abundance, but bacterial growth rates are not affected. Also, the probability of viral infection is reduced by diluting the bacteria. In contrast, even 0.6-μm filtration alone would not affect bacterial phages. In one experiment, Coveney and Wetzel[2] found no difference between TdR factors measured by the two approaches in a lake, but comparisons in other environments are needed.

It is preferable but not essential that bacterial growth rates during the conversion factor experiments equal rates in undisturbed samples. What is critical is that the relationship between incorporation of the radiolabeled compound and bacterial biomass production is not altered by sample manipulations. This critical assumption has not been tested directly — one would have to know the conversion factor for the undisturbed sample, which is precisely the point of the entire experiment. Some investigations have shown that different manipulations of the sample affect conversion factors,[2] whereas others have shown that organic enrichments which affect growth rates do not affect conversion factors.[3]

MATERIALS REQUIRED

Conversion factor experiments require the same materials as are needed for measuring thymidine or leucine incorporation rates and microbial abundance (see Chapters 56-58). In addition, the following are needed:

- Large nitrocellulose filters (0.22 μm $\times$ 45 mm or larger diameter)
- Filter apparatus to filter large volumes (>500 ml)
- Large incubation vessel (500 ml or larger)
- Various bottles to hold water during filtration

PROCEDURE

The protocol consists of preparing water with zero grazing activity and then measuring incorporation rates and bacterial abundance over time. Here the dilution method is presented.

Dilution Method

1. Rinse cellulose filters (0.22 μm $\times$ 45 mm, e.g., Millipore) with distilled water and then with sample water, which is discarded before use. After this rinsing step, filter sample water. Follow the clean techniques used for the analysis of dissolved organic matter (DOM).[4]

2. Add 1 part of unfiltered water (the inoculum) to 9 parts of the 0.22-μm filtered water. Incubate in the dark at the *in situ* temperature.

3. Begin withdrawing subsamples (20 to 50 ml) in order to measure incorporation rates and to preserve water which later can be examined for microbial abundance. Note that incor-

poration *rates* are measured over time and that the radiolabeled compound is added to the small subsamples (10 ml), not the large incubation vessel (e.g., 1000 ml). The incubation is stopped after 1 h or some appropriate time. The methods for measuring thymidine and leucine incorporation are described in detail elsewhere. The same methodology used to measure *in situ* incorporation rates should be used in the conversion factor experiment.

4. Step 3 is repeated over time. The sampling will vary depending on how fast the bacteria are growing. Ideally, sampling should continue at least for one doubling of bacterial abundance. For example, if the generation time is about 1 day, then ideally samples should be taken every 2 to 4 h for 24 h. A decrease in bacterial abundance indicates that grazing has begun and time points after this cannot be used to calculate conversion factors.

Calculation of Conversion Factors

Currently, there is no consensus about the best method for calculating conversion factors. The following equations are two alternatives, with the first being used more commonly:

$$\text{Integrative Method: } CF = [N_f - N_0]/\int (\text{TdR dt}) \tag{1}$$

where N_0 and N_f are the initial and final bacterial abundances and $\int (\text{TdR dt})$ is the rate of TdR incorporation integrated over the course of the experiment.

$$\text{Modified Derivative Method: } CF = \mu \, e^B/e^b \tag{2}$$

where μ is the growth rate determined from the change in cell numbers over time and B and b are the y-intercepts of ln(cells) and ln(TdR incorporation rates) vs. time. That is, e^B should be equivalent to N_0, but is a better estimate because it uses more data and is not dependent on a single measurement. A similar argument applies to e^b. Recently, Ducklow et al.[5] termed this method the "modified" derivative method because it is very similar to the equation originally proposed by Kirchman et al.,[8] which has been called the derivative method.[6] The modified derivative method was first proposed by Ducklow and Hill.[7]

NOTES AND COMMENTS

1. The volume of water needed for the experiment will depend on the number of time points. For example, 10 sample points each requiring about 40 ml (3 × 10 ml for incorporation rates, another 10 ml for bacterial counts) would total 400 ml. To minimize bottle effects and changing surface area, ideally only half of the entire volume should be used. Less water may be required during the latter stages of the experiment because the subsample volume can be reduced and microbial abundance and incorporation rates will still be measurable because they are high late in the experiment.
2. Instead of dilution, grazers can be eliminated by filtration although filtration alone would not reduce loss due to viruses. The most appropriate pore size of the filter varies with the environment. For oligotrophic marine systems, filters with 0.6- or 0.8-μm pore sizes are most effective. For eutrophic freshwaters, 1 or even 2 μm may suffice. The small pore size filters remove more grazers but may remove a large and important fraction of the bacterial assemblage. Still another technique is both to filter and to dilute. That is, the inoculum is filtered (0.6 to 2.0 μm) and then diluted with 0.22-μ filtered water. Kirchman[3] diluted the filtered inoculum only 1:1; their conversion factors are similar to those calculated by others using 1:9 dilution. The advantage of this approach is that the initial rates are easier to measure compared to 1:9 diluted inocula, which is especially important in oligotrophic waters. In one lake experiment, Coveney and Wetzel[2] did find that a combination

of 0.8-μm filtration and five-fold dilution gave a higher TdR conversion factor than either treatment alone.

3. Although the experiment does not depend on exactly reproducing the growth conditions and rates of the original aquatic system, clearly differences from the natural system should be minimized. The perturbation that is hardest to avoid is DOM contamination, e.g., cell breakage during filtration. Fuhrman and Bell[4] discuss ways to avoid DOM contamination. However, Kirchman[3] observed that additions of free amino acids increased TdR incorporation rates (the water was not filtered) only after a long lag. This implies that DOM contamination will not affect initial bacterial growth in conversion factor experiments and thus, will not compromise the factors calculated by the modified derivative approach. In fact, Coveney and Wetzel[2] found no significant effect of organic enrichments on TdR conversion factors calculated by the integration approach, although they did observe that phosphate enrichment lead to lower conversion factors. The DOM contamination may affect the integration approach more than the modified derivative approach because the former uses the increase in bacterial biomass and incorporation rates measured throughout the experiment. Thus, contamination and enrichments have more time to affect the calculated conversion factors.

4. When changes in incorporation rates equal changes in bacterial numbers all calculation methods give the same conversion factor. Problems arise when incorporation rates increase faster than can be explained by increases in numbers or biovolume (i.e., biomass). In these cases, the original derivative method[8] gives high factors, whereas the integration method gives low factors because the incorporation rates are in the numerator in the former and in the denominator in the latter.[5]

    We recommend that the modified derivative approach be used, which will give factors between the other two approaches. This approach uses the change in bacterial abundance to estimate unambiguously bacterial production at time zero of the experiment. The initial incorporation rate seems to be the rate closest to that of the *in situ* bacterial assemblage. This can be tested by comparing incorporation rates per cell by the *in situ* bacteria with the initial per cell rate in the conversion factor experiment.

5. One problem with the integration approach is that it is difficult to determine unarbitrarily what incubation period should be used to do the integration. For example, the initial one or two sample points often should not be included because rates and the ratio of rates to bacterial numbers differ greatly from the rest of the experiment.[6] Also, the time period should be long enough to avoid periods of unbalanced growth when incorporation rates are uncoupled from rates of biomass production.

6. Fuhrman and Azam[9] estimated an empirical conversion factor for TdR by dividing experiments with 0.8-μm filtered water into time periods when bacterial growth was detected. A graph was then constructed of TdR incorporation rates vs. changes in cell abundance. The slope of that graph was taken as the conversion factor. The disadvantages of this approach is that it requires several experiments to obtain one conversion factor.

7. Bjørnsen and Kuparinen[10] suggested still another method for calculating conversion factors, which they term the ''cumulative'' method. This is related to the integration method in that it calculates the slope of cell numbers at a time point vs. the incorporation rates integrated to that time point. This slope effectively represents a way of averaging the factors calculated by the integration method over time.

REFERENCES

1. Landry, M. and Hassett, R., Estimating the grazing impact of marine microzooplankton, *Mar. Biol.*, 67, 283, 1982.
2. Coveney, M. F. and Wetzel, R. G., Experimental evaluation of conversion factors for the [³H]thymidine incorporation assay of bacterial secondary productivity, *Appl. Environ. Microbiol.*, 54, 2018, 1988.
3. Kirchman, D. L., Thymidine and leucine incorporation in the subarctic Pacific: measuring bacterial production, *Mar. Ecol. Progr. Ser.*, 82, 301, 1992.
4. Fuhrman, J. A. and Bell, T. M., Biological considerations in the measurement of dissolved free amino acids in seawater and implications for chemical and microbiological studies, *Mar. Ecol. Progr. Ser.*, 25, 13, 1985.
5. Ducklow, H. W., Kirchman, D. L., and Quinby, H. L., Determination of bacterioplankton growth rates during the North Atlantic spring phytoplankton bloom, *Microb. Ecol.*, 24, 125, 1992.
6. Kirchman, D. L. and Hoch, M. P., Bacterial production in the Delaware Bay estuary estimated from thymidine and leucine incorporation rates, *Mar. Ecol. Progr. Ser.*, 45, 169, 1988.
7. Ducklow, H. W. and Hill, S. M., Tritiated thymidine incorporation and the growth of heterotrophic bacteria in warm core rings, *Limnol. Oceanogr.*, 30, 260, 1985.
8. Kirchman, D. L., Ducklow, H. W., and Mitchell, R., Estimates of bacterial growth from changes in uptake rates and biomass, *Appl. Environ. Microbiol.*, 44, 1296, 1982.
9. Fuhrman, J. A. and Azam, F., Bacterioplankton secondary production estimates for coastal waters of British Columbia, Antarctica, and California, *Appl. Environ. Microbiol.*, 39, 1085, 1980.
10. Bjørnsen, P. K. and Kuparinen, J., Determination of bacterioplankton biomass, net production and growth efficiency in the Southern Ocean, *Mar. Ecol. Progr. Ser.*, 71, 185, 1991.

ADDITIONAL REFERENCES

Bell, T. R., An explanation for the variability in the conversion factor deriving bacterial cell production from incorporation of [³H]thymidine, *Limnol. Oceanogr.*, 35, 910, 1990.

Cole, J. J., Caraco, N. F., Strayer, D. L., Ochs, C., and Nolan, S., A detailed organic carbon budget as an ecosystem-level calibration of bacterial respiration in an oligotrophic lake during midsummer, *Limnol. Oceanogr.*, 34, 286, 1989.

Ellenbroek, F. M. and Cappenberg, T. E., DNA synthesis and tritiated thymidine incorporation by heterotrophic freshwater bacteria in continuous culture, *Appl. Environ. Microbiol.*, 57, 1675, 1991.

Reimann, B., Bjørnsen, P. K., Newell, S. Y., and Fallon, R. D., Calculation of cell production of coastal marine bacteria based on measured incorporation of [³H]thymidine, *Limnol. Oceanogr.*, 32, 471, 1987.

Smits, J. D. and Riemann, B., Calculation of cell production from [³H]thymidine incorporation with freshwater bacteria, *Appl. Environ. Microbiol.*, 54, 2213, 1988.

CHAPTER 60

Bacterial Production in Anaerobic Water Columns

Carlos Pedrós-Alió, Josefina García-Cantizano, and Juan I. Calderón

INTRODUCTION

Aquatic anaerobic environments are more widespread than usually thought.[1] Most eutrophic lakes develop anaerobic hypolimnia during stratification. Meromictic lakes have a permanent monimolimnion which is usually anaerobic. Several ocean basins, such as the Black Sea or the Cariaco Trench, are anaerobic. Many coastal lagoons, rivers with saline wedges, and fjords have anaerobic marine waters underneath aerobic fresh water layers. Finally, the digestive tract of many animals and waste disposal systems are other examples of anaerobic aquatic environments. Of course, most sediments in lakes, marshes, and oceans are anaerobic at some depth, but these systems fall outside the scope of the present chapter.

The techniques used to determine bacterial production in anaerobic environments are essentially the same as those for aerobic habitats, but the few studies carried out in the former, plus a general lack of knowledge on the metabolism of heterotrophic anaerobic bacteria, make interpretations difficult. We will review the modifications used in anaerobic environments for both autotrophic and heterotrophic microorganisms. The reader should consult the appropriate chapters for the standard protocols of each technique.

Microbial populations in anaerobic environments live in a world very much different from ours. Thus, sampling procedures must be devised to minimize exposure to the oxygen, light, and temperature conditions prevailing at the surface. Many organisms, such as methanogenic bacteria, are extremely sensitive to oxygen, while others (such as purple sulfur bacteria) can tolerate moderate amounts of oxygen. Ciliates range from the very tolerant to the extremely sensitive to oxygen exposure. Researchers, therefore, should start by carrying out the most careful sampling possible. Experiments can then be performed to check whether less stringent conditions can be used in each particular case.

Anaerobic water columns are the result of microbial activities. Usually, organic matter accumulates in sediments. First, aerobic bacteria oxidize the organic matter to CO_2 consuming the oxygen in the process. Next fermentation also occurs as an intermediate step, but terminal electron transport occurs to a series of heterotrophic guilds of bacteria, resulting in the release of reduced compounds such as Fe^{2+}, NH_4^+, H_2S, and CH_4. When oxygen import is slow (such as in the hypolimnia of lakes), these reduced substances diffuse into the water column. At the oxic-anoxic interface, autotrophic guilds of microorganisms use the reduced compounds as a source of electrons (and sometimes of energy) and reoxidize them. As long as no major turbulence occurs, heterotrophic processes predominate and the concentration of reduced compounds increases with time. This allows for mass balance calculations for the whole anaerobic water mass. On the other hand, estimates of instantaneous rates will have to concentrate at the oxic-anoxic interface for autotrophic processes and at the sediment

0-87371-564-0/93/$0.00 + $.50

surface for heterotrophic processes. Although the latter also occur in the anaerobic water column, most of the activity is concentrated in the sediments.

At the interfaces between oxygen-containing and oxygen-free waters, a gradient of redox conditions exists which causes the layering of microorganisms. Each population stratifies at a certain layer in the gradient according to its oxygen tolerance and energetic metabolism. This fine layering requires extremely careful sampling with special devices (see below), otherwise, some of the most significant activities may be missed. On the other hand, the different metabolisms across the gradient offer the possibility to measure the production of different bacterial guilds separately (such as oxygenic and anoxygenic photosynthesis, chemolithoautotrophy, denitrification, sulfate reduction, methylotrophy, etc.), something which is very difficult to do in aerobic systems.

MATERIALS REQUIRED

Anaerobic Sampling

- Sampling device: Niskin bottles (for oceanography), pump, double cone and hose (preferred system), or syringe system (for sharply stratified waters)
- Opaque coolers
- Sampling bottles: glass biological oxygen demand (BOD) or screw-capped; caps with septa optional
- Optional: oxygen-free nitrogen, syringes, and hosing for nitrogen flushing

Autotrophic Production

Equipment

- Laboratory for radioactivity work
- Desiccator
- Incubators
- Incubation platform
- Liquid scintillation counter
- Icebox

Supplies

- Tightly capped incubation tubes or bottles
- Micropipettes and tips
- Aluminum foil
- 1-ml glass syringe and needle
- Glass fiber or membrane filters
- Plastic syringes
- 25-mm filter holders
- Forceps
- Scintillation vials

Solutions and Reagents

- NaH[^{14}C]CO$_3$ — we use a 100-μCi ml^{-1} (50 to 60 mCi mmol^{-1} sp. act.) stock solution (1 mCi NaH [^{14}C]CO$_3$ in 10 ml of 10 mM HCl-Tris buffer, pH 9.1) in a rubber cap tube; in every experiment the required amount of solution is extracted with a glass syringe through the septum in the stopper

- 1 mM 3-(3′,4′-dichlorophenil)-1,1′-dimethyl urea (DCMU) dissolved in absolute ethanol
- 40% formaldehyde
- Concentrated HCl
- Scintillation cocktail

PROCEDURES

Anaerobic Sampling

Sampling Devices

When designing sampling devices it must be kept in mind that anaerobic environments may contain sulfide, which quickly reacts with many metals. Both the structural parts of the device and paints should be prepared to resist sulfide. Conventional hydrographic bottles, of the Niskin or Van Dorn types, should be avoided whenever possible (i.e., in lakes), since they do not sample fine layers well and create considerable turbulence when emptied. Thus, oxygen may mix in with the water, reacting with reduced compounds and killing sensitive organisms. In deep ocean basins, such as the Black Sea or the Baltic, there seems to be no alternative to Niskin bottles. Authors should try to use the utmost care in these cases.

Preferably a fine sampling system should be used. Two different sampling systems are available. The first is an improved version of the fine-layer sampler described by Jørgensen et al.[2] and is shown in Figure 1A.[3] This conical inlet device, connected to a surface pump, smoothly sucks the organisms toward the tubing, minimizing turbulence and escape responses of the larger organisms such as rotifers and crustacean nauplii and juveniles. The double cone allows pumping large volumes of water from a given depth, with minimal disturbance of the stratification. The sampler is made of three Plexiglass cones: two hollow and one solid (Figure 1A). The lower cone is loaded with lead pellets as ballast. The solid cone fits on a specially carved rim of the lower cone and is glued to it. The upper hollow cone is attached to the lower one through three screws (only two appear in the figure) leaving a 1-cm circumferential aperture between them. The transition from the outer inlet to the small diameter hose is, thus, smoother than in the original design of Jørgensen et al. A screwable connector allows attachment between the cone (by the screw) and the hose (through clamping of the hose to the upper portion of the screw). The hose and screw can be easily removed by unscrewing the cone. This is very convenient for transport or for changing hoses of different lengths for sampling lakes of different depths. Ideally, a peristaltic pump should be used on the boat. Submerged pumps next to the inlet should be avoided because vibrations of the pump may disrupt the fine layering. The device is sustained by a measuring tape. These tapes are nonstretchable and, thus, provide the most precise and convenient way to determine the exact depth at which each sample is taken.

The second sampling system is a modification of the syringe systems devised by Baker et al.[4] and Mitchell and Fuhrman[5] and is shown in Figure 1B. The syringes are filled simultaneously at the desired depth by the action of a vacuum pump connected to the device. A detailed instantaneous image of the vertical distribution of the organisms can be obtained quickly and avoids problems created by surface waves. The model shown in Figure 1B has 33 syringes in a 1-m copper column, but other lengths and spacing among syringes can be used as desired. This system works only down to approximately 2 m before hydrostatic pressure fills the syringes spontaneously (at 3 to 4 m). For such depths, air can be pumped into the system from the surface while it is being lowered to the desired depth. This may work, depending on the syringes used, down to 10 or 15 m. For better performance at depth, however, a messenger-triggered system[6] is recommended.

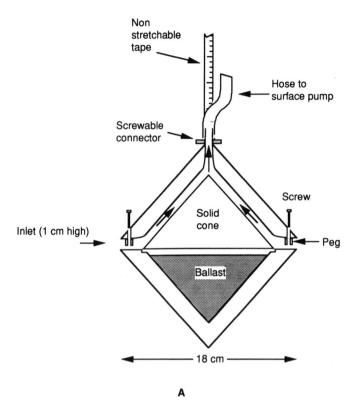

Figure 1.   Devices for sampling sharply stratified water columns: (A) modified double cone and (B) syringe system. (Reproduced with permission from Miracle, R. M., Vicente, E., and Pedrós-Alió, C., *Limnetica*, 8, 59, 1992.)

These systems are lowered to the desired depth with upmost care so as not to disturb the fine stratification of the organisms. A good procedure consists of slowly lowering the device to the desired depth, then very slowly and smoothly moving it sideways (syringe-side first in the case of the syringe system) to sample an undisturbed section of the lake.

Sampling Strategy

Diurnal changes in depth of the chemocline have been described.[7,8] The boundary between O_2 and H_2S moves downward during the day due to production of oxygen by oxygenic photosynthesis and consumption of sulfide by anoxygenic photosynthesis. At night, sulfate reduction increases sulfide concentration, aerobic respiration decreases oxygen concentration, and photosynthesis is inactive. All these processes contribute to the upward displacement of the interface. This has important consequences for production estimates:

1.   In long incubation experiments bottles will be located only temporarily in their original environment. In lakes with nonmotile green bacteria, the plate is fixed but the chemical changes also occur.

2.   During diel cycles, sharply stratified populations will be found at different depths each time they are sampled, thus, the sampling strategy must be different each time in order not to miss any important biomass peak.

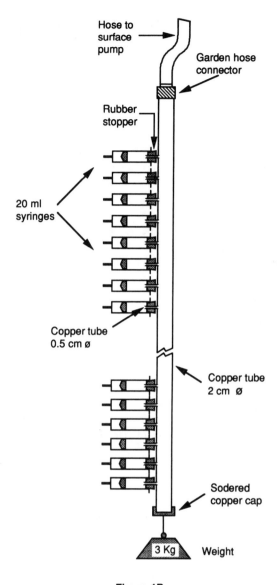

Figure 1B.

3. Layers of maximal activity may not coincide with layers of maximal biomass. In the case of phototrophic microorganisms, for example, layers have been divided into top, middle, and bottom portions.[9,10] The maximal specific activity is at the top, while the maximal biomass is at the middle.

Dispensing Samples into Bottles

Ideally, sample bottles are submerged in opaque coolers or buckets filled with water from the sampling depth in order to keep *in situ* temperatures as long as possible.[15] In this way, water can be pumped into the bottles with minimal disturbance. Sampling bottles should be gently flushed several times with the water before tightly closing them. Many plastic bottles are permeable to gases and, thus, not suitable for anaerobic samples. Some workers flush their bottles with oxygen-free nitrogen before filling them to make sure that no oxygen remains inside.

Since the syringe system is emptied into sampling bottles by pressure, concern may arise about turbulence and mixing with air during this process. However, we have not encountered problems either with oxygen-supersaturated or with sulfide-rich waters retrieved from the metalimnion and emptied at the surface.

Incubations

Stoppered bottles with a septum are a convenient way to carry out incubations, since reactants and fixatives can be added, or samples removed, with syringes without having to open the bottle and expose it to air. Rubber, however, has been shown to affect determinations of some anaerobic processes significantly by adsorbing or releasing methylated sulfur compounds.[11] Conventional glass bottles are a valid alternative. In this case, a needle with nitrogen pressure can be used to maintain the open bottles free of oxygen. Some authors recommend leaving a pea-sized bubble inside the bottles to accommodate pressure changes during incubation. This is especially important during estimates of photosynthesis, since several gases are generated or consumed. The bottles might break with the undesirable result of radioactive chemicals being released in the environment. With purple bacteria the bubble can be simply air, but with more delicate organisms, it should be nitrogen.

When incubating anaerobic samples, workers have to be aware that either the electron donors or acceptors for any given organisms may disappear quite quickly from the bottles, even if growth is slow. In the water column, this consumption is compensated by diffusion and biosynthesis, but in the bottles diffusion is not possible, and biosynthesis may be altered. Thus, incubation times become critical and time courses should be carried out whenever possible. This is very important for populations living at oxic-anoxic interfaces, where high activities can be maintained with low apparent concentrations of substrates, if the latter are constantly supplied by diffusion. Adding substrates before incubation is not a good idea, since they may be inhibitory to the organisms (for example high sulfide may be toxic to purple sulfur bacteria).

Autotrophic Production

The method commonly used for measuring primary production in all environments is based on the ^{14}C-technique, introduced by Steeman-Nielsen in 1952.[12] In anaerobic waters, primary production occurs mainly close to oxic-anoxic interfaces. The simultaneous presence of oxygen, light, and reduced compounds in these interfaces allows the coexistence of several autotrophic guilds: oxygenic phototrophs such as eukaryotic algae and cyanobacteria, anoxygenic phototrophs (some cyanobacteria and sulfur dependent bacteria), as well as different groups of chemolithoautotrophic bacteria. Thus, several modifications and corrections have to be applied to distinguish the participation of the different autotrophic metabolisms present in each particular sample. These involve the use of selective inhibitors, which specifically block one or several of these metabolisms (Tables 1 to 3). The researcher should decide which combination of inhibitors is optimal for each particular system on the basis of microscopical inspection of the microbiota, physicochemical composition of the water, and previous experience with the system. For details and problems with the different inhibitors, the reader should consult the thorough review by Oremland and Capone[13] and references therein.

The following protocol is the one we have following in Lake Ciso. It should be applicable to any anaerobic system with only minor modifications. If desired, further tubes can be added with additional inhibitors to estimate the contribution of the different dark fixers.

Table 1. Commonly Used Inhibitors of Photoautotrophic Processes

| Inhibitor | Light Dependent CO_2 Fixation Process | | |
	Eukaryotic Algae	Cyanobacteria	Anaerobic Sulfur Bacteria
Darkness	+	+	+
DCMU	+	±	−
Cycloheximide	+	−	−
Propanil		+	−
Oxygen	−	±	+
D,L-glyceraldehyde	+	+	+
Carboxypentitol biphosphate	+	+	+

Table 2. Commonly Used Inhibitors of Chemoautotrophic Processes

| Inhibitor | Dark CO_2 Fixation Process | | | |
	Metal Oxidizers	Nitrifiers	Sulfide Oxidizers	Methane Oxidizers
Darkness	−	−	−	−
Nytrapirin (N-serve)	−	+	−	+
Acetylene	−	+	−	+
Chlorate	−	+	−	−
Oxygen	−	−	+	+
Picolinic acid	−	−	−	+
CH_3Fl	−	−	−	+
D,L-glyceraldehyde	+	+	+	+
Carboxypentitol biphosphate	+	+	+	+

Table 3. Commonly Used Inhibitors of Anaerobic Respiratory Processes

| Inhibitor | Respiratory Process | | | |
	Metal Reducers	Denitrifiers	Sulfate Reducers	Methanogens
Acetylene	−	+	−	−
H_2S	−	+	−	−
Oxygen	−	+	+	+
Na-molybdate	−	−	+	−
β-fluorolactate	−	−	+	−
Chloroform	−	−	−	+
BES	−	−	−	+

1. Dispense the sample into eight 20-ml screw-capped tubes following the recommendations given above. In case of sparse populations, larger bottles may have to be used. Pyrex screw-capped bottles, transparent BOD bottles, or bottles capped with serum stoppers have all been used by different authors.

- Two "clear" tubes are filled with sample only, to determine total incorporation.

- Two more tubes are covered with aluminum foil or black electrical tape to determine dark incorporation.

- Two other tubes receive DCMU at a final concentration of 2 μM. This concentration had been previously shown to inhibit oxygenic, but not anoxygenic, photosynthesis in Lake Cisó.[9] The exact concentration of DCMU which inhibits oxygenic photosynthesis without inhibiting anoxygenic photosynthesis has to be determined for each system. Other authors have used concentrations ranging from 10 μM (Reference 14) to 1.5 μM (Reference 15).

- One tube receives 2 ml of formalin (4% final concentration) as killed control.

- The last tube is used to check the possible inhibitory effect of the ethanol used to dissolve DCMU. DCMU is not easily soluble in water, thus, there are two alternatives. In the first alternative, a concentrated stock solution is prepared in ethanol. In this way very small volumes need to be added per sample. The possible inhibitory effect of the ethanol can be checked in additional tubes with ethanol and without DCMU. In the second approach, a very dilute stock solution is prepared in water, then larger volumes have to be added to samples. The last tube, thus, receives an amount of ethanol equivalent to that in the DCMU tubes (0.18-μl ethanol per ml sample). After several experiments in which the ethanol tube was never significantly different from the clear tubes, this control was discontinued. Other parallel experiments showed that preincubation with DCMU was not necessary for proper inhibition of oxygenic photosynthesis, thus, simplifying field procedures.

2. Add 50 μl of the stock ^{14}C-bicarbonate solution to each tube with a micropipette (5 μCi per 20 ml of sample). The tubes are kept in a cooler, in the dark, until deployed (usually around 10 min). If populations are sparse, higher specific activities of CO_2 may have to be added. Researchers should carry out time-course experiments to find out the incubation time giving measurable incorporation without departure from linear uptake. The specific activity gives an additional variable to play with.

3. The set of eight tubes is then placed in custom-made incubators and suspended at the desired depth through a winch mounted on a floating platform.[9] Incubations may last from 2 to 4 h. In the ocean, *in situ* incubations may not be feasible. In this case, incubators can be set up on a board with the *in situ* temperature and light conditions. Extrapolation to the field, however, should be done with caution.

4. The incubators are retrieved and immediately placed in coolers in the dark. The whole contents of each tube are quickly filtered through Whatman GF/F glass fiber filters mounted in filter holders with a syringe. Some authors fix their samples with formalin. This frequently causes faster clogging of the filters and, in our experience, larger variability among replicates. A compromise has to be found between easiness of filtration (which should be as quick as possible) and appropriate retention of all CO_2-fixing microorganisms. We have successfully used Whatman GF/C filters in Lake Cisó, but these may be too large for certain bacteria. In any case, hypolimnetic bacteria tend to be much larger than epilimnetic bacteria.

5. Filters are rinsed with lake water and allowed to dry in the dark until processed. During the drying period, cells on the filters could potentially respire part of the incorporated CO_2, thus leading to underestimates of production. We conducted experiments to determine optimal drying times. In several trials, adding the wet filters directly to vials with the scintillation cocktail gave highly variable counts. After 2- to 24-h drying periods, however, counts were never lower than those in replicates killed with formalin at the beginning of the drying period. Thus, we used overnight drying periods for convenience. Other authors place the wet filters directly in vials with HCl and then strip the acid by bubbling with air.

6. Once dry, the filters are exposed to HCl fumes for 20 min to release precipiated CO_2 and placed in scintillation vials. Then 10 ml of scintillation cocktail are added and the radioactivity counted in a scintillation counter for 1 min. Quench corrections are done by any of the conventional methods.

7. Alkalinity and pH must be determined for each field sample. The available carbon can then be calculated from them. Alternatively, dissolved inorganic carbon (DIC) can be determined directly by gas chromatography (for example by the method of Stainton[16]). Biebl and Pfennig[17] discuss the alternative ways to determine DIC and conditions under which each one is more appropriate. Under most conditions, the calculations from alkalinity and pH seem sufficiently accurate.

8. Calculations:
    The carbon uptake rate ($^{12}C_{as}$ in mg C L^{-1} h^{-1}) can be calculated from the equation (given for example in Vollenweider[18]):

$$^{12}C_{as} = (^{14}C_{as}/^{14}C_{ab}) \times {}^{12}C_{ab} \times 1.06 \tag{1}$$

where: $^{12}C_{ab}$ (mg C L^{-1}) = alkalinity (mEq L^{-1}) × F_{pH}; F_{pH} is a factor depending on temperature and pH.[19,20] In order to determine alkalinity in sulfide-rich waters, sulfide has to be precipitated with some drops of 2% $CdNO_3$ before the assay; $^{14}C_{ab}$ = total dpm added; $^{14}C_{as}$ = assimilated dpm; and 1.06 = isotope discrimination factor. This factor corrects for the fact that $^{14}CO_2$ is taken slightly more slowly than $^{12}CO_2$.

The dpm assimilated by oxygenic photosynthesis are calculated by subtracting dpm incorporated in clear tubes plus DCMU from dpm incorporated in clear tubes. The dpm incorporated by anoxygenic photosynthesis are determined by subtracting the dpm incorporated in dark tubes from dpm incorporated in clear tubes plus DCMU. Finally, the dpm incorporated in the dark are calculated by subtracting dpm incorporated in the killed control from dpm incorporated in dark tubes.

Heterotrophic Production

The techniques for estimating heterotrophic bacterial production in anaerobic waters are the same as those for aerobic waters, except for the sampling and incubation strategies indicated above. Thymidine (TdR) incorporation into macromolecules has been the most widely used technique in anaerobic waters,[21-25] although leucine incorporation into protein, adenine incorporation into DNA, and frequency of dividing cells (FDC) have been used in a few cases[23,26] (also J. J. Cole and M. L. Pace, personal communication). The sulfate uptake into protein technique cannot be used in anaerobic waters with sulfide, since sulfide is the preferred source of sulfur for many microorganisms.

In several cases, maximal heterotrophic activity has been found at the oxic-anoxic interface, regardless of the method used. This has been the case in Lake Oglethorpe[23] using thymidine and leucine incorporation and FDC; in Big Soda Lake, Nevada[24] with the thymidine technique; in Lake Cisó with both the FDC and the thymidine techniques; and in Upton Lake, New York (M. Felip, M. L. Pace and J. J. Cole, personal communication) where the leucine technique was used.

Of these studies, only that by McDonough et al.[23] attempted to address the methodological problems specific to anaerobic waters. These authors compared TdR and leucine uptake and FDC across the vertical profile of Lake Oglethorpe (Georgia, U.S.). While the shape of the profiles was the same for the three techniques, TdR gave lower values, due to the large incorporation of TdR into proteins at depths without oxygen. These results were calculated with conversion factors from the literature. All of the following comments, however, suggest that different conversion factors are likely to be valid for the oxic and anoxic layers of the water column.

Next, we review some of the differences found between aerobic and anaerobic waters with respect to these techniques. Leucine and adenine uptake and FDC have been used so few times that most comments will deal with the TdR uptake technique.

TdR Incorporation into DNA

For a general protocol, consult other chapters in this volume. There are several assumptions that microbial populations should fulfill for this technique to be valid:

1. Only bacteria incorporate TdR at nanomolar concentrations and during short incubation times (less than 1 h). This seems to be true for both aerobic and anaerobic systems. Protists are the only eukaryotic organisms in anaerobic water columns, especially ciliates. We have never observed any label in anaerobic ciliates through ^{3}H-TdR autoradiography.

2. All actively growing heterotrophic bacteria are capable of incorporating exogenous TdR during growth. This is known to be false even for aerobic bacteria.[27] Some anaerobic bacteria are able to take up TdR, since Pollard and Moriarty[25] found that a mixed culture of anaerobic sediment bacteria incorporated TdR at rates compatible with the growth rate estimates from cell increase. McDonough et al.[23] found that aeration of anaerobic samples inhibited TdR incorporation significantly. We have found in Lake Cisó that the percent of bacteria labeled with TdR in anaerobic waters is similar to that in aerobic waters (from 1 to 7%). Therefore, there is no reason to assume a different situation from that in aerobic waters. Other groups of bacteria living in anaerobic habitats, however, do not incorporate TdR. Thus, Pollard and Moriarty[25] could not detect TdR incorporation by acetate-utilizing, sulfate-reducing bacteria. Likewise, Gilmour et al.[28] showed that a collection of marine isolates of sulfate-reducing bacteria could not take up TdR. Three strains of nitrogen-dependent chemolithoautotrophic bacteria and two strains of methylotrophic bacteria were also incapable of TdR uptake.[29] Some autotrophic bacteria which accumulate at oxic-anoxic interfaces seem to take up TdR. Kraffzik and Conrad[30] demonstrated TdR uptake in pure cultures of CO, H_2, and CH_4-oxidizing chemolithotrophic bacteria. We have found that some photolithoautotrophic purple (*Amoebobacter* M3 and *Chromatium minus*) and green *(Chlorobium limicola)* sulfur bacteria isolated from Lake Cisó do take up TdR, although at very low rates. Through autoradiography, we have observed labeled purple bacteria in the field at some winter dates, but not during most of the year. The interpretation of TdR uptake at interfaces, therefore, is very difficult, since both autotrophic and heterotrophic bacteria (but not all) seem able to incorporate it.

3. A constant percentage of cold TCA-insoluble label is in DNA. In several cases, the amount of TdR incorporated into DNA was higher (around 80%) in the aerobic waters than in the anaerobic waters (0 to 50%).[22-24] In anaerobic layers, a high percent of TdR was incorporated into protein.[23,24] The number of cases studied, however, is so small that no firm conclusion can be derived. Researchers should preferably determine the incorporation of TdR into DNA instead of the incorporation into TCA-insoluble material (see Chapter 56).

4. The specific activity of the added TdR is unaffected by endogenous or exogenous TdR. If the isotope dilution approach is used, the results tend to be different in aerobic and anaerobic waters.[23] Lovel and Konopka[22] could not obtain linearity in Crooked Lake (Indiana, U.S.). In Lake Cisó, linearity could not be obtained in anaerobic waters with this approach, despite good adjustment in the aerobic or microaerophilic waters.

5. The uptake of TdR is saturated at the concentration used. In general, saturating concentrations of TdR vary with depth, being different in the aerobic and anaerobic portions of lakes. The saturating TdR concentration has been shown to be different when determined in aerobic and anaerobic waters both in Big Soda Lake[24] and in Lake Cisó.

Other Techniques

Leucine and adenine incorporation and FDC have been used even less often than the TdR uptake technique in anaerobic waters. McDonough et al.[23] compared two of them with TdR uptake and concluded that similar qualitative results could be obtained with the three techniques. However, estimates of bacterial production based on FDC and leucine incorporation were usually significantly higher than estimates based on TdR uptake in the anaerobic waters. The leucine uptake technique has been used in several lakes in New York (J. J. Cole and M. L. Pace, personal communication), where much more repeatable and less variable results were obtained with this technique than with the TdR technique. Our preliminary data from Lake Cisó indicate that similar conversion factors between FDC and growth rate apply to the aerobic and anaerobic waters. Karl and Knauer[26] used adenine incorporation into DNA as an estimate of total microbial production in the Black Sea. They found production in aerobic waters to be about ten times higher than in the sulfide-rich waters. No methodological experiments, however, were carried out to check whether differences existed among the two water masses with respect to the optimal parameters for the technique.

In summary, all techniques may work in anaerobic habitats, but the conversion factors, saturating concentrations, isotope dilution, and other details of the techniques are probably different in the oxic and anoxic layers of the same water body. Thus, they have to be estimated independently. Calculating conversion factors separately for epi-, meta- and hypolimnion involves a lot of work. In Lake Cisó, for example, we had to determine over 180 cell counts in samples from 18 different filtration-dilution cultures (six from each layer) in order to determine one conversion factor for each layer. This is obviously not adequate for routine work, but at least some determinations should be done in each system before calculations of bacterial production can be done with a minimal degree of confidence. Literature conversion factors should be avoided. Special care should be taken with the TdR technique by determining TdR incorporation into DNA. Alternative techniques such as the leucine and adenine uptake and the FDC techniques, should receive more attention in anaerobic habitats. Finally, the use of two techniques simultaneously would add a considerable degree of confidence to estimates of heterotrophic bacterial production in anaerobic habitats.

Activity of Specific Guilds in Gradients

In addition to CO_2 incorporation in the presence of the appropriate inhibitors (see above), some activities can be determined by the appearance of endproducts, the disappearance of substrates, the use of surrogate substrates, or the use of tracers with radioactive or heavy isotopes. Since these are not properly production estimates, however, they will not be considered here. Interested readers should consult Oremland and Capone,[13] Zehnder,[31] and Underhill.[32]

ACKNOWLEDGMENTS

The work of the authors has been supported by DGICYT grant PB87-0183 from the Spanish Ministry of Education and Science.

REFERENCES

1. Jones, J. G., Diversity of freshwater microbiology, *Soc. Gen. Microbiol. Spec. Symp.*, 41, 235, 1987.
2. Jørgensen, B. B., Kuenen, J. G., and Cohen, Y., Microbial transformations of sulfur compounds in a stratified lake (Solar Lake, Sinai), *Limnol. Oceanogr.*, 28, 1075, 1979.
3. Miracle, R. M., Vicente, E., and Pedrós-Alió, C., Biological studies in Spanish meromictic and stratified karstic lakes, *Limnetica*, 8, 59, 1992.
4. Baker, A. L., Baker, K. K., and Tyler, P. A., A family of pneumatically-operated thin layer samplers for replicate sampling of heterogeneous water columns, *Hydrobiologia*, 122, 207, 1985.
5. Mitchell, J. G. and Fuhrman, J. A., Microdistribution of heterotrophic bacteria, *Mar. Ecol. Prog. Ser.*, 53, 141, 1989.
6. Bjørnsen, P. K. and Nielsen, T. G., Decimeter scale heterogeneity in the plankton during a pycnocline bloom of *Gyrodinium aureolum, Mar. Ecol. Prog. Ser.*, 73, 263, 1991.
7. Van Gemerden, H., On the Bacterial Sulfur Cycle of Inland Waters, Ph.D. thesis, University of Leiden, The Netherlands, 1967.
8. Pedrós-Alió, C. and Sala, Ma. M., Microdistribution and diel vertical migration of flagellated vs. gas-vacuolate purple sulfur bacteria in a stratified water body, *Limnol. Oceanogr.*, 35, 1637, 1990.

9. Guerrero, R., Montesinos, E., Pedrós-Alió, C., Esteve, I., Mas, J., Gemerden, H. V., Hofman, P. A. G., and Bakker, J. F., Phototrophic bacteria in two Spanish lakes. Vertical distribution and limiting factors, *Limnol. Oceanogr.*, 30, 919, 1985.

10. Konopka, A., Physiological changes within a metalimnetic layer of *Oscillatoria rubescens, Appl. Environ. Microbiol.*, 40, 681, 1980.

11. Kiene, R. P. and Capone, D. G., Microbial transformations of methylated sulfur compounds in anoxic salt marsh sediments, *Microb. Ecol.*, 15, 275, 1988.

12. Steeman-Nielsen, E., The use of radioactive carbon (^{14}C) for measuring organic production in the sea, *J. Cons. Int. Explor. Mer.*, 18, 117, 1952.

13. Oremland, R. S. and Capone, D. G., Use of "specific" inhibitors in biogeochemistry and microbial ecology, *Adv. Microb. Ecol.*, 10, 285, 1988.

14. Børsheim, K. Y., Gijs Kuenen, J., Gottschall, J., and Dundas, I., Microbial activities and chemical gradients in the chemocline of a meromictic lake in relation to the precision of a sampling procedure, *FEMS Microb. Ecol.*, 31, 337, 1985.

15. Steenbergen, C. L. M. and Van Den Hoven, P., A note on the measurement of production of phototrophic bacteria in deep layer, *Arch. Hydrobiol. Beih., Ergebn. Limnol.*, 34, 249, 1990.

16. Stainton, M. P., A syringe gas-stripping procedure for gas-chromatographic determination of dissolved inorganic and organic carbon in fresh water and carbonates in sediments, *J. Fish. Res. Board Can.*, 30, 1441, 1973.

17. Biebl, H. and Pfennig, N., Anaerobic CO_2 uptake by phototrophic bacteria. A review, *Arch. Hydrobiol. Beih., Ergebn. Limnol.*, 12, 48, 1979.

18. Vollenweider, R. A., *Primary Production in Aquatic Environments*, IBP Handbook 12, 2nd ed., Blackwell Scientific Publications, Oxford, 1974.

19. Almgren, T., Dyrssen, D., and Fonselius, S., Determination of alkalinity and total carbonate, in *Methods of Seawater Analysis, 2nd ed.*, Grasshoff, K., Ehrhardt, M., and Kremling, K., Eds., Verlag Chemie GmbH, Weinheim, Germany, 1983, 99.

20. Parsons, T. R., Maita, Y., and Lalli, C. M., *A Manual of Chemical and Biological Methods for Seawater Analysis*, Pergamon Press, Oxford, U.K., 1984.

21. Bloem, J., Ellenbroek, F. M., Bär-Gillisen, M.-J., and Cappenberg, T. E., Protozoan grazing and bacterial production in stratified Lake Vechten estimated with fluorescently labeled bacteria and by thymidine incorporation, *Appl. Environ. Microbiol.*, 55, 1787, 1989.

22. Lovell, C. R. and Konopka, A., Primary production and bacterial production in two dimictic Indiana lakes, *Appl. Environ. Microbiol.*, 49, 485, 1985.

23. McDonough, R. J., Saunders, R. W., Porter, K. G., and Kirchman, D. L., Depth distribution of bacterial production in a stratified lake with anoxic hypolimnion, *Appl. Environ. Microbiol.*, 52, 992, 1986.

24. Zehr, J. P., Harvey, R. W., Oremland, R. S., Cloern, J. E., and George, L. H., Big Soda Lake (Nevada). 1. Pelagic bacterial heterotrophy and biomass, *Limnol. Oceanogr.*, 32, 781, 1987.

25. Pollard, P. C. and Moriarty, D. J. W., Validity of the tritiated thymidine method for estimating bacterial growth rates: measurement of isotope dilution during DNA synthesis, *Appl. Environ. Microbiol.*, 48, 1076, 1984.

26. Karl, D. M. and Knauer, G. A., Microbial production and particle flux in the upper 350 m of the Black Sea, *Deep-Sea Res.*, 32 (Suppl. 2) S921, 1991.

27. Pedrós-Alió, C. and Newell, S. Y., Microautoradiographic study of thymidine uptake in brackish waters around Sapelo Island, Georgia, U.S.A., *Mar. Ecol. Prog. Ser.*, 55, 83, 1989.

28. Gilmour, C. G., Leavitt, M. E., and Shiaris, M. P., Evidence against incorporation of exogenous thymidine by sulfate-reducing bacteria, *Limnol. Oceanogr.*, 35, 1401, 1990.

29. Johnstone, B. H. and Jones, R. D., A study of the lack of [*methyl*-^{3}H]thymidine uptake and incorporation by chemolithoautotrophic bacteria, *Microb. Ecol.*, 18, 73, 1989.

30. Kraffzik, B. and Conrad, R., Thymidine incorporation into lake water bacterioplankton and pure cultures of chemolithotrophic (CO, H_2) and methanotrophic bacteria, *FEMS Microbiol. Ecol.*, 23, 7, 1991.

31. Zehnder, A. J. B., Ed., *Biology of Anaerobic Microorganisms*, J. Wiley and Sons, New York, 1988.

32. Underhill, S. E., Techniques for studying the microbial ecology of nitrification, *Methods Microbiol.*, 22, 417, 1990.

CHAPTER 61

Production of Heterotrophic Bacteria Inhabiting Marine Snow

Alice L. Alldredge

INTRODUCTION

A significant fraction of the suspended matter in the ocean exists as macroscopic aggregates of organic detritus, living microorganisms, and inorganic particles, known as marine snow. Analogous detrital aggregates also occur in the pelagic zone of lakes.[1] Marine snow is enriched in organic matter and harbors dense microbial communities at concentrations two to five orders of magnitude higher than found in the surrounding water (see Alldredge and Silver).[2] Thus, it can be a significant site of heterotrophic bacterial production in the water column. This production is measured by incubating samples with tritiated thymidine and measuring incorporation of the radiolabel into bacterial DNA.[3,4] Cell growth rate and heterotrophic carbon production are then calculated from thymidine incorporation.

This method is extremely sensitive and simple. Minute quantities of incorporated label may be measured, allowing determination of the bacterial production occurring on single detrital particles or even subsamples of single particles. Sample manipulation is minimal and incubation time short.

All the disadvantages and uncertainties of the thymidine method (see Chapters 56 and 57) also apply when it is used with aggregates. Disadvantages specific to aggregates are: (1) interstitial concentrations of natural dissolved thymidine within aggregates may be high resulting in underestimation of production due to isotope dilution; (2) the particle may act as a diffusional barrier, radiolabel may not penetrate all the interstitial spaces of the particle resulting in underestimation of production; (3) the DNA:cell growth rate conversion factor for bacteria attached to marine snow is assumed to be the same as that for free-living bacteria. Although this may not be appropriate, the conversion factor for attached bacteria has not been determined, and (4) in longer incubations, intense grazing by bacteriovores on aggregates may result in release of label and underestimation of thymidine incorporation.

MATERIALS REQUIRED

Equipment

- Filtration rig for 25-mm filters
- Water bath or environmental chamber set at ambient water temperature
- Liquid scintillation spectrometer

Supplies

- Acid-washed syringes of various sizes (see below under sample collection)
- 10-ml acid-washed test tubes
- 0.2-μm pore size, 25-mm diameter Nuclepore filters
- Scintillation vials
- Pipettes
- 10-ml tissue grinder and pestle (optional)

Solutions

- Stock solution of high specific activity (>50 Ci mmol^{-1}) [methyl-^{3}H] thymidine mixed in filter-sterilized distilled water to a final concentration of 1 mCi ml^{-1}, can be stored at 4°C (do not freeze) for no longer than 1 month[4]
- 10% trichloroacetic acid (TCA) in distilled water, ice-cold
- 5% TCA in distilled water, ice-cold
- Scintillation cocktail, suitable for aqueous samples

PROCEDURES

Sample Collection

Visible detrital aggregates are collected by SCUBA divers at shallow depths (<30 m) and with submersibles at deeper depths. Hamner[5] or Heine[6] should be consulted for safe diving protocols in open water. Aggregates may be collected by SCUBA divers either individually or in pooled, multi-aggregate samples of up to 40 particles each. Pooled samples are useful if additional subsamples for measurement of organic carbon, nitrogen, chlorophyll a, particle dry weight, microbial community composition, etc. are also required, and if mean production rather than variability among particles is of interest.

Undisturbed individual aggregates are collected in acid-washed 1-, 6-, or 20-ml syringe barrels stoppered with a syringe plunger at each end. Then 20 to 50 syringes can be strung together through a hole in one plunger shaft for easy handling underwater (Figure 1). The syringe plungers are not toxic to marine microbes.[7]

Multi-aggregate samples may also be collected using standard disposable syringes. Up to 40 aggregates are drawn through the tip of a sterile, needleless syringe which is then capped underwater with the plastic base of a hypodermic needle (needles are cut off, the bases are ground smooth for safety, and sealed). The total number of aggregates collected is most easily determined if the same number of aggregates is collected in each syringe. An experienced diver can collect about 500 aggregates in a 50 min dive if marine snow is abundant. Aggregates are disrupted slightly when collected by this method and their interstitial microenvironments may be altered.

Aggregate-free surrounding seawater required for background seawater blanks is also collected by divers using 60-ml syringes. These are filled carefully so as to exclude visible particles. Samples should be kept in a cooler during transport back to the ship or laboratory to protect them from light shock and temperature variation.

Aggregates have been collected individually at deeper depths from submersibles using 5-L transparent, mechanically triggered Van Dorn bottle-type samplers or hydraulically operated cylindrical samplers.[8,9] Experiments must be carefully designed to use minimal material since only a few aggregates may be collected per submersible dive.

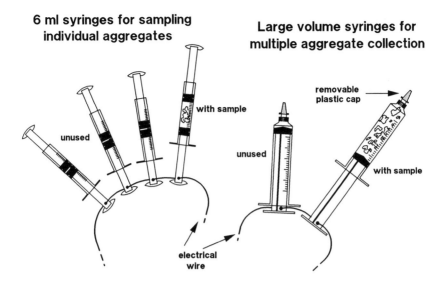

6 ml syringes for sampling individual aggregates

Large volume syringes for multiple aggregate collection

removable plastic cap

with sample

unused

unused

with sample

electrical wire

Figure 1. Examples of syringe samplers used by divers to collect marine snow *in situ*.

Experimental Procedure for Pooled Aggregate Samples

1. If the samples are to be pooled, combine the contents of all the syringes into an acid-washed graduated cylinder. Note the total volume and the number of originally collected particles in the aggregate slurry (usually around 1 aggregate ml^{-1}). Mix the slurry gently but thoroughly and remove 10-ml aliquots to test tubes. Volumes as small as 2 ml may also be used. We incubate three to five replicates of each treatment. Aggregates sink rapidly so the slurry should be mixed before removing each aliquot. The remaining slurry is available for ancillary measurements such as bacterial abundance, aggregate dry weight, etc.

2. Prepare background seawater controls by pipetting 10 ml of surrounding seawater samples into test tubes.

3. Prepare adsorption blanks by adding $HgCl_2$ for a final concentration of 0.2% to three aggregate and three surrounding-seawater samples. Formalin-killed samples tend to yield high adsorption blanks.

4. Add [^{3}H]-thymidine stock to each test tube for a final concentration of 20 nM [^{3}H]-thymidine L^{-1}.

5. Incubate all tubes in the dark at ambient water temperature for a maximum of 45 min.

6. At the end of the incubation, place tubes in crushed ice for 1 min to cool. At this point, subsamples can be taken for bacterial abundance estimates (see below). Then add an equal volume of ice-cold 10% TCA to each tube and extract on ice for 5 to 15 min.

7. Filter samples through 0.2-μm Nuclepore filters. Rinse each filter five times with 1 ml cold 5% TCA and place it in a scintillation vial with 10 ml of aqueous scintillation cocktail. Nuclepore filters give very low blanks. We have found that cellulose-acetate filters give unacceptably high and variable blanks. While 0.45-μm pore size filters may also be used, we use 0.2-μm filters to maximize bacterial retention.

8. Assay each sample for radioactivity by liquid scintillation spectrometry.

Alternate Procedure for Individual Aggregate Samples

1.   If thymidine incorporation of individual aggregates is required, gently pipette each aggregate from its collecting syringe into a test tube and bring the total volume to 4 ml by addition of surrounding, particle-free seawater. The 1-ml automatic pipettes whose disposable tips have been snipped off to widen the bore or straight-sided glass or plastic pipettes (5- to 8-mm inside diameter) are best for transferring aggregates without disruption.

2.   Follow steps 2 through 5 in the section above.

3.   After incubation, pour each sample into a 7- or 15-ml tissue grinder containing a large clearance pestle. Gently disrupt the particle for 10 slow, cavitation-free strokes.

4.   Remove 1 ml of sample and preserve in 2% formalin for later determination of bacterial abundance using standard methods (treat aggregates with pyrophosphate followed by sonication to disassociate particle-bound bacteria as described by Velji and Albright,[10] stain with acridine orange, and count by epifluorescence microscopy as described in Hobbie et al.[11]). Since these samples are radioactive they must be handled very carefully. If the aggregates are large, subsamples may also be removed for determination of aggregate dry weight.[12]

5.   Pour the remaining sample into a test tube and rinse the tissue grinder with 1 ml of filtered seawater. Follow steps 6-8 in the section above.

Calculations

Thymidine Incorporation

Calculate the thymidine incorporation of the aggregates as follows:

1.   We assume that the surrounding seawater comprises 100% of each sample volume. Subtract the mean dpm (disintegrations per minute) per filter of the surrounding seawater samples from each aggregate sample to obtain the dpm of aggregates alone. One must correct for volume if the surrounding seawater and aggregate samples are of different volumes. Accurate determination of aggregate volume is not easy. Since aggregates generally take up less than 10% of the volume of the samples, the assumption that surrounding seawater comprises 100% of each sample only slightly overestimates the production of the background seawater. Moreover, thymidine incorporation of the background seawater is generally one or two orders of magnitude lower than that of the aggregates, so the assumption that seawater comprises all of the sample tends to introduce trivial error. If the aggregates appear to occupy more than about 10% of the sample by volume, however, the volume of surrounding seawater relative to the aggregates may be estimated by allowing the pooled sample to settle in the graduated cylinder for about 30 min in the dark. The volume of the surrounding seawater can also be estimated by examination of aggregate samples while still in their collecting syringes. Counts due to background seawater are then adjusted appropriately.

2.   Subtract the mean dpm of killed surrounding seawater controls from killed aggregate controls to obtain adsorption for aggregates alone.

3.   Finally, subtract the mean dpm due to aggregate adsorption from experimental aggregate samples.

4.   Calculate the number of moles of thymidine incorporated $aggregate^{-1} d^{-1}$ from the formula:

$$\text{moles } aggregate^{-1} d^{-1} = (dpm) (SA^{-1}) (4.5 \times 10^{-13})/(NA) (t) \tag{1}$$

where SA is the specific activity of the thymidine solution in Ci mol^{-1}, 4.5×10^{-13} is the number of curies per dpm, NA is the number of aggregates in the sample, and t is the incubation time in days.

Production

Calculate production from thymidine incorporation as follows:

1. The number of bacteria cells produced aggregate^{-1} d^{-1} is calculated from thymidine incorporation with a conversion factor. This conversion factor is in the range 0.85×10^{18} to 1.75×10^{18} cells produced per mol thymidine incorporated^{-1} for free-living marine bacteria (Riemann, et al.[13]; see also Chapter 59). Conversion factors for bacteria attached to marine snow have not been determined. Thus, it may be best to report a range of production values which reflect the range of conversion factors for free-living bacteria. Freshwater aggregate production may be calculated using a conversion factor of 4.5×10^{18} cells mole thymidine^{-1} determined by Kirchman[14] for attached freshwater bacteria.

2. Bacterial secondary production expressed as carbon is calculated from the following equation:

$$\text{Carbon aggregate}^{-1} \text{ d}^{-1} = (CV) (1.21 \times 10^{-7}) (CP) \qquad (2)$$

where CV is the mean volume of an attached cell in μm^3, CP is production in cells aggregate^{-1} d^{-1}, and 1.21×10^{-7} is the mg carbon μm^{-3} of bacterial cell volume from Watson et al.[15] Bacteria attached to marine snow tend to be significantly larger than free-living forms,[16] resulting in higher carbon production per mol of thymidine incorporated than free-living cells. Since aggregates vary in size, parallel measures of bacterial abundance or particle dry weight can be used to standardize production measurements for comparison. For example, carbon production expressed as carbon mg dry weight of aggregate^{-1} d^{-1} would allow more ready comparisons among particles from different days or locations if aggregate size is highly variable.

NOTES AND COMMENTS

1. An isotope dilution study of coastal marine snow[12] indicates that natural concentrations of thymidine in the interstitial fluid of marine aggregates is on the order of 3 nM. Low natural thymidine levels are also indicated by apparent kinetic saturation of marine snow samples at 2 nM thymidine concentrations.[16] Although the concentration of dissolved thymidine within aggregates probably varies considerably with location and among individual aggregates, we selected 20 nM as an adequate concentration to dilute natural concentrations and inhibit thymidine synthesis by bacterial cells.[17] Thymidine incorporation rates of lake snow have never been measured, so these generalizations apply only to marine aggregates.
2. Alldredge and Gotschalk[12] found that the percent of labeled thymidine incorporated into DNA by bacteria attached to marine snow remained fairly constant at 86 to 89% over the first 45 min of incubation. However, at incubation times longer than 45 min, the distribution of label into RNA and protein increased dramatically. Thus, long incubation times are not recommended. Most conversion factors used to calculate cell growth rates from thymidine incorporation assume at least 80% incorporation into DNA. If aggregates are so small or inactive that short incubations do not yield sufficient incorporation for accurate radiolabel measurements, then sample volumes, rather than incubation times, should be increased.
3. If handling radioactive samples for bacterial counts is objectionable, individual aggregates may also be disrupted and subsampled prior to addition of the radiolabel. Disruption can

be beneficial in that it allows the label to fully penetrate the particle. However, it also destroys any specialized microenvironment which may be maintained within the interstices of the aggregate. Gentle grinding does not result in cell breakage or significantly reduce photosynthesis of algae inhabiting marine snow.[7] Effects of disruption on heterotrophic activity have not been determined, but are also probably small.

4.   Thymidine incorporation rates of marine snow measured with this method range from 10^{-13} to 10^{-12} moles aggregate^{-1} d^{-1} for open ocean and neritic marine snow[12,16] and from 10^{-13} to 10^{-11} moles aggregate^{-1} d^{-1} for deep sea marine snow.[9] Incorporation per cell ranges from 10^{-21} to 10^{-19} moles cell^{-1} d^{-1} in surface waters and from 10^{-20} to 10^{-18} moles cell^{-1} d^{-1} at mesopelagic depths.

REFERENCES

1. Paerl, H. W., Bacterial uptake of dissolved organic matter in relation to detrital aggregation in marine and freshwater systems, *Limnol. Oceanogr.,* 19, 966, 1973.
2. Alldredge, A. L. and Silver, M. W., Characteristics, dynamics and significance of marine snow, *Prog. Oceanogr.,* 20, 41, 1988.
3. Fuhrman, J. A. and Azam, F., Bacterioplankton secondary production estimates for coastal waters of British Columbia, Antarctica and California, *Appl. Environ. Microbiol.,* 39, 1085, 1980.
4. Fuhrman, J. A. and Azam, F., Thymidine incorporation as a measure of heterotrophic bacterioplankton production in marine surface waters: evaluation and field results, *Mar. Biol.,* 66, 109, 1982.
5. Hamner, W., Underwater observations of blue-water plankton: logistics, techniques and safety procedures for divers at sea, *Limnol. Oceanogr.,* 20, 1045, 1975.
6. Heine, J. N., Blue water diving guidelines, California Sea Grant College Program Publication No. T-CSGCP-014, Institute of Marine Resources, University of California, La Jolla, 1986.
7. Gotschalk, C. C., Enhanced primary production and nutrient regeneration within aggregated marine diatoms: implications for mass flocculation of diatom blooms, Masters thesis, University of California, Santa Barbara, 1988, 49.
8. Silver, M. W. and Alldredge, A. L., Bathypelagic marine snow: vertical transport system and deep-sea algal and detrital community, *J. Mar. Res.,* 39, 501, 1981.
9. Alldredge, A. L. and Youngbluth, M. J., The significance of macroscopic aggregates (marine snow) as sites of heterotrophic bacterial production in the mesopelagic zone of the subtropical Atlantic, *Deep-Sea Res.,* 32, 1445, 1985.
10. Velji, M. J. and Albright, L. T., Microscopic enumeration of attached marine bacteria of seawater, marine sediment, fecal matter, and kelp blade samples following pyrophosphate and ultrasound treatment, *Can. J. Microbiol.,* 32, 121, 1986.
11. Hobbie, J. E., Daley, R. J., and Jasper, S., Use of nuclepore filters for counting bacteria by fluorescence microscopy, *Appl. Environ. Microbiol.,* 33, 1225, 1977.
12. Alldredge, A. L. and Gotschalk, C., The relative contribution of marine snow of different origins to biological processes in coastal waters, *Cont. Shelf Res.,* 10, 41, 1990.
13. Riemann, B., Bell, R. T., and Jorgensen, N. O. G., Incorporation of thymidine, adenine, and leucine into natural bacterial assemblages, *Mar. Ecol. Prog. Ser.,* 65, 87, 1990.
14. Kirchman, D., The production of bacteria attached to particles suspended in a freshwater pond, *Limnol. Oceanogr.,* 28, 858, 1983.
15. Watson, S. E., Novitsky, J., Quinby, H. L., and Valois, F. W., Determination of bacterial number and biomass in the marine environment, *Appl. Environ. Microbiol.,* 33, 940, 1977.
16. Alldredge, A. L., Cole, J., and Caron, J. A., Production of heterotrophic bacteria inhabiting organic aggregates (marine snow) from surface waters, *Limnol. Oceanogr.,* 31, 68, 1986.
17. Pollard, P. C. and Moriarty, D. J. W., Validity of the tritiated thymidine method for estimating bacterial growth rates: measurement of isotope dilution during DNA synthesis, *Appl. Environ. Microbiol.,* 48, 1076, 1984.

CHAPTER 62

Bacterial Growth Rates Measured by Pulse Labeling

Paul LaRock and Jung-Ho Hyun

INTRODUCTION

In order to probe rate processes in microbial systems, radioisotopes offer a convenient and extremely sensitive means of quantifying the synthesis of individual cellular components which are related to community growth. Radioisotopes are far more sensitive than chemical assays, can be used on small samples, and provide information on the rate of formation of newly synthesized material in the presence of old or preexisting cellular components. In order to perform successful labeling experiments, one must have some insight as to how the label is used and incorporated by the cell, what molecules to label, and what isotopes to use.

Ideally, one would choose a labeled compound that participates in a particular biosynthetic pathway of interest and has a specific endproduct in which the label may be found. Amino acids (^{3}H-leucine and ^{35}S-methionine) have been used in studies of protein synthesis and ^{3}H-adenine in experiments on nucleic acid production. General labeling, with such compounds as ^{14}C-labeled sugars or inorganic ^{32}PO$_4$, may have application in particular instances, but would require that additional chemical separations be used to isolate compounds of interest. Redalje and Laws,[1] for example, used ^{14}CO$_2$ in algal cultures and then employed chromatographic techniques to isolate labeled chlorophyll and determine its rate of production.

When one adds a radiolabel to a growing culture of bacteria, the rate at which it appears in the population depends on the metabolic role of the target molecule. Compounds that are rapidly consumed, such as amino acids used in the synthesis of protein, will quickly attain a specific activity comparable to that in the medium, usually within a small fraction of a generation time. Macromolecules such as DNA, RNA, or membrane lipids, however, will only accumulate radiolabel as a function of growth, and will require several generations before isotopic uptake and growth have equilibrated, i.e., before the growth rate of the bacteria and the rate of radiolabel incorporation become equal. Such a labeling pattern indicates metabolic stability[2] and that these molecules will not turnover within the time frame of most experiments. We can illustrate this concept by performing a time-course experiment in which we add ^{3}H-adenine to a water sample, remove aliquots periodically, and follow the rate at which the nucleic acids are labeled (Figure 1). Assuming that all of the cells in the sample are actively growing (this is not necessarily the case), we find that after one doubling, 50% of the population has accumulated radiolabel; after two doublings,

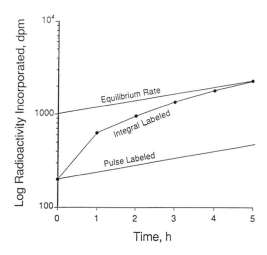

Figure 1.   Macromolecule radiolabeling pattern using integral-labeling and pulse-labeling approaches. In this hypothetical example the culture has a doubling time of one hour, and it requires from four to five generations before the rate of radiolabel incorporation approaches the true culture growth rate (or equilibrium rate). In the pulse-labeling experiment, label incorporation accurately follows growth, although the level of radioactivity is considerably lower. See text for details.

75% of the cells are labeled, and so forth until by the fourth generation 94% of the community has acquired ^{3}H-adenine. From this point on, the increase in cellular radioactivity is equivalent to the growth rate, but prior to isotopic equilibration, radioactivity in the culture appears at the rate that is about 2.5 times faster than culture growth because of the mixed kinetics of growth and isotope incorporation.[3] Even in communities that are declining, radioactivity initially appears to increase.

Another consideration in growth experiments is the manner in which the numbers of bacteria within the community change. Usually we think of microbial growth as an exponential process,[4,5] but models have been developed in which linear increases in cell numbers are used to describe microbial development.[6,7] Which approach is correct, or are there conditions when growth may shift from logarithmic to a linear pattern?

This disparity of growth models, the effects of isotopic equilibrium, and death or decline within a population can be resolved unequivocally by generating the true growth curve of the community and visualizing the actual pattern of development (logarithmic or linear) and whether the population is growing or dying and the effect that various treatments have on growth. To do this, one uses a pulse-labeling technique which is free from errors resulting from lack of isotopic equilibrium, knowledge of specific activity of the isotopic pools, and the shape of the growth curve.

In a pulse-labeling experiment, one takes and maintains a relatively large sample (4 to 20 L) at *in situ* temperature, and periodically withdraws subsamples to measure the increase or decrease in community size as represented by changes in macromolecule synthesis (nucleic acids or phospholipid). The fundamental tenet is that the growth rate remains the same in the isolated material as it was in the environment, at least for some finite period (the pulse-labeling method is capable of detecting changes in growth rate should they occur). In a typical experiment, a radiolabel is added to aliquots of the sample for a brief interval, after which the incubation is terminated, the cellular target that was labeled is extracted, and the incorporated label is counted. A second aliquot is treated in the same fashion at a later time interval, usually one hour after the previous pulse, and then a third aliquot an hour later, and so forth until five to six data points are generated. If the pulse period is kept constant, the amount of isotope incorporated per cell will be the same for each pulse period. The number of cells in the sample, however, will have increased (or decreased) with time, and we have a means of following the rate of production of the cellular component, which is equivalent to the growth (or death) rate of the community.

Most labeling experiments are not done by pulsing, but rather by integral labeling of a sample by adding the radioisotope at the start of an experiment and withdrawing aliquots for analysis over some time period usually specified in hours. This is the method used in the hypothetical time-course experiment described earlier (Figure 1, "Integral Labeled"). As noted earlier, in the early phases of an integral-labeling experiment radiolabel uptake is from two to three times greater than the actual growth rate. In a pulse-labeling experiment, this error is overcome because each aliquot is pulsed for exactly the same time duration so that the isotope incorporated will be a constant fraction of the equilibrium value, i.e., the error is consistent in all samples and thus, cancels itself. the changes that are observed in radioactivity are solely the result of growth (or death) and not the rate at which the isotope appears in the population. As illustrated in Figure 1, the results of a pulse-labeling experiment will exactly parallel the equilibrium rate, although the amount of label incorporated is considerably less. Furthermore, the question of specific activity is of no concern in growth-rate determinations, as the relative proportion of label incorporation is consistent in all pulsed aliquots.

In theory, pulse labeling can be used with any cellular constituent provided the turnover of that constituent is sufficiently longer than the generation period, and the labeled compound can be extracted and purified. Macromolecules fulfill these requirements, and represent approximately 96% of cellular dry weight, 80% of which is protein and RNA, 17 to 18% is lipid and 5 to 6% is DNA.[8] To date we have used tritiated adenine (labels both DNA and RNA), tritiated uridine (labels RNA), ^{125}I-iododeoxyuridine (labels DNA), and inorganic $^{32}PO_4$ (labels phospholipids or membranes) to determine growth rates, all with equivalent results.[3] In fact, it is this ability to label several different markers to corroborate the end result, that makes this approach particularly useful. Interestingly, thymidine does not work and significantly underestimates growth rate.

Three peices of information are obtained in a typical pulse-labeling experiment: (1) the absolute growth (or death) rate of the active community, free of any assumptions or qualifications; (2) the relative size of the growing component; and (3) the shape of the growth curve itself (i.e., is it logarithmic or linear?). The growth-rate constant, μ, (the slope of the growth curve) is most important because it enables one to determine cause-and-effect relationships, such as found in ocean frontal systems, upwelling areas, groundwater or polluted areas (see Note 1). The relative size of the active component is easily determined from the Y-intercept of the growth curve, and the shape of the growth curve reveals whether community development is exponential or linear (linear growth is occasionally observed in coastal environments and groundwater systems which may be interpreted as a growth-limiting condition).

MATERIALS REQUIRED

Equipment

- Water bath or other constant-temperature incubator
- Ice bath/bucket
- Vacuum pump for filtration

- Filtration apparatus: in the course of an elaborate pulse-labeling experiment many hundreds of membrane filters will be used, which necessitates having numerous filter holders. Commercially available manifolds holding either three or six filter units can be used, although we recommend a minimum of six filtering units (this allows the simultaneous filtering of three duplicate pulsed samples). We use all-glass holders accepting 25-mm diameter filters, and having 150-ml funnels (available from Micro Filtration Systems, Dublin, CA).

Supplies

- Cubitainers
- 100-ml tricornered polypropylene beakers
- 125-ml polycarbonate screw-cap bottles
  Membrane filters: as a compromise between collection efficiency and filtration rate, we have chosen to use 0.3-μm pore size cellulosic filters (Millipore Corporation, Bedford, MA) which can be used for either nucleic acid or ^{32}P-phospholipid labeling; polycarbonate filters can only be used for labeling nucleic acids, as $^{32}PO_4$ seems to "stick" to the filters, thus, masking the actual uptake.
- Liquid scintillation vials
- Pasteur pipettes

Solutions

- Appropriate radioisotope
- 10% trichloroacetic acid (TCA)
- Methanol
- Methylene chloride
- Distilled water
- 95% ethanol
- Liquid scintillation fluid

PROCEDURES

General Laboratory Procedures

The following outline details the various steps in carrying out a pulse-labeling experiment. The temporal sequence of events in a typical pulse-labeling experiment is depicted graphically in Figure 2. Various treatments may be performed once the sample is collected depending on the application at hand, but the general pulsing procedure will follow the format given below:

1. Collect sample in a Cubitainer and maintain at *in situ* temperature for the duration of the experiment (see Note 2). Record the time of collection; this is time zero.

2. Remove duplicate 100-ml aliquots (see Note 3), transferring them to 125-ml polycarbonate screw-cap bottles for radioisotope addition.

3. From a stock solution of the appropriate radioisotope (see Note 4), add the label to each of the replicates, and record the time of the pulse (i.e., when the isotope is added).

4. Incubate the pulsed aliquots at the *in situ* temperature for 30 min, then collect the labeled bacteria on a membrane filter (see Note 5).

5. Purify the labeled material using an appropriate extraction technique, as described below.

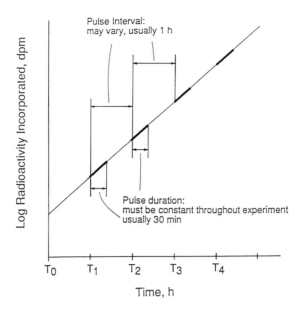

Figure 2. A schematic representation of a pulse-labeling experiment. Note that the pulse periods must be kept constant, although the interval between the pulses may be varied during an experiment as long as the time of pulse is recorded.

6. The radioactive nucleic acids may be precipitated using the technique of Kennell[9] (see Note 6).

   - Ice-cold 10% TCA is first passed through the filter and then excess (about 5 ml) is added to the filter holder with the vacuum off and allowed to stand for 7 to 10 min.

   - The TCA is then drawn through the filter, and is followed by three washes of 95% ethanol (about 5 to 6 ml each) to remove residual TCA and prevent quenching in the subsequent liquid scintillation counting (LSC).

   - The filter is then placed in an LSC vial for counting in the laboratory.

7. Extracting labeled phospholipids is considerably more labor intensive, although the initial measures taken in the field are relatively easy. The technique used is that of McKinley et al.[10] and is a miniaturized version of the Blight and Dyer[11] solvent extraction (see Note 7).

   - The $^{32}PO_4$ labeled cells are collected on cellulosic membrane filters (polycarbonate filters are to be avoided as they yield exceptionally high backgrounds) and placed in an LSC vial to which 1 ml of a mixture of methanol and methylene chloride (2:1) is added.

   - After an hour or so, the solvent is allowed to evaporate, the vials sealed, and returned to the lab for complete extraction.

   - In the laboratory, 7.5 ml of the methanol-methylene chloride solution is added to each LSC vial and allowed to stand for at least 2 h, preferably longer.

   - Next, 2.5 ml of methylene chloride and 2.5 ml of distilled water are added, the mixture shaken, and the phases allowed to separate.

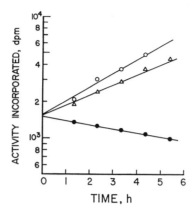

Figure 3.   The effects of mixing water masses of different growth characteristics. Equal volumes of
seawater taken from the continental shelf (open circles) and the Gulf Stream (solid circles)
were mixed (triangles) and the various growth rates measured by pulse labeling. The mixture
acquired the growth characteristics of the parent samples. Note the pulse labeling is the only
method that can assess negative growth rates like that found in the Gulf Stream sample.

- The upper aqueous phase is carefully and completely aspirated off with a Pasteur pipette
  connected to a solvent trap attached to a vacuum source.

- The organic phase is transferred to a clean LSC vial and allowed to air dry. Scintillation
  cocktail is added and the activity counted.

8.   Repeat steps 2 through 7 until the desired number of data points have been generated,
     usually five or six (see Note 8).

9.   Measure the incorporated radioactivity and plot the resultant counts (actually use the
     distintegrations per minute, correcting for quench) on a semi-logarithmic ordinate, using
     the elapsed time between sample collection and pulse additions as the abscissa. Fit a curve
     to the data, noting the slope of the curve (the growth-rate constant), the Y-intercept (see
     Note 9), and the correlation coefficient. Alternatively, one can use any of a number of
     plotting programs to fit the data and determine the respective equation coefficients.

Applications

   The pulse-labeling technique is particularly useful in establishing cause-and-effect relationships that might exist owing to changes in conditions that may arise through the addition
of nutrients and toxicants, or by seasonal events. Some of the salient uses of the pulse-
labeling technique can be best illustrated by specific applications in which the effect of
mixing water masses and bacterial predation were probed. Ocean frontal systems are those
areas between different water masses that are characterized by intense horizontal, chemical,
and thermal gradients. One example is the boundary between the colder, lower salinity,
nutrient-rich shelf water and the Gulf Stream across which some mixing takes place as rings
are formed. To illustrate the effects of mixing two water masses, an experiment was contrived
in which the microbial growth rates in samples of shelf water, a developing Gulf stream
ring, and a 1:1 mixture of the two were determined. The results of this experiment are seen
in Figure 3, in which we see that the shelf water had a positive and rapid growth rate (μ
= 0.28, or a doubling time of 2.48 h). In contrast, the Gulf stream ring had a negative
growth rate (μ = -0.08, or a half time of 8.66 h) and a population that was declining.
An equal mixture of the two waters yielded a curve, which as expected, had growth char-

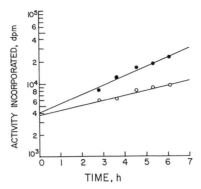

Figure 4. Assessment of bacterial grazing rate by determining the rate of bacterial increase in the presence (open circles) and absence (solid circles) of predators. Details are provided in the text.

acteristics that were an average of the individual water masses ($\mu = 0.20$, or a doubling time of 3.47 h). The significant points to note are that, (1) decaying or declining populations can be followed and (2) the growth rate of any given sample reflects the average development of the community.

With appropriate experimental manipulations, pulse labeling can be used to evaluate bacterial grazing rates. A considerable literature has evolved concerning bacterial consumption by protozoa and the various means used to access the process, usually with epifluorescence microscopy.[12-14] Because pulse labeling measures the net community growth rate, it is possible to design an experiment that measures growth in untreated samples and compares this to the growth rate in a sample from which the grazers had been removed. Since predation is a removal process, it is analogous to the decay curve for the Gulf Stream sample in the previous example, and as such, a sample containing predators should have a lower growth rate than the same sample from which the predators had been removed.

An example of how pulse labeling may be used to determine grazing pressure is given in Figure 4, which was performed in coastal waters off Florida in the Gulf of Mexico. Predators were removed by passing a portion of the sample through a 3.0-μm polycarbonate membrane filter, and then performing pulse-labeling experiments on the treated (no predators) and untreated fractions. Time zero in this instance was when the sample was split and filtered, and the identical Y-intercept for both growth curves indicates that the population in both fractions was equal at the start of the experiment (i.e., filtering did not remove bacteria). The raw sample had a growth-rate constant of 0.14 h^{-1} compared to 0.29 h^{-1} for the predator-free water, indicating predation essentially reduced bacterial production by 50%. For an initial bacterial density of 10^6 cells ml^{-1}, the removal will be about 1.1×10^5 cells ml^{-1} h^{-1}, a figure that agrees quite well with the values of Sherr and Sherr.[14]

NOTES AND COMMENTS

1. The slope of the growth curve is actually the rate constant in the growth equation:

$$N_t = N_0\, e^{\mu t} \qquad (1)$$

where N_t is the number of cells after t hours, N_0 is the number of cells originally present, t is the elapsed time, and μ is the growth-rate constant of the culture, and generally has the unit of reciprocal hours (h^{-1}). Alternatively, the growth-rate constant may be expressed as doubling time by the following expression:

$$\text{Doubling time} = \ln{(2)}/\mu = 0.693/\mu \tag{2}$$

and has the unit of hours if μ is expressed in reciprocal hours.

2. Sufficiently large samples should be taken so that the volume will not change dramatically with the removal of aliquots for pulsing. The most versatile sampling vessels are collapsible polyethylene Cubitainers (Hedwin Corporation, Baltimore, MD) in either the 4- or 20-l size, which fold for convenient shipping, and can be compressed to expel air when dealing with oxygen-sensitive substrates such as sulfide.

3. When dealing with large numbers of samples all requiring measurement of a constant fixed volume, it is more convenient to fill a measuring container to capacity and use this as the aliquot volume rather than measuring out precisely 100 ml each time. This is particularly true in long experiments where fatigue becomes a factor. We have found that 100-ml, tri-cornered polypropylene beakers are extremely uniform and will deliver 120 ml consistently. The additional volume is compensated for in subsequent calculations dealing with radioactivity incorporated per unit volume.

4. Typically, a stock solution of the radioisotope is prepared containing one μCi ml^{-1} and 1 ml of this solution is used to pulse the sample aliquots. Sufficient stock is prepared just before the experiment, and quadruplicate 25 μl are withdrawn and counted to accurately determine the radioactivity that has been added.

5. The pulse duration is that time period during which the subsamples accumulate the radioisotope, and must be kept constant throughout an experiment. While the choice of a pulse time is somewhat arbitrary, we have found 30 min to be most convenient as it allows for a reasonable quantity of label to be accumulated. With longer labeling periods more radioisotope is incorporated, but experiments run longer, and phytoplankton will take up labels such as ^{3}H-adenine which they do not in shorter periods.

6. In theory it is possible to use any cellular constitutent for labeling provided that its turnover time is sufficiently long (i.e., the compound is stable), and that the labeled material can be extracted and purified. Most macromolecules satisfy these demands and constitute approximately 96% of the total cellular dry weight.[8] Nucleic acids are particularly convenient for labeling since their separation chemistry is relatively easy and straightforward. One can use ^{3}H-adenine in pulse labeling, and unlike production estimates, it is unnecessary to differentiate between the RNA and DNA. Radiolabels that are applicable for total nucleic acid labeling are ^{3}H-adenine, ^{3}H-uridine, and ^{125}I-iododeoxyuridine (a DNA-specific analog of thymidine), all of which yield equivalent growth rates. Typically, 1 μCi of activity is added to each aliquot, regardless of its actual volume. Interestingly, ^{3}H-thymidine consistently underestimates growth by a factor of four, and should be avoided in pulse-labeling experiments until the reasons for this varience are understood.

7. The use of ^{32}PO$_4$ to label phospholipid (cell membranes) has also been satisfactorily used in pulse-labeling experiments, and has the advantage of using inorganic chemical species to assess bacterial growth. Because of the longer sample processing time and sort half-life of ^{32}P, we normally add 4 μCi to each pulsed aliquot.

8. Unlike the pulse duration, the time interval between pulses is not critical. Experience with oceanic samples indicates doubling times on the order of three hours, and that once a sample has been collected, it will only undergo two or three doublings before changing growth rate. If one wishes to generate five or six data points, then pulses should be performed hourly. In practice it may be necessary to vary the pulse interval depending on the work load, but a 30 min pulse every hour is a good point of beginning.

9. The relationship of total bacterial number in a sample to the radioactivity incorporated (the Y-intercept) in a given experiment is shown in Figure 5. It is this direct relationship that enables one to use the isotopic uptake as a relative indicator of active biomass. In

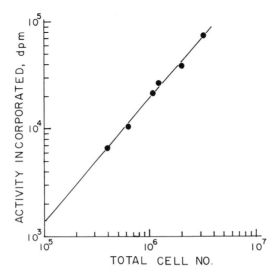

Figure 5.   The relationship of radioactivity incorporated in a pulse-labeling experiment (the Y-intercept of the growth curve) to the total number of actively growing bacteria in the sample.

order for such an assessment to be meaningful, it is necessary to use the same level of radioactivity for each sample, and that the starting point for the resultant growth curves is the time the sample was actually collected, and not the time when laboratory manipulations are begun.

REFERENCES

1. Redalje, D. G. and Laws, E. A., A new method for estimating phytoplankton growth rates and carbon biomass, *Mar. Biol.,* 62, 73, 1981.
2. Neidhardt, F. C., Ingraham, J. L., and Schaechter, M., *Physiology of the Bacterial Cell,* Sinauer Associates, Sunderland, MA, 1990, 506.
3. LaRock, P. A., Schwarz, J. R., and Hofer, K. G., Pulse labelling: a method for measuring microbial growth rates in the ocean, *J. Microbiol. Methods,* 8, 281, 1988.
4. Li, W. K. W., Microbial uptake of radiolabeled substrates: estimates of growth rates from time course measurements, *Appl. Environ. Microbiol.,* 47, 184, 1984.
5. Taylor, C. D. and Jannasch, H. W., Subsampling technique for measuring growth of bacterial cultures under high hydrostatic pressure, *Appl. Environ. Microbiol.,* 32, 355, 1976.
6. Wright, R. T. and Coffin, R. B., Factors affecting bacterioplankton density and productivity in salt marsh estuaries, in *Current Perspectives in Microbial Ecology,* Klug, M. J. and Reddy, C. A., Eds., American Society for Microbiology, Washington, D.C., 1984.
7. Fuhrman, J. A. and Azam, F., Bacterioplankton secondary production estimates for coastal waters of British Columbia, Antarctica and California, *Appl. Environ. Microbiol.,* 39, 1085, 1980.
8. Ingraham, J. L., Maaloe, D., and Neidhart, F. C., *Growth of the Bacterial Cell,* Sinauer Associates, Sunderland, MA, 1983, 435.
9. Kennell, D., Use of filters to separate radioactivity in RNA, DNA and protein, in *Methods in Enzymology Vol. XII, Part A,* Grossman, L. and Moldave, K., Eds., 1967, 686.
10. McKinley, V. L., Federle, T. W., and Vestal, J. R., Effects of petroleum hydrocarbons on plant litter microbiota in an arctic lake, *Appl. Environ. Microbiol.,* 43, 129, 1982.
11. Bligh, E. G. and Dyer, W. J., A rapid method of total lipid extraction and purification, *Can. J. Biochem. Physiol.,* 37, 911, 1959.

12. Fuhrman, J. A. and McManus, G. B., Do bacteria-sized marine eukaryotes consume significant bacterial production?, *Science,* 224, 1257, 1984.

13. Gonzalez, J. M., Sherr, E. B., and Sherr, B. F., Size-selective grazing on bacteria by natural assemblages of estuarine flagellates and cilliates, *Appl. Environ. Microbiol.,* 56, 583, 1990.

14. Sherr, E. B. and Sherr, B. F., High rates of consumption of bacteria by pelagic ciliates, *Nature,* 325, 710, 1987.

CHAPTER 63

Utilization of Amino Acids and Precursors for Amino Acid *De Novo* Synthesis by Planktonic Bacteria

Meinhard Simon

INTRODUCTION

Dissolved free (DFAA) and combined amino acids (DCAA; mostly proteins and peptides) are important substrates for growth of heterotrophic planktonic bacteria since protein comprises 60% of their biomass.[1] If the carbon and nitrogen requirements are not met by DFAA and DCAA *de novo* synthesis from precursors such as ammonium and a carbon source, preferentially carbohydrates, occurs.[2] Uptake of DFAA by bacteria is measured directly by multiplying the turnover rate of ^{3}H-DFAA times the concentration of DFAA as measured by high-performance liquid chromatography (HPLC).[3] Sources of precursors for free amino acids in the intracellular pool are estimated from the mol% distribution[4] and isotope dilution (ID) of intracellular amino acids (Figure 1). (ID of intracellular amino acids is the ratio of the specific activity of amino acids in the water over the specific activity of amino acids in the intracellular pool.) If glutamic acid dominates the pool and the ID of glutamic acid is significantly higher than the ID of the most abundant DFAA, ammonium and a carbon source are the predominant substrates for bacterial growth besides DFAA.[5] If glutamic acid is not dominating the bacterial pool and ID of the most abundant amino acids are equally high, DCAA are the most important substrates for bacterial growth.[6] If ID of all amino acids is low, DFAA are the major substrates for bacterial growth.

Measuring uptake rates of DFAA is attractive because it is a sensitive method by which a major fraction of substrates for growth of planktonic bacteria is often assessed.[7,8] Further, this method allows for the estimation of how fast the DFAA pool is turning over. Determining the mol% distribution and ID of intracellular amino acids in one approach gives a rapid indication of whether ammonium and a carbon source or DCAA are other relevant precursors of amino acids for biosynthetic requirements. This approach, however, yields rather semi-quantitative results than accurate uptake rates of the various amino acid precursors.

A drawback of measuring DFAA uptake rates and intracellular ID of amino acids is that amino acid concentrations are determined by HPLC which is a very sensitive method, though rather sophisticated with an expensive and delicate instrument.[1,3,6,7,9]

0-87371-564-0/93/$0.00 + $.50

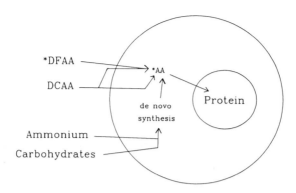

Figure 1.   Schematic diagram of possible pathways of amino acids and precursors for amino acid *de novo* synthesis into the intracellular bacterial pool. Outside the large circle is the extracellular environment of a bacterium and inside is its intracellular pool and the protein fraction. *DFAA, tritiated dissolved free amino acids with a given specific activity, and *AA, tritiated amino acids in the intracellular bacterial pool with a specific activity < that of *DFAA. (From Simon, M. and Rosenstock, B., *Limnol. Oceanogr.*, 37, 1496, 1992. With permission.)

MATERIALS REQUIRED

Equipment

- Filtration manifolds for 25- and 47-mm filters
- Vacuum pump
- Liquid scintillation counter
- HPLC for amino acid analysis after pre-column derivatization with *o*-phthaldialdehyde
- Heating block
- Ice chest
- Fraction collector

Supplies

- Plastic test tubes (10, 20, or 50 ml; polycarbonate or polypropylene)
- Scintillation vials
- 10-ml glass syringe
- 0.2-μm filter kits (low protein binding capacity [polysulfon], e.g., Gelman Acrodisc #4192)
- 0.45-μm 25-mm cellulose nitrate filters (Sartorius or Millipore)
- Nuclepore filter 47-mm (1.0- and 0.2-μm)
- Several 100- to 250-ml flasks (plastic or glass)
- 10- to 20-ml glass vials

Solutions and Reagents

- ³H-amino acid mixture (working solution, see notes)
- Formalin (buffered at pH 7.0)
- 5% ice-cold trichloroacetic acid (TCA)
- 0.2-μm prefiltered sample water
- Ethylacetate
- Scintillation fluor (e.g., Beckman Readysafe)
- HPLC-grade water

PROCEDURES

Uptake Rates of DFAA

1. Take a water sample with a well prerinsed sampler and avoid contaminating it by amino acids (wear gloves since your skin releases amino acids).

2. Transfer the water into a clean bottle. Withdraw 10 ml for HPLC analysis of DFAA.

3. Analysis is best done immediately. For later analysis, prefilter 6 to 8 ml with a 10-ml glass syringe combusted at 550°C and connected to a prefilter kit (0.2-μm pore size; low protein-binding capacity) into a precombusted glass vial. This can be stored frozen at −20°C until analysis.

4. For DFAA uptake measurements withdraw 5 to 50 ml from the sample (depending on the activity) in quadruplicates into plastic test tubes (glass tubes may release heavy metals or adsorb nutrients, particularly in oligotrophic waters).

5. Add ³H-amino acids at tracer amounts with a final concentration of <10% of the ambient DFAA concentration (usually 0.5 to 2 nM ³H-amino acids final concentration is okay).

6. Pipette the same amount into a scintillation vial to measure the added radioactivity.

7. Fix one parallel immediately as a control with buffered formalin (pH 7; 1 to 2% final concentration).

8. Incubate all samples preferentially at *in situ* temperature in the dark for 30 to 60 min depending on the activity of the bacteria. Make sure in pre-experiments that uptake is linear over the incubation time.

9. Stop the incubation by adding buffered formalin for later filtration or by filtering the sample through a 0.45-μm, 25-mm diameter cellulose membrane filter (<0.3 bar vacuum).

10. After filtration, rinse the filter twice with particle-free sample water and put the filter into a scintillation vial.

11. For dissolving the filter add 0.75 to 1 ml of ethylacetate and after 2 min scintillation fluor.

12. Wait several hours to reduce chemiluminescence and count the sample in a liquid scintillation counter.

13. Data evaluation:

$$\text{Turnover rate (h}^{-1}) = (\text{dpm}_{\text{sample}} - \text{dpm}_{\text{control}})/\text{dpm}_{\text{added}} \qquad (1)$$

$$\text{Uptake rate (nmol l}^{-1}\text{ h}^{-1}) = [\text{DFAA}] \times \text{turnover rate} \qquad (2)$$

From the C:N ratio of the amino acids uptake rates can be converted to g C or N which is more convenient for comparison with bacterial production.

Instead of DFAA uptake rates comprising uptake into the intracellular pool and the macromolecules, mostly the protein fraction, DFAA incorporation rates into total bacterial macromolecules can be determined. Therefore, the bacterial pool is extracted and the macromolecules precipitated by 5% ice-cold TCA prior to filtration. This is done most conveniently by filtering the sample onto the filter, chilling it with ice-cold, prefiltered water, turning off the vacuum, and adding 5 ml of 5% ice-cold TCA. After a 5-min extraction the TCA is removed by filtration, the filter rinsed again with TCA and processed further as described above.

Isotope Dilution of Intracellular Amino Acids

1. Be very careful not to contaminate the sample with amino acids. It is best to wear gloves for the whole experiment. Use very clean glassware, if possible pre-combusted at 550°C.

2. Take the sample used also for measuring the DFAA uptake rate and pre-filter it through 0.6- to 1.0 μm Nuclepore filters (47-mm diameter) by very gentle vacuum or best, by gravity. This is to remove nonbacterial organisms which have to be excluded for extracting the intracellular pool of the bacterioplankton. The total volume needed should contain a total of approximately 7×10^8 bacteria.

3. Subdivide the sample into triplicates with equal volumes into plastic or glass bottles so that each subsample contains about 2×10^8 bacteria. This number of bacteria is needed to get measurable amounts of tritium and concentrations of amino acids in the intracellular pool.

4. Add ^{3}H-amino acids at the same final concentration as for the DFAA uptake experiment to each triplicate (0.5 to 2 nM final concentration).

5. Withdraw 5 ml from the samples into a combusted glass vial for later analysis of the specific activity of the added ^{3}H-amino acids.

6. Incubate the samples for about the same time as above at ambient temperature in the dark. Incubation is stopped by rapid filtration of the sample through a 0.2-μm Nuclepore filter (47-mm diameter).

7. For extracting the intracellular pool dip the filter quickly into 3 ml of boiling HPLC-grade water.

8. After a 4-min extraction, place the extracted sample on ice. An extraction time >4 min leads to protein hydrolysis and an amino acid mol% distribution significantly different than that of the intracellular pool. (The mol% distribution of the pool amino acids is a good check for the reliability of the extraction). The extracted samples are ready for analysis or kept frozen at −20°C until analysis.

9. The amino acid concentration and mol% distribution is analyzed by HPLC. The eluted sample is collected in fractions of 1 ml (flow rate 1 ml min^{-1}) with a fraction collector and plastic scintillation vials.

10. After adding scintillation fluor measure the radioactivity of the samples with a liquid scintillation counter.

11. Data evaluation: relate the amino acid concentration (pmol sample^{-1}) to the radioactivity of the respective fraction (dpm or Ci sample^{-1}). Subtract the lowest dpm of all fractions of the HPLC run as a blank from the individual amino acid fractions and calculate the specific activity (SA) of all individual fractions as follows:

$$\text{Specific activity (dpm/mmol)} = (\text{dpm sample}^{-1})/(\text{pmol sample}^{-1}) \qquad (3)$$

$$\text{Intracellular isotope ilution (ID)} = SA_{extra}/SA_{intra} \qquad (4)$$

where SA_{extra} = specific activity of the individual amino acids in the extracellular sample, and SA_{intra} = specific activity of the individual amino acids in the intracellular pool.

12. Interpretation of the results: the mol% distribution of amino acids in the intracellular pool should be intermediate between the mol% distribution of DFAA and the bacterial protein.[6] If ammonium is the major N source, glutamic acid dominates the pool by >35%.[4-6] In this case, the mol% of glutamic acid is not intermediate between that of the DFAA pool and bacterial protein.

The ID can only be >1, ID = 1 means that all tritium of the respective amino acid in the intracellular pool entered via uptake of DFAA, and ID >1 means that the SA of a given amino acid in the intracellular pool has been diluted by an unlabeled fraction which entered the intracellular pool as DCAA or was synthesized *de novo*. If ID of the major amino acids in the DFAA pool is low (i.e., 1 to 3) this means that between 33 and 100% of the amino acids in the intracellular pool entered via uptake of DFAA. If ID of the major amino acids in the DFAA pool is high (>10) and ID of glutamic acid is not substantially higher, this means that the major part of these amino acids (>90%) did not enter via uptake of DFAA but via hydrolysis and coupled uptake of DCAA through uptake systems independent of DFAA uptake. If glutamic acid dominates the intracellular pool and its ID is substantially higher than that of other amino acids, the major source of amino acids in the intracellular pool besides DFAA is *de novo* synthesis from ammonium and a carbon source.

A high ID could also occur if (1) only a small fraction of the planktonic bacteria takes up amino acids (a large fraction is nonresponsive) but the majority of the bacteria keeps up a normal intracellular amino acid pool which is extracted and (2) the ^{3}H-amino acids taken up do not mix with the total intracellular amino acid pool but remain compartmentalized. A low ID of a few ^{3}H-amino acids (1 to 2) added at tracer concentrations, however, ensures that both objections are irrelevant because such low ID factors can only occur when the majority of the bacteria takes up amino acids[10] and no compartmentalization occurs. Further proof is to add the amino acid mixture or a single amino acid at saturating concentrations to maximize uptake of DFAA. This usually results in a low ID.[1,5,6]

NOTES AND COMMENTS

1. ^{3}H-amino acids are purchased as a mixture from Amersham, New England Nuclear, or ICN, or can be made up from individual ^{3}H-amino acids representing the most abundant DFAAs. Prepare a working solution from the stock so that a convenient volume is added (i.e., 20 µl) to the sample. Instead of ^{3}H-amino acids, ^{14}C-amino acids can be used but they have a lower specific activity, resulting in a lower sensitivity. However, respiration can be measured conveniently with ^{14}C-amino acids.[11]

2. 0.45-µm filters have a nominal pore size greater than the smallest planktonic bacteria. However, they retain at least as much radioactivity as 0.2-µm filters but have a better filtration rate. Only in very oligotrophic water with minute bacteria may 0.2-µm filters yield a higher activity.

3. If an HPLC is not available, total DFAA concentration can be determined as primary amines after *o*-phthaldialdehyde derivatization with a fluorometer.[12] Ammonium as a contaminant has to be substracted from the measured concentration, after it is separately determined.

4. The isotope dilution assay is also used to determine ID of ^{3}H-leucine for the conversion factor of bacterial production measurements (Reference 1; see also Chapter 58). The only difference to the protocol described here is that instead of a ^{3}H-amino acid mixture, ^{3}H-leucine at saturating concentration is added to the sample.

REFERENCES

1. Simon, M. and Azam, F., Protein content and protein synthesis rates of planktonic marine bacteria, *Mar. Ecol. Prog. Ser.*, 51, 201, 1989.
2. Kirchman, D. L., Limitation of bacterial growth by dissolved organic matter in the subarctic Pacific, *Mar. Ecol. Prog. Ser.*, 62, 57, 1990.
3. Lindroth, P. and Mopper, K., High performance liquid chromatographic determination of sub-picomole amounts of amino acids by precolumn fluorescence derivatization with *o*-phthaldi-aldehyde, *Anal. Chem.*, 52, 1667, 1979.
4. Tempest, D. W., Meers, J. M., and Brown, C. M., Influence of environment on the content and composition of microbial free amino acid pools, *J. Gen. Microbiol.*, 64, 171, 1970.
5. Simon, M., Isotope dilution of intracellular amino acids as a tracer of carbon and nitrogen sources of planktonic bacteria, *Mar. Ecol. Prog. Ser.*, 74, 295, 1991.
6. Simon, M. and Rosenstock, B., Carbon and nitrogen sources of planktonic bacteria in Lake Constance studied by the composition and isotope dilution of intracellular amino acids, *Limnol. Oceanogr.*, 37, 1496, 1992.
7. Fuhrman, J. A., Dissolved free amino acid cycling in an estuarine outflow plume, *Mar. Ecol. Prog. Ser.*, 66, 197, 1990.
8. Jørgensen, N. O. G., Free amino acids in lakes: concentrations and assimilation rates in relation to phytoplankton and bacterial production, *Limnol. Oceanogr.*, 32, 97, 1987.
9. Fuhrman, J. A. and Bell, T. M., Biological consideration in the measurements of dissolved free amino acids in seawater and implications for chemical and microbiological studies, *Mar. Ecol. Prog. Ser.*, 25, 13, 1985.
10. Simon, M., Improved assessment of bacterial production: combined measurements of protein synthesis via leucine and cell multiplication via thymidine incorporation, *Ergeb. Limnol.*, 34, 151, 1990.
11. Hobbie, J. E. and Crawford, C. C., Respiration corrections for bacterial uptake of dissolved organic compounds in natural waters, *Limnol. Oceanogr.*, 14, 528, 1969.
12. Parsons, T. R., Maita, Y., and Lalli, C. M., *A Manual of Chemical and Biological Methods for Seawater Analysis*, Pergamon Press, Oxford, 1984.

CHAPTER 64

Dialysis Bag Incubation as a Nonradiolabeling Technique to Estimate Bacterioplankton Production *In Situ*

Gerhard J. Herndl, Elisabeth Kaltenböck, and Gerald Müller-Niklas

INTRODUCTION

In situ incubation of water in dialysis bags allows diffusion of dissolved organic material and therefore, incubation of bacterioplankton without substrate limitations over more than 24 h. Size-fractionated and whole water samples can be incubated for up to 1 to 2 d without significant bacterial growth on the surface of the dialysis bags. When using size-fractionated incubations (0.6 to 0.8 μm) one is able to directly estimate bacterial cell production. Comparing the bacterial density at the beginning and the end of the incubation by means of epifluorescence microscopy is a suitable alternative to the conventional radiolabeling techniques to estimate bacterioplankton production.

Advantages of the method are: (1) it is inexpensive and no radiotracers are necessary; (2) increase in bacterial density is followed over a full diel cycle rather than in short-term incubations using radiotracers; this method accounts for the frequently detectable distinct diel periodicity (at least in the euphotic zone) in bacterioplankton activity; (3) it is an *in situ* incubation technique, thus, ''real'' temperature, irradiance, and dissolved organic matter (DOM) fluctuations over diel cycles; (4) no substrate limitation occurs as dialysis bags allow DOM to diffuse across the membrane; (5) using different molecular weight cut-offs the relative importance of various molecular weight fractions of DOM on bacterioplankton growth can be estimated; and (6) no conversion factors are required to estimate bacterioplankton cell production over a day.

Disadvantages of the method: (1) filtration is required prior to incubating the samples in order to minimize grazing by bacteriovores on bacterioplankton; this might lead to a selective removal of fractions of the bacterial community and/or to incomplete removal of small bacteriovores since those are in the same size range as bacterioplankton; (2) bacterial production has to be estimated from increase in cell numbers, which is a time-consuming task; and (3) frequently cells increase in volume during the course of incubation, and therefore, changes in cell volume should be recorded as well.

0-87371-564-0/93/$0.00 + $.50

MATERIALS REQUIRED

Equipment

- Autoclave
- Refrigerator

Supplies

- Dialysis tubing (e.g., Spectra/Por)
- Glass beaker or flask (100 to 1000 ml)
- Suction flask
- Filtration rig for 47-mm filters
- Polycarbonate filters (Nuclepore, 0.6-μm size)
- Frame consisting of an upper and lower rim (0.5-m diameter) about 1 m apart from each other; the two rims connected with three rods; the frame should be made of inert material
- Buoy, rope, and anchor
- Stoppers, strings, syringes to collect incubated water

Solutions

- Concentrated formalin
- 4% formalin in distilled water

PROCEDURES

Dialysis bags are regenerated cellulose tubes containing glycerol and sulfides to plasticize and preserve tubing, respectively. Spectra/Por dialysis tubings manufactured by Spectrum Medical Industries have been successfully used. They are available in various molecular weight cut-offs ranging from 1000 to 50,000 and can be ordered from Fisher Scientific. They should be stored in a desiccator until they are used for the first time.

1. Prior to incubation the dialysis tubes are cut into about 0.5-m tubes and rinsed with hot tap water for 3 h and overnight in cold tap water.

2. Subsequently the tubes should be submerged in distilled water and autoclaved in a suitable flask (100 to 1000 ml depending on the amount of tubes used later) where they can be stored until deployment. Dialysis tubes are sealed on one end using strings prior to autoclaving them.

3. Once the tubes have been rinsed, attention should be paid to keep them wet as otherwise, the pore size might change. Between experimental deployments, dialysis bags should be stored refrigerated in 4% formalin in distilled water.

4. Prior to redeployment, dialysis bags are rinsed thoroughly with cold and hot tap water overnight and autoclaved.

5. Depending on the purpose of the dialysis incubations, water can be prefiltered to reduce bacterivory. We used this technique mainly to estimate bacterial secondary production; for this purpose 0.6-μm filtration through polycarbonate filters (Nuclepore) gave best results for nearshore waters.

6.   Bags are filled with water (about 1 L) without leaving a headspace and tightly closed with strings. The use of stoppers with a small opening which can be stoppered separately to seal the upper end of the dialysis bags facilitates sampling.

7.   Prior to deployment, two 10-ml subsamples are removed from each bag and fixed immediately with concentrated formalin to yield a final concentration of 4% (v/v). These samples are enumerated as described below to determine initial bacterial densities.

8.   The bags are positioned in a frame with their upper and lower ends fixed to the frame. Four to six dialysis bags can easily be fixed to one frame. Using such a device, the bags can be fixed in a distinct depth layer with an anchor and a buoy or the frame can be deployed freely floating in a parcel of water using drifting buoys; deployment period should not exceed 48 h as enhanced wall growth of bacteria have been detected thereafter — usually 24 h will be sufficient for most purposes.

9.   For bacterial production estimates, 10-ml water samples are removed from the dialysis bags in 4- to 6-h intervals and fixed immediately as described above. Store formalin-fixed samples refrigerated but no longer than for 1 week.

10.   Stain sample with acridine orange or DAPI and filter onto a black 25-mm polycarbonate filter (pore size 0.2 μm).

11.   Count bacteria under the epifluorescence microscope and calculate bacterial density ml^{-1}; check also dimension of bacterial cells.

12.   From the increase in bacterial cell number over time one can calculate bacterial cell production and bacterial carbon production if cell numbers are converted into carbon equivalents (20 fg bacterial cell^{-1} is currently the most widely used conversion factor).

13.   By comparing the increase in bacterial density in filtered and unfiltered samples, estimates on flagellate grazing on bacteria or cyanobacteria can be performed if flagellates are enumerated concurrently with bacteria.

NOTES AND COMMENTS

1.   Filtering water for subsequent measurements or experiments should be done with minimum suction pressure; ideally, gravity filtration should be performed whenever possible. Filtration pressure might disrupt phytoplankton cells and consequently, might lead to higher DOM concentrations in the water.[1-4]

2.   Dialysis tubes should hold at least 0.5 L of water; their diameter should be about 3 cm in order to ensure sufficient diffusion rates into the dialysis bags. The frame in which the dialysis bags can be held in position is described in detail in Turley and Lochte[5] and Herndl and Malacic.[6]

3.   Bacterial growth rates can be determined as described in Ducklow and Hill,[7] Wright and Coffin,[8] and Chapter 82. Attention should be paid whether bacterial growth represents an exponential growth curve or a linear; therefore, sampling at 4- to 6-h intervals is recommended. Grazing rates also can be determined as described in Wright and Coffin[8] (see Chapter 82). Check also for the appearance of heterotrophic flagellates as they significantly diminish the bacterial community. Depending on the study site and the temperature regime, heterotrophic flagellates appear in significant numbers after approximately 30 h.

4.   A wide range of factors have been used to convert bacterial cell numbers into carbon equivalents (see Chapter 37).

REFERENCES

1. Ferguson, R. L., Buckley, E. N., and Palumbo, A. V., Response of marine bacterioplankton to differential filtration and confinement, *Appl. Environ. Microbiol.*, 47, 49, 1984.
2. Hilmer, T. and Bate, G. C., Filter types, filtration and post-filtration treatment in phytoplankton production studies, *J. Plankton Res.*, 11, 49, 1989.
3. Johnson, B. D. and Wangersky, P. J., Seawater filtration: particle flow and impaction considerations, *Limnol. Oceanogr.*, 30, 966, 1985.
4. Nagata, T. and Kirchman, D. L., Filtration-induced release of dissolved free amino acids: application to cultures of marine protozoa, *Mar. Ecol. Prog. Ser.*, 68, 1, 1990.
5. Turley, C. M. and Lochte, K., Direct measurement of bacterial productivity in stratified waters close to a front in the Irish Sea, *Mar. Ecol. Prog. Ser.*, 23, 209, 1985.
6. Herndl, G. J. and Malacic, V., Impact of the pycnocline layer on bacterioplankton: diel and spatial variations in microbial parameters in the stratified water column of the Gulf of Trieste (Northern Adriatic Sea), *Mar. Ecol. Prog. Ser.*, 38, 295, 1987.
7. Ducklow, H. W. and Hill, S. M., The growth of heterotrophic bacteria in the surface waters of warm core rings, *Limnol. Oceanogr.*, 30, 239, 1985.
8. Wright, R. T. and Coffin, R. B., Measuring microzooplankton grazing on planktonic marine bacteria by its impact on bacterial production, *Microb. Ecol.*, 10, 137, 1984.

CHAPTER 65

Growth Rates of Natural Populations of Heterotrophic Nanoplankton

George B. McManus

INTRODUCTION

Nonphotosynthetic protists with cell diameters less than 20 μm are known to be important grazers of both phytoplankton and bacteria.[1-5] These organisms are usually referred to as heterotrophic nanoplankton (or sometimes "microflagellates" or "nanoflagellates" because flagellated cells dominate the group numerically, though not always in biomass[6]). To evaluate their position in planktonic food webs, efforts have been made to measure their abundance and growth rates, and to discover their fate in natural waters.[1,7]

Some investigators have measured changes in heterotrophic nanoplankton abundance *in situ* to estimate growth rates, especially to evaluate diel or other variations in population dynamics.[7,8] While this may be possible in some physically well-defined systems, it provides only net growth rate. Since most evidence indicates that heterotrophic nanoplankton populations are closely controlled by grazers, however, net population growth rates may be much smaller than cell division rates. The method described below provides an estimate of growth in the absence of grazing mortality. It requires direct measurement of abundance changes in water samples prescreened to eliminate grazers. Unscreened controls allow the simultaneous estimation of grazing mortality.

Advantages of the method include: (1) growth is observed directly as change in cell abundance rather than in some biochemical surrogate such as ATP or carbon; (2) it is not technically difficult to perform; and (3) much additional information can be obtained from the microscopy (e.g., changes in relative abundance of heterotrophs and autotrophs or average cell size). Disadvantages include: (1) relatively long incubations (about 24 h) run the risk of introducing containment artifacts; (2) it is labor-intensive (about 0.5 to 1 man hours per replicate per time point for the microscopy alone).

Alternative methods, like the use of flow cytometry instead of microscopy for enumeration or the direct estimation of cell division rates from the frequency of dividing cells,[9] have not proven feasible to date.

METHOD PROTOCOL

Equipment

- Temperature-controlled bath or incubator. As with any biological process, control of temperature is important. A flow-through incubator with running water pumped from the sampled environment is inexpensive and usually convenient. Simulated *in situ* light conditions should also be maintained, if possible.

- Epifluorescence microscope.

Supplies

- 10-μm mesh nylon screen
- Incubation containers: 2-L polycarbonate bottles (2 to 4)
- Pipettes for sampling and measuring stains and other reagents
- 20-ml sample vials (10 to 20); scintillation vials of this capacity provide an adequate sample size for nanoplankton abundance estimates, even in oligotrophic environments
- Filtration apparatus for 25-mm filters
- 0.8-μm pore size black polycarbonate membrane filters (plain ones may be stained overnight with 2 g L^{-1} Irgalan black solution in 2% glacial acetic acid); polycarbonate membranes are available in a variety of pore sizes (e.g., 3- or 8-μm), and may be used to subdivide the heterotrophic nanoplankton population more finely[10-12]
- Slides and coverslips
- Nonfluorescing immersion oil

Solutions and Reagents

- 20% (w:v) glutaraldehyde in filtered water: this solution, prepared from a more concentrated solution by diluting with filtered water, will be diluted to a final concentration of 1% in the sample.
- DAPI solution
- Citrate buffer

PROCEDURES

Size Fractionation and Incubation

1. A well-mixed water sample is screened through the 10-μm mesh into two 2-L polycarbonate bottles.

2. Two bottles of unscreened water may be used as controls to estimate net growth rate and mortality.

3. The bottles are placed in the incubator and held under simulated *in situ* conditions for 24 to 48 h. Since the heterotrophic nanoplankton will often be dividing once or more per day, an incubation time of 24 h should usually be enough to observe significant growth in the screened samples, given sampling error and the imprecision inherent in microscopic counts.

4. Although the growth rate could be determined from abundances at two points in time, it is better to sample more frequently, as resources allow, to evaluate the shape of the growth curve (see discussion below). Beginning and end points should be separated by 3 to 5 evenly spaced sampling intervals. At each sampling, place 19 ml in a vial and add 1 ml of the 20% glutaraldehyde solution.

5. Mix gently and store at 4°C. Samples should be kept dark to avoid bleaching chlorophyll from autotrophic nanoplankton and slides should be made within 24 h.

6. To prepare slides, filter 10 ml of preserved sample onto a 0.8-μm pore size black Nuclepore filter, add 1 ml of citrate buffer[16] (pH 4) and 0.1 ml of a 10 μg ml^{-1} DAPI (4',6-diamidino-

2-phenylindole) solution. Stain for 1 min, filter to dryness, and mount in nonfluorescing immersion oil. Slides can be stored for months at $-20°C$.[17]

7. Examine the slide for nanoplankton using epifluorescence microscopy at $1000 \times$ or greater (see also Chapter 26). For DAPI, ultraviolet excitation is required.

Calculation of Growth Rates

For an exponentially growing population, the growth rate is calculated as:

$$k = 1/t * \ln(N_t/N_0) \tag{1}$$

where k is the intrinsic growth rate (dimension of $1/t$), and N_0 and N_t are abundances at initial and final sampling times. If multiple time points are sampled, a linear regression of $\ln(N_t)$ vs. t will have a slope equal to k. An exponential growth model may not always be appropriate, since k is not constant under conditions of density-dependent growth. When that is the case, growth is represented better by the logistic curve[18] which, to a rough approximation, has regions of exponential, linear, and no growth under grazer-free conditions. A linear growth trajectory suggests that the sampled population is near the inflection point of the logistic curve, with growth being controlled by both grazing mortality and food supply. This may be the most common situation in meso- and eutrophic waters.[1,7,19] Thus, if intermediate time points are sampled and growth appears to be linear with time (N_t vs. time is a better fit than $\ln(N_t)$ vs. time), growth rate may be estimated as:

$$k = 1/N_x * dN/dt \tag{2}$$

where N_x is the value of N at the midpoint of the incubation, and dN/dt is the slope of the N vs. time regression.

Productivity (P), or biomass of heterotrophic nanoplankton produced per unit time, is computed as:

$$P = k * B_0 \tag{3}$$

where B_0 is initial biomass, calculated from cell dimensions and a volume-to-weight conversion factor.

An important consideration when reporting k is to make explicit whether "doubling" or "per capita" rate is being specified. Equation 1 computes the per capita growth rate, which is based on the natural logarithms. Doubling rate, d (also referred to as "division rate"), is computed as:

$$d = k/\ln(2) = k/0.693 \tag{4}$$

Growth in unscreened samples presumably represents what is happening biologically to the population *in situ*, without the complicating effects of mixing, advection, and sinking. Any realized increase or decrease results from the imbalance of growth and mortality. The latter is computed as:

$$g = k - 1/t * \ln(N_t/N_0) \tag{5}$$

for the case where growth is exponential, where g is mortality due to grazing, and N is abundance in unscreened samples. When growth is linear with time, g is computed as:

$$g = k - 1/N_x * dN/dt \qquad (6)$$

as in Equation 2. If multiple time points are sampled, g is computed by subtracting the slope of the control regression from that of the screened samples, for both exponential and linear models.

NOTES AND COMMENTS

1. Although the nominal cutoff diameter for nanoplankton is 20 μm, woven nylon mesh of that diameter will pass many larger cells. A 10-μm mesh will pass nearly all of the nanoplankton, except those attached to larger particles, and retain most larger grazers, such as tintinnids and other large ciliates and dinoflagellates. For filtration of large volumes, a two-part Buchner funnel can be fitted with the mesh, which can be removed subsequently for cleaning.

2. If used, controls (unscreened water) should have a full suite of grazers. The smaller the container, the smaller the chance of obtaining a representative population of rarer forms. For example, adult copepods and other mesozooplankters usually occur at abundances of a few individuals per liter, so samples of that size or smaller do not collect them adequately. If these larger grazers are known to have only a negligible impact on nanoplankton, undersampling them may not be critical. This is probably true in most marine systems, where copepods feed mostly on larger cells, but not in lakes containing large populations of *Daphnia* and other cladocerans capable of filtering <10 μm particles efficiently.[13,14] Replicated, clean polycarbonate or polyethylene bottles or disposable polyethylene bags with 1- to 10-L capacity are recommended, the larger the better, both to ensure better representation of the grazer population in controls and to minimize containment artifacts associated with container surfaces.[15] If containers are too small to sample mesozooplankton adequately (i.e., <10 L in most mesotrophic lakes and coastal waters), then controls should be screened through 202-μm mesh to ensure complete exclusion of these larger organisms in all replicates.

3. Any division into size categories is arbitrary. There is a continuum of grazers across the entire size spectrum of planktonic organisms. Even within the nanoplankton, it has been shown that larger flagellates and ciliates eat smaller ones.[2,20-22] Thus, the growth rates calculated by the method outlined here are necessarily somewhat artificial. They probably do not represent the actual growth rates of any taxonomic group, but rather the net growth of all organisms passing the 10-μm screen. Some of these are bacterivores, some herbivores. What is proposed here is merely a standardized method for studying production of a rather heterogeneous group. The method is adaptable to the study of other groups, depending only on the size of the mesh chosen to eliminate larger grazers.[23,24]

4. Some issues have not been discussed here because they have been well-reviewed elsewhere. Furnas,[15] for example, discusses the large body of data on *in situ* growth rates of marine phytoplankton, including results of incubations similar to the ones described here for heterotrophs. Important issues include changes in species composition during incubtions, containment effects, and the alternative of using diffusion or dialysis chambers incubated *in situ*.

5. The technique might also be adaptable for the measurement of growth in "mixotrophic" nanoplankton (organisms that can both photosynthesize and ingest particles). Mixotrophic organisms are occasionally important in lakes, and may be so in marine systems as well.[13,25,26] One possibility would be to add fluorescently labeled bacteria (FLB) to the incubations to "tag" mixotrophs, thus, enabling them to be distinguished from nonphagotrophic phytoplankton. Growth rates of these organisms could then be computed separately. One difficulty is that FLB should be added at a sufficiently high concentration to ensure that every mixotroph will ingest at least one of them during the incubation,

without adding so many that nutrition of the mixotrophs, and hence, their growth rates, are enhanced.[27] The alternative of using nonnutritious latex beads has the risk of failing to label some mixotrophs that may discriminate against artificial food.

6. One final caveat is that this method is intended to measure growth rate in the absence of grazing. In light of recent information regarding viruses in marine plankton,[28-30] it should be emphasized here that other causes of mortality may be operating, even in the absence of grazing. Thus, the growth and productivity estimates measured with this technique are conservative. If anything, potential growth and productivity of heterotrophic nanoplankton is probably greater than what is measured.

ACKNOWLEDGMENTS

I thank Dr. Jeng Chang for suggesting the use of citrate buffer with the DAPI. This makes staining much more economical. Drs. Barry and/or Evelyn Sherr provided helpful comments on the manuscript.

REFERENCES

1. Sherr, B. F., Sherr, E. B., and Newell, S. Y., Abundance and productivity of heterotrophic nanoplankton in Georgia coastal waters, *J. Plankton Res.*, 6, 195, 1984.
2. Goldman, J. C. and Caron, D. A., Experimental studies on an omnivorous microflagellate: implications for grazing and nutrient regeneration in the marine microbial food chain, *Deep-Sea Res.*, 32, 899, 1985.
3. Fenchel, T., The ecology of heterotrophic flagellates, *Adv. Microb. Ecol.*, 9, 57, 1986.
4. McManus, G. B. and Fuhrman, J. A., Clearance of bacteria-sized particles by natural populations of nanoplankton in the Chesapeake Bay outflow plume, *Mar. Ecol. Prog. Ser.*, 42, 199, 1988.
5. Sherr, E. B., Sherr, B. F., and McDaniel, J., Clearance rates of <6 μm fluorescently labeled algae (FLA) by estuarine protozoa: potential grazing impact of flagellates and ciliates, *Mar. Ecol. Prog. Ser.*, 69, 81, 1991.
6. Sherr, E. B., Sherr, B. F., Fallon, R. D., and Newell, S. Y., Small aloricate ciliates as a major component of the marine heterotrophic nanoplankton, *Limnol. Oceanogr.*, 31, 177, 1986.
7. McManus, G. B. and Fuhrman, J. A., Mesoscale and seasonal variability of heterotrophic nanoflagellate abundance in an estuarine outflow plume, *Mar. Ecol. Prog. Ser.*, 61, 207, 1990.
8. Davis, P. G., Caron, D. A., Johnson, P. W., and Sieburth, J. McN., Phototrophic and apochlorotic components of picoplankton and nanoplankton in the North Atlantic: geographic, vertical, seasonal and diel distributions, *Mar. Ecol. Prog. Ser.*, 21, 15, 1985.
9. Sherr, E. B. and Sherr, B. F., Double-staining epifluorescence technique to assess frequency of dividing cells and bacterivory in natural populations of heterotrophic microprotozoa, *Appl. Environ. Microbiol.*, 46, 1388, 1983.
10. Landry, M. R., Haas, L. W., and Fagerness, V. L., Dynamics of microbial plankton communities: experiments in Kaneohe Bay, Hawaii, *Mar. Ecol. Prog. Ser.*, 16, 127, 1984.
11. Andersen, P. and Fenchel, T., Bacterivory by microheterotrophic flagellates in seawater samples, *Limnol. Oceanogr.*, 30, 198, 1985.
12. Rassoulzadegan, F. and Sheldon, R. W., Predator-prey interactions of nanozooplankton and bacteria in an oligotrophic marine environment, *Limnol. Oceanogr.*, 31, 1010, 1986.
13. Pace, M. L., McManus, G. B., and Findlay, S. E. G., Plankton community structure determines the fate of bacterial production in a temperate lake, *Limnol. Oceanogr.*, 35, 795, 1990.
14. Sanders, R. W. and Porter, K. G., Bacterivorous flagellates as food resources for the freshwater crustacean zooplankter *Daphnia ambigua, Limnol. Oceanogr.*, 35, 188, 1990.
15. Furnas, M. J., In situ growth rates of marine phytoplankton: approaches to measurement, community and species growth rates, *J. Plankton Res.*, 12, 1117, 1990.

16. Gerhardt, P., Murray, R. G. E., Costilow, R. N., Nester, E. W., Wood, W. A., Krieg, N. R., and Phillips, G. B., *Manual of Methods for General Bacteriology,* American Society of Microbiology, Washington, D.C., 1981, 68.

17. Bloem, J., Bar-Gilissen, M. B., and Cappenberg, T. E., Fixation, counting and manipulation of heterotrophic nanoflagellates, *Appl. Environ. Microbiol.,* 52, 1266, 1986.

18. Slobodkin, L., *Growth and Regulation of Animal Populations,* Dover, New York, 1980, 234.

19. Wright, R. T., A model for short-term control of the bacterioplankton by substrate and grazing, *Hydrobiologia,* 159, 111, 1988.

20. Rivier, A., Brownlee, D. C., Sheldon, R. W., and Rassoulzadegan, F., Growth of microzooplankton: a comparative study of bactivorous zooflagellates and ciliates, *Mar. Microb. Food Webs,* 1, 51, 1985.

21. Wikner, J. and Hagström, Å., Evidence for a tightly coupled nanoplanktonic predator-prey link regulating the bacterivores in the marine environment, *Mar. Ecol. Prog. Ser.,* 50, 137, 1988.

22. Suttle, C. A., Chan, A. M., Taylor, W. D., and Harrison, P. J., Grazing of planktonic diatoms by microflagellates, *J. Plankton Res.,* 8, 393, 1986.

23. Verity, P. G., Growth rates of natural tintinnid populations in Narragansett Bay, *Mar. Ecol. Prog. Ser.,* 29, 117, 1986.

24. Nielsen, T. G. and Kiørboe, T., Effects of a storm event on the structure of the pelagic food web with special emphasis on planktonic ciliates, *J. Plankton Res.,* 13, 35, 1991.

25. Bird, D. F. and Kalff, J., Bacterial grazing by planktonic lake algae, *Science,* 231, 493, 1986.

26. Estep, K. W., Davis, P. W., Keller, M. D., and Sieburth, J. McN., How important are oceanic nanoflagellates in bacterivory?, *Limnol. Oceanogr.,* 31, 646, 1986.

27. McManus, G. B. and Okubo, A., On the use of surrogate food particles to measure protistan ingestion, *Limnol. Oceanogr.,* 36, 613, 1991.

28. Bergh, O., Børsheim, K. Y., Bratbak, G., and Heldal, M., High abundance of viruses found in aquatic environments, *Nature,* 340, 467, 1989.

29. Proctor, L. M. and Fuhrman, J. A., Viral mortality of marine bacteria and cyanobacteria, *Nature,* 343, 60, 1990.

30. Suttle, C. A., Chan, A. M., and Cottrell, M. T., Infection of phytoplankton by viruses and reduction of primary productivity, *Nature,* 387, 467, 1990.

CHAPTER 66

The Labeled Chlorophyll *a* Technique for Determining Photoautotrophic Carbon Specific Growth Rates and Carbon Biomass

Donald G. Redalje

INTRODUCTION

Assessment of the magnitude and variability, both spatial and temporal, of photoautotrophic community carbon-specific growth rates and carbon biomass in the ocean has been a central focus of oceanographic research over the past two or three decades.[1,2] One of the major problems encountered by researchers has been our inability to isolate components of the photoautotrophic community from other forms of particulate matter found in natural water samples so that we may examine the important rate processes (e.g., the carbon-specific growth rate, μ) and determine carbon biomass (C_p) of solely the phytoplankton.[2-4] Because of our inability to isolate phytoplankton from other particulate matter, the ratio of some measurable parameter to C_p has often been employed to infer the value of C_p. Ratios of parameters such as phytoplankton cell volume (determined by microscopy), or the concentration of chlorophyll *a* (chl *a*), particulate organic carbon (POC) or adenosine triphosphate (ATP), to C_p have been of value.[2,4-6] Knowledge of reasonable values for these ratios and accurate determinations of the parameter of interest can provide a first-order estimate of C_p.[2] Uncertainty in the accuracy of C_p determined in this manner arises from the fact that these ratios are generally species specific, a function of phytoplankton physiological state, and have been shown to exhibit significant diel, seasonal, and spatial variability.[5,7]

Carbon-specific growth rates have generally been calculated using an equation of the form

$$\mu = (1/C_p)(dC_p/dt) \qquad (1)$$

where dC_p/dt is the photosynthetic rate (e.g., derived from the use of the ^{14}C primary production method) and C_p is the carbon biomass estimated using one of the above approaches. In addition, there have been a number of direct and indirect approaches to measuring community and individual species μ.[1] Some of these approaches include techniques which have been based upon diel changes in beam attenuation,[8] a combination of profiles of microscopically determined cell counts with estimates of turbulent diffusion rates,[9] sample dilution incubations,[10] the frequency of dividing cells,[11,12] and pulse labeling of nucleic acids and phospholipids with ^{32}P.[13] Each of these approaches has inherent drawbacks as well as benefits and may, in fact, depend on time-series measurements of biomass estimated by one of the "ratio" techniques described above.

Redalje and Laws[3] introduced a technique which provides accurate estimates of both μ and C_p, based on the incorporation of ^{14}C into chl a. Initial studies with both batch and continuous phytoplankton culture techniques in the laboratory have provided evidence that the labeled chl a technique provides values for both μ and C_p which are comparable with direct measurements of phytoplankton POC and the specific rate of increase in POC over 24 h.[3,4,14] The technique has been applied in the field on numerous occasions, providing reasonable estimates of community μ and C_p.[3,4,14-17] It should be noted that the technique has been extended to include examination of the labeling patterns of taxon-specific pigments in order to provide information on the growth rates of individual components of the phytoplankton community.[18,19] These modifications of the labeled chl a technique will not be addressed here. However, with further study and application, these modifications may prove quite useful to the field scientist.

This method is essentially an extension of the basic ^{14}C technique for measuring primary production.[20] When an alga is incubated in the presence of $H^{14}CO_3^-$, the labeled C is incorporated into 3-phosphoglyceric acid; this compound passes into the Calvin-Benson cycle and eventually label is incorporated into a pool of various small molecular weight intermediates.[21] These labeled compounds subsequently enter the chl a biosynthetic pathway.[22] There is some indication that ^{14}C becomes incorporated into the phytol chain of the chl a molecule more rapidly than it is incorporated into the porphyrin ring structure.[19,23] However, it seems clear that under most environmental conditions of interest to phytoplankton ecologists, chl a will become sufficiently labeled under similar time scales as those most often used for *in situ* or simulated *in situ* ^{14}C productivity experiments.[3,4,14-17] It is assumed that after a sufficiently long incubation, the specific activity of the chl a carbon (R^*_{chl}, dpm $\mu g\ C^{-1}$) will become equal to that of the total phytoplankton carbon pool (R^*_{Cp}, dpm μg C^{-1}).[3] This assumption has been supported by the results from a wide variety of laboratory culture experiments.[3,4,14] Results also indicate that for phytoplankton grown in continuous culture under environmentally realistic laboratory conditions (e.g., NH_4^+-limited growth, 12h/12h light/dark cycle), R^*_{chl} has become equal to R^*_{Cp} over time scales from a few hours to 24 h, depending upon growth conditions.[3,4] It should be noted that under natural photoperiods, phytoplankton carbon incorporation and chlorophyll synthesis may become uncoupled.[4,19] It is, thus, highly recommended that time-course or 24 hour endpoint experiments be used when employing this technique.

Jackson[24] presented a model which describes the impact on productivity measurements as a result of zooplankton grazing within an incubation container. If one were using Equation 1 to determine μ, grazing would remove newly fixed C and lead to an estimate of μ which would be lower than the actual value. However, as pointed out by Welschmeyer and Lorenzen[14] and Laws,[25] one of the principal advantages of the labeled chl a technique is that determinations of μ are free of artifacts due to zooplankton grazing in the incubation container. This is because labeled chl a determinations of μ are based on R^*_{chl}, the specific activity of the chl a-C, which is a ratio (see Equation 3, below), a value that does not depend on the absolute concentration of the chl a within the container.[3] Labeled chl a based determinations of C_p, however, could be subject to grazing related errors.[24,25] Grazing within an incubation container would lead to an underestimation of C_p. This is because C_p is a function of the total amount of ^{14}C incorporated into the phytoplankton carbon pool (see Equation 5, below)[3] and grazing will reduce this value to some degree. Laws[25] has developed a model which can be used to correct field data and provide highly accurate estimates of C_p. These corrections are a function of μ and the fraction of the labeled phytoplankton carbon which is respired during the incubation; in incubations of 1 to 2 doublings in duration, the errors are expected to be on the order of 20% or less.[25]

MATERIALS REQUIRED

Incubation

Equipment

- On-deck incubator or *in situ* incubation apparatus
- Filtration system (manifold, vacuum pump, 25- and 47-mm filter holders, filtration reservoir for liquid radioactive filtrate waste)
- Liquid nitrogen dewar or extended-life refrigerator
- Liquid scintillation (LS) counter
- pH meter or some type of CO_2 analyzer

Supplies

- Polycarbonate bottles
- Pipettors

Pigment Extraction

Equipment

- High-performance liquid chromatography (HPLC) gradient analytical system, including fluorescence or photodiode array detector, peak integrator or chart recorder, and fraction collector
- Tissue grinding apparatus
- Centrifuge or solvent filtration apparatus including vacuum pump and filter holder with capability of collecting filtrate
- LS counter
- Pipettors
- Fume hood

Supplies

- Nitrogen gas cylinder with regulator and tubing capable of producing a concentrated stream of N_2
- 60-ml glass separatory funnels and separatory funnel rack

PROCEDURES

As stated above, the labeled chl *a* technique is essentially an extension of a standard *in situ* or simulated *in situ* [14]C primary production experiment. It is highly recommended that trace metal clean techniques for water sampling and incubations, as outlined by Fitzwater et al.,[26] be employed for this type of experiment, just as one would for standard ^{14}C procedures. The following protocol will include descriptions of the each phase of the method procedure: the incubation and initial sample treatment, the extraction and purification of chl *a*, and the calculation of results.

Sampling and Incubation

1. Water samples are obtained using established, trace metal clean procedures.[26] Incubations should be conducted in acid-cleaned polycarbonate bottles.[26] Because this method requires the isolation and purification of chl *a*, it is recommended that the incubation container be at least 1 L in volume for coastal samples and as much as 4 L for offshore samples. The large sample container volume also reduces the chances for containment related artifacts.[27]

2. The sample is then inoculated with $H^{14}CO_3^-$ (prepared as suggested by Fitzwater et al.[26]). The exact amount of activity added to the sample depends on the phytoplankton biomass. Inoculations of 100 to 150 $\mu Ci L^{-1}$ have generally been sufficient to provide enough ^{14}C incorporated into chl *a* to allow successful application of the technique. It is important that enough ^{14}C activity be incorporated into the purified chl *a* so that the detected activity will be sufficiently above that of background (see below). It is also recommended that a "time-zero blank" (I generally use a 5-ml sample for time-zero samples) be taken as soon as the container is inoculated.[28]

3. Then 5 ml of sample is removed from the incubation container and filtered through a 25-mm Whatman GF/F glass fiber filter.

4. The filter is rinsed with a small volume (about 5 to 10 ml) of filtered seawater, removed from the filter holder, and placed into a LS vial which contains 500 μl of 1 N HCl to remove any residual ^{14}C which is adsorbed onto the filter.[29] These samples should be allowed to stand for at least 24 h prior to adding the appropriate volume of a LS counting solution. We often purge the vial with a stream of N_2 gas to displace any $^{14}CO_2$ contained in the LS vial prior to adding the LS solution.

5. The duration of the incubation is, of course, up to the discretion of the researchers and may depend upon the exact nature of the experimental question being studied. However, it is highly recommended that for accurate determinations of population growth rates, an incubation of 24 h, sunrise to sunrise, be employed.[1,4] Shorter incubations may provide rates which do not approximate the daily population growth rate (e.g., Redalje[4]).

6. One of the required ancillary pieces of information necessary for this procedure is knowledge of the concentration of dissolved CO_2 (DIC) in the water sample. This information can be obtained by following the alkalinity titration procedures listed in Parsons et al.[30] Alternatively, there are several commercially available CO_2 analyzers which will provide this information. In this laboratory we routinely use a Horiba Infrared CO_2 analyzer to determine the concentration of dissolved inorganic carbon.

7. Upon termination of the incubation the following subsampling procedures should be followed.

8. A 500-μl subsample is taken, combined with an equal volume of a 50% (v/v) mixture of ethanolamine and ethanol and placed into an LS vial. This subsample will provide information on the amount of ^{14}C activity added to the incubation container (S*).

9. Upon return to the laboratory, LS solution is added to these subsamples.

10. Next, 2 to 3 replicate subsamples, 50 to 100 ml in volume, are filtered through 25-mm Whatman GF/F glass fiber filters.

11. Each filter is placed into an LS vial which contains 500 μl of 1 N HCl to remove residual $H^{14}CO_3^-$.[29] These subsamples are then treated as described above for the time-zero subsamples. The activity contained in these subsamples will be used to calculate the total amount of ^{14}C incorporated into the phytoplankton (A*, dpm ml^{-1}) and can also be used in the determination of primary production using the standard procedures.[30]

12. The remaining sample is filtered through a 47-mm Whatman GF/F glass fiber filter. More than one filter can be used if the filter becomes clogged and filtration slows excessively.

13. When the sample is completely filtered, each filter is folded into quarters and rolled to fit into a 1- to 2-ml volume cryotube for storage in liquid nitrogen (LN). Samples can be stored for extended periods of time in liquid nitrogen prior to subsequent analysis in the laboratory.

Pigment Extraction and Chromatographic Procedures

In this phase of the procedure description, the techniques used to extract and concentrate chl *a* and subsequently isolate chl *a* from the concentrated extract using HPLC will be presented. It should be noted that it is not necessary for the extraction procedure to be quantitative. The parameter which will be derived, R^*_{chl}, is a ratio of the mass of C in the isolated chl *a* to the ^{14}C activity contained in that isolated chl *a*. It is, thus, necessary that each of these two values be determined quantitatively. Additionally, the extract from a single sample can be separated into several aliquots to provide replicate determinations of R^*_{chl}. The HPLC chl *a* isolation procedures described below may need to be modified depending on the specific HPLC system utilized, the type of detector, and the type of fraction collector employed. In addition, it is recommended, though not necessary, that the investigator consider the use of a photodiode array detector for this procedure. Fluorometric detectors are much more sensitive than absorbance detectors, such as the photodiode array detectors, which is a definite advantage for samples obtained from oligotrophic environments. However, photodiode array detectors have the advantage of allowing the investigator to examine the absorbance spectrum of each peak of interest. These spectra can then be compared with those obtained for pigment standards to determine the purity of each isolated peak.

1. A sample is retrieved from the LN dewar and removed from the cryotube.

2. This filter, with its collected plant material, is then placed into a tissue grinder tube with a small volume of 90% acetone and ground to disrupt the phytoplankton cells using a tissue grinder. Care should be taken so that the total volume of acetone remains relatively small (e.g., 5 to 7 ml).

3. At this point, the filter and cellular debris can be removed from the extract by either centrifugation or by filtration through another glass fiber filter. In this laboratory, refiltration is preferred. The extract is then collected for further treatment using solvent phase separation.[14]

4. This part of the procedure is performed inside a fume hood, under low illumination to prevent any photooxidation of pigments.

    ● The 90% acetone/pigment extract is placed into a 60-ml separatory funnel.

    ● To this, 2 ml of hexane is added.

    ● The mixture is vigorously shaken, vented, and then placed on a separatory funnel rack to allow the phases to separate. The chl *a* will be dissolved in the hexane upper phase. It may be necessary to add a small volume of 10% (m/v) aqueous NaCl to the initial hexane-aqueous-acetone mixture to increase the efficiency of the extraction process.

    ● The lower aqueous-acetone phase is drained off into an LS vial and set aside.

    ● The hexane phase is then drained into another LS vial and set aside as well.

    ● The aqueous-acetone phase is then reintroduced into the separatory funnel and the hexane extraction procedure repeated twice. All of the hexane extracts are pooled in the LS vial.

    ● The pooled hexane extracts are then evaporated to dryness under a stream of N_2 gas and the LS vial capped tightly.

    ● The vial can then be stored in a freezer until further analysis by HPLC.

Table 1. HPLC Gradient and Flow Profiles

Time, minutes	Gradient Profile (% Solvent B)	Flow Profile (ml min^{-1})
0.00	0.00	1.50
0.13 (sample injection)	0.00	1.50
0.50	0.00	1.50
5.50	50.00	2.08
8.00	69.23	2.50
12.00	100.00	2.50
13.00	0.00	2.17
15.00	0.00	1.50
16.00	0.00	1.50
17.00	0.00	0.00

5. Prior to analysis by HPLC, the sample is dissolved in 1 ml of methanol.

6. A 200-µl aliquot of this is mixed with 100 µl of the ion pairing agent defined by Mantoura and Llewellyn[31] (7.7 g ammonium acetate and 1.5 g tetrabutyl ammonium acetate in 100 ml of H$_2$O).

7. This methanol-pigment-ion pairing agent mixture is injected into the HPLC sample loop. In this laboratory we use a Gilson Gradient Analytical HPLC system with an Alltech Absorbosphere HS high carbon loading reverse phase C18 chromatography column (250-mm length, 5-µm particles) and a 200-µl sample loop. The gradient system used for the isolation of chl a involves a 17-min cycle. Following the recommendations of Mantoura and Llewellyn[31] we vary solvent flow rate to maintain a constant pressure in the system of 2000 PSI. The Gradient and flow profiles are presented in Table 1. Mobile phase B is 100% acetone and mobile phase A is 100% methanol. Sample injection occurs 0.13 min into the separation cycle. The chl a peak is detected with a filter fluorometer (Farrand Model A4) equipped with a 10-µl flow cell. The output of the fluorometer is linked to the Gilson Fraction Collector so the isolated chl a peak can be collected. A sample chromatogram shows that the chl a peak is detected 7.3 min into the separation cycle (Figure 1). The peak integrator for the Gilson HPLC system integrates the area of the chl a peak in the output chromatogram. Standards of purified chl a of known mass (either prepared commercially or from previous HPLC runs) are used to derive a standard linear regression relating peak integrator area to chl a mass. In this manner, the chl a mass for unknown samples can be determined. The standard peaks should also be collected to use as blanks for the LS counting procedure described below.

8. The collected chl a peak is then placed into an LS vial and evaporated to dryness under a stream of N$_2$ gas.

9. LS counting solution is then added to the vial and the ^{14}C activity of the isolated chl a is then determined. It is generally necessary to use long LS counting times (e.g., 60 min per sample) to obtain sample activities which are statistically different from unlabeled standard blanks.

Calculations of µ and C$_p$

The procedures described above provide the following parameters necessary for calculating the photoautotrophic µ and C$_p$:

- DIC, the concentration of dissolved CO$_2$ (µg L^{-1});
- S*, the activity of ^{14}C added to the sample (dpm L^{-1});
- A*, the particulate ^{14}C for the experiment (dpm L^{-1});

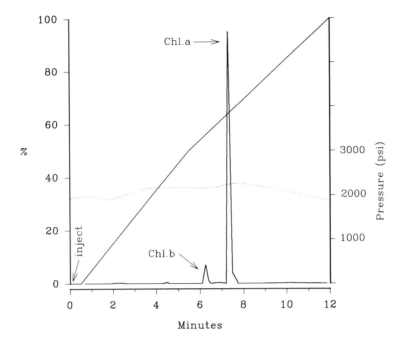

Figure 1. The chromatogram above is the output of the HPLC procedure described in Pigment Extraction and Chromatographic Procedures, which shows the ioslation of chl *a* from other pigments contained in an extract which had been pretreated as also described in Pigment Extraction and Chromatographic Procedures. The left ventrical axis, labeled %, indicates the HPLC gradient (diagonal line) scaled to % Solvent B (100% acetone) content. The chl *a* peak has been scaled to 100% of the left axis length. In this manner, the relative contribution of other pigment peaks can be interpreted relative to that of chl *a*. System pressure (dotted line) is indicated on the right vertical axis.

- M_{chl}, the mass of HPLC chl *a* peak (μg);
- M^*_{chl}, the blank corrected activity of the HPLC chl *a* peak (dpm); and
- t, the length of the incubation (h).

These parameters, including units, will be used in the equations presented below for the calculation of phytoplankton community μ and C_p. The equations are the same or slight modifications of those presented by Redalje and Laws[3] or Welschmeyer and Lorenzen.[14] In addition, one can use the suggestions of Laws[25] to correct C_p for the effects of grazing within the incubation container if appropriate to the specific experiments. These grazing correction equations will not be presented here and the reader is directed to Laws[25] for information regarding experimental conditions which merit use of the suggested corrections (e.g., $\mu > 0.7 \; d^{-1}$, see Introduction).

Calculation of I*

$$I^* \; (dpm \; \mu g \; C^{-1}) = S^*/DIC \qquad (2)$$

Calculation of the C-Specific Activity of chl a

$$R^*_{chl} \; (dpm \; \mu g \; C^{-1}) = M^*_{chl}/(0.74)(M_{chl}) \qquad (3)$$

where 0.74 is the fraction of the molecular weight of chl *a* contributed by C.

Calculation of Photoautotrophic Community μ

$$\mu \ (h^{-1}) = -(1/t)\ln(1 - [1.05 \ R^*_{chl}/I^*]) \tag{4}$$

where 1.05 is a correction for the ^{14}C-isotope discrimination effect.[30] The value of μ can easily be converted to units of d^{-1}.

Calculation of the Photoautotrophic Biomass at the End of the Incubation (C_p) and at the Beginning of the Incubation ($C_{p,o}$)

$$Cp \ (\mu g \ C \ L^{-1}) = A^*/R^*_{chl} \tag{5}$$

Since $C_{p,o}$ can be related to C_p by the equation

$$C_p = C_{p,o} \ e^{\mu t} \tag{6}$$

by rearranging Equation 6, we can determine $C_{p,o}$ using Equation 7:

$$C_{p,o} \ (\mu g \ C \ L^{-1}) = \ln^{-1} (\ln C_p - [\mu t]) \tag{7}$$

NOTES AND COMMENTS

1. There are several potential problems which might be encountered by researchers when the labeled chl *a* technique is applied. It cannot be overemphasized that trace metal clean procedures[26] should be used throughout the incubation part of the technique. Trace metal contamination can result in reduced assimilation of carbon into the phytoplankton and, in turn, into chl *a*. Another obvious concern is that the sample containing the labeled pigments be sufficiently large to assure that the chl *a* can easily be detected during the HPLC phase of the procedure. This concern is in part alleviated by the use of large (e.g., 1 L for nearshore studies and 4 L for offshore research) incubation containers. It is also highly recommended that one carefully consider the use of the grazing correction terms suggested by Laws,[25] especially in oligotrophic ocean waters where microbial/microzooplankton food webs are thought to be most active.[32]

2. Perhaps the most important potential area of concern in the application of the technique is that there is a possibility that labeled, but nonpigmented lipids could coelute with chl *a* during the HPLC procedure. If there is significant ^{14}C activity associated with these lipids, R*$_{chl}$ and thus, μ, would be overestimated to some degree. In addition, C_p would be underestimated. Goericke[19] found that with the HPLC system and gradient program in use in his laboratory, it was necessary to acidify the chl *a*, converting the pigment to phaeophytin *a* (phaeo *a*) before HPLC analysis. The phaeo *a* peak is detected after the chl *a* peak during the HPLC procedure and it does not coelute with nonpigmented lipids under the HPLC conditions used by Goericke.[19] However, in the procedures discussed above, it should be noted that chl *a* does not coelute with nonpigmented lipids. Thus, it is not generally necessary to acidify the pigment extract prior to HPLC analysis. Separation of pigments by HPLC is not always a "straightforward" procedure. There is no substitute for complete familiarization with HPLC theory and instrumentation if the labeled chl *a* technique is to be employed.

3. Another related concern is that some HPLC columns may retain ^{14}C activity which slowly elutes and may contaminate successive samples. If unlabeled chl *a* standards are run at both the beginning and end of a set of samples, it will be possible to detect activity associated with column contamination. Each column manufacturer provides information

on column maintenance and cleaning procedures. It is critical that these procedures be followed. Prolonged column cleaning and rinsing will likely remove any accumulated ^{14}C activity.

4. The pattern and kinetics of pigment labeling is an area of ongoing research. Goericke[19] suggests that a two-step model, where the labeling of a chl a precursor is accounted for in the calculation of μ, may provide a better estimate of μ than the single endpoint model presented above. However, at this time it does not appear that the more complicated model provides a significantly different result than the endpoint model. The continuous culture studies presented by Redalje and Laws[3] and Redalje[4] have shown that after 24-h incubations, determinations of μ derived using the endpoint model, were not significantly different from the steady-state values of μ. However, it is possible that future studies may provide evidence in support of the use of more complicated models.

5. A final point is that the values of μ and C_p determined as described above will describe the photoautotrophic community. Group- or species-specific estimates of μ are also of interest.[18,19] It is possible to address the growth dynamics of various groups of phytoplankton by examining the labeling of taxon-specific pigments, such as fucoxanthin (diatoms), 19′-hexanoyloxyfucoxanthin (prymnesiophytes) and zeaxanthin (cyanobacteria).[18,19] Knowledge of group-specific growth may help us to understand temporal variability in the ocean to a greater extent than community rates because community composition will likely change as a function of changes in environmental conditions. Thus, as our understanding of pigment-specific labeling patterns evolves, these taxon-specific techniques may become valuable to the oceanographic community.

REFERENCES

1. Furnas, M. J., In situ growth rates of marine phytoplankton: approaches to measurement, community and species growth rates, *J. Plankton Res.*, 12, 1117, 1990.
2. Eppley, R. W., Harrison, W. G., Chisholm, S. W., and Stewart, E., Particulate organic matter in surface waters off Southern California and its relationship to phytoplankton, *J. Mar. Res.*, 35, 671, 1977.
3. Redalje, D. G. and Laws, E. A., A new method for estimating phytoplankton growth rates and carbon biomass, *Mar. Biol.*, 62, 73, 1981.
4. Redalje, D. G., Phytoplankton carbon biomass and specific growth rates determined with the labeled chlorophyll a technique, *Mar. Ecol. Prog. Ser.*, 11, 217, 1983.
5. Sinclair, M., Keighan, E., and Jones, I., ATP as a measure of living phytoplankton carbon in estuaries, *J. Fish. Res. Board. Can.*, 36, 180, 1979.
6. Banse, K., Determining the carbon-to-chlorophyll ratio of natural phytoplankton, *Mar. Biol.*, 41, 199, 1977.
7. Hunter, B. L. and Laws, E. A., ATP and chlorophyll *a* as estimators of phytoplankton carbon biomass, *Limnol. Oceanogr.*, 26, 944, 1981.
8. Cullen, J. J., Lewis, M. R., Davis, C. O., and Barber, R. T., Photosynthetic characteristics and estimated growth rates of phytoplankton in the equatorial Pacific, *J. Geophys. Res.*, 97, 639, 1992.
9. Lande, R., Li, W. K. W., Horne, E. P. W., and Wood, A. M., Phytoplankton growth rates estimated from depth profiles of cell concentration and turbulent diffusion, *Deep-Sea Res.*, 36, 1141, 1989.
10. Landry, M. R. and Hassett, R. P., Estimating the grazing impact of marine microzooplankton, *Mar. Biol.*, 67, 283, 1982.
11. Weiler, C. S. and Chisholm, S. W., Phased cell division in natural populations of marine dinoflagellates from shipboard cultures, *J. Exp. Mar. Biol. Ecol.*, 25, 239, 1976.
12. Weiler, C. S., Population structure and in situ division rates of *Ceratium* in oligotrophic waters of the North Pacific Central Gyre, *Limnol. Oceanogr.*, 25, 610, 1980.
13. LaRock, P. A., Schwarz, J. R., and Hofer, K. G., Pulse labelling: a method for measuring microbial growth rates in the ocean, *J. Microbiol. Methods*, 8, 281, 1988.

14. Welschmeyer, N. A. and Lorenzen, C. J., Carbon-14 labeling of phytoplankton carbon and chlorophyll *a* carbon: determination of specific growth rates, *Limnol. Oceanogr.*, 29, 135, 1984.

15. Laws, E. A., Redalje, D. G., Haas, L. W., Bienfang, P. K., Eppley, R. W., Harrison, W. G., Karl, D. M., and Marra, J., High phytoplankton growth and production rates in oligotrophic Hawaiian coastal waters, *Limnol. Oceanogr.*, 29, 1161, 1984.

16. Laws, E. A., DiTullio, G. R., and Redalje, D. G., High phytoplankton growth and production rates in the North Pacific subtropical gyre, *Limnol. Oceanogr.*, 34, 905, 1987.

17. Knauer, G. A., Redalje, D. G., Harrison, W. G., and Karl, D. M., New production at the VERTEX time-series site, *Deep-Sea Res.*, 37, 1121, 1990.

18. Gieskes, W. W. C. and Kraay, G. W., Estimating the carbon-specific growth rate of the major algal species groups in eastern Indonesian waters by the ^{14}C labeling of taxon-specific carotenoids, *Deep-Sea Res.*, 36, 1127, 1989.

19. Goericke, R., Pigments as ecological tracers for the study of the abundance and growth of marine phytoplankton, Doctoral dissertation, Harvard University, Cambridge, MA, 1990, 418.

20. Steemann-Nielsen, E., The use of radioactive carbon (^{14}C) for measuring organic production in the sea, *J. Cons. Perm. Int. Explor. Mer.*, 18, 117, 1952.

21. Calvin, M., Bassham, J. A., Benson, A. A., Lynch, V. H., Quellet, C., Shou, L., Stepka, W., and Tolbert, N. E., Carbon dioxide assimilation in plants, *Symp. Soc. Exp. Biol.*, 5, 284, 1951.

22. Bogorad, L., Chlorophyll biosynthesis, in *Chemistry and Biochemistry of Plant Pigments*, Vol. 1, 2nd ed., Goodwin, T. W., Ed., Academic Press, New York, 1976, 64.

23. Grumbach, K. H., Lichtenthaler, H. K., and Erismann, K. H., Incorporation of ^{14}CO$_2$ in photosynthetic pigments of *Chlorella pyrenoidosa, Planta*, 140, 37, 1978.

24. Jackson, G. A., Zooplankton grazing effects on ^{14}C-based phytoplankton production measurements: a theoretical study, *J. Plankton Res.*, 5, 83, 1983.

25. Laws, E. A., Improved estimates of phytoplankton carbon based on ^{14}C incorporation into chlorophyll *a, J. Theor. Biol.*, 110, 425, 1984.

26. Fitzwater, S. E., Knauer, G. A., and Martin, J. H., Metal contamination and its effects on primary production measurements, *Limnol. Oceanogr.*, 27, 544, 1982.

27. Gieskes, W. W. C., Kraay, G. W., and Baars, M. A., Current ^{14}C methods for measuring primary production: gross underestimates in oceanic waters, *Neth. J. Sea Res.*, 13, 58, 1979.

28. Morris, I., Yentsch, C. M., and Yentsch, C. S., Relationship between light CO$_2$ fixation and dark CO$_2$ fixation by marine algae, *Limnol. Oceanogr.*, 16, 854, 1971.

29. Lean, D. R. S. and Burnison, B. K., An evaluation of errors in the ^{14}C method of primary production measurement, *Limnol. Oceanogr.*, 24, 917, 1979.

30. Parsons, T. R., Maita, Y., and Lalli, C. M., *A Manual of Chemical and Biological Methods for Seawater Analysis*, Pergamon Press, Oxford, U.K., 1984, 173.

31. Mantoura, R. F. C. and Llewellyn, C. A., The rapid determination of algal chlorophyll and carotenoid pigments and their breakdown products in natural waters by reverse-phase high-performance liquid chromatography, *Anal. Chim. Acta*, 151, 297, 1983.

32. Williams, P. J. LeB., Incorporation of microheterotrophic processes into the paradigm of the planktonic food web, *Kiel. Meeres. Sonderh.*, 5, 1, 1981.

CHAPTER 67

Incorporation of $^{14}CO_2$ into Protein as an Estimate of Phytoplankton N-Assimilation and Relative Growth Rate

Giacomo R. DiTullio

INTRODUCTION

The protein labeling method provides estimates of N-assimilation and relative growth rates of phytoplankton populations in the open ocean. The method was devised to circumvent some of the inherent problems associated with the ^{15}N method[1] in oligotrophic ocean waters. Phytoplankton N-assimilation rates in the oligotrophic ocean can be estimated from total ^{14}C uptake and the percentage of ^{14}C incorporated into the protein fraction by assuming that: (1) under N-limitation protein N accounts for 80 to 90% of total cellular N and (2) the N/C ratio of pure protein is constant at 0.30 ± 0.01 (by weight).[2] Following a standard ^{14}C incubation the fraction of labeled protein carbon is then:

$$\text{protein C/total C} = N(0.85)/C(0.30) = 2.8 \ N/C \tag{1}$$

Redalje analyzed seven marine species under 35 distinct N-limited conditions and found a ratio of 2.7 between the fraction of protein carbon and N/C[3], in close agreement to our estimated value of 2.8. The rationale behind the method is that the N/C ratio is directly proportional to the percentage of cellular carbon allocated to protein under N-limitation.[4] In N-limited systems, N/C ratios are linearly correlated with growth rate[5-7] with N/C values approaching the Redfield[8] ratio of 0.15 (by atoms) at maximum relative growth rate.[9]

The N/C assimilation ratio in five taxonomically diverse species of phytoplankton grown in NH_4^+-limited cyclostats was compared to the actual N/C composition ratio measured at 6-h intervals throughout a 12:12 light-dark cycle.[10] It was observed that the N/C assimilation ratio underestimated the true N/C composition ratio during the photoperiod because the specific activity of the protein C was less than the specific activity of the total C. Protein synthesis at night fueled by the C endproducts of carbohydrate respiration resulted in isotopic equilibrium between the protein C and total cellular C pools. The net result was an accurate prediction of the N/C composition ratio from the N/C assimilation ratio after a 24-h incubation.[10]

The main advantages of the protein labeling method are:

1. The estimated N-assimilation rates are specific for phytoplankton (unless dark bacterial uptake of $^{14}CO_2$ is significant, see below).

0-87371-564-0/93/$0.00 + $.50

2. The method measures total N-assimilation from all organic (e.g., urea, amino acids) and inorganic N compounds, hence, a single ^{14}C incubation can provide an estimate of the total autotrophic N-assimilation rate. Moreover, a [15]NO_3^- uptake experiment performed in conjunction with measurement of total N-assimilation from protein labeling can provide a relatively simple estimate for the f-ratio.[11] Traditional estimates of the f-ratio using only [15]NO_3^- and [15]NH_4^+ assimilation will overestimate the true f-ratio when urea and organic nitrogen uptake rates are significant.

3. The N/C assimilation ratio can also provide an estimate for the relative growth rate of the phytoplankton population.[10]

4. The protein labeling method was not significantly affected by grazing processes after a 24-h incubation as determined by dilution experiments.[12]

5. Tracing the pathway of $H^{14}CO_3^-$ assimilation into various metabolic endproducts also can provide valuable information as to the physiological state of natural phytoplankton populations.[13]

The protein labeling method has three major potential disadvantages.

1. The protein labeling method has only been tested under N-limited growth conditions.[10] Under N-sufficient conditions the percentage of cellular N allocated to protein will be somewhat less than 85%.[14] For instance, a cellular N budget in six species of phytoplankton grown under N-deplete and -replete conditions demonstrated that under N-replete conditions 56 ± 13% of total particulate nitrogen (PN) was in the protein fraction, while protein N accounted for 78 ± 14% under N-limited conditions.[14] A significant fraction of the total cellular PN, however, was unaccounted for by the sum of the cellular N compounds in that study. If we assume that half of the unaccounted PN was protein-N then the average percentage of PN represented by protein-N becomes 83 ± 9% and 71 ± 7% in the N-deplete and -replete cultures, respectively. The value of 83% is in close agreement to our estimated 85% under N-limited conditions.[2] Hence, it may also be reasonable to estimate N-assimilation by the protein labeling method in N-rich waters by assuming that protein N constitutes 70% of total cellular N. The protein labeling method should also work in P-limited systems because there is no apparent "luxurious" uptake of N under P-limitation.[7] Further lab studies are required, however, to verify the accuracy of these assumptions.

2. Because protein labeling is an indirect method and depends on [14]CO_2 uptake, all errors associated with the ^{14}C method itself will be propagated to the protein labeling method. The ^{14}C method has many potential problems.[15] Nevertheless, [14]CO_2 assimilation has been shown to be a relatively accurate method for estimating autotrophic oceanic primary production rates.[16] On occasion, however, significant dark uptake of [14]CO_2 has been reported.[17-19] One approach to determine the relative importance of phytoplankton vs. bacterial dark [14]CO_2 uptake is to monitor the change in the chl a carbon-specific activity at night via the chl a labeling method.[20] For instance, in a study performed in the Caribbean Sea and western Atlantic Ocean, 26% of the integral carbon "photosynthetic" production was associated with nocturnal ^{14}C uptake with 76% of this nocturnal uptake being associated with phytoplankton.[21] However, if the relative contribution of bacterial dark [14]CO_2 uptake is significant then autotrophic carbon and nitrogen assimilation rates will be overestimated. The extent and relative importance of dark bacterial CO_2 uptake needs further study.

3. Another potential disadvantage of the method involves the 24-h incubation time required to obtain accurate N-assimilation rates. Cyclostat studies demonstrated that N-assimilation rates as measured by protein labeling were overestimated during the day (because of isotopic disequilibrium) and underestimated at night (because of N-pooling), with the most accurate result obtained after 24-h.[10] Large polycarbonate incubation bottles (>2 L) and clean sampling techniques should minimize the deleterious effects from 24-h incubations.

MATERIALS REQUIRED

Equipment

- Water bath
- Scintillation counter
- Centrifuge

Supplies

- Filtration rig
- Vortexer
- Liquid scintillation counter (LSC) vials
- Pipettors
- Centrifuge tubes
- Disposable glass tubes

Solutions and Reagents

- Chloroform
- Methanol
- Trichloroacetic acid (TCA)
- Sodium hydroxide
- LSC cocktail

PROCEDURES

Solvent Fractionation

Various solvent fractionation schemes have been developed to separate macromolecules but all are basically derived from the early work of Roberts et al.[22] With a few modifications (Figure 1), we have adopted the general fractionation scheme of Li et al.[23]

1. Following a 24-h incubation, aliquots are filtered and frozen at $-20°C$.

2. The solvent fractionation scheme (Figure 1) begins with a Bligh-Dyer lipid extraction.[24] Solvents are added sequentially to facilitate penetration and disruption of the cell membrane.

3. The final mixture of $CHCl_3:MeOH:H_2O$ (10:10:9) is filtered and centrifuged and the $MeOH:H_2O$ hyperphase represents the low molecular weight compounds (e.g., amino acids) and the $CHCl_3$ hypophase represents lipid compounds.

4. The filter residue is then heated at 85°C for 30 min in fresh 5% TCA to precipitate the protein.

At temperatures above 90°C in 5% TCA there may be a release of acid soluble peptides.[25] We have found that the protein ^{14}C activity was 5 to 10% lower after heating at 100°C compared to 85°C (unpublished). The TCA filtrate represents both polysaccharides and nucleic acids, but because of the relatively low isotopic labeling of nucleic acids relative to polysaccharides,[13] this fraction is typically referred to as the polysaccharide fraction. The

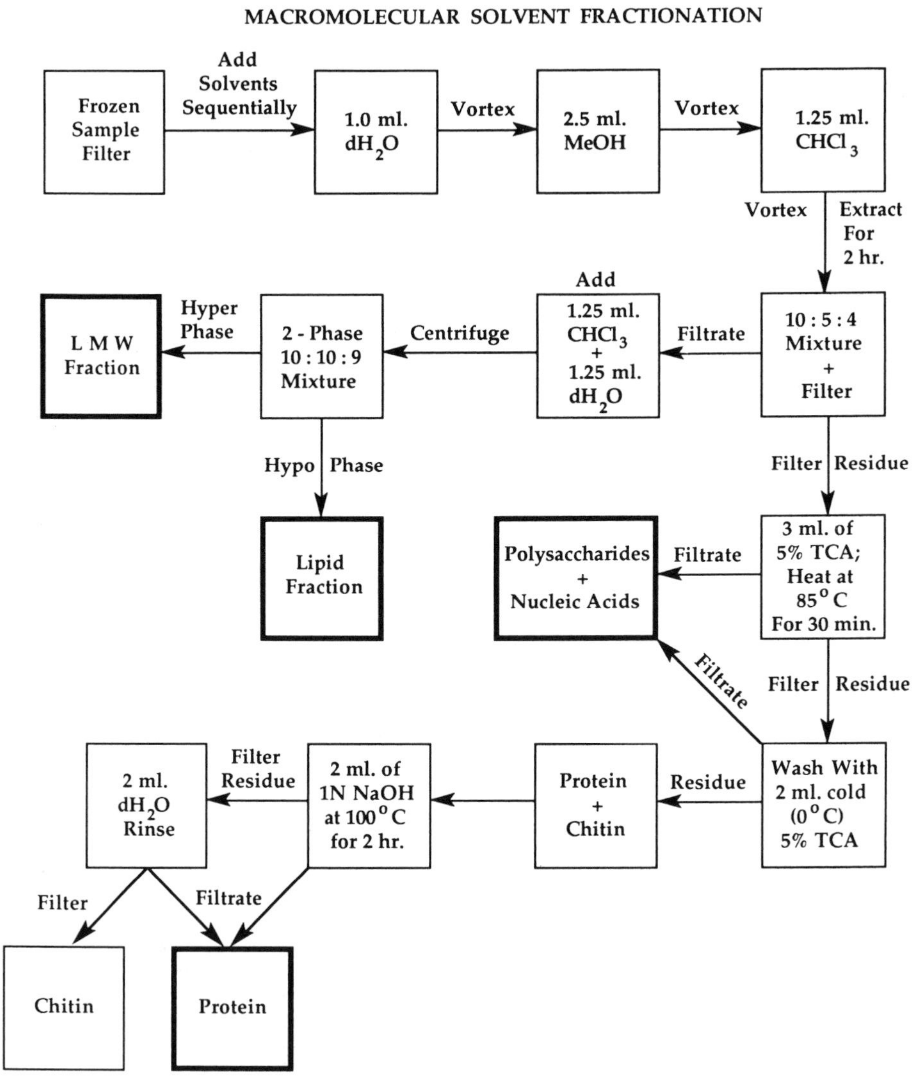

MACROMOLECULAR SOLVENT FRACTIONATION

Figure 1. Flow diagram of solvent extraction procedure to separate protein, polysaccharides, lipids, and low molecular weight (LMW) intermediate compounds. $CHCl_3$ = chloroform, MeOH = methanol, NaOH = sodium hydroxide, TCA = trichloroacetic acid.

solution of 5% TCA is made fresh daily and the stock solution of 100% TCA is kept refrigerated to minimize decomposition.

The TCA-insoluble precipitate is mostly protein.[13,22] A significant fraction (especially in diatoms) of the precipitate, however, may represent the amino sugar polymer, chitin.[26] An alkali hydrolysis (Figure 1) to solubilize the protein will not affect the chitin precipitate on the filter.[26] Hence, the filtrate from the NaOH hydrolysis represents the protein fraction free of chitin interference. The sum of the activities from the four fractions is typically ±5% of the total unfractionated ^{14}C activity.[7,10]

Calculations

N-assimilation rates are calculated using the fraction of ^{14}C incorporated into protein and the total particulate ^{14}C activity.[2] In oligotrophic regions the N-assimilation rate is:

$$\text{mg N m}^{-3}\text{d}^{-1} = (^{14}\text{C-uptake}/2.8) \text{ (protein } ^{14}\text{C/particulate } ^{14}\text{C)} \qquad (2)$$

In high nutrient oceanic waters a factor of 2.3 may be substituted. The relative growth rate of the phytoplankton population as defined by Goldman[27] can be calculated by assuming that the percent protein C (p) varies from 15 to 50% as the relative growth rate (μ_r) of the population approaches unity.[10] Hence, for relative growth rate estimates from protein labeling, $p = 15 + 35\mu_r$.

REFERENCES

1. Harrison, W. G., Nitrogen in the marine environment: use of isotopes, in *Nitrogen in the Marine Environment,* Carpenter, E. J. and Capone, D. G., Eds., Academic Press, New York, 1983, 763.
2. DiTullio, G. R. and Laws, E. A., Estimates of phytoplankton N uptake based on $^{14}CO_2$ incorporation into protein, *Limnol. Oceanogr.,* 28, 177, 1983.
3. Redalje, D. G. and Laws, E. A., The effects of environmental factor on growth and the chemical and biochemical composition of marine diatoms. I. Light and temperature effects, *J. Exp. Mar. Biol. Ecol.,* 68, 59, 1983.
4. Redalje, D. G., The effects of environmental factors on the general patterns of carbon metabolism for marine phytoplankton, Ph.D dissertation, University of Hawaii, 1980, 137.
5. Laws, E. A. and Bannister, T. T., Nutrient- and light-limited growth of *Thalassiosira fluviatilis* in continuous culture, with implications for phytoplankton growth in the ocean, *Limnol. Oceanogr.,* 25, 457, 1980.
6. Terry, K. L., Nitrogen and phosphorus requirements of *Pavlova lutheri* in continuous culture, *Botanica Marina,* 13, 757, 1980.
7. Terry, K. L., Hirata, J., and Laws, E. A., Light-, nitrogen-, and phosphorus-limited growth of Phaeodactylum tricornutum Bohlin Strain TFX-1; chemical composition, carbon partitioning, and the diel periodicity of physiological processes, *J. Exp. Mar. Biol. Ecol.,* 86, 85, 1985.
8. Redfield, A. C., Ketchum, B. H., and Richards, A. F., The influence of organisms on the composition of seawater, in *The Sea,* Vol. II, Hill, M. N., Ed., Interscience, New York, 1963, 26.
9. Goldman, J. C., McCarthy, J. J., and Peavey, D. G., Growth rate influence on the chemical composition of phytoplankton in oceanic waters, *Nature,* 279, 210, 1979.
10. DiTullio, G. R. and Laws, E. A., Diel periodicity of nitrogen and carbon assimilation in five species of marine phytoplankton: accuracy of methodology for predicting N-assimilation rates and N/C composition ratios, *Mar. Ecol. Prog. Ser.,* 32, 123, 1986.
11. Eppley, R. W. and Peterson, B. J., Particulate organic matter flux and planktonic new production in the deep ocean, *Nature,* 282, 677, 1979.
12. DiTullio, G. R., Chemotaxonomic identification and physiological status of phytoplankton in the North Pacific Subtropical Gyre, Ph.D. dissertation, University of Hawaii, 1987, 246.
13. Morris, I., Glover, H. E., and Yentsch, C. S., Products of photosynthesis by marine phytoplankton: the effect of environmental factors on the relative rates of protein synthesis, *Mar. Biol.,* 27, 1, 1974.
14. Dortch, Q., Clayton, J. R., Jr., Thoresen, S. S., and Ahmed, S. I., Species differences in accumulation of nitrogen pools in phytoplankton, *Mar. Biol.,* 81, 237, 1984.
15. Peterson, B. J., Aquatic primary productivity and the ^{14}C-CO_2 method: a history of the productivity problem, *Annu. Rev. Ecol. Syst.,* 11, 359, 1980.
16. Grande, K. D. and others, Primary production in the North Pacific Gyre: a comparison of rates determined by the ^{14}C, O_2, and ^{18}O methods, *Deep-Sea Res.,* 36, 1621, 1989.
17. Li, W. K. W., Experimental approaches to field measurements: methods and interpretation, in *Photosynthetic Picoplankton,* Platt, T. and Li, W. K. W., Eds., Canadian Bulletin of Fisheries and Aquatic Sciences, Vol. 214, 1986, 583.
18. Taguchi, S., Dark fixation of CO_2 in the subtropical North Pacific Ocean and Weddell Sea, *Bull. Plankton Soc. Jpn.,* 30, 115, 1983.

19. Prakash, A., Sheldon, R. W., and Sutcliffe, W. H., Jr., Geographic variation of oceanic ^{14}C dark uptake, *Limnol. Oceanogr.*, 36, 30, 1991.
20. Redalje, D. G. and Laws, E. A., A new method for estimating phytoplankton growth rates and carbon biomass, *Mar. Biol.*, 62, 73, 1981.
21. Taguchi, S., DiTullio, G. R., and Laws, E. A., Physiological characteristics and production of mixed layer and chlorophyll maximum phytoplankton populations in the Caribbean Sea and western Atlantic Ocean, *Deep-Sea Res.*, 35, 1363, 1988.
22. Roberts, R. B., Abelson, P. H., Cowie, D. B., Bolton, E. T., and Britton, R. J., Studies of biosynthesis in *Escherichia coli, Carnegie Inst. Wash. Publ.*, 607, 1955, 521 pp.
23. Li, W. K. W., Glover, H. E., and Morris, I., Physiology of carbon photoassimilation by *Oscillatoria thiebautii* in the Caribbean Sea, *Limnol. Oceanogr.*, 25, 447, 1980.
24. Bligh, E. G. and Dyer, W. G., A rapid method of total lipid extraction and purification, *Can. J. Biochem. Physiol.*, 37, 911, 1959.
25. Marchesi, S. L. and Kennell, D., Magnesium starvation of Aerobacter aerogenes III. Protein metabolism, *J. Bacteriol.*, 93, 357, 1967.
26. Smucker, R. A. and Dawson, R., Products of photosynthesis by marine phytoplankton: chitin in TCA "protein" precipitates, *J. Exp. Mar. Biol. Ecol.*, 104, 143, 1986.
27. Goldman, J., Physiological processes, nutrient availability and the concept of relative growth rate in marine phytoplankton ecology, in *Primary Production in the Sea*, Falkowski, P., Ed., Brookhaven Symp. 31, Plenum Press, New York, 1980, 179.

CHAPTER 68

Membrane-Containing Fungal Mass and Fungal Specific Growth Rate in Natural Samples

Steven Y. Newell

INTRODUCTION

The kingdom Fungi is extraordinarily diverse. The number of present day species of fungi (especially *sensu lato*, including oomycotes[1] and chytridiomycotes[1]) is probably of the same order of magnitude (10^6) as that of insects.[2] One of the niches into which fungi have evolved is decomposition systems based on submerged and intertidal debris of vascular plants.[3-5] Mycelial fungi can clearly be strong competitors for the litter of plants in aquatic environments under natural circumstances.[6,7] An essential competitive advantage of these eukaryotic organo-osmotrophs is their mycelial pervasion of solid substrates, enabling internal digestion of plant macromolecules (Figure 1).

Since the fungal-mycelial network becomes an integral internal portion of the decomposing litter, fungal mass cannot be readily separated and measured by direct microscopy in the same fashion as can bacteria.[8] Although methods are available in the literature for estimating fungal mass by direct microscopy,[9,10] these are very likely to result in underestimates when used with litter, largely due to destruction (by homogenization), persistent retention, and unrecognizability of bits of the fungal-mycelial network.[11]

An alternative assay for fungal mass in decaying litter involves the use of a biochemical index, namely ergosterol.[12] The principal location of ergosterol is within the plasma membranes of fungal cells, so dead, evacuated hyphae containing no cytoplasm would not be expected to contribute to fungal-ergosterol yield, and estimates of fungal mass based on ergosterol are likely to be estimates of living mass.[5] It is only within the kingdom Fungi that ergosterol is the principal membrane sterol, and only two large groups within the kingdom do not have ergosterol as the principal membrane sterol (chytrids and rusts, neither of which are mycelial decomposers).[5]

A distinct advantage has recently been built onto the ergosterol method for estimating fungal mass: one can now also rapidly estimate fungal specific growth rate (μ) as the rate of synthesis of ergosterol.[13] "Production of mycelium in the field . . . usually causes more measurement problems than any other dimension [of fungal presence or activity]. Net production can be assessed by consecutive sampling and summing of positive changes in biomass, but gross production can be obtained only if account is taken of [fungal] biomass losses, caused by such factors as grazing and lysis."[14] There is clear potential for grazing loss of fungal mass in aquatic decomposition systems,[15-17] and preliminary application of

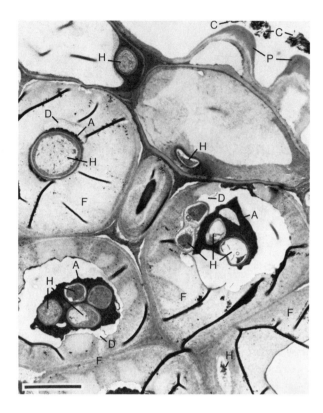

Figure 1. Transmission electron micrograph of a cross-section of a decaying leaf of *Spartina alterniflora* Loisel. (smooth cordgrass). The leaf was fixed immediately upon collection in the natural, standing state.[6] Hyphae (H) of *Phaeosphaeria spartinicola* Leuchtmann have infiltrated and are digesting the highly lignocellulosic fiber cells (F) of the leaf. Other symbols: C, external clay film; P, adaxial papilla; D, digestive erosion of lignocellulose; A, altered lignocellulosic material (+ hyphal sheath?). Bar represents 5 μm. Photo courtesy of David Porter and Wilma Lingle.

the ergosterol-synthesis method (Ac→ERG) resulted in the suggestion that total fungal production in decaying leaves of a marsh grass was 3.5 times net fungal production.[13] The Ac→ERG method involves measurement of the flow of a simple sterol-precursor molecule (acetate, labeled with ^{14}C) into ergosterol as it is synthesized by fungi (see Moriarty[18] and Riemann and Bell[19] for prokaryotic analogs), and subsequent calculation of the rate of fungal-mass formation.

The following recipe for ergosterol analysis is principally derived from Newell et al.[12] and Gessner et al.[20] methodologies founded on Seitz et al.[21,22] The instructions for measurement of acetate incorporation into ergosterol are derived from Newell and Fallon.[13]

MATERIALS REQUIRED

Equipment

- Microwave oven
- Ashing oven
- Refluxing system (multiple glassware units + 80°C water bath)

- Dry bath (40°C, aluminum wells + glass vials to fit snugly)
- Gas or air-distribution drying manifold
- Sonicating cleaning bath
- High-performance liquid chromatographic (HPLC) system, including precision pump, injection valve with sample loop, reverse-phase column, ultraviolet (UV) detector, and integrator/recorder (all parts of flow train compatible with methanol)
- Gas/liquid-tight syringe (500 μl; Teflon plunger) to fit HPLC injection valve
- Two-way valve at HPLC detector exit
- Scintillation counter
- Autopipettes (0.5 to 10 μl; 10 to 100 μl; 100 to 1000 μl; 500 to 5000 μl)

Supplies

- Inert boiling chips
- Glass scintillation vials (20 ml)
- Separatory funnels with glass collection beakers
- Glass vials (20 ml) to snugly fit dry-bath wells
- Teflon filter cartridges (0.45-μm)
- Glass vials (2 ml; Teflon-gasket screw caps)
- Safety apparel for handling toxic liquids and radioisotopes (consult local radiation-safety officials)

Solutions and Reagents

- Solvents (HPLC grade): methanol, ethanol (reagent), pentane (in repipetting containers)
- Scintillation fluor
- Ergosterol (pure, for standards)
- Potassium hydroxide
- Sodium acetate
- [1-^{14}C]sodium acetate

PROCEDURES

1. Collect samples of decaying litter. Rinse away loosely adherent mud, etc. if this is not to be considered part of the litter system (but see Newell et al.[23]). Store samples as briefly as possible, at conditions near those in the field (avoiding extremes of temperature, light, water-content). Remove small subsamples of standard size (area or volume) from each sample (e.g., cork-borer disks from leaves). Divide replicate standard subsamples into two groups (ideally, at least three replicates each), one for determination of organic mass per replicate and one for the ergosterol procedures.

2. Place organic-mass replicates in a microwave oven in open glass dishes and expose to radiation until fully dry (e.g., 6 min "cooking" in a 1000-W oven[24]). Immediately weigh (zeroed balance containing desiccant) and record dry masses. Ash (450°C, 4 h), weigh, and subtract ash mass from dry mass to find organic mass per replicate subsample.

3. Place ergosterol replicates each into a clean, sterile scintillation vial. Add a measured quantity of bacteria-free water (0.2 μm filtered; e.g., 5 ml[13]) to each vial and incubate in darkness 1 h to permit acclimation of microbes to incubation conditions. Avoid violent agitation, which can inhibit fungal activity (see, e.g., Bergbauer and Newell[25]). Add formaldehyde (final concentration, 2%) to adsorption (killed) controls (ideally, at least two replicates) and cap tightly to retain formaldehyde.

4. Add [1-^{14}C]sodium acetate in bacteria-free solution to a final concentration of 5 mM in both living and adsorption-control vials. (Pre-experimentation is required to discover whether degree of participation of radiolabel is maximized at this concentration, and the extent to which expensive radioacetate can be diluted with nonradioacetate without losing adequate final signal in ergosterol.[13]) Incubate with radiolabel for one hour. (Longer incubations may be required for samples from cold environments, e.g., mountain streams.[26] Time-course experimentation may be required to ensure that incubating samples do not show evidence of unnatural incorporation patterns.[13])

5. Terminate radiolabel incorporation by bringing living-replicate vials to 2% formaldehyde. Immediately remove litter subsamples from radiolabel solution, rinse thoroughly, and place them in vials under 5 ml methanol at 4°C in darkness. (Addition of an antioxidant [e.g., butylated hydroxytoluene] may be beneficial during storage.[20])

6. Prepare a series of concentrations of ergosterol procedural standards in 5 ml methanol encompassing the range expected for samples. If practicable, also add standard ergosterol to extra sample replicates (spiked replicates).

7. Transfer methanol and subsamples (from living incubations, adsorption controls, procedural standards, and spiked replicates) to reflux flasks. Rinse sample vials with 5 ml methanol and add rinses to reflux flasks. Add a measured additional quantity of methanol (more than enough to permit full dissolution of ergosterol; ergosterol solubility in methanol, approximately 750 μg·ml^{-1} at 20°C). Reflux at 80°C for 2 h in a fume hood (methanol is toxic and flammable); set temperature of cooling water flowing through outer jackets of reflux units much lower (e.g., 10°C) than the boiling point of methanol (65°C).

8. Dissolve potassium hydroxide (40 mg·ml^{-1}) in reagent ethanol (95:5, ethanol:water; KOH will not dissolve without the water). This solution deteriorates in storage. Add 5 ml of the ethanolic KOH to each reflux flask. (It may be beneficial to remove solids before adding KOH.[12,20]) Add a boiling chip to each flask. Reflux (80°C) 30 min.

9. Transfer liquid from each cooled reflux flask to a separatory funnel.[20] Add 10 ml distilled water to each funnel. Add 10 ml pentane (very flammable!) to each funnel and thoroughly mix (continuous agitation) for 30 s. Allow polar and nonpolar fraction to fully separate, and collect each lower, polar fraction in a separate beaker. Collect the nonpolar, upper fraction in a dry-bath vial. Return each polar fraction from the beaker to the funnel and add another 10 ml pentane. Repeat 30 s agitation and allow to fully separate. Discard lower, polar fraction (including any precipitates [probably proteinaceous or tannic] at the polar/nonpolar interface) into proper radioactive-waste container. Combine second pentane fractions with the first ones in the dry-bath vials.

10. Set dry-bath temperature at 40°C in a fume hood. Fit vials into dry-bath wells and immediately situate a drying-manifold outlet above each vial. (**Do not** allow pentane in vials to boil [36°C boiling point; flash point < − 40°C].) Direct a gentle flow of air into each vial until all pentane has evaporated.

11. Add a carefully measured small quantity (e.g., 1000 μl) of methanol to each dry-bath vial to redissolve neutral lipids. Tightly cap vials and partially submerge in sonic cleaner. Sonicate 5 min. Push each redissolved sample through a Teflon filter cartridge into a 2-ml vial. Immediately cap securely to prevent evaporation.

12. Start flow of eluent (methanol) in HPLC system; allow to equilibrate for 1 h. Set detector at 282 nm wavelength. Rinse syringe and injector valve with methanol (also done between all samples) and inject methanol blank. If baseline is satisfactory, inject pure standard. If standard peak is satisfactory, inject remaining standards and samples. During standard chromatographings, record time of appearance and completion of ergosterol peaks. Use these times, along with recorder output, to indicate when two-way valve should be activated for ergosterol collection immediately after it passes the detector beam. Place a labeled

scintillation vial handy to the two-way valve sample outlet, and use it to collect ergosterol for each sample. Rinse two-way valve with eluent between samples. Balance timing of sample and standard replicates throughout HPLC run, in case of system-operational drift, e.g., in sensitivity. Collect at least two standard ergosterol fractions for determination of background values for [14]C.

13. Add 10 ml scintillation fluor to each vial containing separated ergosterol fraction. Mix thoroughly, wipe clean, and count in a scintillation counter. Subtract counts for adsorption controls from values for no-formaldehyde incubations to find counts due to fungal synthesis of ergosterol from [^{14}C]acetate.

14. Find ergosterol concentration per sample by comparison of areas of sample and standard peaks. Divide by sample organic mass (step 2) to find ergosterol per unit organic mass. Find rate (μ_e, d^{-1}) of ergosterol synthesis based on specific activity (SA) of radioacetate solution (dpm per pmole acetate) used in incubations,[27] quenching (Q, counting efficiency) of sample scintillation count,[28] decimal fraction of sample injected (I), duration of incubation with radiolabel (T, converted to days), amount of ergosterol carbon (pg) per sample (E_c), and pg C per pmol acetate (24). Thus:

$$\mu_e = 24 \cdot CPM_e \cdot Q^{-1} \cdot SA^{-1} \cdot I^{-1} \cdot T^{-1} \cdot E_c^{-1} \qquad (1)$$

NOTES AND COMMENTS

1. A large taxon of eukaryotic, mycelial, zoosporic organo-osmotrophs, sometimes allied systematically with true fungi but more likely to be protoctists evolutionarily quite distinct from eumycotes, the oomycotes,[1] cannot be treated with the ergosterol technique, because its members do not synthesize ergosterol.[5] Also, if there is a high ratio of green microalgal or protozoan mass to fungal mass in samples, then there is the potential for interference by nonfungal ergosterol.[11]

2. See steps 3 and 4. An example sample size[13] for individual [^{14}C]acetate incubations is 6 cm² area (one side) of decaying grass leaves (approximately 70 mg organic mass; 4 weeks postsenescence) in 5 ml of acetate solution (10^8 Bq·mmol^{-1}, specific activity of added acetate, present at 5 mM concentration; 1 Bq = 27 pCi = 1 DPS). After 1.5 h of incubation at 19°C, there were about 500 cpm (adsorption control subtracted; includes background) per 50 μl HPLC injection incorporated into the ergosterol fraction of HPLC output, equal to 10^4 cpm per 6 cm² sample. Ergosterol concentration for this example was about 450 μg·g^{-1} organic mass.[13] For samples like this example, only about 5% of the added radioacetate is taken out of solution, so label exhaustion is not a problem. Adsorption controls were not statistically significantly greater than background.

3. See step 4. Tritiated sodium acetate is considerably less expensive than [^{14}C]acetate, and is likely to be an acceptable alternative radioprecursor molecule.[13]

4. See step 5. Lyophilization may be an equally effective means of preserving samples,[20] especially if samples are instantly frozen.[12]

5. See step 6. Although ergosterol is subject to photoconversion by UV light, ordinary laboratory exposure to indoor lighting will not damage samples or standards.[12] Regardless, storage of samples and stock solutions should be in darkness. But note: double-bond rearrangement can occur if sample clean up is performed by SPE (solid-phase extraction) column chromatography.[5] If SPE clean up is to be used, then standard solutions of ergosterol of concentration as low as those of samples must be used to quantitate potential loss of 282-nm UV absorbance; this loss can be severe, and it may not be proportional to ergosterol concentration.[29]

6. See steps 7 and 8. It may be equally effective to omit the 2-h methanol refluxing, and go straight to the 30-min saponification.[20] If 30-min saponification proves insufficient, longer saponification refluxing can be performed without damaging ergosterol.[12] Martin

et al.[30] have claimed that saponification (which releases esterified ergosterol) is not necessary, but it is well known that the percentage esterified of total ergosterol can be highly variable in fungal mycelia.[31] For the purpose of fungal-mass estimation, the pertinent (and still largely open) question here is: which has the more constant relationship to fungal mass, free ergosterol or free + esterified ergosterol?

7. See step 9. Some substrates (e.g., leaves of red mangrove, *Rhizophora mangle*) may yield large amounts of precipitate at the polar/nonpolar interface. It may be necessary to remove these by centrifugation or filtration (see Martin et al.[30]) if they interfere with efficient collection of the non-polar extractant.

8. See step 12. The opening and closing of the two-way valve may cause backpressure and column bleed, leading to interference with ergosterol peaks. This can be minimized by opening/closing the valve repeatedly prior to beginning the HPLC injections.

9. See step 12. Caution: although ergosterol has a UV absorbance spectrum unique among major sterols,[5] I have found large (relative to ergosterol) peaks in HPLC with 282-nm detection, that had retentivity very similar (relative retention time $\cong 0.9$) to that of ergosterol for both mangrove and sedge leaves (species of *Rhizophora* and *Carex*).[29] These have been confirmed by mass spectrometry[32] to be other than ergosterol, and are probably minor plant sterols with strong upper UV absorbance.

10. See step 13. One should remain alert to the possibility that other molecules potentially containing ^{14}C from acetate could coelute with ergosterol but not be detected because they have no 282-nm absorbance.[13] This can be checked by changing detector wavelength to 210 nm (at which more neutral lipids have absorbance), by using diode-array UV detection to check peak purity, by gas chromatography/mass spectrometry, and/or by running cpm scans of the HPLC output.[33-35] For some types of sample, normal-phase columns and/or different mobile phases may yield better separations of ergosterol than the system described herein (e.g., References 35 and 36).

11. See step 14. The specific activity of acetate may require adjustment to account for the presence of exogenous and endogenous molecules ''competing'' for incorporation into ergosterol.[13] This can be estimated via Moriarty-Pollard dilution plotting,[37] but even with this correction, true specific growth rates may be underestimated,[13] possibly due to inability to induce target fungi to fully replace endogenous sources of ergosterol-precursor molecules with those added. An alternative method of finding fungal growth rates from rates of acetate incorporation into ergosterol is to perform conversion-factor experiments.[13,38] This entails growing the fungal decomposers in simulated natural systems (likely to involve the same degree of isotope dilution) and directly measuring rates of increase in fungal mass and, in parallel, rates of [^{14}C]acetate incorporation into ergosterol.

12. See step 14. Note that it is an integral part of the secondary production strategy of mycelial fungi to grow forward within extending hyphal walls and conserve material by recycling organics (especially nitrogenous material) toward hyphal tips, closing off old hyphal compartments.[39,40] Sterols are susceptible to recycling in eukaryotes.[41] The extent to which this factor might affect interpretation of results of the [^{14}C]acetate method is unknown.

ACKNOWLEDGMENTS

I thank Donna Poppell for word processing and David Porter and Wilma Lingle for providing the electron micrograph in Figure 1. Financial support for my research into fungal-ecological methodology was provided by the U.S. National Science Foundation and the Sapelo Island Research Foundation. This is contribution 700 of the University of Georgia Marine Institute.

REFERENCES

1. Margulis, L., Corliss, J. O., Melkonian, M., and Chapman, D. J., Eds., *Handbook of Protoctista,* Jones and Bartlett, Boston, 1990.
2. Hawksworth, D. L., Presidential address 1990. The fungal dimension of biodiversity: magnitude, significance, and conservation, *Mycol. Res.,* 95, 641, 1991.
3. Suberkropp, K., Aquatic hyphomycete communities, in *The Fungal Community. Its Organization and Role in the Ecosystem,* 2nd ed., Carroll, G. C. and Wicklow, D. T., Eds., Marcel Dekker, New York, 1992, 729.
4. Bärlocher, F., Ed., *The Ecology of Aquatic Hyphomycetes,* Springer, Heidelberg, 1992.
5. Newell, S. Y., Methods for determining biomass and productivity of marine fungi, in *The Isolation and Study of Marine Fungi,* Jones, E. B. G., Ed., Wiley, Chichester, in press.
6. Newell, S. Y., Fallon, R. D., and Miller, J. D., Decomposition and microbial dynamics for standing, naturally positioned leaves of the saltmarsh grass *Spartina alterniflora, Mar. Biol.,* 101, 471, 1989.
7. Gessner, M. O. and Schwoerbel, J., Fungal biomass associated with decaying leaf litter in a stream, *Oecologia,* 87, 602, 1991.
8. Newell, S. Y., Fallon, R. D., and Tabor, P. S., Direct microscopy of natural assemblages, in *Bacteria in Nature,* Vol. 2, Poindexter, J. S. and Leadbetter, E. R., Eds., Plenum, New York, 1986, chap. 1.
9. West, A. W., Specimen preparation, stain type, and extraction and observation procedures as factors in the estimation of soil mycelial lengths and volumes by light microscopy, *Biol. Fertil. Soils,* 7, 88, 1988.
10. Lodge, D. J. and Ingham, E. R., A comparison of agar film techniques for estimating fungal biovolumes in litter and soil, in *Modern Techniques in Soil Ecology,* Crossley, D. A. and Coleman, D. C., Eds., Elsevier, The Hague, in press.
11. Newell, S. Y., Estimating fungal biomass and productivity in decomposing litter, in *The Fungal Community. Its Organization and Role in the Ecosystem,* 2nd ed., Carroll, G. C. and Wicklow, D. T., Eds., Marcel Dekker, New York, 1992, 521.
12. Newell, S. Y., Arsuffi, T. L., and Fallon, R. D., Fundamental procedures for determining ergosterol content of decaying plant material by liquid chromatography, *Appl. Environ. Microbiol.,* 54, 1876, 1988.
13. Newell, S. Y. and Fallon, R. D., Toward a method for measuring instantaneous fungal growth rates in field samples, *Ecology,* 72, 1547, 1991.
14. Frankland, J. C., Ecological methods of observing and quantifying soil fungi, *Trans. Mycol. Soc. Jpn.,* 31, 89, 1990.
15. Bärlocher, F., Newell, S. Y., and Arsuffi, T. L., Digestion of *Spartina alterniflora* Loisel. material with and without fungal constituents by the periwinkle *Littorina irrorata* Say (Mollusca:Gastropoda), *J. Exp. Mar. Biol. Ecol.,* 130, 45, 1989.
16. Bärlocher, F., The role of fungi in the nutrition of stream invertebrates, *Bot. J. Linn. Soc.,* 91, 83, 1985.
17. Kemp, P. F., Newell, S. Y., and Hopkinson, C. S., Importance of grazing on the salt-marsh grass *Spartina alterniflora* to nitrogen turnover in a macrofaunal consumer, *Littorina irrorata,* and to decomposition of standing-dead *Spartina, Mar. Biol.,* 104, 311, 1990.
18. Moriarty, D. J. W., Techniques for estimating bacterial growth rates and production of biomass in aquatic environments, in *Methods in Microbiology,* Vol. 22, Grigorova, R. and Norris, J. K., Eds., Academic, New York, 1990, 211.
19. Riemann, B. and Bell, R. T., Advances in estimating bacterial biomass and growth in aquatic systems, *Arch. Hydrobiol.,* 118, 385, 1990.
20. Gessner, M. O., Bauchrowitz, M. A., and Escautier, M., Extraction and quantification of ergosterol as a measure of fungal biomass in leaf litter, *Microb. Ecol.,* 22, 285, 1991.
21. Seitz, L. M., Mohr, H. E., Burroughs, R., and Sauer, D. B., Ergosterol as an indicator of fungal invasion in grains, *Cereal Chem.,* 54, 1207, 1977.
22. Seitz, L. M., Sauer, D. B., Burroughs, R., Mohr, H. E., and Hubbard, J. D., Ergosterol as a measure of fungal growth, *Phytopathology,* 69, 1202, 1979.

23. Newell, S. Y., Hopkinson, C. S., and Scott, L. A., Patterns of nitrogenase activity (acetylene reduction) associated with standing, decaying shoots of *Spartina alterniflora*, *Est. Coastal Shelf Sci.*, 35, 127, 1992.

24. Newell, S. Y., Arsuffi, T. L., Kemp, P. F., and Scott, L. A., Water potential of standing-dead shoots of an intertidal grass, *Oecologia*, 85, 321, 1991.

25. Bergbauer, M. and Newell, S. Y., Contribution to lignocellulose degradation and DOC formation from a salt marsh macrophyte by the ascomycete *Phaeosphaeria spartinicola*, *FEMS Microbiol. Ecol.*, 86, 341, 1992.

26. Gessner, M. O., personal communication.

27. Coleman, D. C. and Fry, B., Eds., *Carbon Isotope Techniques*, Academic Press, New York, 1991.

28. Long, E. C., Applications of Quench Monitoring by Compton Edge Effect: The "H#". Technical Report 1096-NUC-77-2T, Beckman Instruments, Fullerton, CA, 1977.

29. Newell, S. Y., unpublished data.

30. Martin, F., Delaruelle, C., and Hilbert, J.-L., An improved ergosterol assay to estimate fungal biomass in ectomycorrhizas, *Mycol. Res.*, 94, 1059, 1990.

31. Margalith, P. Z., *Steroid Microbiology*, Thomas, Springfield, IL, 1986.

32. Miller, J. D., Agriculture Canada, personal communication.

33. Görög, S., Ed., *Steroid Analysis in the Pharmaceutical Industry: Hormonal Steroids, Sterols, Vitamins D, Cardiac Glycosides*, Ellis Horwood, Chichester, 1989.

34. Xu, S., Norton, R. A., Crumley, F. G., and Nes, W. D., Comparison of the chromatographic properties of sterols, select additional steroids and triterpenoids, *J. Chromatogr.*, 452, 377, 1988.

35. Peacock, G. A. and Goosey, M. W., Separation of fungal sterols by normal-phase high-performance liquid chromatography, *J. Chromatogr.*, 469, 293, 1989.

36. Schwadorf, K. and Müller, H.-M., Determination of ergosterol in cereals, mixed feed components, and mixed feeds by liquid chromatography, *J. Assoc. Off. Anal. Chem.*, 72, 457, 1989.

37. Moriarty, D. J. W., Measurement of bacterial growth rates in aquatic systems from rates of nucleic acid synthesis, *Adv. Microb. Ecol.*, 9, 245, 1986.

38. Bell, R. T., An explanation for the variability in the conversion factor deriving bacterial cell production from incorporation of [^{3}H]thymidine, *Limnol. Oceanogr.*, 35, 910, 1990.

39. Paustian, K. and Schnürer, J., Fungal growth response to carbon and nitrogen limitation: a theoretical model, *Soil Biol. Biochem.*, 19, 613, 1987.

40. Wessels, J. G. H., Fungal growth and development: a molecular perspective, in *Frontiers in Mycology*, Hawksworth, D. L., Ed., C.A.B. International, Wallingford, U.K., 1991, 27.

41. Voelker, D. R., Organelle biogenesis and intracellular lipid transport in eukaryotes, *Microbiol. Rev.*, 55, 543, 1991.

Section V
Organic Matter Decomposition and Nutrient Regeneration

CHAPTER 69

Radiotracer Approaches for the Study of Plant Polymer Biodegradation

Ronald Benner

INTRODUCTION

Terrestrial and aquatic rooted macrophytes contribute substantial quantities of organic matter to freshwater and estuarine environments. This plant material is chemically complex and is composed primarily of structural polymers, such as cellulose, hemicellulose, and lignin, that are collectively referred to as lignocellulose. Lignocellulosic tissues are resistant to decomposition and are degraded primarily by bacteria and fungi. A variety of approaches have been used to study the microbial transformations and fates of lignocellulosic substrates in aquatic environments, and the use of radiolabeled plant tissues is discussed in this chapter. Plant tissues can be specifically radiolabeled in either the polysaccharide or lignin components, and the labeled materials can be used to determine the rates of microbial mineralization (to CO_2) and solubilization of these components.

The widespread application of radiotracer assays for the study of plant tissue decomposition during the past 15 years has led to a better understanding of the biochemistry and physiology of the decay process as well as of the environmental factors affecting rates of microbial degradation. Radiolabeled preparations, including uniformly labeled tissues, specifically labeled tissues, and purified or synthetic polymers, have been used to investigate the microbial utilization of plant material. Only the preparation, characterization, and use of "natural" substrates will be discussed herein because these preparations are most appropriate for ecological studies. The polymers comprising plant tissues are physically and chemically bound in complex associations, and the microbial degradation of specific chemical components is influenced by the matrix within which the component is embedded.

There are a variety of advantages of the radiotracer approach to the study of the decomposition of lignocellulosic tissues over the more classical approach of following bulk weight loss and changes in concentrations of specific chemical components. The radiotracer approach is highly sensitive allowing for relatively short incubation periods (days-weeks) and small sample sizes. Another important advantage of the radiotracer approach is that the fate of labeled carbon atoms in bulk tissues, specific chemical components, or specific locations within a molecule can be traced regardless of chemical or physical form.

There are, likewise, several disadvantages of the radiotracer approach. The radiolabeled carbon atoms may not be distributed among target macromolecules in plant tissues as intended during preparation, and the investigator must thoroughly characterize the distribution of

0-87371-564-0/93/$0.00 + $.50

radiolabel in the plant tissue. The use of radiolabeled substrates is, for the most part, restricted to the laboratory where it is often not possible to mimic environmental conditions. And finally, the use of radiolabeled preparations provides only limited information about molecular and submolecular changes in chemical composition during decomposition.

MATERIALS REQUIRED

Equipment

- Water bath
- Drying oven
- Freezer or desiccator
- Centrifuge
- Soxhlet apparatus
- Vacuum rotary evaporator
- Ultraviolet (UV) light source
- Liquid scintillation counter
- Analytical balance

Supplies

- Whatman GF/F filters.
- Thin-layer chromatography (TLC): cellulose-coated TLC plates. For simple sugars, solvent system consisting of formic acid:2-butanone:*tert*-butanol:water (15:30:40:15); detector spray reagent composed of diphenylamine (1 g), aniline (1 ml), acetone (50 ml), and phosphoric acid (5 ml). For esterified acids, solvent system consisting of *n*-butanol:acetic acid:water (62:15:23).
- Milk dilution bottles (125 ml) used for incubation. Flexible tubing is attached to the ends of glass tubes in #2 two-hole stoppers. Solid glass plugs are used to seal the tubes during incubation.

Solutions

- Ethanol (95%)
- Ethanol-benzene (1:2, vol/vol)
- Radiolabeled precursors, e.g., cinnamic, ferulic or coumaric acids; see discussion in Procedures section.
- 12 M (72 wt%) sulfuric acid
- Distilled water
- Acid-compatible liquid scintillation cocktail, e.g., Scintiverse II (Fisher Scientific Company)
- CO_2-absorbing liquid scintillation cocktail; a suitable cocktail can be prepared using toluene (1500 ml), methanol (1200 ml), ethanolamine (300 ml), and the scintillator BBOT (6 g)
- Barium hydroxide
- Pepsin solution (4 ml, 1% in 0.1 N HCl)
- 1 N NaOH
- 6 N HCl
- Diethyl ether
- Anhydrous $NaSO_4$
- Nitrogen gas

PROCEDURES

Preparation of Specifically ^{14}C-Labeled Plant Tissues

The basic approach used for preparation of specifically labeled plant tissues is to "feed" a radioactive precursor of the target polymer (usually lignin or polysaccharides) to a living plant so that the plant can incorporate the precursor into newly synthesized polymer within the natural matrix of the plant tissue. Several methods have been used for introducing radiolabeled precursors into living plants,[1] and I have found that the cut stem procedure works for most plant tissues, including grasses, sedges, rushes, and tree branches.[2] I had variable success in obtaining well-labeled leaves and needles using this method with tree branches, but individual leaves can also be radiolabeled using the following procedures.[2] All necessary precautions for the handling and use of radiotracers should be followed during the preparation and use of radiolabeled plant tissues.

1. Herbaceous plants (~50 cm in height) and tree branches (~1 cm in diameter) are wiped clean and cut with a razor blade. Keep the cut end moist to prevent breaking the capillary action that will carry the precursor into the cut stem.

2. Immerse the cut end in 1 to 2 ml of sterile water containing 10 to 50 μCi of precursor in a test tube.

3. Place the plants in an open-air environment with subdued natural lighting, such as a greenhouse, and allow the plant to absorb the labeled solution.

4. Monitor the level of water in the test tube and after the initial labeled solution is absorbed, add water to keep the cut ends submerged until the plants wilt (3 to 6 d).

5. After incorporation of the radiolabeled precursor the plant material is dried at 45 to 50°C. Leaves, needles, and bark should be removed from tree branches prior to drying.

6. After drying, the plant material is ground to pass a 40-mesh sieve (425 μm).

7. Unincorporated precursor and other nonstructural components of the radiolabeled plant tissues are removed by serially extracting the plant material in a Soxhlet apparatus with ethanol (95%), ethanol-benzene (1:2, vol/vol), and water.[3]

8. The extracted material, commonly referred to as lignocellulose, is dried (45 to 50°C) and stored in a sealed container in a freezer or desiccator. The lignocellulose fraction typically accounts for 60 to 90% of the total dry weight of vascular plant tissues.

9. The specific activity (dpm mg^{-1}) of radiolabeled lignocellulse preparations and other particulate samples is best determined by combusting materials (~10 mg) at high temperature (700 to 900°C) and trapping the resulting carbon dioxide in a liquid scintillation cocktail that contains an organic base, such as ethanolamine (see Solutions section, above). I know of two manufacturers of automated instruments for the combustion of radioactive particulate samples, R.J. Harvey Company (Hillsdale, NJ) and Packard Instrument Company (Downers Grove, IL). Typical specific activities for specifically radiolabeled lignocelluloses range from 4000 to 12,000 dpm mg^{-1}.

Radiolabeled Precursors for Lignins and Polysaccharides

Lignin is an aromatic polymer composed of phenylpropane subunits linked by carbon-carbon and ether linkages.[4] Phenylalanine (and tyrosine in herbaceous plants) is the precursor of the three cinnamyl alcohols that are the direct monomeric precursors of lignin.[4] Phenylalanine and tyrosine are available from commercial suppliers in a variety of radiolabeled forms and have been used to specifically radiolabel the lignin component of plant tissues.[2,5] However, these amino acids are also readily incorporated into proteins, and microbial degradation of labeled protein interferes with the determination of rates of lignin degradation.[2] Herbaceous plant tissues typically have much higher protein concentrations than woody tissues, and the labeling of proteins with [^{14}C]phenylalanine and [^{14}C]tyrosine can be a serious problem in these tissues.

Deaminated lignin precursors, such as cinnamic, ferulic, and coumaric acids, are not incorporated into protein[1,2] and are, therefore, the most appropriate precursors for labeling the lignin component of herbaceous plants. However, of these lignin precursors, I could only find cinnamic acid in radiolabeled form [side-chain labeled 3-^{14}C] from a single commercial supplier (Amersham Corporation). There are some fairly simple methods for enzymatically producing many of these lignin precursors from radiolabeled compounds that are commercially available. For example, [U-^{14}C]cinnamic acid can be produced from [U-^{14}C]phenylalanine using the enzyme phenylalanine ammonia-lyase.[6] Lignin precursors that are specifically labeled in the side-chain, ring, or methoxyl group can also be synthesized and used to investigate rates of microbial degradation.[1,7,8]

The polysaccharide components of plant tissues have been specifically radiolabeled using [U-^{14}C]glucose which is readily available from most commercial suppliers.[2,9] During the characterization of glucose-labeled plant tissues (see below), it was found that the label was distributed among a variety of simple sugars that were released upon acid hydrolysis of the plant tissues.[2] Thus, enzymatic interconversions of glucose to other sugars occur during the labeling of plant tissues, resulting in the incorporation of radiolabel in hemicellulose and other polysaccharides as well as cellulose.

Characterization of Specifically ^{14}C-Labeled Lignocelluloses

Each preparation of ^{14}C-labeled lignocellulose must be analyzed by a variety of methods to determine the distribution of label among chemical components. The most basic of these analyses is the crude separation of the polysaccharide and lignin components. The standard procedure used by wood chemists is to hydrolyze the polysaccharides in strong acid leaving lignin (Klason lignin) as a residue.[3] The acid hydrolysis procedure does not provide a quantitative separation of lignin and polysaccharides because some lignin is acid-soluble and components other than lignin, such as certain tannins and cuticular lipids, are insoluble in strong acid.

The distribution of radioactivity among specific sugars released during acid hydrolysis can be determined by separation of the sugars by TLC[10] and quantification of the radioactivity associated with specific sugars. As previously mentioned, the occurrence of radiolabeled protein in [^{14}C-lignin] lignocellulose preparations can interfere with determination of rates of lignin or polysaccharide degradation. The percentage of label associated with proteins in lignin-labeled preparations can be determined using a protease, such as pepsin. Herbaceous plant tissues often contain cinnamic acid derivatives that are esterified with lignin or carbohydrates, and it appears that they are more susceptible to microbial degradation than core lignin.[2] Therefore, the percentage of radiolabel associated with esterified acids also should be determined for [^{14}C-lignin] lignocellulose preparations.

Lignin Residue and Simple Sugars

1.   Radiolabeled lignocellulose (100 mg) is hydrolyzed using 12 M (72 wt%) sulfuric acid (2 ml) at 20°C for 2 h with occasional stirring.

2. The sample is diluted to 1.2 M sulfuric acid with distilled water and heated to 100°C in a water bath for 3 h.

3. The hydrolyzed sample is filtered (tared Whatman GF/F for weight loss determination), washed with warm water, and the combined acid-soluble filtrate is radioassayed.

4. A subsample of the acid soluble filtrate (1 ml) is added directly to an acid-compatible liquid scintillation cocktail (10 ml). The samples are radioassayed using a liquid scintillation counter and appropriate methods for quench correction.

5. The following steps pertain to assay of radioactivity in simple sugars.

6. The filtrate from the acid hydrolysis is neutralized (pH 5.5 to 6.0) with barium hydroxide and the precipitate is removed by centrifugation.

7. The neutralized filtrate is concentrated to a volume of ~5 ml by rotoevaporation under vacuum.

8. The concentrated filtrate from the acid hydrolysis is chromatographed using cellulose-coated TLC plates and a solvent system consisting of formic acid:2-butanone:*tert*-butanol:water (15:30:40:15). Enhanced separation of sugars is achieved using multiple developments. After development, the sugars can be detected using a spray reagent composed of diphenylamine (1 g), aniline (1 ml), acetone (50 ml), and phosphoric acid (5 ml). The spots on the TLC plate corresponding to specific sugars are scraped and radioassayed.

Protein

1. Lignocellulose (100 mg) is digested in a pepsin solution (4 ml, 1% in 0.1 N HCl) at 40°C for 20 h.[3] The weak acid in the pepsin solution can hydrolyze some polysaccharides so the fraction of label solubilized from [^{14}C-polysaccharide] lignocellulose preparations is not necessarily associated with proteins.

2. The pepsin-digested lignocellulose is thoroughly washed with warm water, and the filtrate and particulate fractions are radioassayed.

Esterified Acid Fraction

1. Lignocellulose samples (40 mg) are treated with 10 ml of 1 N NaOH at 20°C for 20 h.

2. The solution is filtered through a tared (for weight loss determination) Whatman GF/F filter and washed well with warm water.

3. The combined filtrates are acidified (pH 2) with 6 N HCl and extracted with diethyl ether (10 ml, twice).

4. A subsample of the combined filtrate (near neutral pH) is saved prior to extraction for radioassay.

5. The combined ether extracts are passed through a column of anhydrous NaSO$_4$ to remove water and dried under a stream of nitrogen.

6. The extract is redissolved in ethanol (5 ml), and a subsample is saved for radioassay.

7. Subsamples of the extract are chromatographed on cellulose TLC plates with *n*-butanol:acetic acid: water (62:15:23).[11]

8. The cinnamyl acid derivatives can be detected by viewing the plates under UV light. The spots corresponding to specific acids are scraped and radioassayed.

Incubation Procedures

A variety of incubation systems for both aerobic[2] and anaerobic[12] conditions can be used for determining the rates of mineralization and solubilization of [^{14}C] lignocellulose preparations. The relatively simple system that I use for maintaining aerobic conditions is convenient for handling up to 80 incubation vessels at a time. Milk dilution bottles (125 ml) are used because they are easily packed.

1. About 10 mg of [^{14}C] lignocellulose is added to each bottle along with 10 to 20 ml of water or a water-sediment inoculum collected from the field.

2. The bottles are aerated every 24 or 48 h for 15 min with filter-sterilized air.

3. The exhaust line from each bottle is connected to a series of two scintillation vials containing CO_2-absorbing liquid scintillation cocktail. The mineralization rate of the [^{14}C] lignocellulose is monitored by radioassaying the CO_2 traps after each aeration. A rack with gangvalves is used to accommodate up to 20 bottles at a time for aeration.

4. Incubations typically last from 10 to 30 d. At the end of the incubation period the contents of the bottle are filtered (0.2-μm pore size), and the filtrate is acidified and radioassayed for acid-stable dissolved organic carbon,[13] (i.e., radioactivity remaining after dissolved inorganic carbon is driven off by acidification).

REFERENCES

1. Crawford, R. L., *Lignin Biodegradation and Transformation,* Wiley-Interscience, New York, 1981.
2. Benner, R., Maccubbin, A. E., and Hodson, R. E., Preparation, characterization, and microbial degradation of specifically radiolabeled [^{14}C]-lignocelluloses from marine and freshwater macrophytes, *Appl. Environ. Microbiol.,* 47, 381, 1984.
3. Browning, B. L., *Methods of Wood Chemistry, Vol. 2,* Interscience Publishers, New York, 1967.
4. Sarkanen, K. V. and Ludwig, C. H., *Lignins: Occurrence, Formation, Structure, and Reactions,* Wiley Interscience, New York, 1971.
5. Crawford, D. L. and Crawford, R. L., Microbial degradation of lignocellulose: the lignin component, *Appl. Environ. Microbiol.,* 31, 714, 1976.
6. Pometto, A. L. and Crawford, D. L., Enzymatic production of the lignin precursor *trans*-[U-^{14}C]cinnamic acid from L-[U-^{14}C]phenylalanine using L-phenylalanine ammonia-lyase, *Enzyme Microb. Technol.,* 3, 73, 1981.
7. Kirk, T. K., Coonors, W. J., Bleam, R. D., Hackett, W. F., and Zeikus, J. G., Preparation and microbial decomposition of synthetic [^{14}C]-lignins, *Proc. Natl. Acad. Sci. U.S.A.,* 72, 2515, 1975.
8. Haider, K. and Trojanowski, J., Decomposition of specifically ^{14}C-labeled phenols and dehydropolymers of coniferyl alcohol as models for lignin degradation by soft and white rot fungi, *Arch. Microbiol.,* 105, 33, 1975.
9. Crawford, D. L., Crawford, R. L., and Pometto, A. L., Preparation of specifically labeled ^{14}C-[Lignin]- and ^{14}C-[Glucan]-lignocelluloses and their decomposition by the microflora of soil, *Appl. Environ. Microbiol.,* 33, 1247, 1977.
10. Vomhof, D. W. and Tucker, T. C., The separation of simple sugars by cellulose thin-layer chromatography, *J. Chromatogr.,* 17, 300, 1965.
11. Hartley, R. D., Carbohydrate esters of ferulic acid as components of cell walls of *Lolium multiflorum, Phytochemistry,* 12, 661, 1973.

12. Benner, R., Maccubbin, A. E., and Hodson, R. E., Anaerobic biodegradation of the lignin and polysaccharide components of lignocellulose and synthetic lignin by sediment microflora, *Appl. Environ. Microbiol.*, 47, 998, 1984.
13. Benner, R., Moran, M. A., and Hodson, R. E., Biogeochemical cycling of lignocellulosic carbon in marine and freshwater ecosystems: relative contributions of procaryotes and eucaryotes, *Limnol. Oceanogr.*, 31, 89, 1986.

CHAPTER 70

Estimating Degradation Rates of Chitin in Aquatic Samples

Michael T. Montgomery and David L. Kirchman

INTRODUCTION

Chitin is a polymer of *N*-acetylglucosamine and rivals cellulose as the most abundant biopolymer in nature. It is found in the exoskeletons of all arthropods and in fungal cell walls. It is also used for buoyancy by common marine diatoms, such as *Thallasiosira* and *Skeletonema*. Because of its high abundance and relatively short turnover time,[1] chitin may be an important source of carbon and nitrogen for heterotrophic bacteria. Chitin degradation can be measured by determining concentrations and turnover rates.

Chitin concentrations in aquatic samples can be estimated by incubating particles with a labeled protein, wheat germ agglutinin (WGA), that binds specifically to the *N*-acetylglucosamine residues that comprise chitin. The amount of label associated with the particles is then related to the chitin concentration in the sample. The assay is relatively simple and rapid, and common aquatic particles, such as clay, bacteria, and cellulose, do not interfere with the chitin measurements.[1]

Turnover rates of chitin can be measured using a substrate analog, methylumbelliferyl-*N*,*N'*-diacetyl-chitobioside (MUF-DC). Analogs have been commonly used to measure extracellular enzyme activity in aquatic samples.[2-4] The chemical bond between the *N*,*N'*-diacetyl-chitobiose and the methylumbelliferyl group is similar to that between the *N*-acetylglucosamine groups of chitin and is hydrolyzed by chitinase. Hydrolysis results in the release of methylumbelliferone which is highly fluorescent and can be measured easily using a fluorometer. The turnover time of chitin in aquatic samples can be estimated from the released fluorescence and the amount of MUF-DC added.

Advantages of the assays include: (1) short assay times (approximately 1 h to measure chitin concentration and as little as 3 h to measure chitin turnover); (2) few steps and reagents; (3) the chitin concentration assay measures chitin that is not sterically hindered from degradation by chitinases and thus, is likely to be measuring chitin that is available for degradation by heterotrophic bacteria; (4) particles commonly found in aquatic samples, such as cellulose, clay (kaolin); (mg per sample), and bacteria (up to 10^7 cells per sample) do not interfere with the chitin concentration assay; and (5) detection of chitinase activity using substrate analogs is sensitive and, after incubation with MUF-DC, fluorescence can be measured later with frozen samples.

Disadvantages of the assay are: (1) the assay for chitin concentrations can only measure chitin that is not sterically hindered from binding with the WGA; (2) the particle size

0-87371-564-0/93/$0.00 + $.50

distribution of chitin used for the standard curve should reflect that of the sample because the assay is surface-area dependent; and (3) hydrolysis of MUF-DC may underestimate chitin turnover because chitinase has a lower affinity for MUF-DC relative to chitin.

MATERIALS REQUIRED

Equipment

- Vacuum pump
- Rotary shaker table
- Scintillation counter
- Fluorometer, <365 nm cut-off primary filter and >460 nm cut-off secondary filter
- Superspeed centrifuge

Supplies

- Polycarbonate filters, 25-mm diameter, pore size = 0.2 μm or greater
- 10 tower filtration manifold
- Centricon-10 protein concentrators (Amicon)
- 20-ml borosilicate glass scintillation vials (Beckman)
- 7-ml plastic and borosilicate scintillation vials

Solutions

- Succinylated WGA (suc-WGA; Vector Labs, Burlingame, CA; or EY Labs, CA)
- Sodium ^{3}H-borohydride (NaB^{3}H$_4$; Amersham)
- 0.2 M borate buffer, pH 8.9
- 0.1 M formaldehyde
- Bovine serum albumin (BSA; Sigma)
  Chitin, purified from crab shell (Sigma)
- 10 mM phosphate or borate buffer, pH 7.4
- Scintillation cocktail
- Methylumbelliferyl-diacetyl-chitobioside (MuF-DC; Sigma)
- Methylumbelliferone (MuF; Sigma)
- Dimethylformamide (Sigma)
- Glycine (50 mM)/ammonium hydroxide (200 mM) solution, pH 10.5

PROCEDURES

Preparation of ^{3}H-WGA

The suc-WGA is labeled by reductive methylation with NaB^{3}H$_4$ according to the method of Tack et al.[5]

1. Dissolve 2 mg of suc-WGA in 200 μl of 0.2 M borate buffer, pH 8.9.

2. Mix with 0.5 mCi of NaB^{3}H$_4$ and 20 μl of 0.1 M formaldehyde.

3. Incubate the mixture on ice for 10 min.

4. Rinse the mixture at least four times using 1 ml of borate buffer each time in Centricon-10 protein concentrators (10,000 MW cut-off; Amicon) with a superspeed centrifuge. This

rinsing procedure removes unreacted B^3H_4 from the mixture. After final rinse, bring final volume to about 1 ml with borate buffer or other appropriate buffer.

Measuring Chitin Concentrations

1. Filter sample with suspended particles onto 0.2-μm (diameter = 47 mm) polycarbonate membrane filters or glass fiber filters.

2. Cut filters with associated particles into ca. 8 pieces and add to 3 ml of 10 mM borate or phosphate buffer (pH 7.4) in a 20-ml borosilicate glass vial. In samples that have high concentrations of clay and silt that may bind WGA nonspecifically, 100 μg of BSA may be added to the sample. The chitin samples are now ready for assay.

3. When measuring 15 μg of chitin or less (typical concentration in 1 L of seawater), add 10 μg of ^{3}H-WGA to the concentrated chitin sample (3 ml) in a 20-ml glass vial.

4. Incubate for 30 min (25°C) on a shaker table (130 rpm).

5. Filter samples onto a 0.2-μm (25-mm pore size) polycarbonate filter.

6. Rinse the sample three times with the borate buffer (5 ml each rinse).

7. Radioassay the filters.

8. A standard curve must be derived by repeating the assay with known amounts of chitin particles (typically 1 to 20 μg per sample). The solution may be buffer or sample water filtered free of chitin particles. The radioactivity associated with the filter is proportional to the amount of chitin in the sample.

Measuring Turnover Rate of Chitin

1. Dissolve MUF-DC and methylumbelliferone standard (MUF) in separate vials with 1 ml of 100% dimethylformamide (final concentration of 5 mM).

2. Add 50 μl of MUF-DC to 4.95 ml of seawater sample (500 μM final MUF-DC concentration in the sample). Conduct assay as soon as possible after collecting the sample. No pretreatment of sample should be necessary.

3. Incubate on the shaker table at *in situ* temperature. The optimal incubation time should be determined, but 2 to 10 h is typical.[6]

4. After incubation, kill the enzymatic reaction with 0.5 ml of the glycine-KOH solution, and then either process immediately or freeze (−20°C) for later analysis.

5. Measure released fluorescence (MUF) using a fluorometer. The amount of released MUF corresponds to the amount of MUF-DC degraded during the incubation time. Concentration of free MUF in the sample is determined using a standard curve of various MUF concentrations and then plotting fluorescence value vs. MUF concentration.

Calculations

The chitin concentration is calculated from the amount of ^{3}H-WGA binding to chitin in the sample and collected on the filter, which is compared with the standard curve.

The turnover rate of chitin is the ratio of MUF-DC degraded (nmol released MUF 1^{-1} h^{-1}) to MUF-DC added to the sample (nmol 1^{-1}). The degradation rate is calculated by multiplying the turnover rate and the chitin concentration in the sample: Degradation rate (μg chitin degraded 1^{-1} h^{-1}) = (MUF released h^{-1}/MUF-DC added) × (μg chitin 1^{-1}).

NOTES AND COMMENTS

1. Polycarbonate filters should be used for measuring chitin concentrations because nonspecific binding of WGA to polycarbonate is lower than to glass fiber filters (0.1 μg vs. 0.3 to 0.5 μg per filter). High nonspecific binding reduces the sensitivity of the assay.

2. Samples for chitin concentrations can be frozen ($-20°$C) if needed for long-term storage. Because formaldehyde (1%) inhibits WGA binding to chitin, sodium azide (0.2%) should be used to inhibit bacterial growth during time-course experiments.

3. Addition of BSA eliminates adsorption of WGA onto nonchitin particles by blocking nonspecific sites that bind any protein.

4. The specific acitvity of the prepared ^{3}H-WGA should be measured (radioactivity per mg WGA) because it is useful for comparing results between batches.

5. In addition to using a standard curve of chitin to calculate chitin concentrations, it may prove useful to calculate concentrations from an internal standard. The assay is run with and without addition of a known amount of chitin. The added chitin should be approximately the same concentration as the sample.

6. Controls for nonspecific binding of WGA to nonchitinous particles include degrading samples with chitinase prior to assay to eliminate chitin from these samples. Another control is to include 10 mM chitotriose (Sigma) to samples prior to adding WGA. The chitrotriose will occupy the chitin-binding site on the WGA, and thus, radioactivity associated with the sample with added chitotriose would be due to nonspecific binding of the WGA. These values for nonspecific binding should be subtracted from those of the sample for the actual measure of WGA that bound specifically to chitin in the sample.

7. Samples for chitin turnover should not be stored prior to addition of the substrate, as doing so may result in an underestimate of enzyme activity. However, once incubated with the MUF-DC, samples may be frozen ($-20°$C) prior to assay for fluorescence.[6] The free MUF liberated by action of the enzyme is stable at $-20°$C for several weeks.[6]

8. More information about chitin turnover can be obtained by measuring the MUF-DC hydrolysis rate with various amounts of added chitin. The turnover rate can be calculated similar to the procedure used by Wright and Hobbie.[7]

REFERENCES

1. Montgomery, M. T., Welschmeyer, N. A., and Kirchman, D. L., A simple assay for chitin: application to sediment trap samples from the subarctic Pacific, *Mar. Ecol. Prog. Ser.*, 64, 301, 1990.

2. Somville, M. and Billen, G., A method for determining exoproteolytic activity in natural waters, *Limnol. Oceanogr.*, 28, 190, 1983.

3. Hoppe, H. G., Significance of exoenzymatic activities in the ecology of brackish water: measurements by means of methylumbelliferyl-substrates, *Mar. Ecol. Prog. Ser.*, 11, 299, 1983.

4. Chrost, R. J., Characterization and significance of β-glucosidase activity in lake water, *Limnol. Oceanogr.*, 34(4), 660, 1989.

5. Tack, B. F., Dean, J., Eilat, D., Lorenz, P. E., and Schechter, A., Tritium labelling of proteins to high specific activity by reductive methylation, *J. Biol. Chem.*, 255, 8842, 1980.

6. Chrost, R. J. and Velimirov, B., Measurement of enzyme kinetics in water samples: effect of freezing and soluble stabilizer, *Mar. Ecol. Prog. Ser.*, 70, 93, 1991.

7. Wright, R. T. and Hobbie, J. E., Uptake of organic solutes in lake water, *Limnol. Oceanogr.*, 10, 22, 1965.

CHAPTER 71

Measurement of Dimethylsulfide (DMS) and Dimethylsulfoniopropionate (DMSP) in Seawater and Estimation of DMS Turnover Rates

Ronald P. Kiene

INTRODUCTION

The flux of the biogenic sulfur gas, dimethyl sulfide (DMS), from natural environments represents a major input of sulfur to the Earth's atmosphere.[1-4] The oceans contribute >50% of the global emission of DMS, therefore, the biogeochemical cycling of DMS in seawater is of great interest. In addition, the flux of DMS has important implications for the chemistry of the troposphere and, possibly, climate regulation.[2,5,6]

Most of the DMS in seawater appears to be derived from the degradation of dimethyl-sulfoniopropionate (DMSP), which is produced in large amounts by some phytoplankton.[7,8] Recent models of DMS cycling in seawater indicate that the distribution and concentration of DMS is governed by complex interactions in the food web involving phytoplankton, zooplankton, protistan grazers, and bacteria.[9-12]

Investigations of DMS and DMSP cycling offer the marine microbial ecologist some unique opportunities to study the interactions of microorganisms in the sea. However, studies of DMS and DMSP biogeochemistry also pose some significant challenges and the *in situ* cycling of DMS and DMSP have only recently been investigated in any detail. This paper is designed to give a description of the basic analytical tools required to measure DMS and DMSP, and to describe a method which has been used to estimate *in situ* rates of DMS turnover.

MATERIALS REQUIRED

Equipment

- Gas chromatograph (GC) with flame photometric detector
- Chromatographic column (examples include Carbopack BHT-100 or Chromosil 330, available from Supelco)
- Recording device (integrator or strip-chart recorder)

0-87371-564-0/93/$0.00 + $.50
© 1993 by Lewis Publishers

Supplies

- Compressed gases (helium or nitrogen carrier gas, air and hydrogen)
- Purge and trap system (see Figure 1)
- Glass syringe (5 to 20 ml)
- Teflon-faced septa
- Microliter syringes (10 to 100 μl)
- Glass fiber filters (Whatman GF/F)
- Incubation bottles (250 to 2000 ml, glass, Teflon, or polycarbonate)

Solutions and Reagents

- Chloroform (high-performance liquid chromatography [HPLC] grade)
- DMS and DMSP standards; Reagent DMSP-HCl can be obtained from Research Plus, Inc., Bayonne, NJ.
- Liquid nitrogen

PROCEDURES

Description of Equipment

DMS concentrations in surface seawater range from <1 to 100 nM, with the world average concentration being only 3.0 nM.[2] The most common approach to measuring DMS in seawater involves the use of a purge-and-trap system combined with a GC and flame photometric detector (FPD). Volatile components, including DMS, are purged from seawater, concentrated in a cold trap (at liquid nitrogen temperatures in this case), subsequently introduced into the GC where they are separated, and finally detected by the FPD. This procedure works exceedingly well for DMS because the FPD is highly ($>10^5$) selective for compounds containing sulfur and also because the FPD is extremely sensitive (<1 pmol of DMS can be detected). These advantages result in relatively simple chromatography and few interferences.

A diagram of the author's purge-and-trap system is shown in Figure 1. Many variations on this general theme are used by different investigators,[13-15] but the general principal is the same in each case. The system in Figure 1 and the procedure described below is designed for analysis of small water samples (2 to 10 ml) which require a minimum of time to purge. With water samples of 2 ml it takes about 5 min to completely process one sample with the author's system. This is considerably shorter than the 10 to 20 min reported by other investigators, who used larger volumes of water (and therefore, require longer sparging times). The capability of running a large number of samples quickly is a great asset when designing and conducting experiments to investigate the microbial ecology of DMS.

The sparging system consists of a glass tube with a porous glass frit on which the water rests (see Figure 1). The water sample is injected through a Teflon-faced septum on a side port above the frit. The septum is held on by an aluminum crimp. Helium (ultrahigh purity) is used as the sparge gas and is introduced under the frit at a rate of 100 ml min^{-1}. Volatile sulfur compounds are stripped from the water sample and carried with the sparge gas through a Nafion dryer (a membrane selectively permeable to water vapor; Perma-Pure, Inc., Toms River, NJ) and into a six-port stainless steel valve (Valco). With the valve in the trap position, the sample gas stream passes through a 30-cm long (3 mm o.d.) Teflon loop immersed in liquid nitrogen. A small amount (0.2 cm^3) of Teflon wool is placed in the

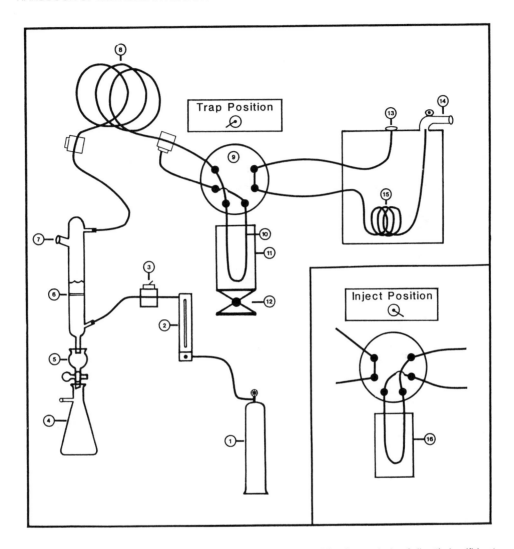

Figure 1. Diagram of a gas trapping and analysis system used for the analysis of dimethyl sulfide: 1, Helium sparge gas; 2, rotometer; 3, toggle valve; 4, vacuum flask; 5, Omnifit two-way valve; 6, porous glass frit; 7, crimp seal port; 8, Nafion dryer; 9, Valco six-port valve; 10, sample loop of 1/8" Teflon tubing; 11, dewar with liquid N2; 12, lab jack; 13, GC injector port; 14, flame photometric detector; 15, analytical column; and 16, dewar with hot (>70°C) water. (Reproduced from Kiene, R. P. and Service, S. K., *Mar. Ecol. Prog. Ser.*, 76, 1, 1991. With permission.)

middle of the Teflon loop to increase the surface area for gas trapping. When the sparging is completed (2.5 min for a 2 ml water sample; larger water samples require longer sparge times), the valve is turned to the inject position which puts the Teflon loop in line with the column in the GC. The loop is then quickly placed in hot (>70°C) water and the carrier gas from the injector port of the GC is used to sweep the sample out of the Teflon loop and into the chromatographic column.

The author uses a Shimadzu GC-9A equipped with a Shimadzu flame photometric detector. Equipment from any number of manufacturers can be used (some investigators build their own systems). Most investigators involved in seawater analysis use packed columns for separating the sulfur gases. The most commonly used are made of Teflon (2 m × 1/8" diameter) and are filled with either Carbopack BHT-100 or Chromosil 330 (both available from Supelco, Inc. Bellefonte, PA). The chromatographic conditions vary depending on the column and complexity of the gas mixture being analyzed. If only DMS is being measured in samples of seawater, we run the GC isothermally (100°C for Carbopack and 60°C for Chromosil). The carrier gas is He at 60 ml min^{-1}. The flame gases (air and hydrogen) are set according to the manufacturer of the FPD being used. Under these conditions, DMS elutes with a retention time of about 1.0 min, and is adequately separated from other sulfur gases. Peak areas are recorded with an integrator. Note that the FPD response is a square function; that is, the square root of the peak area is a linear function of the amount of DMS passing the detector. Some manufacturers incorporate linearizing algorithms in their detector electronics, and some integrators can be programmed to do the calculations automatically.

DMS Analysis

1. DMS primary standards are prepared by carefully weighing small amounts (usually 10 to 100 mg) of DMS into ethylene glycol. Andreae and Barnard[13] used degassed glycol, but in practice, I have not found any problems with using nondegassed glycol provided standards are stored in the dark and are used within 1 month. I make my standards in small (14 ml) serum vials which are filled to within 0.2 ml of capacity with ethylene glycol and sealed with Teflon-faced septa. The weight of the glycol in the bottle is used to accurately determine the volume of the diluent. DMS is then injected through the septum and the weight of added DMS determined. This procedure minimizes both the loss of DMS during the weighing step and uptake of water vapor by the glycol.

2. The primary standard is diluted to the required concentration into known volumes of glycol using microliter syringes. Microliter quantities (5 to 50 µl) of standard are then injected directly into the sparging system which contains 2 ml of presparged distilled water. The standard is then sparged and trapped in the same way as samples. Then DMS standards prepared in glycol can be compared to DMSP standards which are treated with base to give DMS (as described below). This type of comparison between the two different standards yields agreement to within 5%.[15,16]

3. Water samples are obtained by withdrawing a subsample from the sampling or incubation bottles with a Teflon tube (1.5-mm diameter) which is attached to a glass-barreled syringe by means of a luer fitting. To avoid degassing of DMS, care should be taken to avoid agitation of the sample or pulling excessive vacuum in the syringe.

4. The Teflon tube is removed from the syringe and replaced by a filter unit (Gelman), with a 2.5-cm glass fiber filter (Whatman GF/F or Gelman A/E, 0.7- and 1.0-µm nominal retention, respectively) and an attached 22-gauge needle. The volume of water taken up should be approximately 2 ml greater than the amount actually injected into the sparger, to allow for dead volume in the filter unit. The volume in the syringe is carefully adjusted to the desired volume (usually 2 to 10 ml) and the sample is then introduced with gentle pressure via a septum into the gas stripping system.

DMSP Analysis

The equipment needed for DMSP analysis is essentially the same as that for DMS. The only additional requirements are a filter tower and a DMSP standard.

Particulate DMSP

Particulate DMSP (DMSP$_{part}$) is defined as that retained on a glass fiber filter (Whatman GF/F or Gelman A/E, nominal retention of 0.7 and 1.0 µm, respectively). The GF/F is recommended for use in open ocean waters. I have not observed any significant $DMSP_{part}$ in seawater which has been passed through a GF/F, even though most bacteria and some autotrophs may pass these filters.

1. Water samples should be gently mixed before 5 to 20 ml is removed for analysis. The sample is allowed to filter by gravity or with a very low (<1 cmHg) vacuum using a filter tower. The filters are then placed into (14-ml) serum vials. At this point two approaches can be taken, depending on the amount of DMSP present. If the amount of DMSP on the filters is sufficiently high (this amount depends on the sensitivity of the GC), then headspace analysis can be used. This is much simpler and faster (~1 min per sample) than the second alternative.

2. For headspace analysis, 2 ml of 5 M NaOH is added and the bottles quickly capped with Teflon-faced septa. The samples are then allowed to react for at least 6 h; usually 12 to 24 h at 25°C during which time DMSP is quantitatively and stoichiometrically decomposed to DMS and acrylate.[17,18] The DMS evolved into the headspace is measured by removing a subsample of the headspace gas and injecting it directly into the GC column. DMSP standards are prepared by adding known amounts of DMSP to the vials and treating these exactly as the samples. Stock solutions should be stored frozen.

3. A second, more sensitive, approach involves adding a known volume of NaOH to nearly fill the serum vial (I use 13 ml in a 14-ml vial). After sealing the vial with a septum and allowing the reaction to proceed to completion, a 1-ml subsample is then removed and injected into the sparging tube and analyzed for DMS as described above. Precision for both these procedures is about 10%.

Dissolved DMSP

Dissolved DMSP ($DMSP_{diss}$) is defined as that which passes through glass fiber filters. Gentle filtration (preferably by gravity) should be used to avoid disruption of cells and liberation of $DMSP_{diss}$. For $DMSP_{diss}$ analysis, I typically use the filtrate from the $DMSP_{part}$ measurements described above. The water sample should be sparged to ensure that any residual DMS is removed. The water sample can then be analyzed immediately or stored for future analysis. If the sample is to be stored, it is essential that it be placed on ice or frozen immediately. This is because DMSP can degrade even in 0.2-μM filtered seawater.[14,19] Storage of 2 to 3 ml water samples at −20°C in polyethylene Omni Vials (Wheaton) results in no appreciable loss of $DMSP_{diss}$ over several months.

1. For analysis of $DMSP_{diss}$, a 1-ml subsample of water is injected into the sparging tube and 1 ml of 5 M NaOH added to produce DMS from the DMSP.

2. The sample is sparged for 5 min after which time tests have shown that >98% of the DMSP is converted to DMS. The DMS is cryotrapped and quantified as described above. Precision is generally in the range of 5 to 10% but is occasionally >10% for reasons discussed below.

Estimation of *In Situ* DMS Turnover Rates

The microbial removal of DMS is thought to be a major sink for this compound in seawater[10,20,21] and it is, therefore, a major factor governing the amount released to the atmosphere. Only a few estimates of DMS consumption rates have been published.[10] These estimates were made with a chloroform inhibition technique which is described below.

The chloroform method is based on the selective inhibition of DMS consumption, by 500 μM $CHCl_3$. This concentration of $CHCl_3$ appears to be effective at blocking most of the

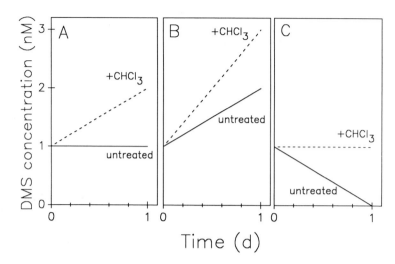

Figure 2. Idealized time courses illustrating the effects of 500 μM CHCl$_3$ on DMS concentrations in seawater. Three cases, which are typical of natural samples, are represented: (A) DMS is in a steady state in untreated samples; (B) net production of DMS occurs in untreated samples; and (C) net consumption of DMS occurs in untreated samples. In each case the chloroform inhibition methods yield a biological consumption rate of 1 nM DMS per day.

DMS consumption,[10,12,22] while having no effect on DMS production from DMSP$_{diss}$.[11,12] In the presence of CHCl$_3$, production of DMS continues, but consumption is blocked. Therefore, the concentration of DMS in the CHCl$_3$-treated seawater increases faster than in untreated water (Figure 2). By making time-course measurements of DMS concentrations, the rate of change of DMS in treated and untreated samples can be determined. The difference between the slope of the CHCl$_3$ treatment and the unamended treatment represents the DMS consumption rate. The method actually measures the production rate of DMS in the absence of biological consumption.

Significant DMS production may or may not occur in a given water mass. It is often observed that DMS concentrations in control samples (untreated) may either remain steady, increase, or decrease during the incubation. This pattern depends on whether there is any net production of DMS. In each case, the slope of DMS vs. time will be greater in samples treated with CHCl$_3$. Figure 2 shows an idealized diagram of these different patterns with the DMS consumption rate being the same in each case. I have performed about 100 chloroform inhibition experiments and in practically every case, DMS concentrations increased faster in the presence of CHCl$_3$. Data from a real application of the technique to estuarine water with a very high DMS consumption rate are shown in Figure 3.

Experiments are carried out by incubating freshly collected water samples at the *in situ* temperature and measuring the concentration of DMS over time. One set of samples is treated with CHCl$_3$ to a level of 500 μM (about 10 μl pure CHCl$_3$ per 250 ml of sample). Incubation vessels should, of course, be clean and nonreactive. I have used glass, Teflon, and polycarbonate bottles with Teflon being the preferred choice. Chloroform reacts with the polycarbonate bottles if droplets are allowed to rest on the bottom. For this reason, separate polycarbonate bottles should be reserved for use with CHCl$_3$, and not used for other purposes.

There are several advantages to the chloroform inhibition technique. First, the procedure is relatively easy, requiring straightforward DMS measurements and duplicate or triplicate water samples for each of the two treatments (total of four or six samples). If very short time courses are required, single samples of each treatment may be used, provided that the

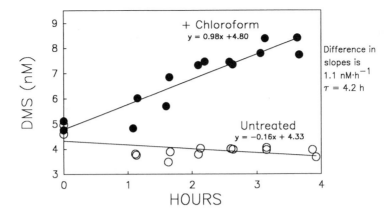

Figure 3. The effects of chloroform (500 μM) on DMS concentrations in estuarine water samples compared with untreated samples. Duplicate sample bottles for each treatment were used and the data for each bottle is plotted. Data are from water obtained from South End Creek, Sapelo Island, GA, July, 1991. The *in situ* temperature was 29°C. τ is the turnover time which is the ambient concentration divided by the consumption rate.

investigator knows from experience that water samples will behave as replicates (to within 10% for DMS) except for the $CHCl_3$ treatment. The second advantage is that reasonably short (<12) incubations are required in most cases. There are some who argue that any incubation in a bottle is too long, thus, this can also be viewed as a disadvantage as well. A third advantage is that no initial perturbation of the DMS concentration needs to be made, as would be the case for ^{14}C-DMS. The DMS concentration will, however, increase over time in the $CHCl_3$ treatment. Another advantage is that the method is sensitive enough to estimate DMS turnover rates as low as 0.5 to 1 nM d^{-1}. The sensitivity of the method depends on the ability of the analytical system to measure significant differences in the slopes of the two treatments within a reasonable incubation time.

The most significant disadvantage of the method is that a toxic "inhibitor" compound must be added to the seawater sample. Although it appears that $CHCl_3$ is selectively inhibitory to DMS consumption and not DMS production, the effects of $CHCl_3$ on the entire biological community have not been tested. An error in the rates determined by the chloroform method would arise if the $CHCl_3$ caused excess release of either $DMSP_{diss}$ or DMS from the $DMSP_{part}$ pool (phytoplankton or grazers). Previous studies[11,24] did not observe any release of DMSP upon addition of $CHCl_3$. However, new findings[22] suggest that 500 μm $CHCl_3$ may cause a small transient increase in $DMSP_{diss}$ concentrations, possibly leading to an overestimate of DMS turnover rates. Thus, the inhibitor approach requires further scrutiny. Another disadvantage of the method is that, while simple, it is labor intensive and time consuming to make rate estimates from a given parcel of water. Consider that each DMS measurement takes at least 5 min. Multiply this by four bottles (two untreated and two with $CHCl_3$) and then by four time points for each treatment and you get 80 min of analysis time. In reality this is even longer, because standard curves need to be run periodically. In order to obtain a depth profile of DMS consumption rates that includes four or five depths, it will require almost constant attention at the GC for an entire day.

Estimation of DMSP Turnover

There are currently no published or generally accepted methods for measuring *in situ* DMSP production or consumption rates. This lack of adequate methodology represents a serious impediment to our further understanding of the microbial ecology and biogeochemistry of DMSP and DMS. As mentioned above, the complexities of the DMS/DMSP cycle and our general lack of understanding of processes involved, coupled with the lack of readily available radioisotopes, have limited progress in this area.

NOTES AND COMMENTS

1. There is some debate in the literature as to whether samples should be filtered prior to DMS analysis.[15] Filtering has the advantages of removing $DMSP_{part}$ which would degrade in the stripper, but it has the drawback that it might cause disruption of cells and release of DMS. The design of the stripper in Figure 1 requires that samples be filtered prior to sparging because $DMSP_{part}$ would accumulate on the glass frit as the water is drained through it. This $DMSP_{part}$ would eventually degrade to DMS and cause unacceptably high and variable blanks. In most cases, it has been observed that filtering does not cause artificially high DMS concentrations in natural samples.[15] In some isolated cases, filtration may be problematic, and this will probably depend on the state of the phytoplankton community in the sample. Intercomparisons of filtered and unfiltered water for DMS analysis as well as multi-investigator intercalibrations both reported in the literature[15] and unpublished (R. P. Kiene, T. S. Bates, P. Matrai, G. Wolfe, and D. Cooper, unpublished data) have shown that DMS measurements are relatively robust toward sample handling and that reasonable agreement among different investigators and analytical systems can be achieved. In contrast to DMS, it should be noted that measurements of DMSP may be subject to filtration artifacts (see below).

2. In the analysis of $DMSP_{diss}$, cases with high variability may be due to release of $DMSP_{diss}$ during the filtration process. This phenomenon is poorly understood, and no published studies have adequately addressed this problem. When used on subsamples from the same water, glass fiber filters generally yield lower $DMSP_{diss}$ concentrations compared to polycarbonate membrane filters (0.2-μm) or cellulose ester filters (0.45-μm), suggesting that they cause the least amount of DMSP release.[25]

3. Although it might seem logical to employ radiotracer methods for DMS turnover measurements, there are some significant limitations associated with isotope methods.[22] First, ^{14}C-DMS is costly and only sparingly available and ^{35}S-DMS is not available at all. Second, the specific activity of ^{14}C-DMS (about 22 mCi mmol^{-1}) is too low to allow tracer additions to seawater with typical DMS concentrations of 1 to 4 nM. Finally, procedures are laborious; there are many potential products of DMS metabolism[26] and special considerations must be taken while working with volatile radioactive compounds. Despite these limitations, the use of ^{14}C-DMS holds some promise, and at least one investigator is making progress with its application.[22,27]

4. It is not yet clear whether samples for DMS turnover should be incubated in the dark or in the ambient light regime. While incubation in the light would seem to be preferable, this adds the possibility that DMS could be photochemically destroyed during the incubation.[28] In theory, the $CHCl_3$ should affect only the biological consumption of DMS and therefore, rates determined in the light or dark should be the same. However, on a recent cruise, the author found that $CHCl_3$ promoted the degradation of DMS in the light,[29] and therefore, it may not be appropriate to incubate $CHCl_3$-treated samples in the light. More work on this topic is necessary.

ACKNOWLEDGMENTS

Support for this research was provided by the National Science Foundation Grant OCE-88-17442 and the University of Georgia Marine Institute. This is contribution #701 of the

University of Georgia Marine Institute. The assistance of S. K. Service and M. Gaylor are acknowledged.

REFERENCES

1. Andreae, M. O., The ocean as a source of atmospheric sulphur compounds, in *The Role of Air Sea Exchange in Geochemical Cycling,* Buat-Menard, Ed., D. Reidel Publishing, Dordrecht, 1986, 331.
2. Andreae, M. O., Ocean-atmosphere interactions in the global biogeochemical sulfur cycle, *Mar. Chem.,* 30, 1, 1990.
3. Erickson, D. J., Ghan, S., and Penner, J., Global ocean-to-atmosphere dimethyl sulfide flux, *J. Geophys. Res.,* 95, 7543, 1990.
4. Bates, T. S., Lamb, B., Guenther, A., Digon, J., and Stoiber, R., Sulfur emissions to the atmosphere from natural sources, *J. Atmos. Chem.,* 14, 315, 1992.
5. Charlson, R. J., Lovelock, J. E., Andreae, M. O., and Warren, S. G., Oceanic phytoplankton, atmospheric sulfur, cloud albedo and climate, *Nature,* 326, 655, 1987.
6. Bates, T. S., Charlson, R. J., and Gammon, R. H., Evidence for the climatic role of marine biogenic sulphur, *Nature,* 329, 319, 1987.
7. Keller, M. D., Bellows, W. K., and Guillard, R. R. L., Dimethyl sulfide production in marine phytoplankton, in *Biogenic Sulfur in the Environment,* Saltzman, E. and Cooper, W. J., Eds., American Chemical Society, Washington, D.C., 1989, 167.
8. Burgermeister, S., Zimmerman, R., Georgii, H., Bingemer, H., Kirst, G., Janssen, M., and Ernst, W., On the origin of dimethylsulfide: relation between chlorophyll, ATP, organismic DMSP, phytoplankton species and DMS distribution in Atlantic surface water and atmosphere, *J. Geophys. Res.,* 95, 20607, 1990.
9. Wakeham, S. G., Howes, B. L., and Dacey, J. W. H., Biogeochemistry of dimethylsulfide in a seasonally stratified coastal salt pond, *Geochim. Cosmochim. Acta,* 51, 1675, 1987.
10. Kiene, R. P. and Bates, T. S., Biological removal of dimethyl sulfide from seawater, *Nature,* 345, 702, 1990.
11. Kiene, R. P. and Service, S. K., Decomposition of dissolved DMSP and DMS in estuarine waters: dependence on temperature and substrate concentration, *Mar. Ecol. Prog. Ser.,* 76, 1, 1991.
12. Kiene, R. P., Dynamics of dimethyl sulfide and dimethylsulfoniopropionate in oceanic water samples, *Mar. Chem.,* 37, 29, 1992.
13. Andreae, M. O. and Barnard, W. R., Determination of trace quantities of dimethyl sulfide in aqueous solutions, *Anal. Chem.,* 55, 608, 1983.
14. Turner, S. M., Malin, G., Bagander, L. E., and Leck, C., Interlaboratory calibration and sample analysis of dimethyl sulphide in water, *Mar. Chem.,* 29, 47, 1990.
15. Turner, S. M., Malin, G., and Liss, P. S., The seasonal variation of dimethyl sulfide and dimethylsulfoniopropionate concentrations in nearshore waters, *Limnol. Oceanogr.,* 33, 364, 1988.
16. Kiene, R. P., unpublished data.
17. White, R. H., Analysis if dimethyl sulfonium compounds in marine algae, *J. Mar. Res.,* 40, 529, 1982.
18. Dacey, H. W. H. and Blough, N., Hydroxide decomposition of DMSP to form DMS, *Geophys. Res. Lett.,* 14, 1246, 1987.
19. Kiene, R. P., unpublished data.
20. Wakeham, S. G. and Dacey, J. W. H., Biogeochemical cycling of dimethyl sulfide in marine environments, in *Biogenic Sulfur in the Environment,* Saltzman, E. S. and Cooper, W. J., Eds., American Chemcial Society, Washington D.C., 1989, 152.
21. Leck, C., Larsson, U., Bagender, L. E., Johansson, S., and Hajdu, S., Dimethylsulfide in the Baltic Sea: Annual variability in relation to biological activity, *J. Geophys. Res.,* 95, C3, 3353, 1990.

22. Wolfe, G. V. and Kiene, R. P., Determination of dimethyl sulfide turnover rates in seawater: comparison of chloroform inhibition and radiotracer methods, submitted for publication.
23. Kiene, R. P., Dimethyl sulfide production from dimethylsulfoniopropionate in coastal seawater samples and bacterial cultures, *Appl. Environ. Microbiol.*, 56, 3292, 1990.
24. Kiene, R. P., unpublished data.
25. Kiene, R. P., Effects of filtration and filter type on the determination of dissolved and particulate DMSP in seawater, in preparation.
26. Taylor, B. F. and Kiene, R. P., Microbial metabolism of dimethyl sulfide, in *Biogenic Sulfur in the Environment*, Saltzman, E. S. and Cooper, W. J., Eds., American Chemical Society, Washington D.C., 1989, 202.
27. Wolfe, G. V., personal communication.
28. Brimblecombe, P. and Shooter, D., Photo-oxidation of dimethylsulphide in aqueous solution, *Mar. Chem.*, 19, 343, 1986.
29. Kiene, R. P., unpublished data.

CHAPTER 72

Sulfate Assimilation by Aquatic Microorganisms

Russell L. Cuhel

INTRODUCTION

Net flux of sulfur from the inorganic mineral nutrient form into biomass, primarily protein, occurs only through the reductive sulfate assimilation pathway of microorganisms (i.e., bacteria, fungi, and algae) and plants. The sulfur content of bulk protein is essentially constant among all organisms, and the predominance of protein synthesis in sulfur metabolism makes it a good tracer of this crucial growth-related process. In addition, reductive sulfate assimilation has become of interest in the global sulfur cycle, since reduced volatile sulfur gases contribute to cloud formation and hence, planetary albedo. Unlike other macronutrient elements (e.g., C, N, P), sulfur is available primarily in one form, sulfate, and its concentration almost never limits growth in any but the most soft-water environments. Hence, problems associated with tracer addition experiments are minimized with respect to sulfur metabolism.

Incubations for sulfate assimilation experiments follow procedures common for almost any rate process measurement involving mixed microbial communities. After inoculation with ^{35}S-sulfate, any of a variety of *in situ*, simulated *in situ*, light gradient, or dark incubation techniques at ambient temperature may be used. Care must be taken to maintain natural light/dark cycles for euphotic-zone microplankton, however, and deeper samples should be dark incubated to avoid contributions from phototrophs which are viable but inactive *in situ*. In freshwaters, incubations as short as 1 to 2 h may be used, but marine samples typically require half-day incubations. Time-course measurements are always recommended. The subcellular fractionation of labeled components of filtered cells yields information pertaining not only to protein synthesis rates but also to the autotrophic vs. heterotrophic contributions to total community protein synthesis.

Advantages of the method are: (1) total microbial community protein synthesis and sulfur metabolism is measured, quantifying the source term for the organic component of the sulfur cycle; (2) growth-saturating environmental concentrations of sulfate result in sulfate assimilation in direct proportion to biomass production requirements, eliminating complications arising from luxury uptake; (3) very small internal pools of sulfur amino acid precursors lead to rapid equilibration of exogenous radioisotopic tracers and hence, quantitative determination of protein synthesis rates; (4) aspects of sulfur metabolism are specific for the photoautotrophic component of the microplankton assemblage, permitting a semi-quantitative interpretation of the relative net protein synthesis rates of phytoplankton vs. heterotrophs; and (5) the small demand for sulfur relative to carbon permits reductive sulfate assimilation

0-87371-564-0/93/$0.00 + $.50
© 1993 by Lewis Publishers

by photoautotrophs at night, and protein synthesis is a major factor in determining the respiration rate of algae at night. Thus, sulfate assimilation complements carbon dioxide fixation measurements over natural day/night cycles.

Disadvantages of the method are: (1) the high sulfate concentration of natural waters relative to growth requirements demands addition of high activities of ^{35}S-sulfate (up to 10 μCi ml^{-1} in oligotrophic marine environments) and hence, makes blank problems more acute; (2) sulfate is assimilated by microbial phototrophs and heterotrophs alike, preventing interpretation of rates specifically in the context of bacterial or algal metabolism; (3) dark incubations are not sufficient to eliminate algal sulfate assimilation, particularly at night; (4) whole cell uptake of sulfate is inadequate for interpreting biosynthetic rates, thus, a moderately involved subcellular fractionation procedure is required to isolate the radioactive tracer in metabolic products of known significance.

MATERIALS REQUIRED

Equipment

- Centrifuge with horizontal rotor capable of 5000 × g
- Water bath capable of 60 and 98°C
- Fume hood for water bath
- Liquid scintillation counter
- Vortex mixer

Supplies

- Repeater pipette or Repipets for reagents
- Micropipettes
- Tissue grinder for glass fiber filters (I use 15-ml heavy duty conical centrifuge tube tissue grinders [BellCo Glass, Vineland, NJ; #1977-20012] in which grinding and all extraction steps are carried out in the same tube, minimizing transfer losses, otherwise 12- to 15-ml conical centrifuge tubes are also required for each sample)
- Pasteur pipettes (9'' with latex bulbs) for withdrawing supernatants, five per sample
- 10-ml graduated cylinders, two per sample (Tekk brand student cylinders without bases are practical because they fit in test tube racks and are wide enough to permit entry of 1 ml automatic pipette tips)
- 12 × 75 mm disposable glass tubes, two per sample
- 16 × 100 mm screw-cap tubes with rubber liners, one per sample
- Liquid scintillation vials (LSV), seven per sample
- Test tube racks, many

Solutions

- 10% trichloroacetic acid (TCA)
- 80% ethanol (ETOH) in water, pH 6.5
- Diethyl ether (DEE)
- 10% TCA containing 0.5% (v:v) n-butanol (TCAB)
- 0.2 N NaOH
- Deionized water (DIW)
- 1 M Tris buffer, pH 12.5
- 1 M Tris buffer, pH 6.5

- 1 M BaCl$_2$
- 50 mM Na$_2$SO$_4$ in 50% ETOH
- Versatile liquid scintillation cocktail (Aquasol, Universol, or similar brand capable of high water content)

PROCEDURES

Incubation of Natural Plankton Assemblages

Sampling and Inoculation

Whenever upper mixed-layer or euphotic-zone populations are analyzed, it is best to sample the water column sufficiently before local dawn to permit dispensing, inoculation, and return to incubation conditions by first light. Sample handling should be undertaken in dim red light if possible. The array of *in situ*, simulated *in situ*, and light gradient methods typically used for carbon dioxide fixation measurements are perfectly suitable for sulfate assimilation studies. The sulfate concentration must be determined for each environment investigated. This is readily obtained by standard ion chromatography techniques for freshwater. In nonestuarine marine samples, the sulfate concentration is directly related to that of chloride (and hence, salinity) through the relationship $[SO_4^{2-}] = [CL^-] * 0.14$ (in weight terms, i.e., mg L^{-1}).[1] Both light and dark bottle incubations are necessary, typically of 250 ml (freshwater) to 500 ml (seawater) volume and containing 1 (freshwater) to 5 to 10 (seawater) μCi of ^{35}SO$_4^{2-}$ per ml. Using sterile technique, working isotope solutions for 1:1000 to 1:10,000 addition (i.e., 10 to 100 mCi ml^{-1}) are prepared from factory stocks (Na$_2$^{35}SO$_4$, >1000 Ci mmol^{-1}; available from virtually all radioisotope suppliers) by dilution with either distilled water or filtered natural water and should be filter-sterilized (0.2-μm syringe filter) before each use. Zero-time blanks (filtered immediately after isotope addition) are the only valid blank. Samples cannot be fixed with preservatives of any kind, but must be filtered at the end of incubation. A typical experiment involving two light and one dark bottle at each of six depths or light intensities and three time points (e.g., midday, full day, and 24 h) requires approximately 15 mCi ^{35}SO$_4^{2-}$. Bought individually, this would cost about $400, however, the cost per millicurie declines exponentially with amount bought at one time. An oceanographic cruise or freshwater season's work (500 mCi) can be undertaken for under $3000 in isotope costs.

Filtration

The high activities encountered in sulfate assimilation tracer studies require special attention to adsorption of unassimilated sulfate onto filters. While the fractionation procedure is effective in removing much of this blank, it is advisable to minimize contamination. The best method involves use of glass fiber filters (25-mm Whatman GF/F is recommended) together with a stainless steel "punch funnel" which excises the periphery of the filter after rinsing (see Figure 1 in Cuhel et al.;[2] may be produced readily in a local machine shop). Alternatively, after filtering the sample and rinsing three times with 10 to 20 ml of filtered natural water, remove the filter tower and rinse the edge of the filter with a circular motion of the squirt bottle for three to four circuits around the edge. Total activity samples (10 to 100 μl) for specific activity determination should be taken from each sample prior to filtering,

PELLET   TREATMENT   SUPERNATANT   FRACTION

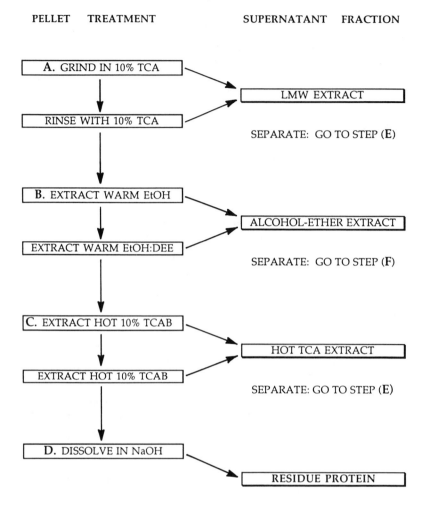

Figure 1.   Flowchart showing various ^{35}S-labeled components of microbial biomass, separated by a serial extraction scheme based on the work of Roberts et al.,[3] with minor modifications for filtered samples.

either adding directly to liquid scintillation cocktail or to an empty vial (sulfate is nonvolatile). Filters should then be frozen at −20°C until processing.

Subcellular Fractionation of Labeled Plankton

The various ^{35}S-labeled components of microbial biomass are separated by a serial extraction scheme based on the work of Roberts et al.[3] with minor modifications for filtered samples (see Figure 1). Each of the derived fractions is amenable to further chemical or chromatographic analysis if desired, and several of these are important in the separation of reductive vs. nonreductive sulfate assimilation.[4] This procedure is adapted for but not specific to ^{35}S-labeled cells: the basic procedure is applicable to any biomass tracer and works equally well with animal tissue[5] as with microplankton. Centrifugation steps may be carried out at room temperature for all steps.

Extraction of Low Molecular Weight (LMW) Metabolites

1. Grind filter to homogeneity in 0.25 to 0.5 ml 10% TCA in conical tissue-grinder style tube, rinsing the pestle four times with 0.5 ml 10% TCA.

2. Chill in refrigerator or ice bath at least 30 min or overnight (it is convenient to do this step the night before the extraction series).

3. Vortex, centrifuge 5000 × g for 20 min. Remove supernatant to a 10-ml graduated cylinder with 9'' Pasteur pipette. Leave pipette in cylinder for next supernatant (rinse).

4. Resuspend pellet in 1 ml 10% TCA, vortex well, and rinse tube sides twice with 0.5 ml 10% TCA. Centrifuge as above and combine supernatant with first extract, discarding pipette afterwards.

5. Record volume of combined extract and vortex cylinder well. Subsample as described below (see section on separation of sulfate from organic sulfur compounds).

Extraction of Lipids and Alcohol-Soluble Protein

1. Resuspend pellet from LMW extract in 1 ml 80% ETOH, pH 6.5, vortex, rinse tube sides twice with 0.5 ml 80% ETOH.

2. Incubate in 60°C water bath for 20 min.

3. Remove from bath, agitate lightly to mix sample, centrifuge 5000 × g for 20 min. Remove supernatant to 16 × 100 mm screw-cap tube, leaving pipette in extract for next extract.

4. Resuspend pellet in 2 ml 80% ETOH:DEE 1:1 (v:v), vortex, rinse tube sides twice with 1 ml 80% ETOH:DEE, 1:1 (v:v).

5. Incubate in 60°C water bath 20 min., agitate lightly to mix sample, centrifuge as above, and combine supernatant with previous extract, discard pipette.

6. During hot-acid hydrolysis (see below), begin separation of lipids from alcohol-soluble protein as described below.

7. If it is necessary to fit the procedure into an 8-h day schedule, this is a convenient stopping point for the first day. Place about two drops (0.1 ml) of TCAB on the pellets, cover the rack of samples with a paper towel, and store in a refrigerator overnight. The operation may resume with the following step the next day with no problems.

Extraction of Polymeric Carbohydrates and Nucleic Acids

1. Resuspend pellet from lipid extract in 0.5 ml 10% TCAB, vortex, and rinse tube sides twice with 1 ml 10% TCAB.

2. Incubate in 97 to 100°C water bath for 20 min.

3. Cool in ice bath until cold to touch, vortex lightly, and centrifuge 5000 × g for 20 min. Remove supernatant to 10-ml graduated cylinder with 9'' Pasteur pipette. Leave pipette in cylinder for next supernatant (rinse).

4. Repeat previous hot-acid extraction step, combining supernatants after centrifugation, discard pipette.

5. Record volume of combined extract and vortex cylinder well. Subsample as described below (see section on separation of sulfate from organic sulfur compounds).

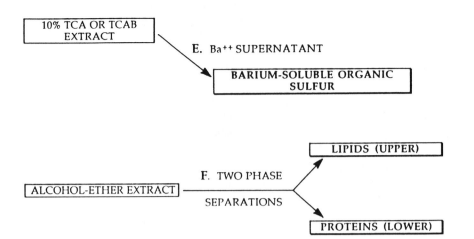

Figure 2. Flowchart showing the separation of (E) inorganic sulfate from organic sulfur compounds and (F) lipids from alcohol-soluble protein.

Residue Protein

1. The remaining material is protein. Add 0.75 ml 0.2 N NaOH, vortex well, and withdraw into LSV.

2. Rinse tube with 1 ml DIW, vortex well, and combine with NaOH solution.

3. Rinse tube with 0.75 ml 1 M Tris pH 6.5 and add to LSV contents. The Tris neutralizes the base, reducing chemiluminescence significantly.

4. Add fluor to make a gel.

Separation of Sulfate from Organic Sulfur Compounds

In previous steps (low molecular weight and polymeric carbohydrates) it is necessary or informative to separate inorganic sulfate from organic sulfur compounds (polysaccharide sulfate esters in the extraction of polymeric carbohydrate step) (see Figure 2E).

1. First, subsample 2 ml of each extract into a 12 × 75 mm tube, add 0.05 ml 50 mM Na_2SO_4 in 50% ETOH, vortex, and while vortexing add 0.25 ml 1 M $BaCl_2$.

2. Continue vortexing until turbidity begins to develop ($BaSO_4$ starts to precipitate in a few seconds), then place samples in a refrigerator for 2 h or more.

3. Meanwhile, withdraw a known volume of the original extracts directly into an LSV (use as much of the remainder as practical).

4. Neutralize the TCA with 0.25 ml of 1 M Tris, pH 12.5, per ml of sample, and add fluor. This is the "total LMW" (in low molecular weight above) or "total hot TCA" (HTCA) (in polymeric carbohydrates above) fraction. The total LMW fraction should contain all of the blank radioactivity (as inorganic sulfate) and hence, is used only as a check on filtration and rinsing effectiveness.

5. After the barium precipitation is complete, agitate the tubes slightly to break up any surface film and centrifuge at about 2500 × g for 10 min (a clinical desk-top centrifuge at full speed is fine).

6. Carefully remove an aliquot (1 to 1.5 ml) of the clear supernatant to an LSV, taking care not to disturb the BaSO$_4$ pellet.

7. Add Tris pH 12.5 and fluor as above. This is the "barium soluble" LMW or HTCA fraction (BaLMW or BaHTCA) and contains organic sulfur compounds.

Separation of Lipids from Alcohol-Soluble Protein

The combined supernatants from (extraction of lipids) above contain both hydrophobic proteins (e.g., membrane proteins) common to all microorganisms, and lipids such as the plant sulfolipid which are restricted to photoautotrophs. These are separated by phase partitioning between DEE and aqueous solution, and should be done in a fume hood (see Figure 2F).

1. To the combined supes in the screw-cap tube add 3.5 ml DEE and vortex well.

2. Add 4 ml DIW and vortex viciously until a uniform emulsion is obtained.

3. Centrifuge the tubes at about 2500 × g for 10 min to separate phases and drive all aqueous solution to the lower phase.

4. Draw off the upper (ether) phase into an LSV.

5. Re-extract the aqueous phase with 3.5 ml DEE, proceeding as above, and combine ether phases in the LSC vial. This is the "lipids" extract.

6. Pour off the aqueous phase into another LSV and rinse the tube with 1.5 ml of 80% ETOH. This is the "alcohol-soluble protein" fraction.

7. Add fluor to both fractions to make a clear fluid.

Liquid Scintillation Counting

Up to 40 samples are conveniently processed at one time (12 to 15 h total) if facilities permit. The preset ^{14}C channel of most liquid scintillation counters may be used for ^{35}S because of the similarity of their emission spectra. Since ^{35}S has a half-life of 87 d, decay correction is usually not necessary as long as (1) the total activity samples are counted along with the fractions, and (2) the whole run takes less than 2 d to count. Calculations (H below) are amenable to spreadsheet format, and to maximize simplicity, it is convenient to count all (n) samples from each fraction in order, i.e., total activities $(1-n)$, total LMW $(1-n)$, BaLMW, alcohol, lipid, total HTCA, BaHTCA, and protein. The statistical terminator of 20 min or 1% error (whichever comes first) is recommended.

Calculations

Determine the specific activity for each sample as dpm ng S^{-1} or dpm nmol^{-1} S using the total activity and the measured sulfate concentration (for weight calculations, in terms of S, not SO$_4^{2-}$). For each fraction, calculate the total radioactivity in the fraction using (dpm in aliquot)(ml extract)(ml counted)$^{-1}$, adjust by (L filtered)$^{-1}$, apply the specific activity term (dpm ngS^{-1})$^{-1}$ to get e.g., ng S L^{-1}, and subtract the blank (zero-time sample) in the same units. For alcohol, lipid, and protein components the whole fraction is counted. For the barium soluble fractions, (ml counted) must be multiplied by 0.8696 to account for added reagents. The results will include (from the perspective of S metabolism):

1. Total LMW: since this fraction contains the unincorporated sulfate blank, it is not involved in the final calculations (but see Notes and Comments).

2. BaLMW: contains free S-amino acids, glutathione and other peptides, vitamins and coenzyme A, as well as other intermediates.

3. Alcohol: contains about 10% of the total cellular protein from membranes.

4. Lipid: contains plant- or algae-produced sulfolipids including sulfoquinovosyl diglyceride (the ubiquitous "plant sulfolipid"[6]) and chlorosulfolipids of more limited distribution.[7]

5. Total HTCA: contains hydrolyzed polysaccharides, RNA, DNA, and some hot-acid soluble protein. Sulfur is not a quantitatively significant component of either RNA, DNA, or polysaccharide, but polysaccharide sulfate esters are relatively common along algae.

6. BaHTCA: contains components of (5) above minus precipitated sulfate from esters.

7. Ester-sulfate: the calculated (i.e., 5 and 6) amount of unreduced sulfur present as polysaccharide sulfate esters.

8. Protein: contains the majority of the cellular protein.

After calculating the assimilation of S into each of the derived fractions listed above, the following summations are appropriate:

- "Total Protein" = 3 + 6 + 8 from above. In weight terms, total protein times 91 equals protein synthesis (protein = 1.1% S by weight).[8]
- "Total Reduced S" = 1 + 3 + 6 + 8 from above. This is the term describing reductive transformation.

In order to effectively resolve the distribution of ^{35}S among the six important fractions, a minimum assimilation of 3000 dpm is required. To maximize isotope use efficiency, added activity needs may be estimated for upper water column samples if the carbon dioxide fixation rate and ambient sulfate concentration are approximately known. Sulfate assimilation tends to be on the order of 1% of carbon assimilation (by weight) under optimal photosynthetic conditions. Total sulfate uptake or incorporation into any particular fraction is calculated from:

$$\text{ng S L}^{-1}\text{h}^{-1} = (\text{UPTAKE})(\text{CONC})(\text{SIZE})^{-1}(\text{ACT})^{-1}(\text{t})^{-1} \tag{1}$$

where UPTAKE is the dpm assimilated per sample, CONC is the available ng SO_4-S per ml, SIZE is the liters incubated per liters filtered, ACT is the final activity of added ^{35}S in dpm ml^{-1}, and t is hours of incubation. By estimating the result and rearranging for ACT, necessary minimum ^{35}S additions can be calculated.

"Total S" = 2 + 3 + 4 + 5 + 8 from above. This reflects the removal of sulfate from the water column in terms of the total S cycle.

Depth profiles may be plotted and integrated to obtain areal sulfate assimilation for any fraction or combination of fractions. Dark samples, while informative, should not be subtracted from light values in any calculation.

NOTES AND COMMENTS

1. At the high activities used for $^{35}SO_4^{2-}$ assimilation studies, contamination with unfixed sulfate is always a concern. A good indication of error will be found in the total LMW

fraction. If this is much higher than the BaLMW, the sample should be regarded with caution. However, even a large background can be eliminated if the first step (LMW) is carefully done.

2. Sulfate uptake by algae is light-stimulated, not light-dependent, and nontrivial dark uptake is to be expected. Furthermore, algae experiencing an adequate daytime light field will usually continue protein synthesis well into or throughout the night.[9-11] Daily areal production estimates should take this into account.

3. Light stimulation is minimal during very short (<2 h) incubations, but becomes more pronounced in half- and full-day incubations of phytoplankton, however, light:dark ratios rarely exceed 10. The light:dark ratio cannot be used as a measure of bacterial activity.

4. Bacteria produce neither sulfur-containing lipids nor ester-sulfate components with very rare exception. Hence, the ratio of residue protein-S:lipid-S is very high (>100) in bacterial cultures and low (about 5 to 10) for algae.[4] This ratio, along with the presence or absence of significant ester sulfate, can be used as an index of bacterial contributions to total community protein synthesis.[12]

5. In most cases the subcellular distribution of ^{35}S is independent of light intensity for mixed natural populations.[12]

REFERENCES

1. Rosenbauer, R. J., Bischoff, J. L., and Seyfried, W. E., Determination of sulfate in seawater and natural brines by ^{133}Ba and membrane dialysis, *Limnol. Oceanogr.*, 24, 393, 1979.

2. Cuhel, R. L., Taylor, C. D., and Jannasch, H. W., Assimilatory sulfur metabolism in marine microorganisms: characteristics and regulation of the sulfate transport system in *Pseudomonas halodurans* and *Alteromonas luteoviolaceus*, *J. Bacteriol.*, 147, 340, 1981.

3. Roberts, R. B., Abelson, P. H., Cowie, D. B., Bolton, E. T., and Britten, R. J., Studies of biosynthesis in *Escherichia coli*, Carnegie Inst. Wash. Publ. No. 607, 1–521, 1963.

4. Cuhel, R. L. and Lean, D. R. S., Protein synthesis by lake plankton measured using in situ carbon dioxide and sulfate assimilation, *Can. J. Fish. Aquat. Sci.*, 44, 2102, 1987.

5. Jannasch, H. W., Cuhel, R. L., Wirsen, C. O., and Taylor, C. D., An approach for in situ studies of deep-sea amphipods and their microbial gut flora, *Deep-Sea Res.*, 27, 867, 1980.

6. Benson, A. A., The plant sulfolipid, *Adv. Lipid Res.*, 1, 387, 1963.

7. Mercer, E. I. and Davies, C. L., Chlorosulpholipids in algae, *Phytochemistry*, 14, 1545, 1975.

8. Jukes, T. H., Holmquist, R., and Moise, H., Amino acid composition of proteins: selection against the genetic code, *Science*, 189, 50, 1975.

9. Cuhel, R. L., Ortner, P. B., and Lean, D. R. S., Night synthesis of protein by algae, *Limnol. Oceanogr.*, 29, 731, 1984.

10. Lancelot, C. and Mathot, S., Biochemical fractionation of primary production by phytoplankton in Belgian coastal waters during short- and long-term incubations with ^{14}C-bicarbonate. I. Mixed diatom population, *Mar. Biol.*, 86, 219, 1985.

11. Morris, I. and Skea, W., Products of photosynthesis in natural populations of marine phytoplankton in the Gulf of Maine, *Mar. Biol.*, 47, 303, 1978.

12. Cuhel, R. L. and Lean, D. R. S., Influence of light intensity, light quality, temperature, and daylength on uptake and assimilation of carbon dioxide and sulfate by lake plankton, *Can. J. Fish. Aquat. Sci.*, 44, 2118, 1987.

CHAPTER 73

Determination of Nitrogenase Activity in Aquatic Samples Using the Acetylene Reduction Procedure

Douglas G. Capone

INTRODUCTION

In many aquatic systems, N_2 fixation is an important component of the nitrogen (N) cycle.[1-3] Our knowledge of the occurrence and importance of N_2 fixation has been greatly fostered over the last two decades by the introduction and application of the acetylene (C_2H_2) reduction assay.

Biological N_2 fixation is the reduction of dinitrogen gas to ammonia with the concurrent evolution of H_2 according to the reaction:[4]

$$N_2 + 8H^+ + 8e \rightarrow 2NH_3 + H_2 \tag{1}$$

This reaction is catalyzed by the enzyme nitrogenase. Nitrogenase is a molybdenum-enzyme complex found in phylogenetically diverse prokaryotes (i.e., eubacteria and archebacteria), of physiologies ranging from aerobic heterotrophs, aerobic and anaerobic phototrophs to strict anaerobes.[4] Among the varied organisms possessing nitrogenase, the enzyme has been highly conserved and is extremely O_2 sensitive even in the strict aerobes. In the absence of Mo, several diazotrophs can synthesize alternative nitrogenases with iron or vanadium.[5]

Nitrogenase can also reduce a relatively broad range of other small molecular weight triply bonded compounds such as azide, cyanide, and C_2H_2.[6,7] C_2H_2 was recognized potentially to be a useful substrate for a biochemical assay for nitrogenase. The reaction for C_2H_2 reduction is:

$$C_2H_2 + 2H^+ + 2e^- \rightarrow C_2H_4 \tag{2}$$

In the presence of C_2H_2 at saturating levels, N_2 fixation is almost wholly inhibited and H_2 evolution is minimal.[8]

The C_2H_2 reduction method, first described by Stewart et al.[9] and Hardy et al.,[10] provided a convenient, very sensitive, relatively simple and inexpensive means to determine nitrogenase activity. Earlier approaches were either not sufficiently sensitive (e.g., increases in cell N) or tedious and not readily accessible to researchers in general (e.g., $^{15}N_2$ fixation, see Reference 11). Many of the aquatic studies using C_2H_2 reduction have been summarized recently.[3,11-13]

Numerous problems and limitations with application of the C_2H_2 reduction method have come to light. The reader is referred to the comprehensive summaries by Taylor,[14] Payne,[15] and Oremland and Capone[16] to obtain a perspective on the potential limitations and the possible relevance of such limitations to particular applications. For instance, in replacing the natural substrate, C_2H_2 inhibits N_2 fixation itself and may, therefore, disrupt N metabolism. Various physiological groups of bacteria such as nitrifiers, methanogens, and hydrocarbon oxidizers are inhibited by C_2H_2. In particular systems, the inhibition of ethylene (C_2H_4) oxidizing bacteria by C_2H_2 can cause accumulation of C_2H_4 in the presence of C_2H_2 not derived from C_2H_2 reduction. C_2H_2 can itself be metabolized aerobically or anaerobically as a growth substrate (see above reviews for details).

The C_2H_2 reduction method provides, by its nature, an indirect estimate of N_2 fixation. Considering the potential problems, within the biological and physical assay system chosen, one should make direct comparisons of C_2H_2 reduction and $^{15}N_2$ fixation to derive empirical conversion factors.[17]

It should be pointed out that the C_2H_2 blockage method for the determination of denitrification has many parallels to the C_2H_2 reduction method.[15,16,18] C_2H_2 is used to block the last step of the denitrification pathway, leading to the accumulation of N_2O. In fact, under certain conditions and with the appropriate detector (electron capture), one may monitor denitrification in the same assays being examined for nitrogenase activity.

MATERIALS REQUIRED

General Needs

- Gas chromatograph (GC; including gases) equipped for flame ionization
- 2-m Porpak R or N 80/100 GC column
- CaC_2, or compressed C_2H_2
- Gas-tight syringe, 100 μL
- C_2H_4 standards

For Planktonic Forms, e.g., *Trichodesmium*

- 10-ml serum vial
- 5 ml 0.45-μm filtered surface seawater
- 10 to 15 colonies of *Trichodesmium*, individually collected by SCUBA or picked from plankton tows with a plastic bacteriological loop and carried through one filtered seawater rinse
- Red silicone rubber septa, crimp sealed type with seals
- 1 ml C_2H_2 (for 20% final concentration)

For Benthic Samples, e.g., Vegetated or Unvegetated Sediment

- 125-ml wide-mouth flask
- 4 cm of 30-cc syringe core (about 15 ml), extruded directly into flask, approximately
- #6 black rubber stopper with recess (Thomas Scientific)
- 15 ml C_2H_2

PROCEDURES

General Approach

Acetylene is added to an enclosed system containing the sample of interest. Over time, samples of the gas or aqueous phase are analyzed for C_2H_4, the product of Mo-nitrogenase reduction of C_2H_2. Two alternative nitrogenases (V and Fe) have been found to reduce C_2H_2 to C_2H_6 and this has been suggested to be diagnostic for these enzymes.[19] Ethylene is most easily quantified by flame ionization gas chromatography (FID-GC).

Depending on the type of experimental material, a range of specific assay approaches are used. For strictly aqueous samples (e.g., micro- or macroplankton, unrooted macrophytes), an aqueous volume containing the sample is placed in a sealed system (e.g., flask) with a gas phase to which C_2H_2 is added to a concentration of about 10 to 20% (v/v). Flett et al.[20] have thoroughly discussed some of the problems and solutions for assaying these specific types of systems.

Sediment-containing systems present a variety of unique difficulties when attempting to apply an assay such as the C_2H_2 reduction procedure (see Capone[11,12] for discussion). The approaches taken are varied, and depend on the research question at hand. To overcome the problem of diffusion of substrate to, and product from, sites of activity, samples may be slurried and incubated with gentle shaking. Alternatively, in order to minimize disruption of natural gradients of O_2 and solutes, assays may employ intact cores to which C_2H_2 saturated water is injected and allowed to incubate, as has been performed from the C_2H_2 blockage assay of denitrification.[18] At the end of the assay period, the core may be segmented, each segment extruded into a vial, fixed, and rapidly sealed and the C_2H_4 formed in each horizon allowed to equilibrate into the gas phase and analyzed. C_2H_2 diffusion within a sediment core is slow (D. G. Capone and J. Sorensen, unpublished data) and its distribution may be inhomogeneous. This is a much more tedious procedure, and thereby limits the extent of replication and/or ability to determine the time course of activity. Perfusion of C_2H_2-saturated water is yet another alternative for sediment systems.[21,22] Acetylene may also be added to benthic chambers to assay activity near the sediment-water column interface.[23]

Assay Vessel

Generally, assays are carried out in sealed gas-tight vessels. Since O_2 concentration may be an important factor in many N_2 fixing systems, glass is generally used with a port sealed with a rubber septa of low gas permeability. Plastics (e.g., polyethylene) are often highly permeable to O_2, and may sorb low molecular weight hydrocarbons such as C_2H_2 and C_2H_4. Some sheet plastics like Saran® wrap, have low permeabilities and have been used for assay vessel construction.[17,24]

For aqueous samples, serum vials are rugged and available in a wide range of volumes. Glassware and seals are available from most laboratory supply houses (e.g., VWR, Fisher, Baxter/American Science Products) as well as chromatography suppliers (e.g., Supelco Alltech). Seals are available as sleeved rubber serum stoppers, or as crimp seal closures in a variety of septa composition. Besides standard red rubber, gray butyl rubber crimp seal closures are less permeable to gases and, in particular, O_2. A clear silicone rubber seal obtained from Alltech was unsuitable because of rapid C_2H_2 and C_2H_4 loss through the stopper. Red silicone was less permeable. An extremely low permeability black butyl stopper for culturing strict anaerobes is available from Bellco. For larger volume assays, flasks and bottles with wider openings may be used. We have sealed larger vessels with the appropriately sized black rubber stopper which has been modified with a central recess to allow easier penetration of a syringe for gas sampling.

Substrate

While other substrate analogs have been used or proposed for nitrogenase assay,[25,26] C_2H_2 is used most widely. C_2H_2 may be obtained commercially as a compressed gas. It is available in various purities and cylinder sizes. Alternatively, C_2H_2 may also be generated in the field by reaction of calcium carbide (CaC_2) with water.[27] To generate C_2H_2 from CaC_2, we routinely employ a small (125-ml) glass side-arm flask containing water with a rubber stopper. A Tygon tube, attached to the side arm, terminates in a needle inserted through the septum of a sealed serum vial and into distilled water (or an H_2SO_4 solution) in the vial. A second needle is inserted through the septa. The reaction is begun by addition of several pellets of CaC_2 to the flask and sealing the flask. C_2H_2 generated in the side arm is forced through the wash in the serum vial. Purified C_2H_2 is obtained by syringe through the septa of the serum vial while C_2H_2 is actively bubbling through the serum vial.

C_2H_4 is generally present in trace quantities, and one needs to correct for this, particularly when nitrogenase activity is low. All preparations also have some level of acetone and phosphine contamniation, and it is generally recommended to remove these by passage of the C_2H_2 through a water or H_2SO_4 wash.[28] H_2 may also be a relevant containment.[28] C_2H_2 prepared from CaC_2 is often as pure as, or purer than, commercial preparations. From either source, C_2H_2 for assays is most easily obtained by syringe with attached needle. For volumes of several milliliters or greater, disposable syringes with a needle generally suffice. Plastic syringes are adequate where O_2 levels are not critical. For systems where O_2 introduction needs to be minimized, glass syringes are advised. For smaller volumes, we use precision gas-tight syringes (e.g., Hamilton). Volumes obtained in syringes may be immediately injected into the assay systems, or be stored in the syringe for brief periods by inserting the needle into a rubber stopper. Larger volumes of C_2H_2 may be stored in gas sampling bags or football bladders for use in the field.

Typically, C_2H_2 is added to a final volume of from 10 to 20% of the gas phase as nitrogenase activity occurring in an aqueous phase in equilibrium with $pC_2H_2 > 0.1$ atm is usually saturated. After injection of C_2H_2, many researchers relieve the overpressurization by pricking the septa with a needle and allowing excess gas to escape. We have generally not found this critical as much of the C_2H_2 goes rapidly into solution. However, one needs to consider the relative volumes of the gas phase and liquid phase and ensure that absolute dissolved levels of C_2H_2 are truly saturating.[20] One may empirically determine the C_2H_2 saturation level of nitrogenase activity for the biological system being examined and within the assay format chosen.

Analyses

After addition of C_2H_2 to the assay system, the production of C_2H_4 from C_2H_2 reduction provides the index of nitrogenase activity. The C_2H_4 formed is most easily quantified by FID-GC. Simple, isothermal instruments are generally sufficient. Relatively small, portable units are available (e.g., Shimadzu Scientific) and may be used in the field. For signal output, a strip chart recorder or integrator is appropriate. We prefer a simple integrator (e.g., HP 3390A) which relieves the operator of the need to attenuate the signal. We have found, for the relatively narrow peaks obtained, somewhat better reproducibility using the peak height mode, rather than area mode, of an integrator.

For the separation of C_2H_2 and C_2H_4, a variety of column-packing materials (and prepacked columns) are available from chromatography suppliers for low molecular weight hydrocarbon separation. Choose a packing from which C_2H_4 is eluted before C_2H_2. We typically use a 2 m $\times$ 1/8'' stainless steel column packed with Porapak R or N, 100/120 mesh. A recent improvement is the Hayesep® series of porous polymers. At carrier flows of 30 ml min^{-1} and oven temperatures of 80°C, C_2H_4 elutes after about 1.0 min, and C_2H_2 at 1.6 min. However, depending on chromatographic conditions, as well as efficiency of packing, column retentivity, and detector sensitivity, elution times may vary. Porapak R and N do not generally provide resolution of C_2H_6 under typical conditions. However, column packings which provide very high resolution of C_2H_4, C_2H_6, and C_2H_2 (in that order) in a relatively brief run (<5 min) and under reasonable chromatographic conditions are available (e.g., Alltech Unibeads®) if C_2H_6 needs to be quantified (e.g., either as an internal standard or as the product of alternative nitrogenase [V] reduction).

For accurate peak quantitation, good resolution between C_2H_4 and C_2H_2 is important. We generally take note of the C_2H_2 peak height in each chromatogram as a semi-quantitative internal standard. While the mass injected may exceed the linear response range of the detector, the peak height response usually remains relatively constant through an experiment. Substantially lower initial values of C_2H_2 among some replicates may indicate insufficient volumes injected. Sudden decreases within a flask from one time point to another may indicate either a septum leak or clogged needle.

Assay systems may be sampled in several ways. Syringe sampling is the simplest and the most typical means of sample collection and introduction into the GC. We use gas-tight syringes (e.g., Hamilton). Alternatively, small disposable "Glass-pak®" (Becton-Dickenson) syringes (0.5 ml with 25-gauge needle) may be used and offer the advantage of allowing one to sample a series of flasks at a common time point and sticking the syringe into a rubber stopper until chromatography. Acetylene and C_2H_4 concentrations are usually stable for several hours.

Subsamples of the gas phase are taken from the assay system through the stopper and directly injected into the GC. Alternatively, if there is no gas phase, a subsample of the liquid phase may be obtained by syringe and injected into an evacuated, sealed vial (e.g., sealed serum vial or Vacutainer®) of larger volume. Vacutainers® (Becton-Dickenson), or similar evacuated blood sampling devices, have been used for some time as convenient gas storage devices.[29] They are available in a variety of volumes. One should be cautious, however, of hydrocarbon contaminants possibly resulting from ethylene oxide sterilization. Many researchers re-evacuate fresh Vacutainers® before use for C_2H_4 storage. One may inject by syringe a volume taken from the experimental flask into the Vacutainer®, or use a double-ended needle to directly sample the experimental flask. Correction for any sample dilution when using an intermediate gas storage vessel (e.g., Vacutainer®) is essential for quantitation. A nonmetabolized internal standard (e.g., C_2H_6 or SF_6) added to the primary assay vessel is useful in this type of application to account for sample dilution, leakage, and nonuniformity in vacuum among sampling vials.

The liquid phase and gas phase are allowed to equilibrate after agitation and the gas phase subsampled and measured chromatographically. The gas phase of assay vials may also be plumbed directly to a sampling valve of the GC and samples of the gas phase drawn into the sample loop and introduced to the chromatograph by valve activation. With any system, the volume of gas sampled and introduced to the GC is arbitrary, but should be consistent. Porpak is noted for its broad sample capacity. Small volumes may limit precision and sensitivity. We typically use 100 µl.

Generally, FID-GCs are highly stable, reliable systems with linear responses over a wide dynamic range. However, detector response can vary over time with variation in carrier and

FID feed gas flow. It is, thus, important to routinely monitor detector response. Upon establishing chromatographic conditions of carrier flow, column temperature, and detector sensitivity, one should confirm linearity of detector response over the range of standard C_2H_4 concentrations which may be encountered. We typically calibrate the detector by injection of a standard volume (100 μl) over a concentration range from 1 to 100 ppm(v). Premixed certified standards may be obtained at various concentrations from chromatography suppliers in small high and low pressure cylinders. Alternatively, one can quite inexpensively make up a range of standards by serial dilution of pure C_2H_4 into sealed glass bottles.[27] After establishing detector linearity, and depending on the stability of the chromatographic system, one should periodically (e.g., hourly, or with each sample series) run several standards within the range of C_2H_4 concentrations found in assays.

Sampling Strategies

When experiments are performed in the field at sites remote from the analytical facilities, assays may be conducted with "endpoint" sampling. That is, assays are set up, amended with C_2H_2, and allowed to incubate for some predetermined period of time after which a sample of the gas phase is removed and stored in a syringe, or in an evacuated gas-tight vessel (e.g., Vacutainer®).

Alternatively, the assay may be fixed in its vessel, and returned to the laboratory for analysis of the gas or aqueous phase for C_2H_4. A variety of metabolic poisons such as $HgCl_2$, formalin, and trichloroacetic acid (TCA), are used to terminate C_2H_2 reduction assays. Consideration should be given to compatability with the method of biomass determination intended, and to the disposal of the poison used. We have found that raising the pH of the aqueous phase to 1 N with NaOH effectively terminates activity in most systems, appropriate for subsequent base hydrolysis for protein analysis. In marine samples, termination with acid causes CO_2 degassing which may compromise the gas-tight seal. Quick freezing is yet another option.

In order to approximate natural rates of activity, it is recommended to keep assays as brief as is practical. Obviously, if induction (e.g., lags) or deceleration in activity occurs during the time frame of an endpoint assay, the estimate of the initial or "*in situ*" rate of C_2H_2 reduction may be over- or underestimated, respectively.

Alternatively, and preferably, one may conduct time-course sampling to determine the variation in nitrogenase activity over time. Such analyses provide insights into phenomena such as nitrogenase induction or inactivation. For endpoint sampling, it is important to establish the time course of nitrogenase activity to validate the assumption of linearity of C_2H_4 production over the assay time frame chosen. If experiments are run proximal to the laboratory, it is relatively simple to obtain gas subsamples and run them immediately. In the field, one may take subsamples from individual assays over a time course, storing those samples until return to the laboratory. Another approach is to initiate many replicates, terminating individual assays over the time course.

We routinely take multiple samples from each assay. This is effectively done in an exponential progression of time (e.g., 0, 0.25, 0.5, 1.0, 2.0 h, etc.). If the multiple samples removed from an individual flasks represent a large cumulative volume, relative to the whole gas phase volume, one may need to correct for sample dilution. Depending on the level of activity expressed, the biomass of diazotrophs assayed, sensitivity of the assay system, etc. the time frame of sampling may be compressed or extended. Zero-time samples are important in order to correct for any background C_2H_4 in the added C_2H_2.

Several levels of controls are appropriate. Samples containing biomass but without C_2H_2 allow detection and correction for any natural C_2H_4 production. A parallel sample to which

C_2H_4 is added at trace levels will demonstrate whether there may be a capacity for C_2H_4 consumption in the system under study. Samples devoid of the assay material but with C_2H_2 provide a check on contaminant diazotrophs (e.g., in assay medium) as well as a benchmark for C_2H_2 concentration and equilibrium in the absence of any consumption. Killed (at zero time) controls are also sometimes used. Potential bottle containment effects are a necessary consideration in any sealed assay system and should be examined.

Calculation of C_2H_2 Reduction

The calculation of C_2H_2 reduction, generally expressed as C_2H_4 production, is relatively straightforward. One may obtain the linear regression coefficients from an C_2H_4 standard curve and use the regression equation to estimate C_2H_4 concentrations in experimental flasks. However, if the detector response of the chromatographic system is highly linear as is typical, one may generally rely on a single standard in the midpoint of the standard curve and calculate the quantity of C_2H_4 formed in an assay vessel at any point in time as:

$$\frac{(\text{pk ht, unk})}{(\text{pk ht, std})} \times [\text{std}] \times (\text{GPV}) \times (\text{SC}) = \text{nmol } C_2H_4 \qquad (3)$$

where: (pk ht, unk) = peak height response for C_2H_4 for standard volume (e.g., 100 μl) sampled from an assay vessel (peak area may be substituted); (pk ht, std) = peak height response for same volume of an appropriate C_2H_4 standard; [std] = concentration of C_2H_4 standard used, expressed in nmol ml^{-1}, as determined by ideal gas law corrected for room temperature and pressure; GPV = gas phase volume, the total volume of the head phase of the assay vessel in ml; and SC = solubility correction for C_2H_4 in aqueous phase.[20]

If the detector response to a particular standard remains relatively constant over the course of a run with no trends in variation, one may potentially average standards over the run. However, we often note a monotonic change in detector response to standards over time and generally use the mean of standards taken most immediately before and after particular time points in each experiment.

For any assay system containing a relatively large aqueous phase, an appreciable fraction of the C_2H_4 formed will be in the aqueous phase. The ratio of the total volume (M) of C_2H_4 formed to the volume transferred to the gaseous phase (X) equals:[20]

$$\text{SC} = M/X = 1 + (\alpha * A/B) \qquad (4)$$

where: α = the Bunsen coefficient for C_2H_4 at the appropriate temperature and salinity; A = the volume of the aqueous phase; and B = the volume of the gas phase (or GPV).

The reciprocal of this value (X/M) represents the fraction of total C_2H_4 formed which is in the gas phase. Hence, the product of the empirically determined moles of C_2H_4 in the gas phase times M/X yield the total moles of C_2H_4 formed in the system.

For endpoint assays, rates of nitrogenase activity may be calculated as:

$$\frac{(\text{nmol } C_2H_4, t_n) - (\text{nmol } C_2H_4, t_o)}{t_n - t_o} = \text{nmol } C_2H_4 \ t^{-1} \qquad (5)$$

where t_n and t_o equal end time and zero time, respectively. For time-course experiments, one may similarly calculate rates between each sampling point. Where rates of C_2H_4 production are steady, we use linear regression analysis of total nmol formed as a function of time to obtain an average rate (slope) over the time period. One generally normalizes total C_2H_4 formed, or the rate of C_2H_4 production, to an appropriate parameter of biomass (e.g., dry weight sample, protein, chlorophyll) or other appropriate dimension (volume of sample assayed, cross-sectional area, etc.). All these calculations are readily adapted to a PC spreadsheet (e.g., Lotus, Excel) format.

Conversion to N_2 Fixed/Calibration

While the theoretical relationship of mol C_2H_2 reduced per mol N_2 is often given as 3:1 (i.e., the 6 mol of reductant required to reduce 1 mol of N_2 would instead reduce 3 mol C_2H_2), a ratio of 4:1 is probably more appropriate as it considers the hydrogenase reaction of nitrogenase (Equation 1), which is inhibited under C_2H_2-reducing conditions (Equation 2).[4] That is, the 8 mol of reductant required to reduce each mol of N_2 and forming 2 mol of ammonium and 1 mol of H_2 would, under C_2H_2-reducing conditions, reduce 4 mol C_2H_2 yielding 4 mol C_2H_4. While many systems examined display ratios not substantially different from this theoretical ratio,[11] ratios considerably greater have been noted.[31-33] Explanations for such departures have been offered.[11]

Thus, the most appropriate means of converting measured rates of C_2H_2 reduction to N_2 fixed is by use of an empirically determined conversion factor for the system under study. In many cases, appropriate factors may be available in the literature. If the appropriate facilities (isotope ratio mass spectrometer or emission spectrometer) are available, one may undertake such comparisons directly. However, it is beyond the scope of this chapter to provide detailed procedures for such calibrations, although much of the general approach is, of necessity, similar. The reader is referred to Burris[17] and Bergerson[34] for more details on procedures for $^{15}N_2$ fixation studies. Alternatively, collaboration with a research group actively engaged in $^{15}N_2$ fixation studies is recommended. If appropriate conversion factors are not available and direct calibration not immediately feasible, nitrogenase activity data should be presented as measured, i.e., in terms of C_2H_4 produced.

Experimental Conditions/Manipulations

When attempting to estimate *in situ* rates of nitrogenase activity, as is often the objective, many factors need be considered. Depending on the subject of study, O_2 may be a critical factor. For anoxic systems, protection from O_2 activation may be important during assay set up. For phototrophic systems, light is an obvious factor which needs to be accounted for during incubation. Similarly, other biological (e.g., presence or absence of grazers, symbioses), chemical (e.g., salinity, pH, inorganic or organic nutrients, toxic factors), or physical (e.g., temperature, pressure) factors may bear directly on observed rates of nitrogenase activity.[11,13]

Experimental manipulation of many of these factors can provide useful information concerning the physiological composition of natural populations of diazotrophs and the factors limiting activity. The C_2H_2 reduction method readily lends itself to these types of investigations. Comparisons of nitrogenase activity in the presence and absence of light, O_2, and organic or inorganic (e.g., sulfate) substrates may reveal the presence of subgroups of diazotrophs within a sample matrix. Coupled with DNA probes,[35] one may definitively establish the relative importance of particular physiological groups. Specific inhibitors have also been used in C_2H_2 reduction assays to demonstrate the importance of different physiological groups of bacteria.[16] The various factors which regulate the activity and expression of nitrogenase activity within a relatively defined system may also be examined experimentally.

With regard to experimental approaches, several are possible. Multiple experimental levels may be established (with replication as needed), relative to appropriate controls. Amendment with C_2H_2 initiates the experiment. At each experimental level, samples may or may not be preincubated under the treatment. Alternatively, a group of replicate assays may be initiated under standard conditions and, after establishing basal rates of C_2H_2 reduction, a subset of the replicates may be subjected to the experimental manipulation (e.g., substrate or inhibitor addition, change in physical parameter). The latter approach often depends on the extent that the particular biological system under study will retain steady rates of C_2H_2 reduction while contained.

Examples

Planktonic Forms, e.g., Colonial Cyanobacterium, Trichodesmium

Trichodesmium is a planktonic cyanobacterium which often occurs in colonial form. It is an active diazotroph in the upper open ocean of the tropics and subtropics. Assay requirements as used in Capone et al.[36] are given in "Materials Required". Assays are typically conducted in a temperature-controlled water bath "on deck" under natural light conditions with two or three layers of neutral density screening to avoid photoinhibition. Typically, assays are conducted aerobically, although lowering O_2 tension generally enhances activity. C_2H_2 production is usually evident within minutes of assay initiation. Comparisons with $^{15}N_2$ found good agreement ratios of C_2H_2 reduction to N_2 fixed of about 4:1.[37]

Benthic Samples, e.g., Vegetated or Unvegetated Sediments

Relatively undisturbed samples of seagrass roots, rhizomes, and associated sediment are obtained by core.[38-40] For sample collection, we have used aluminum or butyrate cores, as well as larger (20 to 50 cc) plastic syringes with the front portion cut off. Cores are sealed with black rubber stoppers for transport to laboratory for assay . Choice of coring device was made to allow direct extrusion into wide-mouth flasks of a slightly larger diameter.

Upon return to the laboratory, cores are directly extruded into flasks while flushing flask with inert gas (e.g., N_2, Ar) through a long canula into the flask. Flasks are sealed with rubber stoppers as the canula is withdrawn. After sealing, the flask is flushed for 1 to 2 min with inert gas through incurrent and excurrent gassing needles. This gassing procedure can be eliminated for aerobic assay. After the system is sealed, the assay is initiated by addition of 15 ml of C_2H_2 through the stopper. Samples are typically incubated in the dark at *in situ* temperatures. At the end of the assay period, sediment may be dried. C_2H_2 reduction may be variously expressed on dry weight, volume, or cross-sectional area of core.

We have generally found short-term linear rates of C_2H_4 accumulation for sediments assayed in this manner. A liquid phase may (e.g., 15 ml of 0.2-μm filtered water) be included in the flask if soluble substrates or inhibitors are to be added. However, we often observe a slight (0.5 to 2 h) lag phase (i.e., period of accelerating C_2H_4 production rather than a period of no observable production) with slurries. In the calculation of C_2H_4 in solution, one should consider the porosity of sediment sample, rather than the total volume which may include inert particles.

ACKNOWLEDGMENTS

The author wishes to thank Barrie Taylor, John Gallon, Judy O'Neil, Veronica Miller, and Kelly Cunningham for reviewing and commenting on the manuscript. I wish to also acknowledge the sustained support of the National Science Foundation Division of Ocean Sciences in my studies of marine N_2 fixation.

REFERENCES

1. Capone, D. G., Aspects of the marine nitrogen cycle with relevance to the dynamics of nitrous and nitric oxide, in *Microbial Production and Consumption of Greenhouse Gases: Methane, Nitrogen Oxides, and Halomethanes,* Rogers, J. E. and Whitman, W. B., Eds., American Society of Microbiology, 1991, 255.
2. Capone, D. G. and Carpenter, E. J., Nitrogen fixation in the marine environment, *Science,* 217, 1140, 1982.
3. Howarth, R. W., Marino, R., Lane, J., and Cole, J. J., Nitrogen fixation in freshwater, estuarine, and marine ecosystems. 1. Rates and importance, *Limnol. Oceanogr.,* 33, 669, 1988.
4. Postgate, J. R., *The Fundamentals of Nitrogen Fixation,* Cambridge University Press, London, 1982.
5. Pau, R. N., Nitrogenase without molybdenum, *Trends Biochem. Sci.,* 14, 183, 1989.
6. Dilworth, M. J., Acetylene-reduction by nitrogen-fixing preparations from *Clostridium pasteurianum, Biochim. Biophys. Acta,* 127, 285, 1966.
7. Schollhorn, R. and Burris, R., Acetylene as a competitive inhibitor of N_2 fixation, *Proc. Natl. Acad. Sci. U.S.A.,* 58, 213, 1967.
8. Rivera-Ortiz, J. M. and Burris, R. H., Interactions among substrates and inhibitors of nitrogenase, *J. Bacteriol.,* 123, 537, 1975.
9. Stewart, W. D. P., Fitzgerald, G. P., and Burris, R. H., *In situ* studies on N_2 fixation, using the acetylene reduction technique, *Proc. Natl. Acad. Sci. U.S.A.,* 58, 2071, 1967.
10. Hardy, R. W. F., Holsten, R. D., Jackson, E. K., and Burns, R. C., The acetylene-ethylene assay for N_2 fixation: laboratory and field evaluation, *Plant Physiol.,* 43, 1185, 1968.
11. Capone, D. G., Benthic nitrogen fixation, in *Nitrogen Cycling in Coastal Marine Environments,* Blackburn, T. H. and Sorenson, J., SCOPE Series, Vol. 33, J. Wiley and Sons, New York, 1988, 85.
12. Capone, D. G., Benthic nitrogen fixation, in *Nitrogen in the Marine Environment,* Carpenter, E. J. and Capone, D. G., Eds., Academic Press, 1983, 105.
13. Paerl, H. W., Physiological ecology and regulation of N_2 fixation in natural waters, in *Advances in Microbial Ecology, Vol. 11,* Marshall, K. C., Ed., Plenum, New York, 1990, 305.
14. Taylor, B. F., Assays of microbial nitrogen transformations, in *Nitrogen in the Marine Environment,* Carpenter, E. J. and Capone, D. G., Eds., Academic Press, New York, 1983, 809.
15. Payne, W. J., Influence of acetylene on microbial and enzymatic assays, *J. Microbiol. Methods,* 2, 117, 1984.
16. Oremland, R. S. and Capone, D. G., Use of specific inhibitors in microbial ecological and biogeochemical studies, in *Advances in Microbial Ecology, Vol. 10,* Marshall, K. C., Ed., 1988, 285.
17. Burris, R. H., Methodology, in *The Biology of N_2 Fixation,* Quispel, A., Ed., North-Holland Publishers, Amsterdam, 1974, 9.
18. Sorensen, J., Denitrification rates in a marine sediment as measured by the acetylene inhibition technique, *Appl. Environ. Microbiol.,* 35, 301, 1978.
19. Dilworth, M. J., Eady, R. R., Robson, R. R., and Miller, R. W., Ethane formation from acetylene as a potential test for vanadium nitrogenase in vivo, *Nature,* 327, 167, 1987.
20. Flett, R. J., Hamilton, R. D., and Campbell, N. E. R., Aquatic acetylene-reduction techniques: solution to several problems, *Can. J. Microbiol.,* 22, 43, 1976.

21. Capone, D. G. and Carpenter, E. J., A perfusion method for assaying microbial activities in estuarine sediments: applicability to studies of N_2 fixation by C_2H_2 reduction, *Appl. Environ. Microbiol.*, 43, 1400, 1982.

22. O'Donohue, M. J., MacRae, I. C., and Moriarty, D. J. W., A comparison of methods for determining rates of acetylene reduction (nitrogen fixation) by heterotrophic bacteria in seagrass sediment, *J. Microbiol. Methods,* 13, 171, 1991.

23. Chan, Y.-K. and Knowles, R., Measurement of denitrification in two freshwater sediments by an *in situ* acetylene inhibition method, *Appl. Environ. Microb.*, 37, 1067, 1979.

24. van Kessel, C., Diffusion of gases through plastic bags containing plants being exposed to acetylene or $^{15}N_2$, *Soil Biol. Biochem.*, 4, 493, 1983.

25. Dalton, H. and Whittenbury, R., The acetylene reduction technique as an assay for nitrogenase activity in the methane oxidizing bacterium *Methylococcus capsulatus* strain Bath, *Arch. Microbiol.*, 109, 147, 1976.

26. McKenna, C. E. and Huang, C. W., In vivo reduction of cyclopropene by *Azobacter vinelandii* nitrogenase, *Nature*, 280, 609, 1979.

27. Postgate, J. R., Acetylene reduction test for nitrogen fixation, in *Methods in Microbiology*, Vol. 6B, Morris, J. R. and Ribbons, D. W., Eds., Academic Press, New York, 1972, 343.

28. Hyman, M. R. and Arp, D. J., Quantification and removal of some contaminating gases from acetylene used to study gas-utilizing enzymes and microorganisms, *Appl. Environ. Microb.*, 53, 298, 1987.

29. Schell, D. W. and Alexander, V., Improved incubation and gas sampling technique for nitrogen fixation studies, *Limnol. Oceanogr.*, 15, 961, 1970.

30. Thake, B. and Rawle, P. R., Non-biological production of ethylene in the acetylene reduction assay for nitrogenase, *Arch. Mikrobiol.*, 85, 39, 1972.

31. Peterson, R. B. and Burris, R. H., Conversion of acetylene reduction rates to nitrogen fixation rates in natural populations of blue-green algae, *Anal. Biochem.*, 73, 404, 1976.

32. Graham, B. M., Hamilton, R. D., and Campbell, N. E. R., Comparison of the nitrogen-15 uptake and acetylene reduction methods for estimating the rates of nitrogen fixation by freshwater blue-green algae, *Can. J. Fish. Aquat. Sci.*, 37, 488, 1980.

33. Seitzinger, S. P. and Garber, J. H., Nitrogen fixation and $^{15}N_2$ calibration of the acetylene reduction assay in coastal marine sediments, *Mar. Ecol. Prog. Ser.*, 37, 65, 1987.

34. Bergerson, F. J., Measurement of nitrogen fixation by direct means, in *Methods for Evaluating Biological Nitrogen Fixation*, Bergerson, F. J., Ed., Wiley and Sons, Chichester, 1980, 65.

35. Kirshstein, J. D., Paerl, H. W., and Zehr, J., Amplification, cloning and sequencing of a nifH segment from aquatic microorganisms and natural communities, *Appl. Environ. Microbiol.*, 57, 2645, 1991.

36. Capone, D. G., O'Neil, J. M., Carpenter, E. J., and Zehr, J., Basis for diel variation in nitrogenase activity in the marine planktonic cyanobacterium, Trichodesmium thiebautii, *Appl. Environ. Microbiol.*, 56, 3532, 1990.

37. Scranton, M. I., Novelli, P. C., Michaels, A., Horrigan, S. G., and Carpenter, E. J., Hydrogen production and nitrogen fixation by Oscillatoria thiebautii during in situ incubations, *Limnol. Oceanogr.*, 32, 998, 1987.

38. Capone, D. G., Nitrogen fixation (acetylene reduction) by rhizosphere sediments of the eelgrass, *Zostera marina* L., *Mar. Ecol. Prog. Ser.*, 10, 67, 1982.

39. Capone, D. G. and Taylor, B. F., N_2 (C_2H_2) fixation in the rhizosphere of *Thalassia testudinum*, *Can. J. Microbiol.*, 26, 998, 1980.

40. O'Neil, J. M. and Capone, D. G., Nitrogenase activity in tropical carbonate marine sediments, *Mar. Ecol. Prog. Ser.*, 56, 145, 1989.

CHAPTER 74

Denitrification and Nitrification Rates in Aquatic Sediments

Sybil P. Seitzinger

INTRODUCTION

Denitrification is the reduction by bacteria of nitrite or nitrate to gaseous N (N_2, N_2O, NO). It is generally considered to be an anaerobic process and requires, in addition to nitrite or nitrate, a source of organic carbon. Denitrification occurs in a wide range of both freshwater and marine sediments.[1,2] The removal of fixed N by denitrification is a major sink for N in many aquatic ecosystems, often removing an amount of N equivalent to 40% or more of the inorganic N inputs to aquatic ecosystems.

Sources of nitrate or nitrite (hereafter referred to as nitrate) for denitrification in aquatic sediments include nitrification in the sediments, nitrate in the overlying water, and in some cases, advection of nitrate-contaminated groundwater through the sediments. Nitrification in the sediments can be an important source of nitrate for denitrification in many aquatic sediments. Nitrification, the oxidation of ammonia to nitrite or nitrate, requires a source of free O_2. Thus, nitrification occurs primarily in the surface few millimeters of sediment where O_2 occurs. As a consequence, denitrification coupled to nitrification in the sediments occurs in close proximity to the thin veneer of oxygenated surface sediments. In addition, oxygen and thus, nitrification can occur in the lining of irrigated burrows of animals or near the roots and rhizomes of submerged aquatic plants if they release oxygen, thus, extending the zone of coupled nitrification/denitrification into the deeper sediments.

Denitrification is directly measured as the production of N_2 from vertically intact sediment cores. The cores with an overlying water and gas phase are incubated in gas-tight glass incubation chambers. The water is changed every 1 to 3 d to keep oxygen, nitrate, and ammonia concentrations near field levels. The background concentration of N_2 inside the chambers is reduced from 78% (ambient) to 1 to 3% by incubating the sediments with an overlying water and gas phase flushed with 80% He and 20% O_2, thus, making it possible to detect increases in N_2 concentration due to production by denitrification. Time-series samples of the gas phase are collected over repeated 24-h incubation periods and analyzed for N_2 and O_2 concentration by thermal conductivity detector (TCD) gas chromatography (GC).[3,4]

Sediment-water nitrate and ammonium fluxes are measured concurrently with denitrification rates. Sediment nitrification rates are calculated as the sum of the measured net nitrate flux out of the sediment plus the nitrate required to supply the measured denitrification.

0-87371-564-0/93/$0.00 + $.50
© 1993 by Lewis Publishers

This calculation assumes that nitrate produced in the sediments is either released to the overlying water or is removed by denitrification. The total inorganic N flux from the sediment is calculated as the sum of the net ammonium and nitrate flux plus the denitrification rate. Under steady-state conditions, the total inorganic N flux should equal the rate of organic N mineralization in the sediment.

Advantages of the method are: (1) the end product of denitrification (N_2) is measured directly; (2) inhibitors, such as acetylene which blocks nitrification, are not used; (3) sediment cores are incubated at near ambient field concentrations of oxygen and nitrate which can affect denitrification rates; (4) sediment-water fluxes of nutrients (e.g., ammonia, nitrate, phosphate), oxygen, and trace gases such as N_2O can be measured simultaneously with denitrification rates; (5) sediment cores are vertically intact and contain the associated organisms, thus, profiles of oxygen and nitrate in the sediments are maintained near field concentrations and effects of bioturbation or irrigation by organisms are included; and (6) nitrification rates and (net) organic N mineralization rates are estimated concurrently with denitrification rates.

Disadvantages of the method are: (1) sediment cores must be incubated for a number of days (3 to 10 d) in the laboratory with low-N_2 overlying water before denitrification rates can be measured, this incubation period is required to deplete the N_2 initially dissolved in the pore water; (2) care must be taken during sampling and water changes to ensure that there is no contamination of chambers or samples with room air (N_2); and (3) sediment nitrification rates may be underestimated if there is significant reduction of nitrate to ammonia.

The acetylene block method has been used extensively to measure denitrification in aquatic sediments.[1,5] While it is quick and relatively easy to use, numerous problems have been demonstrated with the acetylene block method, including the finding that acetylene blocks nitrification (e.g., see review by Knowles[6]). A recent comparative methods study has demonstrated that the acetylene block method greatly underestimates denitrification rates in aquatic sediments because it does not measure denitrification coupled to nitrification in the sediments and it only measures a portion of the denitrification due to nitrate diffusing from the overlying water.[7]

MATERIALS REQUIRED

Equipment

- Sample coring device, e.g., plastic core tubes, 7-cm diameter.
- Incubation chambers: the incubation chambers are constructed of Pyrex glass pipe (Figure 1). Two ports (5-mm i.d.) are located within 1 cm of the top of each chamber for water changing and gas phase sampling and are closed with ground glass stopcocks. A Teflon-coated magnetic stir bar suspended in a Teflon housing (Nalgene #6630) is attached with silicon sealant to the inside of each chamber top to stir the water over the sediments and facilitate equilibration of gases between the water and overlying gas phase. The stir bars are driven by water turbine magnetic stirrers which sit on top of each chamber. The gas phase provides a reservoir of oxygen and is relatively easy to sample and analyze for N_2 and O_2 concentration.
- GC (isothermal) equipped with TCD detector and 2 m × 0.318 cm o.d. stainless steel column packed with activated molecular sieve 5A, 45/60 mesh.
- For N_2O studies, GC (isothermal) equipped with [63]Ni electron capture detector and a 2.5 m × 0.318 cm o.d. stainless steel column packed with Poropak Q, 80/100 mesh; the column is operated at 40°C with a CH_4-Ar or N_2 carrier gas flow rate of 27 cm³ min⁻¹.
- Integrator.

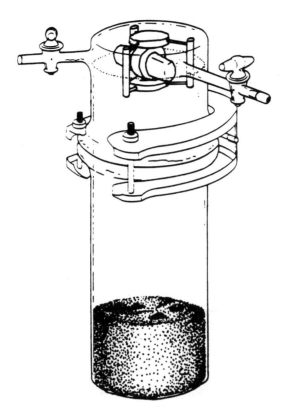

Figure 1. Gas-tight glass incubation chambers for incubating sediment cores for denitrification measurements. Chambers are constructed from two pieces of glass pipe joint (75 mm i.d.). The joint is sealed with a rubber O-ring and held together with a metal clamp. The two ports (5 mm i.d.) in the top section are used for sampling the gas phase and changing the water. The ports are approximately 1 cm from the top and are sealed with ground glass stopcocks (2-mm threaded pressure stopcock with 8-mm side arm). A rubber serum bottle stopper is placed on the port which is used for sampling the gas phase of the chamber. A water turbine magnetic stirrer is placed on the top of the chamber to drive a Teflon-coated magnetic stir bar suspended in a Teflon housing (Nalgene #6630, modified to fit i.d. of chamber) that is attached with silicone sealant to the inside of the chamber top. The approximate heights (cm) of the sediment, water, and gas phases are 4, 15, and 1.3, respectively.

- Gas regulators (4 or 5).
- Magnetic stirrers to drive water stirring system inside chambers (to turn magnetic stirrers inside the chambers use either water-driven magnetic stirrers on top of each chamber or larger magnetic stirrer with chambers positioned around; do not use stirrers on top of chambers that generate heat and thus, increase temperature inside chambers; magnets attached to rods positioned over each chamber that are belt driven using a single constant-speed motor also work well).
- Temperature and light-controlled incubator.
- Spectrophotometer or autoanalyzer for nutrient analyses.

Supplies

For denitrification studies:

- He (ultra high purity) carrier gas
- Gas standard (certified 3% N_2, 20% O_2, balance He)

- He (industrial grade) for sampling procedure
- Gas mixture of 20% O_2 and balance He (use either ready made compressed gas mixture or make with gas mixing valve)
- Gas-tight syringe (250 μl) with 22-gauge, replaceable 4'' needles
- Septa for GC (change daily)
- Serum bottle stoppers (5 mm with 9 mm collar) for chambers (change with each water change)
- Flexible tubing (e.g., Tygon) for water changes, and sampling of chambers and gas standards (1/4'' i.d.; ~50')
- 4-L Erlenmeyer flask or similarly sized container
- Air stones

For nitrification studies:

- Glass fiber filters (Whatman GF/F)
- Filtration setup
- Beakers for control water
- Reagents and standards for ammonia and nitrate analyses

For N_2O studies:

- CH_4-Ar or N_2 (ultra high purity) carrier gas
- Gas-tight syringe (2.0 ml) with 22-gauge, 4'' long replaceable needles

PROCEDURES

Sediment Collection and Incubation

1. Collect sediment cores in the field, preferably by SCUBA divers if the water depth warrants, using plastic core tubes and rubber stoppers. Be careful during coring not to disturb the sediment-water interface. During transport to the laboratory, maintain the cores at ambient field water temperature and keep the overlying water aerated.

2. Transfer the sediment cores (approximately 4-cm deep) to gas-tight glass incubation chambers (Figure 1) by first carefully removing the overlying water, and then removing the stopper from the bottom of the core tubes while they are held directly over the bottom portion of the disassembled glass chambers. Carefully siphon ~5 cm of water over the sediments. Before putting on the chamber top be sure that the O-ring is clean and lightly greased and that the O-ring groove in the chamber top and bottom is clean. Finish adding water to the chambers as described below (see section on water changing) and incubate under the appropriate temperature and light conditions; ensure that the water is continuously stirred.

Denitrification

Water Changing

Because of the high background concentration of N_2 in the atmosphere and dissolved in water, it is generally not possible to detect increases in N_2 concentration due to denitrification. Therefore, the background concentration of N_2 in the chambers is lowered to 1 to 3% by using water and an overlying gas phase which is sparged with a mixture of 80% He and 20% O_2.

1. Sparge water in a large Erlenmeyer flask for 30 min with 80% He and 20% O_2.

2. Then, while continuously sparging the water, siphon the water into each chamber through one of the ports. Stir the water inside the chamber while changing; use two chamber-volumes of water; let overflow water exit from opposite port into a waste container; fill the chamber completely with water during the water change and ensure that no room air is trapped inside the chamber.

3. After two chamber-volumes have flowed through, lower the water level to create an ~1 cm gas phase by slowly flowing He/O_2 gas mixture through the chamber ports. Collect and measure the volume of this water to determine the exact volume of gas phase in each chamber. This water can also be analyzed to determine initial concentrations of nitrate and ammonia.

4. Continue to flush the gas phase with He/O_2 for 3 to 5 min.

5. Close chambers by placing a rubber serum bottle stopper on the exit port and closing the stopcock on the input port; do this simultaneously to prevent excess pressure buildup inside chamber. Do not fold collar back on serum bottle stopper.

6. Lower pressure inside chamber to ~0.9 atm by connecting a vacuum pump to the input port. With pump set to 0.9 atm, quickly open and close stopcock on input port. (The pressure is decreased so that no sample is lost from the sampling syringe when the needle is withdrawn from the chamber; see section on gas sampling below.)

7. Set chambers in an environmental room or incubator under the desired temperature and light conditions; ensure that the water is continuously stirred.

8. Change the water every 1 to 3 d throughout the experiment with He/O_2 flushed water to maintain nitrate and ammonia concentrations near field levels; frequency of water change depends on the rate of sediment-water nitrate and ammonia fluxes and ambient water concentrations.

9. Collect initial (see above) and final water samples for each water change during the denitrification measurement period for nitrate and ammonia analysis. These samples are used to calculate sediment-water nutrient fluxes, nitrification rates, and net organic N mineralization rates as described below.

Initially the flux of N_2 out of the sediments is due to a combination of N_2 production due to denitrification and re-equilibration of N_2 originally dissolved in the pore water with the low-N_2 overlying water (hereafter referred to as N_2 diffusive flux). The time required to deplete the N_2 initially dissolved in the pore water so that the measured N_2 flux is due to denitrification (plus any small background diffusive flux) depends on a number of factors including the depth of sediment, bioturbation, and frequency of flushing of the gas phase. For example, it generally takes approximately 7 to 10 d of incubation with low-N_2 overlying water before the diffusive N_2 flux (5 cm deep core) is less than 25 μg-at N m^{-2} h^{-1};[4,8,9] the gas phase is flushed with He/O_2 two to three times daily for 3 to 5 min during that time. The time to deplete the diffusive flux may vary with investigator and experimental setup and should be checked, particularly when a new user is setting up the method. The point at which it is considered that the N_2 flux is due to denitrification is when consecutive N_2 measurements no longer give significantly decreasing rates. (Also see section on controls.)

Nowicki[10] developed the following modification which decreases the time required before denitrification rates can be obtained to approximately 3 to 4 d. Nowicki measures the pore water diffusive N_2 flux in a parallel set of cores that are incubated under anaerobic conditions with no nitrate in the overlying water, thus, eliminating all sources of nitrate and therefore, stopping denitrification. The measured N_2 diffusive flux is then subtracted from the total

N_2 flux (measured concurrently using cores incubated as described above) to obtain the denitrification rate.

Gas Sampling

After the pore water N_2 has equilibrated with the low-N_2 overlying water, changes in N_2 concentration in the chambers are measured over repeated 24-h intervals and denitrification rates are calculated. Two to four separate flux measurements are generally made on each core.

Care must be taken during sampling to avoid contamination of chambers or samples with room air. The following procedure has been repeatedly tested and demonstrated to produce no atmospheric N_2 contamination of samples or chambers. Beginning approximately 4–6 h after the water and/or gas phase is changed and the pressure is reduced (a sufficient time to allow for equilibration of the overlying water and gas phases), collect initial gas samples from the chambers and immediately analyze for N_2 and O_2 concentration as follows. (Note: the pressure inside the chambers is initially reduced to slightly below room pressure as described above [see section on water changing] to ensure that none of the gas sample is lost from the syringe as it is withdrawn from the chamber. We have encountered contamination problems whenever syringes with shut-off valves are used and therefore, we do not recommend them. Gardner et al.[11] have modified the chamber design with vertical ports on the chamber top which makes it possible to use a gas sampling valve for directly withdrawing samples from the chambers and injecting them into the GC. They do not reduce the pressure inside their chambers.)

1. Flush the gas-tight sampling syringe six times with He before each sample is taken. Then with the needle tip continuously inside the He flushing line, flush the outside of the septa on the sampling port of the chamber.

2. Open the stopcock on the sampling port and insert the needle through the center of the septa and into the central area of the chamber.

3. Flush the syringe with sample six times, draw a 75-μl sample, and then while continuing to flush the septa area with He, slowly withdraw the needle through the septa. The outside of the septa must be continuously and vigorously flushed with He as the tip is withdrawn from the septa to ensure that no atmospheric N_2 is drawn into the syringe as the pressure in the syringe equalizes with the atmospheric pressure.

4. Continuously flush the needle tip and inject the entire sample immediately into the GC. N_2 and O_2 in the samples are analyzed using a GC equipped with a TCD (25 cm^3 min^{-1} He carrier gas flow rate; 2 m × 0.318 cm o.d. stainless steel column packed with activated molecular sieve 5A, 45/60 mesh, operated at 45°C).

5. Calibrate the detector response to N_2 and O_2 by analyzing variable volume injections (e.g., 75 μl, 100 μl, 150 μl) of a certified gas standard (3% N_2, 21% O_2, balance He) at the beginning of each day and at intervals throughout the day. Follow similar procedures as for chamber sampling: flush syringe six times with gas standard and continuously flush needle tip and GC sample port with gas standard or He. Coefficients of variation for repeated injections of standards should be 2% or less and of samples should be 10% or less. Typically, the lower limit of detection of denitrification rates is approximately 25 μg-at m^{-2} h^{-1}.

Calculation of Denitrification Rates and Sediment Oxygen Consumption

Calculate the sediment-water fluxes of N_2 from the difference in concentration between time series samples taken between water changes or gas-phase flushes. After calculating the

N_2 concentration in the chamber gas phase (μg-at N ml^{-1}), calculate the total N_2 (N_{2tot}) in the gas and water phase of each chamber knowing the gas-phase volume, water volume, and solubility coefficient for N_2 at the appropriate temperature and salinity.[12] Calculate the denitrification rate (DNF) based on the time between samples (T) and surface area of sediment (A) as follows:

$$\text{DNF } (\mu\text{g-at N m}^{-2}\text{ h}^{-1}) = (\text{final N}_{2\text{ tot}} - \text{initial N}_{2\text{ tot}})/(T \cdot A) \qquad (1)$$

Calculate O_2 fluxes using the above procedures with O_2 concentrations and solubility coefficients.[12]

Controls

Chambers in which only water is incubated serve as controls to demonstrate that changes in N_2 concentration are due to biological production in the sediments and not due to contamination of chambers or samples with room air. The water should be changed and chambers treated and sampled exactly as performed with the chambers containing sediment cores. These water controls can also be used to determine the exact time required in your chambers to equilibrate the gas and water phases after changing the water or reflushing the gas phase (i.e., the time at which the initial gas sample can be taken following a water change). In our chambers with 0.75 L of water which is continuously stirred and a 60 ml gas phase, it takes 4 to 6 h as noted above. Determine this time for your chambers and stirring system by sampling the gas phase approximately every hour following a water change until there is no significant difference in N_2 concentration.

As noted above (see section on water changing), the number of days it takes to equilibrate the N_2 initially dissolved in the pore waters with the low-N_2 overlying water is determined as the point at which there is no significant decrease in consecutive N_2 flux rates. As an additional check to ensure that the measured fluxes are due to denitrification, the following measurements are frequently used. After denitrification measurements are completed, all sources of nitrate and nitrite (overlying water and nitrification in sediments) are eliminated from the chambers containing sediments. This is accomplished by using overlying water that is nitrate and nitrite free and flushing the water and gas phase with O_2-free He. The chambers are then treated and sampled as described above for the denitrification measurements. If the nitrate concentration in the water during the denitrification measurements was high, it may take 1 to 2 d after nitrate-free water is added to deplete the nitrate in the pore waters. Additionally, if the denitrification rates were high, there may be some N_2 buildup in the pore waters which may take a couple of days to diffuse out. However, after that time, the N_2 flux should decrease to zero.

Nitrification and Sediment-Water Nutrient Fluxes

Measurements

Sediment-water nutrient fluxes are measured on the same cores used for denitrification measurements.

1. Collect initial water samples from the chambers after the water is changed over a core but before the chamber is closed for N_2 flux measurements.

2. Collect final samples approximately 24 h later, after the final gas sample is collected.

3. Determine the water volume inside each chamber by measuring total water height during an incubation and knowing the i.d. of the chambers.

4. Filter the samples through prerinsed glass fiber filters (Whatman GF/F) and analyze for nitrate and nitrite[15] and ammonia.[16]

Changes in nutrient concentrations attributable to water column processes are measured by incubating water without sediment. These controls can be incubated in glass beakers that are similar in size to the sediment incubation chambers. Incubate and sample the controls as described for the sediment chambers.

Calculations

Calculate the sediment-water flux of nutrients based on the initial and final concentration in the chamber water ($CORE_i$, $CORE_f$) and control water ($CNTRL_i$, $CNTRL_f$), the volume of water over the cores (VOL), surface area of sediment (A), and incubation time (T) as follows:

$$Flux = [((CORE_f - CORE_i) - (CNTRL_f - CNTRL_i)) \cdot VOL]/(T \cdot A) \qquad (2)$$

Sediment nitrification rates are then calculated as the sum of the net flux of nitrate across the sediment-water interface plus the rate of nitrate production that would be necessary to supply the measured denitrification rate (DNF from Equation 1). Note that this calculation is valid regardless of whether there is a net flux of nitrate into or out of the sediments; just be sure and keep the sign of the nitrate flux negative if there is a net uptake into the sediments. If there is significant reduction of nitrate to ammonia in the sediments,[17,18] this method will underestimate nitrification rates.

Net organic N mineralization rates are calculated as the sum of the sediment-water inorganic N fluxes (ammonia + nitrate and nitrite + N_2).

N_2O fluxes

N_2O is a trace gas in the atmosphere that can be both produced and consumed during denitrification and produced during nitrification.[1] The net sediment-water flux of N_2O can be relatively easily measured using the same setup as described above for N_2 production measurements. Time series samples (1.5 ml) of the gas phase in the chambers are analyzed for N_2O concentration by direct injection of duplicate samples into a GC equipped with a [63]Ni electron capture detector (340°C). Variable volume injections of a certified standard gas mixture containing 350 ppb N_2O in synthetic air are used to calibrate the detector. Because of the high sensitivity of electron capture detection (ECD) gas chromatography to N_2O, differences in N_2O concentration inside the chambers can normally be measured on samples collected 1 to 2 h apart. In addition, because ambient N_2O concentrations are low, N_2O fluxes can be measured when the sediments are initially placed in the chambers. The solubility coefficient of N_2O as a function of water temperature and salinity[19] is used to calculate the total N_2O in the gas and water phases in the chambers. The sediment-water flux of N_2O is then calculated according to Equation 1.

NOTES AND COMMENTS

1. Considerable care must be taken to ensure that the samples or incubation chambers are not contaminated with room air. The water changing and sampling procedures outlined above have repeatedly been tested and demonstrated to provide contamination-free samples. It typically takes 3 to 5 d of practice before an individual consistently obtains

contamination-free samples. We have tried chambers made of various plastics and found that only glass provides the high degree of gas tightness required.

2. O_2 levels inside the incubation chambers should be kept near field concentrations, since oxygen concentration can affect denitrification rates directly as well as indirectly by affecting nitrification rates. O_2 levels are maintained near the desired level by regulating the frequency of water changing and gas phase flushing.

3. The denitrification method described has been used primarily with unvegetated sediments.[4,8,9,11,13] However, the method has also been used with sediments containing benthic algae and seagrasses and in wetland sediments.[14] We used larger chambers (20 cm diameter and 1 m high) when working with tall seagrasses. If sediments are vegetated and incubated in the light, cover the outside of the chamber bottom around the sediment core with foil to keep the outside of the core in the dark.

REFERENCES

1. Knowles, R., Denitrification, *Microbiol. Rev.*, 46, 43, 1982.

2. Seitzinger, S. P., Denitrification in freshwater and coastal marine ecosystems: ecological and geochemical significance, *Limnol. Oceanogr.*, 33, 702, 1988.

3. Seitzinger, S. P., Nixon, S. W., Pilson, M. E. Q., and Burke, S., Denitrification and N_2O production in nearshore marine sediments, *Geochim. Cosmochim. Acta*, 44, 1853, 1980.

4. Seitzinger, S. P., Nitrogen biogeochemistry in an unpolluted estuary: the importance of benthic denitrification, *Mar. Ecol. Progr. Ser.*, 37, 65, 1987.

5. Balderston, W. L., Sherr, B., and Payne, W. J., Blockage by acetylene of nitrous oxide reduction in *Pseudomonas perfectomarinus*, *Appl. Environ. Microbiol.*, 31, 504, 1976.

6. Knowles, R., Acetylene inhibition technique: development, advantages and potential problems, in *Denitrification in Soil and Sediment*, Revsbech, N. P. and Sørensen, J., Eds., FEMS Symp. No. 56, Plenum Press, New York, 1990, 151.

7. Seitzinger, S. P., Nielsen, L. P., Caffrey, J., and Christensen, P. B., Denitrification measurements in aquatic sediments: a comparison of three methods, submitted.

8. Seitzinger, S. P., Nixon, S. W., and Pilson, M. E. Q., Denitrification and nitrous oxide production in a coastal marine ecosystem, *Limnol. Oceanogr.*, 29, 73, 1984.

9. Nowicki, B. L., Estuaries as nutrient traps and transformers: evidence from mesocosm experiments, Ph.D. thesis, University of Rhode Island, Graduate School Oceanography, Kingston, 1990.

10. Nowicki, B., personal communication.

11. Gardner, W. S., Nalepa, T. F., and Malczyk, J. M., Nitrogen mineralization and denitrification in Lake Michigan sediments, *Limnol. Oceanogr.*, 32, 1226, 1987.

12. Weiss, R. F., The solubility of nitrogen, oxygen and argon in water and seawater, *Deep-Sea Res.*, 17, 721, 1970.

13. Nowicki, B. L. and Oviatt, C. A., Are estuaries traps for anthropogenic nutrients? Evidence from estuarine mesocosms, *Mar. Ecol. Progr. Ser.*, 66, 131, 1990.

14. Seitzinger, S. P. and DeKorsey, R., unpublished data.

15. Armstrong, F. A., Stearns, C. R., and Strickland, J. D., The measurement of upwelling and subsequent biological processes by means of the Technicon autoanalyzer and associated equipment, *Deep-Sea Res.*, 14, 381, 1967.

16. Solorzano, L., Determination of ammonia in natural waters by the phenylhypochlorite method, *Limnol. Oceanogr.*, 14, 799, 1969.

17. Nishio, T., Koike, I., and Hatorri, A., Denitrification, nitrate reduction and oxygen consumption in coastal and estuarine sediments, *Appl. Environ. Microbiol.*, 43, 648, 1982.

18. Binnerup, S. J., Jensen, K., Revsbech, N. P., Jensen, M. L., and Sørensen, J., Denitrification, dissimilatory reduction of nitrate to ammonium and nitrification in bioturbated estuarine sediments as measured with ^{15}N and microsensor techniques, *Appl. Environ. Microbiol.*, 58, 303, 1992.

19. Weiss, R. F. and Price, B. A., Nitrous oxide solubility in water and seawater, *Mar. Chem.*, 8, 347, 1980.

CHAPTER 75

Turnover of $^{15}NH_4^+$ Tracer in Sediments

T. H. Blackburn

INTRODUCTION

The main purpose in experimenting with the addition of $^{15}NH_4^+$ to sediment is to obtain information on the dynamics of the NH_4^+ pool in sediments. Temporal changes in this pool reflect net uptake or net production of NH_4^+ and this information can often be obtained without recourse to $^{15}NH_4^+$ methodology. Net production of NH_4^+ is most commonly observed and reflects the excess production of NH_4^+ from the breakdown of particulate organic nitrogen (PON). Negative changes in the pool size indicate that uptake or disappearance of NH_4^+ has occurred. Most investigations of the nitrogen cycle in sediments are concerned with the determination of the extent to which the sediment is a source of mineralized nitrogen to the overlying water. When this is the purpose, net rates of change in the NH_4^+ pool are the measurements of most interest. If the situation demands a more complete understanding of the processes which result in these net changes, then $^{15}NH_4^+$ methods are very helpful.

The disappearance of $^{15}NH_4^+$, during the course of an incubation, indicates that the label has gone into another pool. This disappearance can be due to incorporation into new biomass or by oxidation to a different inorganic species, e.g., NO_3^-. Anaerobic incubations, in which NO_3^- cannot be produced, limit the disappearance of label into the biomass of anaerobic organisms, mainly bacteria. Aerobic incubations allow nitrification to occur and label can be incorporated into all cells, aerobic at the sediment surface and anaerobic below the oxic zone. This disappearance of label, e.g., into root biomass, can be used to deduce the amount of ammonium available for uptake into a plant, in situations where this is difficult to measure by more direct means. This type of situation will be discussed later.

In anaerobic incubations, if there is an independent estimate of the rate of NH_4^+ incorporation into microbial biomass, it is possible to calculate the probable C:N ratio of the PON, which is undergoing hydrolysis and degradation. It should be remembered, however, that the bacterial population is probably in a steady state; the production of new biomass is matched by the degradation of an equal amount of decomposing bacterial cells. This means that the NH_4^+ results from the degradation of both these cells and from the background PON. Correction should be made for this cellular contribution in order to obtain the C:N ratio of the PON. Incubations are generally so short, that no recycling of ^{15}N label occurs. This is a restriction which must be met in all experiments.

A major problem arises after the addition of $^{15}NH_4^+$ to sediment. A variable amount of the label immediately disappears and cannot be easily extracted with KCl or one of the other salts, which are commonly used to free adsorbed NH_4^+ from sediment particles. Allowance can be made for this disappearance, but unfortunately some of the $^{15}NH_4^+$ can reappear during the incubation. The kinetics of these processes are complex and difficult to model.

MATERIALS REQUIRED

Equipment

- Carlo Erba NA Nitrogen Analyzer
- Anasira mass spectrometer (VG Isogas, England)

Supplies

- Plexiglass tubes for sampling, 2.6 and 4.2 cm in diameter
- N_2 gas
- 500-ml beakers
- Syringes with ends cut off
- Glass centrifuge tubes
- Glass scintillation vials, 20 ml

Solutions and Reagents

- Anoxic seawater
- Chloroform
- 4 or 2 M KCl
- $^{15}NH_4^+$, about 99% ^{15}N
- 50% NaOH
- 5 N H_2SO_4
- N-free Al_2O_3

PROCEDURES

Incubations

Anoxic Incubation of $^{15}NH_4^+$ in Mixed Sediments

1. Sediment is collected in a suitable Plexiglass tube (e.g., 4.2 cm in diameter). The sediment is extruded and sectioned into 2-cm slices. The slices are placed in 500-ml beakers, under a N_2 gas stream.

2. Anoxic seawater (20 ml) containing $\sim$ 2000 nmol $^{15}NH_4^+$ is added to the sediment.

3. The labeled, stirred sediment is dispensed anoxically in 5 ml amounts, using an open-ended syringe, to a series of weighed glass centrifuge tubes. The tubes are stoppered under N_2, weighed, and incubated at the *in situ* temperature for some days (0, 2, and 5, depending on activity of the sediment).

4. The incubation is stopped by adding some drops of chloroform and 1.0 ml 4 M KCl, followed by shaking at room temperature for 2 h, to liberate the adsorbed NH_4^+.

5. The tubes are centrifuged for 15 min at 3000 $\times$ g. The supernatants are decanted and measured for NH_4^+ and ^{15}N-content. The total NH_4^+ content per cubic centimeter of the

original sediment is calculated from the water content, specific gravity, volume of seawater, and KCl which are used. Intial NH_4^+ concentrations may be calculated by subtracting the quantity of label added.

A modification of the above method, based on subsample volumes of sediment rather than weights, has some advantages. After slicing the sediment core, 5-ml volumes of the sediment are removed with a cut-off syringe and are introduced anaerobically into centrifuge tubes. An appropriate quantity of $^{15}NH_4^+$ is injected into the 5-ml subsample. At time intervals, 5 to 10 ml of 2.0 *M* KCl are added to the sediment samples in the centrifuge tubes to extract the exchangeable NH_4^+ pool.

Oxic Incubations of $^{15}NH_4^+$ Injected into Sediment Cores

The purpose is to incubate sediment under conditions which will not result in the death of the aerobic flora and fauna.

1. Sediment cores (2.6 cm in diameter) are injected with $^{15}NH_4^+$ (100 μl 0.01 *M* 99% ^{15}N) at 2-cm intervals. All overlying water is removed from the surface of the cores, to ensure that label is not lost by diffusion upward. It is usually necessary to inject 20 cores, because of sediment heterogeneity.

2. Half the cores are processed at ~12 h and the other half 36 h after injection. The cores are sectioned at 2-cm intervals, each section being extracted with 10.0 or 20.0 ml of 2.0 *M* KCl, depending on the clay content of the sediment.

Measurement of $^{15}NH_4^+$

Irrespective of the method of incubation, both NH_4^+ and ^{15}N content of the samples must be measured. The former is most conveniently assayed by an autoanalyzer using salicylic acid.[1] The ^{15}N content in the NH_4^+ pool is usually measured by emission spectrometry or mass spectrometry. Both methods require that the $^{15}NH_4^+$ be concentrated before measurement. This may be achieved by diffusion of NH_4^+ into a glass capillary for subsequent processing before measurement by emission spectrometry.[2] This methodology is less commonly used now, because of the availability of relatively cheap mass spectrometers. Therefore, the unpublished procedure for sample preparation for mass spectrometry will be described in more detail.

Mass spectrometers require that a certain minimum quantity of N must be available for an accurate N-isotopic determination to be made. For example, the mass spectrometer may require that 3 μmol N are introduced to ensure that pressure effects do not bias the ^{15}N measurement. Typically, the KCl extracts derived from the incubations, described above, might contain NH_4^+ at a concentration of 100 μ*M* (3 μmol per 30 ml). The procedure, which is used for the preparation of the sample, consists of the diffusion of the NH_4^+ from a volume of <5.0 ml.

1. Acidified Al_2O_3 is prepared by pipetting 10.0 ml of 5.0 N H_2SO_4 to 10.0 g of N-free Al_2O_3, which is then dried at 80°C. A portion of the dried powder (50 mg containing 250 μmol H^+) is added to a 4.0 × 6.0 mm aluminum capsule (Mikro Kemi AB, Sweden), which is then suspended below the rubber stopper (Figure 1).

2. Then 1 ml of sample is added to 1.0 ml of 3.0 m*M* natural abundance (i.e., non-^{15}N-enriched) NH_4^+, in a glass scintillation vial which is used as the diffusion chamber (Figure 1). The mixed samples plus carrier are made alkaline by the addition of 1.0 ml of 50% NaOH.

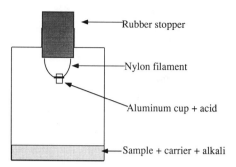

Figure 1.   Diffusion chamber. The chamber consists of a glass scintillation vial, as described in the text.

3.   The liberated NH_3 diffuses to the acidified Al_2O_3. The vial is closed by the rubber stopper, with attached cup, immediately after the addition of the alkali to the sample. Gentle agitation of the closed vial at 55°C for 24 h gives >95% transfer of the $^{15}NH_4^+$ to the acidified Al_2O_3.

4.   The capsule is carefully closed and introduced into a Carlo Erba NA Nitrogen Analyzer coupled to an Anasira mass spectrometer (VG Isogas, England) with a PC-controlled interface. The NH_4^+ is converted to N_2, whose $^{14}N^{14}N$, $^{14}N^{15}N$, and $^{15}N^{15}N$ content is determined and $\%^{15}N$ calculated.[3] Standards of the same NH_4^+ content and ^{15}N enrichment are run in place of every fifth sample. The $\%^{15}N$ of the original sample, without carrier is calculated from:

$$N_m = (N \cdot A_o + N_c \cdot A_c)/(A_o \cdot A_c) \qquad (1)$$

where N_m is the $\%^{15}N$ in the mixed sample plus carrier, N is the $\%^{15}N$ in the original extract, A_o is the quantity of NH_4^+ in the original extract (nmoles), N_c is the $\%^{15}N$ in the carrier (3000 nmol), and A_c is the $\%^{15}N$ in the carrier (0.366). On rearrangement this becomes:

$$N = (N_m \cdot A_o + N_m \cdot A_c - N_c \cdot A_c)/A_o \qquad (2)$$

The concentration of NH_4^+ in the original sample is calculated from:

$$C = (V_s \cdot P + V_k)C_k/V_s \qquad (3)$$

where   C is the concentration of NH_4^+ in the original sediment (nmol cm^{-3}), V_s is the volume of sediment in the sample (typically 10 cm^3), P is the porosity (ml cm^{-3}), V_k is the volume of KCl used in the extraction (typically 10 ml), and C_k is the concentration of NH_4^+ in the KCl extract (μM).

A linear regression is made of the NH_4^+ concentrations against time to calculate the rate of net NH_4^+ production. The slope of this plot is the rate (d − i), where d is the gross rate of NH_4^+ production and i is the rate of NH_4^+ disappearance, uptake and oxidation (Figure 2A). If there is a change in the NH_4^+ concentration during the incubation, this change must be used to calculate the rate d:

$$\ln (N_t - N_c) = \ln (N_o - N_c) - [d/(d - i)] \ln(C_t/C_o) \qquad (4)$$

where $(N - N_c)$ is the excess $\%^{15}N$ in the sample at time t, $(N_o - N_c)$ is the excess $\%^{15}N$ in the sample at time zero, C_t is the concentration of NH_4^+ at time t and C_o is the concentration of NH_4^+ at time zero. A plot of $\ln(C_t/C_o)$ against $\ln (N_t$

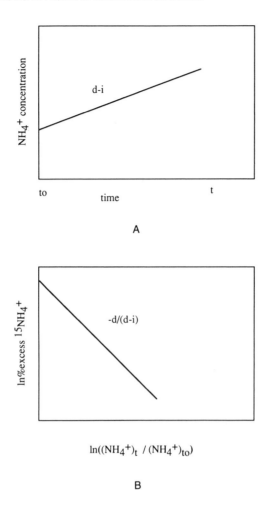

Figure 2. (A) Plot of NH_4^+ concentration against time. The slope of the plot is the net rate of NH_4^+ production. (B) Plot of ln(% excess $^{15}NH_4^+$) against the log of the final NH_4^+ concentration divided by the starting concentration. The slope is minus the gross NH_4^+ production rate divided by the net rate of NH_4^+ production.

$- N_c$) gives a slope of $d/(d - i)$, as seen in Figure 2B. The rate, d, is calculated from $- d/(d - i)$ and the measured $(d - i)$. Generally, this procedure can be used to process data from all types of incubations. The procedure of Blackburn and Henriksen[4] can be used in the situation where no net change in NH_4^+ concentration occurs during the incubation.

Interpretation of Results

One of the worst problems with the $^{15}NH_4^+$ tracer method is the disappearance of $^{15}NH_4^+$ immediately after addition to the sediment and its subsequent reappearance during the incubation. This occurs most frequently where sediment mineralization is relatively low and where the NH_4^+ pool is high. It is impossible that $^{15}NH_4^+$ should increase due to biological activity; the basis of the calculation requires that $^{15}NH_4^+$ content remains constant or decreases with time. The problem is greatest in silty sediments. Goeyens et al.[5] found a complete recovery of added label, where a dilute sediment slurry was used. Dilute slurries, however, have the disadvantage that coupled processes in the sediment become uncoupled, and the incubation conditions are a little unnatural. It must be concluded that the $^{15}NH_4^+$ tracer procedure has serious limitations when applied to many sediment types. The procedure, however, has many advantages when applied to the sediments in the rhizosphere of aquatic plants.

The dynamics of nitrogen cycling in the rhizosphere of aquatic plants is interesting because, in many situations, the sediment appears to be the major source of nitrogen for the plants. The incubations can be of slurries of sediment, having first removed the roots,[6,7] or $^{15}NH_4^+$ can be injected into cores taken in the rhizosphere. In the latter situation, excised roots can continue to take up NH_4^+ or may contribute to the production of NH_4^+ if the roots die and decompose during the incubation. Very often, the sediments which are under aquatic angiosperms, contain very low concentrations of NH_4^+, but this does not necessarily indicate that this NH_4^+ pool is unimportant to the plant. On the contrary, this pool can be very dynamic and turnover many times during a day. The dynamics of such pools can be best explored by using $^{15}NH_4^+$ methodology.

$^{15}NH_4^+$ methodology can also be used to explore the probable C:N ratio of the substrate undergoing degradation and the rate of bacterial biomass synthesis.[8]

REFERENCES

1. Bower, C. E. and Holm-Hansen, T., A salicylate-hypochlorite method for determining ammonia in seawater, *Can. J. Fish. Aquat. Sci.,* 37, 794, 1980.
2. Blackburn, T. H., Method for measuring rates of NH_4^+ turnover in anoxic marine sediments, using a $^{15}NH_4^+$ dilution technique, *Appl. Environ. Microbiol.,* 37, 760, 1979.
3. Fiedler, R. and Proksch, G., The determination of nitrogen-15 by emission and mass spectrometry in biochemical analysis: a review, *Anal. Chim. Acta,* 78, 1, 1975.
4. Blackburn, T. H. and Henriksen, K., Nitrogen recycling in different types of sediments from Danish waters, *Limnol. Oceanogr.,* 28, 477, 1983.
5. Goeyens, L., De Vries, R. T. P., Bakker, J. F., and Helder, W., An experiment on the relative importance of denitrification, nitrate reduction and ammonification in coastal marine sediment, *Neth. J. Sea Res.,* 21, 171, 1987.
6. Iizumi, H., Hattori, A., and McRoy, C. P., Ammonium regeneration and assimilation in eelgrass (*Zostera marina*) beds, *Mar. Biol.,* 66, 59, 1982.
7. Bowden, W. B., A nitrogen-15 isotope dilution study of ammonium production and consumption in a marsh sediment, *Limnol. Oceanogr.,* 29, 1004, 1984.
8. Blackburn, T. H., Nitrogen cycle in marine sediments, *Ophelia,* 26, 65, 1986.

CHAPTER 76

Microbial Cycling of Inorganic and Organic Phosphorus in the Water Column

James W. Ammerman

INTRODUCTION

Rates of microbial cycling of inorganic and organic phosphate (P) compounds in aquatic environments can be measured by radioisotope and fluorescence methods. These measurements include uptake and regeneration of inorganic orthophosphate (Pi), enzymatic hydrolysis of specific components of the dissolved organic P (DOP), and uptake of the released Pi (coupled uptake). These processes are included in a simplified diagram of the P cycle shown in Figure 1. Alkaline phosphatase (AP) is the best-studied microbial cell-surface enzyme, or "ectoenzyme",[1] involved in DOP breakdown, though 5′-nucleotidase and others may also be important.[2] This issue is beyond the scope of this paper, see Ammerman and Azam[3] for a detailed discussion. Radioisotope methods for Pi uptake and regeneration have been used for some time,[4] as have fluorescent measurements of AP activity.[5] Radioisotope methods for measuring enzymatic hydrolysis of organic phosphate and uptake of the released Pi are relatively recent,[3,6] although "uptake" of organic phosphates, presumably of the hydrolyzed Pi, was measured earlier.[7] Since both algae and bacteria are responsible for these processes, size-fractionation methods must be used to separate them (see below). Methods for the measurement of Pi, DOP, and particulate phosphate are not discussed here but can be found in published literature.[8-12] Most of these methods can be adapted for the use of 10-ml samples.

To measure uptake of Pi, radiolabeled Pi is added to a water sample and after a suitable incubation time, the plankton are collected on a filter, and the radioactive P on the filter is counted. Pi regeneration is determined by measuring the ratio of radiolabeled to chemical Pi in the filtrate of the above Pi uptake samples over time. To measure enzymatic hydrolysis of organic P, a radiolabeled organic P compound (e.g., ATP) is added to a water sample and incubated. The microbes hydrolyze the Pi from ATP and take up some or all of it. The hydrolyzed Pi taken up by the microbes is collected on a filter, and the released Pi which was not taken up is separated from the unhydrolyzed substrate and counted. This assumes that most DOP species must be dephosphorylated and the P taken up as Pi;[2] this has been shown in natural waters by Pi inhibition of ATP "uptake" but not hydrolysis.[3] The fluorescent technique for AP uses an organic phosphate analog which has little fluorescence when derivatized with Pi but is highly fluorescent when not derivatized. The assay simply follows the increase in fluorescence over time as more of the Pi is hydrolyzed by the enzyme.

The advantages of the isotopic techniques are: (1) very high sensitivity, because labeled substrates are available at tracer or near-tracer concentrations so adding label does not increase

0-87371-564-0/93/$0.00 + $.50

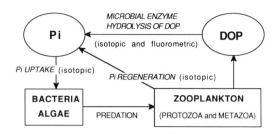

Figure 1. Simplified aquatic phosphorus cycle with the rate processes discussed in the text shown in italics, followed by the type of method(s) used to assay them. An example of the simplification is that Pi regeneration also includes microbial hydrolysis of DOP and is not due solely to zooplankton. Abbreviations: Pi, orthophosphate and DOP, dissolved organic phosphate. (Modified from Ammerman, J. W., *Microbial Enzymes in Aquatic Environments*, Chróst, R. J., Ed., Springer-Verlag, 1991, 165.)

the overall substrate concentrations; (2) labeled substrates are actual substrates and not analogs; (3) P-labeled organics allow the fate of the released Pi to be followed after hydrolysis; and (4) small samples can be rapidly batch-processed using multiplace filtration rigs. The assays below have been optimized for 10-ml samples processed in multiplace filtration rigs, but in many cases even smaller samples could be used. Disadvantages of isotopic techniques are: (1) radiation requires proper disposal, and for ^{32}P may also require shielding to prevent a safety hazard; (2) samples cannot be stored for very long prior to counting, due to the short half-lives of the isotopes; and (3) samples must be incubated in closed containers with the possibility of "bottle effects".

Phosphorus compounds can be labeled with either ^{32}P or ^{33}P (Amersham Corporation, Arlington Heights, IL; Dupont Company, NEN Research Products, Boston, MA; or ICN Biomedicals, Inc., Costa Mesa, CA), though the latter is less common (Dupont Company, NEN Research Products). ^{33}P is also more expensive, but has the advantages of lower energy (comparable to ^{14}C) and a longer half-life (25 d). ^{32}P is inexpensive (especially when purchased under a university contract with one of the above companies), even for most labeled organic compounds, though the selection is largely limited to nucleotides. However, ^{32}P also has a short half-life (14 d) which means stock solutions are short-lived. The higher energy of ^{32}P requires shielding for quantities of 1 mCi (37 MBq) or larger, but this high energy also results in very high counting efficiencies and easy detection of any contamination with a radiation meter (see below). ^{32}P counts with nearly 100% efficiency when cocktail is added and requires no quench curve. It can also be counted by Cerenkov radiation at lower efficiencies without cocktail. ^{33}P has an energy similar to ^{14}C and requires a quench curve for optimum dpm determination.

The fluorometric method for AP and other cell-surface enzymes has the following advantages: (1) high sensitivity, in the nanomolar range, though not as high as isotopic methods and usually above tracer concentrations; (2) no radiation is used, so no special disposal or safety procedures are required; (3) samples can be frozen following incubation for later analysis, at least according to a recent study;[13] and (4) with certain fluorescent substrates, time-course samples can be measured continuously, directly in the fluorometer. The disadvantages of the fluorometric techniques are: (1) they are useful for DOP hydrolysis studies only; (2) the substrates are analogs, not true substrates, so the rates of hydrolysis may not reflect actual rates; (3) the released Pi is not labeled so its fate subsequent to hydrolysis cannot be followed; (4) as with the isotopic methods, samples must be incubated in closed containers with the possibility of bottle effects; and (5) the samples have to be processed individually, rather than in batch mode.

The emphasis in all the protocols below, both for isotopic and fluorometric methods, is on small volume, replicated, simple assays, which can be quickly and conveniently done aboard ship or in a small field lab. Time-course assays are recommended in all protocols for the initial samples from a new environment, at least until optimal incubation times have been established. However, since this limits the total number of samples that can be processed, the only assay that is always run as a time course is the fluorescent AP assay, since the initial reading can vary with the background fluorescence of the water and substrate. Other assays are run as single end point assays once reasonable incubation times are established. Incubation times for all assays are also kept to the minimum time possible, in order to maximize throughput as well as limiting substrate depletion and bottle effects. Short incubations should also minimize the problem of isotope dilution of labeled substrate.[14] Therefore, most incubations are conducted in a temperature-controlled environment in the lab, which allows easy access to samples and confines the use of potentially hazardous isotopes like ^{32}P. Since Pi uptake is sometimes stimulated by light,[15,16] however, ^{32}Pi incubations longer than an hour should be carried out under natural light when possible.

I also favor tracer methods over kinetic assays, in order to estimate actual *in situ* substrate turnover times rather than potential activities. (A few kinetics experiments are important in each environment, however, to establish kinetic parameters). Only the fluorescent AP assay is not a tracer assay, although it may approximate one if the added substrate concentration is 100 nM or less. The rates determined by the above methods can be expressed as turnover rates, turnover times (the reciprocal), or flux rates. Flux rates (turnover rate $\times$ substrate concentration) are ultimately preferable, though turnover rates or times are more easily understood. Flux rates, however, require knowledge of the *in situ* substrate concentrations so except for total Pi and DOP fluxes, they are usually unobtainable. Pi and DOP are measurable in natural waters, but there are few good methods for specific components of the DOP, such as dissolved ATP (see discussion in Ammerman and Azam[17]). Even the chemical Pi method includes some fraction of the DOP, so Pi flux calculations should be treated with caution.[18]

Results of time-course assays are calculated by determining the slope of activity vs. time, usually determined with a least squares fit to a linear regression line, though other regression models (logarithmic, exponential) can also be used. Since the rates of most hydrolysis and uptake processes decline over a time course, it is not always clear how many and which time points to include in a regression. Many argue that initial rates are correct (see arguments for short incubations above) while others maintain that initial rates are unrealistically high (e.g., represent exchange rather than uptake), and use later values.[19] I prefer initial rates, and often perform time-course calculations with the first two time points (e.g., fluorescent AP assays). Single endpoint measurements of isotopic activities are equivalent to a two point time course with the initial point equal to zero.) Equations for single end point assays are given in the text and a figure below and references are included for the time-course calculations.

MATERIALS REQUIRED

DOP Hydrolysis

Equipment

- 10- or 12-place filtration rig for 25-mm filters (Hoefer Scientific Instruments, San Francisco, CA; Millipore Corporation), the Hoefer filtration rig is better for collecting filtrates and has individual vacuum controls for each filter tower
- Micropipettes
- Cooling/heating/circulating water bath
- Foam or other inexpensive cooler with heating/cooling coils attached to circulator

- Liquid scintillation counter
- Radiation mixer
- Vortex mixer

Supplies

- HCl-washed activated charcoal (Sigma Chemical Company, St. Louis, MO)
- 7-ml plastic scintillation vials
- 15-ml polystyrene disposable tubes with caps, sterile preferred
- 25-mm, 0.2-μm Nuclepore filters or equivalent
- 47-mm Nuclepore filters of various pore sizes for size fractionation
- 25-mm, 0.45-μm Millipore filters (HA) or equivalent
- Wash bottle with filtered sample water of appropriate salinity for rinsing filters

Solutions and Reagents

- [γ-^{32}P]ATP (Amersham Corporation; Dupont Company, NEN Research Products; or ICN Biomedicals, Inc.): the gamma form is preferable because the only likely labeled species are [γ-^{32}P]ATP and ^{32}Pi; specific activities between 100 and 1000 Ci mmol^{-1} (3.7 to 37 TBq mmol^{-1}) are best, in order to provide sufficient sensitivity without significant radiochemical breakdown
- 0.03 N H$_2$SO$_4$
- Formalin solution for making killed controls, or alternatively a hot plate, beaker, and screw-capped 50-ml culture tubes for making boiled controls

Pi Uptake and Regeneration

Additional Solutions and Supplies

- ^{32}Pi or ^{33}Pi in aqueous solution (Amersham Corporation, Dupont Company, NEN Research Products; or ICN Biomedicals, Inc.). ^{32}Pi-specific activity should be around 9000 Ci mmol^{-1} (333 TBq mmol^{-1}), or carrier-free (no unlabeled Pi molecules). ^{33}Pi-specific activity will be lower. Prepare and store ^{32}Pi stock solutions the same way as [γ-^{32}P]ATP stocks (see section on procedures).
- 25-mm Nucleopore filters of various pore sizes are used for size fractionation.

Fluorometric Methods for AP

Additional Equipment

- A fluorometer equipped for measurements at the 4-methylumbelliferone excitation wavelength of 365 nm and emission wavelength of 455 nm. This requires: (1) a spectrofluorometer set at the above wavelengths; (2) a filter fluorometer (Turner model 111, Sequoia-Turner Corporation, Mountain View, CA; Turner Designs Model 10, Turner Designs, Mountain View, CA; or similar) with a 4-W F4T5, near-UV fluorescent lamp (GTE Sylvania F4T5/350 BL or equivalent) and a Kodak (Wratten) color specification 7-60 excitation filter and 2A and 5-60 emission filters; (3) or a Hoefer TKO 100 mini-fluorometer which is made for only these wavelengths. Alternatively, a fluorometer equipped for measurements at the 3-O-methylfluorescein excitation wavelength of 490 nm and emission wavelength of 520 nm can be used. This requires a spectrofluorometer, or a filter fluorometer with a 4-W F4T5 fluorescent lamp and a Kodak (Wratten) color specification 47B excitation filter and a 2A-12 combination emission filters.

- Fluorometer cuvettes: square cuvettes with four clear slides for a spectrofluorometer; or 13 mm × 100 mm disposable glass culture tubes for a filter fluorometer.

Additional Solutions and Reagents

- 4-methylumbelliferyl phosphate (MUF-P; Sigma Chemical Company). Store solid in the dark at −20°C or below. make a 100-mM concentrated stock solution by dissolving 25.6 mg MUF-P in 1 ml 2-methoxyethanol using a vortex mixer to dissolve. Store in the freezer. It should be good for up to a week. Working stocks (100 μM) should be made fresh daily by diluting this concentrated stock with water.
- 4-methylumbelliferone, sodium salt (MUF; Sigma Chemical Company). Solid can be stored at room temperature. Concentrated and working stocks are made and stored as for 4-MUF-P, but use 19.8 mg for concentrated stock.
- 3-O-methylfluorescein phosphate (Sigma Chemical Company). Store stocks as for MUF-P. For 100 mM concentrated stock dissolve 52.5 mg in 1 ml of water. Working stock should be diluted as for MUF-P.
- 3-O-methylfluorescein (Sigma Chemical Company). Store stocks as for MUF-P. For 100-mM concentrated stock, dissolve 34.6 mg in 1 ml of absolute methanol. Working stock should be diluted as for MUF-P.
- 50 mM borate buffer (pH 10.8), 0.22-μm membrane filtered.
- 2-methoxyethanol (Aldrich Chemical Company, Milwaukee, WI).

PROCEDURES

DOP Hydrolysis

This method is discussed first because it is the most complex of the isotopic methods and the other methods are largely a subset of it. ATP is used as a model organic-P compound in this case though other labeled organic-P compounds have also been used (see below). As discussed above, various microbial ectoenzymes, including AP and 5′-nucleotidase, may be responsible for this hydrolysis. This issue is discussed in detail elsewhere.[3] ATP is a substrate for both of the above enzymes. To avoid possible phosphate or trace-metal contamination, all phosphate-related analyses should be carried out in new disposable plasticware, or else plastic containers which have been acid washed (10% HCl) and rinsed with deionized water. All reagents should also be made up in deionized water unless otherwise specified.

The stocks and working stocks of [γ-^{32}P]ATP should be stored at −20°C or less. Before use, stock solutions should be diluted to 10^5 to 10^6 dpm or less per 10-μl volume and filtered through a 0.2-μm filter attached to a syringe. These working stocks should be frozen in small aliquots in acid-washed plastic scintillation vials; each vial should contain a day's worth of isotope (using 10-μl stock per 10-ml sample) to avoid repeated freezing and thawing. The final chemical concentration of [γ-^{32}P]ATP after 10 μl of working stock is added to a 10-ml water sample should be less than 1 nM.

The procedure for measuring ATP hydrolysis is outlined in the flow chart in Figure 2 and should be carried out as soon as possible after water samples are brought on deck. If a delay of analysis for a few hours is necessary, then samples can be maintained in acid-cleaned plastic bottles in an opaque cooler away from sunlight or other heat or light sources.

1. Aliquot duplicate 10-ml subsamples from each water sample into 15-ml tubes and incubate in the circulating water bath at an average *in situ* temperature for the water samples being processed. Incubating samples in a test tube rack in an attached cooler, rather than the

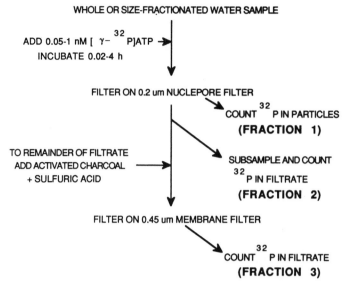

WHOLE OR SIZE-FRACTIONATED WATER SAMPLE

ADD 0.05-1 nM [$\gamma-^{32}$P]ATP
INCUBATE 0.02-4 h

FILTER ON 0.2 um NUCLEPORE FILTER

COUNT ^{32}P IN PARTICLES
(FRACTION 1)

TO REMAINDER OF FILTRATE
ADD ACTIVATED CHARCOAL
+ SULFURIC ACID

SUBSAMPLE AND COUNT
^{32}P IN FILTRATE
(FRACTION 2)

FILTER ON 0.45 um MEMBRANE FILTER

COUNT ^{32}P IN FILTRATE
(FRACTION 3)

A. ATP HYDROLYSIS (%/h) = (COUNTS IN FRACTIONS 1+3/COUNTS IN FRACTIONS 1+2) x 1/h

B. UPTAKE OF Pi RELEASED BY ATP HYDROLYSIS (%/h) =
(COUNTS IN FRACTION 1/COUNTS IN FRACTIONS 1 + 2) x 1/h

C. % UPTAKE OF Pi RELEASED BY ATP HYDROLYSIS = B/A x 100

Figure 2. Flow chart of the dissolved organic phosphate hydrolysis method, using radiolabeled ATP as a model dissolved organic phosphate compound. Note that the turnover rates calculated in A and B must be multiplied by 100 to be expressed as percentages. (Modified from Ammerman, J. W. and Azam, F., *Limnol. Oceanogr.*, 36, 1427, 1991.)

water bath itself, usually provides more room for samples and prevents possible radioactive contamination of the water bath.

2. Inoculate tubes with 10 µl of isotope each as quickly as possible (a repeating pipette is helpful for this), and note the time of inoculation.

3. After incubation, filter each 10-ml subsample onto a 25-mm, 0.2-µm Nuclepore (or equivalent) filter in a 10-place filtration rig and note the time of filtration. Collect the filtrate from each tube in another set of 10-ml tubes which were loaded into the filtration rig prior to filtration.

4. Remove the tubes containing the filtrates, *prior* to rinsing the filters.

5. Rinse each filter three times from the wash bottle containing filtered sample or similar water, adding enough water each time to completely cover the filter. (If the samples vary significantly in salinity, then each sample should be rinsed with filtered water of an appropriate salinity to avoid bursting the cells.)

6. Place the filters (Fraction 1 in Figure 2) in scintillation vials (the 7-ml size is sufficient) with enough scintillation cocktail to cover the filter, vortex, and count in a liquid scintillation counter. (It is normally unnecessary to dissolve the filters to count ^{32}P.)

7. Next, remove 1 ml of the filtrate (Fraction 2 in Figure 2) with a micropipette and count with 4 to 5 ml added cocktail (enough to avoid a two-phase solution) in a 7-ml scintillation vial.

8. To the remaining filtrate add about 20 mg of activated charcoal and 1 ml of 0.03 N H_2SO_4 and mix. The charcoal and acid may be added together as a slurry as long as the slurry is well-mixed.

9. Filter the charcoal-containing filtrates a second time, this time using filters which will remove all the charcoal (25-mm, 0.45-μm Millipore HA are among the easiest to handle).

10. Finally, count 1 ml of this filtrate (Fraction 3 in Figure 2) in the same way you counted Fraction 2.

11. Calculate the rates of ATP hydrolysis and uptake of released Pi (coupled uptake) using the equations in Figure 2 in a microcomputer spreadsheet. Control rates are calculated similarly and subtracted from the experimental rates. The percent uptake of Pi released by ATP hydrolysis (Figure 2, Equation C) should not exceed 100% and is usually determined by the ambient Pi concentration.[3]

The above ATP-hydrolysis method has been used in numerous coastal and estuarine studies by this lab.[3,6,17,20] Other investigators have also used this method to measure [γ-^{32}P] ATP hydrolysis and coupled uptake of released Pi in mesocosms in the Baltic Sea[21] and in lakes.[22] Another recent study in freshwater measured uptake of ^{32}Pi from [α-^{32}P]ATP with a modification of this method, but did not measure hydrolysis.[23] Other studies have used glucose-6-^{32}P to measure uptake of organic-P,[19,24] but again hydrolysis was not measured. Glucose-6-^{32}P is a good AP substrate, but unlike ATP, it is not a 5′-nucleotidase substrate. Unfortunately, glucose-6-^{32}P is not available commercially, it must be custom-synthesized by one of the companies listed above[19] or in the investigator's lab.[24] We have also recently used ^{32}P-pyrophosphate (^{32}PPi) to measure AP since, like glucose-6-^{32}P, it is an AP substrate but not a 5′-nucleotidase substrate.[25] The advantage over glucose-6-^{32}P, however is that ^{32}PPi is commercially available at specific activities of up to 60 Ci mmol^{-1} (2.2 TBq mmol^{-1}; Dupont Company, NEN Research Products). The assay for ^{32}PPi hydrolysis is identical to the ATP hydrolysis assay discussed above, except that thin-layer chromatography (TLC),[26] is used to separate ^{32}Pi from unreacted ^{32}PPi.

Pi Uptake and Regeneration

Microbial Pi uptake (Figure 2) is measured by substituting ^{32}Pi for [γ-^{32}P]ATP in the above assay and eliminating the second filtration. It is still useful but not essential to collect the original filtrate, however, it allows precise calculation of the amount of isotope added to each sample and greatly lessens the radioactive contamination of the filter rig. If larger samples are incubated and Pi turnover is slow, then Pi regeneration rates can be determined as well.

Start the experiment by adding 10 μl of ^{32}Pi to a 10-ml subsample and follow the procedure outlined in Figure 2 and discussed above. Size fractionation can be done by prefiltration as for ATP hydrolysis (see Notes and Comments section), or postfiltration of whole-water samples directly onto Nuclepore filters of appropriate pore size following incubation. The experiment is complete after the filtrate (Fraction 2 in Figure 2) is subsampled and counted. The remainder of the filtrate is discarded in the liquid radioactive waste. Pi uptake rates are calculated with Equation B in Figure 2, and the Pi uptake rate in the controls is subtracted from the rates in the experimental samples. Time-course experiments may be particularly important with Pi uptake because turnover times may be only a few minutes in P-deficient environments. Studies have used this method[20] or the very similar method of following the time course of ^{32}Pi depletion in the filtrate to calculate the uptake rate constant.[4,27] The equation for time-course determinations of turnover rate is given by Lean et al.[27] Detailed methods for measuring Pi uptake kinetics are presented in Bentzen and Taylor.[28]

Numerous ammonium regeneration studies have been done by isotope dilution methods in a variety of aquatic environments (see references in Glibert et al.[29]). A few studies have applied similar methods to studies of Pi regeneration,[14,19,30,31] but they are complex and work best in environments with long Pi turnover times (greater than 24 h). The method is briefly discussed here for the sake of completeness, but it cannot be recommended without further study. Large transparent plastic bottles (250 ml or larger) are inoculated with ^{32}Pi and incubated under simulated *in situ* conditions. Both ^{32}P and chemical Pi concentrations are measured in the 10-ml filtrates and the result is a time course of the Pi specific activity. From the equations given in Harrison[30] and Harrison and Harris,[14] these time-course data are used to calculate a regeneration rate, assuming that it is constant.

Some fraction of the ^{32}P in the filtrates in these regeneration experiments may be ^{32}P-labeled organics released by organisms and not ^{32}Pi. If this fraction is significant, the isobutanol extraction method for Pi[11] must be used to separate ^{32}Pi from the ^{32}P-labeled organics.[31] This method would require much larger volume filtrates and details are beyond the scope of this paper.

Fluorometric Methods for AP

Unlike the above assays, I always measure AP as a time course (see Introduction).

1. Start the assay by adding 10 μl of MUF-P to 10-ml samples in replicate 15-ml tubes to yield a final substrate concentration of 100 nM (use 1 μM if activity is low). (Size fractionation must be done by prefiltration as for ATP hydrolysis; see Notes and Comments.) Mix the sample and immediately remove a 3-ml subsample into a clean fluorometer cuvette. (If more than three time points are needed, incubate larger samples in Whirl-pak bags or plastic bottles.)

2. Add 1 ml of 50 mM borate buffer (pH 10.8) to the subsample in the cuvette and mix, giving a final pH above 10.

3. Wipe off the outside of the cuvette, to prevent interference, and place the cuvette in the fluorometer (previously zeroed on distilled water plus buffer) for an initial reading. The fluorescent breakdown product of the AP reaction, 4-methylumbelliferone (MUF), is most fluorescent at pH values of 10 or higher.[32]

4. Remove and similarly process other 3-ml aliquots at periodic intervals appropriate to the amount of activity. As with the above assays, necessary incubation times can range from a few minutes to several hours. A pair of formalin-killed or boiled controls should also be run through the entire procedure with each set of samples and treated like a regular sample (see Notes and Comments).

5. The turnover time of MUF-P (at 100 nM final concentration) by AP activity is calculated with the following equation:

    turnover time (h) = fluorescence of 100 nM MUF standard (in fluorescence units)/slope of fluorescence time course (in fluorescence units/h).

    The reciprocal would give turnover rate, comparable to Figure 2. The slope is usually determined from just the first two time points, to measure initial rates (see Introduction). Control samples are generally not subtracted but are used to ensure that there is no significant substrate hydrolysis due to nonbiological processes. To calibrate the fluorometer, one or more standard MUF solutions (between 1 and 100 nM as appropriate) are measured just like the samples, after mixing 3 ml of standard solution with 1 ml of borate buffer.

NOTES AND COMMENTS

DOP Hydrolysis

1. See Procedures, step 1.

   - For size fractionation experiments, all fractions are prefiltered through 47-mm Nuclepore filters prior to aliquoting into the 15-ml tubes; postfiltration cannot be used for enzyme activity measurements. The <0.6 μm fractions and larger size fractions can be filtered by gravity alone, and smaller fractions including <0.2 μm (cell-free) fractions are filtered by gentle vacuum (less than 25 cm Hg).
   - A pair of formalin-killed or boiled controls should be included with each set of water samples. Samples with widely varying particle loads, such as along an estuarine salinity gradient, should have a pair of controls for every water sample. Formalin-killed controls should be killed with 0.2 ml formalin added to each 10-ml control at least 15 min before substrate is added (boiled controls are boiled in culture tubes for 10 min and then cooled before aliquoting into 15-ml tubes). Controls are then processed like regular samples.

2. See Procedures, step 2. Appropriate incubation times vary widely in different environments and range from 1 min to several hours. This can best be determined by running time courses on the initial samples. Incubate time-course samples in a series of individual 15 ml tubes, or in a larger plastic bottle, and then remove 10-ml subsamples. (Whirl-pak bags are convenient sterile, disposal containers but are not recommended for use with ^{32}P because of the chance of a spill).

3. See Procedures, step 3. Nuclepore polycarbonate filters may retain less isotope in control samples and act more like sieves than membrane filters. However, membrane filters such as Millipore HA (0.45-μm pore size) and GS (0.22-μm pore size) are also acceptable and are easier to handle.

4. A radiation meter can be used to make preliminary counts on ^{32}P samples, close enough to optimize incubation times, if a scintillation counter is unavailable on the ship or at the field site. Final counts should be done with a liquid scintillation counter, however. Single-sample, bench-top ^{32}P counters are also available (Bioscan, Washington, D.C.) but are too expensive for occasional use. I recommend one of the newer nontoxic or "ecological" cocktails (e.g., EcoLite and Ecolume, ICN Biomedicals, Inc.; Formula 989, Dupont Company, NEN Research Products; Ready Safe, Beckman Instruments, Inc., Fullerton, CA) for both safety reasons and ease of disposal.

5. Working stocks of [γ-^{32}P]ATP should be periodically checked for chemical hydrolysis of ^{32}Pi. This can be done by treating a small aliquot of the working stock with activated charcoal and dilute sulfuric acid and filtering it, as outlined in Figure 2. Control samples in ATP hydrolysis experiments also provide a convenient check on the integrity of the [γ-^{32}P]ATP working stock and can be used to correct for chemical hydrolysis. Even fresh stocks of [γ-^{32}P]ATP typically have about a 5% background of ^{32}Pi. Experiments conducted with working stocks with significant backgrounds of ^{32}Pi may exhibit ^{32}Pi uptake rates which are artifactually higher than 100% of the [γ-^{32}P]ATP hydrolysis rate.

Pi Uptake and Regeneration

1. The Pi regeneration experiments described above require high precision Pi measurements, probably ±5%, in order to provide good estimates of regeneration rates.

2. Multiplace filter rigs, which are used with radioactive P isotopes, need to be cleaned frequently to prevent a buildup of isotope. Metal-containing units, like the Hoefer model,

are especially prone to this problem, particularly older units. I use 25% phosphoric acid, followed by copious rinsing with distilled water. Plastic units, like the Millipore one, may become stained with charcoal, particularly on the gaskets. This can be lessened, but not eliminated, by frequent cleaning.

Fluorometric Methods for AP

1. Since fluorometers are very sensitive, any particulate or other contamination can lead to spuriously high readings. The fluorometer cuvettes should be thoroughly rinsed between samples.

2. The borate buffer may need to be periodically refiltered to remove particles.

3. Numerous AP studies have been done using MUF-P;[1,20,33,34] it is currently the preferred fluorometric substrate for environmental AP studies. Some of these studies[1,33] use a kinetic approach with saturating substrate concentrations for the AP assay, rather than the "near-tracer" approach used here. The kinetic approach permits the calculation of the kinetic parameters K_m and V_{max}, but provides only a maximum potential activity rather than a true *in situ* turnover rate. The maximal activity is most useful when normalized to biomass.

4. AP is not the only enzyme which can be assayed with a MUF-derivatized substrate. Numerous MUF-derivatives are available for assaying other ectoenzymes,[1] especially sugar-degrading enzymes and peptidases (Sigma Chemical Company). Since all are MUF-derivatives, these other assays use the same wavelength settings on the fluorometer as the AP assay.

5. Probably the first fluorescent substrate used for environmental AP assays was 3-*O*-methylfluorescein phosphate.[5] This substrate is another fluorescent analog which functions much like MUF-P. It has a similar sensitivity and is assayed in the same way, though the excitation and emission wavelengths are different (see above). The disadvantages of 3-*O*-methylfluorescein phosphate are that it costs more than MUF-P (Sigma Chemical Company) and there are few other enzymes substrates available as 3-*O*-methylfluorescein derivatives. Its one major advantage is that no addition of buffer is required to raise the pH before measuring the fluorescence. This means that continuous AP assays can be run by incubating the samples directly in the fluorometer, as long as the cuvette holder is temperature-controlled. Such continuous assays are useful for measuring very accurate time courses of activity, though they limit the total number of samples that can be run.

REFERENCES

1. Chróst, R. J., Microbial ectoenzymes in aquatic environments, in *Aquatic Microbial Ecology: Biochemical and Molecular Approaches,* Overbeck, J. and Chróst, R. J., Eds., Springer-Verlag, 1990, 47.

2. Lugtenberg, B., The pho regulon in *Escherichia coli,* in *Phosphate Metabolism and Cellular Regulation in Microorganisms,* Torriani-Gorini, A., Rothman, F. G., Silver, S., Wright, A., and Yagil, E., Eds., American Society for Microbiology, 1987, 1.

3. Ammerman, J. W. and Azam, F., Bacterial 5′-nucleotidase activity in estuarine and coastal marine waters: characterization of enzyme activity, *Limnol. Oceanogr.,* 36, 1427, 1991.

4. Cembella, A. D., Antia, N. J., and Harrison, P. J., The utilization of inorganic and organic phosphorus compounds as nutrients by eukaryotic microalgae: a multidisciplinary perspective: Part 2, *CRC Crit. Rev. Microbiol.,* 11, 13, 1984.

5. Perry, M. J., Alkaline phosphatase activity in subtropical Central North Pacific waters using a sensitive fluorometric method, *Mar. Biol.,* 15, 113, 1972.

6. Ammerman, J. W. and Azam, F., Bacterial 5′-nucleotidase in aquatic ecosystems: a novel mechanism of phosphorus regeneration, *Science,* 227, 1338, 1985.

7. Taft, J. L., Loftus, M. E., and Taylor, W. R., Phosphate uptake from phosphomonoesters by phytoplankton in the Chesapeake Bay, *Limnol. Oceanogr.,* 22, 1012, 1977.

8. Clesceri, L. S., Greenberg, A. E., and Trussell, R. R., *Standard Methods for the Examination of Water and Wastewater,* American Public Health Association, 1989.

9. Parsons, T. R., Maita, Y., and Lalli, C. M., *A Manual of Chemical and Biological Methods for Seawater Analysis,* Pergamon Press, 1984.

10. Ridal, J. J. and Moore, R. M., A re-examination of the measurement of dissolved organic phosphorus in seawater, *Mar. Chem.,* 29, 19, 1990.

11. Strickland, J. D. H. and Parsons, T. R., *A Practical Handbook of Seawater Analysis, 2nd. ed.,* Fish. Res. Board Canada, 1972.

12. Solorzano, L. and Sharp, J. H., Determination of total dissolved phosphorus and particulate phosphorus in natural waters, *Limnol. Oceanogr.,* 25, 754, 1980.

13. Chróst, R. J. and Velimirov, B., Measurement of enzyme kinetics in water samples: effect of freezing and soluble stabilizer, *Mar. Ecol. Prog. Ser.,* 70, 93, 1991.

14. Harrison, W. G. and Harris, L. R., Isotope-dilution and its effects on measurements of nitrogen and phosphorus uptake by oceanic microplankton, *Mar. Ecol. Prog. Ser.,* 27, 253, 1986.

15. Nalewajko, C. and Lee, K., Light stimulation of phosphate uptake in marine phytoplankton, *Mar. Biol.,* 74, 9, 1983.

16. Cembella, A. D., Antia, N. J., and Harrison, P. J., The utilization of inorganic and organic phosphorus compounds as nutrients by eukaryotic microalgae: a multidisciplinary perspective: Part 1, *CRC Crit. Rev. Microbiol.,* 10, 317, 1984.

17. Ammerman, J. W. and Azam, F., Bacterial 5'-nucleotidase activity in estuarine and coastal marine waters: role in phosphorus regeneration, *Limnol. Oceanogr.,* 36, 1437, 1991.

18. Lean, D. R. S. and White, E., Chemical and radiotracer measurements of phosphorus uptake by lake plankton, *Can. J. Fish. Aquat. Sci.,* 40, 147, 1983.

19. Taft, J. L., Taylor, W. R., and McCarthy, J. J., Uptake and release of phosphorus by phytoplankton in the Chesapeake Bay estuary, USA, *Mar. Biol.,* 33, 21, 1975.

20. Ammerman, J. W., Role of ecto-phosphohydrolases in phosphorus regeneration in estuarine and coastal ecosystems, in *Microbial Enzymes in Aquatic Environments,* Chróst, R. J., Ed., Springer-Verlag, 1991, 165.

21. Tamminen, T., Dissolved organic phosphorus regeneration by bacterioplankton: 5'-nucleotidase activity and subsequent phosphate uptake in a mesocosm enrichment experiment, *Mar. Ecol. Prog. Ser.,* 58, 89, 1989.

22. Cotner, J. B. and Wetzel, R. G., 5'-Nucleotidase activity in a eutrophic lake and an oligotrophic lake, *Appl. Environ. Microbiol.,* 57, 1306, 1991.

23. Bentzen, E. and Taylor, W. D., Estimating organic P utilization by freshwater plankton using [^{32}P]ATP, *J. Plankton Res.,* 13, 1223, 1991.

24. Heath, R. T. and Edinger, A. C., Uptake of ^{32}P-phosphoryl from glucose-6-phosphate by plankton in an acid bog lake, *Verh. Int. Verein. Limnol.,* 24, 210, 1990.

25. Ammerman, J. W. and Beifuss, M. J., A new tracer alkaline phosphatase assay for field use, in preparation.

26. Chróst, R. J., Phosphorus and microplankton development in a eutrophic lake, *Acta Microbiol. Polon.,* 37, 205, 1988.

27. Lean, D. R. S., Abbott, A. A., and Pick, F. R., Phosphorus deficiency of Lake Ontario plankton, *Can. J. Fish Aquat. Sci.,* 44, 2069, 1987.

28. Bentzen, E. and Taylor, W. D., Estimating Michaelis-Menten parameters and lake water phosphate by the Rigler bioassay: importance of fitting technique, plankton size, and substrate range, *Can. J. Fish. Aquat. Sci.,* 48, 73, 1991.

29. Glibert, P. A., Lipschultz, F., McCarthy, J. J., and Altabet, M. A., Isotope dilution models of uptake and remineralization of ammonium by marine plankton, *Limnol. Oceanogr.,* 27, 639, 1982.

30. Harrison, W. G., Uptake and recycling of soluble reactive phosphorus by marine microplankton, *Mar. Ecol. Prog. Ser.,* 10, 127, 1983.

31. Smith, R. E. H., Harrison, W. G., and Harris, L., Phosphorus exchange in marine microplankton communities near Hawaii, *Mar. Biol.,* 86, 75, 1985.

32. Chróst, R. J. and Krambeck, H. J., Fluorescence correction for measurements of enzyme activity in natural waters using methylumbelliferyl-substrates, *Arch. Hydrobiol.,* 106, 79, 1986.

33. Chróst, R. J. and Overbeck, J., Kinetics of alkaline phosphatase activity and phosphorus availability for phytoplankton and bacterioplankton in Lake Plußee (North German eutrophic lake), *Microb. Ecol.,* 13, 229, 1987.

34. Hoppe, H.-G., Significance of exoenzymatic activities in the ecology of brackish water: measurements by means of methylumbelliferyl-substrates, *Mar. Ecol. Prog. Ser.,* 11, 299, 1983.

Section VI
Food Webs and Trophic Interactions

CHAPTER 77

Approaches for Measuring Stable Carbon and Nitrogen Isotopes in Bacteria

Richard B. Coffin and Luis A. Cifuentes

INTRODUCTION

Stable isotopes have been used successfully over the past three decades to trace geochemical cycles and pathways of elemental transfer through aquatic food chains. This technique, however, has only recently been used to examine aquatic microbial roles in elemental cycling. The major obstacle to measuring stable isotope compositions in bacteria has been the concentration of enough bacteria and separation of bacteria from other microorganisms and bacterial-sized particles. In the past three years, direct and indirect approaches have been used to measure stable carbon and nitrogen isotope compositions and sources of bacterial carbon and nitrogen.[1-3]

The bioassay incubation[2] is the simplest approach for examining stable isotope ratios in bacteria. Water samples are filtered through 0.2-μm pore size filters to remove particles, inoculated with a 1.0% volume of indigenous bacteria, and incubated for 24 to 48 h until bacteria produce enough biomass for stable isotopic analysis. This method relies on the observation that bacteria grown on organic matter with a known stable carbon isotope value ($\delta^{13}C$) produce particulate biomass with a similar value.[2-6] For example, indigenous bacteria grown on oak tree leachate with a $\delta^{13}C$ of -28.8% produced biomass with values of -25.9%, while the same bacterial assemblage grown on *Spartina* leachate with $\delta^{13}C$ of -13.3% resulted in bacteria with values of -14.1%.[2] With this approach, some sources of dissolved organic matter that are used by bacteria can be identified.

To measure stable carbon and nitrogen ratios *in situ*, nucleic acids are used as a biomarker for bacteria.[3] For this approach, particles <1.0 μm are concentrated with tangential flow filtration followed by extraction of nucleic acids from these concentrates. Preliminary studies have shown that nucleic acids can be used as a biomarker for bacterial cells by demonstrating $\delta^{13}C$ isotopic fidelity between nucleic acids extracted from bacteria, the whole bacterial cell, and the substrate on which bacteria were grown.[3,6] Furthermore, Coffin et al.[3] demonstrated that enough nucleic acids for stable isotope analysis could be extracted from <1.0 μm particles concentrated from 50 to 100 L of estuarine water in approximately 2 h. To confirm that the source of the nucleic acids was indeed bacterial, oligonucleotide probes, targeted for 16S rRNA were used to demonstrate that greater than 90% of the nucleic acid extracted from 1-μm filtrates was from bacteria.[3]

An alternative approach is to use $\delta^{13}C$ analysis of dissolved inorganic carbon (DIC) to examine the source of organic matter that is respired by bacteria. Water or soil samples are enclosed in respirometer flasks and respired CO_2 is trapped in sodium hydroxide and analyzed for $\delta^{13}C$ composition. While this approach requires further testing to determine the degree of fidelity between respired CO_2, bacteria, and the substrate source, preliminary results suggest substrate sources with largely different $\delta^{13}C$ can be differentiated. For example, this technique has recently been used to compare the relative degradation of *Spartina* and crude oil in a salt marsh that was damaged by an oil spill. Results indicated that the largest proportion of bacterial substrate was provided by organic matter from *Spartina*.[7] This approach is most suited to samples from complex environments where it is difficult to isolate bacteria and extract nucleic acids, or is not feasible to isolate bacteria for bioassay experiments.

MATERIALS REQUIRED

Bioassay Incubations

Equipment

- Peristaltic pump
- Filter cartridge holder

Supplies

- 0.2-μm cartridge filters
- 47-mm 1.0-μm filters and filtration apparatus
- 47-mm filtration apparatus
- Incubation containers
- 25-mm GF/F filters and filtration apparatus

Nucleic Acids Used as an *In Situ* Biomarker

Equipment

- Peristaltic pump
- Tangential flow filtration apparatus
- High-speed centrifuge
- Vacuum oven
- Freeze dryer
- Spectrophotometer
- Bead beater
- Filter cartridge holder

Supplies

- Glass beads, baked overnight at 450°C
- 1.0-μm cartridge filters
- Centrifuge tubes, 30 ml

Solutions and Reagents

- $NaClO_4$ 1.5 M
- NaCl 15 mM

- Na_2HPO_4 1.5 mM (pH 7.0)
- Chloroform
- Ethanol 95%

Stable Isotope Analyses

Equipment

- Vacuum line ($<10^{-5}$ torr)
- Torch
- Programmable furnace
- Triple-collector isotope ratio mass spectrometer (CO_2)
- Dual-collector isotope ratio mass spectrometer (N_2)

Supplies

- Quartz tubing (8 mm i.d., 10 mm o.d.)
- Dewar flasks
- Sample bulbs or Pyrex tubing (CO_2)
- Sample bulbs w/silica gel (N_2)

Solutions and Reagents

- Liquid nitrogen
- CuO
- Cu
- Dry ice/ethanol

PROCEDURES

Bioassay Incubation

1. To obtain bacteria for stable carbon or nitrogen isotope analysis with the bioassay approach,[2] water samples are passed through 0.2-μm cartridge filters (Millipore Milligard) to remove bacterial sized particles (refer to Figure 1).

2. Filtered water samples are inoculated with a 1.0% volume of the same water sample that has been passed through a 1.0-μm filter. Generally, 20-L bioassay experiments are incubated in 5 gal cubitainers (Cole Palmer) in the dark.

3. Samples for bacterial abundance (AODC[8] method) are taken every 12 h, for 24 to 48 h. The length of incubation depends on the bacterial growth rate. Incubations must be sufficiently long to produce enough bacterial biomass for isotopic analysis. Depending on the quality of the mass spectrometer, a minimum of 100 to 200 μg bacterial carbon is required.

4. Bacteria are concentrated by filtration on to GF/F filters (Whatman) that have been baked at 500°C for 1 h. The GF/F filters do not concentrate all of the bacteria, however, the filters are more efficient with higher particulate loading. If water volumes that are passed through filters are kept as large as practical considering time constraints, up to 70% bacterial retention is possible. Filters are stored frozen at −20°C in plastic 47-mm Petri dishes until they can be prepared for analysis. In preparation for analysis, filters are thawed and dried in an oven at 60°C. To fit the dried filters into quartz combustion tubes, they are

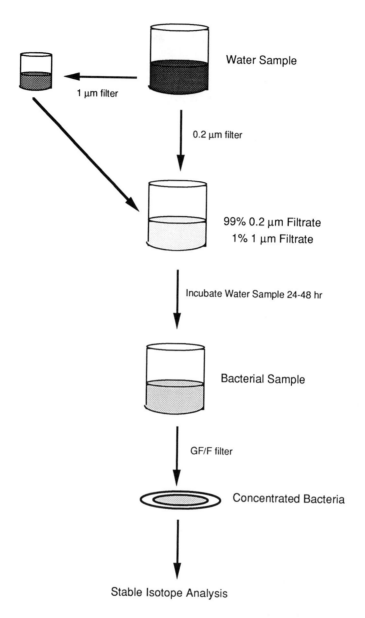

Figure 1. Flow diagram of protocol for bioassay incubation experiments.

ground to a fine powder with a mortar and pestle. Combustion of the filters is described below.

Nucleic Acids Used as an *In Situ* Biomarker

1.   Water samples are prefiltered through 1.0-μm cartridge filters (Millipore Polygard) to remove large particles and organisms (Figure 2). This step may be omitted for samples in which particle-attached bacteria are a significant contribution to the total bacterial biomass. The size of the water sample that should be prefiltered depends on the bacterial biomass. In estuarine water samples from 25 to 100 L water are necessary, while in coastal waters 200 to 300 L are necessary.

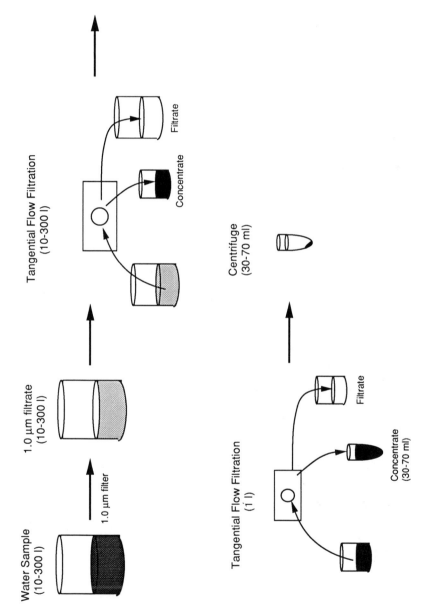

Figure 2. Flow diagram for the concentration of bacteria by tangential flow filtration.

2.   After the prefiltration, <1 μm particles are concentrated with tangential flow filtration (Figure 2). Both Millipore Pellicon and Amicon tangential flow concentration systems equipped with 10^5 Da filter cartridges have been used. For the Millipore system, three cassettes totaling 15 ft^2 filter area are used to concentrate bacteria. The Amicon system is equipped with one cartridge (Hollow Fiber Cartridge) with a total surface area of 4.9 ft^2. Particles in water samples are concentrated to approximately a 1-L volume depending on the dead volume of the system.

3.   Sample volumes are reduced further using smaller tangential flow filtration systems; such as a Pellicon Minitan system equipped with four cassettes or an Amicon Hollow Fiber cartridge. With the smaller tangential flow systems, water volumes are reduced to 30 to 70 ml.

4.   A final concentration is completed by centrifugation using a SS34 centrifuge head at 20,000 rpm for 1 h. If samples cannot be centrifuged immediately, they may be frozen or possibly preserved with 2% HgCl$_2$ (5 ml/100 ml sample). The effect of HgCl$_2$ on isotope ratios of extracted nucleic acids has not yet been thoroughly examined but is not expected to be significant.

Nucleic Acid Extractions

Nucleic acids are extracted using a modification of a method that has been described by Blair et al.[6]

1.   To lyse bacteria, pellets are suspended in 5 to 10 ml of 1.5 M NaClO$_4$ and shaken for 15 min (refer to Figure 3).

2.   To ensure efficient lysis of cells, 2 g of glass beads (baked 2 h at 500°C) is added to bacterial suspensions and cells are blended in a bead beater (BioSpec Products, Bartlesville, OK) for 1 min.

3.   Lysates are extracted three times in polypropylene centrifuge tubes with an equal volume of chloroform. For each extraction, the aqueous phase is separated from the intermediate cell debris/protein layer and chloroform using an inverted pipette. After three extractions, precipitated protein should not be visible at the interface. If large amounts of protein continue to precipitate, extractions should be repeated.

4.   Nucleic acids are precipitated from the lysates by adding 0.2 M NaCl (final concentration) and twice the lysate volume of 95% cold ethanol and chilling the ethanol/lysate solution in a -70°C freezer for 10 min.

5.   Precipitated nucleic acids are concentrated by centrifugation at 17,000 $\times$ g for 30 min.

6.   Nucleic acid pellets are washed with 70% cold ethanol (-20°C). Excess ethanol is removed by placing samples in a vacuum oven at room temperature for 30 min.

7.   The purity and concentration of the extracted nucleic acid is determined using ultraviolet (UV) spectrophotometry at 260 and 280 nm.[9] The ratio of UV absorption at 260 to 280 nm is used to indicate the purity of the extracted nucleic acids. Ratios of 1.8 indicate pure DNA and 2.0 pure RNA.

8.   Finally, nucleic acids are freeze dried in preparation for stable isotope analysis.

16S rRNA-Targeted Oligonucleotide Probes

Extraction of nucleic acids from a concentration of diverse microorganisms necessitates a confirmation that the nucleic acids are, in fact, from the targeted microbial assemblage.

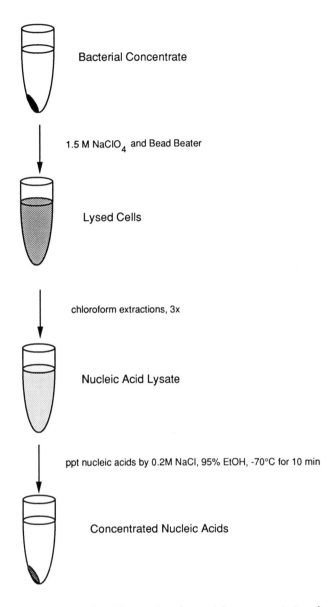

Figure 3. Flow diagram for nucleic acid extractions from <1.0-μm concentrates of seawater samples.

Hybridizations of extracted nucleic acids with universal- and kingdom-level oligonucleotide probes targeted to 16S rRNA may be used to probe the proportions of eubacteria and eukaryotes that comprise microbial biomass.[3,10-11] Oligonucleotide probes described in Coffin et al.[3] do not address the contribution of nucleic acids from cyanobacteria. Recently, Giovannoni et al.[12] described oligonucleotide probes that are targeted for 16S rRNA from cyanobacteria. These probes may be used to check samples for contributions from cyanobacteria and perhaps, owing to their cyanobacterial origin, prochlorophytes.[13] Methods for the use of 16S rRNA-targeted oligonucleotide probes are described elsewhere in this volume. If it is not possible to set up this method in a laboratory, enumeration of phototrophic organisms in cell concentrates using epifluorescence microscopy to examine the relative amount of autofluorescing cells and heterotrophic bacteria will suffice.

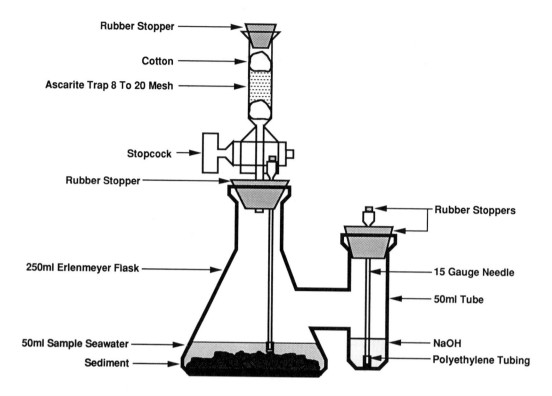

Figure 4.   Diagram of biometer flask used to trap CO_2 for stable carbon isotope analysis.

Respired CO_2 as a Tracer of Bacterial Substrate

1.   Biometer flasks (Belco) are fitted with stoppers that contain sample ports for water samples and CO_2 traps (Figure 4). Sample ports are fixed with stoppered luer-lok connectors so that subsamples may be taken with syringes. In addition, the stopper above the water sample inlet is fixed with an ascarite CO_2 trap that is closed to the sample by a stopcock. The stopcock is opened to the atmosphere when water or NaOH subsamples are taken, thus, CO_2-free volumes of air replace the sample volume drawn from flasks.

2.   Water and/or sediment samples are placed in the bottom of the flask and 10 ml NaOH (0.5 M) is placed in the bottom of the side arm. The NaOH used to trap CO_2 is boiled for 2 h and stored in capped serum bottles which are sealed and flushed with nitrogen prior to being filled with sodium hydroxide.

3.   Biometer flasks are incubated at the desired temperature with gentle shaking. Subsamples may be taken from the sample flask or the sodium hydroxide in the side arm. Sodium hydroxide subsamples are injected into capped serum bottles that have been flushed with nitrogen and stored for isotope analyses of $\delta^{13}C$ of the trapped CO_2.

4.   Dissolved CO_2 concentrations are measured by gas chromatography (GC); (Hewlett-Packard 5890 Series II/3, equipped with an automatic head space analyzer, and a 20 m $\times$ 0.53 mm Pore Plot Q fused silica column. CO_2 is reduced in the presence of Ni-catalyst to CH_4, and CH_4 concentrations are measured with a flame ionization detector).

Stable Carbon and Nitrogen Isotope Analyses

1.   Bioassay concentrates and nucleic acids are converted to CO_2 and N_2 gas by a modified Dumas combustion procedure.[14] Sufficient sample to produce >50 μg N_2 and >100 μg

CO_2 is placed in a 24-cm quartz tube (8-mm o.d.; 6-mm i.d.; preheated to 650°C for 2 h) with 2 g of Cu and 4 g of CuO (preheated to 650°C for 2 h). The tube is then evacuated to $<10^{-5}$ torr and sealed. In turn, quartz tubes are heated to 900°C at a rate of 450°C h^{-1}, kept at 900°C for 2 h, and cooled to room temperature at a rate of 60°C h^{-1}. The slow cooling cycle ensures that any oxides of nitrogen are reduced to N_2.

2. Cryogenic distillation — following sample combustion, CO_2 and N_2 gas are separated by cryogenic distillation. The quartz tube containing the combustion products is scored and attached to the vacuum line with a Cajon connector. The vacuum line contains a liquid nitrogen trap that retains condensible gases. The N_2 gas, which does not condense at liquid nitrogen temperatures, is transferred to a small sample bulb containing silica gel and fitted with a high-vacuum valve. At liquid nitrogen temperatures, N_2 gas quantitatively absorbs onto the silica gel. When warmed to room temperatures, N_2 gas quantitatively desorbs so that it can be directly introduced into the mass spectrometer from the sample bulb. After N_2 gas is recovered in the sample bulb, the CO_2 is allowed to sublimate and then transferred through an H_2O trap (ethanol, dry ice slush) to either a sample bulb or to a 6-mm Pyrex tube, which is sealed. It is important that CO_2 samples be free of H_2O before introduction into the mass spectrometer.

3. Purging respired inorganic carbon — samples from the respirometer experiments are prepared for isotopic analysis by a modification of the *in vacuo* acidification and purging technique described in Grossman.[15] A 30-ml flask containing 1 ml of concentrated phosphoric acid is connected to the vacuum line with a suitable high-vacuum joint. Between 5 and 8 ml of the 0.5 M NaOH solution containing the trapped CO_2 is injected through a septum into the flask. The released CO_2 gas is cryogenically distilled through a series of water traps and stored either in a sample bulb or in a 6-mm, sealed Pyrex tube. As stated above, it is imperative that CO_2 samples be free of H_2O before introduction into the mass spectrometer.

4. Isotopic analysis — the CO_2 gas can be analyzed on an isotope ratio mass spectrometer fitted with a triple collector that can detect mass 44, 45, and 46. A dual-collector instrument set to detect mass 28 and 29 is sufficient to analyze N_2 gas. Results are reported in standard delta (δ) notation:

$$\delta^hX = 1000 * [((^hX/^lX)_{SAM}/(^hX/^lX)_{STD}) - 1] \qquad (1)$$

where X is either carbon or nitrogen, h is the heavier isotope, l is the lighter isotope, SAM is the sample, and STD is the standard. The recognized standard for carbon is PeeDee Belemnite (NBS-1) and that for nitrogen is atmospheric N_2. In general, the reproducibility of the $\delta^{13}C$ measurement is between ± 0.1 and ± 0.2 Σ. For $\delta^{15}N$, the precision is often between ± 0.2 and $\pm 0.3\Sigma$.

NOTES AND COMMENTS

1. Stable isotopes can be used now to trace sources of carbon and nitrogen for aquatic bacteria. Continued use of the techniques described in this chapter will improve our understanding of bacterial processes, particularly particle cycling, and their influence on isotope biogeochemistry. When methods were not available to examine components of particulate organic matter (POM) in planktonic environments, stable isotope studies grouped all particulate material into one class. This practice was defended by postulating little differences in stable isotope ratios of biotic components in POM (e.g., Altabet and McCarthy[16]) Comparing $\delta^{13}C$ and $\delta^{15}N$ of bacteria and suspended particulate matter from the Gulf of Mexico, however, revealed that bacteria were consistently depleted in ^{13}C and enriched in ^{15}N relative to particulate carbon and nitrogen.[17] More work is needed to

document the range of stable isotope values across components of POM and the factors that influence the observed similarities and differences. Three different approaches for measuring stable isotope ratios in bacteria have been outlined. The simplest is to grow bacteria in incubation experiments. This approach is based on the observation that the isotopic ratio of the accumulated biomass is similar to that of the substrate. Thus, when numerous sources of organic matter (e.g., algae, seagrasses, vascular plant detritus), which are isotopically distinct, are available to planktonic bacteria, it is possible to determine the relative contribution of these sources to bacterial production. Because filtration of particles from seawater eliminates phytoplankton, zooplankton excretion or grazing on phytoplankton and hydrolysis of organic debris, the isotopic composition of bacteria grown in bioassay experiments only provides measures of "potential" substrate sources. Thus, this approach may seriously underestimate the importance of particulate sources of organic matter for bacteria. The extent to which the significance of these sources may be underestimated may be a function of the coupling between bacteria and substrate sources and of the turnover time of pools of organic matter derived from particulates (i.e., algae).

2.   Incubation time is an important variable in the bioassay experiments. Isotopic changes occurring after bacteria growing in batch cultures reach a stationary phase have not been thoroughly documented. Recent experiments have compared the $\delta^{13}C$ of bacteria grown with identical carbon, but varying nitrogen concentrations. When grown on lower nitrogen concentrations, bacteria produced biomass that was depleted in ^{13}C (R. B. Coffin and L. A. Cifuentes, unpublished data). These results may indicate that the isotopic composition of bacteria is related to the nutritional status. For these reasons, bioassay incubations should be monitored for abundance and incubation time and terminated as the cells reach stationary phase growth.

3.   In environments where bacteria depend on planktonic sources of organic matter, the *in situ* isotopic measurement may be more applicable than the bioassay incubation. Nucleic acids used as a biomarker for bacteria provide an *in situ* method to study bacterial stable isotope compositions. While this approach is more labor intensive than the bioassay method, no other method is available that provides direct, *in situ* information on the sources of bacterial substrate. However, some technical difficulties with nucleic acid extractions often make it difficult to concentrate enough, pure nucleic acid for isotopic analysis. A practical consideration is the amount of water that must be concentrated to obtain sufficient concentrations of nucleic acids. Generally, 100 μg C and 50 μg N are sufficient for isotope analysis. Assuming nucleic acids are 45% carbon and approximately 20% nitrogen, a minimum of 250 μg nucleic acid is required. Nucleic acid concentrations in bacteria concentrated from estuarine samples ranged from 6 to 35 fg cell^{-1} and concentration of bacteria with tangential flow filtration is generally 40% efficient.[3] With these values, a conservative estimate of the number of bacteria required for stable isotope analysis of nucleic acids is 10^{11}. Assuming bacterial concentrations in estuaries ranging between 10^9 and 10^{10} L^{-1}, between 10 and 100 L water samples are required; in coastal and open ocean waters substantially larger waters samples may be necessary. Generally, it takes 2 h to concentrate 50-L water samples, therefore, fewer samples can be processed when larger water samples are required. Another technical difficulty with nucleic acid extractions is obtaining pure extracts. In brackish and freshwaters with high concentrations of humic substances, nucleic acids extracted from <1 μm concentrates are occasionally stained brown. Although humic acids coextracting with nucleic acids may not be a significant source of bacterial substrate, they could bias the stable isotope measurement. To address this potential contamination, experiments were conducted by adding 5 mg L^{-1} concentrations of humic acid standards (Sigma Chemical Corporation) to *Pseudomonas aeruginosa* cultures. The color of the extracted nucleic acids from cultures receiving the humic acid was brown compared with the control samples without humic acid, however, the $\delta^{13}C$ of the samples was not different (J. Morin and R. Coffin, unpublished data). While these preliminary results are encouraging, we continue to study the potential for contaminants to coextract with nucleic acids in aquatic ecosystems with high concentrations of humic matter.

4. The stable isotope ratio of bacteria growing on mixtures of organic matter from different sources could build biomass that resembles an unrelated carbon source. For example, bacterial biomass produced from growth on dissolved carbon from terrestrial sources and a *Spartina* marsh could have a stable isotope value similar to that of phytoplankton. Mixed substrate sources have been delineated in particulate organic matter and in larger estuarine organisms with multiple isotope tracers.[18] A similar approach may be useful for delineating carbon and nitrogen sources for bacteria. Alternatively, combining isotopic analyses of both extracted nucleic acids and bioassay incubations may be useful for defining complex sources of bacterial substrate. This approach studies the effect of removing planktonic sources (bioassay incubation), which is not the case in the *in situ* (nucleic acid) measurement.

5. The primary emphasis of published stable isotope studies on bacteria has been carbon. In contrast, nitrogen has not been examined as thoroughly. Paralleling the application of stable isotopes to particulate organic matter, the use of stable nitrogen isotopes as a tracer in bacteria may be more complex than that of stable carbon isotopes. While bacterial fractionation of stable carbon isotopes in the uptake of dissolved organic carbon (DOC) is small,[2] the isotopic discrimination between nitrogen source and bacteria can be substantially greater. For example, the range for *Vibrio harveyii* grown on varying initial concentrations of NH_4^+ (23 μM to 23 mM) was -3 to $-14‰$, and with ammonium concentrations of 180 μM the maximum was $-27‰$.[19] Bacteria use two different, concentration-dependent enzyme systems to assimilate ammonium. If ammonia diffuses across the cell membrane, it is assimilated by glutamate dehydrogenase. When active ammonium transport occurs, ammonium assimilation is accomplished by glutamine synthetase. The transition between ammonium assimilating enzymes occurs between 0.1 and 1 mM NH_4^+.[19] Glutamine synthetase is generally expected to be the major mechanism for ammonium assimilation by bacteria in aquatic ecosystems because of low NH_4^+ concentrations. Further confounding the study of stable nitrogen isotopes in bacteria are the varied sources of nitrogen for bacteria, including ammonium, dissolved amino acids, and nitrate.[20-26] Generally amino acids are preferred nitrogen sources over inorganic nitrogen,[27,28] however, bacteria may account for 5 to 60% of the total ammonium uptake in aquatic environments. Therefore, when amino acids concentrations are low, ammonium may be a significant proportion of the total heterotrophic nitrogen uptake. Clearly the combination of varying Δ_n for ammonium, coupled with bacterial uptake of multiple nitrogen sources, may render the identification of nitrogen sources difficult.

6. Finally, there are environments (highly turbid waters, sediments, pore waters, and soils) where it is not possible at this time to use either the bioassay incubation or the nucleic acid technique described earlier. For these cases, we have initiated research on the potential for using CO_2 to trace the source of organic matter which bacteria use for energy. This approach is based on the observation that there is little isotopic discrimination between the carbon substrate and the respired CO_2. To conduct these experiments, the bacterial assemblage and substrates are removed from their original environments and incubated under conditions designed to accelerate respiration. Thus, results from this type of experiment must be interpreted with caution until a larger database is established.

7. Mention of trade names does not constitute product endorsement of recommendation for use by the U.S. EPA.

REFERENCES

1. Coffin, R. B., Cifuentes, L. A., and Kovack, J., Stable nitrogen isotope analysis of bacterioplankton, *Appl. Environ. Microb.*, in preparation.
2. Coffin, R. B., Fry, B., Peterson, B. J., and Wright, R. T., Carbon isotopic compositions of estuarine bacteria, *Limnol. Oceanogr.*, 34, 1305, 1989.

3. Coffin, R. B., Velinsky, D. J., Devereux, R., Price, W. A., and Cifuentes, L. A., Stable carbon isotope analysis of nucleic acids to trace sources of dissolved substrates used by estuarine bacteria, *Appl. Environ. Microbiol.,* 56, 2012, 1990.

4. Abelson, P. H. and Hoering, T. C., Carbon isotope fractionation in formation of amino acids by photosynthetic organisms, *Proc. Natl. Acad. Sci. U.S.A.,* 47, 623, 1961.

5. Monson, K. D. and Hayes, J. M., Carbon isotopic fractionation in the biosynthesis of bacterial fatty acids. Ozonolysis of unsaturated fatty acids as a means of determining the intramolecular distribution of carbon isotopes, *Geochim. Cosmochim. Acta,* 46, 139, 1982.

6. Blair, N. et al., Carbon isotopic fractionation in heterotrophic microbial metabolism, *Appl. Environ. Microbiol.,* 50, 996, 1985.

7. Shelton, M. E., Mueller, J. G., Coffin, R. B., Cifuentes, L. A., and Pritchard, P. H., Laboratory and field evaluation of bioremediation measures applied to a salt marsh contaminated with crude oil, *Appl. Environ. Microb.,* submitted.

8. Hobbie, J. E., Daley, R. J., and Jasper, S., Use of Nuclepore filters for counting bacteria by fluorescence microscopy, *Appl. Environ. Microb.,* 33, 1225, 1977.

9. Maniatis, T., Fritsch, E. F., and Sambrook, J., Molecular cloning. A laboratory manual, Cold Spring Harbor Laboratory, Cold Spring Habor, New York, 1982, 109.

10. Giovannoni, S. J., DeLong, E. F., Olsen, G. J., and Pace, N. R., Phylogenetic group-specific oligodeoxynucleotide probes for identification of single microbial cells, *J. Bacteriol.,* 170, 720, 1988.

11. Pace, N. R., Stahl, D. A., Lane, D. J., and Olsen, G. J., The use of rRNA sequences to characterize natural microbial populations, *Adv. Microbial. Ecol.,* 9, 1, 1986.

12. Giovannoni, S. J., Turner, S., Olsen, G. J., Barns, S., Lane, D. J., and Pace, N. R., Evolutionary relationships among cyanobacteria and green chloroplasts, *J. Bacteriol.,* 170, 3584, 1988.

13. Chisolm, S. W., Olsen, R. J., Zettler, E. R., Goericke, R., Waterbury, J. B., and Welshmeyer, N. A., A novel free-living prochlorophyte abundant in the oceanic euphotic zone, *Nature,* 334, 340, 1988.

14. Macko, S. A., Stable nitrogen isotope ratios as tracers of organic geochemical processes, Ph.D. thesis, University of Texas at Austin, 1981, 181.

15. Grossman, E. O., Carbon isotopic fractionation in live benthic foraminifera — comparison with inorganic precipitate studies, *Geochim. Cosmochim. Acta,* 48, 1505, 1984.

16. Altabet, M. A. and McCarthy, J. J., Temporal and spatial variations in the natural abundance of 15N in PON from a warm core ring, *Deep-Sea Res.,* 7, 755, 1985.

17. Cifuentes, L. A. and Coffin, R. B., Stable isotopes as tracers of microbial processes, published abstract in 91st ASM General Meeting, Dallas, Texas, 1991.

18. Peterson, B. J., Howarth, R. W., and Garrit, R. H., Multiple stable isotopes used to trace the flow of organic matter in estuarine food webs, *Science,* 277, 1361, 1985.

19. Hoch, M., Ammonium uptake by heterotrophic marine bacteria: environmental control and nitrogen isotope biogeochemistry, Dissertation, University of Delaware, 1991.

20. Crawford, C. C., Hobbie, J. E., and Webb, K. L., The utilization of dissolved free amino acids by estuarine microorganisms, *Ecology,* 55, 551, 1974.

21. Hollibaugh, J. T., Carruthers, A. B., Fuhrman, J. A., and Azam, F., Cycling of organic nitrogen in marine plankton communities studied in enclosed water columns, *Mar. Biol.,* 59, 15, 1980.

22. Hollibaugh, J. T. and Azam, F., Microbial degradation of dissolved proteins in seawater, *Limnol. Oceanogr.,* 28, 1104, 1983.

23. Wheeler, P. A. and Kirchman, D., Utilization of inorganic and organic nitrogen by bacteria in marine systems, *Limnol. Oceanogr.,* 31, 998, 1988.

24. Coffin, R. B., Bacterial uptake of dissolved free and combined amino acids in estuarine waters, *Limnol. Oceanogr.,* 34, 531, 1989.

25. Horrigan, S. G., Hagström, A., Koike, I., and Azam, F., Inorganic nitrogen utilization by assemblages of marine bacteria in seawater culture, *Mar. Ecol. Prog. Ser.,* 50, 147, 1988.

26. Kirchman, D. L., Suzuki, Y., Garside, G., and Ducklow, H. W., High turnover rates of dissolved organic carbon during a spring bloom, *Nature,* 352, 612, 1991.

27. Keil, R. G. and Kirchman, D. L., Contribution of dissolved free amino acids and ammonium to the nitrogen requirements of heterotrophic bacterioplankton, *Mar. Ecol. Prog. Ser.*, 73, 1, 1991.
28. Kirchman, D. L., Kiel, R. G., and Wheeler, P. A., The effect of amino acids on ammonium utilization and regeneration by heterotrophic bacteria in the subarctic Pacific, *Deep-Sea Res.*, 36, 1763, 1989.

CHAPTER 78

Bacterial Sinking Losses

Carlos Pedrós-Alió and Jordi Mas

INTRODUCTION

Sedimentation is usually overlooked as a loss factor for bacteria in planktonic environments due to the extremely low sinking rates predicted by theory for organisms of their size range. Thus, typical bacterioplankton cells with diameters between 0.5 and 1 μm, and with a buoyant density of 1.08 would sink, according to Stokes' Law, at rates close to 0.1 cm per day, which will usually result in negligible losses.

However, because the sinking speed increases with the square of the particle radius, there are some cases in which sedimentation is large enough to be considered as an important loss factor. Examples are colonial cyanobacteria and large-sized purple bacteria, bacteria attached to particles, bacteria on aggregates such as marine snow, bacteria in fecal pellets, or flocculates at waste disposal plants. In general, whenever bacteria are found in particles many times their individual size, they will experience significant losses by sedimentation. Thus, measurement of sedimentary fluxes is a must when trying to assess the factors affecting the dynamics of bacterioplankton.

Determining sedimentation losses invariably requires the utilization of some kind of sediment trap.[1,2] Sediment traps allow quantification of the number of bacteria that sediment out of the water column during a certain period of time.

Sediment traps consist of an open receptacle with a known "catching" area, which is deployed at a certain depth during a given period of time. The concentration (N) of bacteria inside the trap increases with time according to the following equation:

$$dN/dt = J \cdot A/V \tag{1}$$

where A is the catching area (cm^2), V is the volume of the trap (cm^3), and J is the sedimentary flux (cells cm^{-2} d^{-1}). Integrating from t = n to t = 0 and rearranging Equation 1 we obtain:

$$J = (N_t - N_0)/(t_n - t_0) \cdot (V/A) \tag{2}$$

in which J can be calculated from the concentration of cells in the trap at the beginning (N$_0$) and at the end (N$_t$) of the deployment period (t$_n$ − t$_0$).

The water column can be divided into two compartments separated by the mouth of the trap. What is measured is the flux of bacteria from the upper to the lower compartment across the sedimentation boundary (the mouth of the trap). The interpretation given to this

0-87371-564-0/93/$0.00 + $.50
© 1993 by Lewis Publishers

sedimentary flux depends, to a great extent, on the characteristics of the upper compartment. When the upper compartment is stratified and structured in layers, J provides an estimate of the sinking losses in the layers close to the trap boundary. When, on the contrary, the upper compartment is well mixed and has a homogeneous composition, J gives an estimate of the overall sedimentary loss for the whole compartment.

One of the main disadvantages of the technique is that a number is always obtained, but it is usually impossible to assess its validity. Factors such as the design of the traps, the occurrence of growth or decomposition within the traps, or overtrapping of bacterioplankton under turbulent conditions can result in totally senseless and misleading results when not properly considered. Devices such as funnels and baffles, which have been demonstrated to impair trap performance under laboratory conditions, are repeatedly used. In particular, fluxes for the whole ocean are often calculated from such traps without further cautioning about possible biases due to trap design.

Sedimentation can proceed at very high (several m h^{-1}) or very low rates (a few cm d^{-1}) in environments embracing a rather variable set of conditions, from highly turbulent to quiescent. For this reason, no single trap design and procedure for measuring sedimentation can be presented that is universally valid. Two different trap designs, which constitute opposite extremes in design complexity, are presented as examples in the next section. The last section reviews several commonly found problems and their possible solutions. The reader will have to decide, on the basis of the specific problem at hand, which trap design and which corrections should be used in each case.

INSTRUMENTATION: TRAP DESIGN

Several aspects have to be taken into account when designing a sediment trap. The shape of the trap and the trap's aspect ratio (the ratio between length and width), its ruggedness, resistance to tilting by waves and turbulence, resistance to corrosion, or other forms of chemical attack (i.e., sulfide), are all factors strongly affecting the performance of the trap. Both gross over- or underestimations of sedimentary fluxes are possible depending on the sedimentary environment. The behavior of several trap designs has been thoroughly analyzed in laboratory[3-8] and field experiments.[9,10] The general conclusion from these studies is that while bottle-shaped traps have a tendency to overcollect and funnel-shaped traps have a tendency to undercollect particles, cylindrical traps with diameters between 5 and 20 cm and aspect ratios larger than 5 (for small lakes and nonturbulent environments) or larger than 10 (for turbulent environments such as large lakes or the sea) provide the best choice. Baffled traps have often been used to improve trapping efficiency although, in general, baffles do not seem to be needed if the aspect ratio of the trap is properly chosen.[4] Moreover, baffling seems to be detrimental in that, depending on the overall characteristics of the trap and of the sedimenting particles, trapping efficiency can actually decrease while increasing the variability of the measurements.[6] There is no standard trap design that can be universally recommended. Different water bodies may require quite different designs. For example, traps deployed in the open ocean, down to thousands of meters, necessarily encounter different kinds of problems than traps in small ponds. Next, we describe two very different trap designs.

Deep Moored Traps

Several automated multisample traps have been designed for deployment in the oceans.[2,11] These traps allow continuous monitoring during prolonged periods of time. They reduce

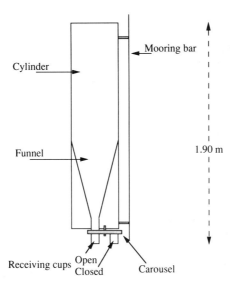

Figure 1. Schematic drawing of an automated multisample sediment trap used in oceanography. (Simplified drawing based on Heussner et al.[2])

ship costs by relieving the worker of frequent visits to the trap. Moreover, they become moored monitoring stations for several instruments such as transmissometers, fluorometers, current meters, thermistors, etc. Thus, an extremely powerful data set can be gathered at a reasonable expense. The PPS 3 time-series sediment trap will be described.[2] The general appearance of this trap can be seen in Figure 1.

This trap consists of a cylindrical upper portion and a funnel-shaped lower portion. The mouth area has to be big enough to provide enough material for analysis within the desired deployment periods. In the lower part of the trap, a rotating plate has seven holes. Six receiving cups are fitted to these holes and the seventh is left open for trap flushing. A battery-powered motor rotates the plate to place a receiving cup under the single eccentric hole in the stationary plate (Figure 1). This is done at preprogrammed time periods.

The traps are fixed at the desired depths to a mooring line with a heavy weight at the bottom end and underwater floats at the upper end. Floats are distributed throughout the line to provide the desired tension so the traps do not tilt much under the impulse of internal waves, fish bumping into them, or currents. These floats also ensure that if the line is broken accidentally two buoyant line portions result.

Each receiving cup is under the eccentric hole for a programmed amount of time (1 h to 45 d), and the whole array is recovered after all the cups have been used. Once on board, the supernatant of each cup is removed by pipette and stored. The particulate sample is photographed and analyzed visually. Next, swimmers (zooplankton or even fish who swim into the trap rather than sinking into it) are removed by sieving the sample through 1-mm nylon mesh and subsequent removal of smaller organisms under a dissecting microscope with fine tweezers. Subsamples are then taken for different analyses with a special device to ensure representative subsamples.[2]

Shallow Moored Traps

A rather simple trap system based on the same recommendations from laboratory studies has been successfully used in the measurement of sedimentation of phytoplankton and

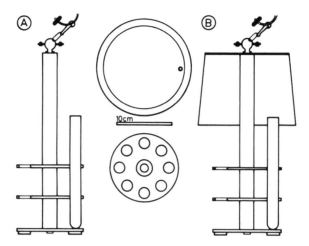

Figure 2.   Schematic drawing of the uncovered (A) and covered (B) sediment traps used in a small pond
(reproduced from Pedrós-Alió, C., Mas, J., Gasol, J. Ma., and Guerrero, R., *J. Plankton Res.,*
11, 87, 1989, by permission of Oxford University Press). A scale bar is provided in the central
part of the figure, where top views of the covered (upper drawing) and uncovered (lower
drawing) traps are shown.

phototrophic bacterioplankton in Lake Cisó, Spain,[12,13] a very small pond 20 m in diameter
and 8 m deep. The lake is protected from strong winds by the surrounding trees and bushes
and, thus, surface waves are negligible as well as violent turbulence. Fish are absent from
the lake. The traps are made from plates and hollow cylinders of PVC (Figure 2). Three
plates are set at regular intervals along a central axis. The two upper plates have eight holes,
through which eight glass tubes (16 × 200 mm, aspect ratio 12.5) can be placed easily and
stably in each trap (Figure 2A). The traps hang from a metal frame kept afloat by several
buoys. The platform consists of a central square of galvanized iron stabilized by four arms
of the same material, with two plastic carboys on each arm providing flotation. The carboys
are partially filled with water to dampen lateral and vertical surface water movements. The
ends of each arm are tightly tied to anchors or to the shore. Movement of the traps due to
surface waves is, thus, negligible. To allow for *in situ* incubations or deployment of sediment
traps, two small winches are mounted on each arm of the platform.

We used tubes of 16-mm mouth diameter and 12.5 aspect ratio for the following reasons:
the mouth diameter is slightly lower than the minimum recommended value, but such values
were calculated for algae, specifically diatoms, which often have diameters around 50 or
70 μm. Thus, 16-mm diameter tubes seemed sufficient for bacteria-sized particles and even
for small algae. These dimensions minimized the entrance of leaves or other large particles
which could have introduced unwanted material in the traps. The aspect ratio was within
the upper part of the range given by Blomqvist and Håkanson,[5] because resuspension of the
material in the trap was more likely for small particles such as bacteria.

When deployed, the tubes are carefully filled with water from the depth at which they
are going to be placed. The whole trap is then hung from one of the winches on the platform
and lowered to the desired depth. After a given exposure period, the upper 5 to 10 ml of
water from all the tubes is removed with a hand pump before collecting the contents for
analysis. The contents of the eight tubes of each trap are pooled into an opaque BOD bottle,
a common procedure in sediment trap methodology.[4,14] One covered and one uncovered trap
per depth were shown to be sufficient for statistically reliable results.

PROCEDURES

Growth and Decomposition

Because the actual measurement underlying the estimation of sedimentary fluxes is a change in cell concentration inside the trap during the deployment period, any processes other than sedimentation which affect this concentration will lead to erroneous fluxes. Decomposition of bacteria inside the trap, grazing by predatory organisms, or even growth of bacteria inside the trap are factors to be taken into account. Several approaches have been used to deal with these problems.

1. Shortening the deployment period decreases the impact of these factors, but it also makes the accurate evaluation of the increase in cell concentration more difficult, because fewer cells accumulate in the trap.

2. Different fixatives and preservatives have been used in an attempt to prevent the alteration of trapped material. While some of them seem to be useful in preserving the samples, many are unable to stop growth, even stimulating growth of specialized groups of microorganisms[4] or, as in the case of organic fixatives, interfering with analyses of organic compounds.[1] Utilization of fixatives presents the additional problem that, during long-term deployments, diffusion can decrease its concentration well below the effective value. This phenomenon has been circumvented by the addition of 10% NaCl which creates a diffusion barrier[15] but can, on the other hand, introduce osmotic alterations in the collected materials.[1]

3. Alternatively one can determine growth and decomposition simultaneously, and correct sedimentary fluxes.

   - Lysis by decomposition and/or predation will be important to a variable extent and depending considerably on the organisms and on whether the traps are deployed in an oxic or an anoxic environment. This means that the specific rate of cell decay (k_d) has to be simultaneously measured under conditions as close as possible to those found in the field in such a way that a correction can later be applied to data gathered from the traps. Because sedimentation and decomposition occur at the same time, the overall change of concentration inside the trap can be described as:

$$dN/dt = J(A/V) - k_d N \tag{3}$$

   which after integration and isolation of J, yields:

$$J = [-k_d(N_t - N_0 e^{-kd\ t})/(e^{-kd\ t} - 1)] \cdot V/A \tag{4}$$

   where sedimentary flux can be calculated from the increase in cell concentration ($N_t - N_0$) and the specific rate of decomposition. Equation 4 must be used cautiously, especially in cases when $N_t - N_0$ is very small (but might not differ significantly from zero) and k_d is large. In these cases, the built-in correction in Equation 4 could lead to the conclusion that sedimentation is important, even when it was not significant.

   - Growth in the traps is not important when measuring sedimentation of phototrophic organisms in the aphotic zone. However, some control should be applied in other cases. Ducklow et al.[1] used thymidine incorporation and increase in cell abundance during 1-d incubations of the collected material under *in situ* conditions to estimate the specific growth rate (μ) of bacteria in the traps. These authors applied this specific growth rate to correct for growth in the traps during the deployment period. They used an approach essentially equivalent to that outline above for decomposition. We will express their correction using the same terms as above for the sake of consistency. Thus, variation of cell concentration in the traps when growth occurs is defined by the following equation:

$$dN/dt = J(A/V) + \mu N \tag{5}$$

which after integration and isolation of J, yields:

$$J = [\mu(N_t - N_0 e^{\mu t})/(e^{\mu t} - 1)] \cdot V/A \tag{6}$$

- When growth and decay occur simultaneously, a similar equation can be derived which corrects for both processes:

$$J = [(\mu - k_d)(N_t - N_0 e^{(\mu - kd)t})/(e^{(\mu - kd)t} - 1)] \cdot V/A \tag{7}$$

Mathematical treatments such as those expressed in Equations 4, 6, and 7 are based on the assumption that growth, losses, or both, proceed at a constant rate during the period of deployment. While this is probably true for short deployment periods in environments subject to little environmental fluctuations, it should be carefully tested when measurements involve either long periods, environments subject to strong fluctuations (where growth and decay can be highly variable), or in organisms in which growth or decay proceed at a rate so high that changes in biomass due to these processes clearly exceed those due to sedimentation.

Turbulence

Even when there is no sedimentation, turbulence can cause some suspended material to enter the traps. In cases when sinking rates are very high, this probably will have little impact on the accuracy of the observations. However, when sinking rates are low, turbulence can eventually contribute more than sedimentation to the material found in the traps. This has been shown in a number of measurements performed by Pedrós-Alió et al.[13] in Lake Cisó, during periods of both high and low thermal stability of the lake, which corresponded to conditions of low and high turbulence. To correct for turbulent overtrapping, an additional set of sedimentation traps was used, covered with a plastic cap to avoid direct sedimentation of suspended material inside the trap (Figure 2B). A correction for turbulent trapping is then applied as follows:

$$J = [(({}^uN_t - {}^uN_0)/t) \cdot ({}^uV/{}^uA)] - [({}^cN_t - {}^cN_0)/t) \cdot ({}^cV/{}^cA)] \tag{8}$$

where the superscripts indicate whether the values of initial and final cell concentration (N_0 and N_t) and the volume and catching area (V and A) correspond to the covered (c) or uncovered (u) traps. In the case that decomposition and/or growth occurred, the corrections outlined above should also be applied. The design of the covered traps shown in Figure 2B may not provide an absolute correction but, as shown in Pedrós-Alió et al.,[13] it is far superior to the use of conventional traps alone.

Resuspension of Bottom Sediments

It is sometimes desirable to deploy the traps close to the sediment surface, in order to have a more realistic estimate of the rate at which suspended particles reach the bottom. In turbulent environments, however, the bottom sediment can be resuspended and collected in the traps, thus, yielding erroneously high estimates. Certain environments are especially prone to resuspension of sediments: bottom slopes with an inclination higher than 5%, river mouth areas, shallow areas, or areas with large effective fetch relative to the water depth.[5] In these cases, some correction has to be applied. Some authors estimate resuspension by following some tracer typical of sedimentary material in the traps. If the C:N ratios[16] or the organic matter content[17] of sedimentary and sinking materials are different, some correction can be applied to the trap material (which presumably will show an intermediate C:N ratio or organic matter content). These corrections, however, involve many assumptions about the dynamics of both the sediment and trap materials and the results are not very convincing. Moreover, if sedimentation of bacterial cells rather than organic carbon or other compounds is desired, the latter approach is no longer valid and, as far as we know, the only way to correct for trapping of resuspended material is the utilization of covered traps. The proper design of these covered traps, however, needs further study under carefully controlled conditions.

ACKNOWLEDGMENTS

We thank T. C. Granata for criticism of the manuscript. The work of C. P.-A. has been supported by DGICYT grant PB87-0183 from the Spanish Ministry of Education and Science.

REFERENCES

1. Ducklow, H. W., Kirchman, D. L., and Rowe, G. T., Production and vertical flux of attached bacteria in the New York Bight as studied with floating sediment traps, *Appl. Environ. Microbiol.*, 43, 769, 1982.
2. Heusnner, S., Ratti, C., and Carbonne, J., The PPS 3 time-series sediment trap and the trap sample processing techniques used during the ECOMARGE experiment, *Cont. Shelf Res.*, 10, 943, 1990.
3. Hargrave, B. T. and Burns, N. M., Assessment of sediment trap collection efficiency, *Limnol. Oceanogr.*, 24, 1124, 1979.
4. Bloesch, J. and Burns, N. M., A critical review of sediment trap technique, *Schweiz. Z. Hydrol.*, 42, 15, 1980.
5. Blomqvist, S. and Håkanson, L., A review on sediment traps in aquatic environments, *Arch. Hydrobiol.*, 91, 101, 1981.
6. Butman, C. A., Sediment trap biases in turbulent flows: results from a laboratory flume study, *J. Mar. Res.*, 44, 645, 1986.
7. Butman, C. A., Grant, W. D., and Stolzenbach, K. D., Predictions of sediment trap biases in turbulent flows: a theoretical analysis based on observations from the literature, *J. Mar. Res.*, 44, 601, 1986.
8. Gardner, W. D., Sediment trap dynamics and calibration: a laboratory evaluation, *J. Mar. Res.*, 38, 17, 1980.
9. Blomqvist, S. and Kofoed, C., Sediment trapping — a subaquatic *in situ* experiment, *Limnol. Oceanogr.*, 26, 585, 1981.
10. Gardner, W. D., Field assessment of sediment traps, *J. Mar. Res.*, 38, 41, 1980.
11. Honjo, S. and Doherty, K. W., Large aperture time-series sediment traps; design objectives, construction and application, *Deep-Sea Res.*, 35, 133, 1988.
12. Mas, J., Pedrós-Alió, C., and Guerrero, R., In situ specific loss and growth rates of purple sulfur bacteria in Lake Cisó, *FEMS Microb. Ecol.*, 73, 271, 1990.
13. Pedrós-Alió, C., Mas, J., Gasol, J. Ma., and Guerrero, R., Sinking speeds of free-living phototrophic bacteria determined with covered and uncovered traps, *J. Plankton Res.*, 11, 887, 1989.

14. Rathke, D. E., Bloesch, J., Burns, N. M., and Rosa, F., Settling fluxes in Lake Erie (Canada) measured by traps and settling chambers, *Int. Verein. Theor. Angew. Limnol. Verh.*, 21, 383, 1981.

15. Fellows, D. A., Karl, D. M., and Knauer, G. A., Large particle fluxes and the vertical transport of living carbon in the upper 1500 m of the Northeast Pacific Ocean, *Deep-Sea Res.*, 28, 921, 1981.

16. Steele, J. H. and Baird, I. E., Sedimentation of organic matter in a Scottish sea loch, *Mem. Ist. Ital Idrobiol.*, 29(Suppl.), 73, 1972.

17. Gasith, A., Tripton sedimentation in eutrophic lakes — simple correction for the resuspended matter, *Verh. Int. Verein. Limnol.*, 19, 116, 1975.

CHAPTER 79

Methods for the Observation and Use in Feeding Experiments of Microbial Exopolymers

Alan W. Decho

INTRODUCTION

Bacterial and microalgal exopolymers coat a wide range of particulate material. They are coincidently ingested by consumers during the active ingestion of microbial cells and other particulates. Exopolymers range from the tight capsules surrounding individual microbial cells to the more loosely defined slime biofilms on the surfaces of sediment particles and detritus. In addition, exopolymers comprise a portion of the high molecular weight DOM in solution and much of the "amorphous" material of water-column aggregates (see Decho[1] for review). These secretions, however, often evade detection using light or electron microscopy. This is because of their tendency to collapse or be removed during most conventional preparation procedures. Therefore, the contribution of exopolymers as a potential food source and in other marine processes is not readily apparent.

Exopolymers represent an abundant source of potentially labile carbon for microbial consumers. The ingestion and absorption efficiencies (AE) of bacterial and microalgal exopolymers by microbial consumers may be estimated by first isolating high activity ^{14}C-labeled exopolymers from microbial cells grown in culture. The exopolymers are then bound to the surfaces of particulates normally ingested by the consumer. These exopolymer-coated particles are added to natural sediments or seawater during a brief "pulse". Their ingestion, retention, and egestion is then followed over time.

Since microbial consumers often key on particles during feeding, such particles are ideal to coat with radiolabeled exopolymers for use in feeding experiments. These can range from individual microbial cells (or similar sized beads) to sediment particles and aggregates. Certain protozoan consumers may also ingest exopolymers while in solution.[2]

Methods are presented here for: (1) qualitatively observing exopolymers on natural particles which minimize the deformation and losses commonly associated with conventional preparation techniques for light and electron microscopy and (2) conducting feeding experiments to determine the ingestion and absorption efficiencies of exopolymers by animals.

Advantages of the methods are that: (1) the exopolymers can be added to natural sediments or seawater, and thus, grazing experiments are conducted in the presence of natural food sources; (2) for short term incubations, the radiolabel will be primarily associated with the exopolymer; and (3) the exopolymer coatings surrounding the particles will mimic (depending on their concentration and composition) the biofilms associated with natural particles.

Disadvantages of the methods are that: (1) since the exopolymers are grown in laboratory culture, their composition (which may affect their labile nature) may differ from the variety of exopolymers which occur under natural conditions and (2) the exopolymers, when added in purified form, will not contain the wide range of adsorbed compounds which they accumulate under natural conditions. Such adsorbed compounds (i.e., humic material, various DOM compounds, metals, etc.) may influence how readily the exopolymers (and the particle which it coats) are ingested by a given consumer.

MATERIALS REQUIRED FOR EXOPOLYMERS IN NATURAL SAMPLES

(Note: Mention of brandname products does not constitute an endorsement by the U.S. Geological Survey).

Exopolymers in Natural Samples

Equipment

- SEM (scanning electron microscope; cold-stage type)
- TEM (transmission electron microscope)
- Ultramicrotome (for TEM) with diamond or glass knives
- Light microscope
- Sputter-Cryo coat apparatus (for SEM)

Supplies

- SEM specimen mount stubs
- Stub transfer device
- Colloidal silver paint (for mounting specimens on stubs)
- TEM film grids (carbon-stabilized
- Dewar flask (for short-term storage of liquid nitrogen)
- Glass microscope slides with coverslips
- Immersion oil (type B) for light microscopy
- Forceps

Solutions, Reagents, and Media

- Sodium cacodylate buffer (0.1 M), pH 7.2 (Polysciences Inc.)
- Ruthenium red stain, concentration = 0.15% w/v to be added to glutaraldehyde (see TEM below)
- Polycationic ferritin (working solution = 10 mg ml^{-1} in 0.1 M cacodylate buffer)
- Electron microscope grade glutaraldehyde (4%) in sodium cacodylate buffer (0.1 M, pH 7.2) containing 0.15% (w/v) ruthenium red and 3% NaCl.
- Aqueous uranyl acetate (2% w/v)
- Agar (4%) in sterile-filtered (0.2 μm) seawater
- Alcian blue 8GX stain (1% alcian blue in 3% acetic acid; pH 2.5); filter (0.2 μm) before use)
- India ink solution (shake well, do not dilute)
- Osmium tetroxide (2%) in 0.1 M sodium cacodylate buffer (pH 7.2)
- Spurr's low viscosity embedding medium
- Liquid nitrogen ($-170°C$)

Exopolymers for Feeding Experiments

Equipment

- Liquid scintillation counter
- Gamma counter
- High-speed centrifuge
- Microscope
- Freeze drier (lyophilizer)
- Freezer (preferably super-cold at $-70°C$)
- Shaker table

Supplies

- Filtration apparatus (0.2 μm)
- 50-ml plastic centrifuge tubes
- Tissue solubilizer, with low quenching (e.g., ScintiGest, Fisher Company)
- Scintillation vials (5- or 20-ml size)
- Radioisotopes: [$^{14}C(U)$] D-glucose, $NaH^{14}Co_3$, $^{51}CrCl_2$ (optional)
- Fluorescent beads (0.4-μm diameter, Polysciences Company; optional)
- Fluorescent paint pigments (Radiant Paint Company, Riverside, CA)
- Pipettes and micropipettes
- Disposable 1.0-ml syringe with needle (for injecting acid and phenylethylamine)
- Whatman No. 1 chromatography paper (cut into strips and folded)
- Glass vials with fitted injection ports (for respiration)

Solutions, Reagents, and Media

- Seawater
- Formalin (concentration 37%)
- Scintillation cocktail (aqueous)
- DIFCO marine medium 2216 (Sigma Chemical Company)
- D-glucose
- Ethanol or methanol
- Phenylethylamine (Sigma Chemical Company)
- H_2SO_4 (concentration)
- f/2 media (or other suitable culture medium for microalgae)
- EDTA (ethylenediaminetetraacetic acid; 8 mM stock solution in 0.2-μm filtered SW)
- Containers for feeding incubations (type of container depends on experimental animal used)
- Dialysis tubing (presoak and rinse in deionized water 30 min before use)

PROCEDURES

Exopolymers in Natural Samples

Light Microscopy

1. Alcian Blue (pH 2.5):
    For the rapid visual observation of biofilm slime exopolymers and amorphous aggregates material using light microscopy, the stain alcian blue (1% alcian blue in 3% acetic acid, approximately pH 2.5) is most efficient and imparts minimal deformation.[3]

- The sample (approximately 100 to 200 μl) to be examined is placed on a glass slide with cover slip, and mixed with a small amount (25 to 50 μl) of alcian blue solution. Staining is very rapid (<1 min).

- The sample is then observed by light microscopy.

2. India ink (wet-film method):
   For closer observations and measurements of individual capsule exopolymers surrounding cells, wet preparations using India ink[4] provide the best visualization to determine the presence or absence of most capsules (it should be noted that very thin capsules are only detectable using electron microscopy).

   - Staining capsules using India ink involves adding a small amount (15 to 30 μl) of India ink solution to a small amount of concentrated sample (i.e., 50 μl) on a glass slide with a cover slip. Microbial cultures may be preconcentrated by centrifugation for best results.

   - Remove excess liquid with blotting paper. The ink film should be about the same thickness as the capsulated bacteria to be observed. If the film is too thin, however, the capsules and cells will be flattened and distorted. Several attempts may be necessary to achieve the proper thickness.

   - The cells and capsules are then observed using oil-immersion light microscopy.

Scanning Electron Microscope

Microbial capsules and slime are destroyed or severely deformed during most preparations for electron microscopy. Such procedures typically require both fixation and dehydration of the samples to be observed. Fixation results in selective loss of some material during washing. Also, since exopolymers are highly hydrated gels (99% water), dehydration steps result in further loss of polymeric material and in significant deformation of the physical structure of the remaining polymer.[5] Methods are presented below which will reduce such losses and deformations, allowing observations of relatively intact exopolymers.

The most nondestructive technique used to observe biofilms relatively intact involves the Sputter-Cryo technique. This allows the examination of highly hydrated biological materials at liquid nitrogen temperatures (i.e., −170°C) without the use of any chemical fixatives.[6] The technique permits observations relatively free from artifacts due to drying or fixation.

1. Preparation of the sample involves removing most of the excess water and carefully mounting the specimen (e.g., sand grain or aggregate) while still wet on an SEM stub with colloidal silver paint.

2. The stub (plus its attached specimen) is quickly quench-cooled by placing it in a nitrogen slush (approximately −200°C) for several minutes using a stub-transfer device.

3. The specimen is moved via the transfer device into the working chamber of an electron microscope Sputter-Cryo apparatus maintained at −170°C. The specimens are then etched for approximately 2 min at 10 mA and then sputter-coated with gold for approximately 7 min at 35 mA.

4. The stubs are moved, via the transfer device, onto the cold-stage of an SEM and observed at approximately 30 kV accelerating voltage.[7,8]

Transmission Electron Microscope

1. About 2 g of sample is placed in a 4% glutaraldehyde solution (4% glutaraldehyde in 0.1 M sodium cacodylate buffer) containing 0.15% (w/v) ruthenium red and 3% sodium chloride, pH 7.2. Fixation is carried out for 2 h at 20°C.

2. The fixative is removed by centrifugation or gentle filtration, depending on the nature of the sample and washed in cacodylate buffer.

3. The sample is placed in agar (4%) at 50°C, cooled, then cut into cubes (1 mm), and washed five times in cacodylate buffer.

4. The agar samples are postfixed in 2% (w/v) osmium tetroxide in 0.1 M sodium cacodylate (pH 7.2) for 2h.

5. The samples are thoroughly rinsed using deionized water, then dehydrated in an ethanol series (30, 50, 70, 90, and 100% for 20 min each) followed by propylene oxide and flat embedding in Spurr's low viscosity resin, then allowed to harden (for 72 h).

6. Ultrathin sections are cut using an ultramicrotome fitted with a diamond knife. Sections are collected on carbon-stabilized grids, poststained with uranyl acetate (2%) for 1 h, then examined using a transmission electron microscope at an accelerating voltage of 60 to 100 kV.

Preparation of Exopolymers for Feeding Experiments

In general, two physical types of microbial exopolymers, capsule and slime, can be differentiated. These two types of exopolymers can be effectively isolated from a given culture using differential centrifugation or by the growth phase in which the polymers are isolated.

Bacteria

1. Cultures of a bacterial isolate should be grown at a constant temperature (e.g., 22°C) on a mechanical shaker table in DIFCO marine medium 2216 or a medium specific for that isolate. The medium may be slightly enriched with D-glucose (0.5 to 2.0% w/v) to produce a more polysaccharide-rich exopolymer.

2. Incubate the cultures to the appropriate growth phase in which exopolymer production commences. Then add a suitable ^{14}C exopolymer precursor such as ^{14}C D-glucose. The precursor is added to cultures during the time when exopolymer production is occurring in order to maximize incorporation of the label into the exopolymer. The [U]-^{14}C-glucose, used as a carbon marker, will uniformly label both polysaccharide and protein fractions of the exopolymers. ^{14}C-L-lysine may be used to specifically label the protein fractions of some polymers.

The following steps (3 through 6a) are for capsule exopolymers only.

3a. Capsule exopolymers may be labeled using [U]-^{14}C-D-glucose (1 to 3 mCi 50 ml^{-1} culture), which is added to cultures during early log-phase of growth or just before capsule production is initiated. This will provide maximal incorporation efficiency of ^{14}C-label into the capsule exopolymers.

4a. The cultures are harvested during late log-phase of growth or just after capsule production has peaked. Cells are first fixed in formalin (0.5% final concentration).

5a.   The cells and their capsule-exopolymers are initially separated from more loosely associated polymers released into the media by a single centrifugation (12,000 × g; 15 min). This initial supernatant is discarded.

6a.   The pelleted cells (and their closely associated capsule exopolymers) are collected and resuspended in filter-sterilized (0.2 μm) seawater, then subjected to repeated (3 to 4 times) higher speed centrifugations (21,000 × g; 30 min), resuspending the pellet each time in the same supernatant. This will generally provide enough shear force to remove a large portion of the capsule exopolymer from the cells. The exopolymer remains suspended in the supernatant, while the cells are sedimented in the pellet.

For slime exopolymers, substitute the following steps (3 through 6b).

3b.   [U]-^{14}C-D-glucose (1 to 3 mCi 50 ml^{-1}) is added to cultures just as maximal slime production begins, usually during late log-phase of growth.

4b.   Cells are fixed in formalin (0.5% final concentration).

5b.   The cultures are harvested during late stationary growth phase.

6b.   Slime exopolymers are extracted by centrifugation at 15,000 × g for 15 min.

For either exopolymer type, continue with the following steps.

7.   Exopolymers in the supernatant are precipitated in cold (4°C) ethanol or methanol (70% final concentration) for 8 h, and purified according to the methods of Kennedy and Sutherland.[9] If the polymers are not abundant in the supernatant, the supernatant can be lyophilized to about 10% of the original volume in order to concentrate the polymers. Then the polymers will readily precipitate in cold alcohol and can be centrifuged to a pellet.

8.   The pelleted polymer is redissolved in a small volume (e.g., 10 ml) of deionized water. The cold-alcohol precipitation steps are then repeated two more times to remove ethanol-soluble small molecular weight compounds associated with the polymers. The exopolymer is finally dialyzed to remove any residual alcohol, then lyophilized to a dry white fibrous matrix.

9.   The activity of the ^{14}C-exopolymer is measured in triplicate (using liquid scintillation counting [LSC]) by dissolving a known amount (e.g., 50 μg) of dry polymer in 500 μl deionized water (in a scintillation vial), adding 5 to 10 ml aqueous scintillation cocktail, then counting by LSC. The activity is expressed as dpm ^{14}C per microgram dry exopolymer. Purified freeze-dried exopolymers can be stored for long periods at −70°C until use.

Diatoms or Microalgae

1.   A marine diatom isolated or obtained in near-axenic culture from a culture collection is cultured in f/2 medium[10] or another suitable medium. The cultures are grown at constant temperature (e.g., 20°C) on a 12 h alternating light-dark cycle (illumination = 200 μEinsteins).

2.   During late log-phase of growth, 1 to 3 mCi 50 ml^{-1} of NaH^{14}Co$_3$ is added as an exopolymer precursor to the cultures, then grown under constant illumination.

3.   Diatom exopolymers are most abundantly extracted during the stationary phase of growth according to methods modified from Huntsman and Sloneker,[11] after limitation of either N, P, or Si has occurred.

- Loose slime exopolymer (in solution within the medium) is easily removed by a single centrifugation (10,000 × g for 10 min) and collected in the supernatant.

- Cells and their more tightly bound capsule exopolymers will be sedimented in the pellet. To separate capsule exopolymers from cells, the addition of EDTA (4 mM; final concentration) for 1 h (slowly stirred), will help facilitate the solubilization of the tightly bound exopolymers. Removal of the exopolymer from cell surfaces is accomplished using repeated (3 to 4 times) centrifugation, resuspending each time in the same supernatant.

4. The supernatant (containing the soluble exopolymer) is reduced to 10% of its original volume by lyophilization, then precipitated in cold ethanol and subsequently purified as described above. After concentration, precipitation, and purification (as described above), the freeze-dried exopolymer is stored ($-70°C$) until use. Activities are determined as described above.

Binding of Exopolymers to Particulate Material

1. Approximately 1 to 10 mg dry ^{14}C-exopolymer is added per g sediment. The exopolymers will adsorb to sediments relatively rapidly (i.e., 10 to 15 min) in the presence of seawater.

2. The sediments are lightly centrifuged (just enough to gently pellet the sediment particles!) to remove the nonadsorbed exopolymer.

3. Inert tracers, such as ^{51}Cr or fluorescent paint pigments, may also be added as a convenient visual marker or inert radiolabel for the label, provided the animal used is not selective in its ingestion or digestion of the tracer (see Chapter 85).

4. The activity of the sediment-exopolymer mixture is then monitored using LSC for ^{14}C (gamma counting if ^{51}Cr is additionally used).

Short-Term Feeding Experiments

Feeding experiments used to determine absorption efficiencies can be conducted according to a mass balance approach (described here) or by using a ^{14}C:^{51}Cr ratio approach (see Chapter 85) if the animals digestive architecture permits.

Basic Procedures

Once the labeled exopolymer has been bound to the particles of choice (i.e., sediment, beads, etc.), it is added to an enclosed microcosm containing natural food sources, sediments (if benthic animals are used), seawater, and the consumer animals.

1. The bound-labeled exopolymers are added for a brief time period (i.e., hot-feed pulse).

2. Once the hot-feed period is over, the animals are removed from the microcosm, briefly washed in filtered (0.2 μm) seawater, then placed in "cold-feed microcosms which are identical to the hot-feed microcosms but contain no added label. The animals will continue to feed naturally until their gut has been cleared of labeled exopolymer material.

3. Feces should be collected during this cold-feed period and ^{14}C-activities measured by LSC.

4. Once the gut has been cleared of labeled material, the animal is sacrificed, its tissues solubilized in a suitable agent with low quenching (e.g., ScintiGest, Fisher Company), and the ^{14}C-activity of the tissues counted by LSC.

Respiration Losses of $^{14}CO_2$

Losses of ^{14}C via respiration of $^{14}CO_2$ during the cold-feed period should also be measured using replicate parallel samples,[12] especially if the cold-feed period is long (i.e., several hours). The percent of ^{14}C-respired is added to the %AE.

1. To measure $^{14}CO_2$ respiration, animals previously fed ^{14}C-labeled exopolymer are placed in enclosed glass vials containing unlabeled food, and fitted with injection ports and suspended folded chromatography paper (Whatman No. 1). Control vials, containing no animals but having identical conditions, should also be employed.

2. At the conclusion of the incubation time (generally the length of the cold-feed period), the vials are injected with a small amount of concentrated acid (H_2SO_4) to cease biological activities and result in a final pH 2 (the volume of acid will depend on the conditions present in the vials and should be determined from preliminary experiments). This will volatilize any CO_2 in solution.

3. Then phenylethylamine (approximately 250 µl) is carefully added to saturate the suspended chromatography paper. The vials are gently shaken on a shaker table for 2 to 4 h (or a sufficient amount of time to capture the CO_2).

4. After this time the chromatography paper is retrieved, placed in a scintillation vial, and cocktail added. The animals are retrived, washed, solubilized, then prepared for LSC in order to determine the amount of ^{14}C-label remaining in the animal. The sample(s) are then counted by LSC.

5. The percent of ingested ^{14}C which has been respired by the animals during the given time period, can then be determined by the equation:

$$\%\ ^{14}\text{C-Respired} = \text{dpm}\ ^{14}CO_2 \text{dpm}^{-1}\ \text{Ingested} \tag{1}$$

Calculation of Percent Assimilation Efficiency

The general design for a feeding experiment is:

$$\text{Hot-feed} \rightarrow \text{wash} \rightarrow \text{cold-feed} \rightarrow \text{animal (gut-clear)}.$$

1. Measure:

    - $^{14}CO_2$ respiration loss
    - ^{14}C feces loss
    - ^{14}C in animal tissues

2. The percent absorption efficiency (%AE) is then calculated as:

$$\%\text{AE} = 1 - (\%\ ^{14}\text{C Fecal}/\%\ ^{14}\text{C Ingested}) \tag{2}$$

where

$$\%\ ^{14}\text{C Ingested} = (^{14}\text{C Tissues}) + (^{14}\text{C Respiration loss}) + (^{14}\text{C Fecal}) \tag{3}$$

NOTES AND COMMENTS

1. Alcian blue stains complex carbohydrates rich in carboxyl (CHO) polyanions and/or sulfated groups, which abundantly reside on acidic mucopolysaccharides. The samples may

be optionally fixed in neutral buffered formalin for 24 to 48 h, so that the nuclei and RNA associated with cells are not stained. However, formalin will result in some deformation of the polymer.

2.   India ink will not penetrate the capsule. The capsule, therefore, appears as a clear light zone between the refractile cell outline and the dark ink background.[4] India ink can be used to effectively measure capsule size and shape.[3] Capsule radial widths of 0.2 to 0.5 μm are the smallest that can be recognized with certainty using this technique. Naked bacteria, having no capsules, are not easily separated from the ink background.

3.   It should be noted that since TEM preparations involve fixations and dehydrations of samples, some shrinkage and modification of the physical structure of the exopolymers is inevitable. Looser slime exopolymers, found in aggregates and biofilms, will be more susceptible to deformation than the more resilent capsule exopolymers surrounding individual cells. Since glutaraldehyde fixes proteins strongly, but not carbohydrates or lipid material, the ruthenium red[13] is additionally used to stain acidic polysaccharides. Polycationic ferric (final concentration 1.0 mg ml^{-1}) may be effectively substituted in place of ruthenium red as a stain for acidic polysaccharides.[14]

4.   Slime exopolymers are loosely associated with cells or copiously released into the surrounding medium. These exopolymers are generally produced during the stationary phase and can easily be separated from the cells by a single, high-speed centrifugation. Capsule exopolymers closely surround cells, are less abundant, and are not easily removed by a single centrifugation. For many bacteria, capsule exopolymers are produced during mid to late log-phase of growth. The patterns of capsule and slime production should be determined for each bacterial strain and medium condition to be used. For example, capsule and slime exopolymers may be produced at the same time or sequentially. Such patterns can be quickly determined by staining with alcian blue or India ink (see below) to determine the nature and integrity of the capsules and slime exopolymer produced.

5.   The culture conditions (i.e., media composition, physiological state of cells, etc.) in which microbial cells are grown will influence the composition of the resulting exopolymers. Not all bacteria and diatoms produce exopolymers under all conditions. For each microbial strain, it is, therefore, important to determine during which growth phase and culture conditions exopolymers are produced, and in what abundances. This will simplify later extractions and increase subsequent yields. Many microbial species produce exopolymers most abundantly during the stationary phase of growth, when a given nutrient(s) (i.e., N, P, Si, etc.) has become limiting.

6.   The activity of exopolymer should ideally be as high as is possible and practical. A higher activity polymer will facilitate greater resolution and more repeatable results in subsequent feeding experiments. The activity will depend on: (1) the ratio of labeled vs. unlabeled precursor (e.g., ^{14}C-glucose vs. glucose) in the medium during the time in which the polymer is being produced by the cells (the higher the ratios the greater the resulting polymer activities) and (2) the time in which the labeled precursor is added (i.e., it should be added just as polymer production begins).

7.   Since the adsorption of exopolymers to the surfaces of particles occurs primarily through ionic attractions, they can be adsorbed to a wide variety of particles. How well the polymer is adsorbed will depend on its specific composition and the ionic concentration (i.e., salinity) of the system. Increasing ionic concentration will generally enhance binding. Other more specific methods for the covalent binding of polymers to beads,[15] or the creation of artificial aggregates, are also available.[16]

8.   In feeding experiments, the length of time for the hot-feed period, determined from preliminary experiments, should be long enough to allow a sufficient amount of labeled exopolymer to be ingested, but short enough to minimize recycling of the label (i.e., respiration as ^{14}CO$_2$, coprophagy, etc.). The length of time needed for the cold-feed period can also be determined from preliminary experiments, or can be determined concurrently if the labeled exopolymer was mixed with fluorescent paint pigments. Since gut-passage rates are dependent upon feeding rates, it is important to incubate the animals under conditions (i.e., food, temperature, salinity) close to those of their ambient environment.

9. Respiration of $^{14}CO_2$ may be highly variable between animals, therefore, an adequate number of replicate animals should be used in estimating respiratory losses. Also, the trapping efficiency of the respiration system should be determined from preliminary experiments. This is determined by directly adding a known amount of $NaH^{14}CO_3$ to the vials containing sterile seawater, then following the abovementioned procedure (i.e., acidification, etc.) Percent $^{14}CO_2$ recovery = dpm recovered dpm^{-1} added.

10. Acidification of high carbonate sediments will quickly oversaturate the phenylethylamine strips with CO_2, resulting in a reduced recovery of $^{14}CO_2$.

REFERENCES

1. Decho, A. W., Microbial exopolymer secretions in ocean environments: their role(s) in food webs and marine processes, *Oceanogr. Mar. Biol. Annu. Rev.*, 28, 73, 1990.

2. Sherr, E. B., Direct use of high molecular weight polysaccharides by heterotrophic flagellates, *Nature (London)*, 335, 348, 1988.

3. Roth, I. L., Physical structure of surface carbohydrates, in *Surface Carbohydrates of the Prokaryote Cell*, Sutherland, I., Ed., Academic Press, New York, 1977, 5.

4. Duguid, J. P., The demonstration of bacterial capsules and slime, *J. Pathol. Bacteriol.*, 63, 673, 1951.

5. Sutherland, I. W., Bacterial exopolysaccharides, *Adv. Microbiol. Physiol.*, 8, 143, 1972.

6. Robards, A. W. and Crosby, P., A comprehensive freezing, fracturing and coating system for low-temperature scanning electron microscopy, *SEM*, 2, 325, 1979.

7. Richards, S. R. and Turner, R. J., A comparative study of techniques for the examination of biofilms by scanning electron microscopy, *Water Res.*, 18, 767, 1984.

8. Howgrave-Graham, A. R. and Wallis, F. M., Preparation techniques for the electron microscopy of granular sludge from an anaerobic digester, *Lett. Appl. Microbiol.*, 13, 87, 1991.

9. Kennedy, A. F. D. and Sutherland, I. W., Analysis of bacterial exopolysaccharides, *Biotechnol. Appl. Biochem.*, 9, 12, 1987.

10. Guillard, R. R. L. and Ryther, J. H., Studies on marine planktonic diatoms. I. *Cyclotella nana* Hustedt and *Detonula confervacea* (Cleve) Gran, *Can. J. Microbiol.*, 8, 229, 1962.

11. Huntsman, S. A. and Sloneker, J. H., An exocellular polysaccharide from the diatom *Gomphonema olivaceum*, *J. Phycology*, 7, 261, 1971.

12. Hobbie, J. E. and Crawford, C. C., Respiration corrections for bacterial uptake of dissolved organic compounds in natural waters, *Limnol. Oceanogr.*, 14, 528, 1969.

13. Springer, E. L. and Roth, I. L., The ultrastructure of the capsules of *Diplococcus pneumoniae* and *Klebsiella pneumonia* stained with ruthenium red, *J. Gen. Microbiol.*, 74, 21, 1973.

14. Jacques, M. and Foiry, B., Comparison of different cationic probes for electron microscopic visualization of bacterial polyanionic capsular material, *Curr. Microbiol.*, 18, 313, 1989.

15. Decho, A. W. and Moriarty, D. J. W., Bacterial exopolymer utilization by a harpacticoid copepod: a methodology and results, *Limnol. Oceanogr.*, 35, 1039, 1990.

16. Shanks, A. L. and Edmondson, E. W., Laboratory-made artificial marine snow: a biological model of the real thing, *Mar. Biol.*, 101, 463, 1989.

CHAPTER 80

Protistan Grazing Rates via Uptake of Fluorescently Labeled Prey

Evelyn B. Sherr and Barry F. Sherr

INTRODUCTION

Rates of grazing on bacteria or phytoplankton by protists in natural waters may be estimated by evaluating the average rate of particle ingestion by specific components of the protistan assemblage in a water sample, and then multiplying the cell-specific ingestion rate by the total number of protists in each component, e.g., flagellates, ciliates, identified morphological types, or size fractions. Using fluorescently labeled particles as analogs of live prey in order to determine cell-specific uptake rates is becoming a standard approach. This method depends on use of an epifluorescence microscope outfitted with both ultraviolet light (DAPI) and blue light (AO, FITC, DTAF, chlorophyll autofluorescence) filter sets (see Chapter 26).

Two types of fluorescently labeled prey (FLP) analogs have been used: (1) inert plastic microspheres (FLM),[1-4] which can be coated with protein[5] and (2) heat-killed bacteria[6-9] or clonal phytoplankton cultures.[10,11] The usual experimental design for this method is one in which short-term, cell-specific uptake rates are determined via quantifying the average number of prey particles within protistan cells over a time course ranging from 10 to 60 min. (Since fluorescently labeled bacteria (FLB) and algae (FLA) are digested by the protists, an alternative approach is possible using these prey particles, in which rate of disappearance of FLB or FLA is followed over longer time periods, 12 to 24 h or more.)

Advantages of the method are: (1) grazing rates are determined via short time-course experiments, minimizing the potential problems associated with long time-course experiments; (2) there is minimum manipulation of the sample, i.e., no size screening, dilution, or addition of chemicals is necessary; (3) the specific components of the protistan assemblage which are the most active consumers of the type of FLP used are directly identified; and (4) aspects of the physiology and behavior of individual species/morphological types of protists within a natural microbial assemblage can be investigated, e.g., rates of ingestion/digestion of prey, prey size preference, feeding response to variations in prey density, switching from one type of prey to another, and variation in grazing rate in response to changes in environmental parameters.

Disadvantages of the method are: (1) selective grazing by protists, in which the added FLP analogs are ingested at either higher or lower rates than are natural prey cells, may occur, especially when FLM are used; (2) adding FLP may change the natural abundance of that type of prey sufficiently to cause a change in protistan grazing rate; (3) only grazing by identified components of the protistan assemblage is quantified, therefore, the value for total protistan grazing rate may be an underestimate of the true rate; and (4) the method requires microscopic inspection of numerous subsamples, therefore, a considerable amount of time and labor is required to arrive at each grazing rate value.

MATERIALS REQUIRED

Preparation of Fluorescently Labeled Prey

Equipment

- Epifluorescence microscope
- High-speed centrifuge
- Sonicator with microtip
- Water bath set at 60°C
- Filtration rig for 25-mm filters
- Micropipettes

Supplies

- 50-ml plastic centrifuge tubes
- 7-ml plastic vials for storing aliquots of the FLB or FLA preparation

Solutions and Cultures

- DTAF (Sigma Chemical Company, St. Louis, MO).
- 0.05 M Na_2HPO_4-NaCl solution, adjusted to pH 9 (we use the same salt content as the medium from which the prey cells were obtained, or for freshwater organisms, 0.85% NaCL, i.e., physiological saline).
- 0.02 M Na_4 pyrophosphate (PP_i)-NaCl solution (0.89 g in 100 ml$_{-1}$ of deionized water, with NaCl added to equal the salinity of the medium of the prey cells, or 0.85% NaCl for freshwater organisms).
- FLA preparation only: 0.02 M PP_i-NaCl solution made up with 10% dimethyl sulfoxide (DMSO).
- Concentrated cultures of prey organism: phytoplankton FLA preparations have been made to date using only single-species isolates of algae or coccoid cyanobacteria grown in nutrient-rich media. We recommend that the algal cultures be axenic. The following phytoplankton species have been successfully used to make FLA:[10,11] *Synechococcus* sp., *Nannochloris atomis*, *Nannochloris* sp., *Chlorella capsulata*, *Chlorella* sp., *Thalassiosira pseudonana*, and *Prorocentrum mariae-lebouria*. Rublee and Gallegos[10] reported that a diatom with long spines, some freshwater algae, and several phytoflagellates did not stain well. Phytoplankton cultures may be obtained from the Provasoli-Guillard Center for Culture of Marine Phytoplankton, Bigelow Laboratory for Ocean Sciences, West Boothbay Harbor, ME. Bacteria may be obtained in two ways: (1) growing up cultures of mixed species assemblages, e.g., 0.8-μm filtered water amended with 1 to 2 mg L^{-1} yeast extract incubated for 1 to 2 d, or of bacterial isolates grown in nutrient medium to a density of $>10^7$ ml^{-1} or (2) concentrating natural assemblages of bacterioplankton present in 10 to 20 L of 0.8-μm filtered water using, e.g., a hollow fiber filter with a nominal molecular size cut-off of 0.1 μm.[12]

Fluorescently Labeled Prey Short-Term Uptake Experiments

Equipment

- Sonicator with microtip
- Temperature-controlled incubator
- Epifluorescence microscope equipped with UV and blue light filter sets
- Filtration rig for 25-mm filters
- Micropipettes

Supplies

- 0.2-μm polycarbonate filters
- 0.2- and 0.8-μm black-stained polycarbonate filters
- Containers for the experimental samples — we routinely use Whirl-pak bags, soaked overnight in 10% HCl, copiously rinsed, which are placed into 1-L beakers half filled with water from the sample to be used, and at the same temperature as the sample water

Solutions and Cultures

- FLP preparation, diluted from the stock to about 10^8 FLP ml^{-1} for bacterial-sized particles, 10^6 FLP ml^{-1} for algal-sized particles
- DAPI solution, 1 mg ml^{-1}
- Acridine orange solution for AODC
- Fixative for protistan samples: we use a final concentration of 0.5% alkaline Lugol's solution, followed immediately by 2% borate buffered formalin and a drop of 3% sodium thiosulfate to clear the Lugol color

PROCEDURES

Preparation of Fluorescently Labeled Prey

Fluorescently Labeled Microspheres

Fluorescently labeled microspheres are plastic beads impregnated with a fluorescent stain. They can be obtained in sizes ranging from 0.05 to 90 μm in diameter, and brightly fluorescent in blue, yellow-green, or red wavelengths, from Polysciences, Inc., Warrington, PA, or from Interfacial Dynamics Corporation, Portland, OR. The FLM cost about $70 to $100 for a vial of 2 to 10 ml of 10^4 FLM ml^{-1} (large particles) to 10^{12} FLM ml^{-1} (small particles). As an analog for suspended bacteria, the usual FLM chosen is Polysciences cat. no. 17152, 0.5-μm YG (yellow-green fluorescence). (A 1-ml trial size of this FLM is available.) The FLM of 1.0, 2.0, 3.0, and 6.0 μm diameters could be used as analog for small size classes of phytoplankton.

In order to use FLM in grazing-rate assays, they must first be diluted from the stock vial, which dispenses the FLM in deionized water via drops from the tip. The FLM may be diluted into 1 to 5 ml of deionized water in a small plastic vial. Using a calibrated micropipette, aliquots of the diluted stock FLM are then added to 2 ml of fresh or salt water and collected onto a 0.2-μm membrane filter for quantitative enumeration via epifluorescence microscopy. The density ml^{-1} of FLM in the diluted stock is used to calculate how much of the diluted stock to add to yield the desired density of FLM in the grazing rate experiments.

One problem with FLM is clumping of the microspheres. To enhance dispersal, Pace and Bailiff[5] aged aliquots of bacterial-sized microspheres in bovine serum albumin (5 mg ml^{-1}) for 24 h before using the FLM in uptake experiments. Coating the FLM with protein may also make them somewhat more palatable to phagotrophic protists, as many protists do show active selection against FLM.

Fluorescently Labeled Bacteria or Algae

We developed a method for staining bacteria with the yellow-green fluorescing dye, 5-(4,6-dichlorotriazin-2-yl) aminofluorescein (DTAF), which binds to protein.[6] The staining procedure was subsequently adapted to produce FLA from monospecific cultures of algae.[10] Our method involves heat-killing the prey cells in order to uniformly stain them. The procedure for preparing FLB and FLA follows.

1. The bacterial or phytoplankton cells should be harvested during log-phase growth, as cell clumping often increases after the stationary phase is reached.

2. Harvest bacterial or phytoplankton cells by centrifugation. We centrifuge 25 ml of medium per 50-ml tube at 22,000 $\times$ g for 12 to 20 min for bacteria, 800 $\times$ g for a shorter time period for phytoplankton.

3. Suspend pellets in 10 ml of phosphate-saline buffer, pH 9.

4. Add 2 mg of DTAF to the cell suspension in the phosphate-saline buffer, and incubate in the 60°C water bath for 2 h. We incubate the suspension in one of the 50-ml centrifuge tubes.

5. After the incubation, centrifuge the cells down, decant the DTAF solution, and wash and centrifuge three times with the phosphate-buffered saline.

6. After the final wash, resuspend the cells in 20 ml of the PP$_i$-saline buffer (use PP$_i$ buffer with DMSO for FLA). (If FLB is prepared from a very dense bacterial culture, it may be necessary to sonicate the suspension and/or filter through a 3.0-μm membrane filter to remove large clumps.)

7. Vortex the FLB or FLA suspension, and pipette 1 or 2 ml aliquots into 7-ml plastic vials.

8. Cap tightly and store frozen. It is possible to freeze-dry both FLB and FLA for shipping or handling at room temperature. We store freeze-dried preparations in dessicant at 5°C, and then rehydrate with 1 to 2 ml of deionized water, sonicate, and keep frozen until used in experiments.

9. Determine the concentration of FLB or FLA in the preparation by pipetting a small aliquot into 2 ml of PP$_i$ buffer, sonicating at 30-W power level for two several second bursts, and collecting onto a 0.2-μm membrane filter for enumeration via epifluorescence microscopy (see Chapter 26).

Fluorescently Labeled Prey Short-Term Uptake Experiments

The optimum experimental design for FLP grazing assays is to add a sufficient concentration of FLP to result in easily quantifiable ingestion rates, but not to greatly increase standing stock abundance of the prey of interest. Based on past experience, the optimum concentration for FLB is about 1 to 2 $\times$ 10^5 FLB ml^{-1} for ciliate bacterivory, and about 1 to 2 $\times$ 10^6 FLB ml^{-1} for flagellate bacterivory; for FLA, the optimum concentration appears to be >500 to 1000 FLA ml^{-1}. These FLP densities may be a fraction of the total bacterial or phytoplankton standing stock in mesotrophic to eutrophic waters, but are equal to or greater than live prey abundances in oligotrophic waters. Adding lower concentrations of FLB or FLA results in very low values of FLP cell^{-1}, thus, ingestion rates are difficult to determine with any precision. For oligotrophic systems, short-term FLP uptake assays may be used to qualitatively determine what components of the protistan assemblage are capable of grazing particular types of prey, while longer time-course experiments, in which disappearance rates of FLB or FLA added at lower concentrations are followed, may be the only reasonable way to accurately analyze overall grazing rate using FLP.

1. Gently pour 50 to 400 ml of sample water into individual, rinsed Whirl-pak bags (or other container).

2. Place the bags into the beakers filled with sample water of the same temperature, and then place into an incubator set at *in situ* temperature. Let sit undisturbed for at least 30 min to allow the microbial assemblage to recover from handling shock.

3. For assays using bacterial-sized FLP, determine *in situ* bacterial abundance via AODC in order to estimate the amount of FLB to add. A good ball park range is adding FLB to equal about 5% of total bacterioplankton for ciliate uptake (but not less than 1×10^5 ml^{-1}), and about 30% of total bacterioplankton for flagellate uptake (but not less than 1×10^6 ml^{-1}). The volume of FLP diluted stock should be less than 1% of the total volume of sample.

4. Briefly sonicate the FLP diluted stock at 30-W power level for two several second bursts, and vortex to uniformly mix. (Note: repeated cycles of freezing and sonication can lead to fragmentation of FLB).

5. Add the FLP and quickly but gently mix into the sample to create a uniform suspension. This may be accomplished by "massaging" the Whirl-pak bag, or by swirling the sample with the end of a pipette.

6. At selected time intervals, withdraw 10- to 20-ml subsamples and immediately preserve in individual containers. Store in the dark at 5°C until epifluorescence analysis.

7. For selected subsamples, filter aliquots onto 0.2-µm plain membrane filters and enumerate the FLP to determine the actual amount added.

8. Stain subsamples with concentrated DAPI (see Chapter 26), filter onto 0.8-µm black membrane filters, and examine via epifluorescence microscopy.

9. Locate individual protistan cells using the UV filter set, and then switch to the blue light filter set to inspect each cell for ingestion of FLP. FLB and FLA often appear brighter inside protistan cells than they do outside on the filter. Focus up and down to be sure that all ingested FLP are counted inside larger cells. For each subsample, determine total number of protists of interest, and total number of FLP ingested by the protistan cells enumerated in the sample.

10. Plot FLP/protistan cell vs. time. The usual relation will show a linear increase of FLB/cell for some period of time, after which FLP/cell begins to level off due to digestion/egestion.[12] A regression of the linear portion of the uptake curve wil yield a rate of FLP uptake by the protists of interest (e.g., flagellates or ciliates). Divide the FLP uptake rate (FLP/cell/h) by the concentration of FLP per nl or µl (10^6 FLP ml^{-1} = 1 FLP nl^{-1}). This will be an estimate of the hourly per cell clearance rate of protists for the type of prey for which the FLP was used as an analog. Multiply the per cell clearance rates by the total number of protists ml^{-1} to obtain an assemblage clearance rate. The assemblage

clearance rates can be multiplied by the abundance of the *in situ* prey of interest to estimate prey ingestion rates by the protists, and by 24 to obtain a daily clearance or ingestion rate.

NOTES AND COMMENTS

1.   Some types of protists have been found in several studies to actively select against FLM compared to FLB or FLA, thus, FLM should be used with caution.[6,11,13]
2.   Sieracki et al.[14] found that preservation of samples with certain fixatives (e.g., formalin alone) resulted in egestion of ingested particles by flagellates. Use of either the Lugol-formalin-thiosulfate method, or v/v addition of ice-cold 4% glutaraldehyde is recommended.[7]
3.   Although at least two studies have addressed the question of whether protists graze FLB at rates similar to those for *in situ* bacterioplankton, and have found no strong evidence to the contrary,[7,8] more work needs to be done to address the selectivity question both for FLB and FLA. Recently, a method for preparing live, fluorescing bacteria was published by Landry et al.[16] These authors reported that a phagotrophic flagellate, *Paraphysomonas vestita*, discriminated between the live preparations and FLB prepared as above. We compared these two preparations with natural assemblages of estuarine protists and found no significant discrimination against FLB.[7] Clearly, more work needs to be done on this potential problem.
4.   Both flagellates and ciliates have been found to graze larger-sized bacteria at higher rates than smaller sized bacteria.[8,15] Thus, attention should be paid to the cell size of FLB compared to the average cell size of suspended bacteria in the experiments.
5.   In some systems, notably freshwater, a significant amount of bacterivory may be due to phytoflagellates.[4]
6.   FLP uptake experiments can be used to determine various aspects of the physiology and behavior of specific types of protists in natural microbial assemblages, e.g., rates of prey digestion[12] and functional feeding response to variations in prey density.[11]

REFERENCES

1. Borsheim, K. Y., Clearance rates of bacteria-sized particles by freshwater ciliates, measured with mono-disperse fluorescent latex beads, *Oecologia,* 63, 286, 1984.
2. McManus, G. B. and Fuhrman, J. A., Bacterivory in seawater studied with the use of inert fluorescent particles, *Limnol. Oceanogr.,* 31, 420, 1986.
3. McManus, G. B. and Fuhrman, J. A., Clearance of bacteria-sized particles by natural populations of nanoplankton in the Chesapeake Bay outflow plume, *Mar. Ecol. Prog. Ser.,* 42, 199, 1988.
4. Bird, D. F. and Kalff, J., Algal phagotrophy: regulating factors and importance relative to photosynthesis in Dinobryon (Chrysophyceae), *Limnol. Oceanogr.,* 32, 277, 1987.
5. Pace, M. L. and Baliff, M. D., An evaluation of a fluorescent microsphere technique for measuring grazing rates of phagotrophic microorganisms, *Mar. Ecol. Prog. Ser.,* 40, 185, 1987.
6. Sherr, B. F., Sherr, E. B., and Fallon, R. D., Use of monodispersed, fluorescently labeled bacteria to estimate in situ protozoan bacterivory, *Appl. Environ. Microbiol.,* 53, 958, 1987.
7. Sherr, B. F., Sherr, E. B., and Pedros-Alio, C., Simultaneous measurement of bacterioplankton production and protozoan bacterivory in estuarine water, *Mar. Ecol. Prog. Ser.,* 54, 209, 1989.
8. Bloem, J., Starink, M., Bar-Gilissen, M.-J. B., and Cappenberg, T. E., Protozoan grazing, bacterial activity, and mineralization in two-stage continuous cultures, *Appl. Environ. Microbiol.,* 54, 3113, 1988.
9. Chrzanowski, T. H. and Simek, K., Prey-size selection by freshwater flagellated protozoa, *Limnol. Oceanogr.,* 35, 1429, 1990.

10. Rublee, P. A. and Gallegos, C. L., Use of fluorescently labelled algae (FLA) to estimate microzooplankton grazing, *Mar. Ecol. Prog. Ser.,* 51, 221, 1989.

11. Sherr, E. B., Sherr, B. F., and McDaniel, J., Clearance rates of <6 μm fluorescently labeled algae (FLA) by estuarine protozoa: potential grazing impact of flagellates and ciliates, *Mar. Ecol. Prog. Ser.,* 69, 81, 1991.

12. Sherr, B. F., Sherr, E. B., and Rassoulzadegan, F., Rates of digestion of bacteria by marine phagotrophic protozoa: temperature dependence, *Appl. Environ. Microbiol.,* 54, 1091, 1988.

13. Nygaard, K., Borsheim, K. Y., and Thingstad, T. F., Grazing rates on bacteria by marine heterotrophic microflagellates compared to uptake rates of bacterial-sized monodisperse fluorescent latex beads, *Mar. Ecol. Prog. Ser.,* 44, 159, 1988.

14. Sieracki, M. E., Haas, L. W., Caron, D. A., and Lessard, E. J., The effect of fixation on particle retention by microflagellates: underestimation of grazing rates, *Mar. Ecol. Prog. Ser.,* 38, 251, 1987.

15. Gonzalez, J. M., Sherr, E. B., and Sherr, B. F., Size-selective grazing on bacteria by natural assemblages of estuarine flagellates and ciliates, *Appl. Environ. Microbiol.,* 56, 583, 1990.

16. Landry, M. R., Lehner-Fournier, J. M., Sundstrom, J. A., Fagerness, V. L., and Selph, K. E., Discrimination between living and heat-killed prey by a marine zooflagellate, *Paraphysomonas vestita* (Stokes), *J. Exp. Mar. Biol. Ecol.,* 146, 139, 1991.

CHAPTER **81**

Grazing Rate of Bacterioplankton via Turnover of Genetically Marked Minicells

Johan Wikner

INTRODUCTION

The grazing rate of the bacterioplankton community may be estimated by determining the rate of disappearance of genetically labeled minicells, added in low concentrations to natural water samples. The change in minicell concentration is monitored at time intervals, during 4-h time-course experiments, based on a recapture technique (MiniCap). The turnover rate of minicells is subsequently calculated to grazing rate of natural bacteria via the bacteria to minicell ratio in the sample.

"Minicells" are small cells with a size distribution similar to marine bacteria, that bud off from bacterial strains with an aberrant cell division pattern.[1] Minicells possess a bacterial surface, which corresponds to that of the natural-sized cells (Gram negative in the case of *Escherichia coli*-derived minicells). Minicells cannot divide due to the lack of chromosome, but may harbor plasmids. Choosing a plasmid coding for a highly expressed protein enables quantification of minicells via the radioactively labeled marker protein.[2] Thus, the method relies on the ability of the minicell to adequately mimic bacterioplankton, its inability to grow, and its specific genetic marker.

Advantages of the method are: (1) grazers encounter a natural bacterial surface with a realistic size distribution, minimizing bias due to selectivity; (2) a detection limit allowing estimates of grazing rates at low concentrations of bacteria (down to 0.5×10^6 ml^{-1}); (3) lack of manipulations which could rupture normal trophic interactions; (4) a relatively short incubation time; (5) direct estimate of total community grazing; and (6) the ability to follow the fate of the minicell isotope.

Disadvantages of the method are: (1) the lack of an estimated selection factor of the minicell; (2) preparation of labeled minicells and processing of the samples requires approximately one week before and after the field sampling; and (3) the laboratory must be equipped to use radiolabeled compounds and carry out electrophoresis.

Field sampling can be readily carried out aboard a research vessel, and samples can be stored and processed later in the laboratory. Therefore, this method allows one to perform sampling extensive enough to study the ecological effect(s) of bactivores on the bacterial community. Despite the apparently laborious protocol to prepare genetically marked minicells, the minicell technique has been used for the majority of diel and annual estimates of grazing, as compared with other grazing methods, at the time of the edition of this chapter.[3,4] This is partly due to the relatively simple field protocol and analysis of samples, and the fact that each batch of radioactively labeled minicells lasts for approximately two months, essentially limited by their specific activity, due to the decay of the [^{35}S]-isotope.

0-87371-564-0/93/$0.00 + $.50
© 1993 by Lewis Publishers

For simplicity of use I have included detailed procedures and solution preparation previously listed in methodology textbooks of bacteriology and molecular biology (e.g., Sambrook et al.[5]), but with any modifications developed at the Department of Microbiology, Umeå University.

MATERIALS REQUIRED

Equipment

- Eppendorf centrifuge
- Cold centrifuge
- Eppendorf incubator or water bath at 95°C
- Freezer, −20°C
- Oven (60°C)
- Equipment for vertical slab gel electrophoresis, including stable power supply (500 V)
- Incubator with shaker at 37°C
- Adjustable micropipettes (2 to 10, 10 to 100, and 100 to 1000 μl)
- Epifluorescence microscope (100× objective)
- Scintillation counter
- Slab gel dryer
- Spectrophotometer (600 nm)
- Swing-out rotor for 60-ml tubes (type Sorvall HB-4)
- Tilted angel rotors for 250- and 60-ml tubes (type Sorvall GSA and SS-34)
- Vortexer
- Vacuum pump (400 mmHg)

Supplies

- 0.2- and 0.6- to 12-μm pore size polycarbonate filters
- 2.5-L clear polycarbonate bottles
- Filter holders with support for 47-mm polycarbonate filters
- Pasteur pipettes
- For grazing experiments, two strains of labeled minicells with different marker proteins, prepared as described
- Pyrex glass centrifuge (15 ml) tubes with rubber adaptor
- 60- and 250-ml centrifuge tubes
- Sterile tips
- Cold 20-ml syringes
- X-ray film (e.g., Kodak X-omat XAR 5, 20.3 × 25.4 cm) and darkroom equipment for developing X-ray film
- Eppendorf tubes, 1.5 ml

Solutions and Reagents

- Antibiotics (500 μg ml^{-1} ampicillin for /pBR322 strain, 36 μg ml^{-1} chloramphenicol for /pACYC 184 strain)
- Sterile 0.9% NaCl in distilled water
- 37% Formaldehyde
- [^{35}S]-methionine: 15.3 mCi ml^{-1}, 1,220 Ci mmol^{-1}; Amersham Corporation, Arlington Heights, IL
- Scintillation fluid: Hisafe III, no. 81000437, Pharmacia LKB
- Methanol 45% v/v and acetic acid 5% v/v

- Tissue solubilizer: Biolute-S, Zinsser Analytic, no. 1310200
- LB, 5x (Luria broth): warm until dissolved and adjust to pH 7.4 with NaOH. autoclave to sterilize

Trypton	42 g
Yeast extract	21 g
NaCl	42 g
dist. H$_2$O	840 ml

- LB-medium (Luria broth): bring following to 500 ml with distilled H$_2$O.

LB 5×	100 ml
E + B$_1$, 50×	10 ml
20% glucose	5 ml

- Medium E + vitamin B$_1$, 50×: Dissolve following chemicals in distilled water in the order listed. Autoclave, then add 1.5 ml of thiamin (B$_1$, 10 mg ml^{-1}, sterile filtered) when temperature <41°C.

dist. H$_2$O	670 ml
MgSO$_4$·7H$_2$O	10 g
Citric acid · 1 H$_2$O	100 g
K$_2$HPO$_4$, anhydrous	500 g
NaHNH$_4$PO$_4$ · H$_2$O	175 g

- LB agar plates, 3 per strain with LB-medium and 0.7% agar.
- LB + methionine: To 15 ml LB-medium, add 0.1 ml of 2% L-methionine.
- Buffered Saline Gelatine (BSG) 10×: Mix following ingredients. Dissolve the gelatine by warming to 100°C in a water bath. Store in refrigerator.

NaCl	0.85% w/v
KH$_2$PO$_4$	0.03% w/v
Na$_2$HPO$_4$	0.06% w/v
Gelatine	100 µg ml^{-1}

- Bacto Methionine Assay Medium (MAM), Difco B423: Dissolve 5.25 g powder in 50 ml distilled water. Heat to boiling for 2 to 3 min. Precipitate should be evenly distributed before transfer.
- Mineral salt solution 56/2 5×: This medium has to be prepared from the following compounds in solution to avoid precipitation.

1 M Na$_2$HPO$_4$	300 ml
1 M KH$_2$PO$_4$	200 ml
10% MgSO$_4$	5 ml
10% (NH$_4$)$_2$SO$_4$	100 ml
1.5% Ca(NO$_3$)$_2$·4 H$_2$O	5 ml
0.05% FeSO$_4$·7 H$_2$O	5 ml
dist. H$_2$O	385 ml

- Minicell medium (MM): Autoclave this solution for 10 min at no higher than 1.0 kg cm^{-2} pressure and 121°C. Let cool below 41°C and add 0.015 ml of vitamin B$_1$ solution (5 mg ml^{-1} stock solution sterile filtered.)

Mineral salt solution 56/2 5×	20 ml
Methionine assay medium	50 ml
Glucose 20% w/v	1 ml
Proline 4% w/v	0.35 ml
Dist. water	30 ml

- Sucrose gradients: Make two sucrose gradients per strain, one in a 60-ml centrifuge tube and one in the 15-ml glass centrifuge tube. Put 32 ml of 1× BSG with 20% w/v of sucrose in the 60-ml centrifuge tube, and 13 ml in the glass tube. Freeze at −20°C until completely frozen. Allow to thaw overnight at 4°C in a cold room (not refrigerator). Keep on ice until used.

- Sample buffer: 0.065 *M* Tris hydrochloride (pH 6.8), 2% Sodium dodecyl sulfate, 9% v/v glycerol, 5% v/v 2-mercaptoethanol, 0.004% w/v Bromphenol blue.
- Polyacrylamide separation gel, 15%, 20 cm long and 1 mm thick

45% w/v acrylamide	9.9 ml
1.7% w/v bis-acrylamide	7.1 ml
dist. H$_2$O	5.3 ml
1.5 *M* Tris HCL, pH 8.8	7.5 ml

Put under 400 mmHg of vacuum for 10 min. Add 75 μl 12% ammonium persulfate (Sigma), 9 μl TEMED (Sigma), and 300 μl 10% SDS (Sigma). When poured onto the glass plates, add 1 ml of distilled water on top of the surface to inhibit direct exchange with air. The gel will start to polymerize and be solid after 1 h at room temperature. Excess water on top of the solid gel is removed with qualitative filter paper before casting the stacking gel.

- Polyacrylamide stacking gel, 5%

45% acrylamide	1.7 ml
1.7% *bis*-acrylamide	1.25 ml
dist. H$_2$O	4.5 ml
0.25 *M* Tris HCL pH 6.8	7.5 ml

Put under 400 mmHg of vacuum for 10 min. Add 36 μl 12% ammonium persulfate (Sigma), 30 μl TEMED (Sigma), and 150 μl 10% SDS (Sigma)

- Running buffer 10×:

Tris base	30 g
Glycine	144 g
Sodium dodecyl sulfate	10 g
add dist. H$_2$O up to	1000 ml

PROCEDURES

Preparation of Minicells

Two strains of the minicell producing *E. coli* M2141 (*pro, lac, str^r, minA, minB*) were constructed by introducing the plasmids pBR322, (Am^r, Tc^r) and pACYC 184 (Cm^r, Tc^r). The *E. coli* M2141 strain can be generally found at most molecular laboratories or culture collections, and the desired plasmids transformed into this strain. The special feature of these plasmids are the high expression of the β-lactamase (pBR322) and chloramphenicol acethyltransferase (pACYC 184), respectively. These two proteins dominate the protein assemblage, as observed on autoradiograms, subsequent radiolabeling, and electrophoretic separation. The β-lactamase and chloramphenicol acethyltransferase, therefore, function as markers for the different types of minicells when radioactively labeled. It is assumed that in the protocol below the two strains with plasmids are readily available (see original Reference 6).

1. Grow strains overnight in 10 ml of LB containing the appropriate antibiotic (500 μg ml^{-1} ampicillin for /pBR322 strain, 36 μg ml^{-1} chloramphenicol for /pACYC 184 strain).

2. Inoculate the 10 ml overnight culture into 400 ml of fresh LB medium prewarmed to 37°C containing appropriate antibiotics as above. Let grow at 37°C, with shaking, for approximately 5 h until an optical density measured at 600 nm (OD$_{600}$) of 0.8 is reached.

3. Centrifuge 10 min in 250-ml centrifuge tubes at 27,000 × g and 4°C.

4. Resuspend cells in 10 ml of $1\times$ BSG.

5. Vortex the 250-ml bucket while disrupting the pellet with a 10-ml pipette. Suck the suspension in and out of the pipette five times. Transfer to a 60-ml centrifuge tube.

6. Pellet the cells 10 min at $20{,}000 \times$ g and 4°C.

7. Resuspend the cells in 2 ml $1\times$ BSG using a vortexer and a 5-ml pipette for at least 60 s.

8. Layer the cell suspension on an ice-cold sucrose gradient in 60-ml centrifuge tubes, by slowly pouring the suspension along the side of the tube.

9. Centrifuge in a cold swing-out rotor (e.g., Sorvall HB-4) at $4100 \times$ g for 20 min at 4°C.

10. A cloud of minicells should be seen in the middle of the tube when holding the tube against a lightbulb. Dislocate the syringe plug, plunge the needle into the middle of the cloud, and suck the cloud into the syringe (approximately 12 ml, leave the upper and lower part of the cloud). Squeeze the minicells vigorously into 60-ml centrifuge tubes with 10 ml of $1\times$ BSG.

11. Pellet the cells at $20{,}000 \times$ g and 4°C for 13 min.

12. Clean used syringes with 70% ethanol. Discard supernatant and resuspend the pellet in 1 ml $1\times$ BSG. Use vortexer and a 1-ml pipette for at least 60 s.

13. Layer the minicells on a sucrose gradient in the 15-ml Pyrex glass centrifuge tube.

14. Centrifuge 20 min at $4100 \times$ g and 4°C with rubber adaptor.

15. Take out the minicell cloud from the middle of the tube as before and squeeze it into a 60-ml centrifuge tube with 10 ml of MM.

16. Centrifuge 13 min at $20{,}000 \times$ g and 4°C.

17. Discard the supernatant and resuspend the pellet in 2 ml of MM by vortexing at least 60 s (no pipette). Add 10 ml of MM and mix carefully.

18. Remove 0.1 ml of minicell suspension from each strain and dilute 10^1, 10^2, and 10^3 times in 0.9% NaCl. Disperse on LA plates to give viable counts of contaminating normal cells. Measure the OD_{600} in a sterile (70% ethanol washed) cuvette. Return the suspension to the tube.

19. Centrifuge 13 min at $20{,}000$ g and 4°C.

20. Resuspend the pellet in 2 ml of MM by vortexing for 30 s. Add an appropriate volume of MM to give a final OD_{600} of 0.25 (cf. OD_{600} measurement above).

21. Dilute 0.05-ml sample 100-fold in 0.9% NaCl and preserve in formaldehyde (final concentration 1.5%), until determination of the minicell concentration according to section on analysis of minicell batch.

22. Store in refrigerator until sample preparation.

Labeling of Minicells

High specific activity of the minicells is crucial for obtaining a sensitive enough detection limit in the field experiments. Labeling should be done immediately after the preparation, as minicells quickly lose their protein synthesis capacity, especially after storage in a refrigerator or freezer. Specific activity of the minicells should be above 5×10^{-4} cpm cell^{-1}, and ideally above 1×10^{-3} cpm cell^{-1}. In my experience the following protocol gives generally high, but *not* fully reproducible, specific activity of the minicells.

1. Take 5 ml of minicell suspension and put it into an Erlenmeyer flask and incubate the flask for 60 min at 37°C.

2. Following incubation, add 40 µl of [^{35}S]-methionine (15.3 mCi ml^{-1}). Incubate for an additional 15 min at 37°C.

3. Chase the protein synthesis with 2 ml of prewarm LB + methionine for 5 min at 37°C (i.e., allow polymerization of stalled polypeptides to the stop codon).

4. Spin down cells for 13 min at 20,000 × g in 15-ml Pyrex glass tubes with rubber adaptor. Discard the highly radioactive supernatant in an appropriate container.

5. Resuspend the pellet in 1 ml of LB + methionine. Transfer the suspension to 1.5-ml Eppendorf tubes and centrifuge 4 min at 12,000 × g.

6. Resuspend pellet in same volume of MM as before the labeling.

7. Transfer 150 µl of suspension to a new Eppendorf tube. Spin down the 150 µl suspension for 4 min at 12,000 × g. Freeze the remainder of the labeled suspension in 2-ml volumes with 15% v/v glycerol at −70°C (preferably screw-cap tubes).

8. Discard the supernatant of the 150-µl sample with a drawn Pasteur pipette and add 25 µl of sample buffer to pellet.

9. Vortex 15 s and heat for 2 min to 95°C.

10. Spin down debris for 2 min at 12,00 × g in the Eppendorf centrifuge.

11. Load onto a 15% polyacrylamide gel for separation of proteins and subsequent quantification of marker protein. Samples at this stage may be frozen at −20°C.

Analysis of Minicell Batch

1. Gel electrophoresis: For complete acrylamide gel electrophoresis technique see appropriate manuals. I use a straight 15% polyacrylamide separation gel of 20-cm length made with 1.5 M Tris HCl buffer, pH 8.8. A 5% polyacrylamide stacking gel 5 cm long, made from 0.25 M Tris HCl buffer (pH 6.8) precedes the separation gel. Running buffer is used according to the recipe in the Solutions section above. Up to 4 gels with 20 5-mm wide slots in each have been run simultaneously. Gels are run at 140 mAh (<30 mA) in a Bio-Rad Protean cell II, usually overnight with cooling applied. This allows the blue saple buffer to migrate to the bottom of the gel, leaving marker proteins in the center of the gel.

    No staining of the gel is necessary since migration rates of the marker proteins are known. However, gels are routinely washed for 1 h in 45% methanol + 5% acetic acid before drying to fix the proteins. Ensure that the orientation of the gel is known at this stage (e.g., by cutting the gel corners). Gels are dried in a Bio-Rad slab gel dryer for 1 h at 50°C with Whatman chromatography paper as support, and then an additional 30 min with heat turned off.

2. Autoradiography:

    ● The dried gel on Whatman chromatography paper support is mounted flat on a glass plate.

    ● One sheet of Kodak X-omat XAR 5,8 × 10 min is fixed on top of the gel and holes made with a needle through the film and gel to allow realignment after developing the film.

    ● Films are exposed at room temperature for 5 to 10 d (for field samples, 1 d for analysis of fresh batch) depending on the age of the batch (i.e., specific activity of the minicells).

3. Radioactivity in proteins:

   ● The dry, developed film is realigned with the gel and holes punched around the gel bands to mark their position on the gel.

   ● Film is removed and protein bands cut out with a pair of scissors and put in 5-ml scintillation vials.

   ● Tissue solubilizer and scintillation liquid is added according to manufacturers recommendations.

   ● Samples are counted in a scintillation counter having an appropriate quench correction curve installed for calculation of dpm, or at least a quench compensation function to adjust samples to the same relative H-number (quench).

4. Direct counts of minicells: Minicells are still effectively bacteria and thus, can be counted in the epifluorescence microscope, according to standard procedures outlined elsewhere in this volume. I normally end up with 10^9 minicells ml^{-1} in a batch following this protocol.

5. Viable counts: Colony-forming units are determined. Normal-sized cells should not exceed 10^5 cells ml^{-1}, since they are diluted 10^4-fold in the field experiments. During a 4-h incubation (and likely longer), interference from normal-sized *E. coli* cells will, thereby, be negligible in the field experiments.

6. 10% Trichloracetic acid (TCA) precipitable material: If accumulation of the [^{35}S]-isotope is to be followed into size fractions (by differential filtration), I have used the radioactivity in TCA-precipitable material cell^{-1} to convert dpm's ml^{-1} to cells ml^{-1}. Then 10 μl of labeled minicell solution is added to 3 ml ice-cold 10% TCA and precipitated for 1 h in ice. The precipitate is harvested on 0.22-μm cellulose acetate-nitrate filters (Millipore), washed three times with ice-cold 5% TCA and counted in the scintillation counter.

7. Cell size: The minicell population has a normal distribution with a size spectrum between 0.1 and 0.9 μm in diameter, averaging 0.5 μm (0.10 μm^3 biovolume). The size distribution of the minicells is reproducible in my hands. However, the size distribution of minicells should be first determined in the initial batches when setting up this method at a new laboratory.

8. Storage of minicells: Labeled minicells can be stored in 15% v/v of glycerol at $-70°C$ for two months or more. Even when obtaining good specific activity in the labeling procedure, however, the decay rate of the [^{35}S]-isotope limits the application of the minicells to field experiments after this time period. For shorter time storage (days) minicells may be stored in a $-20°C$ freezer or a refrigerator at 0 to 4°C; in the latter case without glycerol. The potential risk when storing in nonfrozen Minicell medium is some growth of the normal-sized cells. Repeated freezing and thawing should be avoided since this may rupture minicells, leading to leakage of the marker protein into the medium.

Measurement of Turnover of Minicells in Seawater Samples

This method assumes that all grazing bactivores will accept minicells as normal food particles, subsequently ingest and digest the minicells. The amount of remaining undigested minicells is measured and the disappearance of minicells can be followed during a time-course experiment. Minicells are added to a water sample at 10% (normally 1 to 2 × 10^5 minicells ml^{-1}) of the abundance of natural bacteria, thus, minimizing a protozoan response to increase in prey abundance. The sample is gently mixed and minicells harvested at appropriate time intervals by filtration. At this stage minicells, harboring a marker protein with a different molecular weight, are added as an internal standard to monitor losses during the subsequent processing of the sample. The cells in the sample are concentrated, lysed, and the sample is then fractionated by electrophoresis. Samples may be stored frozen prior to electrophoresis.

Addition of Minicells

1. Thaw an amount of labeled minicells (/pACYC 184 type) required to give a final minicell concentration of 10% of the abundance of the natural bacterial assemblage (approximately 10^5 minicells ml^{-1}).

2. In addition, thaw minicells of the /pBR322 type to give a dpm of radioactive marker protein equivalent to a 200-ml sample of pACYC 184 minicells (typically this will be 2 $\times$ 10^7 pBR322-type minicells per subsample). This is the stock suspension of pBR322-type minicells. Remember to take the dilution by the glycerol into account in your calculations. The pBR322-type minicells are used as an internal standard added to samples taken during incubations.

3. Pellet the minicells for 5 min at 12,000 $\times$ g at room temperature.

4. Discard the supernatant and resuspend pACYC 184-type minicells in the original volume (excluding glycerol volume) of 0.2-μm filtered seawater.

5. Vortex at least 1 min and suck the suspension several times with a micropipette tip to carefully disperse the minicells before addition to the water sample.

6. Resuspend pBR322-type minicells in 0.9% NaCl to yield a 100-fold dilution compared to the concentration in the stock suspension in step 2. Disperse cells carefully as above (step 5). This diluton results in lowered variability between samples, probably because of the better dispersion of minicells.

7. The pACYC 184-type minicells are first added to approximately 100 ml of the water sample to be examined, and then the remainder of the sample may be added to ensure good mixing. I recommend that at least six subsamples are included to give an estimate of the minicell turnover rate. As long as the disappearance of minicells can be approximated as linear, triplicate endpoint samples may be used. This requires a total sample volume of 1200 ml.

8. Slowly turn the spiked water samples upside down 50 times and let the sample rest 15 min before taking the time-zero sample.

Sampling of Minicells

Four-hour incubations have been employed routinely to estimate grazing rates by the minicell technique. This minimizes artifacts due to the enclosure of the sample. Samples may be taken as triplicates in the beginning and end of the time course, or distributed over the 4-h time period to verify linearity. In an annual study a 24-h incubation was employed to allow for diel variability in predation, while minimizing labor and cost.[3] Although the best daily estimate is preferably done by repeated 4-h incubations,[4] rates determined from 24-h incubations are observed to be of the same order of magnitude as those obtained by repeated sampling.

1. Add 20 ml of sample to a 250-ml filter holder with a 0.2-μm polycarbonate filter mounted. Add pBR322-type minicells as the internal standard, in the amount as outlined above. Then add the remainder of the sample (180 ml).

2. Apply 400 mmHg of vacuum and suck until the filter surface is dry. Disconnect the vacuum and then add 1.5 ml of 0.9% NaCl. With a Pasteur pipette, flush the filter surface 30 times while keeping the filter rig tilted to one side to wash down the minicells. Keeping the filter holder still tilted, transfer the sample to an Eppendorf tube.

3. Pellet the minicells 5 min at 12,000 × g.

4. Discard the supernatant with a drawn Pasteur pipette.

5. Add 25 μl of sample buffer. Vortex 1 min at low speed.

6. Heat the sample at 95°C for 2 min.

7. At this stage samples may be frozen ($-20°C$) or kept ice cold until the electrophoretic separation.

8. Just prior to the electrophoretic separation, samples are centrifuged 2 min to spin down debris. The samples are then subjected to electrophoresis and further processed as outlined above (see section on analysis of minicell batch).

9. Calculations:
    The scintillation count will give the amount of marker protein of each type in each sample. The concentration of pACYC 184 minicells in the water sample at each time point (N_{mc}) is calculated as:

$$N_{mc} = (\beta/\beta_x)(\alpha_x/SA_\alpha) \qquad (1)$$

where β = amount of βLA protein (pBR322-type minicells) added (dpm); β_x = amount of βLA protein measured in the sample (dpm); α_x = concentration of CAT protein in the sample (dpm ml^{-1}); and SA_α = specific activity of the pACYC 184-type minicell (dpm cell^{-1}).

Plotting N_{mc} vs. time produces graphs like in Figure 1. The slope of the least squares linear regression gives an estimate of the minicell turnover rate (S_{mc}). Occasionally samples clearly diverge from a subjective slope like in the middle panel and can be omitted. At grazing rates lower than 10^4 bacteria ml^{-1} h^{-1}, experimental variability is normally too high to allow slopes statistically different ($p < 0.05$) from zero.

To calculate the rate of grazing of marine bacteria from the minicell turnover rate, I have assumed that the minicell is representative of the average bacterial prey (i.e., no selection). Furthermore, since bactivores (e.g., flagellates and ciliates) appear to opeate at lower than maximum ingestion rates,[7] added minicells are assumed to be ingested in addition to marine bacteria (as opposed to "instead of"). Therefore, the turnover rate of minicells (S_{mc}) is calculated to the grazing rate of bacteria (Gz_b) by multiplication with the reciprocal of the minicell to bacteria quotient (as opposed to percentage) according to:

$$Gz_b = S_{mc}(N_b/N_{mc}) \qquad (2)$$

where N_b = initial abundance of natural bacteria in the sample and N_{mc} = initial abundance of minicells in the sample, or in 24-h, time-course experiments the estimated minicell concentration after 10 h.

Although the flagellates and ciliates exposed to the minicells in the laboratory have been able to use them as a food source, and have demonstrated ingestion rates similar to these found when grazing marine bacteria,[2] selective grazing still remains the largest potential source of error in this method. Size is one factor that can cause selective grazing.[8] As long as the size distribution of minicells mimics that of the natural bacterial community, then selection for or against the minicell is mainly due to the differing surface properties of the

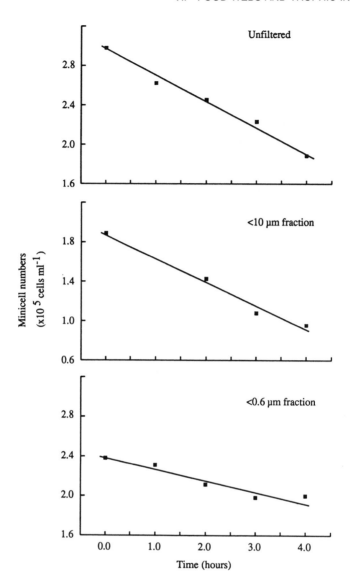

Figure 1. Change in minicell concentration over a 4-h incubation period in natural and size fractionated water samples. The least-squares linear regression lines had slope coefficient, 95% confidence interval, coefficient of determination and significance level as follows. Unfiltered sample: $-0.26 \pm 0.057 \times 10^5$ cells ml^{-1} h^{-1} ($r^2 = 0.985$, $p = 0.001$); <10 μm fraction: $-0.24 \pm 0.099 \times 10^5$ cells ml^{-1} h^{-1} ($r^2 = 0.983$, $p = 0.010$); and <0.6 μm fraction: $-0.11 \pm 0.064 \times 10^5$ cells ml^{-1} h^{-1} ($r^2 = 0.911$, $p = 0.012$). Bacterial concentration in the <0.6 μm fraction was a few percent of that in the unfiltered sample, resulting in low grazing rate of bacteria in this sample, despite significant turnover of minicells. Grazing in the <0.2 μm fraction was below detection limits (10^3 minicell ml^{-1} h^{-1}).

prey bacteria and minicells. Finally, potential selection for or against the minicell appear to be constant in different environments, as the community clearance rate (i.e., capture efficiency) shows low variability between water masses.[3]

Disappearance of minicells stained with 5-(4,6-dichlorotriazin-2-yl) aminofluorescein (DTAF), as monitored by microscopy, was employed by Pace et al.[9] to give estimates of total community grazing. It should be mentioned, however, the dye or staining procedure might change the surface properties of the minicell and thus selectivity. Additionally, longer

(>15 h) incubations had to be used to monitor significant changes in minicell abundance by this technique, as compared to when a genetic marker is used for quantification.

The DTAF-stained minicells were also used by Pace et al.[9] to estimate ingestion rates of groups of phagotrophs, by recording the increase in number of ingested minicells per protozoa with time.

Obviously, the minicell technique will not monitor bacterial mortality due to processes with specific requirements on surface properties, like viral or bacterial attack on bacterial prey.

Incorporation of Bacterial Biomass into Size Fractions

This method may be used in conjunction with the grazing-rate assay outlined above to record the rate of accumulation of the minicell label into size fractions. This will give an estimate of the rate of accumulation of bacterial biomass into different size fractions, give the approximate sizes of important bactivores, and estimate the transfer efficiency of bacterial biomass within the food web.[4]

1. Mix labeled minicells carefully in the water sample at a final concentration of 2×10^5 ml^{-1}. Preferably they are dispersed in a 100-ml volume of water sample prior to the addition of the rest of the sample.

2. At selected time intervals, pour 100 ml of sample into a glass bottle with formalin, for a final concentration of 4% formaldehyde.

3. Filter through a polycarbonate filter using only gravity or mild vacuum (<50 mmHg). I have used 1, 3, and 10-μm pore sizes. Actual pore size should be chosen according to the size structure of the microbial community in the sample and the experimental objective.

4. Fold the filter and put into scintillation vials and dry at 60°C for 1 h. Add scintillation liquid and determine amount of radioactivity with a scintillation counter.

5. Calculations:
   For incubations shorter than 4 h, incorporation of natural bacterial biomass (I_b) into a size fraction can be recalculated from amount of [^{35}S]-isotope incorporated, specific activity of TCA-precipitated minicells (SA_{tca}, dpm cell^{-1}), and the natural bacteria:minicell ratio according to:

$$I_b = (I_x/SA_{tca})(N_b/N_{mc}) \tag{3}$$

where I_x = amount of incorporated [^{35}S]-isotope per volume (dpm ml^{-1}); N_b = initial bacterial concentration (cells ml^{-1}); and N_{mc} = initial minicell concentration (cells ml^{-1}).

When following the incorporation of the [^{35}S]-isotope for longer time periods (>4 h), it is necessary to make consecutive adjustments of the minicell concentration, due to the grazing of minicells. Simultaneous measurement of the grazing rate of minicells is necessary, so that the actual minicell concentration N_{mc} can be used in Equation 3. Thus, N_{mc} will decrease during the experiment if grazing occurs, and I_b is affected accordingly.

It is assumed that the isotope used is representative of the minicell biomass, and that the minicell biomass is assimilated with a similar efficiency as that of natural bacteria. Calculating growth efficiency of bactivores from these types of data and simultaneous grazing rates, have given values close to growth efficiencies of flagellates found in laboratory experiments.[4]

ACKNOWLEDGMENTS

I am grateful to Professor Hagström for introducing me to aquatic microbial ecology, and for financial and scientific support during this project. Dr. W. Cochlan carefully read and provided valuable comments on the final version of the manuscript.

REFERENCES

1. Frazer, A. C. and Curtiss, R., Production, properties and utility of bacterial minicells, *Curr. Top. Microbiol. Immunol.*, 69, 1, 1975.
2. Wikner, J., Andersson, A., Normark, S., and Hagström, Å., Use of genetically marked minicells as a probe in measurement of predation on bacteria in aquatic environments, *Appl. Environ. Microbiol.*, 52, 4, 1986.
3. Wikner, J. and Hagström, Å., Annual study of bacterioplankton community dynamics, *Limnol. Oceanogr.*, 36, 1313, 1991.
4. Wikner, J., Rassoulzadegan, F., and Hagström, Å., Periodic bacterivore activity balances bacterial growth in the marine environment, *Limnol. Oceanogr.*, 35, 313, 1990.
5. Sambrook, J., Fritsch, E. F., and Maniatis, T., *Molecular Cloning: A Laboratory Manual, 2nd ed.*, Vol. 1, 2, and 3, Cold Spring Harbor Laboratory, Cold Spring Harbor, New York, 1989, 1625.
6. Kennedy, N., Beutin, N., and Achtman, M., Conjugation proteins encoded by the F sex factor, *Nature*, 270, 580, 1977.
7. Fenchel, T., Ecology of heterotrophic microflagellates. II. Bioenergetics and growth, *Mar. Ecol. Prog. Ser.*, 8, 225, 1982.
8. Andersson, A., Larsson, U., and Hagström, Å., Size-selective grazing by a microflagellate on pelagic bacteria, *Mar. Ecol. Prog. Ser.*, 33, 51, 1986.
9. Pace, M. L., McManus, G. B., and Findlay, S. E. G., Planktonic community structures determines the fate of bacterial production in a temperate lake, *Limnol. Oceanogr.*, 35, 795, 1990.

CHAPTER 82

Estimating Rates of Growth and Grazing Mortality of Phytoplankton by the Dilution Method

Michael R. Landry

INTRODUCTION

Dilution of natural microbial plankton communities with particle-free water from the same source can be used, under appropriate conditions, to estimate instantaneous rates of growth and grazing mortality of phytoplankton.[1,2] The basic principle of the approach is that an "observable" variable, the net growth rate of phytoplankton, should change as the density of primary consumers is manipulated by dilution, thereby reducing grazing pressure. The analytical framework for the method uses differences in measured net growth rates at different dilution levels to uncouple average growth and mortality rates over a defined incubation period.

The dilution method is based on the exponential model of population growth in which the net observed rate of growth (r) for a time interval (t) is the difference between instantaneous rates of population growth (μ) and mortality (m);

$$r = \ln (P_t/P_0)/t = \mu - m \tag{1}$$

where P_0 and P_t are, respectively, measured initial and time t estimates of population abundance. Since the rate of mortality, m (time^{-1}), can be expressed as the product of the individual clearance rate (F = volume swept clear of prey grazer^{-1} time^{-1}) and the time-averaged density of grazers ($\langle D \rangle$ = grazers volume^{-1}), Equation 1 can be rewritten as

$$r = \ln (P_t/P_0)/t = \mu - F \cdot \langle D \rangle \tag{2}$$

where

$$\langle D \rangle = (D_t - D_0)/(\ln D_t/D_0) \tag{3}$$

after Frost.[3] Thus, for constant μ and F (see below), net growth rate of phytoplankton varies linearly as a function of grazer density, with intercept μ and slope $-F$. In practice, one generally wants to know the rate of grazing mortality ($m_n = F \cdot D_n$) at the natural initial density of grazers ($D_{n,0}$); hence Equation 2 becomes

0-87371-564-0/93/$0.00 + $.50
© 1993 by Lewis Publishers

$$r = \mu - m_n \cdot \langle D \rangle / D_{n,0} \qquad (4)$$

Accordingly, grazing mortality is the slope of the relationship between net phytoplankton growth rate and grazer abundance normalized to natural density. This derivation is slightly different in terminology and detail from previous representations[1,2] in order to clarify where critical assumptions effect the analyses and to emphasize the importance of measuring changes in grazer density during dilution experiments. This latter point was not developed when the dilution approach was first proposed,[1] and its incorporation into subsequent analyses[2] appears to have been largely overlooked in practice, although it has received some attention from a theoretical perspective.[4] Most applications of the dilution approach mistakenly use the initial dilution factor, $D_0/D_{n,0}$, in lieu of $\langle D \rangle / D_{n,0}$ in Equation 4.

The preceding theoretical development involves three restrictive assumptions. First, the exponential model is assumed to apply in the broad sense. This allows that growth and mortality rates may vary on short time scales but provides a framework for computing average rates over incubation periods on the order of a day. Second, the mean specific growth rate (μ) of phytoplankton is assumed to be density independent (i.e., unaffected by dilution). To satisfy this assumption, dissolved nutrients must remain nonlimiting, or equally limiting, to growth at all dilutions during the experimental incubation. Third, the average clearance rate of individual microzooplankton is assumed to be constant at all dilutions. In addition to these three explicit assumptions, most users fail to measure or account for changes in the abundances of grazers during experimental incubations; hence, they implicitly assume that such changes are negiligible for all dilutions.

Assumptions two and three can be relaxed under certain conditions. Gallegos[4] extended the method to account for situations in which consumer grazing rates were saturated at ambient conditions and nonlinearly related to food concentration in diluted fractions. His elaboration of the dilution protocol, the "three-point" method, measures the growth rate of phytoplankton at very high levels of dilution and can be used to assess the functional relationship between microzooplankton grazing rate and phytoplankton concentration. Andersen et al.[5] developed the method to study the sources of nutrients (internal stores, external pools, and recycling) for phytoplankton growth. They dealt specifically with limiting nitrogen and phosphorus relative to incubations with excess nutrients, but their theory is applicable generally to nutrient-limited conditions.

The major advantages of the dilution approach are: (1) it provides rate estimates for both growth and grazing mortality of phytoplankton in a single experiment; (2) it involves minimal handling and physical disruption of the organisms (e.g., relative to size-fractionation); and (3) it allows simultaneous analysis of the dynamics of different components of the phytoplankton community.

Disadvantages of the method are: (1) it can require a considerable effort in setup and sample analysis to run one experiment successfully; (2) the assumptions are difficult to test routinely; (3) relative growth rates in the diluted samples may be differentially affected by contaminants or enhancements in the added filtered water; (4) "threshold" feeding at high dilutions could depress grazing rates in more dilute fractions below levels expected by dilution alone; (5) it requires relatively long incubations and may yield ambiguous or imprecise rate estimates; and (6) the required estimates of average grazer densities may be difficult or impossible to assess if different categories of potential consumers change in relative abundance during the incubation period.

PROCEDURES

Scale and Scope

There is no unique or optimal design for a dilution experiment. Procedures and equipment must be adapted to the system under study and the scope and scale of the problem as defined by the investigator. Relevant parameters include the relative richness of the system, tradeoffs in the desired level of precision of individual experimental results vs. replicated experiments, time and equipment constraints on volume of water filtered, and the sample volume required to assess population abundance according to the chosen analytical technique(s).

There are a number of ways to assess population abundances or concentrations of phytoplankton. Fluorometrically determined chlorophyll a, a standard method for measuring total phytoplankton concentration, has the advantages of being relatively simple and rapid, yet sensitive and precise. These qualities make chlorophyll an ideal variable for indexing initial dilution ratios (see section on subsampling and incubation strategies) and giving early indications of experimental results.[1,4-9] The major advantage of the dilution approach in studying the dynamics of communities of organisms is not fully realized, however, unless appropriate methods are used to distinguish component populations. High-pressure liquid chromatography (HPLC), for example, can be used to measure the relative abundances of major phytoplankton groups by their unique accessory pigments.[10,11] Microscopy or flow cytometry, on the other hand, can be used to separate populations numerically in terms of specific taxa, size, morphological characteristics, and autofluorescence properties.[2,12-14] In many ways, microscopic techniques are the most revealing means of assessing population changes in conjunction with dilution experiments, but they are also the most time consuming and least precise.

Experimental containers should be as large as practical to minimize surface area effects during the incubation and to accommodate all of the desired subsampling. HPLC analyses of oligotrophic ocean water, for instance, may set a minimum bottle size of 2 to 4 L. Polycarbonate bottles are recommended for most applications.[11] Large volume requirements reduce the effectiveness and generally preclude the use of dialysis sacs or diffusion cages to minimize differential effects of nutrient availability at the different dilution levels.[1,2] Where nutrient limitation is anticipated to be a problem, nutrients should be added equally to each bottle in an amount sufficient to support one or two doublings of phytoplankton biomass in the most concentrated sample.[5] When this is done, additional bottles of undiluted water without added nutrients should be run as controls.[1,5] Gifford[6] suggested that a fraction of the oligotrichous ciliates could be lost when nutrients are added, but did not provide evidence supporting this claim. While it is likely that very high levels could be detrimental to delicate organisms, it is not clear that reasonable additions within the range of natural concentrations will increase grazer mortality.

Only two dilution levels are required to solve Equation 4 for μ and m_n, the two unknowns.[2] However, most applications of the dilution method involve incubations of replicated experimental bottles containing four to six dilution levels. These are usually distributed more or less uniformly in the range from undiluted water to 5:1 or 9:1 mixtures of filtered to unfiltered seawater. In addition, Gallegos[4] has made a compelling case for using two highly diluted conditions (95 to 99% filtered water) to estimate phytoplankton growth rates when ambient phytoplankton concentrations are high enough to saturate grazer feeding rates. This is likely to be necessary for applications in eutrophic, coastal environments, but it is not viewed as essential or practical in the open ocean where phytoplankton concentrations are three to four orders of magnitude less than Gallegos' study site.[4]

Filtration Procedures

The objective of the filtration step is to produce, as efficiently as possible, a desired volume of water, free of the populations of interest and free of contaminants that might either enhance or depress growth rates relative to ambient water. The most demanding conditions are those in the open ocean where the dominant phytoplankton are extremely tiny and where any contamination of the filtered water may be significant relative to background levels of dissolved substrates. The recommended filters for this application are large (147 mm), 0.2-μm, "Triton-free" methyl cellulose filters (MicronSep Inc., Westboro, MA). Such filters have faster flow rates and higher retention efficiencies than comparably sized 0.2-μm polycarbonate membranes.[15-17] While they filter water slower than Whatman GF/F glass fiber filters or polycarbonate membranes with nominal pores ≥1.0 μm, they are less likely to break cells and elevate concentrations of dissolved substrates in the filtrate.[18] Where very large volumes of water must be filtered and where time is a more important constraint than cost, Gelman filter capsules may be an alternative to the use of flat filters. These pleated methyl cellulose filters have approximately an order of magnitude more filter surface area than the 147-mm filter circles and are sold (40 to $50 each) sterile in sealed polycarbonate containers, eliminating the need for and maintenance of separate filter holders.

Regardless of the filter type selected, it is essential that the entire filtration system be made from nontoxic plastic or Teflon-coated materials. Tubing should be silicone rubber.[19] Anything that will contact the water during the filtration operation (i.e., filters, filter holders, tubing, containers) should at least be well soaked with 10% HCl and thoroughly rinsed with Milli-Q (or equivalent) deionized water before use and protected in plastic bags between uses. Rigorous "clean" procedures require numerous cleaning and rinse steps,[20] the wearing of plastic gloves throughout the experimental setup, and a filtration system closed to the external environment. When large water (e.g., 30-L Go-Flo) bottles are used to collect water on shipboard, water can be gravity filtered directly from the bottle through the filter holder and into a carbuoy.

Preparation of the Dilution Series

The most important aspect of this phase of the experimental setup is to blend natural and filtered water to the desired dilution levels and to draw initial subsamples with minimal negative impact on the planktonic community. Ciliated protozoa are particularly susceptible to destruction by vigorous filling and mixing procedures.[6,21] The gentlest approach is to add measured volumes of filtered water to the experimental containers and then fill the remaining volume with freshly collected natural water. The free end of tubing used to transfer water from the collection device (e.g., water bottle) to the experimental container should at all times be submerged below the water level to minimize bubbling. An alternative procedure is to premix dilution levels in larger containers (e.g., carbuoys) before transferring to the smaller experimental containers. This procedure allows better replicability and easier subsampling of initial dilution levels (from the residual in mixing containers), and it is essential if the experimental containers hold a variable volume (e.g., nonrigid diffusion sacs or Whirl-Pak bags). The additional transfer step, however, increases that probability that delicate protozoans will be damaged.

Subsampling and Incubation Strategies

In order to determine rates of change of autotrophic populations and mean densities of grazer populations during dilution experiments, the dilution conditions must be subsampled at least at the beginning and end of experiments; intermediate time points may also be desirable under some circumstances (e.g., if growth and grazing rates are expected to be exceptionally high or to determine day-night differences in rates). There is technically no limit to the number or volume of initial subsamples that can be taken for each dilution level when (1) the subsamples are drawn from large mixing containers used to fill experimental bottles or (2) additional replicate experimental bottles are prepared and randomly sacrificed. It should be recognized, however, that relatively precise estimates of initial abundances are needed to minimize the error propagated through calculations of population rates of change. Generally, there are practical limits to the number of initial samples that one might want to process in order to estimate several variables of interest within acceptable levels of precision (e.g., <5% standard error). For population estimators requiring relatively costly or time-consuming analyses, e.g., HPLC pigments or microscopic counts, initial subsampling should focus (at least duplicate, generally triplicate) on abundances in the undiluted ambient water. Initial concentrations in the dilution series can be determined from these estimates of natural abundances and dilution ratios derived from (1) accurately measured mixing volumes or (2) a single variable, such as fluorometrically measured chlorophyll a, that can be subsampled from each bottle and quantified easily and precisely. Subsamples for all parameters of interest should be taken from each bottle at the end of the experiment. In addition, filtered water used to set up the experiment should be incubated with the experimental dilutions and subsampled at the beginning and end to account for differential growth of any organisms that may have passed through the filter.[17]

Experimental bottles should be incubated under conditions approaching, as closely as possible, those in the ambient environment. *In situ* incubations are optimal (essential for experiments involving diffusion containers) and the only way to ensure that the experimental containers experience natural temperature and light conditions. Where the bottom depth is shallow and the tidal amplitude low, bottles can be attached at appropriate depths to a taut line connecting a surface or subsurface float to an anchor.[2] For most deep water or open ocean work, *in situ* incubations involve attaching the bottles to a weighted line below a free-drifting surface float. Mesh bags of monofilament netting, attached to rings in the line above and below the depth of interest, work well for securing one to several bottles in energetic environments.[11] Individual bags of soft nylon mesh are effective in attaching and protecting dialysis tubing and diffusion chambers on *in situ* arrays.[1] When *in situ* incubations are not feasible, alternative water bath incubation systems should simulate ambient light quantity and spectral quality and control water temperature (within 1°C) to avoid significantly enhancing or retarding growth and grazing rates.

Analysis and Interpretation

Analyzed by standard linear regression techniques,[1,5] the relationship between observed growth rate of phytoplankton and relative grazer density should yield a negative slope corresponding to the magnitude of natural grazing mortality, m_n. In situations where excess nutrients are added to experimental bottles, the intercept of the regression line with the growth rate axis may or may not represent the true specific growth rate of phytoplankton in nature. Thus it is essential, to the extent that any closed-bottle incubation might provide results consistent with nature, to also have incubated undiluted samples without added nutrients. The specific growth rate of phytoplankton, calculated from these samples as the sum of the estimated grazing mortality and the net growth of phytoplankton without added nutrients, is interpreted as the rate of phytoplankton growth under nutrient-limited field conditions.[5]

The mere existence of a significant negative linear relationship in the analysis of a dilution experiment does not ensure that the results are correct. For instance, phytoplankton growth rates could be greater at the higher dilutions levels relative to less dilute mixtures due to differential nutrient availability.[1] In such circumstances, the slope of the regression relationship would be exaggerated and grazing mortality, m_n, overestimated. Conversely, underestimates (possibly even negative estimates if the slope of the regression is positive) of m_n could occur if the filtered water contained contaminants that negatively affected phytoplankton growth. Negative or nonlinear experimental results may also arise due to complex cycling of nutrients between internal and external pools unless nutrients are available in excess of phytoplankton needs.[5] In addition, if initial dilution levels, rather than the corrected relative density of grazers, is used as the independent variable in the regression analysis, grazing rate could either be significantly under- or overestimated depending, respectively, on whether grazers declined or increased during the incubation.[4]

For applications in environments with high phytoplankton concentrations, ingestion rates of consumers may be saturated at natural food concentrations, resulting in nonlinear relationships between net phytoplankton growth and grazer density which will lead, in turn, to underestimates of grazing mortality. In such cases, the three-point method of Gallegos[4] can be used to compute phytoplankton growth from two highly diluted samples. The underlying relationship between microzooplankton grazing and phytoplankton concentration can then be elucidated by using this growth rate estimate as a known term for computing grazing mortalities from the net growth rates observed at each dilution level.

Relation to Other Methods

Despite its restrictive assumptions and potential for misinterpretation, the dilution technique provides information about the dynamics of whole phytoplankton communities that other methods cannot. Selective metabolic inhibitors, for instance, are useful for separating growth and grazing rates on prokaryotic populations,[22-24] but are clearly limited in what they can do to resolve the interactions of eukaryotic phytoplankton and consumers. Campbell and Carpenter[12] demonstrated that dilution and inhibitor methods yield comparable estimates of grazing on the cyanobacteria *Synechococcus*; however, dilution was less sensitive, i.e., more likely to give statistically insignificant estimates, when *Synechococcus* densities were low. Neither dilution nor inhibitor approaches are capable of resolving grazing behaviors of specific microzooplankton taxa or of providing rate estimates on time scales as short as can be accomplished using fluorescently labeled prey.[25,26] On the other hand, given the great diversity in size and morphology of phytoplankton and the possibility that some grazers can distinguish nonliving analogs of prey from the real thing,[27] it would seem that labeled prey are best suited for studying the details of grazing on specific populations and impractical as a routine approach at the community level. There is no perfect method for estimating the grazing rates of microzooplankton under all conditions. That is why it is important to regard the above methods not as alternatives, but as complementary experimental techniques — each to be used in a manner consistent with its strengths and each testing, and hopefully supporting, the others in the areas in which they overlap.

Although the present paper only advocates dilution for studying growth and grazing mortality of phytoplankton (i.e., photosynthetic autotrophs), the method has been applied, with positive reports, to studies of heterotrophic bacteria.[2,13,28] The principle of the approach remains the same for this application. However, the interpretation of results is complicated by the fact that the source of energy for growth is not light, which can be provided equally at all dilution levels, but organic substrates provided indirectly by other organisms. Cell breakage during filtration can artificially enhance substrate levels at high dilutions,[18] and dilution itself may alter "nuturing" relationships among organisms in the microbial community.[29] These effects cannot be easily remedied by adding organic substrates in excess, because substrate availability dramatically affects bacterial size as well as growth rate[30] and because the microzooplankton grazing impact on bacteria is sensitive to prey size.[31,32]

REFERENCES

1. Landry, M. R. and Hassett, R. P., Estimating the grazing impact of marine micro-zooplankton, *Mar. Biol.,* 67, 283, 1982.
2. Landry, M. R., Haas, L. W., and Fagerness, V. L., Dynamics of microbial plankton communities: experiments in Kaneohe Bay, Hawaii, *Mar. Ecol. Prog. Ser.,* 16, 127, 1984.
3. Frost, B. W., Effects of size and concentration of food particles on the feeding behavior of the marine planktonic copepod *Calanus pacificus, Limnol. Oceanogr.,* 17, 805, 1972.
4. Gallegos, C. L., Microzooplankton grazing on phytoplankton in the Rhode River, Maryland: nonlinear feeding kinetics, *Mar. Ecol. Prog. Ser.,* 57, 23, 1989.
5. Andersen, T., Schartau, A. K. L., and Paasche, E., Quantifying external and internal nitrogen and phosphorous pools, as well as nitrogen and phosphorus supplied through remineralization, in coastal marine plankton by means of a dilution technique, *Mar. Ecol. Prog. Ser.,* 69, 67, 1991.
6. Gifford, D. J., Impact of grazing by microzooplankton in the Northwest Arm of Halifax Harbour, Nova Scotia, *Mar. Ecol. Prog. Ser.,* 47, 249, 1988.
7. Paranjape, M. A., Grazing by microzooplankton in the eastern Canadian arctic in summer 1983, *Mar. Ecol. Prog. Ser.,* 40, 239, 1987.
8. Paranjape, M. A., Microzooplankton herbivory on the Grand Bank (Newfoundland, Canada): a seasonal study, *Mar. Biol.,* 107, 321, 1990.
9. Taylor, G. T. and Haberstroh, P. R., Microzooplankton grazing and planktonic production in the Bransfield Strait observed during the RACER program, *Antarct. J.,* 23, 126, 1988.
10. Burkill, P. H., Mantoura, R. F. C., Llewellyn, C. A., and Owens, N. J. P., Microzooplankton grazing and selectivity of phytoplankton in coastal waters, *Mar. Biol.,* 93, 581, 1987.
11. Strom, S. L. and Welschmeyer, N. A., Pigment-specific rates of phytoplankton growth and microzooplankton grazing in the open subarctic Pacific Ocean, *Limnol. Oceanogr.,* 36, 50, 1991.
12. Campbell, L. and Carpenter, E. J., Estimating the grazing pressure of heterotrophic nanoplankton on *Synechococcus* spp. using the sea water dilution and selective inhibitor techniques, *Mar. Ecol. Prog. Ser.,* 33, 121, 1986.
13. Tremaine, S. C. and Mills, A. L., Tests of the critical assumptions of the dilution method for estimating bacterivory by microeucaryotes, *Appl. Environ. Microbiol.,* 53, 2914, 1987.
14. Weisse, T. and Scheffel-Möser, U., Growth and grazing loss rates in single-celled Phaeocystis sp. (Prymnesiophyscae), *Mar. Biol.,* 106, 153, 1990.
15. Li, W. K. W. and Dickie, P. M., Growth of bacteria in seawater filtered through 0.2 μm Nucleopore membranes: implications for dilution experiments, *Mar. Ecol. Prog. Ser.,* 26, 245, 1985.
16. Stockner, J. G., Klut, M. E., and Cochlan, W. P., Leaky filters: a warning to aquatic ecologists, *Can. J. Fish. Aquat. Sci.,* 47, 16, 1989.
17. Li, W. K. W., Particles in "particle-free" seawater: growth of ultraplankton and implication for dilution experiments, *Can. J. Fish. Aquat. Sci.,* 47, 1258, 1990.
18. Fuhrman, J. A. and Bell, T. M., Biological considerations in the measurement of dissolved free amino acids in seawater and implications for chemical and microbiological studies, *Mar. Ecol. Prog. Ser.,* 25, 13, 1985.
19. Price, N. M., Harrison, P. J., Landry, M. R., Azam, F., and Hall, K. J. F., Toxic effects of latex and Tygon tubing on marine phytoplankton, zooplankton and bacteria, *Mar. Ecol. Prog. Ser.,* 34, 41, 1986.

20. Fitzwater, S. E., Knauer, G. A., and Martin, J. H., Metal contamination and its effect on primary production measurements, *Limnol. Oceanogr.*, 27, 544, 1982.

21. Gifford, D. J., Laboratory culture of marine planktonic oligotrichs (Ciliophora, Oligotrichida), *Mar. Ecol. Prog. Ser.*, 23, 257, 1985.

22. Newell, S. Y., Sherr, B. F., Sherr, E. B., and Fallon, R. D., Bacterial response to presence of eukaryote inhibitors in water from a coastal marine environment, *Mar. Environ. Res.*, 10, 147, 1983.

23. Fuhrman, J. A. and McManus, G. B., Do bacteria-sized eukaryotes consume significant bacterial production?, *Science,* 224, 1257, 1984.

24. Sherr, B. F., Sherr, E. B., Andrew, T. L., Fallon, R. D., and Newell, S. Y., Trophic interactions between heterotrophic Protozoa and bacterioplankton in estuarine water analyzed with selective metabolic inhibitors, *Mar. Ecol. Prog. Ser.*, 32, 169, 1986.

25. Sherr, B. F., Sherr, E. B., and Fallon, R. D., Use of monodispersed, fluorescently-labeled bacteria to estimate in situ protozoan bacterivory, *Appl. Environ. Microbiol.*, 53, 958, 1987.

26. Rublee, P. A. and Gallegos, C. L., Use of fluorescently labelled algae (FLA) to estimate microzooplankton grazing, *Mar. Ecol. Prog. Ser.*, 51, 221, 1989.

27. Landry, M. R., Lehner-Fournier, J. M., Sundstrom, J. A., Fagerness, V. L., and Selph, K. E., Discrimination between living and heat-killed prey by a marine zooflagellate, *Paraphysomonas vestita, J. Exp. Mar. Biol. Ecol.*, 146, 139, 1991.

28. Ducklow, H. W. and Hill, S. M., The growth of heterotrophic bacteria in the surface waters of warm core rings, *Limnol. Oceanogr.*, 30, 239, 1985.

29. Sieburth, J. McN. and Davis, P. G., The role of heterotrophic nanoplankton in the grazing and nurturing of planktonic bacteria in the Sargasso and Caribbean Seas, *Ann. Inst. Oceanogr., Paris,* 58(Suppl.), 285, 1982.

30. Ammerman, J. W., Fuhrman, J. A., Hagstrom, Å., and Azam, F., Bacterioplankton growth in seawater. I. Growth kinetics and cellular characteristics in seawater cultures, *Mar. Ecol. Prog. Ser.*, 18, 31, 1984.

31. Gonzales, J. M., Sherr, E. B., and Sherr, B. F., Size-selective grazing on bacteria by natural assemblages of estuarine flagellates and ciliates, *Appl. Environ. Microbiol.*, 56, 583, 1990.

32. Monger, B. C. and Landry, M. R., Prey-size dependency of grazing by free-living marine flagellates, *Mar. Ecol. Prog. Ser.*, 74, 239, 1991.

CHAPTER 83

Consumption of Protozoa by Copepods Feeding on Natural Microplankton Assemblages

Dian J. Gifford

INTRODUCTION

Planktonic protozoa function as important trophic intermediaries in pelagic food webs by repackaging small bacterial and algal cells into food items which are accessible to larger consumers such as calanoid copepods. Early laboratory studies using cultured prey documented consumption of Protozoa by calanoid copepods[1,2] and demonstrated the potential importance of these prey in the diets of their consumers. More recent studies employing natural prey assemblages have demonstrated that calanoid copepods consume Protozoa under field, as well as laboratory, conditions and that they do so at rates and in quantities which are physiologically meaningful to the consumer organisms and demographically meaningful to the prey populations.[2-4]

The protocol described below was designed to measure clearance and ingestion rates of marine calanoid copepods on Protozoa by monitoring changes in natural assemblages of microplankton. The target prey organisms are ciliates and heterotrophic dinoflagellates; chlorophyll is measured to evaluate consumption of phytoplankton. The protocol can be adapted for use with freshwater copepods, and it can be used to measure feeding rates of other suspension feeding taxa, such as salps (D. J. Gifford, unpublished data). An obvious advantage of the method is simplicity of design and execution. The major disadvantage is that enumeration and measurement of prey organisms in the microplankton assemblages is extremely labor intensive.

MATERIALS REQUIRED

Equipment

- Dissecting microscope
- Inverted microscope
- Inverted microscope settling columns and counting chambers
- Temperature-controlled plankton wheel. The rotating wheel incubator keeps prey cells suspended homogeneously in the incubation bottles, preventing measurement of spuriously high feeding rates which may occur if the copepods feed on concentrated patches of prey (it should be noted, however, that copepods may well feed on patches of food in nature): a

number of possible designs exist for rotating plankton wheels; for example, they may be electrically powered or water-driven and the devices have in common that they are inevitably custom designed and built.

- Fluorometer
- Filtration manifold
- Plankton net of appropriate mesh

Supplies

- Silicon tubing
- Surface bucket, Go-Flo or Niskin bottles
- Polycarbonate carbuoys
- Parafilm
- Neutral density screening
- Polycarbonate bottles
- Sample bottles
- GF/F Filters
- Forceps
- Fluorometer tubes
- Graduated cylinders (250 to 500 ml)
- Pasteur pipettes
- Conical centrifuge tubes

Solutions

- Fixative (see Notes and Comments section)

PROCEDURES

Field Collection of Microplankton

The natural microplankton assemblage is collected from the environment of choice using clean methods. How "clean" the collection conditions are is dictated by the specific environment: for example, estuarine forms should not be as susceptible to metal, or other toxic contamination as are oceanic forms. A rigorous metal-free cleaning protocol is described by Fitzwater et al.[5] This cleaning method is optimal for seawater collected from oceanic environments, and may be applied less rigorously for collection from nearshore and estuarine environments.

To collect bulk seawater, a clean plastic surface bucket may suffice. It is important to note that even brief exposure to the standard latex tubing supplied on Niskin bottles is extremely toxic to marine organisms.[6] Hence, in oceanic environments, clean Teflon-coated Go-Flo bottles, which do not contact the surface film of the water and do not contain interior metal parts, are preferred. If these are not available, Niskin bottles can be re-rigged with silicon tubing and O-rings and Teflon-coated springs, then cleaned. If Go-Flo or Niskin bottles are used, the water is drained from them through silicon tubing into clean polycarbonate carbuoys. Turbulence destroys some protozoans and should be avoided while draining; the siphon tube should contact the bottom of the receiving vessel so that water is transferred into water rather than into air. If large grazers that might prey on the protozoans are present, the seawater may be siphoned gently through a submerged 200-μm mesh to remove them, although some losses of aloricate ciliates will occur.

Field Collection of Copepods

1. Collect copepods by vertical or horizontal tows of a plankton net. Undamaged copepods, having intact appendages and antennae, are collected using a net with as small a mesh as possible. The small mesh apparently does not abrade the animals, as do larger meshes (C. B. Miller, unpublished observation).

2. Sort copepods under a dissecting microscope.

3. Acclimate copepods to the experimental conditions (bottles) for 24 to 48 h by incubating copepods in polycarbonate bottles containing the natural microplankton assemblage at *in situ* environmental temperature. The bottles are topped up with microplankton assemblage, covered with parafilm to exclude airspace, capped tightly, wrapped in neutral density screening to simulate *in situ* light intensity, and oriented along their vertical axes on a plankton wheel which is rotated slowly. If the experiment is incubated on the deck of a research vessel, temperature is controlled by running seawater through the incubator. If the experiment is done in the laboratory, a temperature-controlled space is needed.

Experiments

Before beginning these experiments, it is helpful to know (1) the rate at which the copepods clear phytoplankton, easily measured by following the disappearance of chlorophyll over 12 to 24 h and (2) the size structure of the microplankton (phytoplankton + microplankton) assemblage. Each experiment is specific to the copepod consumer and the taxonomic and size distribution of the natural microplankton assemblage. As a general rule, suspension-feeding copepods clear large particles at rates higher than those at which they clear small particles. Hence, if the natural assemblage consists primarily of large diatoms, such items are likely to be the primary food consumed; whereas if the natural assemblage consists primarily of small plant cells and larger protozoans, the protozoans are likely to be the primary food consumed.

Light regime is dictated by the behavior of the specific copepods used and by available experimental facilities. Because many copepods feed on a diel basis, the experimental duration should encompass the animals' natural hours of darkness. The entire experiment may be done under conditions of darkness, or ambient light conditions may be used. The latter is a feasible option if the experiment is done in an on-deck incubator. In this case, the bottles and/or incubator are screened with neutral density mesh to simulate *in situ* light intensity.

1. Experimental design — Experiments utilize two treatments: an experimental treatment consisting of the microplankton assemblage incubated in bottles with copepods, and a control treatment in which bottles contain only the microplankton assemblage. Because the microplankton assemblage is incubated without an airspace in order to avoid destructive losses of protozoans, bottles cannot be sampled more than once. Instead, duplicate control bottles are set up, half to be harvested at the beginning of the experiment, the remainder upon termination. Experimental bottles are harvested only upon termination of the experiment. The number of replicates of each treatment is determined by space on the rotating plankton wheels and by the labor required to process the microplankton samples under the inverted microscope (see below). The treatments are set up in polycarbonate bottles. The bottle volume employed depends on the size of the copepods under study. For example, 500-ml bottles are suitable for small copepods such as *Acartia* spp., while 2-L bottles are appropriate for large *Neocalanus* spp.[4] Whatever bottle volume is used, it is necessary to have sufficient volume to count several hundred of the most abundant protozoans in the microplankton assemblage.

2. Collect the microplankton assemblage immediately before setting up the experiment.

3. Mix microplankton assemblage gently with a plastic paddle and siphon it gently through silicon tubing into the experimental vessels, filling the bottles to about 90% of their volume.

4. Add the acclimated copepods to the experimental treatment replicates.

5. Top up all bottles with microplankton assemblage, seal with parafilm, cap as described above, and place on a slowly rotating (~0.5 rpm) plankton wheel.

6. Incubate all bottles on the plankton wheel for 1 to 2 h at ambient environmental temperature, then harvest the initial control treatments. In my experience, most losses of Protozoa due to handling occur immediately; after a period of stabilization, better between-bottle replication is achieved. Collect microplankton and chlorophyll samples from the initial control treatments. A number of samples can be collected from each bottle, depending on the goals of the particular experiment. Typically, the bottles are sampled for microplankton and chlorophyll. The microplankton samples are collected by gently siphoning a subsample of appropriate volume (typically 250 to 500 ml) into a bottle containing fixative (see Notes and Comments section). Chlorophyll samples are collected similarly and analyzed by fluorometry.[7]

7. Harvest the final control treatments and the experimental treatments after 24 h, collecting microplankton and chlorophyll samples.

8. Concentrate the microplankton samples prior to counting. The volume concentrated depends on the abundances of Protozoa in the sample: a minimum of 100 to 200 of the most common forms will eventually be enumerated. If the volume of sample to be settled is 100 ml, settling columns are used. If the volume is greater than 100 ml, I concentrate the sample by the following method.

   - The well-mixed sample is placed in a graduated cylinder and settled for 4 h per cm of chamber height.[8]

   - The top ~75% of fluid is siphoned off through fine-bore tubing.

   - The remaining volume (25 to 50 ml) is transferred into a conical centrifuge tube and settled, as above.

   - The top fluid is removed with a Pasteur pipette and discarded, so that a final volume of 1 to 2 ml remains in the tube.

   - The fluid is pipetted into an inverted microscope counting chamber.

   - The centrifuge tube is rinsed with 1 ml of filtered seawater, which is also pipetted into the counting chamber.

   - The chamber volume is topped up with filtered seawater, and a cover slip placed over it.

9. The sample is enumerated using an inverted microscope at a magnification of 100 to 250×. The contents of the entire chamber are enumerated.

The usual target prey items of the experiment are ciliate and dinoflagellate microzooplankton. If desired, phytoplankton taxa can also be enumerated. Natural microplankton assemblages are generally diverse with respect to species composition and size of phytoplankton and microzooplankton. It is difficult, if not impossible, to identify aloricate ciliates to species without resorting to protargol staining, except in taxa with distinct morphology, such as Laboea or Tontonia. Hence, ciliates are classified on the basis of geometry: spheres, cones, ice cream cones, oblate spheroids, etc. Linear dimensions for calculation of geometric volumes are measured using a calibrated ocular micrometer.[9]

As noted above, the success of the method depends on separating the copepod feeding signal from the counting variation of the microplankton. Statistically significant changes in abundance are resolved only in prey categories containing abundant cells, i.e., in which 100 to 200 cells are counted.[10] If copepod clearance, but not ingestion, rates are required, these prey categories alone are enumerated and measured. If ingestion rates are needed, all protozoan categories are enumerated and measured.

Data Analysis and Interpretation

Carbon Content

Protozoan carbon content is calculated from empirically derived, volume-specific relationships for particular taxa.[11,12] The question of protozoan cell shrinkage resulting from fixation becomes critical at this point.[11] If a fixative that shrinks the cells has been used, the carbon-to-volume calculation may be corrected by the appropriate factor. It should be noted that the Putt and Stoecker[11] conversion factor corrects for shrinkage in 2% acid Lugol's solution. Alternatively, one may not correct for shrinkage and consider the calculated carbon content to be conservative.

Calculation of Copepod Clearance and Ingestion Rates

Per capita rates of clearance and ingestion of the consumer organisms are calculated using a modification of Frost's[13] grazing equations. If only clearance rate is needed, it is calculated from changes in the abundances of only the most abundant prey categories: a rate is calculated for each separate prey category and the mean taken.

If ingestion rates are needed, one makes the simplifying assumption that all prey greater than a certain size are cleared at equal rates.[14-16] The mean clearance rate, F, calculated for the most abundant prey categories is multiplied by the mean prey concentration, $\langle C \rangle$, of all prey categories in the control treatments to give total ingestion.

The prey growth coefficient, k, is calculated by

$$C_2 = C_1 \cdot e^{k(t_2 - t_1)} \tag{1}$$

where C_1 and C_2 are prey concentrations in the control treatments at the beginning (t_1) and end (t_2) of the experiment.

The grazing coefficient, g, is calculated by

$$C_2^* = C_1 \cdot e^{(k-g)(t_2 - t_1)} \tag{2}$$

where C_2^* is the prey concentration in the experimental treatment at the end of the experiment.

The mean prey concentration, $\langle C \rangle$, is calculated by

$$\langle C \rangle = C_1 \, (e^{(k-g)(t_2 - t_1)} - 1)/(t_2 - t_1)(k - g) \tag{3}$$

Clearance rate, F, is calculated by

$$F = Vg/N \tag{4}$$

where V is the volume of the incubation bottle and N is the number of copepods in the bottle.

Ingestion rate, I is calculated by

$$I = F \cdot \langle C \rangle \tag{5}$$

The units of k and g are inverse time, $\langle C \rangle$ may be expressed as numbers or biomass (carbon), F is volume cleared/copepod/time, and I is numbers or biomass consumed/copepod/time.

NOTES AND COMMENTS

1.   The number of copepods added to experimental treatments is determined by the need to separate their feeding signal from the counting variation in the microplankton samples. Because the coefficient of variation of microplankton enumerated in settled samples is on the order of ~20%, the copepods must clear more than this fraction of the bottle volume if a feeding signal is to be discerned. As a practical matter, I attempt to set up experiments so that ~40% of the bottle volume is cleared during an experiment.

2.   Choice of fixatives. These experiments require that absolute numerical abundances of the protozoan prey are measured and that the volumes of the prey items be estimated accurately in order to calculate their carbon content. Unfortunately, there is no perfect fixative. Acid Lugol's solution[17] at a concentration of 10% (v/v) or greater appears to preserve all ciliates (D. J. Gifford, unpublished observation; D. K. Stoecker, unpublished observation) but shrinks most ciliate cells by a factor of ~50% (D. K. Stoecker, unpublished observation). Bouins fixative[18] also appears to preserve virtually all ciliates but also shrinks cell volumes to the same degree as acid Lugol's solution (D. K. Stoecker, unpublished observations). Formaldehyde, whether buffered or not, destroys significant numbers of aloricate ciliates (D. J. Gifford, unpublished observations; D. K. Stoecker, unpublished observations). Such losses can be high (20 to 40%), but are not consistent, and thus, cannot be corrected easily. Because of this, I recommend choosing a fixative which preserves the absolute numbers of ciliates and correcting for shrinkage when calculating carbon content. Acid Lugol's solution and Bouins fixative have the disadvantage that they destroy chlorophyll fluorescence and cannot be used to preserve samples for examination by epifluorescence microscopy. Thus, ciliates and dinoflagellates which are strictly autotrophic, strictly heterotrophic, or mixotrophic cannot be identified unambiguously. If knowledge of the trophic status of the copepod prey is required, a second set of samples can be preserved with an aldehyde (which will involve significant losses of ciliates and athecate dinoflagellates) for examination by epifluorescence microscopy. Alternatively, forms known to hetertrophic (e.g., *Protoperidinium* spp.), autotrophic (e.g., *Myrionecta rubra* (ex *Mesodinium rubrum*), or mixotrophic (e.g., *Laboea strobila; Tontonia* spp.) are enumerated as such in the acid Lugol's samples. Fixatives are discussed further in Chapter 25.

3.   Replication. Enumeration and measurement of protozoan prey using the inverted microscope is extremely laborious. In general, better results are obtained by running fewer replicates and counting more cells in each treatment.

ACKNOWLEDGMENTS

Manuscript preparation was supported by grant number N00014-90-J-1437 from the U.S. Office of Naval Research.

REFERENCES

1. Stoecker, D. K. and Capuzzo, J. M., Predation on Protozoa: its importance to zooplankton, *J. Plankton Res.*, 12, 1990.
2. Gifford, D. J., The protozoan-metazoan trophic link in pelagic ecosystems, *J. Protozool.*, 38, 81, 1991.
3. Gifford, D. J. and Dagg, M. J., Feeding of the estuarine copepod *Acartia tonsa* Dana: carnivory vs. herbivory in natural microplankton assemblages, *Bull. Mar. Sci.*, 43, 458, 1988.
4. Gifford, D. J. and Dagg, M. J., The microzooplankton-mesozooplankton link: consumption of planktonic Protozoa by the calanoid copepods *Acartia tonsa* Dana and *Neocalanus plumchrus* Murukawa, *Mar. Microb. Food Webs*, 5, 161, 1991.
5. Fitzwater, S. E., Knauer, G. A., and Martin, J. H., Metal contamination and its effect on primary production, *Limnol. Oceanogr.*, 27, 544, 1982.
6. Price, N. M., Harrison, P. J., Landry, M. R., Azam, F., and Hall, K. J. F., Toxic effects of latex and tygon tubing on marine phytoplankton, zooplankton and bacteria, *Mar. Ecol. Progr. Ser.*, 34, 41, 1986.
7. Strickland, J. D. H. and Parsons, T. R., *A Practical Handbook of Seawater Analysis*, National Research Council of Canada, Ottawa, Canada, 1972.
8. Hasle, G. R., Using the inverted microscope, *Phytoplankton Manual*, UNESCO, Paris, 1978, 191.
9. Beers, J. R., Reid, F. M. H., and Stewart, G. L., Microplankton of the North Pacific central gyre. Population structure and abundance, June, 1973. *Int. Revue Ges. Hydrobiol.*, 60, 607, 1975.
10. Venrick, E. L., How many cells to count?, *Phytoplankton Manual*, UNESCO, Paris, 1978, 167.
11. Putt, M. and Stoecker, D. K., An experimentally determined carbon:volume ratio for marine "oligotrichous" ciliates from estuarine and coastal waters, *Limnol. Oceanogr.*, 34, 1097, 1989.
12. Lessard, E. J., The trophic role of heterotrophic dinoflagellates in diverse marine environments, *Mar. Microb. Food Webs*, 5, 49, 1991.
13. Frost, B. W., Effects of size and concentration of food particles on the feeding behavior of the marine planktonic copepod *Calanus pacificus, Limnol. Oceanogr.*, 17, 805, 1972.
14. Nival, P. and Nival, S., Particle retention efficiencies of an herbivorous copepod, *Acartia clausi* (adult and copepodite): effects on grazing, *Limnol. Oceanogr.*, 21, 24, 1976.
15. Bartram, W. C., Experimental development of a model for feeding of neritic copepods on phytoplankton, *J. Plankton Res.*, 3, 25, 1980.
16. Frost, B. W., Landry, M. R., and Hassett, R. P., Feeding behavior of the large calanoid copepods *Neocalanus plumchrus* and *Neocalanus cristatus* from the subarctic Pacific Ocean, *Deep-Sea Res.*, 30, 1, 1981.
17. Throndsen, J., Preservation and storage, *Phytoplankton Manual*, UNESCO, Paris, 1978, 69.
18. Lee, J. J., Small, E. B., Lynn, D. H., and Bovee, E. C., Some techniques for collecting, cultivating, and observing protozoa, *Illustrated Guide to the Protozoa*, Society of Protozoologists, Lawrence, Kansas, 1985, 1.

CHAPTER 84

Predation on Planktonic Protists Assessed by Immunochemical Assays

Mark D. Ohman

INTRODUCTION

Measuring the sources and rates of predation on planktonic protists is of interest from a number of perspectives. In studies of protist population dynamics, mortality rates as well as division rates need to be assessed.[1] From a predator's perspective, the abundance (and specific composition of assemblages) of planktonic protists can markedly influence a predator's growth rate. From the point of view of ecosystem studies, the intensity and selectivity of predation on planktonic protists can alter the pathways of material and energy transfer in pelagic food webs.

The present method was developed to measure predation on ciliate protists by planktonic suspension-feeding copepods. This prey-predator interaction is addressed because of the suspected importance of ciliates as a prey resource for planktonic metazoans[2,3,4,5] and the likelihood that ciliates are significant intermediaries in the transfer of microbial loop production to larger suspension feeders. The need for new methods to study predation on planktonic ciliates arises because many ciliates are quite fragile and lack recognizable hard parts. Conventional methods for collecting, concentrating, incubating, and fixing planktonic organisms frequently result in lysis of these cells.[6] This is particularly true for ciliates in the order Oligotrichida and those members of the Choreotrichida that lack a lorica. Despite the particular focus of the method outlined here, the principles of this immunochemical method should also be applicable to other protist-predator studies.

The principle of this method is that antibodies raised against ciliate antigens can be used as a probe to detect the presence of ciliate antigens from intact cells or the gut contents or fecal pellets of a predator. The reaction between antigen and antibody (Ag-Ab) can be visualized in a number of ways. The procedure presented here employs dot blots, using an enzyme-linked conjugate to visualize the intensity of the reaction. In a dot blot procedure the antigen solution (e.g., gut contents of a predator) is applied (blotted) onto nitrocellulose membranes. The immobilized antigen is reacted with a primary antibody (which recognizes the target Ag), then with a secondary antibody (which recognizes the primary Ab) conjugated to a reporter molecule (in this case the enzyme alkaline phosphatase). The Ag-Ab reaction is then visualized by allowing the dot blot to react with a chromogenic substrate solution.

The advantages of this immunochemical method include: (1) *in situ* predation on ciliates can be measured; (2) there is no need to collect, concentrate, incubate, or preserve ciliates, any of which can result in cell lysis and lead to experimental artifacts; (3) immunochemical probes can be developed with different levels of specificity, depending on the question being addressed; and (4) individual predators can be probed, rather than relying on ensemble measures of ingestion.

Disadvantages of the method include: (1) the targeted prey organisms usually must be cultured for production of antibodies; (2) crossreactions must be screened carefully in advance; and (3) the predator digestive tract may have to be dissected and removed to minimize crossreactions between the antibodies and predator tissues.

MATERIALS REQUIRED

Equipment

- Orbital shaker
- Microliter pipette
- Reflectance densitometer
- Dot blot manifold (optional)

Solutions

- Nitrocellulose (0.45 μm; from Schleicher and Schuell)
- Blotto (concentration of merthiolate increased over that in Reference 17):
    - 5% (w/v) nonfat dry milk prepared in buffer (e.g., phosphate-buffered saline, PBS; pH 7.4).
    - 0.01% (v/v) Antifoam A
    - 0.00025% (w/v) Merthiolate
- Primary antibody diluted in Blotto
- Secondary antibody (e.g., alkaline phosphatase conjugated goat anti-rabbit IgG) diluted 1:1,000 in Blotto
- 3% (v/v) normal goat serum, in Blotto
- Stock stain solutions:[18]
    5-bromo-4-chloro-3-indolyl phosphate (BCIP), *p*-toluidine salt (5 mg/ml in dimethylformamide); nitro blue tetrazolium (NBT) (1 mg/ml in either Veronal-acetate or 0.15 M Tris buffer, pH 9.6); and 2 M MgCl$_2$.

PROCEDURES

These procedures assume that the objective is to produce a probe consisting of polyclonal antisera. In predation studies, polyclonal antisera can be advantageous because detection of the presence of prey remains does not depend upon the immunoreactivity of a single epitope. For production of monoclonal antibodies consult the current literature. For the basic elements of immunochemistry, including antibody production and antiserum purification, see Warr[7,8] and Johnstone and Thorpe.[9] Only those aspects unique to the application used here are specified in any detail.

Antigen Preparation

Generally it will be necessary to culture ciliates or other protists to obtain sufficient monospecific antigen for antibody production. As noted below, care must be taken when preparing ciliate antigens for immunization. The specificity of the antisera will be influenced by procedures at this step. In particular:

- Avoid contamination by bacteria and other microorganisms by use of monoxenic cultures and harvesting techniques that retain only the ciliates of interest.

- Avoid inclusion of particulate debris egested by ciliates by harvesting only from fresh cultures.

- Maintain antigen extracts on ice to minimize proteolytic activity; if protease inhibitors are used, ensure that they are used at concentrations that will not adversely affect the host organism producing antibodies (e.g., 1 mM phenylmethyl sulfonyl fluoride, 2 mM EDTA used in Reference 4).

Formaldehyde-fixed cells are commonly used for antibody production. However, some aloricate ciliates do not survive fixation and preservation.[6] In such cases extracts of live or frozen cells should be used as the immunogen.

Particular care must be taken in concentrating and harvesting aloricate ciliates for antibody production because of their fragility. The following procedure has worked effectively in the author's laboratory.

1. Allow cultured ciliates to reduce their prey to background levels.

2. For bacterivorous taxa, harvest ciliates separately from bacteria by differential centrifugation (e.g., 270 $\times$ g for 15 min for *Strombidium*, 1000 $\times$ g for 20 min for *Uronema*, at 4°C).

3. Discard the supernatant and quickly resuspend cells in buffer (e.g., PBS). (For herbivorous or carnivorous ciliates use other methods to concentrate ciliates and to separate them from their prey. Alternatives include gentle filtration, reverse flow filtration, centrifugation, pipetting, or perhaps chemotaxis.)

Often it will be useful to characterize the molecular mass of antigens by sodium dodecyl sulfate-polyacrylamide gel electrophoresis (SDS-PAGE) and to characterize the specific antigens recognized by the antibodies on Western blots.[9]

Antibody Production

Several different immunization protocols and mammalian hosts can be used for antibody production.[9] The approach of Vaitukaitis[10] requires small quantities of antigen, injected intradermally at multiple sites in New Zealand white rabbits. It has been used successfully to produce antibodies to yolk protein from larvae of the northern anchovy.[11] A modified version of the protocol permitted Ohman et al.[4] to produce sensitive antisera using just 90 μg of protein extracted from oligotrich ciliates as the immunogen. In the protocol used by Ohman et al.,[4] 6 to 8 intradermal injections were made in Freund's complete adjuvant on day 0. This was followed by multiple intramuscular booster injections using Freund's incomplete adjuvant on day 21. Additional booster injections followed at 14-d intervals, until the antiserum titer was acceptably high.

It is advisable to have adequate antigenic material available for titer checks and for calibrating the immunoassays. Two or more rabbits should be immunized with any immunization protocol to allow for genetic and physiological differences in the animals' immune response to particular antigens. Pre-immune serum should be withdrawn to serve in control assays. The titer of test bleeds and the specificity of the antibodies can be assessed by dot blot assays (below).

Purification of Antibodies

The resulting crude antisera can be used for immunoassays, but purification can lead to greater specificity. Also, it is sometimes desirable to remove the targeted immunoglobulins (usually IgG) from serum albumin, proteases, other proteins, carbohydrates, and other serum constituents. A variety of techniques can be used to accomplish this purification, including ion exchange chromatography, high-performance liquid chromatography (HPLC), and ligand affinity columns (see References 9, 12, and 13). Perhaps the simplest is the use of Protein A or Protein G affinity columns. Specific purifications may be required for special applications or where undesired crossreactions occur.

Immunoassay

The reaction between antibodies and the prey antigens from a predator's gut contents is visualized by dot blots. General discussions of dot blotting techniques can be found in References 14, 15, and 16. Specific applications can be found in References 4 and 11. The procedure entails, in order: homogenization of gut contents in a buffer and application onto nitrocellulose; reaction with primary antibody; reaction with secondary antibody conjugated to alkaline phosphatase; incubation with chromogenic substrate; and quantification of the reaction intensity.

An additional dot blot series should be incubated with pre-immune serum as a control for false positives. This is particularly important when assaying antibody titer and testing for crossreactions. If the reaction is to be quantified by densitometry, a standard of known reactivity with the primary antibody should be blotted along with each batch of samples to ensure consistent reaction intensity. The original immunogen serves well for this purpose.

1. Immerse nitrocellulose in distilled water for 5 min and air dry.
2. Blot gut contents onto nitrocellulose and air dry.
3. Block with Blotto for 30 min.
4. Bind primary Ab for 60 min.
5. Wash with Blotto for 10 min ($2 \times$).
6. Block with 3% normal goat serum for 20 min.
7. Bind secondary Ab for 60 min.
8. Wash with Blotto for 10 min ($2 \times$).
9. Add alkaline phosphatase substrate for 20 min. (From stock solutions combine immediately before use in the following proportions: 9.0 ml buffer, 1.0 ml NBT, 0.10 ml BCIP, and 0.02 ml $MgCl_2$.)
10. Stop reaction with deionized or distilled water rinse ($3 \times$).
11. Record results. A qualitative scale can be established for the intensity of precipitated dye product.
12. Air dry, then read with reflection densitometer.

Quantification

The presence or absence of a significant reaction is sufficient information for some predator-prey studies. In this case, dot blot intensity can be evaluated visually and compared with the reaction blank obtained from assay of the gut contents of starved predators. The detection limit can be determined from serial dilution and dot blot assay of known quantities of antigen.

To quantify the amount of antigenic material in a predator's gut and estimate ingestion rates further requires: (1) calibration of densitometric measures of the intensity of dot blots;

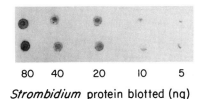

80   40   20   10   5

Strombidium protein blotted (ng)

Figure 1. Dot blots of extracts of the ciliate *Strombidium* sp. reacted with anti-*Strombidium* serum. Serial dilutions (left to right) were blotted in duplicate (top and bottom rows).

(2) knowledge of changes in immunoreactivity of antigens with time during digestion; and (3) knowledge of the gut residence time of antigenic material. Changes in immunoreactivity (2) and gut passage (3) may or may not occur simultaneously. If early digestion alters epitopes sufficiently to change the immunoreactivity of the ingested prey prior to complete assimilation, then the timing of these two processes needs to be resolved separately. Feller and Ferguson[19] point out additional concerns when quantifying immunochemical estimates of predator gut contents.

The intensity of the dot blot is proportional to the amount of antigenic material present, provided antibodies and chromogenic substrate are available in excess. With an appropriate calibration curve, the reaction intensity can be expressed in terms of the quantity of the original antigen (e.g., mass of protein if the primary antigenic substance is proteinaceous). A calibration curve is prepared by serial dilution of a known antigen which is blotted onto nitrocellulose and taken through the assay procedure in the immunoassay section (Figure 1). When the reaction is read by reflectance densitometry the response is typically curvilinear (rectilinear on a semi-log plot, Figure 2). Because of this log-linear relation, the assay must be calibrated carefully over the range of antigenic material found in samples. Dot blots should be uniform in shape and surface area for densitometry; blotting templates or manifolds are helpful for this purpose.

Areas for Further Development

The Ag-Ab reaction can be read in a number of ways. The primary antibody can be reacted with a secondary antibody coupled to a variety of reporter molecules (enzymes,[20] fluorochromes, immunogold, or radiotracers). For some purposes different reporters may be more advantageous than that presented here. For example, immunofluorescence may be useful for visualizing the sites of prey assimilation in a predator's digestive tract, or for enumerating intact ciliate cells by flow cytometry. Further enhancement of the sensitivity of the dot blot assay may come from the use of chemiluminescence (rather than histochemical stains) to visualize the reaction.

The reaction can also be carried out in solution by ELISA (enzyme-linked immunosorbent assay) techniques rather than in a solid phase bound to nitrocellulose. ELISAs have the advantage of greater sensitivity than dot blots and the availability of automated procedures for plate washing and spectrophotometric or fluorometric reading of results. However, care must be taken to select microtiter plates that bind the antigens (or antibodies) of interest with high affinity. Two advantages of the solid-phase dot blot approach over ELISA are the excellent protein-binding characteristics of nitrocellulose and the permanent record of the dot blot that is obtained.

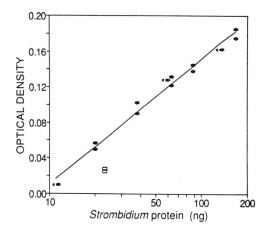

Figure 2. Relationship between the optical density of immunochemical dot blots and the quantity of ciliate protein. Extracts of the ciliate *Strombidium* sp. were reacted with anti-*Strombidium* serum that had been immunoadsorbed against extracts of larval anchovy to eliminate cross-reacting antibodies (see text). The two points indicated by open squares were excluded from the regression. (Reprinted from Ohman, M. D., Theilacker, G. H., and Kaupp, S. E., *Biol. Bull.*, 181, 500, 1991. With permission of *The Biological Bulletin*.)

NOTES AND COMMENTS

1. Immunoassay. Assay can be done at 21 or 37°C, but the temperature must be consistent for all assays. Incubation times should be optimized for the temperature and antibodies employed. Agitate on orbital shaker throughout incubations. Use a dot blot template if applying numerous samples to a single piece of nitrocellulose. Blot samples in duplicate or triplicate, provided sufficient gut material is available. Control dot blots should be carried out using pre-immune serum as a control for false positives.

2. Cross reactions. It is essential to screen antisera carefully for crossreactivity with non-homologous antigens prior to use in predation studies. Predator tissues should also be screened for crossreactions with the antibodies. To minimize such reactions, the gut contents of the predator can be dissected away from predator tissues onto a microscope slide, homogenized in a few microliters of a suitable buffer, and blotted onto nitrocellulose. Undesired crossreactions can sometimes be eliminated. For example, Ohman et al.[4] used an affinity column to eliminate a strong, unexpected crossreaction between the antiserum raised against ciliates and extracts of teleost larvae. This procedure entailed, first, binding soluble extracts of fish larvae to Sepharose and packing the beads into a microaffinity column. The antiserum was then immunoadsorbed against the immobilized larval extracts until the eluted antiserum had virtually no remaining crossreactivity with the teleost tissues. The antiserum used for the calibration relation in Figure 2 had been immunoadsorbed in this manner.

3. Ethical concerns. Alternatives to the production of antibodies in mammalian hosts are being discussed,[21] but these are not practical options at present. Since mammalian hosts remain necessary, the activity should not be carried out frivolously. The user should ensure that an immunochemical approach is the best method available for the question at hand prior to the immunization of experimental animals. Also, proper facilities and protocols are needed for animal care. If these facilities are not available, it is possible to have custom antibodies produced by an experienced commercial laboratory.

ACKNOWLEDGMENTS

I wish to thank G. H. Theilacker, R. J. Feller, and an anonymous referee for comments on the manuscript. Supported by NSF OCE-9019639.

REFERENCES

1. Ohman, M. D. and Snyder, R. A., Growth kinetics of the omnivorous oligotrich ciliate *Strombidium* sp., *Limnol. Oceanogr.*, 36, 922, 1991.
2. Sherr, E. B., Sherr, B. F., and Paffenhöfer, G.-A., Phagotrophic protozoa as food for metazoans: a "missing" trophic link in marine pelagic food webs?, *Mar. Microb. Food Webs*, 1, 61, 1986.
3. Stoecker, D. K. and Capuzzo, J. M., Predation on Protozoa: its importance to zooplankton, *J. Plank. Res.*, 12, 891, 1990.
4. Ohman, M. D., Theilacker, G. H., and Kaupp, S. E., Immunochemical detection of predation on ciliate protists by larvae of the northern anchovy (*Engraulis mordax*), *Biol. Bull.*, 181, 500, 1991.
5. Gifford, D. J., The protozoan-metazoan trophic link in pelagic ecosystems, *J. Protozool.*, 38, 81, 1991.
6. Snyder, R. A. and Ohman, M. D., Description of a new species of Strombidinopsidae (Ciliophora: Choreotrichida) from coastal waters of southern California, U.S.A., *Trans. Am. Micro. Soc.*, 110, 237, 1991.
7. Warr, G. W., Preparation of antigens and principles of immunization, in *Antibody as a tool. The Applications of Immunochemistry,* Marchalonis, J. J. and Warr, G. W., Eds., John Wiley, New York, 1982, chap. 2.
8. Warr, G. W., Purification of antibodies, in *Antibody as a Tool. The Applications of Immunochemistry,* Marchalonis, J. J. and Warr, G. W., Eds., John Wiley, New York, 1982, chap. 3.
9. Johnstone, A. and Thorpe, R., *Immunochemistry in Practice,* 2nd ed., Blackwell Scientific, Oxford, 1987, 1.
10. Vaitukaitis, J. L., Production of antisera with small doses of immunogen: multiple intradermal injections, *Methods Enzymol.*, 73, 46, 1981.
11. Theilacker, G. H., Kimball, A. S., and Trimmer, J. S., Use of an ELISPOT immunoassay to detect euphausiid predation on larval anchovy, *Mar. Ecol. Prog. Ser.*, 30, 127, 1986.
12. Anon., *Affinity Chromatography. Principles and Methods,* Pharmacia LKB Biotechnology, Ref. 50-01-020, Uppsala, Sweden, 1988, 1.
13. Anon., *The Art of Antibody Purification,* W. R. Grace & Co., Pub. No. 868, Danvers, MA, 1989, 1.
14. Monroe, D., The solid-phase enzyme-linked immunospot assay. Current and potential applications, *BioTechniques,* 3, 222, 1985.
15. Hawkes, R., The dot immunobinding assay, *Methods Enzymol.*, 121, 484, 1986.
16. Garfin, D. E. and Bers, G., Basic aspects of protein blotting, in *Protein Blotting: Methodology, Research, and Diagnostic Applications,* Baldo, B. A. and Tovey, E. R., Eds., S. Karger, Basel, 1989, chap. 2.
17. Johnson, D. A., Gautsch, J. W., Sportsman, J. R., and Elder, J. H., Improved technique utilizing nonfat dry milk for analysis of proteins and nucleic acids transferred to nitrocellulose, *Gene Anal. Technol.*, 1, 3, 1984.
18. Blake, M. S., Johnston, K. H., Russell-Jones, G. J., and Gotschlich, E. C., A rapid, sensitive method for detection of alkaline phosphatase-conjugated anti-antibody on western blots, *Anal. Biochem.*, 136, 175, 1984.
19. Feller, R.J. and Ferguson, R. B., Quantifying stomach contents using immunoassays: a critique, in *Immunochemical Approaches to Coastal, Estuarine and Oceanographic Questions,* Lecture notes on coastal and estuarine studies, Vol. 25, Yentsch, C. M., Mague, F. C., and Horan, P. K., Eds., Springer-Verlag, New York, 1988, 295.
20. Gershoni, J. M., Protein blotting: a manual, *Methods Biochem. Anal.*, 33, 1, 1988.
21. Winter, G. and Milstein, C., Man-made antibodies, *Nature,* 349, 293, 1991.

CHAPTER 85

Absorption of Microbes by Benthic Macrofauna by the ¹⁴C:⁵¹Cr Dual-Labeling Method

Glenn Lopez

INTRODUCTION

The purpose of this paper is to provide a protocol for estimating absorption of sediment-associated microorganisms by deposit-feeding animals. Benthic deposit feeders utilize sedimentary bacteria, microbial products, and in shallower waters, microalgae, along with detritus as part of their complex diet.[1] Microorganisms provide essential compounds that may limit animal growth, such as protein, amino and fatty acids, and sterols. Thus, there has been extensive investigation of selective ingestion and absorption of bacteria and microalgae. Radiotracer techniques have been used in many of these studies because they offer improved sensitivity in measuring absorption in small animals. The ability to specifically label sedimentary microalgae and bacteria with appropriate metabolic substrates is equally important in the acceptance of radiotracer methods.[2-5] In addition, ¹⁴C labeling of food allows measurement of allocation and respiration of the absorbed food.[2,3] ¹⁴C-labeled foods can be additionally labeled with an inert radiotracer, such as ⁵¹Cr, thereby allowing estimation of ¹⁴C absorption by the change in ¹⁴C:⁵¹Cr upon gut passage.[4-6] Biological membranes are nearly impermeable to the trivalent form of chromium, so ⁵¹Cr serves as an unabsorbed marker, a radioactive equivalent to the ash content in the ash ratio method.[7] The ¹⁴C:⁵¹Cr dual-tracer method combines the advantages of indicator methods (no quantitative sampling of feces) and radiotracer methods.[8]

Lopez and Cheng[5,9] modified the Calow and Fletcher[6] dual-tracer method to additionally measure ingestion selectivity. This approach was based on the assumption that neither ⁵¹Cr or the inorganic fraction of sediment (mineral grains) is absorbed during gut passage, so the ⁵¹Cr/ash ratio in the feces could be used to estimate the amount of inorganic matter ingested.

The ideal radiolabeling technique would label only the target microbial group (i.e., heterotrophic bacteria), the label would be homogeneously distributed within each cell and among different cells of the target group, and the labeled cells would not change in specific activity during the time course of the experiment. These traits are irreconcilable because metabolic activity of uniformly labeled cells will cause an efflux of the label. The best compromise is to use an intermediate labeling period, and to limit the duration of the experiment.

0-87371-564-0/93/$0.00 + $.50
© 1993 by Lewis Publishers

MATERIALS REQUIRED

Equipment

- Centrifuge
- Two-channel liquid scintillation counter (LSC)
- Gamma counter
- Analytical balance
- Micro- and macropipettes
- Filtration gear for 47- and 25-mm filters

Supplies

- 50-ml polycarbonate centrifuge tubes with gasketed screw caps

Solutions and Reagents

- $^{51}CrCl_3$ stock solution: we normally use high specific activity (400 to 1200 Ci g^{-1}) $^{51}CrCl_3$ in 0.5 M HCl (Dupont-NEN), diluting it in 0.5 M HCl to 1 mCi ml^{-1}; ^{14}C-mixed amino acids (i.e., U-^{14}C protein hydrolysate, Amersham); for labeling heterotrophs; and ^{14}C-sodium bicarbonate (aqueous solution, pH 9.5 or crystalline solid); for labeling microalgae.
- Filtered seawater (0.2 μm)
- 0.5 M NaOH solution
- Sediment: most deposit feeders live in and preferentially ingest the finer sediment fractions (<63 μm). If sediment is poorly sorted, it may be preferable to sieve, reserving the finer fraction. Allow this fraction to settle in filtered seawater until overlying water is clear. The labeling method can be adapted for all sediment types.

PROCEDURES

Preparation of Labeled Sediment and Microbes

The first step in radiolabeling sediment is to decide on the specific activity desired for a particular experiment. This is based on the size of the smallest sample; in our work, it is an individual fecal pellet. The method described below for "normal" specific activity has been used to study groups of small (few millimeters in length) macrofauna (e.g., nuculid bivalves[5]) or to study larger animals such as amphipods.[10] We have recently taken to labeling smaller volumes of sediment to much higher specific activity so to be able to follow activity in individual, small animals (G. Lopez, unpublished observation).

Sediment to be dual-labeled should be first labeled with ^{51}Cr because it takes longer than microbial labeling.

Sediment is then labeled with a ^{14}C substrate for either microalgae or heterotrophs. We radiolabel microalgae by using a relatively short incubation with ^{14}C bicarbonate in the light. Longer incubation times would label the algae uniformly, but would result in change in algal concentration, and would label much nonalgal material through recycling. Light/dark comparisons of ^{14}C uptake by surface sediments have shown that dark uptake is usually no more than a few percent of light uptake.[11] The activity of the added ^{14}C bicarbonate can vary from 5 μCi ^{14}C cm^{-3} settled sediment (for normal activity) to 30 to 50 μCi ^{14}C added to three drops settled sediment in 5-ml filtered seawater (for very high specific activity). Heterotrophic microorganisms in the sediment, presumably mostly bacteria, are labeled by the addition of metabolic substrates to the ^{51}Cr-labeled sediment. We have used uniformly labeled ^{14}C glucose, acetate, and mixed amino acids with good results.

1. For ^{51}Cr labeling to a normal specific activity, add 1 to 5 cm^3 settled sediment to a centrifuge tube.

2. Add filtered seawater to bring volume to 30 to 40 ml.

3. Add equal volumes of the NaOH and ^{51}Cr solutions, adding the NaOH first.

4. Cap the centrifuge tube, mix the sediment suspension, and place in darkness on a shaking table for 24 h. ^{51}Cr will then have equilibrated, and the sediment is ready for microbial labeling. ^{51}Cr-specific activity should be in the range of 5 to 10 × 10^4 dpm mg^{-1} dry sediment.[5] For high specific activity, replace each cm^3 settled sediment with 2 to 10 drops settled sediment.

5. Microbial labeling:

   ● Labeling of microalgae: For high activity, we use crystalline ^{14}C sodium bicarbonate. This first should be tested for contaminants by acidifying (<pH 3) a ^{14}C-bicarbonate solution of known activity, followed by ^{14}CO$_2$ trapping in a base.[3] Added activity and trapped activity should agree. If they do not agree, the crystalline ^{14}C bicarbonate may be contaminated with nonacid volatile compounds. Add the preferred activity to the ^{51}Cr-labeled sediment, recap the centrifuge tube, and incubate for 4 to 6 h under fluorescent light (200 to 250 μE m^{-2} s^{-1}). At this point, the sediment is ready for centrifuge/rinsings.

   ● Labeling of heterotrophs: we vary the ^{14}C activity from 2 μCi cm^{-3} settled sediment to 50 μCi to a few drops settled sediment. Incubate the tube in the dark for 1 to 3 h.

6. Cleanup of dual-labeled sediment; Unincorporated radioisotopes are separated from the dual-labeled sediment by centrifugation or filtration.

   ● Muds and sandy muds should be centrifuged for at least 10 min at 2500 × g. Decant the radioactive supernate (discard as high activity wastewater), and resuspend the sediment in filtered seawater. Repeat five times, then bring up to original volume of the settled sediment.

   ● Filtration may be preferred when labeling small volumes of sediment. Filter labeled sediment onto a polycarbonate filter (0.2 μm), rinse with filtered seawater, and resuspend the sediment in seawater.

7. Marking the dual-labeled sediment with fluorescent particles makes absorption experiments much easier to conduct. We use particles produced for fluorescent paint and ink manufacture (courtesy of Radiant Color, Richmond, CA), but other fluorescent particles are suitable (e.g., Polysciences). The Radiant particles are small (2 to 7 μm, specific gravity 1.4), inert, brightly fluorescent in both daylight and blue light, and are readily ingested by deposit feeders. The particles are prepared by soaking in seawater until thoroughly wetted. A few drops of settled particles are added for each cm^3 sediment.

Absorption Experiments

Design

We employ a pulse-chase design in dual-tracer absorption experiments; animals are allowed to feed for a short period on labeled sediment (less than one gut filling time, determined previously), and then transferred to unlabeled sediment for egestion of labeled feces. The fluorescent particles mark the labeled feces, aiding in their collection. Many animals produce compact pellets, which can be collected by pipette or sieving. Loose feces can also be collected by pipette; unlabeled sediment that is inadvertently mixed with the labeled feces will not affect the radioisotope ratio, though it will increase quenching of liquid scintillants. Put each sample directly into a scintillation vial. If they are not to be counted immediately, freeze them. Pellet collection continues until the pulse of radioactive sediment has passed completely through the animal. This can take anywhere from a few minutes to a few hours. Then the animal is collected, put into a scintillation vial, and frozen before counting. (The activity in the collected animal is not used in calculating absorption efficiency, but it is a good check.)

A separate group of animals are collected for analysis immediately after ingestion of labeled sediment; this ingestion group is used to account for a change in isotope ratio due to selective ingestion. The experiment is conducted in a dish of appropriate size; we commonly use 6- or 12-well multiwell dishes. Ideally, animals should be run individually rather than in groups.

The details for conducting these experiments depend upon the animal; for mobile surface feeders (e.g., gastropods), animals are placed in labeled sediment for several minutes, then they are sieved out, cleaned of adhering particles, and transferred to unlabeled sediment.[12] For tubicolous surface feeders (e.g., spionid polychaetes), labeled sediment is placed near an animal already feeding on unlabeled sediment. It is allowed to feed for a few minutes, then the labeled sediment is removed by pipette, and unlabeled sediment is layered over (G. Lopez, unpublished observations). Subsurface feeders (e.g., nuculid bivalves) can be placed directly into labeled sediment for ingestion.[5]

A critical assumption of the dual-tracer method is that ^{51}Cr passes through the gut unabsorbed.[6] In fact, most animals studied absorb a small fraction of ingested ^{51}Cr.[4-6] Therefore, measurement of ^{51}Cr absorption is required. This is best done by feeding animals a pulse of ^{51}Cr-labeled sediment, immediately counting the live animals in a gamma counter, putting animals on unlabeled sediment, then repeatedly count the animals and the collected feces in a gamma counter. From the loss of activity from an animal, and the corresponding activity increase in its feces, ^{51}Cr absorption can be estimated. In most cases, animals absorb less than 5% ^{51}Cr, with little variation within a given species. Therefore, once $^{14}C/^{51}Cr$ ratios are corrected for ^{51}Cr absorption, little error due to ^{51}Cr absorption is propagated to the calculation of absorption efficiency.

Sample Preparation and Counting

1. At the beginning of an experiment, three to five subsamples of the labeled sediment are filtered onto preweighed 25-mm, 0.2-μm polycarbonate filters. Sediment samples should be kept small (<50 mg) to minimize quenching of the liquid scintillant. Fold filter in half with sample inside, and put into a low background LSC vial. The vial should be glass if tissue solubilizer must be used (see below). We use minivials that fit into 17-mm diameter tube holders for the gamma counter. Do not yet add cocktail.

2. Count ^{51}Cr activity of samples on a gamma counter. We prefer a true well counter because of its better counting efficiency and lower background. With a gamma counter (NaI crystal) optimized for ^{51}Cr counting, approximately 5% counting efficiency can be expected. Run appropriate blanks and standards with each set of samples. Correct sample counts for ^{51}Cr decay ($t_{1/2} = 27.7$ d).

3. The samples can now be prepared for LSC of ^{14}C. If necessary, add tissue solubilizer to samples. For small animals (few milligrams) and individual fecal pellets, we simply add

a few milliliters of scintillant, which acts as a reasonable solubilizer over the course of a few days. If solubilizer was used, use an organic scintillant (e.g., Amersham OCS) to avoid chemiluminescence. Samples may also require addition of acid to neutralize the tissue solubilizer; follow instructions for solubilizer use. Otherwise, use a water compatible cocktail (e.g., Amersham PCS).

4. Count in an LSC. To correct for ^{51}Cr crossover in ^{14}C channel, use a ^{3}H-^{14}C dual-channel program standardized with ^{51}Cr instead of ^{3}H in the lower energy channel. As with all dual-channel programs, the external standards ratio method must be used. ^{51}Cr counting efficiency in the higher energy ^{14}C channel rarely exceeds 1%.

If a gamma counter is not available, it is possible to estimate ^{51}Cr activity in the tritium window of an LSC, but this approach is rarely satisfactory because the Auger electrons emitted by ^{51}Cr are of such low energy (4 KeV) that they are easily quenched. A preferable alternative is to count the sample in full ^{14}C channel, then recount a couple weeks later, calculating ^{51}Cr activity from the reduction in cpm.[10]

Calculation of Absorption Efficiency and Ingestion Selectivity

All sample counts should now be expressed in dpm. To calculate the ^{14}C/^{51}Cr ratio in the feces, sum the activities of each isotope in all the pellets, then take the ratio of the sum. The isotope ratio of the animals collected immediately after they ingested labeled sediment (ingestion group) is estimated in the same way.

Absorption efficiency is estimated as

$$100 * [1 - (^{14}C:^{51}Cr, feces/^{14}C:^{51}Cr, ingested)] \tag{1}$$

The summed ^{51}Cr in the feces should be corrected for ^{51}Cr absorption. For example, if ^{51}Cr absorption was 2%, then the summed ^{51}Cr would be corrected by multiplying by 1.02.

A measure of selective ingestion of the ^{14}C-labeled cells can be calculated by dividing ^{14}C/^{51}Cr ingested by ^{14}C/^{51}Cr in the offered sediment.[10] Other measures of ingestion selectivity require additional measurements.[9,10]

NOTES AND COMMENTS

1. ^{51}Cr is a reasonable choice for the unabsorbed isotope because it is relatively inexpensive, easy to count and dispose (short half-life), and is poorly absorbable. One of its main disadvantages is that it cannot be easily combined with ^{3}H for dual labeling. Alternative gamma-emitting isotopes may be better suited for multiple labeling approaches. ^{241}Am is more particle reactive and less absorbable than ^{51}Cr, although it also interferes with some beta emitters.[13,14] ^{60}Co has been used for measuring absorption of metals.[15]

2. An advantage of the pulse-chase experimental approach is that it allows compartmental analysis of turnover in complex guts.[12] Many deposit feeders have relatively complex guts consisting of several compartments (e.g., in bivalves, the stomach, the style sac, and the diverticula). By tracking the time course of egestion of a labeled meal, it is possible to estimate the number, size, and turnover times of functional compartments.[12,16]

REFERENCES

1. Lopez, G. L. and Levinton, J. S., Ecology of deposit-feeding animals in marine sediments, *Q. Rev. Biol.*, 62, 235, 1987.
2. Kofoed, L. H., The feeding biology of *Hydrobia ventrosa* (Montague). I. The assimilation of different components of food, *J. Exp. Mar. Biol. Ecol.*, 19, 233, 1975.
3. Kofoed, L. H., The feeding biology of *Hydrobia ventrosa* (Montague). II. Allocation of the components of the carbon budget and the significance of the secretion of dissolved organic material, *J. Exp. Mar. Biol. Ecol.*, 19, 243, 1975.
4. Cammen, L. M., The significance of microbial carbon in the nutrition of the deposit-feeding polychaete *Nereis succinea, Mar. Biol.*, 61, 9, 1980.
5. Lopez, G. R. and Cheng, I.-J., Synoptic measurements of ingestion rate, ingestion selectivity, and absorption efficiency of natural foods in the deposit-feeding molluscs *Nucula annulata* (Bivalvia) and *Hydrobia totteni* (Gastropoda), *Mar. Ecol. Prog. Ser.*, 11, 55, 1983.
6. Calow, P. and Fletcher, C., A new radiotracer technique involving ^{14}C and ^{51}Cr for estimating the assimilation efficiency of aquatic primary producers, *Oecologia*, 9, 155, 1972.
7. Conover, R. J., Assimilation of organic matter by zooplankton, *Limnol. Oceanogr.*, 11, 338, 1966.
8. Lopez, G., Tantichodok, P., and Cheng, I.-J., Radiotracer methods for determining utilization of sedimentary organic matter by deposit feeders, *Ecology of Marine Deposit Feeders*, Lopez, G., Taghon, G., and Levinton, J., Eds., Springer-Verlag, Heidelberg, 1989, chap. 7.
9. Lopez, G. R. and Cheng, I.-J., Ingestion selectivity of sedimentary organic matter by the deposit-feeder *Nucula annulata* (Bivalvia: Nuculidae), *Mar. Ecol. Prog. Ser.*, 8, 279, 1982.
10. Lopez, G. and Elmgren, R., Feeding depths and organic absorption for the deposit-feeding benthic amphipods *Pontoporeia affinis* and *Pontoporeia femorata, Limnol. Oceanogr.*, 34, 982, 1989.
11. Lopez, G. R., The availability of attached microorganisms as food for some marine benthic deposit-feeding molluscs, with notes on microbial detachment due to the crystalline style, *Marine Benthic Dynamics*, Tenore, K. R. and Coull, B. C., Eds., University South Carolina Press, Columbia, 1980, 387.
12. Kofoed, L., Forbes, V., and Lopez, G., Time-dependent absorption in deposit feeders, *Ecology of Marine Deposit Feeders*, Lopez, G., Taghon, G., and Levinton, J., Eds., Springer-Verlag, Heidelberg, 1989, chap. 6.
13. Fisher, N. S. and Reinfelder, J. R., Assimilation of selenium in the marine copepod *Acartia tonsa* studied with a radiotracer ratio method, *Mar. Ecol. Prog. Ser.*, 70, 157, 1991.
14. Reinfelder, J. R. and Fisher, N. S., The assimilation of elements ingested by marine copepods, *Science*, 251, 794, 1991.
15. Weeks, J. M. and Rainbow, P. S., A dual-labelling technique to measure the relative assimilation efficiencies of invertebrates taking up trace metals from food, *Functional Ecol.*, 4, 711, 1990.
16. Bricelj, V. M., Bass, A. E., and Lopez, G. R., Absorption and gut passage time of microalgae in a suspension feeder: an evaluation of the $^{51}Cr:^{14}C$ twin tracer technique, *Mar. Ecol. Prog. Ser.*, 17, 57, 1984.

CHAPTER 86

Radioisotope Technique to Quantify *In Situ* Microbivory by Meiofauna in Sediments

Paul A. Montagna

INTRODUCTION

There are three size claseses of benthic organisms corresponding to bacteria, meiofauna, and macrofauna.[1,2] Microalgae[1,3] and protozoans (e.g., Foraminifera)[4] are usually in the low end of the size range for meiofauna. Theoretical arguments, based on the size spectrum, have been made that trophic relationships exist among the benthic taxonomic groups.[3,5] Experimental evidence has also established trophic relationships among the taxonomic groups. Meiofauna eat bacteria[6] and macrofauna eat meiofauna.[7,8] Benthic detritivores also consume bacteria associated with detritus.[9] Meiofauna and macrofauna also eat microalgae.[6] However, the presence of sediment, small size of most benthic invertebrates, and large differences among sizes of benthic groups makes trophic studies difficult to perform. It is particularly difficult to make quantitative measurements of rates of microbivory by meiofauna.

Radiolabeled substrates have proven to be useful to study material flow in benthic food webs.[6] There are two different ways to employ radioactive tracers to measure the flow of organic material in food webs via feeding. Microbial prey are either prelabeled[10] or labeled while they are being grazed.[11] Both techniques have advantages and disadvantages that limit their usefulness to either laboratory or field studies. The prelabeling technique requires growing microbial prey with a radioactive tracer, introducing the labeled microbe to the existing microbial community and determining the specific activity of the food source. Such conditions are best achieved in laboratory studies. The synoptic labeling technique is more amenable to *in situ* studies, because only the radioactive tracer is introduced and microbial uptake of the tracer and invertebrate grazing can be measured at the same time.

The synoptic, *in situ*, labeling technique to measure invertebrate grazing on microbial prey has many advantages over other techniques. It is easy to employ in the field and incubation times are short. Grazing on autotrophic microalgae and heterotrophic bacteria can be measured in the same experiment using dual-labels. The relative importance of dissolved organic matter uptake (DOM) and grazing to the grazer can be assessed by inhibiting either prey label uptake or grazer metabolic processes. Microbial production and grazing impact can also be measured simultaneously. The materials required are inexpensive.

There are also several disadvantages to using this technique. There are a large number of control experiments necessary to determine that the technique is working and to correct label uptake for nongrazing processes. Postincubation analyses are labor intensive. Protists require

0-87371-564-0/93/$0.00 + $.50
© 1993 by Lewis Publishers

different handling techniques than metazoans, so they are either ignored or difficult to deal with. Labeling kinetics and labeling specificity of prey must be checked. Sediment chemistry can affect behavior or kinetics of the label. The labeling rate of prey and grazer must be sufficient to ensure accurate counting of the radioactivity.

MATERIALS REQUIRED

Equipment

- Sediment sampling devices, e.g., cores and stoppers
- Water column sampling devices
- Incubation chamber with lights
- Vortex mixer
- Scintillation counter
- Dissecting microscope
- Warming plate or drying oven

Supplies

- Tweezers
- 63-μm sieves
- Wash bottles
- Filter devices
- Funnel
- Radioactive waste containers
- Micropipettes
- Pipettes
- Aluminum foil
- 50-ml clear plastic centrifuge tubes
- 0.2-μm filters
- Scintillation vials
- Petri dishes

Solutions and reagents

- Tracer solution stock solution: 5 μCi of ^{3}HTdR and 5 μCi of $H^{14}CO_3^-$ is needed per sample, make in either sterile sea or freshwater
- Bacterial inhibitor solution: double saturated solution of nalidixic acid (200 μg ml^{-1}) plus 5'-deoxythymidine (2 μg ml^{-1}) in seawater
- Scintillation cocktail for aqueous solutions
- Tissue solubilizer
- Scintillation cocktail compatible with solubilizer

METHOD PROTOCOL

The method described below is a dual-labeling technique for measuring grazing on bacteria and microalgae by meiofauna at the same time. A single label can be used to investigate only bacteria or microalgae as prey. The invertebrate grazers can be small (e.g., meiofauna) or large (e.g., macrofauna). Species specific studies can be performed by only sorting out the taxa of interest. Although I use the term grazing, ingestion by filter feeders and deposit feeders is also measured.

In situ grazing studies require controls, because meiofauna may absorb DOM,[12] or adsorb dissolved inorganic carbon (DIC).[13] A properly designed *in situ* meiofaunal grazing experiment must consider and correct for isotopic uptake by microbes, meiofauna, and macrofauna via nongrazing processes. This requires parallel incubations to obtain experimental and control values.

Field Incubations

Meiofaunal grazing rates on bacteria and microalgae are measured by incubating sediment with two radiolabeled substrates: [methyl-^{3}H]-thymidine (^{3}HTdR) and ^{14}C bicarbonate (H^{14}CO$_3^-$).[12] Bacteria are labeled with the thymidine via DNA synthesis, and microalgae are labeled with the bicarbonate via photosynthesis. Incubation must take place in an appropriate chamber with respect to temperature and ambient light conditions. Appropriate geochemical horizons of the sediment must be used. It is thought that thymidine incorporation is basically an aerobic process,[14] therefore, reducing sediment should not be used if thymidine is used to label bacteria. Sediment samples can be incubated in whole intact cores or in slurries. Slurries made by adding label contained in a small volume of water (e.g., 50 μl) to the sediment surface (e.g., a whole 9.6-cm^2 core) and stirring the top 1 cm do not work well.[15] I define a slurry as an approximately 1:1 mix of aerobic sediment to sea water,[16] but I have also used 1.7:1 mixes.[12]

Inhibitor controls are used to correct for nongrazing label uptake by live meiofauna.[12,13] These processes include, but are not limited to, adsorption, absorption, or drinking of the label. A saturated solution of nalidixic acid plus 5'-deoxythymidine can be added to a sediment slurry sample to inhibit prokaryotic incorporation of ^{3}HTdR.[12,17] Control samples can also be incubated in the dark to inhibit photosynthetic fixation of ^{14}CO$_2$. The object of the inhibitor-control treatment is to determine the amount of uptake of ^{3}H or ^{14}C that is not due to feeding on microbes by meiofauna. Meiofauna are feeding during the control experiments, but microbial uptake of label is inhibited by the treatment.

Killed controls can be used to distinguish between label uptake due to absorption and adsorption by meiofauna in the control treatments. When formalin is added to a sample, the amount of label taken up by meiofauna is due to nongrazing and nonliving processes, i.e., adsorption. The difference between inhibitor-control uptake and killed-control uptake is absorption.

Other control and verification experiments may be necessary. If a microbial inhibitor is used in control experiments, then the effectiveness of the inhibitor should be checked. Time-course experiments of label uptake by the microbes should be used to establish that label uptake is linear over the period of time selected for the incubation. The model used to calculate the grazing rates also assumes that label uptake by the grazer is hyperbolic, and that the incubation period is short enough so that label recycling does not occur. These two assumptions should also be verified.

1. Collect sediment and water samples. I use 60-cm^3 syringes with the bottoms cut off to collect sediment samples. If the density of meiofauna in the study area is low, then use larger cores. For each grazing rate measurement you will need one core each for the experimental, microbial inhibitor-control, and killed-control incubations. If you plan 3 replicates, then you need a total of nine cores per station or experimental unit.

2. Place the top 2 cm of sediment (12 cm^3) in the centrifuge tube. I find the 2-cm depth horizon useful for a variety of sandy and muddy sediments, but the exact amount should

be adjusted for local conditions. I incubate samples in 60-cm³ clear centrifuge tubes, because they are cheap, sterile, disposable, and easy to handle. You can also use plastic Petri dishes, or any suitable container. Alternatively, you can incubate the whole core. However, if you intend to measure microalgal grazing, the container must let light pass through it.

3. Make experimental solutions:

- Filter-sterilize the station seawater. Add 5 μCi of ³HTdR and 5 μCi of $H^{14}CO_3^-$ from the stock solution to sterile station water so that the final volume is 12 ml. The seawater label solutions will be used for the live-grazing measurements.

- Inhibitor-control solutions are made with label added to the inhibitor solutions, so that a total of 12 ml contains 5 μCi of ³HTdR and 5 μCi of $H^{14}CO_3^-$ and 200 μg ml^{-1} nalidixic acid plus 2 μg ml^{-1} 5′-deoxythymidine. The solubility in water at 23°C is 100 μg ml^{-1} for nalidixic acid and 1 μg ml^{-1} 5′-deoxythymidine. The inhibitors are effective only when the solution is within 40 to 80 of saturation.[12] Since I make slurries that are about 1:1 mixtures of water and sediment sample, the saturation of the inhibitor mixture could be reduced to 50%. Therefore, I double the amount of inhibitor to make stock solutions and use a vortex machine to suspend the undissolved crystals and dilute it with station water. When diluted, the inhibitor goes into solution. A killed-control solution is made by adding label and formalin to the seawater, so that you have a labeled solution containing 4% formalin.

4. Incubate the samples. Samples should be incubated at *in situ* temperature using a water bath or an incubator. The incubation time has to be long enough for label to accumulate in the grazer, but not too long so that the label will be recycled from the grazer. I have found that 1 to 2 h is a good length of time. The best time can be determined by performing a time-series experiment on just the live-grazing treatment. If you are measuring grazing on microalgae, then you must incubate the samples under light so that photosynthesis is occurring. Conditions during the incubation should be designed so that optimal label uptake rates by the microbes occur. Therefore, you should incubate at light levels that are above saturation, but below the point of photoinhibition. This should be in the range of 100 to 300 μE·m^{-2}·s^{-1}. Control samples should be incubated in the dark. I wrap the centrifuge tubes in aluminum foil.

5. Terminate the incubation. The live-grazing incubations and the inhibitor-incubations are terminated by adding 4% formalin. I add enough formalin to bring the total volume within the centrifuge tube to 50.0 ml. The formalin-incubation should be terminated by adding sterilized station water to a volume of 50.0 ml; this will maintain the similar final concentrations of about 2% formalin in all treatments.

6. Withdraw a 1.0-ml subsample from the slurry to determine the uptake by the microbes. This is physically difficult to perform. I use disposable 1-ml pipettes that have large bores. I sometimes use larger pipettes if the sediment grain size is large, e.g., sand. Use a vortex machine to suspend and mix the sediment, and subsample as the tube is being vortexed. The subsample is filtered onto a 0.2-μm Millipore filter and rinsed three times with filtered seawater to estimate uptake of $H^{14}CO_3^-$ by microalgae and ³HTdR by bacteria. The filters can also be rinsed with 1% HCl to remove excess bicarbonate radioactivity.[18] The subsample is dispersed and suspended in 5-ml distilled water and 15-ml Insta-Gel for dual-label liquid scintillation counting.

7. Harvest grazers. Meiofauna are separated from sediments by diluting samples with sea water, swirling to suspend the animals, and decanting them and the supernatant onto 63-μm Nitex sieves. This procedure can generate a lot of radioactive wastewater. I collect the waste by sieving over a funnel stuck into a 5-gal jug. Typically, a field experiment where 27 replicate cores are sieved will require a 5-gal waste bottle. Meiofauna collected

on the sieve are then rinsed into jars and kept in refrigerated 2% formalin until sorting. I have also preserved meiofauna samples with 5% glutaraldehyde in cacodylate buffer, a preservative used for electron microscopy.

Laboratory Extraction

I have been concerned that labeled grazers preserved in seawater might leak label. Label levels do not change over a 7-d period.[6] You should be careful not to take so many samples that you may not be able to sort them all in a reasonable amount of time. I try to complete all sorting within 2 d.

1. Sort meiofauna using a dissecting microscope. Sort meiofauna by species, taxa, or other grouping into scintillation vials containing 1 ml distilled water. The water is needed so that you can get the meiofauna off your tweezers. *You must be very careful not to get detritus into the scintillation vials.* Detritus will contain very high counts of labeled bacteria.

    You may want to normalize the grazing rates to the abundance, biomas, or taxonomic content of the samples. You should record whatever data is necessary for this during the sorting procedure. Alternatively, you can make independent measurements of community structure.

2. Dry meiofauna at 60°C for a 24-h period.

3. Determine radioactivity.

    ● Add 100 μl of tissue solubilizer to each sample.

    ● After 24 h, add a scintillation cocktail that is compatible with the solubilizer. For example, I have used Packard Soluene® 350, which contains ammonium hydroxide, with Packard Hionic-Flour™, which has a high capacity for alkaline and salt solutions.

    ● Count samples by dual-label liquid scintillation spectrophotometry.

Calculation of Grazing Rates: Model and Statistics

Meiofaunal grazing rates on bacteria and microalgae are calculated using a model proposed by Daro.[11] The meiofaunal grazing rate (G) is the proportion of material flowing from the donor (or prey) compartment to the recipient (or grazer) compartment per unit time. The model assumes that label is in excess and that during the incubation period label does not become limiting, prey uptake is linear, and grazer label recycling is zero. In this case, G is expressed in units of h^{-1} and is calculated as follows:[6,12]

$$G = 2F/t \qquad (1)$$

$$F = M/B \qquad (2)$$

where F is the fraction of label uptake in the grazer, M (in this case meiofauna), relative to the prey, B (bacteria or microalgae), at time, t in hours. M and B are in units of disintegrations per minute (dpm). Since $2/t$ is a constant, variability in G is due to variability in F, i.e., M and B.

When controls are performed, the mathematics are tricky. For example, suppose you perform three replicate grazing experiments and three replicate controls. Which control value do you subtract from each experimental value? It is tempting to calculate the grazing fraction for each core (F$_i$) using the mean control value and subtracting it from each experimental value as follows:

$$F_i = \frac{(M_i - \overline{M}_c)}{(B_i - \overline{B}_c)} \tag{3}$$

where each replicate core generates a grazer value M$_i$ and a prey value B$_i$, and the average control value is $\overline{M}_c$ for grazers and $\overline{B}_c$ for prey. Equation 3 can generate one value for each replicate, and these values can be used in statistical analyses to find differences among stations or other treatment effects in an experiment. There is a problem with using this approach. Each replicate value is now a function of two sources of variability: the variability due to the grazing replicate value and the variability due to the error associated with the estimate of the average control value. An analysis of variance would underestimate variability, since control variability is not included in the analysis. If control values are larger than individual experimental values, then the grazing rate will be a negative number. You can also argue that the grazing rate should be set to zero when negative values are calculated, but this will lead to an overestimate of the grazing rate.

We are really performing two separate experiments to estimate the experimental and control values. Grazing rates are then estimated as a function of these two estimates. Therefore, the mean and variance of the grazing rates are also a function of the means and variances of the two values.[19] Using this approach the mean grazing fraction ($\overline{F}$) is calculated as:

$$\overline{F} = \frac{(\overline{M} - \overline{M}_c)}{(\overline{B} - \overline{B}_c)} \tag{4}$$

and the variance of the fraction (S$_F^2$) is calculated as:[12]

$$S_F^2 = \frac{(S_M^2 + S_{M_c}^2) + \overline{F}^2 \times (S_B^2 + S_{B_c}^2)}{(\overline{B} - \overline{B}_c)^2} \tag{5}$$

where S$_M^2$ and S$_{M_c}^2$ are the variance estimates of the meiofauna experimental and control uptake values, and S$_B^2$ and S$_{B_c}^2$ are the variance estimates for the microbial experimental and control uptake values. For each grazing rate sample mean, 95% confidence intervals can be constructed using this variance estimate to determine if there are significant differences among the means at the 5% level.

NOTES ON DIFFICULTIES

1. Choosing labels. Labels for prey should have high specificity, high labeling rates, and low recycling rates (i.e., label is not mineralized by prey). The label should not interfere with grazer metabolism or be taken up by the grazer. If the grazer also incorporates the label via nongrazing processes then inhibitor controls will be required. If prey incorporation rates are low, then high activities will be necessary.

2. Uptake of ³HTdR has been used for estimating both bacterial production and the transfer of bacterial carbon to higher trophic levels.[20] The use of ³HTdR for food-web studies is attractive because it is taken up and incorporated into macromolecules exclusively by prokaryotes.[21] However, thymidine is apparently not incorporated by either chemolithotrophic[22] or sulfate-reducing[14] bacteria.

3. Low labeling rates may also be a problem with bacteria. Uptake of either [^{3}H]-acetate or ^{3}HTdR could be detected by microautoradiography in only approximately 2% of the bacteria in samples from sandy intertidal sediments.[23] Only 10% of the bacteria in Halifax Harbor muddy sediments are actively growing.[24] If grazers are selecting only growing bacteria, then we will be overestimating grazing rates by as much as an order of magnitude when using thymidine. Lacking information about selection of bacteria by grazers, we must assume that grazers are not selecting the labeled bacteria over the unlabeled bacteria. Other compounds may be better suited for labeling bacteria. About 25% of the bacteria were labeled with [U-^{14}C]-glucose in intertidal mudflats.[25] This only appears to be a potential problem for bacteria. Labeling rates of microphytobenthos with bicarbonate are high. Uptake of [^{14}C]-bicarbonate was detected by microautoradiography in up to 75% of the microalgae in sandy sediments.[23]

4. Delivering labels. Rapid and uniform prey labeling is difficult. Four different techniques have been used: direct "injection" below the sediment surface,[26] sediment pore-water "replacement",[27] "slurries" of sediment and seawater,[12] and "addition" to the overlying water.[6] Each of these methods has advantages and disadvantages, and each method has been used in a wide variety of microbial studies. Direct "injection" into whole, intact, sediment cores is the preferred method.[15,23,28] This technique preserves the vertical distribution of geochemical electron acceptors, does not disturb the grazers behavior, and is least likely to perturb microbial activity. Replacement of pore water would disrupt the vertical distribution of oxygen and sulfide gradients and would alter metabolism of both aerobic and anaerobic heterotrophs. Replacement is also limited to sandy sediments. It could be difficult to achieve uniform labeling with additions of label, since diffusion of label into the media could permit microbes to be labeled at different rates or not at all. However, the top few millimeters of sediment contains the most active bacteria,[24] all of the photosynthetically active microalgae (G. F. Blanchard and P. A. Montagna, unpublished data), and most of the meiofauna, indicating that additions should be given more consideration in the future. Slurries can achieve uniform labeling at the expense of sediment disturbance. For example, bicarbonate uptake by microalgae in sandy sediments decreased from 75 to 50% in slurries.[23] Slurries can also alter meiofauna behavior and change the feeding rate by disrupting microbial-meiofaunal spatial relationships.[15]

5. Leaking labels. The time-consuming task of sorting samples means that label could diffuse out of the grazer before the sample is counted. This would lead to underestimates of grazing rates. Label leakage from meiofauna was not detected during daily sampling over the first seven days after preservation with 2% formalin solution, but label leakage could have occurred during the first day.[6] Label loss by meiofauna has received too little study, but several studies have been performed on zooplankton. In the worst case, cladocerans lost half of the ^{14}C within 1 h when preserved in 4% formalin at room temperature after feeding on algae.[29] Loss rates in cladocerans can vary from study to study, but 50% of the loss always occurs during the first 24 h and is reasonably stable thereafter, leading Mourelatos to conclude that "sorting of animals should be delayed until losses have stabilized . . . and a correction factor of $2\times$ be applied".[29] Rapid freezing on dry ice can also improve retention of label by zooplankton over chemical preservatives.[30] Freezing samples in formalin and sorting within 2 h of thawing yielded consistent loss rates, indicating a correction factor applied to this method may be the best technique.[29] Lugol's iodine has good preservative properties for minimizing loss, reaching stable values of 38% loss after 1 h.[31] It is clear that all meiobenthic grazing rates measured to date are conservative measures, since label loss must have occurred in most studies.

6. High control values. If during inhibitor-control experiments, you cannot inhibit prey uptake of label then you will get high control values for grazers. This can be circumvented by performing separate experiments to determine grazer label uptake kinetics in sterile seawater. This will allow you to calculate the label uptake by the grazer due to absorption based on the label concentration during the grazing experiment. This problem may exist just with labeling bacteria. It appears that label uptake by bacteria associated with the

cuticle of meiofauna can be responsible for most of the meiofauna uptake.[12] Using autoradiography, it was found that uptake of [^{14}C]-acetate by the harpacticoid, *Zausodes arenicolus*, was entirely by epicuticular bacteria, but [^{14}C]-bicarbonate uptake was exclusively due to grazing on photoautotrophs.[32]

7. Chemical interferences. Sediment pore water contains high concentrations of DOM and inorganic nutrients, and clay particles that are often electrically charged. This chemical milieu promotes the label being bound chemically by adsorption processes. If this happens then not all label introduced to the incubation will be available for uptake by the microbes. Therefore, the concentration of available label will be decreased. The same processes also bind the inhibitors in control experiments leading to less inhibition than predicted. In my experience this problem is especially acute in organic rich muds, and in estuarine sediments near river mouths.[16]

8. Dual-label counting. ^{14}C has a higher energy level than ^{3}H. If you use too much ^{14}C relative to ^{3}H, then the ^{3}H signal can be overwhelmed by the ^{14}C signal. I always try to use at least equal specific activities of ^{14}C and ^{3}H. This problem is compounded, if uptake by bacteria is generally lower than uptake by microalgae.[23] If it is possible, use higher amounts of ^{3}H.

9. Statistical. Since the grazing rate is a function of four estimates (microbial experimental and control values and meiofaunal experimental and control values) the "ball park" is enormous. Each estimate contributes a source of variability. It is imperative to obtain replication for each source of variability. Using Equation 4 instead of 3 will at least allow you to calculate the true magnitude of the variance of the grazing rate.

10. Time. The incubation time should be kept short, so that label is not recycled by the grazers. If you expect recycling of label, then Equation 1 is not valid. The method measures ingestion, not assimilation. There is a need for more information on meiofauna gut evacuation rates and passage times.

REFERENCES

1. Schwinghamer, P., Characteristic size distributions of integral benthic communities, *Can. J. Fish. Aquat. Sci.*, 38, 1255, 1981.
2. Warwick, R. M. and Joint, I. R., The size distribution of organisms in the Celtic Sea: from bacteria to Metazoa, *Oecologia*, 73, 185, 1987.
3. Schwinghamer, P., Generating ecological hypotheses from biomass spectra using causal analysis: a benthic example, *Mar. Ecol. Prog. Ser.*, 13, 151, 1983.
4. Gerlach, S. A., Hahn, A. E., and Schrage, M., Size spectra of benthic biomass and metabolism, *Mar. Ecol. Prog. Ser.*, 26, 161, 1985.
5. Warwick, R. M., Species size distributions in marine benthic communities, *Oecologia*, 61, 32, 1984.
6. Montagna, P. A., In situ measurement of meiobenthic grazing rates on sediment bacteria and edaphic diatoms, *Mar. Ecol. Prog. Ser.*, 18, 119, 1984.
7. Watzin, M. C., Interactions among temporary and permanent meiofauna: observations on the feeding and behavior of selected taxa, *Biol. Bull.*, 169, 397, 1985.
8. Pihl, L., Food selection and consumption of mobile epibenthic fauna in shallow marine areas, *Mar. Ecol. Prog. Ser.*, 22, 169, 1985.
9. Findlay, S. and Tenore, K., Nitrogen source for a detritivore: detritus substrate versus associated microbes, *Science*, 218, 371, 1982.
10. Haney, J. F., An in situ method for measurement of zooplankton grazing rates, *Limnol. Oceanogr.*, 16, 970, 1971.
11. Daro, M. H., A simplified ^{14}C method for grazing measurements on natural planktonic populations, *Helgol. Wiss. Merresunters.*, 31, 241, 1978.
12. Montagna, P. A. and Bauer, J. E., Partitioning radiolabeled thymidine uptake by bacteria and meiofauna using metabolic blocks and poisons in benthic feeding studies, *Mar. Biol.*, 98, 101, 1988.

13. Montagna, P. A., Live controls for radioisotope food chain experiments using meiofauna, *Mar. Ecol. Prog. Ser.,* 12, 43, 1983.

14. Gilmour, C. C., Leavitt, M. E., and Shiaris, M. P., Evidence against incorporation of exogenous thymidine by sulfate-reducing bacteria, *Limnol. Oceanogr.,* 35, 1401, 1990.

15. Carman, K. R., Dobbs, F. C., and Guckert, J. B., Comparison of three techniques for administering radiolabeled substrates to sediments for trophic studies: uptake of label by harpacticoid copepods, *Mar. Biol.,* 102, 119, 1989.

16. Montagna, P. A. and Yoon, W. B., The effect of freshwater inflow on meiofaunal consumption of sediment bacteria and microphytobenthos in San Antonio, Bay, Texas, U.S.A., *Estuar. Coast. Shelf Sci.,* 33, 529, 1991.

17. Findlay, S., Meyer, J. L., and Smith, P. J., Significance of bacterial biomass in the nutrition of a freshwater isopod (*Lirceus* sp.), *Oecologia,* 63, 38, 1984.

18. Blanchard, G. F., Measurement of meiofauna grazing rates on microphytobenthos: is primary production a limiting factor?, *J. Exp. Mar. Biol. Ecol.,* 147, 37, 1991.

19. Kempthorne, O. and Allmaras, R. R., Errors in observation, in *Methods in Soil Analysis, Part 1, Physical and Mineralogical Properties, including Statistics of Measurement and Sampling,* Black, C. A., Ed., American Society of Agronomy, Madison, WI, 1965, 1.

20. Hollibaugh, J. T., Furhman, J. A., and Azam, F., Radioactively labeling of natural assemblages of bacterioplankton for use in trophic studies, *Limnol. Oceanogr.,* 25, 172, 1980.

21. Moriarty, D. J. W. and Pollard, P. C., Diel variation of bacterial productivity in seagrass (*Zostera capricorni*) beds measured by the rate of thymidine incorporation into DNA, *Mar. Biol.,* 72, 165, 1982.

22. Johnstone, B. H. and Jones, R. D., A study on the lack of [methyl-^{3}] thymidine uptake and incorporation by chemolithotrophic bacteria, *Microb. Ecol.,* 18, 73, 1989.

23. Carman, K. R., Radioactive labeling of a natural assemblage of marine sedimentary bacteria and microalgae for trophic studies: an autoradiographic study, *Microb. Ecol.,* 19, 279, 1990.

24. Novitsky, J. A., Microbial growth rates and biomass production in a marine sediment: evidence for a very active but mostly nongrowing community, *Appl. Environ. Microbiol.,* 53, 2368, 1989.

25. Juniper, S. K., Stimulation of bacterial activity by a deposit feeder in two New Zealand intertidal inlets, *Bull. Mar. Sci.,* 31, 691, 1981.

26. Carman, K. R. and Thistle, D., Microbial food partitioning by three species of benthic copepods, *Mar. Biol.,* 88, 143, 1985.

27. Meyer-Reil, L.-A. and Faubel, A., Uptake of organic matter by meiofauna organisms and interrelationships with bacteria, *Mar. Ecol. Prog. Ser.,* 3:251, 1980.

28. Dobbs, F. C., Guckert, J. B., and Carman, K. R., Comparison of three techniques for administering radiolabeled substrates to sediments for trophic studies: incorporation by microbes, *Microb. Ecol.,* 17, 237, 1989.

29. Mourelatos, S., A new technique for long preservation of ^{14}C-labelled cladocerans, *Hydrobiologia,* 190, 147, 1990.

30. Sierszen, M. E. and Watras, C. J., Rapid Freeze preservation minimizes radioisotope leakage from zooplankton in feeding experiments, *J. Plankton Res.,* 9, 945, 1987.

31. Holtby, L. B. and Knoechel, R., Zooplankton filtering rates: error due to loss of radioisotopic label in chemically preserved samples, *Limnol. Oceanogr.,* 26, 774, 1981.

32. Carman, K., Mechanisms of uptake of radioactive labels by meiobenthic copepods during grazing experiments, *Mar. Ecol. Prog. Ser.,* 68, 71, 1990.

Index

INDEX